Viktor Sadovnichy • Askar Akaev •
Ilya Ilyin • Sergey Malkov • Leonid Grinin •
Andrey Korotayev
Editors

Reconsidering the Limits to Growth

A Report to the Russian Association of the Club of Rome

Springer

Editors
Viktor Sadovnichy
Lomonosov Moscow State University
Moscow, Russia

Askar Akaev
Lomonosov Moscow State University
Moscow, Russia

Ilya Ilyin
Lomonosov Moscow State University
Moscow, Russia

Sergey Malkov
Institute of Economics, Russian Academy of Sciences
Moscow, Russia

Leonid Grinin
Institute of Oriental Studies, Russian Academy of Sciences
Moscow, Russia

Andrey Korotayev
HSE University
Moscow, Russia

ISSN 2522-0985 ISSN 2522-0993 (electronic)
World-Systems Evolution and Global Futures
ISBN 978-3-031-34998-0 ISBN 978-3-031-34999-7 (eBook)
https://doi.org/10.1007/978-3-031-34999-7

This Springer imprint is published by the registered company Springer Nature Switzerland AG
The registered company address is: Gewerbestrasse 11, 6330 Cham, Switzerland

World-Systems Evolution and Global Futures

This series seeks to promote understanding of large-scale and long-term processes of social change, in particular the many facets and implications of globalization. It critically explores the factors that affect the historical formation and current evolution of social systems, on both the regional and global level. Processes and factors that are examined include economies, technologies, geopolitics, institutions, conflicts, demographic trends, climate change, global culture, social movements, global inequalities, etc. Building on world-systems analysis, the series addresses topics such as globalization from historical and comparative perspectives, trends in global inequalities, core-periphery relations and the rise and fall of hegemonic core states, transnational institutions, and the long-term energy transition. This ambitious interdisciplinary and international series presents cutting-edge research by social scientists who study whole human systems and is relevant for all readers interested in systems approaches to the emerging world society, especially historians, political scientists, economists, sociologists, geographers and anthropologists.

This book series is indexed in Scopus.

All titles in this series are peer-reviewed.

Foreword: On to the Future

The world is currently a dangerous place. Will it become more or less dangerous in the future? Science fiction novels and movies vary between utopian landscapes where there is no poverty, famine, or unemployment and dystopian hellholes where people fight over small quantities of gasoline or water. The real future probably lies somewhere in between these ends of the future continuum. We have a very binary choice in considering what that future will look like. Either we ignore it altogether and kick the can down the road for future generations to figure it out or we can use data, imagination, and creativity we currently possess to try forecasting what the future might look like. Whether or not we do anything about our forecasts, the future should at least be less of a shock than if we emulate ostriches and ignore it altogether.

This book is put together by a stellar group of analysts working in Moscow. They follow in what is now a fifty year plus tradition in evaluating the limits to growth in the world. When the Club of Rome issued its first forecast in 1972, the emphasis was on the possibility of overpopulation and deficits in carrying capacity, but one of its most amazing features was that it saw the world as a system of interacting parts. Up to that time, most forecasting had had more modest ambition, looking at singular processes and/or less than planetary scope. The world is now even more integrated than it was in the early 1970s, and the need for world modeling and forecasting is all the more crucial because what happens in faraway places is more likely to impact throughout the system than once might have been the case. For instance, the flapping wings of butterflies in Brazil may not lead to tornados in Oklahoma, but destroying their habitat in the Brazilian jungle diminishes a major forested area that plays a significant role in reducing worldwide greenhouse emission.

Yet the first thing to recognize about world forecasting is that the world is a heterogenous place. The Middle East's future will not resemble closely the future of South America. Future European problems will not look the same as future South Asian problems. The world is composed of a number of different regions, each possessing distinctive mixes of resources, capacities, and prospects. For many purposes, it may not be prudent to attempt modeling the full extent of regional differences. But, at the very least, we will need to recognize the differences between the Global North and South. Northern futures, in general, look very different than Southern futures. There may be ways to bridge that inherent gap so that the North and South can converge on mutually rewarding solutions to different kinds of problems. We will have to wait and see.

In the meantime, the "Club of Moscow" response to the Club of Rome chooses to look at the interactions among aging, capitalism, inequality, climate change, resources, technology, and world order. This list does not exhaust all of the future topics that might be examined, but it certainly represents an ambitious undertaking. Differences between the Global South and North quickly emerge. A number of Southern states still have high birth rates while Northern states have populations who are proportionally older than once was the case. Profound effects in terms of economic productivity, employment, societal welfare, and perhaps even finding enough military personnel to engage in war ensue.

Climate change is likely to be more devastating in the South than in the North. Both groups will experience chaotic weather patterns, heat waves, and flooding associated with rising seas. But warming (and drying) impacts will probably be greater in already arid and warm areas such as South Asia, the Middle East, and the Sahel. Food and water crises will increase mortality rates in these zones. Climate migrants may have no place to go. Affluent areas can build machines to process sea water into potable water. Impoverished and land-locked areas cannot. Food can be purchased on the world market, but one has to be able to export demanded commodities to pay for imports. If the amount of land devoted to agricultural production actually diminishes, the competition for, and the pricing of, food will escalate as well. The thawing of frozen tundra does not necessarily yield good soil for growing things.

The differences continue. The South needs technology desperately. The North worries about how to control or even suppress some of the new technology that it has. The North wants access to Southern resources that are needed by the new technology. The South seems eager to sell raw materials, but questions of fair pricing and environmental degradation persist. The North wishes to hang on to the tattered world order that emerged after World War II. The South prefers selective or wholesale changes in order arrangements that speak to rules that discriminate against developing and lower income areas.

The Moscow Report is surprisingly optimistic that these problems, or most of them, can be resolved readily. I say surprisingly only because the problems are serious and most politicians are reluctant to expend resources on future problems—that is, on someone else's watch. Yet, where older limits to growth analyses were fixated on the dilemmas of excessive population growth, the Moscow Club sees that as largely a dead issue (outside Africa). "Depopulation" will be the new demographic focus as societies fail to produce enough children to replace their current populations—even in places where fertility rates are still very high. At the same time, these societies will be becoming greyer as older segments of the population expand. Technological change will work to counteract these problems even as it extends life expectancy considerably. An artificial intelligence-based Cybernetic Revolution by the mid-twenty-first century will bring radical change to economic structures and incentives. Capitalism will gradually become less consumer and profit oriented. Sustainable, long-term development will be prioritized instead. Resource acquisition will be less characterized by intense competition. Global warming will be managed through a reliance on nuclear, hydrogen, and thermonuclear energy sources. Hegemonic-centric world order (whether presided over by the USA or a successor) will give way to a more integrated, supranational coordination of problem-solving.

Earlier, I noted that contemplation of the future tends to fall on a continuum ranging from relative utopian to sheer dystopian. The Moscow Club foresees movement toward the relative utopian end of the continuum. They recognize such a transition will not be easy. The transition will be uneven and it will be resisted, but in their telling demographic imperatives and technological change will win out eventually—even if a new global system is not seen as fully emerging until as late as 2130. The limits to growth will be supplanted, or at least coordinated, by management and supranational governance.

None of the people reading this are likely to be around to test the accuracy of these predictions. The authors are safe in that regard. However, the main point of forecasting is not whether we anticipate how the future will look exactly. The main point is to project trends in play now forward so that we do not simply wait to see what "fate" has in store for us. What are the alternative paths to the future? How can we tell which trajectory is most likely? How can we facilitate the preferred paths and limit the non-preferred tracks? One thing is clear. Demographic dynamics and technological change, however powerful shapers they are likely to be, will not suffice to transform the world. Human agency will need to help the movement toward more utopian and less dystopian choices. Forecasts, like the one provided by the Moscow Club, are needed to inform us about possibilities and probabilities. We will not make it through any fundamental transformation unscathed, but at least we will have a better sense of what might lead to what along the way.

Indiana University
Bloomington, IN, USA
March 2023

William R. Thompson

Acknowledgments

First and foremost, we would like to express our gratitude to Co-Presidents of the Club of Rome Sandrine Dixson-Declève and Mamphela Ramphele, its Vice-President Carlos Alvarez Pereira, and Executive Committee, as well as three anonymous referees whom they commissioned to write reviews of the first draft of our report. All their recommendations have been taken into account, which helped us very much with the preparation of its final version.

Our depest gratitude also goes to the leaders of the Russian Association of the Club of Rome Sergej Kamionsky and Alexander Ageev for their continuous support.

A very important role was also played by the discussions of the preliminary versions of our report that were held in Lomonosov Moscow State University. We would like to thank our university colleagues Dmitry Kavtaradze (Senior Research Professor of the Department of General Biology and Hydrobiology), Valery Petrosyan (Professor of the Department of Organic Chemestry), Andrey Smurov (Director of the Earth Science Museum), and Valery Snakin (Sector Head of the Earth Science Museum, Head of the Laboratory for Landscape Ecology at the Institute of Basic Biological Problems of the Russian Academy of Sciences). Their invaluable comments helped us very much in our work on the present report.

This volume is an output of a research project implemented with support by the Russian Science Foundation (Project No. 20-61-46004).

About This Book

One of the landmark scientific works that gave impetus to understanding the changed trends was the famous first report to the Club of Rome *The Limits to Growth* (by Donella and Dennis Meadows, Jorgen Randers, and William Behrens III) which has played a prominent role in understanding the path of humanity towards the future. During the past fifty years both the concept and understanding of these limits have significantly changed. We believe that at present the evolution of the World System has approached a new critical milestone. The World System and humanity as a whole are currently moving into a fundamentally new phase of historical development, when the old economic and social technologies no longer work as efficiently as before or even begin to function counterproductively, which leads the World System into a systemic crisis. We believe that there is a transition of human society to a new phase state, the shape of which has not yet been determined. Therefore, we believe that new approaches are needed both to the analysis of the global situation and to forecasts.

The main goals of the book are:

- To give an analysis of the changes with which the World System has come to the present, based on an integrated approach that includes the world-systems, historical and evolutionary approaches, mathematical modeling, and a systematic view of society, in which changes in one subsystem cause transformations in others
- To define the main vectors of transformations of the World System
- To make a detailed forecast of the development of all the main subsystems of society and the World System, while presenting three or four horizons of changes (from short-term to ultra-long-term up to 100 years)
- To present different development scenarios and make recommendations on how to switch to the most favorable development scenario

The title of our report, *Reconsidering the Limits to Growth*, certainly echoes the famous *The Limits to Growth*. This is no coincidence, as 2022 marked the fiftieth anniversary of this report, which continues to influence minds. However, in this report, the concept of limits is understood more broadly than usual, implying situations when such limits become less important. To put it more precisely, those limits are serious but solvable challenges which can and should be solved.

Keywords: global futures, forecasting, World System, modeling, global aging, global warming, energy transition, Africa, environment, technological revolution, evolution, historical trends, social change.

Contents

Introduction: Hoping for the Future 1
Victor Sadovnichy, Askar Akaev, Ilya Ilyin, Sergey Malkov,
Leonid Grinin, Yury Sayamov, and Andrey Korotayev

Part I Looking into the Past and the Future

Macrohistorical Approach 17
Victor Sadovnichy, Askar Akaev, Ilya Ilyin, Sergey Malkov,
Leonid Grinin, Ivan Aleshkovski, and Andrey Korotayev

The Future Society and the Transition to It 35
Viktor Sadovnichy, Askar Akaev, Ilya Ilyin, Sergey Malkov,
Leonid Grinin, Aleksey Andreev, and Andrey Korotayev

Part II Problems, Forecasts, Solutions (Climate, Ecology, Demography, Aging)

Climate and Energy: Energy Transition Scenarios and Global Temperature Changes Based on Current Technologies and Trends 53
Askar Akaev and Olga Davydova

Ecology: Life in the "Unstable Biosphere" 71
Natalia Kovaleva

Demography: Toward Optimization of Demographic Processes 97
Andrey Korotayev, Sergey Malkov, and Jameelah Musieva

Global Aging: An Integral Problem of the Future. How to Turn a Problem into a Development Driver? 117
Leonid Grinin, Anton Grinin, and Andrey Korotayev

Part III Problems, Forecasts, Solutions (Technology, Economics, Socio-Political Development)

Technologies: Limitless Possibilities and Effective Control 139
Leonid Grinin and Anton Grinin

Economics: Optimizing Growth 155
Leonid Grinin, Anton Grinin, and Sergey Malkov

Sociopolitical Transformations: A Difficult Path to Cybernetic Society 169
Leonid Grinin, Anton Grinin, and Sergey Malkov

Future Political Change. Toward a more Efficient World Order 191
Leonid Grinin, Anton Grinin, and Andrey Korotayev

High-Income and Low-Income Countries. Toward a Common Goal at Different Speeds 207
Leonid Grinin, Sergey Malkov, and Andrey Korotayev

Africa: The Continent of the Future. Challenges and Opportunities 225
Leonid Grinin and Andrey Korotayev

Part IV Modeling Social Self-Organization and Historical Dynamics – from Agrarian to Cybernetic W-Society

Modeling Social Self-Organization and Historical Dynamics. An Overview 241
Victor Sadovnichy, Askar Akaev, Ilya Ilyin, Sergey Malkov, Leonid Grinin, Vasily Popov, and Andrey Korotayev

Modeling Social Self-Organization and Historical Dynamics. A General Approach 253
Askar Akaev, Sergey Malkov, Stanislav Bilyuga, Artemy Malkov, Jameelah Musieva, and Andrey Korotayev

Modeling Social Self-Organization and Historical Dynamics. Agrarian Society 309
Sergey Malkov, Natalia Kovaleva, Leonid Grinin, and Andrey Korotayev

Modeling Social Self-Organization and Historical Dynamics. Industrial Society 337
Askar Akaev, Sergey Malkov, Olga Davydova, Natalia Kovaleva, Artemy Malkov, Leonid Grinin, and Andrey Korotayev

Modeling Social Self-organization and Historical Dynamics: Global Phase Transitions 387
Sergey Malkov, Leonid Grinin, Anton Grinin, Jameelah Musieva, and Andrey Korotayev

Modeling Social Self-Organization and Historical Dynamics: Modern Society and a Look into the Global Future. Cybernetic W-Society 419
Askar Akaev, Sergey Malkov, Leonid Grinin, Stanislav Bilyuga, Olga Davydova, Anton Grinin, Natalia Kovaleva, Artemy Malkov, Jameelah Musieva, and Andrey Korotayev

Modeling Social Self-Organization and Historical Dynamics: Africa's Futures 461
Andrey Korotayev, Sergey Shulgin, Vadim Ustyuzhanin, Julia Zinkina, and Leonid Grinin

Analyzing Social Self-Organization and Historical Dynamics. Future Cybernetic W-Society: Sociopolitical Aspects 491
Leonid Grinin and Anton Grinin

Analyzing Social Self-Organization and Historical Dynamics: Life Quality Index 521
Sergey Malkov, Stanislav Bilyuga, and Jameelah Musieva

Conclusion: Reconsidering the Limits – Suggestions (Come On!) 531
Victor Sadovnichy, Askar Akaev, Ilya Ilyin, Sergey Malkov, Leonid Grinin, Dmitry Ivanov, and Andrey Korotayev

Introduction: Hoping for the Future

Victor Sadovnichy, Askar Akaev, Ilya Ilyin, Sergey Malkov, Leonid Grinin, Yury Sayamov, and Andrey Korotayev

Abstract

In this introductory chapter, Sadovnichy et al. offer a preliminary analysis of the previous reports to the Club of Rome. This report is the result of more than 10 years of work on modeling and forecasting world dynamics and it reflects the views of Russian scientists on the future of global development. The main goals of the book are to do a preliminary work on the following tasks: (1) To give an analysis of changes through which the World System has come to its present state, based on an integrated approach (that incorporates the world-systems,

V. Sadovnichy · I. Ilyin · Y. Sayamov
Lomonosov Moscow State University, Moscow, Russia
e-mail: info@rector.msu.ru; dekanat@fgp.msu.ru

A. Akaev
Lomonosov Moscow State University, Moscow, Russia

HSE University, Moscow, Russia

S. Malkov
Lomonosov Moscow State University, Moscow, Russia

HSE University, Moscow, Russia

Institute of Economics, Russian Academy of Sciences, Moscow, Russia
e-mail: s@malkov.org

L. Grinin
HSE University, Moscow, Russia

Institute of Oriental Studies, Russian Academy of Sciences, Moscow, Russia

A. Korotayev (✉)
Lomonosov Moscow State University, Moscow, Russia

HSE University, Moscow, Russia

Institute for African Studies, Russian Academy of Sciences, Moscow, Russia

V. Sadovnichy et al. (eds.), *Reconsidering the Limits to Growth*, World-Systems Evolution and Global Futures, https://doi.org/10.1007/978-3-031-34999-7_1

historical and evolutionary approaches), on mathematical modeling, as well as on a systematic view of society, in which changes in one subsystem cause transformations in others; (2) To define the main vectors of transformations of the World System; (3) To make a detailed forecast of the development of all the main subsystems of society and the World System, while presenting three or four horizons of changes (from short-term to ultra-long-term up to a hundred years); (4) To present different development scenarios and make recommendations on how to switch to the most favorable development scenario. Since the publication in 1972 of the first report "The Limits to Growth" (by Donella and Dennis Meadows, Jorgen Randers and William Behrens III), which first drew attention of the world community to the problem of limitations to the possibilities of a secure world development, the views of the Club of Rome experts on the factors that negatively affect the dynamics and nature of global development have significantly evolved. If in 1972 they spoke about the exponential growth of the world population as the main factor in depletion of natural resources, then in later reports, amid stabilization of global demographic growth, this problem started to be explained by the imperfection of the capitalist economy and the inability of modern states to make decisions on self-limitation in favor of long-term results. This introductory chapter analyzes the evolution of views of the Club of Rome experts that was influenced by noticeable transformations taking place in the world, which determines the need for careful reflection and work on forecasts for future development. So, during the past 50 years, both the concept and understanding of these limits have significantly changed. Sadovnichy et al. believe that, at present, the evolution of the World System has reached a new critical milestone. The World System and humanity as a whole are currently moving into a fundamentally new phase of historical development when the old economic and social technologies no longer work as efficiently as before or even begin to function counterproductively, which leads the World System into a systemic crisis. There is a transition of human society to a new phase, the shape of which has not yet been determined. Therefore, Sadovnichy et al. insist that new approaches are needed both to the analysis of the global situation and to forecasts. The title of this report to the Club of Rome, "Reconsidering the Limits", certainly echoes the famous "The Limits to Growth". We would like to state clearly that our title should not give a misleading impression of where the thinking of the Club of Rome is today, especially in connection to the 50th anniversary of the famous report. What is more, we entirely recognize the presence of objective limits to growth. However, these limits are different in different periods, and besides the borders of limits are not frozen. So they might be expanded significantly as a result of special planning and efforts. In some respect, our report can be regarded as preliminary research attempting to understand how we can expand the limits in the future. This is no coincidence, as 2022 marked the fiftieth anniversary of this report, which continues to influence minds. However, in this report, the concept of limits is understood more broadly than usual, implying situations when such limits become less important. This introductory chapter spells out the methodological assumptions of the report. Particular attention is paid to the point that the changes taking place in the world are fundamental and long-term; therefore, they should be considered in a broad historical context (and not just as a continuation of the trends of the twentieth century). Special attention is paid to the explicit consideration of social and political factors and their influence on other factors, which makes it possible to consider the adaptability of the social system to external changes (sociopolitical factors are considered as key to ensuring adaptability). Attention is drawn to the adaptability of the social system to external changes, which is why the study is also aimed at modeling the basic processes of interaction between various spheres of life at the respective phases of historical development. The chapter ends with a description of how the present volume is organized.

One of the cornerstone scientific works that gave impetus to understanding changed trends was the famous first report to the Club of Rome *The Limits to Growth* (by Donella and Dennis Meadows, Jorgen Randers and William Behrens III), which played a prominent role in understanding the path of humanity toward the future. This report is titled *Reconsidering the Limits to Growth*, which is a transparent reference to the aforementioned famous report which in 2022 marked its fiftieth anniversary. *The Limits to Growth* immediately caught worldwide attention. It was a pioneering work to pose the problem of limited opportunities for global demographic and economic growth, which proceeded at extremely high rates in the middle of the twentieth century. Of particular significance is the fact that the analysis of future problems was carried out on the basis of the results of forecast calculations carried out using the World3 mathematical model. This model, specifically created for the task, describes the long-term dynamics of global indicators such as the world population, the level of industrial production, the amount of nonrenewable natural resources, industrial environmental pollution, etc. Forecast calculations showed that if humankind had proceeded to develop inertially, a catastrophe would have inevitably occurred due to resource depletion, aggravation of environmental problems, food shortages, etc.

The results of the study got widely recognized; ecology, climate, and other topics became subjects of active discussion both at the global and national governments. Gradually, these topics began to significantly influence political and economic decision-making.

In the subsequent 50 years, several dozen reports to the Club of Rome have been published on various topics related to the problems of world development. Let us briefly describe the most significant ones, such as those issued for the anniversaries of the formation of the Club of Rome, or for the anniversaries of "The Limits to Growth" publication. This short review will show how the logic of the reports has gradually transformed.

1972, "The Limits to Growth" report (Meadows et al., 1972) states that the exponential growth of demographic and economic characteristics observed in the world since the beginning of the twentieth century is a natural and regular process. However, in the second half of the twentieth century, this growth would become a problem as it would lead to the inevitable depletion of resources and a global crisis in the 2020s. Therefore, humankind is facing the most important task of stopping population growth if it wants to avoid the looming catastrophe.

1979, "No limits to learning" report (Botkin et al., 1979): the authors dwell on the need to revise the current teaching methods, which means the formation of a new approach to life that should not be limited to the traditional scheme of secondary and higher education. The gap between the complexity of the world and the ability of humans to navigate in the new conditions of life leads to those difficulties of humankind, which are the subject of special attention on the part of the Club of Rome. In order to successfully solve global problems and eliminate crisis situations, it is necessary, according to the authors of the report, to radically revise existing teaching methods. It is important to introduce new forms of education based on innovative learning that orients a person to foresight and actively participate in creating her or his future and, thus, a better future for humanity.

1989, "Beyond the Limits to Growth" report (Pestel, 1989) calls to think systematically, to move from unmanaged quantitative to "organic" growth and development, and to create a world government (since nation-states are unable to coordinate and solve long-term tasks collectively).

2004, "Limits to Growth: The 30-Year Update" report (Meadows et al., 2004) considers the second phase of the demographic transition (declining birth rates in high-income countries), but states that exponential demographic growth continues regardless because mortality decreases along with decreasing fertility (accordingly, the gap between fertility and mortality persists). Social block is lacking in the World3 model, and all macro-characteristics are averaged (disregarding the large social stratification and regional differences in the world). Therefore, the World3 model produces overly optimistic forecasts. To avoid unfavorable paths of development, "new values" are needed, according to the authors of the report: "Individualism and shortsightedness are the greatest problems of the current social system, we think, and the deepest cause of unsustainability. Love and compassion institutionalized in collective solutions is the better alternative" (Meadows et al., 2004: 281). However, the report does not elaborate how exactly "individualism and shortsightedness" should be replaced with "love and participation" in people's minds, and only expresses hopes that this is possible.

2012, "2052: A Global Forecast for the Next Forty Years" (Randers, 2012) contains a conclusion about the imminent deceleration of population growth (below eight billion people) and a further drop in the population of the Earth—not due to an increase in mortality (as "The Limits to Growth" forecasted), but rather due to a change in the family model (transition from a large family model to a model of family with few or no children). At the same time, despite the deceleration of demographic growth, "going beyond" is in full swing due to the ongoing increase in world GDP. The reasons for this include, among other things, the shortsightedness of governments and businesses (for which economic growth is an end goal in itself), and the persistence of poverty due to uneven income distribution. It is concluded that capitalism as a social system is to blame, but the question "what should be done?" remains unanswered.

2018, "Come On! Capitalism, Short-termism, Population and the Destruction of the Planet" (von Weizsaecker et al., 2018) is dedicated to the half-century anniversary of the Club of Rome. This report states that the world has changed dramatically, has become closed, capitalism has ceased to be effective.[1] It proposes to change the value system and the system of education.[2]

2018, "Transformation is feasible" (Randers et al., 2018) forecasts that fertility will decline and population will stabilize, but at the same time no possible option for global dynamics can be satisfactory in the absence of changes in the social system. The necessary changes in the social system are discussed (redistribution of income from the super-rich to the poor, long-term decision-making, rejection of "democracy" focused on short-term conditions), while stating that capitalism is not capable of this. It is stated that it is necessary to change the World3 model to the World4 model, where a social block will be present (in the World3 model, it was lacking).

2021, "Learning New Ways of Becoming Human" (Álvarez Pereira, 2021) criticizes capitalism and individualism that hinder the transition to global sustainable development. This report regards the need to abandon Eurocentrism: "we have to go beyond the supposed 'universality' of our own perspective to accept that we can learn from other civilizational paradigms (from Africa, Asia, indigenous cultures)" (Álvarez Pereira, 2021: 15).

Thus, the views of the Club of Rome experts have seriously evolved during the 50 years that have passed since the publication of "The Limits to Growth". Initially (Meadows et al., 1972), the main problem was seen in the exponential growth of the world population and the inevitable depletion of natural resources related to it (therefore, the main recommendation was to urgently bring down fertility rates). Later (Meadows et al., 2004), the demographic transition and deceleration of demographic growth (Randers, 2012) were the focus of attention. The forecasts remained pessimistic, but rather due to disbelief in the ability of capitalism and parliamentary democracies to make decisions that require self-restraint and are aimed at achieving long-term (rather than momentary) results (von Weizsäcker et al., 2018; Randers et al., 2018). That is, the general more or less pessimistic tone of long-term forecasts has persisted, but while at first (50 years ago) unstoppable demographic growth was seen as the main negative factor, now the focus has shifted to the imperfection of the socio-economic and political system in most countries of the world, the inability of governments to solve long-term problems.

The transformation of the tone and logic of reports to the Club of Rome over the past 50 years reflects the epoch-making changes taking place in the world. These changes need to be thoroughly thought about in order to identify their underlying causes; they require an answer to the question: what should we expect in the future?

Our report presents some considerations about the changes taking place in the world, based both on the results of these works and numerous other works of scholars from different countries as well as more than 10 years of combined work of our collective.[3] We would like to emphasize that in our report, we spell out our

[1] For our view on the future of capitalism see Grinin et al. (2022).

[2] For our analysis of this report see Grinin and Grinin (2021).

[3] Under the guidance of Viktor Sadovnichy and Askar Akaev at the Lomonosov Moscow State University has been modeling and forecasting world dynamics (see, e.g., Sadovnichy et al., 2012,

own view not only on the mainstream subjects and problems such as climate, ecology, demography, inequality, etc., but also on those issues that do not receive fair attention being nevertheless extremely important—such as aging, different pathways of high income and low-income countries toward the common human goals, specific problems and ways of development in African societies, a vision of future society and different scenarios of transition to it. Such approaches also help to overcome the limitation of the Eurocentric view, which is fairly criticized (e.g., Das et al., 2001; Hwang, 2020; Álvarez Pereira, 2021: 15; Ramphele, 2021).

The main goals of the book are to do preliminary work on the following tasks: (1) To give an analysis of changes through which the World System has come to its present state, based on an integrated approach (that incorporates the world systems, historical and evolutionary approaches), on mathematical modeling, as well as on a systematic view of society, in which changes in one subsystem cause transformations in others; (2) To define the main vectors of transformations of the World System; (3) To make forecasts of the development of all the main subsystems of the society and the World System, while presenting three or four horizons of changes (from short-term to ultra-long-term up to 100 years); (4) To present different development scenarios and make recommendations on how to switch to the most favorable development scenario.

Our research is based on the following methodological assumptions:

- we use a wide range of methodological and research approaches, including the system-dynamic approach. We consider a variety of drivers and factors that influence and may influence the development of the situation in the present and future, including such bottom-up drivers as mindshifts and awareness. However, we proceed from the fact that the strength of different factors manifests itself differently at different time distances. And if mindshifts and awareness are very influential in the short-term period, then in the long-term period, the role of such factors as technology, demographic trends or climate increases.

We have tried to consider different spheres of society's life, along with changes in them and to make forecasts in three projections: short-term, medium-term and long-term. Each of the projections gives its own picture. In some cases, we pay special attention to long-term trends for the following reasons: (a) they often receive insufficient attention; (b) they are more suitable for mathematical modeling; (c) it is at such long-term trends that the places of trend reversals are more clearly visible, which are more difficult to notice at short-term distances.

- to understand a number of trends, the analysis of technological development is of particular importance, especially in the long run. At the same time, technological development proceeds unevenly, with leaps (taking the form of "technological

---

2014, 2017; references to research work of other members of this group are provided in corresponding chapters; see also Grinin et al., 2017a, b, 2020, 2021a, b).

revolutions") occurring from time to time, which contributes to the unevenness of the historical process (see Grinin, 2007; Grinin & Grinin, 2015, 2016; Grinin & Korotayev, 2015; Korotayev, 2020);

- the most important requirement for research is to use an integrated approach, that is, to take into account the mutual influence of all the most important factors (ecology, climate, technology, demography, economy, social sphere, politics) that affect macrohistorical development. At the same time, we take into account the fact that different regions of the world develop at different rates; thus, approaches to solving modern problems depend on that;
- special attention is paid to the consideration of social and political factors and their influence on other factors, which makes it possible to recognize the adaptability of the social system to external changes (sociopolitical factors are considered as key to ensuring adaptability). Adaptations get reinforced at the level of social institutions. Institutions that prove effective over long-time horizons form civilizations as sustainable modes of social self-organization that have passed the test of time;
- in accordance with that, the objective of research and modeling is the basic processes (including social and political ones) that determine features of interaction of various spheres of life at their respective phases of historical development. Accordingly, modeling is carried out at a high level of aggregation; the purpose of modeling is not so much to determine the specific values of variables as to reveal the logic of their long-term dynamics. It is reasonable to present the results of modeling not in the form of graphs that reflect the change in the values of certain variables over time within specific scenario conditions, but rather with the help of *phase portraits*. Analysis of the *phase portraits* makes it possible to identify "order parameters" (i.e., key factors on which the type of phase portrait depends) and characteristics of the sustainability of economic and social development at the respective phases of historical development (based on analysis of the conditions under which a radical transformation of the phase portrait occurs).

In our report, the concept of "limits" is presented in a broad sense, sometimes, they should be interpreted as powerful challenges to which, however, answers can be given that can reduce their negative effect.[4] The limits that, in our opinion, can be reconsidered are indicated in the concluding chapter of the report.[5]

The structure of our report is as follows. It consists of an Introduction, three parts, Conclusion, and Mathematical Appendices. The names of these parts speak for themselves. Part I ("Looking into the Past and the Future") contains two chapters,

[4]Especially, if there is a serious ideological breakthrough (see, e.g., Álvarez Pereira, 2016). As is noted by Rodolfo A. Fiorini, Carlos Álvarez Pereira et al. (2020: 56), "individuals have within themselves vast resources for self-understanding and for changing their self-conceptions, basic attitudes and self-directed behavior, and these resources can be tapped if a climate of facilitative psychological conditions is provided".

[5]See chapter "Conclusion. Reconsidering The Limits. Suggestions (Come on!)" (Sadovnichy et al., 2023c, this volume).

which outline the general approach of our team. Chapter "Macrohistorical Approach" (Sadovnichy et al., 2023a, this volume) provides the macrohistorical background, whereas the other chapter of that part presents some assessments of modern processes arising against that background, as well as some forecasts [see chapter "Future Society and Transition to It" (Sadovnichy et al., 2023b, this volume)]. The main content of the report is presented in Parts II and III, which together consist of 10 chapters. These chapters cover all the main subsystems of modern and future society.

Part II—"Problems, Forecasts, Solutions (Climate, Ecology, Demography, Aging)"—consists of four chapters. The first two chapters are devoted to the extremely acute problem of climate and ecology conservation. The energy policies of states and the world as a whole are now inextricably linked with the problem of climate and global warming. Energy aspects are analyzed in chapter "Climate and Energy. Scenarios of Energy Transition and Global Temperature Changes Based on Current Technologies and Trends" (Akaev & Davydova, 2023, this volume). The mathematical model underlying our scenarios for energy transition and changes in global temperature is described in the mathematical appendix to this report in chapter "Modeling Social Self-Organization and Historical Dynamics. A General Approach" (Akaev et al., 2023a). Pressing ecological problems of our planet are discussed in chapter "Ecology. Life in the "Unstable Biosphere"" (Kovaleva, 2023, this volume).

Chapters "Demography. Toward Optimization of Demographic Processes" (Korotayev et al., 2023a, this volume) and "Global Aging – an Integral Problem of the Future. How to Turn a Problem into a Development Driver?" (Grinin et al., 2023d, this volume) regard the problems of demography in general, as well as aging, perhaps the most important process of the present and the future that can have a tremendous impact on all aspects of human life.

Part III "Problems, Forecasts, Solutions (Technology, Economics, Sociopolitical Development)" consists of six chapters devoted to the analysis of modern and future technologies, forecasts of economic development, as well as sociopolitical development, including an analysis of the possible impact that aging can have on the sociopolitical structure of the future society.[6] We would like to mention again that of particular importance are the chapters devoted to the analysis of differences in the development of advanced and catching-up countries,[7] with a particular focus on

[6]See chapter "Technology. Limitless Possibilities, Effective Control" (Grinin & Grinin, 2023b, this volume), chapter "Economics. Optimizing Growth" (Grinin et al., 2023c, this volume), chapter "Socio-Political Transformations. A Difficult Path to Cybernetic Society" (Grinin et al., 2023e, this volume), chapter "Future Political Change. Toward a More Efficient World Order" (Grinin et al., 2023b, this volume), chapter "High Income and Low Income Countries. Toward a Common Goal at Different Speeds" (Grinin et al., 2023a, this volume), and chapter "Africa – the Continent of the Future. Challenges and Opportunities" (Grinin & Korotayev, 2023, this volume).

[7]Chapter "High Income and Low Income Countries. Toward a Common Goal at Different Speeds" (Grinin et al., 2023a, this volume).

African countries.[8] This gave us the opportunity to refine forecasts and recommendations on optimizing the development of different groups of countries in terms of their economy, demography, and attitude to climate.

We aimed to title the chapters in Parts II and III in a "self-explanatory" way so that the main idea could already be obvious from the title. But these are chapters of different authors, and sometimes different authors offer different scenarios for the development of society in the future, since, obviously, the future is not predetermined, it depends on many factors, including our own efforts.

In chapter "Conclusion. Reconsidering The Limits. Suggestions (Come on!)" (Sadovnichy et al., 2023c, this volume) we sum up the limits (challenges) that we are talking about, briefly summarize how the humankind might be able to reconsider them. We focus on the fundamental idea that the more actively we fight for the future, the more likely it will be closer to our ideal, although, of course, we must simultaneously remain realistic.

There are also eight methodological/mathematical chapters united under common title "Modeling Social Self-Organization and Historical Dynamics"[9] and an introductory chapter to them.[10] They cover certain issues in more detail and show the mathematical tools that we use. They demonstrate how we see combining of the world-systems, historical and evolutionary approaches, a systematic view of society, and mathematical modeling within a single research program.

References

Akaev, A., & Davydova, O. (2023). Climate and energy. Scenarios of energy transition and global temperature changes based on current technologies and trends. In V. Sadovnichy et al. (Eds.), *Reconsidering the limits to growth. A report to the Russian Association of the Club of Rome* (pp. 53–70). Springer. https://doi.org/10.1007/978-3-031-34999-7_4

[8] Chapter "Africa – the Continent of the Future. Challenges and Opportunities" (Grinin & Korotayev, 2023, this volume).

[9] See chapter "Modeling Social Self-Organization and Historical Dynamics. A General Approach" (Akaev et al., 2023a, this volume), chapter "Modeling Social Self-Organization and Historical Dynamics. Agrarian Society" (Malkov et al., 2023c, this volume), chapter "Modeling Social Self-Organization and Historical Dynamics. Industrial Society" (Akaev et al., 2023b, this volume), chapter "Modeling Social Self-Organization and Historical Dynamics. Global Phase Transitions" (Malkov et al., 2023b, this volume), chapter "Modeling Social Self-Organization and Historical Dynamics. Modern Society and Problems of Global Transition" (Akaev et al., 2023d, this volume), chapter "Modeling Social Self-Organization and Historical Dynamics. Modern Society and a Look into the Global Futures: Cybernetic W-Society" (Akaev et al., 2023c, this volume), chapter "Modeling Social Self-Organization and Historical Dynamics. Africa's Futures" (Korotayev et al., 2023b, this volume), chapter "Analyzing Social Self-Organization and Historical Dynamics. Future Cybernetic W-Society: Socio-Political Aspect" (Grinin & Grinin, 2023a, this volume), and chapter "Modeling Social Self-Organization and Historical Dynamics. Life Quality Index" (Malkov et al., 2023a, this volume).

[10] See chapter "Modeling Social Self-Organization and Historical Dynamics. An Overview" (Sadovnichy et al., 2023d, this volume).

Akaev, A., Malkov, S., Bilyuga, S., Malkov, A., Musieva, J., & Korotayev, A. (2023a). Modeling social self-organization and historical dynamics. A general approach. In V. Sadovnichy et al. (Eds.), *Reconsidering the limits to growth. A report to the Russian Association of the Club of Rome* (pp. 253–307). Springer. https://doi.org/10.1007/978-3-031-34999-7_15

Akaev, A., Malkov, S., Davydova, O., Kovaleva, N., Malkov, A., Grinin, L., & Korotayev, A. (2023b). Modeling social self-organization and historical dynamics. Industrial society. In V. Sadovnichy et al. (Eds.), *Reconsidering the limits to growth. A report to the Russian Association of the Club of Rome* (pp. 337–385). Springer. https://doi.org/10.1007/978-3-031-34999-7_17

Akaev, A., Malkov, S., Grinin, L., Bilyuga, S., Davydova, O., Grinin, A., Kovaleva, N., Malkov, A., Musieva, J., & Korotayev, A. (2023c). Modeling social self-organization and historical dynamics. Modern society and a look into the global futures: Cybernetic W-society. In V. Sadovnichy et al. (Eds.), *Reconsidering the limits to growth. A report to the Russian Association of the Club of Rome* (pp. 419–460). Springer. https://doi.org/10.1007/978-3-031-34999-7_19

Akaev, A., Malkov, S., Grinin, L., Bilyuga, S., Davydova, O., Grinin, A., Kovaleva, N., Malkov, A., Musieva, J., & Korotayev, A. (2023d). Modeling social self-organization and historical dynamics. Modern society and problems of global transition. In V. Sadovnichy et al. (Eds.), *Overcoming the limits to growth. A report to the Club of Rome* (pp. 461–490). Springer. https://doi.org/10.1007/978-3-031-34999-7_20

Álvarez Pereira, C. (2016). Towards a society of living: Provocations on economy and economics by a layman and entrepreneur. *Eruditio, e-Journal of the World Academy of Art & Science, 2*(2), 72–101.

Álvarez Pereira, C. (2021). *Learning new ways of becoming human. A report to The Club of Rome*. The Club of Rome.

Botkin, J., Elmandjra, M., & Malitza, M. (1979). *No limits to learning*. Pergamon Press.

Das, V., Kleinman, A., Lock, M. M., Ramphele, M., & Reynolds, P. (Eds.). (2001). *Remaking a world: Violence, social suffering, and recovery*. University of California Press.

Fiorini, R. A., Álvarez Pereira, C., Jacobs, G., Kiniger-Passigli, D., Zucconi, A., Nešković, N., & Cocchiarella, L. (2020). Global transformative leadership in the 21st century: A science, engineering, technology integrated and strategic perspective. *Cadmus, 4*(2), 56–87.

Grinin, L. (2007). Production revolutions and periodization of history: A comparative and theoretic-mathematical approach. *Social Evolution & History, 6*(2), 11–55.

Grinin, L., & Grinin, A. (2015). *Ot rubil do nanorobotov. Mir na puti k epokhe samoupravlyayemykh sistem (istoriya tekhnologiy i opisaniye ikh budushchego)*. Uchitel.

Grinin, L., & Grinin, A. (2016). *The cybernetic revolution and the forthcoming epoch of self-regulating systems*. Uchitel.

Grinin, A., & Grinin, L. (2021). Analyzing the global problems of the twenty-first century. A review and forecast based on the report to the Club of Rome 'come on!'. *Journal of Globalization Studies, 12*(2), 181–195. https://doi.org/10.30884/jogs/2021.02.10

Grinin, L., & Grinin, A. (2023a). Analyzing social self-organization and historical dynamics. Future cybernetic W-society: Sociopolitical aspects. In V. Sadovnichy et al. (Eds.), *Reconsidering the limits to growth. A report to the Russian Association of the Club of Rome* (pp. 491–519). Springer. https://doi.org/10.1007/978-3-031-34999-7_21

Grinin, L., & Grinin, A. (2023b). Technology. Limitless possibilities, effective control. In V. Sadovnichy et al. (Eds.), *Reconsidering the limits to growth. A report to the Russian Association of the Club of Rome* (pp. 139–154). Springer. https://doi.org/10.1007/978-3-031-34999-7_8

Grinin, L., & Korotayev, A. (2015). *Great divergence and great convergence. A global perspective*. Springer. https://doi.org/10.1007/978-3-319-17780-9

Grinin, L., & Korotayev, A. (2023). Africa – The continent of the future. Challenges and opportunities. In V. Sadovnichy et al. (Eds.), *Reconsidering the limits to growth. A report to the Russian Association of the Club of Rome* (pp. 225–238). Springer. https://doi.org/10.1007/978-3-031-34999-7_13

Grinin, L., Grinin, A., & Korotayev, A. (2017a). The MANBRIC-Technologies in the Forthcoming Technological Revolution. In T. Devezas, J. Leitão, & A. Sarygulov (Eds.), *Industry 4.0 – Entrepreneurship and structural change in the new digital landscape: What is coming on along with the fourth industrial revolution* (pp. 243–261). Springer. https://doi.org/10.1007/978-3-319-49604-7_13

Grinin, L., Grinin, A., & Korotayev, A. (2017b). Forthcoming Kondratieff wave, cybernetic revolution, and global ageing. *Technological Forecasting and Social Change, 115*, 52–68. https://doi.org/10.1016/j.techfore.2016.09.017

Grinin, L., Grinin, A., & Korotayev, A. (2020). A quantitative analysis of worldwide long-term technology growth: From 40,000 BCE to the early 22nd century. *Technological Forecasting and Social Change, 155*, 1–15. https://doi.org/10.1016/j.techfore.2020.119955

Grinin, L., Grinin, A., & Korotayev, A. (2021a). COVID-19 pandemic as a trigger for the acceleration of the cybernetic revolution, transition from E-government to E-state, and change in social relations. *Technological Forecasting and Social Change, 175*, 1–17. https://doi.org/10.1016/j.techfore.2021.121348

Grinin, L., Grinin, A., & Korotayev, A. (2021b). Global trends and forecasts of the 21st century. *World Futures, 77*(5), 335–370. https://doi.org/10.1080/02604027.2021.1949939

Grinin, L., Malkov, S., Grinin, A., & Korotayev, A. (2022). Will capitalism die? Reflections on the capitalism of the past, present and future. *Sotsiologicheskiy Zhurnal = Sociological Journal, 28*(2), 100–130. https://doi.org/10.19181/socjour.2022.28.2.8988

Grinin, L., Malkov, S., & Korotayev, A. (2023a). High income and low income countries. Toward a common goal at different speeds. In V. Sadovnichy et al. (Eds.), *Reconsidering the limits to growth. A report to the Russian Association of the Club of Rome* (pp. 207–224). Springer. https://doi.org/10.1007/978-3-031-34999-7_12

Grinin, L., Grinin, A., & Korotayev, A. (2023b). Future political change. Toward a more efficient world order. In V. Sadovnichy et al. (Eds.), *Reconsidering the limits to growth. A report to the Russian Association of the Club of Rome* (pp. 191–206). Springer. https://doi.org/10.1007/978-3-031-34999-7_11

Grinin, L., Grinin, A., & Malkov, S. (2023c). Economics. Optimizing growth. In V. Sadovnichy et al. (Eds.), *Reconsidering the limits to growth. A report to the Russian Association of the Club of Rome* (pp. 155–168). Springer. https://doi.org/10.1007/978-3-031-349999-7_9

Grinin, L., Grinin, A., & Korotayev, A. (2023d). Global aging – An integral problem of the future. How to turn a problem into a development driver? In V. Sadovnichy et al. (Eds.), *Reconsidering the limits to growth. A report to the Russian Association of the Club of Rome* (pp. 117–135). Springer. https://doi.org/10.1007/978-3-031-34999-7_7

Grinin, L., Grinin, A., & Malkov, S. (2023e). Socio-political transformations. A difficult path to cybernetic society. In V. Sadovnichy et al. (Eds.), *Reconsidering the limits to growth. A report to the Russian Association of the Club of Rome* (pp. 169–189). Springer. https://doi.org/10.1007/978-3-031-34999-7_10

Hwang, K. K. (2020). Escape from Kantian Eurocentric bias in cross-cultural psychology. *Culture & Psychology, 26*(4), 863–878. https://doi.org/10.1177/1354067X20922145

Korotayev, A. (2020). The 21st century singularity in the big history perspective. A re-analysis. In A. Korotayev & D. LePoire (Eds.), *The 21st century singularity and global futures. A big history perspective* (pp. 19–75). Springer. https://doi.org/10.1007/978-3-030-33730-8_2

Korotayev, A., Malkov, S., & Musieva, J. (2023a). Demography. Toward optimization of demographic processes. In V. Sadovnichy et al. (Eds.), *Reconsidering the limits to growth. A report to the Russian Association of the Club of Rome* (pp. 97–116). Springer. https://doi.org/10.1007/978-3-031-34999-7_6

Korotayev, A., Shulgin, S., Ustyuzhanin, V., Zinkina, J., & Grinin, L. (2023b). Modeling social self-organization and historical dynamics. Africa's futures. In V. Sadovnichy et al. (Eds.),

Reconsidering the limits to growth. A report to the Russian Association of the Club of Rome (pp. 461–490). Springer. https://doi.org/10.1007/978-3-031-34999-7_20

Kovaleva, N. (2023). Ecology. Life in the "unstable biosphere". In V. Sadovnichy et al. (Eds.), *Reconsidering the limits to growth. A report to the Russian Association of the Club of Rome* (pp. 71–95). Springer. https://doi.org/10.1007/978-3-031-34999-7_5

Malkov, S., Bilyuga, S., & Musieva, J. (2023a). Analyzing social self-organization and historical dynamics. Life quality index. In V. Sadovnichy et al. (Eds.), *Reconsidering the limits to growth. A report to the Russian Association of the Club of Rome* (pp. 521–530). Springer. https://doi.org/10.1007/978-3-031-34999-7_22

Malkov, S., Grinin, L., Grinin, A., Musieva, J., & Korotayev, A. (2023b). Modeling social self-organization and historical dynamics. Global phase transitions. In V. Sadovnichy et al. (Eds.), *Reconsidering the limits to growth. A report to the Russian Association of the Club of Rome* (pp. 387–417). Springer. https://doi.org/10.1007/978-3-031-34999-7_18

Malkov, S., Kovaleva, N., Grinin, L., & Korotayev, A. (2023c). Modeling social self-organization and historical dynamics. Agrarian society. In V. Sadovnichy et al. (Eds.), *Reconsidering the limits to growth. A report to the Russian Association of the Club of Rome* (pp. 309–335). Springer. https://doi.org/10.1007/978-3-031-34999-7_16

Meadows, D. H., Meadows, D. L., Randers, J., & Behrens, W. W. (1972). *The limits to growth: A report for the Club of Rome's project on the predicament of mankind*. Universe Books.

Meadows, D. H., Randers, J., & Meadows, D. L. (2004). *Limits to growth: The 30-year update. A report to The Club of Rome*. Earthscan.

Pestel, E. (1989). *Beyond the limit to growth. A report to The Club of Rome*. The Club of Rome.

Ramphele, M. (2021). Mamphela Ramphele critiques Johan Fourie's 'A long walk to economic Freedom'. *News24, 19.05.2021*. https://www.news24.com/news24/books/review-mamphela-ramphele-critiques-johan-fouries-a-long-walk-to-economic-freedom-20210519

Randers, J. (2012). *2052. A global forecast for the next forty years. A report to the club of Rome*. Chelsea Green Publishing.

Randers, J., Rockström, J., Stoknes, P. E., Golüke, U., Collste, D., & Cornell, S. (2018). *Transformation is feasible. A report to The Club of Rome*. Stockholm Resilience Centre.

Sadovnichy, V., Akaev, A., Korotayev, A., & Malkov, S. (2012). *Modelirovaniye i prognozirovaniye mirovoy dinamiki*. ISPI RAN.

Sadovnichy, V., Akaev, A., Korotayev, A., & Malkov, S. (2014). *Kompleksnoye modelirovaniye i prognozirovaniye razvitiya stran BRIKS v kontekste mirovoy di-namiki*. Nauka.

Sadovnichy, V., Akaev, A., Korotayev, A., Malkov, S., & Sokolov, V. (2017). *Analiz i modelirovaniye mirovoy i stranovoy dinamiki*. LENAND.

Sadovnichy, V., Akaev, A., Ilyin, I., Malkov, S., Grinin, L., Aleshkovski, I., & Korotayev, A. (2023a). Macrohistorical approach. In V. Sadovnichy et al., (Eds.) *Reconsidering the limits to growth. A report to the Russian Association of the Club of Rome* (pp. 17–33). Springer. https://doi.org/10.1007/978-3-031-34999-7_2

Sadovnichy, V., Akaev, A., Ilyin, I., Malkov, S., Grinin, L., Andreev, A., & Korotayev, A. (2023b). The future society and the transition to it. In V. Sadovnichy et al., (Eds.) *Reconsidering the limits to growth. A report to the Russian Association of the Club of Rome* (pp. 35–50). Springer. https://doi.org/10.1007/978-3-031-34999-7_3

Sadovnichy, V., Akaev, A., Ilyin, I., Malkov, S., Grinin, L., Ivanov, D., & Korotayev, A. (2023c). Conclusion. Reconsidering the limits – Suggestions (come on!). In V. Sadovnichy et al., (Eds.) *Reconsidering the limits to growth. A report to the Russian Association of the Club of Rome* (pp. 531–524). Springer. https://doi.org/10.1007/978-3-031-34999-7_23

Sadovnichy, V., Akaev, A., Ilyin, I., Malkov, S., Grinin, L., Popov, V., & Korotayev, A. (2023d). Modeling social selforganization and historical dynamics. An overview. In V. Sadovnichy et al. (Eds.), *Reconsidering the limits to growth. A report to the Russian Association of the Club of Rome* (pp. 253–307). Springer. https://doi.org/10.1007/978-3-031-34999-7_15

von Weizsäcker, E. U., & Wijkman, A. (2018). *Come on! Capitalism, short-termism, population and the destruction of the planet*. Springer.

Viktor Sadovnichy is a Doctor of Physics and Mathematics, Professor, Full Member of the Russian Academy of Sciences, and Rector of Lomonosov Moscow State University since 1992. Fields of scientific research: mathematics, mechanics, and computer science. Sadovnichy's research deals with the theory of differential operators, functional analysis, and mathematical modeling of long-term social and economic dynamics. He has obtained important theoretical and applied results related to the issues of dynamic simulation of controlled flights, in particular, of spacecrafts and aircrafts. This work formed the basis for the creation of the first simulator of zero-gravity in the Earth conditions in the world.

Askar Akaev is the First President of the Kyrgyz Republic and Foreign Full Member of the Russian Academy of Sciences. He works as Chief Research Professor at the Prigogine Institute for Mathematical Investigations of Complex Systems at Lomonosov Moscow State University. He is also the Academic Supervisor of the Centre for Stability and Risk Analysis at HSE University, Moscow, Russia. Askar Akaev has authored and co-authored over 300 scientific publications, including such monographs as *Coherent Optical Computers* (1977), *Modeling and Forecasting the World Dynamics* (2012), *Quality of Education, Effectiveness of R&D, and Economic Growth. Quantitative Analysis and Mathematical Modeling* (2016), etc.

Ilya Ilyin is a well-known Russian researcher, Head of the Department of Global Studies, Dean of the Faculty of Global Studies at Lomonosov Moscow State University, and author of more than 350 academic papers. He is the Chairman of the Presidium of the Society for Global Studies, Chairman of the Russian Committee and Vice President of the Intergovernmental Council for the UNESCO Program "Management of Social transformations – MOST", honorary worker in the field of state youth policy and education, editor-in-chief of the journal *Bulletin of Moscow University.* Series 27: *Global Studies and Geopolitics*.

Sergey Malkov heads the Center for Long-Term Forecasting and Strategic Planning at Lomonosov Moscow State University; he also works as a Senior Research Professor at the Institute of Economics of the Russian Academy of Sciences and at the Centre for Stability and Risk Analysis of HSE University in Moscow. He is the author of over 300 scholarly publications, including such monographs as *Social self-organization and historical process: possibilities of mathematical modeling* (URSS, 2009) and *The twenty-first Century Singularity and Global Futures. A Big History Perspective* (Springer, 2020, together with David J. LePoire and others).

Leonid Grinin is a Senior Research Professor at the Laboratory for Monitoring of Sociopolitical Destabilization Risks at HSE University in Moscow, as well as a Senior Research Professor of the Institute of Oriental Studies, Russian Academy of Sciences. He is the author of more than 600 publications, including more than 33 monographs in Russian, English, Spanish, German, and Chinese. His current research focuses on comparative political studies, the theory of revolution, political anthropology, global economy, global history, historical sociology, and futures studies.

Yury Sayamov is a Scientist and Diplomat, Ambassador, Doctor of Historical Sciences, Professor and Head of the UNESCO Chair on Global Problems and Emerging Social and Ethical Challenges at the Faculty of Global Studies of the Lomonosov Moscow State University, a Full Member of the Club of Rome, International Global Research Academy, Russian Academy of Natural Sciences, Russian Academy of Ecology, Geopolitical Club of Nice and other academic institutions.

Andrey Korotayev heads the Centre for Stability and Risk Analysis at HSE University in Moscow, Russia. He is also a Senior Research Professor at the Institute of Oriental Studies and Institute for African Studies, Russian Academy of Sciences, as well as at the Faculty of Global Studies of Lomonosov Moscow State University and the Russian Presidential Academy of National Economy and Public Administration. He has authored or co-authored over 650 scholarly publications.

Part I

Looking into the Past and the Future

Macrohistorical Approach

Victor Sadovnichy, Askar Akaev, Ilya Ilyin, Sergey Malkov, Leonid Grinin, Ivan Aleshkovski, and Andrey Korotayev

Abstract

This chapter outlines a methodological approach, on the basis of which the report further analyzes the processes that have been taking place in the world in recent decades. Sadovnichy et al. call this approach macrohistorical. It is based on the notion that the ongoing processes are considered in a broad historical context, covering hundreds and thousands of years. Sadovnichy et al. consider it necessary to avoid "historical myopia" and subjectivity in the interpretation of current events, to identify historical patterns of what is happening and to try to look

V. Sadovnichy · I. Ilyin · I. Aleshkovski
Lomonosov Moscow State University, Moscow, Russia
e-mail: info@rector.msu.ru; dekanat@fgp.msu.ru; aleshkovski@fgp.msu.ru

A. Akaev
Lomonosov Moscow State University, Moscow, Russia

HSE University, Moscow, Russia

S. Malkov
Lomonosov Moscow State University, Moscow, Russia

HSE University, Moscow, Russia

Institute of Economics, Russian Academy of Sciences, Moscow, Russia
e-mail: s@malkov.org

L. Grinin
HSE University, Moscow, Russia

Institute of Oriental Studies, Russian Academy of Sciences, Moscow, Russia

A. Korotayev (✉)
Lomonosov Moscow State University, Moscow, Russia

HSE University, Moscow, Russia

Institute of Oriental Studies, Russian Academy of Sciences, Moscow, Russia

V. Sadovnichy et al. (eds.), *Reconsidering the Limits to Growth*, World-Systems Evolution and Global Futures, https://doi.org/10.1007/978-3-031-34999-7_2

into the future with an unbiased view. Sadovnichy et al. emphasize the need for a comprehensive analysis that takes into account not only the long-term historical trends, but also the interaction of all important factors, driving forces, and spheres (ecology, climate, technology, demography, economy, sociosphere, politics, ideology, culture, etc.) that affect macrohistorical development and try to identify specific influences of each of them in a certain period. It is necessary to take into account macrohistorical and evolutional paradigms, patterns, and trends. Though no factors can be considered absolutely dominant, some spheres can be regarded as more significant with respect to their influence; therefore, changes within them are more likely to affect other spheres rather than the other way around. In this respect, within the framework of this approach, an important role in the history of humankind is played by technological development, which does not occur evenly, but in spurts (taking the form of "technological revolutions"). The introduction of new technologies entails serious economic changes, expressed, in particular, in labor and land productivity growth, and in expansion of the resource base. This, in turn, contributes to population growth. The latter, and especially the increase in population density, makes inevitable changes in the social structure of society, and, consequently, changes in the political sphere. At the same time, economic activity simultaneously affects the external environment (ecology), and that, in turn, can affect nature and climate. The purpose of the macrohistorical approach is to link all these mutual influences into a single picture; its analysis makes it possible to understand the processes taking place in the world and the prospects for their further development. The importance of the historical approach to the analysis of the present and the future means that we should look at the current situation within the framework of not just historical but macrohistorical development. The chapter presents some of the results of such analysis.

1 Macrohistorical Approach: Introduction

In this chapter, we would like to outline a methodological approach, on the basis of which we further analyze the processes that have been taking place in the world in recent decades. The authors call this approach macrohistorical. Its essence lies in the assumption that the ongoing processes are considered in a broad historical context, covering hundreds and thousands of years. The authors consider it necessary to avoid "historical myopia" and subjectivity in the interpretation of current events, to identify the historical pattern of what is happening, and to try to look into the future with an unbiased view. The study of long-term tendencies allows to identify zones of exhaustion of previous trends where their reversals become possible, that is, to detect major turning points. At present, the World System has found itself within one of such bifurcation zones. It is easier to understand such bifurcations and possible directions of further evolution on the basis of long-term macrohistorical analysis. We would like to emphasize that we use a macrohistorical approach not in order to extrapolate long-term historical trends but rather to detect when, where, and how those long-term trends might be reversed. It is impossible not to add here that, at such critical stages and moments of macrotrend reversals, the use of the ideas of Erich Jantsch (1980), Ilya Prigogine (1997, 2017), Prigogine & Stengers (1984, 2018) and Hermann Haken (1978, 1980, 1983, 1984, 2006a, 2006b) about self-organization, nonlinear dynamics, dissipative structures, nonequilibrium states and bifurcation points become especially productive.

At the bifurcation points, the system is in such an unstable state that even a relatively weak impact at the right time and place can change the further development trajectory. As the World System today is at a bifurcation point, it is important to understand in which direction we should apply our efforts in order to enter an optimal channel of evolution. Thus, we think that the most effective way is to study long-term historical trends, and present critical points and short-term and middle-term possibilities to identify the optimum direction of our actions. The study of long-term historical trends, research of present critical points, and investigation of short-term and middle-term possibilities do not exclude each other, but it is necessary to analyze them together.

We emphasize the need for a comprehensive analysis that takes into account not only the long-term historical trends but also the interaction of all important factors, driving forces, and spheres (ecology, climate, technology, demography, economy, culture, sociosphere, politics, ideology, culture, etc.), as well as relations around redistribution, stratification, inequality, injustice and so on, that affect macrohistorical development and try to identify specific influences of each of them in a certain period. In addition, it is necessary to take into account macrohistorical and evolutional regularities, patterns, and trends. The macrohistorical approach aims to link all these mutual influences into a single picture; its analysis makes it possible to understand the processes taking place in the world and the prospects for their further development. The chapter presents some of the results of the analysis.

Moreover, we would like to emphasize the importance of the historical approach to analyzing the present and the future. This means that the changes taking place in the world are fundamental and long-term; therefore, they should be considered in a broad historical context (and not just as a continuation of the trends of the twentieth century). Therefore, we should look at the current situation within the framework of not just historical but macrohistorical development. Many fundamental trends take a long time to emerge and develop, so considering their historical evolution and dynamics is absolutely necessary. At the same time, it is important to understand that the development of a society is a very complex result of the mutual influence of different subsystems of society.[1]

---

[1]For examples of macrohistorical approach (see Braudel, 1973, 1982; Wallerstein, 1974, 2004; Gellner, 1988; Frank & Gills, 1993; Modelski & Thompson, 1996; Arrighi & Silver, 1999; Diakonoff, 1999; Denemark et al., 2000; Korotayev, 2004, 2008, 2020; Korotayev et al., 2006; Modelski et al., 2007; Akaev et al., 2012, 2017; Akaev & Sadovnichiy, 2016; Ilyin & Ursul, 2012; Korotayev & Malkov, 2012; McNeill, 2013; Grinin 2012, 2022; Grinin & Grinin, 2015, 2016, 2023c; Grinin & Korotayev, 2009, 2015, 2018; Chase-Dunn & Hall, 1997; Chase-Dunn & Anderson, 2005; Chase-Dunn et al. 2010; Gills & Thompson, 2012; Chase-Dunn & Lerro, 2016; Turchin, 2018; Zinkina et al. 2019; Malkov & Davydova, 2021; Rodrigue, 2021).

Naturally, it is impossible to talk about dominant and influenced subsystems for all cases, since the strength of the direct and reverse influence depends on numerous circumstances in each specific case. However, in general, it seems that technological development can be considered the key factor in the historical development of humankind. Technologies do not develop evenly but experience certain leaps ("technological revolutions"[2]). The introduction of new technologies entails serious economic changes, expressed, in particular, in the growth of labor and/or land productivity and in the expansion of the resource base. This in turn contributes to population growth. Population growth and especially the increase in its density makes changes in the social structure of society inevitable, and, consequently, changes in the political sphere, redistribution, and consumption. At the same time, economic activity simultaneously affects the external environment (ecology), and that, in turn, can affect nature and climate.

These powerful technological shifts are caused by deep crises periodically occurring in human history that, in turn, is caused by fundamental circumstances (limited natural resources, social institutions, demographic pressure, climate change, etc.). During crises, society's demand for change and innovation rises sharply in search of coping with a difficult situation. New technologies are one of the possible responses to the challenges and limitations of crises. All these changes occur at different rates, which complicates the overall picture. Therefore, the macrohistorical approach is especially important here.

Let us consider very briefly what the macrohistorical analysis by subsystems shows (this is described in more detail in Parts II and III, as well as in mathematical chapters of this volume[3]).

---

[2] See chapter "Technology. Limitless Possibilities, Effective Control" (Grinin & Grinin, 2023b, this volume) for detail; see also Grinin (2006, 2007), Grinin and Grinin (2015), Grinin and Korotayev (2015), Grinin et al. (2017, 2020a, 2020b, 2021).

[3] See chapter "Modeling Social Self-Organization and Historical Dynamics. An Overview" (Sadovnichy et al., 2023, this volume), chapter "Modeling Social Self-Organization and Historical Dynamics. A General Approach" (Akaev et al., 2023a, this volume), chapter "Modeling Social Self-Organization and Historical Dynamics. Agrarian Society" (Malkov et al., 2023c, this volume), chapter "Modeling Social Self-Organization and Historical Dynamics. Industrial Society" (Akaev et al., 2023b, this volume), chapter "Modeling Social Self-Organization and Historical Dynamics. Modern Society and a Look into the Global Future: Cybernetic W-Society" (Akaev et al., 2023c, this volume), chapter "Modeling Social Self-Organization and Historical Dynamics. Global Phase Transitions" (Malkov et al., 2023b, this volume), chapter "Modeling Social Self-Organization and Historical Dynamics. Modern Society and a Look into the Global Future: Cybernetic W-Society" (Grinin et al., 2023d, this volume), chapter "Modeling Social Self-Organization and Historical Dynamics. Africa's Futures" (Korotayev et al., 2023b, this volume), chapter "Analyzing Social Self-Organization and Historical Dynamics. Future Cybernetic W-Society: Socio-Political Aspects" (Grinin & Grinin, 2023a), and chapter "Modeling Social Self-Organization and Historical Dynamics. Life Quality Index" (Malkov et al., 2023a, this volume).

2 Natural and Climatic Factors

Climate change is a natural process that has been cyclical over millions of years. The socio-economic homeostasis was periodically disrupted by natural climatic crises that sharply increased the demand for innovations and initiated technological revolutions that changed the sociopolitical landscape [the urban revolution (Childe, 1952), the 'Axial Age' (Jaspers, 1953), the industrial revolution (Cipolla, 1976; Allen, 2009, 2011; Goldstone, 2009; Grinin & Korotayev, 2015)] and contributed to the transition to the next phase of historical development. On the one hand, there is an obvious downward trend in the dependence of humans and society on the natural environment and its limitations. This trend is due to the growth of productive forces that allow society to receive more and more energy, resources, and other benefits from nature. On the other hand, this trend has been increasingly fraught with deep problems in recent decades.[4] Not only the dependence of society on nature has become apparent again, but also the dependence of nature on society has become obvious. There is a pressing issue of finding a way to overcome the problems that have arisen: should we look for a technological way out or is it necessary to change the very way of life of the humankind?

Currently, the climate has changed significantly, and global warming is progressing; the temperature of the world's ocean is increasing, glaciers are melting, and natural disasters have become more frequent and more intense. According to the UN, in 2019, the average global temperature had already risen by 1.1 °C compared to preindustrial levels. Research shows that two-thirds of global warming during that period was caused by an increase in the concentration of CO_2 in the Earth's atmosphere, which is largely a direct consequence of anthropogenic activity. If CO_2 emissions are not curbed by the middle of the century, global warming will exceed the level of 2 °C, which will bring catastrophic consequences (see, e.g., Randers, 2012; Wijkman & Skånberg, 2017; Bardi, 2017, Randers et al., 2018; von Weizsäcker & Wijkman, 2018; Gills & Morgan, 2022).[5]

3 Technological Subsystem

From the macrohistorical perspective, technical equipment of humankind has constantly been increasing, paralleled by the ever-deeper use of natural resources, an increase in labor productivity, the lengthening of technological chains, and an increase in the importance of technology in the general production system (Grinin, 2006, 2007; Grinin & Grinin, 2015, 2016; Grinin & Korotayev, 2015; Grinin et al., 2020a, 2020b; Korotayev, 2020). Human labor has been increasingly replaced with

[4] See chapter "Ecology. Life in the "Unstable Biosphere"" (Kovaleva, 2023, this volume).

[5] See chapter "Climate and Energy. Scenarios of Energy Transition and Global Temperature Changes Based on Current Technologies and Trends" (Akaev & Davydova, 2023, this volume) for more details.

technology—first, manual labor, followed by high-qualified physical labor, mental labor, high-qualified intellectual labor, and so on. This trend is evident today, and in many ways, it has accelerated. In particular, the rapid growth of technology makes a number of professions obsolete and useless, raising the fear that artificial intelligence (AI) and robots will make many people superfluous in production. Indeed, one can expect increasing amounts of people to become redundant due to the introduction of new technologies. However, at the same time, the problem of labor shortages is aggravated today due to the downward demographic trends, global aging, and depopulation that has already begun in a number of countries.[6] Thus, the current historical situation is special, since, in more and more countries, the growth in levels of production and labor productivity does not cause population growth, in contrast to all previous historical eras.

Importantly, technological development has always been extremely uneven. Epochs of slow change (technological stagnation) are interspersed with periods of rapid technological growth. The most important production (technological) revolutions in history included agrarian, industrial, and cybernetic (e.g., Grinin et al., 2020a, 2020b[7]). These revolutions have typical similarities. In particular, the agrarian and industrial revolutions had three phases, each of which led to a powerful growth of technology and production (which, in turn, led to population growth and changes in the structure of society). Based on that, it is possible to make forecasts about new technological waves in the coming decades.[8]

However, there is some dissonance between expectations from technologies and risks from the introduction of them in reality, and these risks are increasing. We stress it in several chapters of this volume, paying attention in particular to the relationship between power structures and the framing of technological development. The information society has experienced and continues to experience conflicts and tensions and has been forced to introduce many regulations.[9] A number of works have already been devoted to the analysis of various aspects of this problem in the present and the future (e.g., Westin, 1966; Ashman et al., 2014; Cecere et al., 2015; Moustaka et al., 2019; Schwartz, 1999; Solove, 2008; Brammer et al., 2020; Alharbi 2020). There is also a growing danger of total digital control over human behavior

[6] See chapter "Demography. Toward the optimization of demographic processes" for more details (Korotayev et al., 2023a, this volume) and chapter "Global Aging – an Integral Problem of the Future. How to Turn a Problem into a Development Driver?" (Grinin et al., 2023b, this volume) for more detail.

[7] On the Agrarian or Neolithic Revolution see, e.g., Childe (1934, 1944, 1948), Reed (1977), Harris and Hillman (1989); on the Industrial Revolution see, e.g., Allen (2009, 2011) and references above; on the Cybernetic Revolution see, e.g., Bernal (1965), Sylvester and Klotz (1983), Rifkin (2011), Schwab (2016), Grinin and Grinin (2015, 2016) and references above.

[8] See chapter "Technology: limitless possibilities and effective control" (Grinin & Grinin, 2023b, this volume).

[9] Of particular importance is the growing contradiction between the values of privacy and constitutional rights to keep private information in secret, on the one hand, and increasing information transparency, which allows tracking every person, on the other hand.

(through social networks, gadgets, and other systems with artificial intelligence and cognitive technologies); people become "transparent" [about negative consequences of this, see chapter "Technology. Limitless Possibilities and Effective Control" (Grinin & Grinin, 2023b, this volume)]. In chapter "Analyzing Social Self-Organization and Historical Dynamics. Future Cybernetic W-Society: Sociopolitical Aspects" (Grinin & Grinin, 2023a, this volume), we provide an idea about socio-technical self-regulating systems (SSSs), which, with the help of artificial intelligence, will regulate various social and administrative relations. SSSs perform social and administrative (i.e., control, verification, distribution, security, checking, identification, issue of documents, rating, and other functions) using a set of technologies with little or no involvement of officials and specialists. We also believe that the widespread introduction of SSSs can accelerate the change in social relations and increase confrontation in society because the use of self-regulating systems in governance will lead to increased technological and political control. This is all the more dangerous because, although such systems have not yet become widespread in private life, they are likely to become omnipresent in not so distant future.

4 Economic Subsystem

In the macrohistorical perspective, the volume of production and surplus products has been increasing (on the basis of which society could become more complex). In the industrial age, growth of production and surplus products overtook population growth (Goldstone, 2009; Allen, 2009, 2011; Grinin & Korotayev, 2015). This resulted in an almost constant rise in the standard of living of the population which, in turn, led to a change in the structure of the economy from a subsistence basis to a market one (e.g., Berger, 1986). Accordingly, the desire to expand markets and stimulate consumer demand increased. The ideology of increasing consumption led to the formation of a consumer society. As a result, wastefulness has increased, and the problems of environmental pollution and resource depletion have become more acute.[10] Now, the limits to consumerism as the main driver of the economy are visible, and the role of population growth in aggregate demand has been decreasing due to population aging and falling birth rates.[11] These problems, however, are less pronounced in low-income countries which have not yet reached a sufficient level of consumption. The challenge is to change the consumption pattern of society and

[10] See chapter "Climate and Energy. Scenarios of Energy Transition and Global Temperature Changes Based on Current Technologies and Trends" (Akaev & Davydova, 2023, this volume), chapter "Ecology. Life in the "Unstable Biosphere"" (Kovaleva, 2023, this volume), and chapter "Economics. Optimizing Growth" (Grinin et al., 2023c, this volume) for more detail.

[11] See chapter "Demography. Toward Optimization of Demographic Processes" for more details (Korotayev et al., 2023a, this volume), chapter "Global Aging – an Integral Problem of the Future. How to Turn a Problem into a Development Driver?" (Grinin et al., 2023b, this volume), and chapter "Socio-Political Transformations. A Difficult Path to Cybernetic Society" (Grinin et al., 2023e, this volume) for more details.

individuals in high-income countries, while taking into account the underconsumption in low-income countries and the problem of climate change.[12]

At the same time, because of globalization, the world has become connected and unified. The "Great Divergence'" [the initial breakaway of the industrialized countries from the countries of the World System periphery (e.g., Goldstone, 2009, 2021; Pomeranz, 2021)] was replaced by the "Great Convergence" (e.g., Akaev, 2015; Grinin & Korotayev, 2015; Korotayev et al., 2015). The problems of convergence will play an increasingly important role, the most serious problem being the integration of Africa into world development.[13]

5 Demographic Subsystem

The general macrohistorical trend observed until the eighteenth century was constant population growth, albeit slow and with large fluctuations, especially for individual societies. During the industrial period, population growth accelerated enormously, reaching record rates in the second half of the twentieth century. Starting from the fourth millennium BC, we also observe urbanization growth, its trend generally resembling that of population growth. Since the eighteenth century, general (but especially infant) mortality has decreased, and life expectancy has increased (e.g., Armengaud, 1976; Chesnais, 1992; Kapitza, 2006; Dyson, 2010; Podlazov, 2017).[14] The subsequent (with a substantial time lag) reduction in birth rates were closely related to changes in the status and education of women, an increase in investments in the development of children, the emergence and development of state policy regarding family and children (education, health care, sports, etc.) (e.g., Singh & Casterline, 1985; Caldwell et al., 2006; Reher, 2011; Livi-Bacci, 2012).

Crucially, the process of global aging is intensifying. We believe (and this is one of the main ideas of our report) that global aging is among the most important trends and limitations of the twenty-first century. Its role will increase, so research on all the other problems must consider the process of global aging.[15]

[12] See chapter "High Income and Low Income Countries. Toward a Common Goal at Different Speeds" (Grinin & Korotayev, 2023b, this volume) and chapter "Africa – the Continent of the Future. Challenges and Opportunities" (Grinin & Korotayev, 2023a, this volume) for more detail.

[13] For detail about future economic changings, problems of economic growth, consumerism etc. see chapter "Economics. Optimizing Growth" (Grinin et al., 2023c, this volume).

[14] On demographic trends see chapter "Demography. Toward Optimization of Demographic Processes" (Korotayev et al., 2023a, this volume), chapter "Global Aging – an Integral Problem of the Future. How to Turn a Problem into a Development Driver?" (Grinin et al., 2023b, this volume), and chapter "Africa – the Continent of the Future. Challenges and Opportunities" (Grinin & Korotayev, 2023a, this volume) for more detail.

[15] See chapter "Global Aging – an Integral Problem of the Future. How to Turn a Problem into a Development Driver?" (Grinin et al., 2023b, this volume), chapter "Socio-Political Transformations. A Difficult Path to Cybernetic Society" (Grinin et al., 2023e, this volume), and chapter "Modeling Social Self-Organization and Historical Dynamics. Global Phase Transitions" (Malkov et al., 2023b, this volume) for more detail.

6 Social Subsystem

From the macrohistorical perspective, the social structure of society has been getting more complex and increasingly stratified. Notably, the foundations of stratification changed more than once due to the development of the political structure, production and surplus product growth, division of labor, an increase in the number of professions, and the development of culture. Human and social values are changing markedly (see, e.g., Aleshkovski et al., 2020; Zinkina et al., 2022; Korotayev et al., 2023c). It is important to note the presence of a constant dichotomy of collectivism and individualism in the development of social processes, along with a long-term trend toward an increase in the importance of individuality (the value of an individual). However, in recent decades, this trend has already acquired certain negative features, so there is a growing need to strengthen the role of collectivism in social interactions (see, e.g., Wallerstein et al., 2013; Andersen, 2020).

It is important to note that, in the eras of slow technological development, limited resource base, and serious external threats (such conditions were typical, for example, for agrarian societies of the Middle Ages), the emerging social institutions shifted toward stronger paternalism-focused on collective survival and subordination of individual interests to the public (the so-called X-type social structures). In the conditions of strong economic growth and expansion of the resource base (for example, due to the emergence of new technologies), social institutions transformed in the direction of encouraging and stimulating economic activity and competition, individualism (the so-called Y-type social structures).[16]

We believe that the transition from X-structures to Y-structures and vice versa is a process of complex restructuring of society, accompanied by sociopolitical crises and revolutions, including the ones in the ideological sphere. New ideologies/religions legitimize the ongoing transition and the principles of organization of the new society (e.g., Korotayev, 2004). Thus, each transition had its own ideological revolution, without which it would not have been possible.

7 Political Subsystem

The general macrohistorical trend of development is the growing complexity of the political structure of societies. This led to state formation (Claessen & Skalník, 1978; Grinin et al., 2004; Turchin et al., 2022). With time, the state that is a complex political subsystem, became leading in some respects, determining to a large extent the development of the other societal subsystems. Interaction between civilizations and tribal peripheries created huge empires and changed the world (see, e.g., Korotayev et al., 2021, 2022; Wilkinson, 2022). In complex agrarian societies

[16]See chapter "Socio-Political Transformations. A Difficult Path to Cybernetic Society" for more details (Grinin et al., 2023e) and chapter "Modeling Social Self-Organization and Historical Dynamics. Global Phase Transitions" (Malkov et al., 2023b, this volume).

political sphere supported social stratification, whereas only a small minority of people had political rights. However, since the Modern Age, the number of people with political rights has gradually increased (along with the growth of prosperity, ongoing urbanization, the spread of education, and other processes), so the number of participants in political processes has grown remarkably. This trend led to the emergence of nation-states and nationalism (Gellner, 1984; Grinin, 2008). However, at the same time, there is a trend that increases the ability of the most influential minority to limit the actual rights of the majority in the political sphere. This is especially associated with the monopolization of the mass media and the growth of artificial intelligence capabilities. The transition to a more sustainable society is impossible without greater de facto equality in the political sphere and the pluralism of opinions.[17]

References

Akaev, A. (2015). *Ot epokhi Velikoy divergentsii k epokhe Velikoy konvergentsii: Matematicheskoye modelirovaniye i prognozirovaniye dolgosrochnogo tekhnologicheskogo i ekonomicheskogo razvitiya mirovoy dinamiki*. LENAND/URSS.

Akaev, A., & Davydova, O. (2023). Climate and energy. Scenarios of energy transition and global temperature changes based on current technologies and trends. In V. Sadovnichy et al. (Eds.), *Reconsidering the limits to growth. A report to the Russian Association of the Club of Rome* (pp. 53–70). Springer. https://doi.org/10.1007/978-3-031-34999-7_4

Akaev, A. A., & Sadovnichiy, V. A. (2016). A closed dynamic model to describe and calculate the Kondratiev long wave of economic development. *Herald of the Russian Academy of Sciences, 86*(5), 371–383. https://doi.org/10.1134/S1019331616050014

Akaev, A., Sadovnichy, V., & Korotayev, A. (2012). On the dynamics of the world demographic transition and financial-economic crises forecasts. *The European Physical Journal Special Topics, 205*(1), 355–373. https://doi.org/10.1140/epjst/e2012-01578-2

Akaev, A., Korotayev, A., Issaev, L., & Zinkina, J. (2017). Technological development and protest waves: Arab spring as a trigger of the global phase transition? *Technological Forecasting and Social Change, 116*, 316–321. https://doi.org/10.1016/j.techfore.2016.08.009

Akaev, A., Malkov, S., Bilyuga, S., Malkov, A., Musieva, J., & Korotayev, A. (2023a). Modeling social self-organization and historical dynamics. A general approach. In V. Sadovnichy et al. (Eds.), *Reconsidering the limits to growth. A report to the Russian Association of the Club of Rome* (pp. 253–307). Springer. https://doi.org/10.1007/978-3-031-34999-7_15

Akaev, A., Malkov, S., Davydova, O., Kovaleva, N., Malkov, A., Grinin, L., & Korotayev, A. (2023b). Modeling social self-organization and historical dynamics. Industrial society. In V. Sadovnichy et al. (Eds.), *Reconsidering the limits to growth. A report to the Russian Association of the Club of Rome* (pp. 337–385). Springer. https://doi.org/10.1007/978-3-031-34999-7_17

Akaev, A., Malkov, S., Grinin, L., Bilyuga, S., Davydova, O., Grinin, A., Kovaleva, N., Malkov, A., Musieva, J., & Korotayev, A. (2023c). Modeling social self-organization and historical dynamics. Modern society and a look into the global futures: Cybernetic W-society. In V. Sadovnichy et al. (Eds.), *Reconsidering the limits to growth. A report to the Russian*

[17] See chapter "Future Political Change. Toward a More Efficient World Order" (Grinin et al., 2023a, in this volume) and chapter "Socio-Political Transformations. A Difficult Path to Cybernetic Society" (Grinin et al., 2023e, in this volume) for more detail.

Association of the Club of Rome (pp. 419–460). Springer. https://doi.org/10.1007/978-3-030-34999-7_19
Aleshkovski, I., Gasprishvili, A., & Smakotina, N. (2020). Global values in the context of civilizational dialogue. *Journal of Globalization Studies, 11*(1), 73–85. https://doi.org/10.30884/jogs/2020.01.05
Alharbi, F. S. (2020). Dealing with data breaches amidst changes in technology. *International Journal of Computer Science and Security (IJCSS), 14*(3), 108–115.
Allen, R. C. (2009). *The British industrial revolution in global perspective*. Cambridge University Press.
Allen, R. C. (2011). *Global economic history: A very short introduction*. Oxford University Press.
Andersen, L. R. (2020). *Bildung – keep growing. A report to the Club of Rome*. Nordic Bildung.
Armengaud, A. (1976). Population in Europe 1700–1914. In C. M. Cipolla (Ed.), *The industrial revolution. 1700–1914* (pp. 22–76). Harvester.
Arrighi, G., & Silver, B. J. (1999). *Chaos and governance in the modern world system*. University of Minnesota Press.
Ashman, H., Brailsford, T., Cristea, A. I., Sheng, Q. Z., Stewart, C., Toms, E. G., & Wade, V. (2014). The ethical and social implications of personalization technologies for e-learning. *Information & Management, 51*(6), 819–832. https://doi.org/10.1016/j.im.2014.04.003
Bardi, U. (2017). *The Seneca effect. Why growth is slow but collapse is rapid. A report to the Club of Rome*. Springer.
Berger, P. L. (1986). *The capitalist revolution*. Basic Books.
Bernal, J. D. (1965). *Science in history* (3rd ed.). Hawthorn Books.
Brammer, S., Branicki, L., & Linnenluecke, M. (2020). COVID-19, societalization and the future of business in society. *Science of the Total Environment, 34*(4), 2–7. https://doi.org/10.5465/amp.2019.0053
Braudel, F. (1973). *Capitalism and material life, 1400–1800*. Harper and Row.
Braudel, F. (1982). *Civilization and capitalism, 15th – 18th century*. Harper and Row.
Caldwell, J. C., Caldwell, B. K., Caldwell, P., McDonald, P. F., & Schindlmayr, T. (2006). *Demographic transition theory*. Springer. https://doi.org/10.1007/978-1-4020-4498-4
Cecere, G., Le Guel, F., & Soulié, N. (2015). Perceived internet privacy concerns on social networks in Europe. *Technological Forecasting and Social Change, 96*, 277–287. https://doi.org/10.1016/j.techfore.2015.01.021
Chase-Dunn, C., & Anderson, E. (Eds.). (2005). *The historical evolution of world-systems*. Springer. https://doi.org/10.1057/9781403980526
Chase-Dunn, C., & Hall, T. D. (1997). *Rise and demise: Comparing world-systems*. Westview Press.
Chase-Dunn, C., & Lerro, B. (2016). *Social change: Globalization from the Stone Age to the present*. Routledge.
Chase-Dunn, C., Niemeyer, R., Alvarez, A., Inoue, H., & Love, J. (2010). Cycles of rise and fall, upsweeps and collapses: Changes in the scale of settlements and polities since the Bronze Age. In L. Grinin, P. Herrmann, A. Korotayev, & A. Tausch (Eds.), *History and mathematics: Processes and models of global dynamics* (pp. 64–91). Uchitel.
Chesnais, J. C. (1992). *The demographic transition: Stages, patterns, and economic implications*. Clarendon Press.
Childe, V. G. (1934). *New light on the most ancient east*. Routledge & Kegan.
Childe, V. G. (1944). *Progress and archaeology*. Watts.
Childe, V. G. (1948). *What happened in history*. Penguin Books.
Childe, V. G. (1952). *New light on the most ancient east* (4th ed.). Routledge & Paul.
Cipolla, C. M. (Ed.). (1976). *The industrial revolution, 1700–1914*. Harvester Press.
Claessen, H. J. M., & Skalník, P. (Eds.). (1978). *The early state*. Mouton.
Denemark, R. A., Friedman, J., Gills, B. K., & Modelski, G. (2000). *World system history: The social science of long-term change*. Routledge.
Diakonoff, I. (1999). *The paths of history*. Cambridge University Press.

Dyson, T. (2010). *Population and development. The demographic transition*. Zed Books.
Frank, A. G., & Gills, B. K. (Eds.). (1993). *The world system: Five hundred years, or five thousand*. Routledge.
Gellner, E. (1984). *Nations and nationalism*. Blackwell.
Gellner, E. (1988). *Plough, sword and book. The structure of human history*. The University of Chicago Press.
Gills, B., & Morgan, J. (Eds.). (2022). *Economics and climate emergency*. Routledge.
Gills, B. K., & Thompson, W. (Eds.). (2012). *Globalization and global history*. Routledge.
Goldstone, J. A. (2009). *Why Europe? The rise of the west in world history. 1500–1850*. McGraw-Hill.
Goldstone, J. A. (2021). Dating the great divergence. *Journal of Global History, 16*(2), 266–285. https://doi.org/10.1017/S1740022820000406
Grinin, L. (2006). *Proizvoditel'nyye sily i istoricheskiy protsess*. URSS.
Grinin, L. (2007). Production revolutions and periodization of history: A comparative and theoretic-mathematical approach. *Social Evolution & History, 6*(2), 11–55.
Grinin, L. (2008). Early state, developed state, mature state: Statehood evolutionary sequence. *Social Evolution & History, 7*(1), 67–81.
Grinin, L. (2012). *Macrohistory and globalization*. Uchitel.
Grinin, L. (2022). On revolutionary waves since the 16th century. In J. A. Goldstone, L. Grinin, & A. Korotayev (Eds.), *Handbook of revolutions in the 21st century: The new waves of revolutions, and the causes and effects of disruptive political change* (pp. 389–411). Springer. https://doi.org/10.1007/978-3-030-86468-2_13
Grinin, L., & Grinin, A. (2015). *Ot rubil do nanorobotov. Mir na puti k epokhe samoupravlyayemykh sistem (istoriya tekhnologiy i opisaniye ikh budushchego)*. Uchitel.
Grinin, L., & Grinin, A. (2016). *The cybernetic revolution and the forthcoming epoch of self-regulating systems*. Uchitel.
Grinin, L., & Grinin, A. (2023a). Analyzing social self-organization and historical dynamics. Future cybernetic W-society: Sociopolitical aspects. In V. Sadovnichy et al. (Eds.), *Reconsidering the limits to growth. A report to the Russian Association of the Club of Rome* (pp. 491–519). Springer. https://doi.org/10.1007/978-3-031-34999-7_21
Grinin, L., & Grinin, A. (2023b). Technology. Limitless possibilities, effective control. In V. Sadovnichy et al. (Eds.), *Reconsidering the limits to growth. A report to the Russian Association of the Club of Rome* (pp. 139–154). Springer. https://doi.org/10.1007/978-3-031-34999-7_8
Grinin, L., & Grinin, A. (2023c). Historical materialism: Does the concept have a future? *Social Evolution & History, 22*(1), 143–178. https://doi.org/10.30884/seh/2023.01.07
Grinin, L., & Korotayev, A. (2009). *Social macroevolution. Genesis and transformations of the world system*. Librocom/URSS.
Grinin, L., & Korotayev, A. (2015). *Great divergence and great convergence. A global perspective*. Springer. https://doi.org/10.1007/978-3-319-17780-9
Grinin, L., & Korotayev, A. (2018). Origins of globalization in the framework of the Afroeurasian world-system history. In T. D. Hall (Ed.), *Comparing globalizations. Historical and world-systems approaches* (pp. 37–70). Springer. https://doi.org/10.1007/978-3-319-68219-8_3
Grinin, L., & Korotayev, A. (2023a). Africa – The continent of the future. Challenges and opportunities. In V. Sadovnichy et al. (Eds.), *Reconsidering the limits to growth. A report to the Russian Association of the Club of Rome* (pp. 225–238). Springer. https://doi.org/10.1007/978-3-031-34999-7_13
Grinin, L., & Korotayev, A. (2023b). High income and low income countries. Toward a common goal at different speeds. In V. Sadovnichy et al. (Eds.), *Reconsidering the limits to growth. A report to the Russian Association of the Club of Rome* (pp. 207–224). Springer. https://doi.org/10.1007/978-3-031-34999-7_12
Grinin, L. E., Carneiro, R. L., Bondarenko, D. M., Kradin, N. N., & Korotayev, A. V. (Eds.). (2004). *The early state, its alternatives and analogues*. Uchitel.

Grinin, L., Grinin, A., & Korotayev, A. (2017). Forthcoming Kondratieff wave, cybernetic revolution, and global ageing. *Technological Forecasting and Social Change, 115*, 52–68. https://doi.org/10.1016/j.techfore.2016.09.017

Grinin, L., Grinin, A., & Korotayev, A. (2020a). A quantitative analysis of worldwide long-term technology growth: From 40,000 BCE to the early 22nd century. *Technological Forecasting and Social Change, 155*, 1–15. https://doi.org/10.1016/j.techfore.2020.119955

Grinin, L., Grinin, A., & Korotayev, A. (2020b). Dynamics of technological growth rate and the forthcoming singularity. In A. Korotayev & D. LePoire (Eds.), *The 21st century singularity and global futures. A big history perspective* (pp. 287–344). Springer. https://doi.org/10.1007/978-3-030-33730-8_14

Grinin, L., Grinin, A., & Korotayev, A. (2021). Global trends and forecasts of the 21st century. *World Futures, 77*(5), 335–370. https://doi.org/10.1080/02604027.2021.1949939

Grinin, L., Grinin, A., & Korotayev, A. (2023a). Future political change. Toward a more efficient world order. In V. Sadovnichy et al. (Eds.), *Reconsidering the limits to growth. A report to the Russian Association of the Club of Rome* (pp. 191–206). Springer. https://doi.org/10.1007/978-3-031-34999-7_11

Grinin, L., Grinin, A., & Korotayev, A. (2023b). Global aging – An integral problem of the future. How to turn a problem into a development driver? In V. Sadovnichy et al. (Eds.), *Reconsidering the limits to growth. A report to the Russian Association of the Club of Rome* (pp. 117–135). Springer. https://doi.org/10.1007/978-3-031-34999-7_7

Grinin, L., Grinin, A., & Malkov, S. (2023c). Economics. Optimizing growth. In V. Sadovnichy et al. (Eds.), *Reconsidering the limits to growth. A report to the Russian Association of the Club of Rome* (pp. 155–168). Springer. https://doi.org/10.1007/978-3-031-34999-7_9

Grinin, L., Grinin, A., & Malkov, S. (2023d). Modeling social self-organization and historical dynamics. Global futures and cybernetic W-society. In V. Sadovnichy et al. (Eds.), *Reconsidering the limits to growth. A report to the Russian Association of the Club of Rome* (pp. 491–519). Springer. https://doi.org/10.1007/978-3-031-34999-7_21

Grinin, L., Grinin, A., & Malkov, S. (2023e). Socio-political transformations. A difficult path to cybernetic society. In V. Sadovnichy et al. (Eds.), *Reconsidering the limits to growth. A report to the Russian Association of the Club of Rome* (pp. 169–189). Springer. https://doi.org/10.1007/978-3-031-34999-7_10

Haken, H. (1978). *Synergetics: An introduction: Nonequilibrium phase transitions and self-organization in physics, chemistry, and biology*. Springer.

Haken, H. (1980). Synergetics. *Naturwissenschaften, 67*(3), 121–128.

Haken, H. (1983). *Advanced synergetics: Instability hierarchies of self-organizing systems and devices*. Springer.

Haken, H. (1984). *The science of structure: Synergetics*. Van Nostrand Reinhold.

Haken, H. (2006a). *Information and self-organization: A macroscopic approach to complex systems*. Springer.

Haken, H. (2006b). *Synergetik in der Psychologie Selbstorganisation verstehen und gestalten*. Hogrefe.

Harris, D., & Hillman, G. (Eds.). (1989). *An evolutionary continuum of people – Plant interaction. Foraging and farming. The evolution of plant exploitation*. Unwin Hyman.

Ilyin, I., & Ursul, A. (2012). Globalistics: New investigative trends in science. *Globalistics and Globalization Studies, 1*, 107–118.

Jantsch, E. (1980). *The self-organizing universe: Scientific and human implications of the emerging paradigm of evolution*. Pergamon.

Jaspers, K. (1953). *The origin and goal of history*. Yale University Press.

Kapitza, S. (2006). *Global population blow-up and after. Report to the Club of Rome*. Global Marshall Plan Initiative.

Korotayev, A. (2004). *World religions and social evolution of the old world Oikumene civilizations: A cross-cultural perspective*. Edwin Mellen Press.

Korotayev, A. (2008). Compact mathematical models of world system development, and how they can help us to clarify our understanding of globalization processes. In G. Modelski, T. Devezas, & W. R. Thompson (Eds.), *Globalization as evolutionary process: Modeling global change* (pp. 133–160). Routledge. https://doi.org/10.4324/9780203937297

Korotayev, A. (2020). The 21st century singularity in the big history perspective. A re-analysis. In A. Korotayev & D. LePoire (Eds.), *The 21st century singularity and global futures. A big history perspective* (pp. 19–75). Springer. https://doi.org/10.1007/978-3-030-33730-8_2

Korotayev, A., & Malkov, S. (2012). Mathematical models of the world-system development. In S. Babones & C. Chase-Dunn (Eds.), *Routledge handbook of world-systems analysis* (pp. 158–161). Routledge. https://doi.org/10.4324/9780203863428.ch4_5

Korotayev, A., Malkov, A., & Khaltourina, D. (2006). *Introduction to social macrodynamics: Compact macromodels of the World System growth*. KomKniga/URSS. https://elibrary.ru/item.asp?id=19676519

Korotayev, A., Goldstone, J. A., & Zinkina, J. (2015). Phases of global demographic transition correlate with phases of the great divergence and great convergence. *Technological Forecasting and Social Change, 95*, 163–169. https://doi.org/10.1016/j.techfore.2015.01.017

Korotayev, A., Grinin, L., & Grinin, A. (2021). Mathematical model of interaction between civilization center and tribal periphery: A description. *Social Evolution & History, 20*(2), 50–78. https://doi.org/10.30884/seh/2021.02.03

Korotayev, A., Grinin, L., & Grinin, A. (2022). Mathematical model of interaction between civilization center and tribal periphery: An analysis. *Social Evolution & History, 21*(1), 67–100. https://doi.org/10.30884/seh/2022.01.03

Korotayev, A., Malkov, S., & Musieva, J. (2023a). Demography. Toward optimization of demographic processes. In V. Sadovnichy et al. (Eds.), *Reconsidering the limits to growth. A report to the Russian Association of the Club of Rome* (pp. 97–116s). Springer. https://doi.org/10.1007/978-3-031-34999-7_6

Korotayev, A., Shulgin, S., Ustyuzhanin, V., Zinkina, J., & Grinin, L. (2023b). Modeling social self-organization and historical dynamics. Africa's futures. In V. Sadovnichy et al. (Eds.), *Reconsidering the limits to growth. A report to the Russian Association of the Club of Rome* (pp. 461–490). Springer. https://doi.org/10.1007/978-3-031-34999-7_20

Korotayev, A., Butovskaya, M., Shulgin, S., & Zinkina, J. (2023c). How can the global ageing affect the global value system? An evolutionary perspective. *Social Evolution & History, 22*(1), 57–76. https://doi.org/10.30884/seh/2023.01.03

Kovaleva, N. (2023). Ecology. Life in the "unstable biosphere". In V. Sadovnichy et al. (Eds.), *Reconsidering the limits to growth. A report to the Russian Association of the Club of Rome* (pp. 71–95). Springer. https://doi.org/10.1007/978-3-031-34999-7_5

Livi-Bacci, M. (2012). *A concise history of world population*. Wiley-Blackwell.

Malkov, S., & Davydova, O. (2021). Modernization as a global process: The experience of mathematical modeling. *Computer Research and Modeling, 13*(4), 859–873. https://doi.org/10.20537/2076-7633-2021-13-4-859-873

Malkov, S., Bilyuga, S., & Musieva, J. (2023a). Modeling social self-organization and historical dynamics. Life quality index. In V. Sadovnichy et al. (Eds.), *Reconsidering the limits to growth. A report to the Russian Association of the Club of Rome* (pp. 521–530). Springer. https://doi.org/10.1007/978-3-031-34999-7_22

Malkov, S., Grinin, L., Grinin, A., Musieva, J., & Korotayev, A. (2023b). Modeling social self-organization and historical dynamics. Global phase transitions. In V. Sadovnichy et al. (Eds.), *Reconsidering the limits to growth. A report to the Russian Association of the Club of Rome* (pp. 387–417). Springer. https://doi.org/10.1007/978-3-031-34999-7_18

Malkov, S., Kovaleva, N., Grinin, L., & Korotayev, A. (2023c). Modeling social self-organization and historical dynamics. Agrarian society. In V. Sadovnichy et al. (Eds.), *Reconsidering the limits to growth. A report to the Russian Association of the Club of Rome* (pp. 309–335). Springer. https://doi.org/10.1007/978-3-031-34999-7_16

McNeill, W. H. (2013). *The pursuit of power: Technology, armed force, and society since AD 1000*. University of Chicago Press.
Modelski, G., & Thompson, W. R. (1996). *Leading sectors and world powers: The coevolution of global politics and economics*. University of South Carolina Press.
Modelski, G., Devezas, T., & Thompson, W. R. (Eds.). (2007). *Globalization as evolutionary process: Modeling global change*. Routledge.
Moustaka, V., Theodosiou, Z., Vakali, A., Kounoudes, A., & Anthopoulos, L. G. (2019). Enhancing social networking in smart cities: Privacy and security borderlines. *Technological Forecasting and Social Change, 142*, 285–300. https://doi.org/10.1016/j.techfore.2018.10.026
Podlazov, A. (2017). A theory of the global demographic process. *Herald of the Russian Academy of Sciences, 87*(3), 256–266. https://doi.org/10.1134/S1019331617030054
Pomeranz, K. (2021). *The great divergence. China, Europe, and the making of the modern world economy* (2nd ed.). Princeton University Press. https://doi.org/10.1515/9780691217192
Prigogine, I. (1997). *The end of certainty*. Free Press.
Prigogine, I. (2017). *Non-equilibrium statistical mechanics (Dover books on physics)*. Dover Publications.
Prigogine, I., & Stengers, I. (1984). *Order out of chaos: Man's new dialogue with nature*. Bantam.
Prigogine, I., & Stengers, I. (2018). *Order out of chaos: Man's new dialogue with nature* (2nd ed.). Verso Books.
Randers, J. (2012). *2052. A global forecast for the next forty years. A report to the club of Rome*. Chelsea Green Publishing.
Randers, J., Rockström, J., Stoknes, P. E., Golüke, U., Collste, D., & Cornell, S. (2018). *Transformation is feasible. A report to the Club of Rome*. Stockholm Resilience Centre.
Reed, C. A. (Ed.). (1977). *The origins of agriculture*. Mouton.
Reher, D. (2011). Economic and social implications of the demographic transition. *Population and Development Review, 37*, 11–33. https://doi.org/10.1111/j.1728-4457.2011.00376.x
Rifkin, J. (2011). *The third industrial revolution. How lateral power is transforming energy, the economy, and the world*. Palgrave Macmillan.
Rodrigue, B. H. (2021). Disaster's offspring: Catastrophe, narrative, and survival in global history. *Journal of Globalization Studies, 12*(1), 159–171. https://doi.org/10.30884/jogs/2021.01.09
Sadovnichy, V., Akaev, A., Ilyin, I., Malkov, S., Grinin, L., Popov, V., & Korotayev, A. (2023). Modeling social self-organization and historical dynamics. An overview. In V. Sadovnichy et al. (Eds.), *Reconsidering the limits to growth. A report to the Russian Association of the Club of Rome* (pp. 241–251). Springer. https://doi.org/10.1007/978-3-031-34999-7_14
Schwab, K. (2016). *The fourth industrial revolution*. World Economic Forum.
Schwartz, P. M. (1999). Internet privacy and the state. *Connecticut Law Review, 32*, 815–829.
Singh, S., & Casterline, J. (1985). The socio-economic determinants of fertility. In J. Cleland & J. Hobcraft (Eds.), *Reproductive change in developing countries. Insights from world fertility survey* (pp. 199–222). Oxford University Press.
Solove, D. J. (2008). *Understanding privacy*. Harvard University Press.
Sylvester, E., & Klotz, L. C. (1983). *The gene age: Genetic engineering and the next industrial revolution*. Scribner.
Turchin, P. (2018). *Historical dynamics: Why states rise and fall* (2nd ed.). Princeton University Press.
Turchin, P., Whitehouse, H., Gavrilets, S., Hoyer, D., François, P., Bennett, J. S., Feeney, K. C., Peregrine, P., Feinman, G., Korotayev, A., Kradin, N., Levine, J., Reddish, J., Cioni, E., Wacziarg, R., Mendel-Gleason, G., & Benam, M. (2022). Disentangling the evolutionary drivers of social complexity: A comprehensive test of hypotheses. *Science Advances, 8*(25), 1–18. https://doi.org/10.1126/sciadv.abn3517
von Weizsäcker, E. U., & Wijkman, A. (2018). *Come on! Capitalism, short-termism, population and the destruction of the planet. A report to the Club of Rome*. Springer. https://doi.org/10.1007/978-1-4939-7419-1
Wallerstein, I. (1974). *The modern world-system*. Academic Press.

Wallerstein, I. (2004). *World-systems analysis: An introduction*. Duke University Press.
Wallerstein, I., Collins, R., Mann, M., Derluguian, G., & Calhoun, C. (2013). *Does capitalism have a future?* Oxford University Press.
Westin, A. F. (1966). Science, privacy, and freedom: Issues and proposals for the 1970's. Part I. the current impact of surveillance on privacy. *Columbia Law Review, 66*(6), 1003–1050.
Wijkman, A., & Skånberg, K. (2017). *The circular economy and benefits for society. Jobs and climate clear winners in an economy based on renewable energy and resource efficiency. A report to the Club of Rome*. Club of Rome.
Wilkinson, D. (2022). The power configurations of the central civilization/world system in the eighth century. *Social Evolution & History, 21*(1), 33–64. https://doi.org/10.30884/seh/2022.01.02
Zinkina, J., Christian, D., Grinin, L., Ilyin, I., Andreev, A., Aleshkovski, I., Shulgin, S., & Korotayev, A. (2019). *A big history of globalization: The emergence of a global world system*. Springer. https://doi.org/10.1007/978-3-030-05707-7
Zinkina, J., Ustiuzhanin, V., & Shulgin, S. (2022). Social macroevolution and changes in the human value systems. How has modernization affected human values. *Social Evolution & History, 21*(2), 91–105. https://doi.org/10.30884/seh/2022.02.06

Victor Sadovnichy is a Doctor of Physics and Mathematics, Professor, full member of the Russian Academy of Sciences, and rector of Lomonosov Moscow State University since 1992. Field of scientific research: mathematics, mechanics, and computer science. Sadovnichy's research deals with the theory of differential operators and functional analysis, as well as the mathematical modeling of long-term social and economic dynamics. He has obtained important theoretical and applied results related to the issues of dynamic simulation of controlled flights, in particular, of spacecraft and aircraft. This work formed the basis for creation of the first simulator of zero-gravity in the Earth conditions in the world.

Askar Akaev is the First President of the Kyrgyz Republic and Foreign Full Member of the Russian Academy of Sciences. He works as Chief Research Professor in the Prigogine Institute for Mathematical Investigations of Complex Systems at Lomonosov Moscow State University. He is also the Academic Supervisor of Centre for Stability and Risk Analysis at HSE University, Moscow, Russia. Askar Akaev has authored and co-authored over 300 scientific publications, including such monographs as *Coherent Optical Computers* (1977), *Optical Methods of Information Processing* (1983), *Holographic Memory* (1996), *Selected Lectures on Optical Computers* (1996), *Economic Transition from the Point of View of a Physicist. A Mathematical Model of Emerging Economy* (2000), *Modeling and Forecasting the World Dynamics* (2012), *Quality of Education, Effectiveness of R&D, and Economic Growth. Quantitative Analysis and Mathematical Modeling* (2016), etc.

Ilya Ilyin is a well-known Russian researcher, Head of the Department of Global Studies, Dean of the Faculty of Global Studies at Lomonosov Moscow State University, author of more than 350 academic papers. He is the Chairman of the Presidium of the Society for Global Studies, President of the International Public Organization for the Promotion of Public Diplomacy, Scientific, Educational and Youth Cooperation "Eurasian Commonwealth", Chief Scientific Secretary of the International Academy of Global Studies, Chairman of the Russian Committee and Vice President of the Intergovernmental Council for the UNESCO Program "Management of Social transformations – MOST", honorary worker in the field of state youth policy and education, editor-in-chief of the journal *Bulletin of Moscow University*. Series 27: *Global Studies and Geopolitics*.

Sergey Malkov heads the Center for Long-Term Forecasting and Strategic Planning at Lomonosov Moscow State University; he also works as a Senior Research Professor at Centre for Stability and Risk Analysis at HSE University in Moscow, as well as at the Institute of Economics,

Russian Academy of Sciences. He is the author of over 300 scholarly publications, including such monographs as *Social self-organization and historical process: possibilities of mathematical modeling* (URSS, 2009), *Modeling and forecasting of world dynamics* (ISPI RAS, 2012, together with V. A. Sadovnichy and others), *A new theory of the reproduction of capital: development and practical application* (Nestor-Istoriya, 2016, together with V. I. Mayevsky and others), *The twenty-first Century Singularity and Global Futures. A Big History Perspective* (Springer, 2020, together with David J. LePoire and others).

Leonid Grinin is a Senior Research Professor at Centre for Stability and Risk Analysis at HSE University in Moscow, as well as Senior Research Professor of the Institute of Oriental Studies, Russian Academy of Sciences. He is co-editor of the international journals *Social Evolution & History* and the *Journal of Globalization Studies*, as well as co-editor of the international yearbooks *Evolution*, *History & Mathematics*, and *Kondratieff Waves*. He is the author of more than 600 publications, including more than 33 monographs in Russian, English, German, Spanish and Chinese. His current research focuses on comparative political studies, theory of revolution, political anthropology, global economy, global history, historical sociology, and futures studies.

Ivan Aleshkovski is an Associate Professor and Deputy Dean of the Faculty of Global Studies, Lomonosov Moscow State University, Russia. His academic interests include globalistics, international migration flows, migration policy, economic and demographic development of Russia and the world. He has authored more than 150 journal articles and contributions.

Andrey Korotayev heads the Centre for Stability and Risk Analysis at HSE University in Moscow, Russia. He is also a Senior Research Professor at the Institute of Oriental Studies and Institute for African Studies, Russian Academy of Sciences, as well as at the Faculty of Global Studies of Lomonosov Moscow State University and the Russian Presidential Academy of National Economy and Public Administration. He has authored or co-authored over 650 scholarly publications.

The Future Society and the Transition to It

Viktor Sadovnichy, Askar Akaev, Ilya Ilyin, Sergey Malkov, Leonid Grinin, Aleksey Andreev, and Andrey Korotayev

Abstract

This chapter analyzes the current historical situation and the prospects for its further development. It is evident that drastic changes are currently taking place in the world, reflecting a phase transition in historical development, namely, the transition from the late industrial phase to the cybernetic phase, features of which are largely determined by the ongoing Cybernetic Revolution. One could also talk about a transition from an economically organized World System to a politically organized one. Sadovnichy et al. consider the most important features of the

V. Sadovnichy · I. Ilyin · A. Andreev
Lomonosov Moscow State University, Moscow, Russia
e-mail: info@rector.msu.ru; dekanat@fgp.msu.ru; andreev@fgp.msu.ru

A. Akaev
Lomonosov Moscow State University, Moscow, Russia

HSE University, Moscow, Russia

S. Malkov
Lomonosov Moscow State University, Moscow, Russia

HSE University, Moscow, Russia

Institute of Economics, Russian Academy of Sciences, Moscow, Russia
e-mail: s@malkov.org

L. Grinin
HSE University, Moscow, Russia

Institute of Oriental Studies, Russian Academy of Sciences, Moscow, Russia

A. Korotayev (✉)
Lomonosov Moscow State University, Moscow, Russia

HSE University, Moscow, Russia

Institute of Oriental Studies, Russian Academy of Sciences, Moscow, Russia

V. Sadovnichy et al. (eds.), *Reconsidering the Limits to Growth*, World-Systems Evolution and Global Futures, https://doi.org/10.1007/978-3-031-34999-7_3

ongoing changes and their impact on various spheres of society's life, taking into account various scenarios of the future global evolution. In particular, Sadovnichy et al. discuss the possible transition from a liberal market capitalist society to a model of society with an equal distribution (but still preserving private initiative and the spirit of enterprise). This global transition is facilitated by the limitation of extensive growth opportunities due to objective reasons. Modern transition is also connected with the achievement of a high level of technological development, which objectively demands the transformation of social, political, and international relations to correspond to the level of technology. There are some contradictions with respect to technological development. On the one hand, the benefits of new technology implementation are significant for solving many modern problems in providing a higher quality of life, since for the first time in history, it is becoming possible to satisfy the material needs of the majority of the world's population. On the other hand, the development of technological innovations leads to the formation of socio-technical self-regulating systems based on artificial intelligence, the introduction of which also carries with it serious risks, and their scale remains unknown. At the moment, the rate of demographic growth is declining, alleviating the problem of overpopulation of the Earth. However, the high-income countries and some of the low-income ones will experience depopulation, while on the contrary, in most Sub-Saharan African countries, fast population growth will continue for a few decades. Therefore, the optimization of demographic processes is an extremely important task. It is very important that the development of medical technologies leads to an increase in life expectancy, as well as to a change in the demographic structure of society, i.e., to global aging, which is leading to changing social and labor force structure. Therefore, ensuring the quality of life of the aging population will present a critical problem. Aging at the same time gradually reduces the acuteness of material problems and their role in motivating people's activities, while at the same time increasing the role of ideological factors. In this regard, it makes sense to talk about the formation of a new type of society (cybernetic W-society), since the ongoing phase transition will affect all the foundations of life and the organization of society. The transition to a new model of society is inevitably interconnected with global political transformations, involving the formation of a new world order in which new principles of interaction between states will appear with a gradual rejection of competitive confrontation in favor of mutually beneficial cooperation.

1 Perspective and Brief Forecasts

The current historical situation is a transition from the late industrial phase to the cybernetic phase of global evolution,[1] from the economically organized World System to the politically organized one.[2] Let us outline the important features of the transition.

First, the world has become much more unified in the informational, transportation, and economic aspects. Existentially, the principles of harmonious co-existence must be developed. We consider rollbacks and fluctuations along this path, periods of de-globalization, and disruption of established relations (similar to what is happening in the world today in connection with the events in Ukraine) to be inevitable on such a difficult path. And we believe that in overcoming those temporary states of de-globalization, important principles for stabilizing relations for the future will be developed.[3] Various political upheavals and forms of destabilization, including revolutions, will also play a significant role in the processes of transition to the politically organized World System and the harmonization of relations between countries, blocs, and leaders (see, e.g., Grinin, 2022; Goldstone et al., 2022; Grinin et al., 2021b).

Second, since the world has become more unified, possibilities for extensive growth have largely been exhausted. Therefore, a transition to new models of society and the world order is inevitable (possibly, from a liberal market economy to a society with a higher level of distributive functions that preserve important advantages of the previous system).

Third, in modern society, technological innovation has become an integral part of life. Development is moving toward the formation of socio-technical self-regulating systems (based on artificial intelligence), which will significantly change the process of regulating social life.[4] However, it is not yet completely clear what consequences such technological development will lead to (since a continuous stream of innovations inevitably reduces the resilience of society and generates unforeseen risks). The fact is that the introduction of digital technologies and artificial intelligence has led to an explosive growth of information about a person's life (through

[1] See chapter "Technology. Limitless Possibilities, Effective Control" (Grinin & Grinin, 2023b, this volume).

[2] See chapter "Economics. Optimizing Growth" (Grinin et al., 2023c), chapter "Socio-Political Transformations. A Difficult Path to Cybernetic Society" (Grinin et al., 2023d, this volume), and chapter "Analyzing Social Self-Organization and Historical Dynamics. Future Cybernetic W-Society: Socio-Political Aspects" (Grinin & Grinin, 2023a, this volume).

[3] See chapter "Future Political Change. Toward a More Efficient World Order" (Grinin et al., 2023a, this volume) and chapter "Socio-Political Transformations. A Difficult Path to Cybernetic Society" (Grinin et al., 2023d, this volume) for more detail (see also Grinin, 2016; Grinin et al., 2016a, b, 2021b).

[4] See chapter "Technology. Limitless Possibilities, Effective Control" (Grinin & Grinin, 2023b, this volume) and chapter "Modeling Social Self-Organization and Historical Dynamics. Global Phase Transitions" (Malkov et al., 2023a, this volume).

social networks, electronic devices, etc.). People have become 'transparent'; their information can be exploited without their consent, violating their constitutional rights and threatening total control over their activities and behavior. A number of works have already been devoted to the analysis of this problem's various aspects in the present and the future (e.g., Westin, 1966; Schwartz, 1999; Solove, 2008; Ashman et al., 2014; Cecere et al., 2015; Moustaka et al., 2019; Brammer et al., 2020; Alharbi, 2020).

The dependence of a person on a computer (on the algorithms embedded in it) might lead to a decrease in the need for personal experience (the lack of personal experience is compensated by the use of artificial intelligence), and to a decrease in the ability to make independent, responsible decisions.

Fourth, for the first time in the history of humankind, the development of technology has made it possible to satisfy the main material needs of the overwhelming majority of the population in high-income countries. As a result, material problems could lose their power as a motivator of activity—indeed, a growing number of people do not want to work constantly, often change occupations, and seek not so much high earnings as self-expression. A certain motivational vacuum emerges. This exacerbates the general ideological vacuum, making it more difficult to solve ecological problems of climate change, as well as issues of people's self-identification. However, we must not forget that a very large part of humankind in low-income countries (and potentially a significant part of new generations) is still very far from the real satisfaction of basic needs, including education.[5]

Fifth, the development of medical technologies has significantly increased life expectancy. Along with a significant decline in birth rate, this, for the first time in history, led to the phenomenon of global aging, a radical deformation of the demographic structures: the society of the young is turning into a society of the elderly.[6]

Sixth, while earlier technological transformations changed nature, now human beings themselves have become the object of changes (biotechnology, genetic engineering, artificial organs, etc.), including the human consciousness (psychotropic substances, cognitive technologies, neuropsychological impact, etc.). We are

[5] See chapter "Socio-Political Transformations. A Difficult Path to Cybernetic Society". (Grinin et al., 2023d, this volume) and chapter "High Income and Low Income Countries. Toward a Common Goal at Different Speeds" (Grinin et al., 2023e, this volume).

[6] See chapter "Demography. Toward Optimization of Demographic Processes" (Korotayev et al., 2023a, this volume), chapter "Global Aging – an Integral Problem of the Future. How to Turn a Problem into a Development Driver?" (Grinin et al., 2023b), chapter "Socio-Political Transformations. A Difficult Path to Cybernetic Society" (Grinin et al., 2023d, this volume), chapter "Modeling Social Self-Organization and Historical Dynamics. Global Phase Transitions" (Malkov et al., 2023a, this volume), and chapter "Modeling Social Self-Organization and Historical Dynamics. Modern Society and a Look into the Global Futures: Cybernetic W-Society" (Akaev et al., 2023a) (see also, e.g., Jackson et al., 2008; Maestas et al., 2016; Goldstone 2015; Lee et al. 2021; McDaniel & Zimmer, 2016; Grinin & Korotayev, 2016; Grinin et al., 2017b; Bengtson, 2018; Fichtner, 2018; Barsukov, 2019; Korotayev et al., 2019, 2021, 2023c; Mitchell & Walker, 2020; UN Population Division, 2020; Whittington et al., 2020; Jackson, 2021; Medici, 2021).

talking about an invasion into the biological nature of humans, which was previously unthinkable (see, e.g., Grinin & Grinin, 2015, 2016, 2020; Bohan, 2020).

Seventh, the human society is moving to a new type of its organization. The macrohistorical analysis [see chapter "Macrohistorical Approach" (Sadovnichy et al., 2023, this volume)] suggests identifying two very important types of society in history. X-type social structures are typical for societies with stronger paternalistic institutions focused on collective survival and subordination of individual interests to the public (e.g., Kirdina, 2014). These are traditional ancient and medieval states and communist and authoritarian societies of the Modern period. Y-type social structures are typical for societies with social institutions aimed at encouraging and stimulating economic activity and competition, individualism. These are liberal market countries.[7]

However, now the situation differs significantly from the conditions under which the former X- and Y-societies flourished and were effective.[8] A chain of economic, social, and political crises and tensions within the last 15 years (not to mention climate and ecology degradation) attest to that (see Grinin et al., 2016a, b; Grinin & Korotayev, 2012, 2021; Grinin et al., 2021b). So, it makes sense to talk about the formation of a new type of society (we will preliminarily call it "cybernetic W-society"), since the restructuring will affect all the foundations of life and organization of society.[9]

2 Preliminary Conclusions

The analysis above suggests the following preliminary conclusions that could help to formulate forecast scenarios.

Cybernetic Revolution An important role in determining the trajectory of further development and the formation of the basic features of the future society will be played by the development of technologies within the ongoing Cybernetic Revolution. This revolution is called Cybernetic after the science of Cybernetics. We base

[7] For more detail on X-type and Y-type societies see chapter "Economics. Optimizing Growth" (Grinin et al., 2023c), chapter "Socio-Political Transformations. A Difficult Path to Cybernetic Society" (Grinin et al., 2023d, this volume), and chapter "Modeling Social Self-Organization and Historical Dynamics. Modern Society and a Look into the Global Futures: Cybernetic W-Society" (Akaev et al., 2023a, this volume).

[8] See chapter "Modeling Social Self-Organization and Historical Dynamics. Agrarian Society" (Malkov et al., 2023b, this volume), chapter "Modeling Social Self-Organization and Historical Dynamics. Industrial Society" (Akaev et al., 2023b, this volume), and chapter "Modeling Social Self-Organization and Historical Dynamics. Global Phase Transitions" (Malkov et al., 2023a, this volume).

[9] See chapter "Socio-Political Transformations. A Difficult Path to Cybernetic Society" (Grinin et al., 2023d, this volume) and chapter "Modeling Social Self-Organization and Historical Dynamics. Modern Society and a Look into the Global Futures: Cybernetic W-Society" (Akaev et al., 2023a, this volume).

our analysis of such systems on ideas from its field, which studies control various complex regulated systems (see, e.g., Wiener, 1948; Ashby, 1956; Beer, 1959, 1994; von Foerster & Zopf, 1962; Umpleby & Dent, 1999; Tesler, 2004).

The Cybernetic Revolution technologies will lead to a wide diffusion of self-regulating systems that are able not only to work but also to make decisions and correct goals with minimal or no human participation. At the same time, the complexity and functionality of such systems will grow dramatically (on the basis of the widespread use of artificial intelligence, robotics, and other smart systems); they will be used in production, biotechnology, medicine, genetic engineering, in the functioning of waste-free production and self-cleaning (involving bio- and nanotechnologies), as well as in social and administrative areas.[10]

The future of liberal capitalism The era of capitalism, based on liberal market principles and focused on maximizing consumption and profits, has come to the point beyond which capitalism will transform toward even more regulation than today, even at the international level (e.g., Wallerstein et al., 2013; Schwab & Malleret, 2020; Schwab & Vanham, 2021; Grinin et al., 2022; see also Fücks, 2013; Randers, 2012; Randers et al., 2018; Rifkin, 2011; Efimovich, 2020). The question of pace and directions of this transformation, as well as its manifestations in different groups of countries (high-income, low-income, and former communist) and regions, is very important and presents one of the main intrigues of the development of the world in the twenty-first century.[11] However, it is clear that there will be no socialism (communism) in its classic form either, since it was a product of the outgoing industrial era that emerged as an internal alternative to capitalism but also focused on "satisfying growing material needs". At the same time, a significant part of modern ideology in the world and in high-income countries is based on the "left" views and continues to shift further to the left. Therefore, the ideas and innovations of a socialist orientation in the future society will probably play an important role, but the extent of that role is still questionable. However, within such a "leftist" scenario of global development, it will be extremely important to preserve the foundations of market principles, as well as strong guarantees of citizens' rights and a balance of social interests.

Obviously, the development of individualism, consumerism, and capitalism, in general, encounters numerous problems. Therefore, and also because of the increasing influence of artificial intelligence and technology on the potential to control behavior, we think that the future society as a whole will be a society with more

[10] See chapter "Technology. Limitless Possibilities and Effective Control" (Grinin & Grinin, 2023b, this volume) and chapter "Modeling Social Self-Organization and Historical Dynamics. Modern Society and a Look into the Global Futures: Cybernetic W-Society" (Akaev et al., 2023a, this volume) (see also Grinin, 2007; Grinin & Grinin, 2016; Grinin et al., 2017a, 2020, 2021a).

[11] See chapter "Economics. Optimizing Growth" (Grinin et al., 2023c, this volume), chapter "Future Political Change. Toward a More Efficient World Order" (Grinin et al., 2023a, this volume), and chapter "High Income and Low Income Countries. Toward a Common Goal at Different Speeds" (Grinin et al., 2023e, this volume).

overregulation and more control than what is present in modern liberal societies. However, it is very important to limit this control and protect the main civil liberties.

Optimization of demographic processes The relative rate of world population growth has been decelerating, and in the twenty-first century, it will almost completely stabilize. Notably, in no way does this stabilization mean optimization of demographic processes; on the contrary, it reflects their radicalization. The high-income countries and some of the low income ones will experience depopulation, while most Sub-Saharan African countries, on the contrary, will experience explosive population growth.[12] Therefore, the optimization of demographic processes is an extremely important task.[13] However, it is important to note that the global population is stabilizing primarily due to internal causes (declining birth rates), and not external (environmental, etc.) constraints (which was the leitmotif of the early reports to the Club of Rome) (see, e.g., Kapitza, 2006; Dyson, 2010; Reher, 2011; Randers, 2012; Livi-Bacci, 2012; Korotayev et al., 2006, 2015; Podlazov, 2017; Randers et al., 2018; Korotayev, 2020; Lesthaeghe, 2020; Vollset et al., 2020; Harper, 2020). It is still unclear at what level stabilization will occur; and various paths are possible. In any case, it will be possible to feed this population with the help of modern technologies. Therefore, the problem of overpopulation is not the key one (except for Tropical Africa[14]). On the other hand, ensuring the quality of life of the aging population will present a critical problem.[15]

Environmental problems A more serious problem is the environmental one (see, e.g., Randers, 2012; Wijkman & Skånberg, 2017; Randers et al., 2018; von Weizsäcker & Wijkman, 2018). It affects the quality of life of both current and future generations. Careful monitoring and proactive measures are needed to prevent the exacerbation of environmental problems.[16]

---

[12] See chapter "Demography. Toward Optimization of Demographic Processes" (Korotayev et al., 2023a, this volume), chapter "Africa – the Continent of the Future. Challenges and Opportunities" (Grinin & Korotayev, 2023, this volume), and chapter "Modeling Social Self-Organization and Historical Dynamics. Africa's Futures" (Korotayev et al., 2023b) (see also, e.g., Alkema et al., 2011; Raftery et al., 2012; Zinkina & Korotayev, 2014a, b; Korotayev et al., 2016, 2022; Podlazov, 2017; Lutz et al., 2018; Kebede et al., 2019; Schoumaker, 2019; UN Population Division, 2019, 2022; May & Rotenberg, 2020; Vollset et al., 2020; Wittgenstein Centre, 2022).

[13] See chapter "Demography. Toward Optimization of Demographic Processes" (Korotayev et al., 2023a, this volume).

[14] See chapter "Africa – the Continent of the Future. Challenges and Opportunities" (Grinin & Korotayev, 2023, this volume) and chapter "Modeling Social Self-Organization and Historical Dynamics. Africa's Futures" (Korotayev et al., 2023b) (see also, e.g., Korotayev & Zinkina, 2015; Korotayev et al., 2016; Korotayev et al., 2022).

[15] See chapter "Global Aging – an Integral Problem of the Future. How to Turn a Problem into a Development Driver?" (Grinin et al., 2023b, this volume) and chapter "Socio-Political Transformations. A Difficult Path to Cybernetic Society" (Grinin et al., 2023d, this volume).

[16] See chapter "Ecology. Life in the "Unstable Biosphere"" (Kovaleva, 2023, this volume).

Global warming Even more serious is the problem of climate change,[17] although, of course, it is very closely related to the environmental one. Global warming, as mentioned in chapter "Climate and Energy. Scenarios of Energy Transition and Global Temperature Changes Based on Current Technologies and Trends" (Akaev & Davydova, 2023, this volume), is caused to a large extent by anthropogenic activity and proceeds at an unprecedented rate. Global temperatures have already risen by more than 1 °C above preindustrial levels. If carbon dioxide emissions are not radically curbed in two or three decades, humanity will not be able to keep the warming at the level of 2 °C (which is required by the Paris Agreement) (see, e.g., Randers, 2012; Randers et al., 2018; Xu et al., 2018; Diffenbaugh & Burke, 2019; Al-Ghussain, 2019; Choi et al., 2020; Masson-Delmotte et al., 2021; Ortiz-Bobea et al., 2021; Zandalinas et al. 2021; Gills & Morgan, 2022).

New world order Another critically important problem is that the former world order based on the U.S. hegemony has begun to weaken noticeably. As a result, political turbulence, conflicts, wars, and other destructive processes are intensifying in the world. This, along with other factors (such as the coronavirus pandemic), at first has slowed down and then reversed the processes of globalization, with negative effects on the economy (see, e.g., Schwab & Malleret, 2020; Schwab & Vanham, 2021; Grinin & Korotayev, 2021; Grinin & Grinin, 2021; Grinin et al., 2021b). It is obvious that the World System needs a new world order. We believe that the new order will be neither unipolar nor bipolar; perhaps for quite a long time, it will be an interaction of major global players (countries and their alliances). The new world order will require a fairly stable balance of forces and interests, new models of supranational governance, and coordination of world processes. This, in turn, requires new principles that would reduce competitive confrontation and stimulate cooperation.[18]

3 A Brief Forecast of Future Societal Variations

We would like to end this chapter with a brief forecast[19] of the possible directions in which future societies might develop.

[17] See chapter "Climate and Energy. Scenarios of Energy Transition and Global Temperature Changes Based on Current Technologies and Trends" (Akaev & Davydova, 2023, this volume) (see also, e.g., Randers, 2012; Wijkman & Skånberg, 2017; Randers et al., 2018; von Weizsäcker & Wijkman, 2018).

[18] See chapter "Future Political Change. Toward a More Efficient World Order" (Grinin et al., 2023a, this volume) (see also Grinin, 2016; Grinin et al., 2016a, b, 2021b).

[19] See chapter "Socio-Political Transformations. A Difficult Path to Cybernetic Society" (Grinin et al., 2023d, this volume) and chapter "Analyzing Social Self-Organization and Historical Dynamics. Future Cybernetic W-Society: Socio-Political Aspects" (Grinin & Grinin, 2023a, this volume) for detail.

In the long term, the range of possible changes is quite wide, more precisely, there will be a continuum of variations of the future social structure and order in different societies, whereas two development poles will compete:

1. The formation of a hierarchical estate (caste) totalitarian society controlled by an explicit or implicit, formal or informal global elite, which would control resource and information flows. Digital technologies would be used for total control over the behavior of the population and control over people's mentality.
2. The formation of a "world-organism"—a society where, against the background of professional specialization, there is no discrimination; and governance, based on network technologies, is carried out in the interests of all social strata. The society uses technologies to control the management elite formed on the basis of meritocratic principles. An important role in such a society would be played by ideology (in the form of ethical norms and self-restraints that suppress antagonistic competition), which establishes a system of rules for social interaction and a collective workout of consensual decisions. Competitive relations are channeled into the sphere of sports, science and technology, and creative professions in the nonantagonistic form. These options are two extreme alternatives. In their pure form, they may not be implemented.

The actual global development is most likely to go somewhere between those two polarities. We would like to say that it would be preferable if we could move to a future society that is based on cooperation rather than competition and, at the same time, preserves privacy, individuality, and initiative of its citizens.

Against this background, the key task is to form a new system of values, a new ideology that should replace both totalitarian and liberal doctrines in order to ensure movement closer to the second path of development. This ideology could take the form of an ethical system that establishes the primacy of collectivist values and social solidarity. The transition from the modern X- and Y-societies to the future cybernetic W-society would be a human-made process. It is crucial to intensify efforts in that direction since this is an issue of designing the future.[20]

References

Akaev, A., & Davydova, O. (2023). Climate and energy. Scenarios of energy transition and global temperature changes based on current technologies and trends. In V. Sadovnichy et al. (Eds.), *Reconsidering the limits to growth. A report to the Russian Association of the Club of Rome* (pp. 53–70). Springer. https://doi.org/10.1007/978-3-031-34999-7_4

Akaev, A., Malkov, S., Grinin, L., Bilyuga, S., Davydova, O., Grinin, A., Kovaleva, N., Malkov, A., Musieva, J., & Korotayev, A. (2023a). Modeling social self-organization and historical dynamics. Modern society and a look into the global futures: Cybernetic W-society. In

[20]See chapter "Socio-Political Transformations. A Difficult Path to Cybernetic Society" (Grinin et al., 2023d, this volume).

V. Sadovnichy et al. (Eds.), *Reconsidering the limits to growth. A report to the Russian Association of the Club of Rome* (pp. 419–460). Springer. https://doi.org/10.1007/978-3-031-34999-7_19

Akaev, A., Malkov, S., Davydova, O., Kovaleva, N., Malkov A., & Korotayev, A. (2023b). Modeling social self-organization and historical dynamics. Industrial society. In V. Sadovnichy et al. (Eds.), *Reconsidering the limits to growth. A report to the Russian Association of the Club of Rome* (pp. 337–385). Springer. https://doi.org/10.1007/978-3-031-34999-7_17

Al-Ghussain, L. (2019). Global warming: Review on driving forces and mitigation. *Environmental Progress & Sustainable Energy, 38*(1), 13–21. https://doi.org/10.1002/ep.13041

Alharbi, F. S. (2020). Dealing with data breaches amidst changes in technology. *International Journal of Computer Science and Security (IJCSS), 14*(3), 108–115.

Alkema, L., Raftery, A. E., Gerland, P., Clark, S. J., Pelletier, F., Buettner, T., & Heilig, G. K. (2011). Probabilistic projections of the total fertility rate for all countries. *Demography, 48*(3), 815–839. https://doi.org/10.1007/s13524-011-0040-5

Ashby, R. (1956). *An introduction to cybernetics*. Chapman and Hall.

Ashman, H., Brailsford, T., Cristea, A. I., Sheng, Q. Z., Stewart, C., Toms, E. G., & Wade, V. (2014). The ethical and social implications of personalization technologies for e-learning. *Information & Management, 51*(6), 819–832. https://doi.org/10.1016/j.im.2014.04.003

Barsukov, V. N. (2019). From the demographic dividend to population ageing: World trends in the system-wide transition. *Economic and Social Changes: Facts, Trends, Forecast, 12*(4), 167–182.

Beer, S. (1959). *Cybernetics and management*. English Universities Press.

Beer, S. (1994). *Decision and control: The meaning of operational research and management cybernetics*. Wiley.

Bengtson, V. (Ed.). (2018). *Global aging and challenges to families*. Routledge.

Bohan, E. (2020). Threshold 9: Big history as a roadmap for the future. In A. Korotayev & D. LePoire (Eds.), *The 21st century singularity and global futures. A big history perspective* (pp. 267–286). Springer. https://doi.org/10.1007/978-3-030-33730-8_13

Brammer, S., Branicki, L., & Linnenluecke, M. (2020). COVID-19, societalization and the future of business in society. *Science of the Total Environment, 34*(4), 2–7. https://doi.org/10.5465/amp.2019.0053

Cecere, G., Le Guel, F., & Soulié, N. (2015). Perceived internet privacy concerns on social networks in Europe. *Technological Forecasting and Social Change, 96*, 277–287. https://doi.org/10.1016/j.techfore.2015.01.021

Choi, D., Gao, Z., & Jiang, W. (2020). Attention to global warming. *The Review of Financial Studies, 33*(3), 1112–1145.

Diffenbaugh, N. S., & Burke, M. (2019). Global warming has increased global economic inequality. *Proceedings of the National Academy of Sciences, 116*(20), 9808–9813. https://doi.org/10.1073/pnas.1816020116

Dyson, T. (2010). *Population and development. The demographic transition*. Zed Books.

Efimovich, A. (2020). Negative rates and other financial technologies in modern economic reality on the world-system scale. *Journal of Globalization Studies, 11*(2), 63–76. https://doi.org/10.30884/jogs/2020.02.05

Fichtner, J. J. (2018). Global aging and public finance. *Business Economics, 53*(2), 72–78.

Fücks, R. (2013). *Intelligent Wachsen. Die Grüne Revolution*. Carl Hanser Verlag.

Gills, B., & Morgan, J. (Eds.). (2022). *Economics and climate emergency*. Routledge.

Goldstone, J. A. (2015). Population ageing and global economic growth. In J. A. Goldstone, L. Grinin, & A. Korotayev (Eds.), *History & mathematics: Demography & ageing* (pp. 147–155). Uchitel.

Goldstone, J. A., Grinin, L., & Korotayev, A. (2022). Conclusion. How many revolutions will we see in the 21st century? In J. A. Goldstone, L. Grinin, & A. Korotayev (Eds.), *Handbook of revolutions in the 21st century: The new waves of revolutions, and the causes and effects of disruptive political change* (pp. 1033–1057). Springer. https://doi.org/10.1007/978-3-030-86468-2_41

Grinin, L. (2007). Production revolutions and periodization of history: A comparative and theoretic-mathematical approach. *Social Evolution & History, 6*(2), 11–55.

Grinin, L. (2016). The new world order and philosophy. In A. N. Chumakov & W. C. Gay (Eds.), *Between past orthodoxies and the future of globalization. Contemporary philosophical problems* (pp. 143–156). Brill/Rodopi.

Grinin, L. (2022). Revolutions of the 21st century as a factor of the world system reconfiguration. In J. A. Goldstone, L. Grinin, & A. Korotayev (Eds.), *Handbook of revolutions in the 21st century: The new waves of revolutions, and the causes and effects of disruptive political change* (pp. 973–996). Springer. https://doi.org/10.1007/978-3-030-86468-2_38

Grinin, L., & Grinin, A. (2015). *Ot rubil do nanorobotov. Mir na puti k epokhe samoupravlyayemykh sistem (istoriya tekhnologiy i opisaniye ikh budushchego)*. Uchitel.

Grinin, L., & Grinin, A. (2016). *The cybernetic revolution and the forthcoming epoch of self-regulating systems*. Uchitel.

Grinin, A., & Grinin, L. (2020). Crossing the threshold of cyborgization. *Journal of Big History, 4*(3), 54–65. https://doi.org/10.22339/jbh.v4i3.4330

Grinin, A., & Grinin, L. (2021). Analyzing the global problems of the twenty-first century. A review and forecast based on the report to the Club of Rome 'Come On!'. *Journal of Globalization Studies, 12*(2), 181–195. https://doi.org/10.30884/jogs/2021.02.10

Grinin, L., & Grinin, A. (2023a). Analyzing social self-organization and historical dynamics. Future cybernetic W-society: Sociopolitical aspects. In V. Sadovnichy et al. (Eds.), *Reconsidering the limits to growth. A report to the Russian Association of the Club of Rome* (pp. 491–519). Springer. https://doi.org/10.1007/978-3-031-34999-7_21

Grinin, L., & Grinin, A. (2023b). Technology. Limitless possibilities and effective control. In V. Sadovnichy et al. (Eds.), *Reconsidering the limits to growth. A report to the Russian Association of the Club of Rome* (pp. 139–154). Springer. https://doi.org/10.1007/978-3-031-34999-7_8

Grinin, L., & Korotayev, A. (2012). Does "Arab spring" mean the beginning of world system reconfiguration? *World Futures, 68*(7), 471–505. https://doi.org/10.1080/02604027.2012.697836

Grinin, L., & Korotayev, A. (2016). Global population ageing, the sixth Kondratieff wave, and the global financial system. *Journal of Globalization Studies, 7*(2), 11–31.

Grinin, L., & Korotayev, A. (2021). Seven weaknesses of the U.S., Donald Trump, and the future of American hegemony. *World Futures, 77*(1), 23–54. https://doi.org/10.1080/02604027.2020.1801309

Grinin, L., & Korotayev, A. (2023). Africa: The continent of the future. Challenges and opportunities. In V. Sadovnichy et al. (Eds.), *Reconsidering the limits to growth. A report to the Russian Association of the Club of Rome* (pp. 225–238). Springer. https://doi.org/10.1007/978-3-031-34999-7_13

Grinin, L., Ilyin, I., & Andreev, A. (2016a). World order in the past, present, and future. *Social Evolution and History, 15*(1), 58–84.

Grinin, L., Korotayev, A., & Tausch, A. (2016b). *Economic cycles, crises, and the global periphery*. Springer. https://doi.org/10.1007/978-3-319-41262-7

Grinin, L., Grinin, A., & Korotayev, A. (2017a). The MANBRIC-technologies in the forthcoming technological revolution. In T. Devezas, J. Leitão, & A. Sarygulov (Eds.), *Industry 4.0 – Entrepreneurship and structural change in the new digital landscape: What is coming on along with the fourth industrial revolution* (pp. 243–261). Springer. https://doi.org/10.1007/978-3-319-49604-7_13

Grinin, L., Grinin, A., & Korotayev, A. (2017b). Forthcoming Kondratieff wave, cybernetic revolution, and global ageing. *Technological Forecasting and Social Change, 115*, 52–68. https://doi.org/10.1016/j.techfore.2016.09.017

Grinin, L., Grinin, A., & Korotayev, A. (2020). A quantitative analysis of worldwide long-term technology growth: From 40,000 BCE to the early 22nd century. *Technological Forecasting and Social Change, 155*, 1–15. https://doi.org/10.1016/j.techfore.2020.119955

Grinin, L., Grinin, A., & Korotayev, A. (2021a). COVID-19 pandemic as a trigger for the acceleration of the cybernetic revolution, transition from E-government to E-state, and change in social relations. *Technological Forecasting and Social Change, 175*, 1–17. https://doi.org/10.1016/j.techfore.2021.121348

Grinin, L., Grinin, A., & Korotayev, A. (2021b). Global trends and forecasts of the 21st century. *World Futures, 77*(5), 335–370. https://doi.org/10.1080/02604027.2021.1949939

Grinin, L., Malkov, S., Grinin, A., & Korotayev, A. (2022). Will capitalism die? Reflections on the capitalism of the past, present and future. *Sotsiologicheskiy Zhurnal = Sociological Journal, 28*(2), 100–130. https://doi.org/10.19181/socjour.2022.28.2.8988

Grinin, L., Grinin, A., & Korotayev, A. (2023a). Future political change. Toward a more efficient world order. In V. Sadovnichy et al. (Eds.), *Reconsidering the limits to growth. A report to the Russian Association of the Club of Rome* (pp. 191–206). Springer. https://doi.org/10.1007/978-3-031-34999-7_11

Grinin, L., Grinin, A., & Korotayev, A. (2023b). Global aging – An integral problem of the future. How to turn a problem into a development driver? In V. Sadovnichy et al. (Eds.), *Reconsidering the limits to growth. A report to the Russian Association of the Club of Rome* (pp. 117–135). Springer. https://doi.org/10.1007/978-3-031-34999-7_7

Grinin, L., Grinin, A., & Malkov, S. (2023c). Economics: Optimizing growth. In V. Sadovnichy et al. (Eds.), *Reconsidering the limits to growth. A report to the Russian Association of the Club of Rome* (pp. 155–168). Springer. https://doi.org/10.1007/978-3-031-34999-7_9

Grinin, L., Grinin, A., & Malkov, S. (2023d). Socio-political transformations. A difficult path to cybernetic society. In V. Sadovnichy et al. (Eds.), *Reconsidering the limits to growth. A report to the Russian Association of the Club of Rome* (pp. 169–189). Springer. https://doi.org/10.1007/978-3-031-34999-7_10

Grinin, L., Malkov, S., & Korotayev, A. (2023e). High income and low income countries. Toward a common goal at different speeds. In V. Sadovnichy et al. (Eds.), *Reconsidering the limits to growth. A report to the Russian Association of the Club of Rome* (pp. 207–224). Springer. https://doi.org/10.1007/978-3-031-34999-7_12

Harper, A. (2020). Fresh look at the global history. A review of the twenty-first century singularity and global futures: A big history perspective. *Journal of Globalization Studies, 11*(1), 134–141. https://doi.org/10.30884/jogs/2020.01.09

Jackson, R. (2021). *The shape of things to come. The macro challenges of population aging*. Concord Coalition & the Global Aging Institute.

Jackson, R., Howe, N., Strauss, R., & Nakasima, K. (2008). *The greying of the great powers. Demography and geopolitics in the 21st century*. Center for Strategic & International Studies.

Kapitza, S. (2006). *Global population blow-up and after. Report to the Club of Rome*. Global Marshall Plan Initiative.

Kebede, E., Goujon, A., & Lutz, W. (2019). Stalls in Africa's decline partly result from disruptions in female education. *Proceedings of the National Academy of Sciences, 116*(8), 2891–2896. https://doi.org/10.1073/pnas.1717288116

Kirdina, S. (2014). Institutional matrices theory, or X&Y theory: The main provisions and applications. *Journal of Institutional Studies, 6*(3), 13–33.

Korotayev, A. (2020). The 21st century singularity in the big history perspective. A re-analysis. In A. Korotayev & D. LePoire (Eds.), *The 21st century singularity and global futures. A big history perspective* (pp. 19–75). Springer. https://doi.org/10.1007/978-3-030-33730-8_2

Korotayev, A., & Zinkina, J. (2015). East Africa in the Malthusian trap? *Journal of Low Income Societies, 31*(3), 1–36. https://doi.org/10.1177/0169796X15590322

Korotayev, A., Malkov, A., & Khaltourina, D. (2006). *Introduction to social macrodynamics: Compact macromodels of the world system growth*. KomKniga/URSS.

Korotayev, A., Goldstone, J. A., & Zinkina, J. (2015). Phases of global demographic transition correlate with phases of the great divergence and great convergence. *Technological Forecasting and Social Change, 95*, 163–169. https://doi.org/10.1016/j.techfore.2015.01.017

Korotayev, A., Zinkina, J., Goldstone, J., & Shulgin, S. (2016). Explaining current fertility dynamics in tropical Africa from an anthropological perspective: A cross-cultural investigation. *Cross-Cultural Research, 50*(3), 251–280. https://doi.org/10.1177/1069397116644158

Korotayev, A., Novikov, K., & Shulgin, S. (2019). Values of the elderly people in an aging world. *Sociology of Power, 31*(1), 114–142. https://doi.org/10.22394/2074-0492-2019-1-114-142

Korotayev, A., Butovskaya, M., Shulgin, S., & Zinkina, J. (2021). Vliyaniye global'nogo stareniya na global'nuyu sistemu tsennostey. *Vek Globalizatsii, 4*, 69–80. https://doi.org/10.30884/vglob/2021.04.05

Korotayev, A., Shulgin, S., Zinkina, J., & Slav, M. (2022). Estimates of the possible economic effect of the demographic dividend for sub-Saharan Africa for the period up to 2036. *Vostok (Oriens), 2*, 108–123. https://doi.org/10.31857/S086919080019128-7

Korotayev, A., Malkov, S., & Musieva, J. (2023a). Demography: Toward optimization of demographic processes. In V. Sadovnichy et al. (Eds.), *Reconsidering the limits to growth. A report to the Russian Association of the Club of Rome* (pp. 97–116). Springer. https://doi.org/10.1007/978-3-031-34999-7_6

Korotayev, A., Shulgin, S., Ustyuzhanin, V., Zinkina, J., & Grinin, L. (2023b). Modeling social self-organization and historical dynamics. Africa's futures. In V. Sadovnichy et al. (Eds.), *Reconsidering the limits to growth. A report to the Russian Association of the Club of Rome* (pp. 461–490). Springer. https://doi.org/10.1007/978-3-031-34999-7_20

Korotayev, A., Butovskaya, M., Shulgin, S., & Zinkina, J. (2023c). How can the global ageing affect the global value system? An evolutionary perspective. *Social Evolution & History, 22*(1), 57–76. https://doi.org/10.30884/seh/2023.01.03

Kovaleva, N. (2023). Ecology. Life in the "unstable biosphere". In V. Sadovnichy et al. (Eds.), *Reconsidering the limits to growth. A report to the Russian Association of the Club of Rome* (pp. 71–95). Springer. https://doi.org/10.1007/978-3-031-34999-7_5

Lee, J., Phillips, D., Wilkens, J., & Gateway to Global Aging Data Team. (2021). Gateway to global aging data: Resources for cross-national comparisons of family, social environment, and healthy aging. *The Journals of Gerontology: Series B, 76*(Suppl 1), S5–S16. https://doi.org/10.1093/geronb/gbab050

Lesthaeghe, R. (2020). The second demographic transition, 1986–2020: Sub-replacement fertility and rising cohabitation – A global update. *Genus, 76*(1), 1–38. https://doi.org/10.1186/s41118-020-00077-4

Livi-Bacci, M. (2012). *A concise history of world population*. Wiley-Blackwell.

Lutz, W., Goujon, A., Kc, S., Stonawski, M., & Stilianakis, N. (2018). *Demographic and human capital scenarios for the 21st century: 2018 assessment for 201 countries*. Publications Office of the European Union.

Maestas, N., Mullen, K. J., & Powell, D. (2016). *The effect of population aging on economic growth, the labor force and productivity*. NBER.

Malkov, S., Grinin, L., Grinin, A., Musieva, J., & Korotayev, A. (2023a). Modeling social self-organization and historical dynamics. Global phase transitions. In V. Sadovnichy et al. (Eds.), *Reconsidering the limits to growth. A report to the Russian Association of the Club of Rome* (pp. 387–417). Springer. https://doi.org/10.1007/978-3-031-34999-7_18

Malkov, S., Kovaleva, N., Grinin, L., & Korotayev, A. (2023b). Modeling social self-organization and historical dynamics. Agrarian society. In V. Sadovnichy et al. (Eds.), *Reconsidering the limits to growth. A report to the Russian Association of the Club of Rome* (pp. 309–335). Springer. https://doi.org/10.1007/978-3-031-34999-7_16

Masson-Delmotte, V., Zhai, P., Pirani, A., Connors, S. L., Péan, C., Berger, S., Caud, N., Chen, Y., Goldfarb, L., Gomis, M. I., Huang, M., Leitzell, K., Lonnoy, E., Matthews, J. B. R., Maycock, T. K., Waterfield, T., Yelekçi, O., Yu R., & Zhou B. (Eds.). (2021). *Summary for Policymakers. Climate Change 2021: The physical science basis. Contribution of Working Group I to the Sixth*

Assessment Report of the Intergovernmental Panel on Climate Change. Cambridge University Press.

May, J. F., & Rotenberg, S. (2020). A call for better integrated policies to accelerate the fertility decline in sub-Saharan Africa. *Studies in Family Planning, 51*(2), 193–204. https://doi.org/10.1111/sifp.12118

McDaniel, S. A., & Zimmer, Z. (Eds.). (2016). *Global ageing in the twenty-first century: Challenges, opportunities and implications*. Routledge.

Medici, A. C. (2021). *Health sector challenges and policies in the context of ageing populations*. United Nations, Department of Economics and Social Affairs, Population Division.

Mitchell, E., & Walker, R. (2020). Global ageing: Successes, challenges and opportunities. *British Journal of Hospital Medicine, 81*(2), 1–9. https://doi.org/10.12968/hmed.2019.0377

Moustaka, V., Theodosiou, Z., Vakali, A., Kounoudes, A., & Anthopoulos, L. G. (2019). Enhancing social networking in smart cities: Privacy and security borderlines. *Technological Forecasting and Social Change, 142*, 285–300. https://doi.org/10.1016/j.techfore.2018.10.026

Ortiz-Bobea, A., Ault, T. R., Carrillo, C. M., et al. (2021). Anthropogenic climate change has slowed global agricultural productivity growth. *Nature. Climate Change, 11*, 306–312. https://doi.org/10.1038/s41558-021-01000-1

Podlazov, A. (2017). A theory of the global demographic process. *Herald of the Russian Academy of Sciences, 87*(3), 256–266. https://doi.org/10.1134/S1019331617030054

Raftery, A. E., Li, N., Ševčíková, H., Gerland, P., & Heilig, G. K. (2012). Bayesian probabilistic population projections for all countries. *Proceedings of the National Academy of Sciences, 109*(35), 13915–13921. https://doi.org/10.1073/pnas.1211452109

Randers, J. (2012). *2052. A global forecast for the next forty years. A report to the Club of Rome*. Chelsea Green Publishing.

Randers, J., Rockström, J., Stoknes, P. E., Golüke, U., Collste, D., & Cornell, S. (2018). *Transformation is feasible. A report to the Club of Rome*. Stockholm Resilience Centre.

Reher, D. (2011). Economic and social implications of the demographic transition. *Population and Development Review, 37*, 11–33. https://doi.org/10.1111/j.1728-4457.2011.00376.x

Rifkin, J. (2011). *The third industrial revolution. How lateral power is transforming energy, the economy, and the world*. Palgrave Macmillan.

Sadovnichy, V., Akaev, A., Ilyin, I., Malkov, S., Grinin, L., Aleshkovski, I., & Korotayev, A. (2023). Macrohistorical approach. In V. Sadovnichy et al. (Eds.), *Reconsidering the limits to growth. A report to the Russian Association of the Club of Rome* (pp. 17–33). Springer. https://doi.org/10.1007/978-3-031-34999-7_2

Schoumaker, B. (2019). Stalls in fertility transitions in sub-Saharan Africa: Revisiting the evidence. *Studies in Family Planning, 50*(3), 257–278. https://doi.org/10.1111/sifp.12098

Schwab, K., & Malleret, T. (2020). *COVID-19: The great reset*. Forum Publishing.

Schwab, K., & Vanham, P. (2021). *Stakeholder capitalism*. Wiley.

Schwartz, P. M. (1999). Internet privacy and the state. *Connecticut Law Review, 32*, 815–829.

Solove, D. J. (2008). *Understanding privacy*. Harvard University Press.

Tesler, G. S. (2004). *Novaya kibernetika*. Logos.

Umpleby, S. A., & Dent, E. B. (1999). The origins and purposes of several traditions in systems theory and cybernetics. *Cybernetics and Systems, 30*, 79–103.

UN Population Division. (2019). *World population prospects 2019*. United Nations.

UN Population Division. (2020). *World population ageing 2020*. United Nations.

UN Population Division. (2022). *United Nations population division database*. United Nations. http://www.un.org/esa/population

Vollset, S. E., Goren, E., Yuan, C.-W., Cao, J., Smith, A. E., Hsiao, T., Murray, C. J. L., et al. (2020). Fertility, mortality, migration, and population scenarios for 195 countries and territories from 2017 to 2100: A forecasting analysis for the global burden of disease study. *The Lancet, 396*(10258), 1285–1306. https://doi.org/10.1016/S0140-6736(20)30677-2

von Foerster, H., & Zopf, G. W. (1962). Principles of self-organization. In M. C. Yovits & S. Cameron (Eds.), *Self-organization* (pp. 31–50). Pergamon Press.

von Weizsäcker, E. U., & Wijkman, A. (2018). *Come on! Capitalism, short-termism, population and the destruction of the planet. A report to the Club of Rome*. Springer. https://doi.org/10.1007/978-1-4939-7419-1

Wallerstein, I., Collins, R., Mann, M., Derluguian, G., & Calhoun, C. (2013). *Does capitalism have a future?* Oxford University Press.

Westin, A. F. (1966). Science, privacy, and freedom: Issues and proposals for the 1970's. Part I. the current impact of surveillance on privacy. *Columbia Law Review, 66*(6), 1003–1050.

Whittington, F. J., Kunkel, S. R., & de Medeiros, K. (Eds.). (2020). *Global aging: Comparative perspectives on aging and the life course*. Springer.

Wiener, N. (1948). *Cybernetics, or control and communication in the animal and the machine*. MIT Press.

Wijkman, A., & Skånberg, K. (2017). *The circular economy and benefits for society. Jobs and climate clear winners in an economy based on renewable energy and resource efficiency. A report to the Club of Rome*. Club of Rome.

Wittgenstein Centre. (2022). *Wittgenstein Centre Human Capital Data Explorer*. http://dataexplorer.wittgensteincentre.org/wcde-v2/

Xu, Y., Ramanathan, V., & Victor, D. G. (2018). Global warming will happen faster than we think. *Nature, 564*(7734), 30–32.

Zandalinas, S. I., Fritschi, F. B., & Mittler, R. (2021). Global warming, climate change, and environmental pollution: Recipe for a multifactorial stress combination disaster. *Trends in Plant Science, 26*(6), 588–599. https://doi.org/10.1016/j.tplants.2021.02.011

Zinkina, J., & Korotayev, A. (2014a). Explosive population growth in tropical Africa: Crucial omission in development forecasts (emerging risks and way out). *World Futures, 70*(4), 271–305. https://doi.org/10.1080/02604027.2014.894868

Zinkina, J., & Korotayev, A. (2014b). Projecting Mozambique's demographic futures. *Journal of Futures Studies, 19*(2), 21–40.

Viktor Sadovnichy is a Doctor of Physics and Mathematics, Professor, Full Member of the Russian Academy of Sciences, and Rector of Lomonosov Moscow State University since 1992. Fields of scientific research: mathematics, mechanics, and computer science. Sadovnichy's research deals with the theory of differential operators, functional analysis, and mathematical modeling of long-term social and economic dynamics. He has obtained important theoretical and applied results related to the issues of dynamic simulation of controlled flights, in particular, of spacecraft and aircraft. This work formed the basis for the creation of the first simulator of zero-gravity in the Earth conditions in the world.

Askar Akaev is the First President of the Kyrgyz Republic and Foreign Full Member of the Russian Academy of Sciences. He works as Chief Research Professor in the Prigogine Institute for Mathematical Investigations of Complex Systems at Lomonosov Moscow State University. He is also the Academic Supervisor of the Centre for Stability and Risk Analysis at HSE University, Moscow, Russia. Askar Akaev has authored and co-authored over 300 scientific publications.

Ilya Ilyin is a well-known Russian researcher, Head of the Department of Global Studies, and Dean of the Faculty of Global Studies at Lomonosov Moscow State University, author of more than 350 academic papers. He is the Chairman of the Presidium of the Society for Global Studies, President of the International Public Organization for the Promotion of Public Diplomacy, Scientific, Educational and Youth Cooperation "Eurasian Commonwealth", Chief Scientific Secretary of the International Academy of Global Studies, Chairman of the Russian Committee and Vice President of the Intergovernmental Council for the UNESCO Program "Management of Social transformations – MOST", honorary worker in the field of state youth policy and education, editor-in-chief of the Bulletin of Moscow University. Series 27: Global Studies and Geopolitics.

Sergey Malkov heads the Center for Long-Term Forecasting and Strategic Planning at the Lomonosov Moscow State University and works as a Senior Research Professor at the Centre for Stability and Risk Analysis at HSE University in Moscow, as well as at the Institute of Economics, Russian Academy of Sciences. He is the author of over 300 scholarly publications.

Leonid Grinin is a Senior Research Professor at the Centre for Stability and Risk Analysis at HSE University in Moscow and a Senior Research Professor at the Institute of Oriental Studies, Russian Academy of Sciences. He is co-editor of the international journals *Social Evolution & History* and the *Journal of Globalization Studies* and the international yearbooks *Evolution, History & Mathematics*, and *Kondratieff Waves*. He is the author of more than 600 publications, including more than 33 monographs in Russian, English, German, Spanish and Chinese.

Aleksey Andreev is a Candidate of Biological Sciences, Associate Professor of the Faculty of Global Studies at Lomonosov Moscow State University. He is a specialist in the field of global socio-natural processes, forecasting the development of science and technology, and economics of education—author of over 100 scientific publications.

Andrey Korotayev heads the Centre for Stability and Risk Analysis at HSE University in Moscow, Russia. He is also a Senior Research Professor at the Institute of Oriental Studies and Institute for African Studies, Russian Academy of Sciences, as well as at the Faculty of Global Studies of Lomonosov Moscow State University and the Russian Presidential Academy of National Economy and Public Administration. He has authored or co-authored over 650 scholarly publications.

Part II

Problems, Forecasts, Solutions (Climate, Ecology, Demography, Aging)

Climate and Energy: Energy Transition Scenarios and Global Temperature Changes Based on Current Technologies and Trends

Askar Akaev and Olga Davydova

Abstract

The chapter is devoted to the analysis of climate change issues and the transition to renewable energy sources. The features of the current climate situation are associated with a general increase in the average global temperature as a result of an extremely high concentration of carbon dioxide (CO_2) in the atmosphere, the amount of which is increasing and posing a threat to the stability of the global ecological system as a whole. Taking into consideration the fact that the main share of CO_2 emissions is accounted for by energy consumption (which experienced over the entire timeline of history transitions from one type of energy resources to another—from biomass to coal, from coal to oil, and from oil to natural gas), Akaev and Davydova analyze the possibilities of transitioning to renewable energy sources (RES) forecasted to take place by the second half of the twenty-first century. They carry out mathematical modeling of this transition with various scenarios for the future of the fuel and energy balance in the twenty-first century. For this, Akaev and Davydova have developed a specialized mathematical model that takes into account current trends in energy consumption based on the data from the largest energy companies and international organizations in the energy sector, such as BP, Equinor, Shell, International Energy Agency (IEA), International Renewable Energy Agency (IRENA), and others. Three scenarios for the increase in the average global temperature of the surface atmosphere in the twenty-first century are proposed: the conservative scenario, the ambitious scenario, and the Net Zero scenario. The conservative scenario assumes that government policies, technologies, and social preferences continue to evolve in the same

A. Akaev
Lomonosov Moscow State University, Moscow, Russia

HSE University, Moscow, Russia

O. Davydova (✉)
Lomonosov Moscow State University, Moscow, Russia

V. Sadovnichy et al. (eds.), *Reconsidering the Limits to Growth*, World-Systems Evolution and Global Futures, https://doi.org/10.1007/978-3-031-34999-7_4

way as in the recent past. The ambitious scenario envisages the introduction of measures leading to a significant reduction in carbon emissions from energy use, which in turn makes it possible to limit the increase in global temperature in the twenty-first century. The Net Zero scenario, which the authors consider the optimal one, assumes that the measures proposed in the ambitious scenario are complemented and reinforced by significant changes in the behavior and preferences of society. The chapter details modern energy-efficient technologies and methods of using renewable energy sources, the implementation of which is envisaged in the framework of the optimal Net Zero scenario.

1 Current Situation

Since 1850, the concentration of carbon dioxide in the atmosphere has increased significantly, from typical for the preindustrial period 280 ppm and observed for hundreds of years, to 421 ppm at present, i.e., more than 50%, which significantly contributed to the climate warming (NOAA, 2022). Two-thirds of global warming has been caused by the increase in the concentration of CO_2 in the Earth's atmosphere in the nineteenth and twentieth centuries. These figures are substantiated by physical measurements over the past 150 years of both the average world temperature of the surface atmosphere and the growth of CO_2 concentration in the Earth's atmosphere (Pachauri & Meyer, 2014; Masson-Delmotte et al., 2021). Carbon dioxide is constantly accumulating in the Earth's atmosphere, and its emissions are currently at their peak (British Petroleum, 2021; IEA, 2020a, b, 2021a, b, c). According to the World Meteorological Organization, the average global temperature in 2021 was 1.2 °C higher than the preindustrial level of +14 °C, which is a direct consequence of anthropogenic activity (WMO, 2021). Climate change has become one of the most pressing issues for humanity. The effects of climate change are everywhere.[1] Global Ocean temperatures are rising, glaciers are melting, sea levels are rising (NOAA, 2020), and extreme weather is becoming more and more severe and destructive. Droughts, wildfires, floods, and hurricanes have become more frequent. In 2020 alone, more than 415 natural disasters occurred worldwide (Statista, 2022). Natural disasters kill an average of 60,000 people a year worldwide (Ritchie & Roser, 2021). The rate of degradation of arable land is 30–35 times higher than historical rates,[2] more territories are desertified, and crop yields are decreasing (World Bank Group, 2014; Ortiz-Bobea et al., 2021) due to the depletion of water reserves among other things. According to Welthungerhilfe, 811 million people were affected by hunger in 2020 (Welthungerhilfe, 2021), which could intensify

[1] This issue has been in the focus of attention of many recent reports to The Club of Rome (see, e.g., Randers, 2012; Maxton et al., 2016; Wijkman & Skånberg, 2017; Randers et al., 2018; von Weizsäcker & Wijkman, 2018; Berg, 2019). Of course, the problem of climate change is one of the most important in the modern ideology. Therefore, this issue can be combined with very different ideological aspects (see, e.g., Garske, 2020; Kiss, 2020)

[2] See Chapter "Ecology. Life in the "unstable biosphere"" (Kovaleva, 2023, this volume).

regional tensions and exacerbate existing conflicts. Climatologists have shown that if CO_2 emissions are not reduced by two or three times by the middle of the century, then warming cannot be kept at the level of 2 °C, and by the end of the twenty-first century, it will exceed 3–4 °C, which will lead to catastrophic consequences (Masson-Delmotte et al., 2021), so urgent action is needed.

In 2015, the Paris Agreement was adopted, the goal of which was to keep the warming at 1.5–2 °C (United Nations, 2016). To achieve this goal, according to estimations of the International Panel on Climate Change, it is necessary to reduce energy emissions of greenhouse gases into the atmosphere by three times compared to 2019 emissions (33.3 Gt) by about 2050 (1.5 °C) or by 2070 (2 °C) (Masson-Delmotte et al., 2018).

Carbon dioxide emissions can be natural or anthropogenic. Natural emissions come from the oceans, during volcanic activity, natural fires, and in the process of decay of organic materials. Previously, such emissions were absorbed through natural processes, and balance was maintained. However, as a result of anthropogenic activities (burning of fossil fuels, deforestation, agriculture, etc.), the balance has been disrupted.[3]

Currently, about 15–20% of CO_2 emissions are generated by the "agriculture, forestry and land use sector" (Our World in Data, 2020; World Resources Institute, 2021). Soils developed by man became sources of carbon dioxide. An increase in global temperature causes more intense release of carbon dioxide from soils. Every year, about 60 petagrams of CO_2 enter the atmosphere from soils due to "breathing" (Our World in Data, 2020).[4]

Of course, global levels of industrial CO_2 emissions into the atmosphere are significantly affected by energy consumption [which has already reached 14 billion tons of oil equivalent per year (British Petroleum, 2021)], as well as the structure of the global fuel and energy balance. The contribution of energy consumption to global CO_2 emissions today exceeds 73% (Our World in Data, 2020). Throughout the historical period, the structure of energy consumption has been constantly changing, and there have been so-called "energy transitions" to new models, from the predominant use of one resource to another. Three types of such transitions are known in history—from biomass to coal, from coal to oil, and from oil to natural gas. At the moment, the shares of energy sources in world energy consumption are distributed as follows: oil—31.2%, natural gas—24.7%, coal—27.2%, nuclear energy—4.3%, hydropower—6.9%, renewable energy sources—5.7% (British Petroleum, 2021). In the twenty-first century, the fourth energy transition to renewable energy sources is forecasted.

The great energy transition from the use of currently dominant fossil hydrocarbons to predominantly using renewable energy sources (RES), when the share of RES in the total energy balance exceeds 40%, may take place in the 2060s.

[3] Ibidem.

[4] Ibidem.

2 Possible Scenarios for Further Development of the Situation

In order to forecast the upcoming energy transition and to choose an optimal scenario for the development of the fuel and energy balance in the twenty-first century, a specialized mathematical model has been developed.[5] To develop and verify the model, the current trends in energy consumption were studied, and statistical data on energy consumption and the fuel and energy balance provided by the following organizations were analyzed: BP (British Petroleum, 2021, 2020), International Energy Agency (IEA) (IEA, 2020b, 2021c), International Renewable Energy Agency (IRENA) (IRENA, 2020, 2021), Intergovernmental Panel on Climate Change (IPCC) (Pachauri & Meyer, 2014; Masson-Delmotte et al., 2018, 2021), World Nuclear Association (WNA, 2020), World Energy Council (WEC) (WEC, 2019), Organization of the Petroleum Exporting Countries (OPEC) (OPEC, 2021), Equinor (Equinor, 2020, 2021), Greenpeace (Greenpeace, 2015), DNV GL (DNV, 2020, 2021), Shell (Shell, 2013, 2018), REN 21 (REN21, 2019, 2021), ExxonMobil (ExxonMobil, 2019, 2021) and others.

The proposed mathematical model allows forecasting changes in the average global temperature of the surface atmosphere in the twenty-first century (Akaev & Davydova, 2020, 2021a, b) in accordance with the following calculation mechanism:

- calculation of various population growth scenarios [see chapter "Demography. Toward the Optimization of Demographic Processes" (Korotayev et al., 2023a, this volume), as well as, e.g., (Akaev & Sadovnichy, 2010; Akaev et al., 2012; Kapitza, 2006)];
- calculation of scenarios for energy demand dynamics (Akaev, 2012, 2014a, b);
- forecast of the energy consumption structure by types of energy sources (coal, oil, gas, renewable energy sources, nuclear energy, hydropower) (Akaev & Davydova, 2021a, 2021b);
- calculation of the dynamics of CO_2 emissions into the atmosphere during the combustion of hydrocarbon fuels, taking into account structural changes in the consumption of organic fossil fuels (coal, oil, gas), as well as the use of carbon capture and storage technologies;
- calculation of the dynamics of CO_2 accumulation in the atmosphere, taking into account nonproductive CO_2 emissions (due to deforestation and soil erosion) and the absorption of part of the emissions by oceans and terrestrial ecosystems;
- calculation of the change in the average global temperature of the surface atmosphere based on the Tarko technique, which relates the dynamics of the deviation of the average global temperature to an increase in the dynamics of carbon (carbon dioxide) accumulation in the Earth's atmosphere.

[5] See Chapter "Modeling social self-organization and historical dynamics. Global phase transitions" (Malkov et al., 2023).

Taking into account the statistical data from 1960 to 2021 and the trends observed in recent years in the field of energy consumption, as well as the implemented energy-efficient technologies, three scenarios of the increase in the average global temperature of the surface atmosphere in the twenty-first century have been specified and calculated, including the conservative scenario, the ambitious scenario, and the Net Zero scenario. The conservative scenario assumes that government policies, technologies, and social preferences continue to evolve in the same way as in the recent past. The ambitious scenario envisages the introduction of measures that would lead to a significant reduction in carbon emissions from energy use, which in turn makes it possible to limit the increase in global temperature in the twenty-first century.[6] The Net Zero scenario assumes that the measures proposed in the ambitious scenario are complemented and reinforced by significant changes in the behavior and preferences of society. The dynamics of changes in the structure of the global fuel and energy balance for the twenty-first century under the conservative scenario, the ambitious scenario, and the Net Zero scenario is shown in Figs. 1, 2 and 3:

With the help of the developed model, the optimal scenario (the Net Zero scenario) has been found, the implementation of which will meet the requirements of the Paris Climate Agreement to keep global warming at the level of 1.5–2 °C compared to the preindustrial level. The Net Zero scenario assumes (Akaev & Davydova, 2021a, 2021b):

- the use of energy-efficient technologies (cp. Randers et al., 2018);
- wide use of hydrogen as an energy carrier;
- further development of renewable energy sources;
- widespread use of chemical technology for the capture, sequestration, and storage of carbon dioxide.

There are many energy-efficient technologies available that can help to reduce carbon emissions [see chapter "Technology. Limitless Possibilities and Effective Control" (Grinin & Grinin, 2023, this volume)]. An example of significant savings in energy consumption is the widespread use of intelligent digital technologies for energy management. The data collected by "smart sensors" is key for the energy consumption system, the operation of which is optimized by intelligent digital devices through adjusting supply and demand in real time.

[6]On the global development scenarios in economic, social, and political dimensions that are relevant for the issue of preventing catastrophic climatic changes see Chapter "Economics. Optimizing Growth" (Grinin et al., 2023c, this volume), Chapter "Socio-Political Transformations. A Difficult Path to Cybernetic Society" (Grinin et al., 2023d, this volume); Chapter "High Income and Low Income Countries. Toward a Common goal at Different Speeds" (Grinin et al., 2023b, this volume); Chapter "Future Political Change. Toward a More Efficient World Order" (Grinin et al., 2023a, in this volume), Chapter "Africa – the Continent of the Future. Challenges and Opportunities" (Grinin & Korotayev, 2023, this volume), and Chapter "Modeling Social Self-Organization and Historical Dynamics. Africa's Futures" (Korotayev et al., 2023b, this volume).

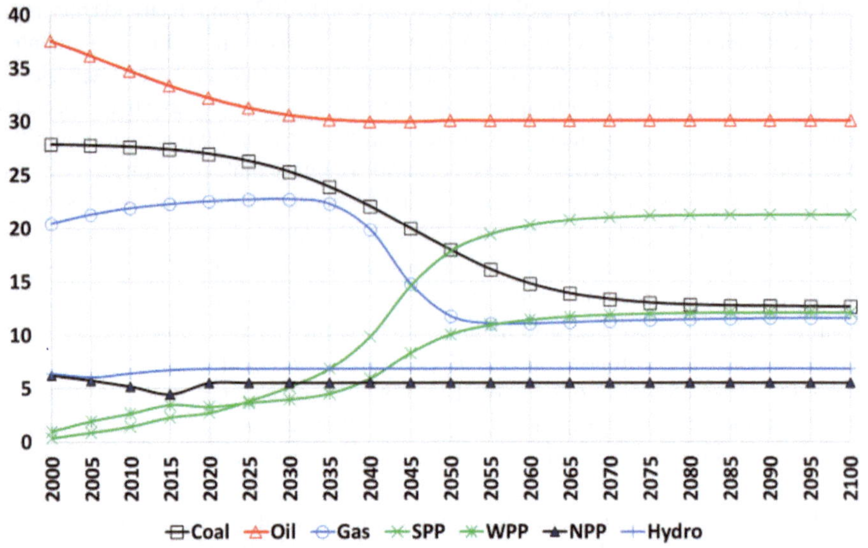

Fig. 1 Dynamics of changes in the structure of the global fuel and energy balance for the twenty-first century (for Coal, Oil, Gas, Solar Power Plants (SPP), Wind Power Plants (WPP), Nuclear Power Plants (NPP), and Hydro energy) under the conservative scenario

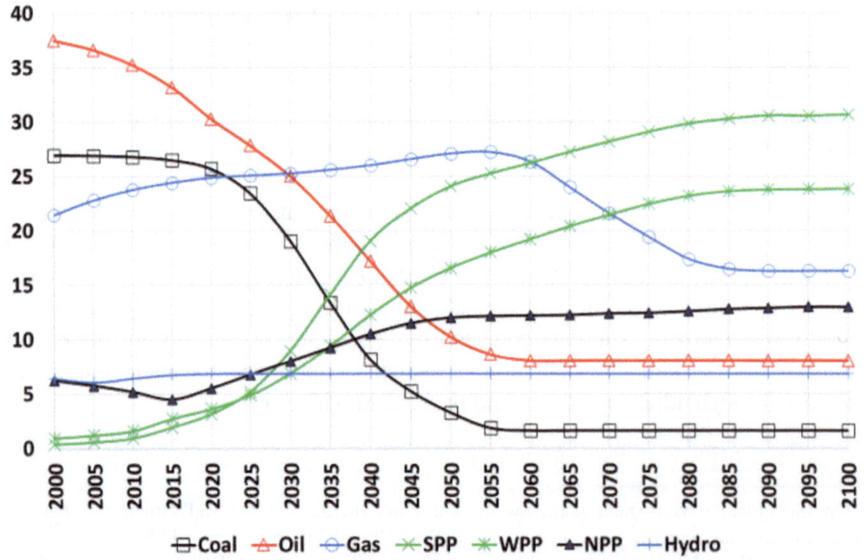

Fig. 2 Dynamics of changes in the structure of the global fuel and energy balance for the twenty-first century under the ambitious scenario

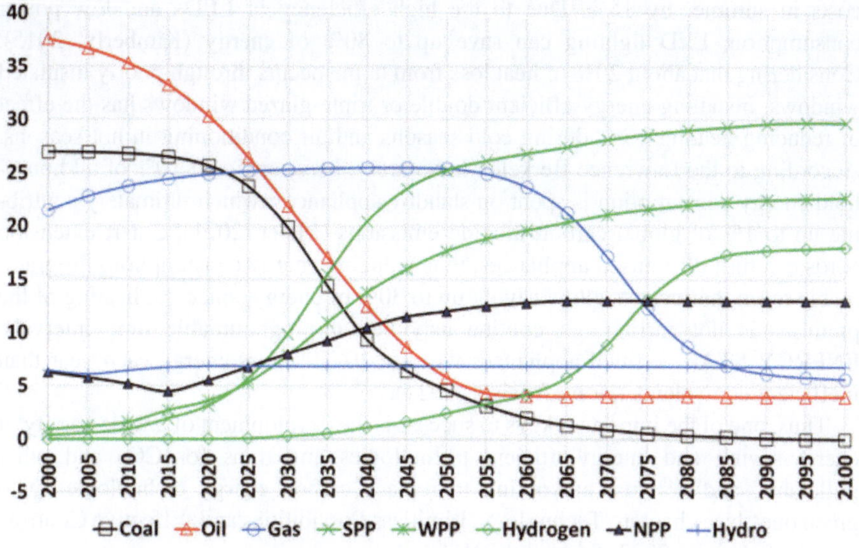

Fig. 3 Dynamics of changes in the structure of the global fuel and energy balance for the twenty-first century under the Net Zero scenario

"Smart grids" based on data from producers and consumers make it possible to synchronize supply and demand in real time in an optimal way. "Smart grids" can regulate the flow of electricity from one region to another, taking into account prevailing weather conditions.

Currently, "smart energy consuming devices" are also being developed. The consumer installs equipment to optimize the modes of acquisition of electric energy based on the needs and load of the system. Thus, it becomes possible for the consumer not only to receive energy, but also to give it to the network, making profit.

Currently, 26% of global CO_2 emissions are generated by the transport sector (IEA, 2019). Vehicles driven by electric motors can solve the problem of such emissions. Hybrid electric vehicles have already achieved 65% greater fuel efficiency than gasoline-powered vehicles (Vorrath, 2015). Gasoline-powered vehicles consume four times more energy than fully electric vehicles (Vorrath, 2015; Energy Efficiency Day, 2020; Virta, 2021). In 2019, more than two million new electric vehicles were sold worldwide (IEA, 2021a), but this number needs to be increased significantly.

The use of various energy-efficient technologies for households should also be expanded. For example, thanks to solar panels, annual greenhouse gas emissions can be reduced by more than 1350 kg within a single household. Using clean, renewable solar energy to power a million homes can reduce carbon dioxide emissions by 4.3 million tons per year (NCAT, 2021). Wind generators can be used to generate direct or alternating current with its further conversion into heat by means of heat pumps in order to heat buildings and heat water. "Cold Roofs" help reduce air conditioning costs in summer by 15%. Due to the high efficiency of LEDs and low power consumption, LED lighting can save up to 80% of energy (Kimberly, 2015). Considering that about 20% of heat loss from home occurs through poorly insulated windows, installing energy-efficient double or triple-glazed windows has the effect of reducing heating costs during cold seasons and air conditioning in hot seasons. According to the Lawrence Berkeley Laboratory, between 5 and 10% of all household energy consumption is spent on standby appliances, which ultimately contribute up to 1% of global carbon dioxide emissions (Meier, 2021). Smart extension cords can turn off unused appliances, thus reducing energy consumption. Programmable room thermostats allow saving up to 30% of energy, since the heating of the premises is not carried out continuously but at programmable time intervals. ENERGY STAR-certified appliances will use 10–50% less energy each year than inefficient equivalents (HomeAdvisor, 2021).

Thus, one of the important keys to success is the development of a wide variety of energy-saving and energy-efficient technologies and traps for CO_2 and other pollutants, and not just attempting to replace carbon energy technologies with green ones [see chapter "Technology. Limitless Possibilities and, Effective Control" (Grinin & Grinin, 2023, this volume)].

Although hydrogen is widely used today, it is far from reaching its full potential as an energy source. A lot of research is required to provide cheap and sustainable clean hydrogen energy derived only from renewable energy sources (hydropower, solar panels, or wind farms) without emissions or at least using carbon capture and storage (CCS) systems, which help avoid carbon emissions into the atmosphere. Green hydrogen is produced by electrolysis, where electricity is generated only from sources with zero carbon content. This technology can significantly reduce CO_2 emissions, but it is currently too expensive (IRENA, 2019). In 2015, the production cost of 1 kg of green hydrogen was $6 (Casey, 2021). Over the past five years, the cost of producing green hydrogen has dropped to $3 per kilogram (S&P Global, 2021). For comparison, 1 kg of gray hydrogen produced from carbon sources costs $1.80, and blue hydrogen (using CCS technology) costs $2.40. The US Department of Energy expects that in 2025 the cost of producing green hydrogen will drop to $2 per kilogram, in which case green hydrogen can become competitive with other nonrenewable sources (EIA US, 2021). The European Union, Japan, South Korea, Australia, the Netherlands, Norway, Chile, and Canada have already developed their hydrogen strategies. The European Union has set itself the goal of increasing the capacity of electrolyzers to 6 GW by 2024 and to 40 GW by 2030 (Sonal, 2021).

Today, the share of green hydrogen is less than 1% of the total hydrogen produced. Exponential growth (up to 60% per year) in green hydrogen production is expected in the coming decades (Holbrook, 2021). In general, most reports assume that hydrogen produced from renewable energy sources only will account for 10–25% of energy consumption in the twenty-first century (Buli, 2021; Flowers, 2020; Scott, 2020). Calculations based on the developed model show that green hydrogen will reach 18% of energy consumption in the twenty-first century (see Fig. 4).

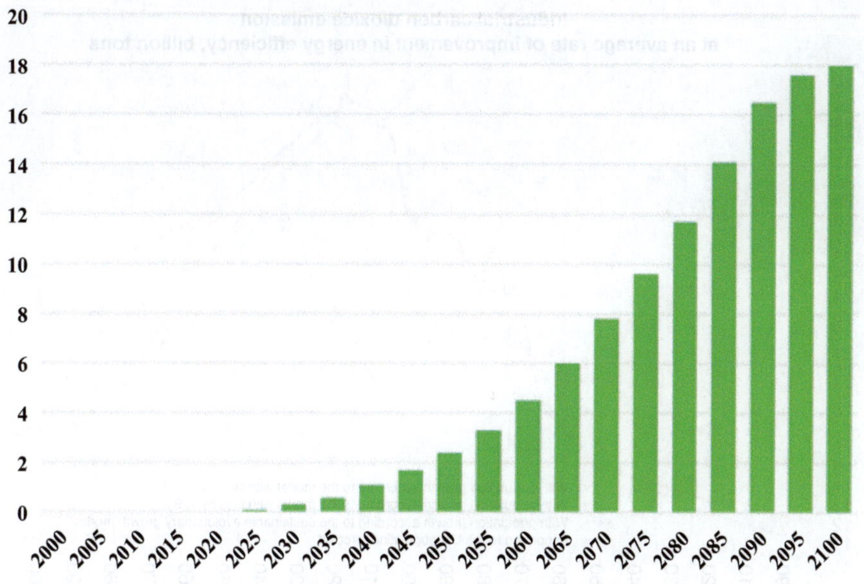

Fig. 4 The share of green hydrogen in the structure of the global fuel and energy balance for the twenty-first century under the Net Zero scenario (calculated by authors)

Also, widespread use of chemical technology is required to capture, sequestrate, and store carbon dioxide, both in the process of burning hydrocarbons in power plants and directly from the atmosphere, which is reflected in the calculations by the model. However, this path is hindered by the high cost of technology.

Evidently, significant investments are required to successfully implement the Net Zero scenario. Unfortunately, investment flows are currently directed toward those areas that bring maximum profit, rather than those that benefit society in the long term (Randers, 2012; Maxton et al., 2016). Only when there is, compelling evidence of damage is the money spent on reducing the negative effects of climate change that could have been prevented.

The results of forecast calculations of the dynamics of the reduction of anthropogenic carbon dioxide emissions into the atmosphere in the twenty-first century under the conservative scenario, the ambitious scenario, and the Net Zero scenario are presented in Figs. 5, 6 and 7:

The results of the projection of the deviation of the average global temperature of the surface atmosphere dynamics in the twenty-first century under the conservative scenario, the ambitious scenario, and the Net Zero scenario are presented in Figs. 8, 9 and 10:

Calculations show that if the conservative scenario is implemented, global warming will reach 2 °C, while if the ambitious scenario is implemented, it will be 1.8 °C. The Net Zero scenario will keep global warming at 1.7 °C (Akaev & Davydova, 2021a, 2021b).

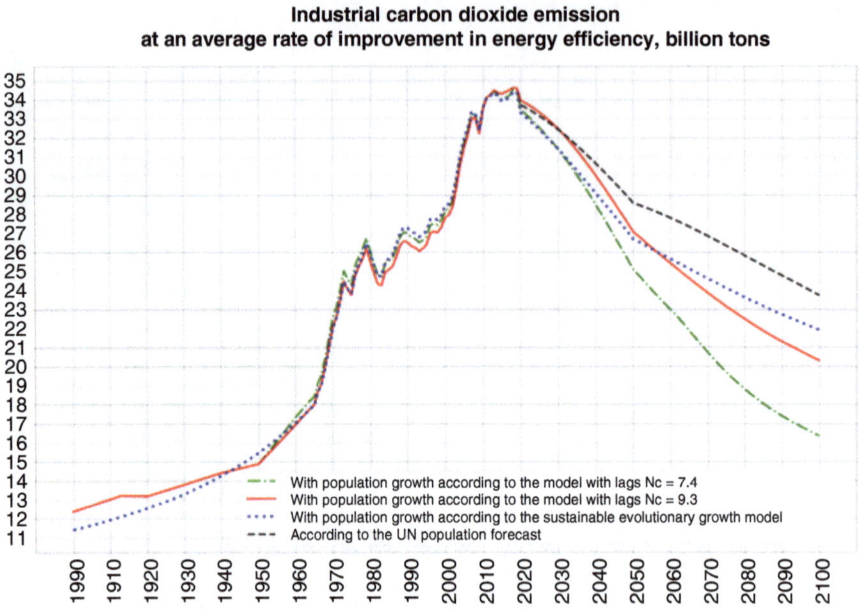

Fig. 5 The dynamics of the reduction of anthropogenic carbon dioxide (CO_2) emissions into the atmosphere in the twenty-first century under the conservative scenario of energy transition

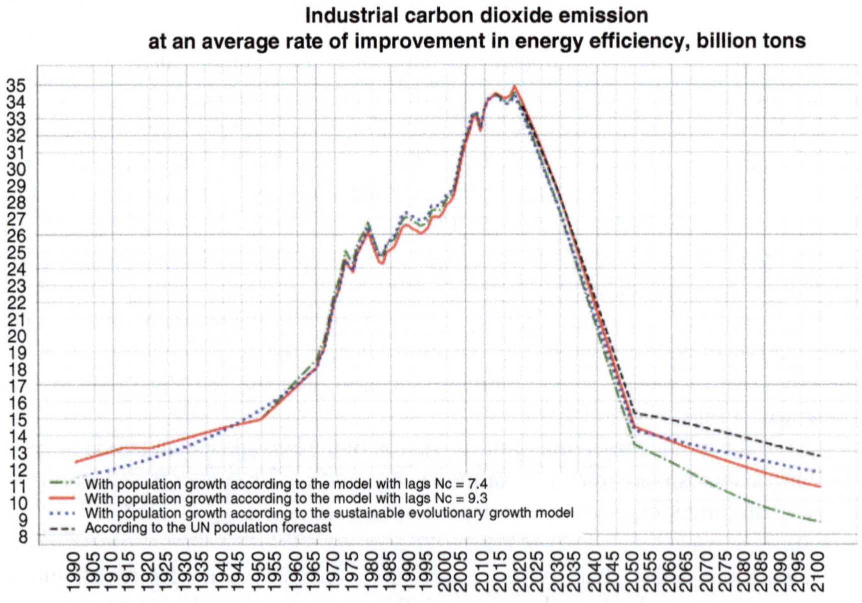

Fig. 6 The dynamics of the reduction of anthropogenic carbon dioxide (CO_2) emissions into the atmosphere in the twenty-first century under the ambitious scenario of energy transition

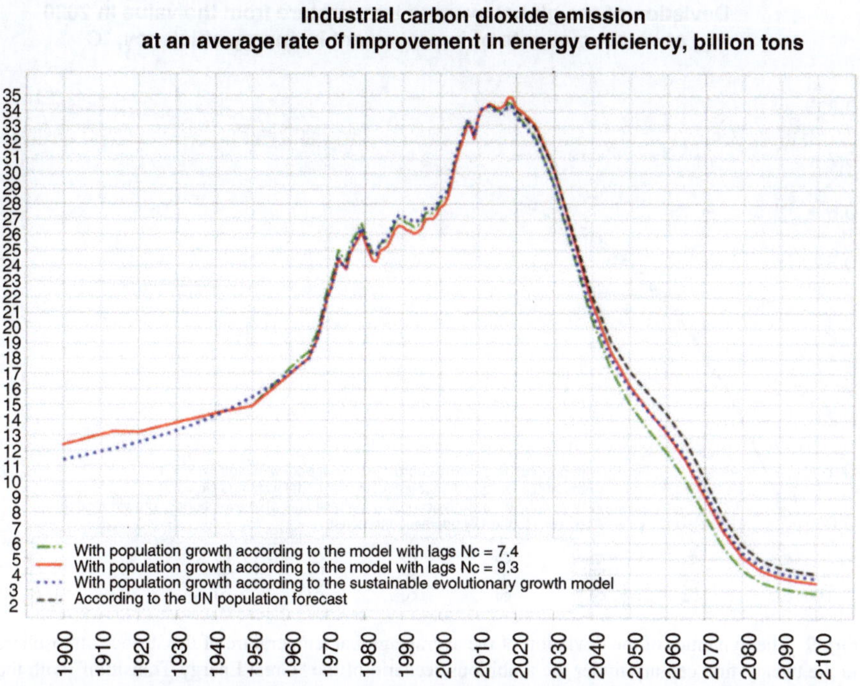

Fig. 7 The dynamics of the reduction of anthropogenic carbon dioxide (CO_2) emissions into the atmosphere in the twenty-first century under the Net Zero scenario of the energy transition with the use of hydrogen and CCS technology for capture and storage of some part of CO_2

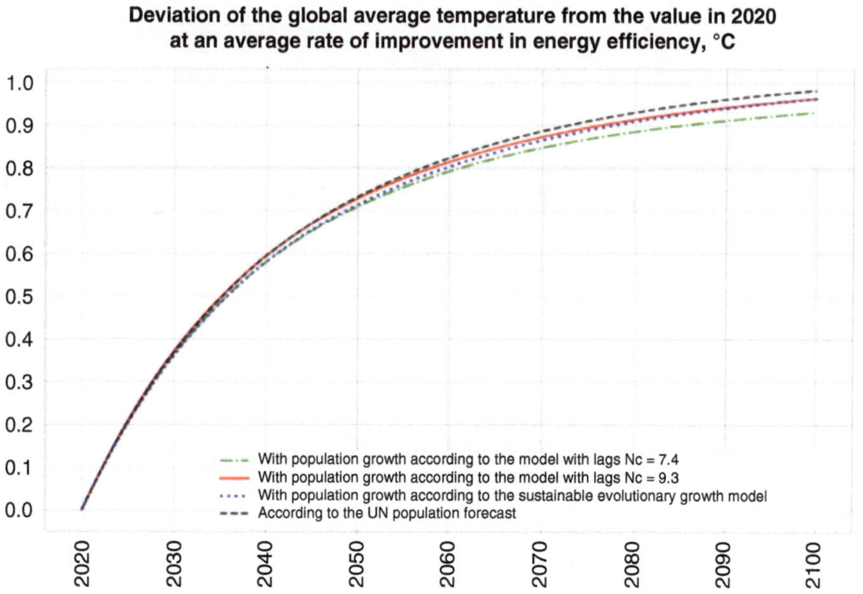

Fig. 8 The dynamics of the deviation of the average global temperature of the surface atmosphere in the twenty-first century under the conservative scenario of energy transition with the use of CCS technology in coal energy

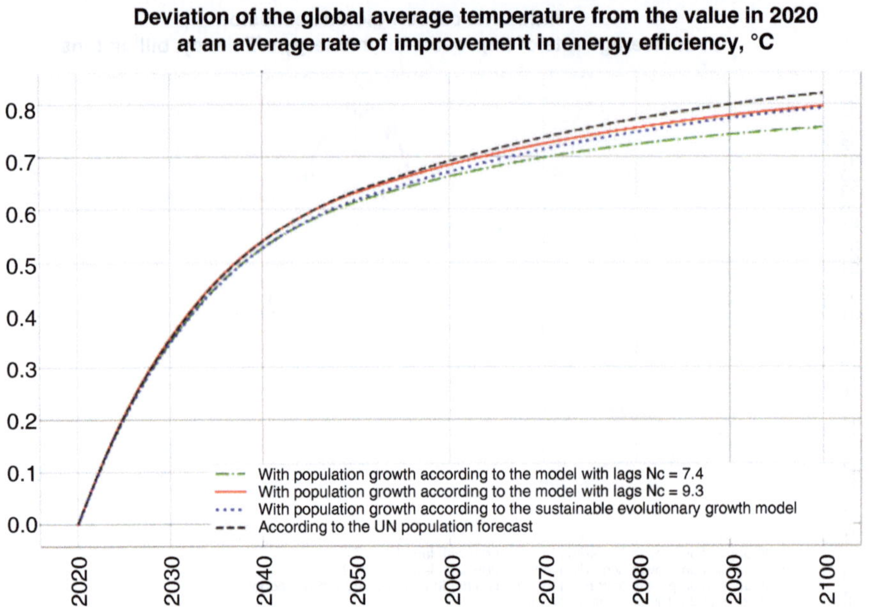

Fig. 9 The dynamics of the deviation of the average global temperature of the surface atmosphere in the twenty-first century under the ambitious scenario of the "Great Energy Transition" with the use of CCS technology in coal energy

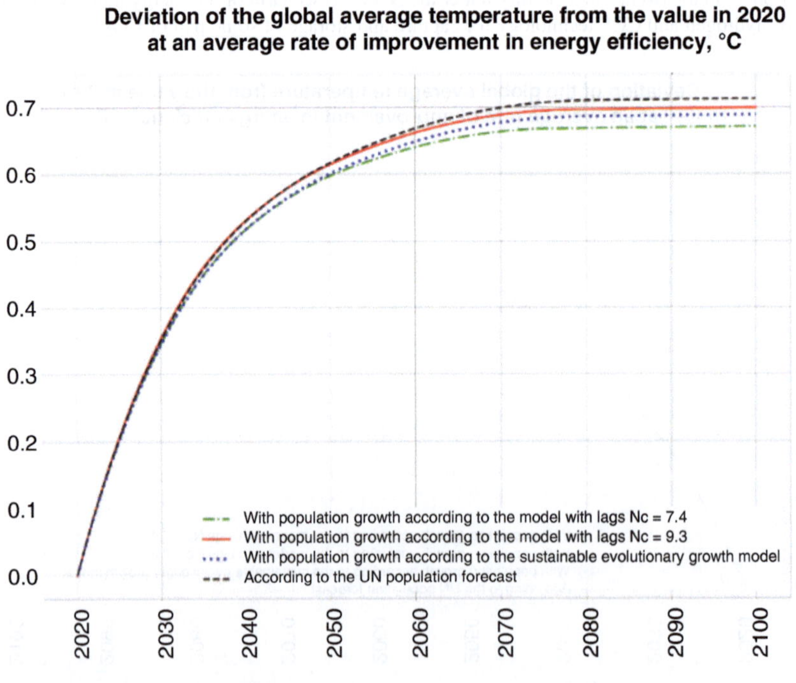

Fig. 10 The dynamics of the deviation of the average global temperature of the surface atmosphere in the twenty-first century under the Net Zero scenario of energy transition with the use of CCS technology in coal energy and the use of hydrogen

Since the conservative scenario assumes that by 2050 the share of fossil energy sources will remain at about 60–65%, this will lead to unacceptable CO_2 emissions. This is far from the current ambitious goal of reducing emissions by 50–80% by 2050. Furthermore, under the conservative scenario, more frequent extreme weather events will be observed in the coming decades. In many places, floods that previously occurred once a century will occur much more frequently by 2050, possibly annually. When sea levels rise by several meters, about 30% of the land, which is a densely populated territory, will be flooded. According to a study conducted by scientists from the Potsdam Institute for the Study of Climate Change, by 2100, due to the melting of continental ice, the level of the World Ocean may rise by 0.75–1.5 m (PIK, 2013). As a result, in 100 years, Venice will go underwater, followed in another 50 years by Amsterdam, Hamburg, Los Angeles, St. Petersburg, and other cities.

The climate crisis can be avoided if the world acts decisively and collectively and takes measures necessary to reduce carbon emissions. In particular, the global community needs to redirect investment flows from the most profitable solutions to those that will benefit society in the long term. Annual global investment in climate averaged $632 billion per year during 2019 and 2020 (Burg, 2021). For comparison, global GDP in 2020 amounted to 85 trillion US dollars, meaning that, at the moment, global investment in climate is less than 1% of the global GDP. Calculations show that successful implementation of the Net Zero scenario requires significant investment in the development of renewable energy sources. The Net Zero scenario assumes an increase in the share of renewable energy sources in the fuel and energy balance to 50–60% by 2050 from the current 6%. Thus, it is necessary to expand solar and wind energy use more significantly. Investments are also needed to increase the use of energy-efficient technologies, including smart digital technologies and energy-efficient technologies for homes. The study and improvement of chemical technology for capturing, sequestration, and storing carbon dioxide also require funding. In addition, investments should be directed to the development of hydrogen, nuclear, and thermonuclear energy. Part of the investment can be obtained through increased taxes on CO_2 emissions, which in turn will contribute to solving the problem of global warming. Calculations show that the implementation of the above measures will meet the requirements of the Paris Climate Agreement to keep global warming at 1.5–2 °C compared to preindustrial levels.

References

Akaev, A. (2012). Stabilization of the planetary climate in the twenty-first century by transition to a new paradigm of energy consumption. *Doklady Earth Sciences, 446*(2), 1180–1184.

Akaev, A. (2014a). Energeticheskaya model' ekonomicheskogo rosta V.N. Pokrovskogo v usloviyakh novoy paradigmy energopotrebleniya. *Vestnik Instituta ekonomiki RAN, 3*, 12–33.

Akaev, A. (2014b). The stabilization of Earth's climate in the 21st century by the stabilization of per capita consumption. In *The Oxford handbook of the macroeconomics of global warming* (pp. 499–554) Oxford University Press.

Akaev, A., & Davydova, O. (2020). The Paris agreement on climate is coming into force: Will the great energy transition take place? *Herald of the Russian Academy of Sciences, 90*(5), 588–599.

Akaev, A., & Davydova, O. (2021a). A mathematical description of selected energy transition scenarios in the 21st century, intended to realize the main goals of the Paris climate agreement. *Energies, 14*(9), 1–28. https://doi.org/10.3390/en14092558

Akaev, A., & Davydova, O. (2021b). Mathematical description of energy transition scenarios based on the latest technologies and trends. *Energies, 14*(24), 1–25. https://doi.org/10.3390/en14248360

Akaev, A., & Sadovnichy, V. (2010). Mathematical model of population dynamics with the world population size stabilizing about a stationary level. *Doklady Mathematics, 82*(3), 978–981.

Akaev, A., Sadovnichy, V., & Korotayev, A. (2012). On the dynamics of the world demographic transition and financial-economic crises forecasts. *The European Physical Journal Special Topics, 205*(1), 355–373. https://doi.org/10.1140/epjst/e2012-01578-2

Berg, C. (2019). *Sustainable action: Overcoming the barriers. A report to the Club of Rome*. Routledge.

BP. (2020). *Energy outlook 2020 edition.* British Petroleum. https://www.bp.com/content/dam/bp/business-sites/en/global/corporate/pdfs/energy-economics/energy-outlook/bp-energy-outlook-2020.pdf

BP. (2021). *Statistical review of world energy.* British Petroleum. https://www.bp.com/en/global/corporate/energy-economics/statistical-review-of-world-energy.html

Buli, N. (2021). *Green hydrogen to account for 20% of European power demand by 2050—Statkraft.* https://www.reuters.com/business/energy/green-hydrogen-account-20-european-power-demand-by-2050-statkraft-2021-10-21/

Burg, N. (2021). *Who funds the fight against climate change?* Bank of the West. https://meansandmatters.bankofthewest.com/article/sustainable-living/taking-action/who-funds-the-fight-against-climate-change/

Casey, J. P. (2021). *Will China do for hydrogen what it did for solar power?* Future Power Technology, Inc. https://www.power-technology.com/features/will-china-do-for-hydrogen-what-it-did-for-solar-power/

DNV. (2020). *Energy transition outlook 2020.* Det Norske Veritas. https://download.dnvgl.com/eto-2020-download

DNV. (2021). *Energy transition outlook 2021.* Det Norske Veritas. https://eto.dnv.com/2020/index.html

EIA US. (2021). *Hydrogen Explained.* https://www.eia.gov/energyexplained/hydrogen/

Energy Efficiency Day. (2020). *Electric vehicles: An efficient choice for transportation and the grid.* https://energyefficiencyday.org/electric-vehicles-an-efficient-choice-for-transportation-and-the-grid/

Equinor. (2020). *Energy perspectives 2020.* https://www.equinor.com/en/sustainability/energy-perspectives.html

Equinor. (2021). *Energy perspectives 2021.* https://www.equinor.com/en/sustainability/energy-perspectives.html

ExxonMobil. (2019). Outlook for Energy: A perspective to 2040. https://corporate.exxonmobil.com/-/media/Global/Files/outlook-for-energy/2019-Outlook-for-Energy_v4.pdf

ExxonMobil. (2021). *Outlook for Energy.* https://corporate.exxonmobil.com/Energy-and-innovation/Outlook-for-Energy

Flowers, S. (2020). *Future energy—Green hydrogen, could it be a pillar of decarbonization.* https://www.woodmac.com/news/the-edge/future-energy-green-hydrogen/

Garske, J. P. (2020). Climate Change, Malthusian Catastrophe, and a Global Rule of Law. *Journal of Globalization Studies, 11*(1), 3–15. https://doi.org/10.30884/jogs/10.30884/jogs/2020.01.00

Greenpeace. (2015). *Energy revolution.* https://wayback.archive-it.org/9650/20200212151746/http:/p3-raw.greenpeace.org/international/en/publications/Campaign-reports/Climate-Reports/Energy-Revolution-2015/

Grinin, L., & Grinin, A. (2023). Technology. Limitless possibilities and effective control. In V. Sadovnichy et al. (Eds.), *Reconsidering the limits to growth. A report to the Russian Association of the Club of Rome* (pp. 139–154). Springer. https://doi.org/10.1007/978-3-031-34999-7_8

Grinin, L., & Korotayev, A. (2023). Africa – The continent of the future. Challenges and opportunities. In V. Sadovnichy et al. (Eds.), *Reconsidering the limits to growth. A report to the Russian Association of the Club of Rome* (pp. 225–238). Springer. https://doi.org/10.1007/978-3-031-34999-7_13

Grinin, L., Grinin, A., & Korotayev, A. (2023a). Future political change. Toward a more efficient world order. In V. Sadovnichy et al. (Eds.), *Reconsidering the limits to growth. A report to the Russian Association of the Club of Rome* (pp. 191–206). Springer. https://doi.org/10.1007/978-3-031-34999-7_11

Grinin, L., Malkov, S., & Korotayev, A. (2023b). High income and low income countries. Toward a common goal at different speeds. In V. Sadovnichy et al. (Eds.), *Reconsidering the limits to growth. A report to the Russian Association of the Club of Rome* (pp. 207–224). Springer. https://doi.org/10.1007/978-3-031-34999-7_12

Grinin, L., Grinin A., & Malkov, S. (2023c). Economics. Optimizing growth. In V. Sadovnichy et al. (Eds.), *Reconsidering the limits to growth. A report to the Russian Association of the Club of Rome* (pp. 155–168). Springer. https://doi.org/10.1007/978-3-031-34999-7_9

Grinin, L., Grinin, A., & Malkov, S. (2023d). Socio-political transformations. A difficult path to cybernetic society. In V. Sadovnichy et al. (Eds.), *Reconsidering the limits to growth. A report to the Russian Association of the Club of Rome* (pp. 169–189). Springer. https://doi.org/10.1007/978-3-031-34999-7_10

Gruppa Vsemirnogo banka. (2014). *Ubav'te teplo. Litsom k litsu s novoy klimaticheskoy normoy.* https://www.google.ru/url?sa=t&rct=j&q=&esrc=s&source=web&cd=&ved=2ahUKEwi898KI05X2AhUXuosKHYfoBfcQFnoECCgQAQ&url=http%3A%2F%2Fdocuments1.worldbank.org%2Fcurated%2Fen%2F318441468238479152%2Fpdf%2F927040v10Russi0s0010ES0with0embargo.pdf&usg=AOvVaw1bfxac1-KkQHKX5-KYb12O

Holbrook, E. (2021). *EnergyCAP, study says global green hydrogen production to skyrocket 57% to 2030.* https://www.environmentalleader.com/2021/01/study-says-global-green-hydrogen-production-to-skyrocket-57-to-2030/

HomeAdvisor. (2021). *Pros, cons and costs: Energy star appliances.* HomeAdvisor. https://www.homeadvisor.com/r/energy-star-appliances/

IEA. (2019). *Global CO2 emissions by sector. IEA.* https://www.iea.org/data-and-statistics/charts/global-co2-emissions-by-sector-2019

IEA. (2020a). *Energy related CO2 emissions, 1990–2019.* IEA https://www.iea.org/data-and-statistics/charts/energy-related-co2-emissions-1990-2019

IEA. (2020b). *World energy outlook 2020.* IEA. https://www.iea.org/reports/world-energy-outlook-2020

IEA. (2021a). *Global Energy Review 2021.* IEA. https://www.iea.org/reports/global-energy-review-2021/co2-emissions.

IEA. (2021b). *Global EV Outlook* 2020. *Entering the Decade of Electric Drive?* IEA. https://iea.blob.core.windows.net/assets/af46e012-18c2-44d6-becd-bad21fa844fd/Global_EV_Outlook_2020.pdf

IEA. (2021c). *World energy outlook 2021.* IEA. https://www.iea.org/reports/world-energy-outlook-2021

IRENA. (2019). *Hydrogen: A renewable energy perspective.* International Renewable Energy Agency.

IRENA. (2020). *Global renewables outlook: Energy transformation 2050.* International Renewable Energy Agency.

IRENA. (2021). *World energy transitions outlook: 1.5°C pathway.* International Renewable Energy Agency.

Kapitza, S. (2006). *Global population blow-up and after. Report to the Club of Rome*. Global Marshall Plan Initiative.

Kimberly, J. (2015). *How an LED uses so much less energy*. Consumer Reports. https://www.consumerreports.org/lightbulbs/why-an-led-uses-so-little-energy-/

Kiss, E. (2020). The Present Dialogue between Nature and Society or the Mutuality of Nature Preservation and Society Protection. *Journal of Globalization Studies, 11*(1), 86–101. https://doi.org/10.30884/jogs/2020.01.0010.30884/jogs/2020.01.06

Korotayev, A., Malkov, S., & Musieva, J. (2023a). Demography. Toward the optimization of demographic processes. In V. Sadovnichy et al. (Eds.), *Reconsidering the limits to growth. A report to the Russian Association of the Club of Rome* (pp. 97–116). Springer. https://doi.org/10.1007/978-3-031-34999-7_6

Korotayev, A., Shulgin, S., Ustyuzhanin, V., Zinkina, J., & Grinin, L. (2023b). Modeling social self-organization and historical dynamics. Africa's futures. In V. Sadovnichy et al. (Eds.), *Reconsidering the limits to growth. A report to the Russian Association of the Club of Rome* (pp. 491–519). Springer. https://doi.org/10.1007/978-3-031-34999-7_21

Kovaleva, N. (2023). Ecology. Life in the "unstable biosphere". In V. Sadovnichy et al. (Eds.), *Reconsidering the limits to growth. A report to the Russian Association of the Club of Rome* (pp. 71–95). Springer. https://doi.org/10.1007/978-3-031-34999-7_5

Malkov, S., Grinin, L., Grinin, A., Musieva, J., & Korotayev, A. (2023). Modeling social self-organization and historical dynamics. Global phase transitions. In V. Sadovnichy et al., (Eds.) *Reconsidering the limits to growth. A report to the Russian Association of the Club of Rome* (pp. 419–460). Springer. https://doi.org/10.1007/978-3-031-34999-7_19

Masson-Delmotte, V., Zhai, P., Pörtner, H.-O., Roberts, D., Skea, J., Shukla, P.R., Pirani, A., Moufouma-Okia, W., Péan, C., Pidcock, R., Connors, S., Matthews, J.B.R., Chen, Y., Zhou, X., Gomis, M.I., Lonnoy, E., Maycock, T., Tignor, M., & Waterfield, T. (2018). *Global warming of 1.5°C. An IPCC special report on the impacts of global warming of 1.5°C above pre-industrial levels and related global greenhouse gas emission pathways, in the context of strengthening the global response to the threat of climate change, sustainable development, and efforts to eradicate poverty*. IPCC.

Masson-Delmotte, V., Zhai, P., Pirani, A., Connors, S. L., Péan, C., Berger, S., Caud, N., Chen, Y., Goldfarb, L., Gomis, M. I., Huang, M., Leitzell, K., Lonnoy, E., Matthews, J. B. R., Maycock, T. K., Waterfield, T., Yelekçi, O., Yu, R., & Zhou, B. (2021). *Summary for Policymakers. Climate Change 2021: The physical science basis. Contribution of Working Group I to the Sixth Assessment Report of the Intergovernmental Panel on Climate Change*. Cambridge University Press.

Maxton, G., Randers, J., & Suzuki, D. (2016). *Reinventing prosperity. A report to the Club of Rome*. Greystone Books.

Meier, A. (2021) *Should the next standby power target be 0-watt?* Lawrence Berkeley National Laboratory. https://escholarship.org/content/qt566951pn/qt566951pn.pdf

NCAT. (2021). *Solar energy*. The National Center for Appropriate Technology. https://www.ncat.org/solar-energy/

NOAA. (2020). *Climate Change: Global Sea Level*. https://www.climate.gov/news-features/understanding-climate/climate-change-global-sea-level

NOAA. (2022). *Trends in Atmospheric Carbon Dioxide, Mauna Loa CO2 weekly mean and historical comparisons*. https://gml.noaa.gov/ccgg/trends/data.html

OPEC. (2021). *World oil outlook 2021*. The Organization of the Petroleum Exporting Countries.

Ortiz-Bobea, A., Ault, T. R., Carrillo, C. M., et al. (2021). Anthropogenic climate change has slowed global agricultural productivity growth. *Nature. Climate Change, 11*, 306–312. https://doi.org/10.1038/s41558-021-01000-1

Our World in Data. (2020). *Emissions by sector*. https://ourworldindata.org/emissions-by-sector

Pachauri, R. K., & Meyer, L. A. (2014). *Climate change 2014: Synthesis report. Contribution of working groups I, II and III to the fifth assessment report of the intergovernmental panel on climate change*. IPCC.

PIK. (2013). *Expert assessment: Sea-level rise could exceed one meter in this century*. Potsdam Institute for Climate Impact Research. https://www.pik-potsdam.de/en/news/latest-news/archive/2013/expert-assessment-sea-level-rise-could-exceed-one-meter-in-this-century

Randers, J. (2012). *2052. A global forecast for the next forty years. A report to the club of Rome*. Chelsea Green Publishing.

Randers, J., Rockström, J., Stoknes, P. E., Golüke, U., Collste, D., & Cornell, S. (2018). *Transformation is feasible. A report to The Club of Rome*. Stockholm Resilience Centre.

REN21. (2019). *Global transition perspectives to renewable energy*. Global report "Renewable energy 2019". REN21. www.ren21.net/GSR

REN21. (2021). *Renewables 2021 Global Status Report*. https://www.ren21.net/reports/global-status-report/

Ritchie, H., & Roser, M. (2021). *Natural disasters. Our world in data*. https://ourworldindata.org/natural-disasters

S&P Global. (2021). *Experts explain why green hydrogen costs have fallen and will keep falling*. https://www.spglobal.com/marketintelligence/en/news-insights/latest-news-headlines/experts-explain-why-green-hydrogen-costs-have-fallen-and-will-keep-falling-63037203

Scott, M. (2020) *Green hydrogen. The fuel of the future, set for 50-fold expansion*. Forbes. https://www.forbes.com/sites/mikescott/2020/12/14/green-hydrogen-the-fuel-of-the-future-set-for-50-fold-expansion/?sh=6f5485306df3

Shell. (2013). *New lens scenarios A shift in perspective for a world in transition*. https://www.google.ru/url?sa=t&rct=j&q=&esrc=s&source=web&cd=&ved=2ahUKEwih_7mr-4ftAhVMr4sKHT_tBAEQFjABegQIAx-AC&url=https%3A%2F%2Fwww.shell.com%2Fcontent%2Fdam%2Froyaldutchshell%2Fdocuments%2Fcorporate%2Fscenarios-newdoc.pdf&usg=AOvVaw2yKUrs_ZGNsFk9icrtWoM1

Shell. (2018). *Sky Meeting the goals of the Paris agreement*. https://www.shell.com/promos/business-customers-promos/download-latest-scenario-sky/_jcr_content.stream/1530643931055/eca19f7fc0d20adbe830d3b0b27bcc9ef72198f5/shell-scenario-sky.pdf

Sonal P. (2021). Countries roll out green hydrogen strategies. *Electrolyzer Targets Power*. https://www.powermag.com/countries-roll-out-green-hydrogen-strategies-electrolyzer-targets/

Statista. (2022). *Annual number of natural disaster events globally from 2000 to 2020*. https://www.statista.com/statistics/510959/number-of-natural-disasters-events-globally/

United Nations. (2016). *Paris agreement*. United Nations. https://unfccc.int/process/conferences/pastconferences/paris-climate-change-conference-november-2015/paris-agreement

Virta. (2021). *Myth buster: Electric vehicles will overload the power grid*. https://www.virta.global/blog/myth-buster-electric-vehicles-will-overload-the-power-grid

von Weizsäcker, E. U., & Wijkman, A. (2018). *Come on!* Springer. https://doi.org/10.1007/978-1-4939-7419-1

Vorrath, S. (2015). *Top 10 technologies to double the energy efficiency, deliver zero emissions*. Renew Economy. https://reneweconomy.com.au/top-10-technologies-to-double-energy-efficiency-deliver-zero-emissions-65210/

WEC. (2019). *World energy scenarios 2019*. World Energy Council. https://www.worldenergy.org/assets/downloads/2019_Scenarios_Full_Report.pdf

Welthungerhilfe. (2021). *Global Hunger Index*. https://www.welthungerhilfe.org/hunger/global-hunger-index/

Wijkman, A., & Skånberg, K. (2017). *The circular economy and benefits for society. Jobs and climate clear winners in an economy based on renewable energy and resource efficiency. A report to the Club of Rome*. Club of Rome.

WMO. (2021). *State of the global climate 2020*. World Meteorological Organization. https://library.wmo.int/index.php?lvl=notice_display&id=21880#.Yf79XfhRVD9

WNA. (2020). *The nuclear fuel report: Expanded summary – Global scenarios for demand and supply availability 2019–2040*, Report No. 2020/005. World Nuclear Association. https://world-nuclear.org/getmedia/b488c502-baf9-4142-8d12-42bab97593c3/nuclear-fuel-report-2019-expanded-summary-final.pdf.aspx

World Resources Institute. (2021). *Climate Watch.* https://www.climatewatchdata.org/key-visualizations?topic=sectoral_emissions&visualization=4

Askar Akaev is the First President of the Kyrgyz Republic and Foreign Full Member of the Russian Academy of Sciences. He works as Chief Research Professor in the Prigogine Institute for Mathematical Investigations of Complex Systems at Lomonosov Moscow State University. He is also the Academic Supervisor of the Centre for Stability and Risk Analysis at HSE University, Moscow, Russia. Askar Akaev has authored and co-authored over 300 scientific publications, including such monographs as *Coherent Optical Computers* (1977), *Optical Methods of Information Processing* (1983), *Holographic Memory* (1996), *Selected Lectures on Optical Computers* (1996), *Economic Transition from the Point of View of a Physicist. A Mathematical Model of Emerging Economy* (2000), *Modeling and Forecasting the World Dynamics* (2012), *Quality of Education, Effectiveness of R&D, and Economic Growth. Quantitative Analysis and Mathematical Modeling* (2016), etc.

Olga Davydova is a specialist in data analysis, development of mathematical models, implementation of models in Python, identification of parameters, and verification of models. She has co-authored more than 15 academic articles including such papers as "Resistance to Socio-Political Instability as an Indicator of the Country's Successful Development" (2015), "Model for Assessing and Forecasting the Development of Human Capital" (2016), "Forecast of the Dynamics of World Imports by Commodity Groups" (2018), "The Paris Agreement on Climate Is Coming into Force: Will the Great Energy Transition Take Place?" (2020). Olga Davydova has participated in numerous research projects, including the ones on the development of mathematical methods for forecasting global and country socio-economic development, mathematical modeling of promising trade and transport corridors, modeling the destabilization of socio-political systems, developing a model of a country's monetary circulation, developing a mathematical model of the energy transition and research in the climate sphere, assessing and forecasting the dynamics of human capital at the federal and regional levels.

Ecology: Life in the "Unstable Biosphere"

Natalia Kovaleva

Abstract

In this chapter, Kovaleva analyzes the modern ecological situation, characterized by an imbalance in the circulation of matter and energy, reduction of the area of the active functioning of the biosphere and biodiversity. Biosphere parameters are approaching acceptable limits of changes, the transition through which entails a loss of system stability and its destruction. Humankind, as thousands of years ago, obtains almost all food as a result of the use of soil energy in agriculture and animal husbandry. These circumstances are forcing many countries to expand the plowing of land by reducing the area of forests, meadows, and pastures. Meanwhile, the share of arable soils on the planet and the productivity of biocenoses are declining. Replacing natural ecosystems with anthropogenic ones leads to the openness of the cycles of elements, primarily biophilic ones—carbon, nitrogen, phosphorus—and the removal of elements from biological cycles, which in turn reduces the stability of the biosphere as a whole. Urbanization, which implies a loss of highly fertile agricultural land, deforestation, and a significant reduction in biological diversity, seriously impacts global ecology. In this chapter, Kovaleva considers the problems of negative anthropogenic impact both within the framework of the global crisis and within the framework of local crisis conditions. She analyzes the experience of implementing global and local measures aimed at the minimization or elimination of anthropogenic damage to the environment, as well as available theoretical and methodological scientific research from various countries aimed at solving the identified problems. Kovaleva outlines priorities in resolving those problems in the framework of developing a science-based strategy for sustainable development and suggests measures to improve the environment and rehabilitate contaminated territories. The author maintains that the next few decades will be of decisive importance for the formation and

N. Kovaleva (✉)
Soil Science Faculty, Lomonosov Moscow State University, Moscow, Russia

V. Sadovnichy et al. (eds.), *Reconsidering the Limits to Growth*, World-Systems Evolution and Global Futures, https://doi.org/10.1007/978-3-031-34999-7_5

implementation of a new global agenda aimed at ensuring the survival of humankind.

1 Current State

The natural environment around us in its natural state is a balanced system (akin to a living organism), in which individual groups of organisms, including humans, interact with each other and with the surrounding abiotic components without losing the quasi-static equilibrium. Such a state of the system is considered stable. However, phases of sustainable development of the biosphere were interrupted throughout the geological history of landscapes by natural disasters (volcanic eruptions, floods, falls of very large meteorites, etc.). Small biological and large geological cycles of elements and energy cycles secure the connections between functioning organisms and the abiotic environment.

In the course of its development, humankind has been influencing natural systems, changing their capacity, as well as the directions and speeds of natural processes, cycles of matter, and energy flows. Quasi-equilibrium is disturbed by changing parameters, but not immediately. Each of the parameters is associated with other ones and has the capacity (the ability of the environment to adsorb foreign substances, excess energy, etc., without changing its state) to compensate for the changes in neighboring parameters. This buffer reaction continues until the capacities of all parameters are exhausted. This property of the system is the theoretical basis for the concept of acceptable limits of change or growth limits (Rozanov, 1984).

Acceptable limits of changes are the minimal and maximal critical values of the parameters of the state of the environment, within which it is stable and not destroyed. In its interaction with human beings, the environment can go through a number of critical points in its development:

1. Normal state of the environment is an ecologically balanced natural state corresponding to the totality of natural conditions and natural values of parameters.
2. Abnormal (disturbed) state is a state in which one or more parameters reach values that deviate from the background characteristics of a given area, or some properties of the environment are disrupted, for example, its carrying capacity. In an abnormal state, the environment does not yet lose its systemic integrity but acquires the characteristics of an ecologically unbalanced system and may have a harmful effect on people or not satisfy their needs.
3. Critical state of the environment is characterized by the parameters of the state approaching acceptable limits to change, the transition through which entails a loss of stability of the system.
4. Destruction of the environment (crisis) is a state in which the environment becomes unsuitable for human habitation or use as a resource.

Table 1 Areas with natural ecosystems disturbed to varying extents on the Earth's continents

Continents	Area, km^2	Nondisturbed territories, %	Partially disturbed territories, %	Disturbed territories, %
Europe	8,759,321	15.6	19.6	64.9
Asia	53,311,557	43.5	27.0	29.5
Africa	33,985,316	48.9	35.8	15.4
North America	26,179,907	56.3	18.8	24.9
South America	20,120,346	62.5	22.5	15.1
Australia	8,487,262	62.3	25.8	12.0
Antarctica	13,208,983	100.0	0	0
Land area, total	162,052,691	51.9	24.2	23.9
All land area without ice, rock, and bare surfaces	134,904,471	27.0	36.7	36.3

Data source: Hannah et al. (1994)

The current ecological situation is characterized by the following:

- planetary circulation of matter and energy is out of balance (the biomass of land has decreased, while the biomass of the ocean has increased; technogenic pollution of the biosphere, etc.);
- progressive expansion of the area of life was followed by a narrowing of the active functioning of the biosphere (mining, wars, construction);
- biodiversity reduction (by 8–10% for vascular plants);
- weakening of the buffering capacity and the ability to self-clean;
- environmental disasters reaching continental and global levels as a result of natural-anthropogenic resonance;
- qualitative decrease in the viability of the environment;
- the "human-land" system changing to "nomad-resource" system; reduction of resources;
- urbanization has led to the emergence of entire regions that differ in their characteristics from the natural zones in which they are located. All cities in the world are similar to each other in environmental properties;
- most natural ecosystems of Europe, China, India, Central Asia, Turkey, and the Russian forest-steppe have completely disappeared (see Table 1).

Global crisis is to be distinguished from local crisis conditions. The latter in their totality, do not always lead to global crisis due to the compensatory functions and great natural diversity. Meanwhile, natural climatic and landscape zoning is the most important theoretical basis for finding ways out of the current ecological crisis and the only basis for the sustainability of the planetary geobiosystem. That is why Vasily Dokuchaev, the Russian geologist and geographer, who discovered the law of geographical zoning and the founder of soil science, became the teacher of Vladimir Vernadsky, the founder of ecology (Chesnokov, 2013). After all, the doctrine of the zones of nature is the antipode of the law for the struggle for existence, exaggerated

Table 2 Carbon stored by the biome

Biome	Gigatons of C
Tropical and subtropical forests	547.8
Tropical and subtropical grasslands, savannas, shrubs	285.3
Deserts and dry bushes	178.0
Steppe, savannas, and shrubs	183.7
Temperate forests	314.9
Boreal forests	384.2
Tundra	155.4

Data source: Trumper et al. (2009)

by some of Darwin's followers, because animals, plants, microorganisms, and the landscape with its inherent features of soils, topography, climate, geology coexist in every natural zone, in every biogeocenosis, interdependently and under conditions of self-regulation (Dokuchayev, 1899).

The total size of the territory needed to ensure the vital activity of one person and to assimilate waste products varies from 5 to 12 hectares, of which 1 to 2 hectares are territories with destroyed landscapes (Klige, 2006). Consequently, with a population of 8 billion people, 60 to 120 million km^2 of developed land is required, along with 300 to 600 million km^2 of natural ecosystems for waste assimilation. The Earth's land surface is only 149 million km^2, and the area of the entire surface is 510 million km^2.

A significant part of artificial and natural ecosystems is degrading and faces additional risks due to climate change and the reduction of biological diversity. Between 1998 and 2013, about 20% of the world's vegetated land showed a consistent downward trend in productivity, with 20% of agricultural land, 16% of forest land, 19% of grassland, and 27% of rangeland showing a clear downward trend. These trends are especially worrisome given the increased demand for agricultural and livestock products requiring intensive land use. Highly productive plant formations are being replaced with unproductive and low-value ones. Broad-leaved forests and feather grass steppes have been completely destroyed. Anthropogenic badland makes up about 10% of the land surface, and the productivity of biocenoses is significantly reduced by 10–15%; therefore, the carbon deposition capacity of the planet also decreases (Table 2).

Critical state of landscape that is approaching its critical threshold (Rozanov, 1984; Golubev, 1999; Randers, 2012; Reymers, 1994, etc.) is caused by well-known factors: (1) degradation of soil productivity as a result of irrational use; (2) reduction of natural vegetation; (3) pollution of landscapes by-products of economic activities. The Human Ecological Footprint, as noted in one of the well-known reports to the Club of Rome, continues to increase following the increase in the Human Development Index and reaches maximum values in "rich" countries (von Weizsäcker & Wijkman, 2018).

In connection with rapid population growth, the development of industry, construction of cities, and transport routes, fertile soils have been increasingly alienated

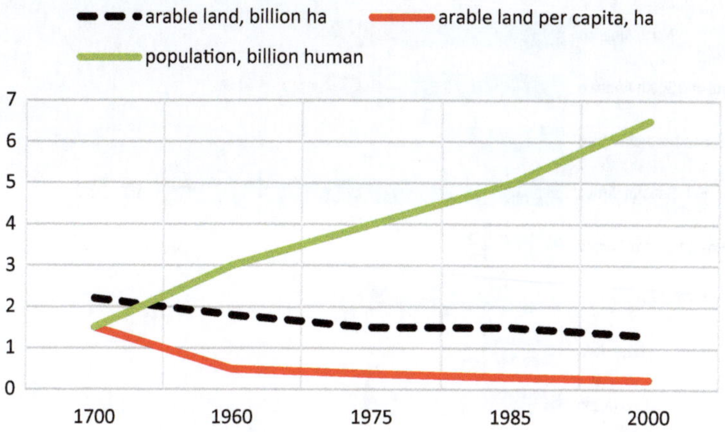

Fig. 1 Dynamics of arable land per capita. Source: Klige (2006), Anthropogenic transformation (2011)

from agricultural circulation, and the rate of anthropogenic degradation is on the rise. These circumstances are forcing many countries to expand the plowing of land by reducing the area of forests, meadows, and pastures.

Meanwhile, humanity still receives more than 98–99% of food (by weight), including 87% of protein (Dobrovolsky, 1997), from the use of fertile soil in agriculture and animal husbandry. While during the first half of the twentieth century, the area of arable land per capita decreased in the world from 0.23 to 0.12 hectares, by 2005, it amounted to only 0.07 hectares (Fig. 1) (Klige, 2006). And this is already a critical value, which will be very difficult to compensate for by further increase in soil fertility. It is time to understand that fertile soil is the ultimate natural resource, no less significant for human life than clean air and freshwater!

Meanwhile, there are very few arable soils on Earth—they comprise 22% of the earth's land area (Table 1), which is about 3.2 billion hectares (Rozanov & Rozanov, 1990). Of these, 1.5 billion hectares are currently under arable land, while 1.7 billion hectares remain unplowed—these are mostly soils that are insufficiently fertile and difficult to develop. In addition, according to a number of researchers, the total area of land that can be used for arable farming should not exceed 2.7 billion hectares. Then the reserve of arable land will be reduced to 1.2 billion hectares (Kovalev & Kovaleva, 2020). This reserve is located mainly in the countries of the tropical belt under moist deciduous forests with red-colored acidic and leached ferrallitic soils, as well as in tropical and subtropical savannahs with alkaline soils. In Russia, after the plowing of virgin lands, there is no reserve of fertile, easily developed soils left (Fig. 2).

The Netherlands Environmental Assessment Agency (PBL), in collaboration with Wageningen University, Utrecht University, and the Joint Research Center of the European Commission, has modeled possible scenarios for social development, taking into account the magnitude of global changes in land use and land conditions

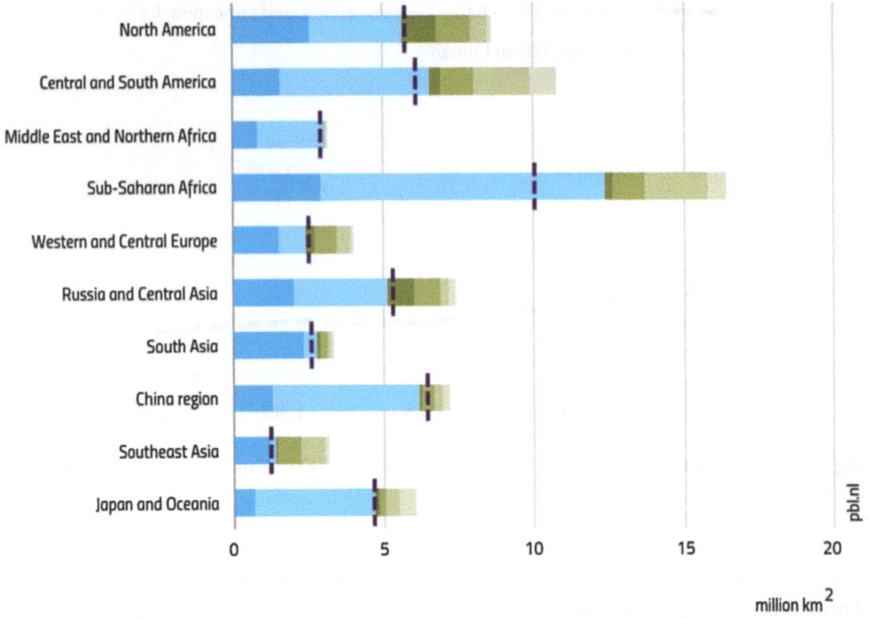

Fig. 2 Assessment of the land fund potential. Source: UN Convention to Combat Desertification (2017: 112)

up to 2050. The "Middle of the Road" scenario (SSP2) is characterized by the continuation of current trends (as usual); the Green Sustainability Scenario (SSP1) represents a more equal and prosperous world committed to sustainable development; and the Fragmentation Scenario (FSP3) reflects a divided world with low economic growth, rapid population growth, and limited environmental concerns (Van Vuuren et al., 2014).

These scenarios are developed using the IMAGE.5 model by applying quantitative projections for population levels (Lutz et al., 2014), urbanization (Jiang & O'Neill, 2015), and economic development (Dellink et al., 2015), as well as by quantifying model parameters that reflect scenarios as described above. The outcomes of the scenarios cover the energy system, food production, land use, greenhouse gas emissions, climate change, biodiversity, and impacts on water and soil properties. Modeling of changes in soil properties, biological diversity, and hydrological systems is supplemented with S-World models (Stoorvogel et al., 2017a, b), GLOBIO model (www.globio.info), and PCR GLOB WB model (Sutanudjaja et al., 2014; De Graaf et al., 2014), respectively.

According to the Dutch PBL/IMAGE model, two out of three scenarios assume an increase in land use for agricultural purposes: it is expected that approximately 50% (in SSP3) and 80% (in SSP2) of this increase will occur in lands with low or average productivity. In contrast, under the WSP1 scenario, global agricultural land area will decrease as a result of a combination of low population growth, an

Table 3 Degree of soil degradation

Degradation type	Area, million ha	%
Washout and destruction by water erosion	1094	56
Dispersion and destruction by wind erosion	548	28
Chemical degradation (nutrient depletion, salinization, pollution, acidification)	239	12
Physical degradation (overconsolidation, waterlogging, subsidence)	83	4
TOTAL	1964	100
Degree of soil degradation		
Weak	749	38
Moderate	911	46
Strong	296	15
Very strong	9	1
TOTAL	1964	100

Data source: Dobrovolsky (1997)

increased focus on sustainable consumption and production (e.g., reduced meat consumption and food waste), and more efficient crop and livestock systems. In Europe and Russia, which contain much of the world's most fertile land, even highly productive land will be subject to land use change or will not be used. The largest expansion of agricultural land is in Sub-Saharan Africa due to high population growth[1] and an increase in demand for food and animal feed that cannot be met only by increasing the efficiency of agricultural production.

At the same time, about 8 million hectares are annually withdrawn from agricultural use due to alienation for other economic needs and about seven million as a result of various soil degradation processes (Table 3). Thus, humankind loses about 15 million hectares of productive land every year. Meanwhile, the process of soil degradation is happening at an exponentially increasing rate: over the past 50 years, it increased 30 times compared to the historical average during the Holocene (Klige, 2006) (Fig. 3).

However, one of the underlying causes of ecological crisis is the underestimation of degradation processes at the level of the soil cover. It is not only food security that is connected with the use of soil resources but also the stability of the entire landscape of the Earth (Table 4). To overcome this problem, the concept of the ecological functions of soils was developed at Moscow State University (Dobrovolsky & Nikitin, 1990).

Anthropogenic ecosystems are aimed at achieving maximum productivity, while "unproductive" natural ecosystems in terms of productivity are set up for maintaining the maximum stability of the biosphere, since, unlike agrocenoses,

[1] See chapter "Africa – the Continent of the Future. Challenges and Opportunities" (Grinin & Korotayev, 2023a, this volume) and chapter "Modeling Social Self-Organization and Historical Dynamics. Africa's Futures" (Korotayev et al., 2023b, this volume) for detail.

Fig. 3 The rate of productive land loss in the last 10,000 years. Source: Klige (2006)

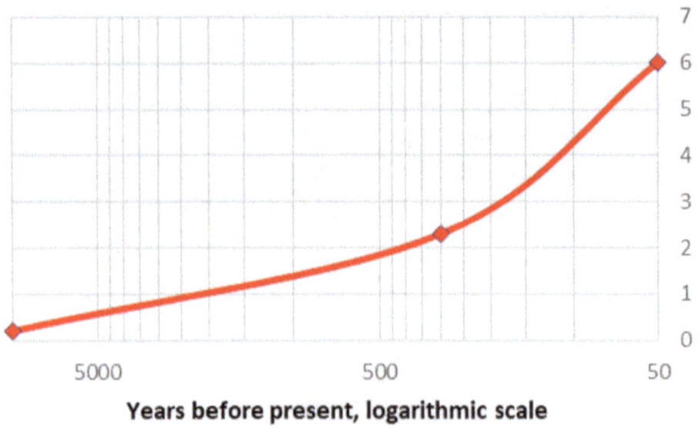

Table 4 Global functions of soils (pedospheres)

Lithospheric	Hydrospheric	Atmospheric	General biospheric
Biochemical transformation of the upper layers of the lithosphere	Transformation of surface water into groundwater	Absorption and reflection of solar radiation	Habitat, accumulation of energy and biophilic substances
Source of matter for the formation of rocks and minerals	Participation in the formation of river runoff	Regulation of atmospheric moisture circulation	Connecting link in the biological and geological circulation of substances
Transfer of accumulated solar energy to the deep parts of the lithosphere	Factor of bioproductivity of water bodies due to the introduction of soil compounds	Source of solid matter and organisms entering the atmosphere	Factor of biological diversity and evolution of organisms
Protection of the lithosphere from excessive erosion and the condition for its normal development	Sorption barrier protecting water areas from pollution	Regulation of the composition and gas regime of the atmosphere	Factor of stability of the functioning of the biosphere

Source: Dobrovolsky (1997)

they are capable of self-regulation by maintaining the biological cycle of substances.[2] Replacing natural ecosystems with anthropogenic ones leads to disconnection of the cycles of elements, primarily biophilic ones, such as carbon, nitrogen, phosphorus, and to the removal of elements from the biological cycles, which, in turn, reduces the stability of the biosphere as a whole (Fig. 4).

[2]On ecological situation and environmental change in the agrarian era see also chapter "Modeling Social Self-Organization and Historical Dynamics. Agrarian Society" (Malkov et al., 2023, this volume); for the industrial epoch see chapter "Modeling Social Self-Organization and Historical Dynamics. Industrial Society" (Akaev et al., 2023a, this volume). About ecological positions of different countries in the Arctic region see Kazanin, 2020.

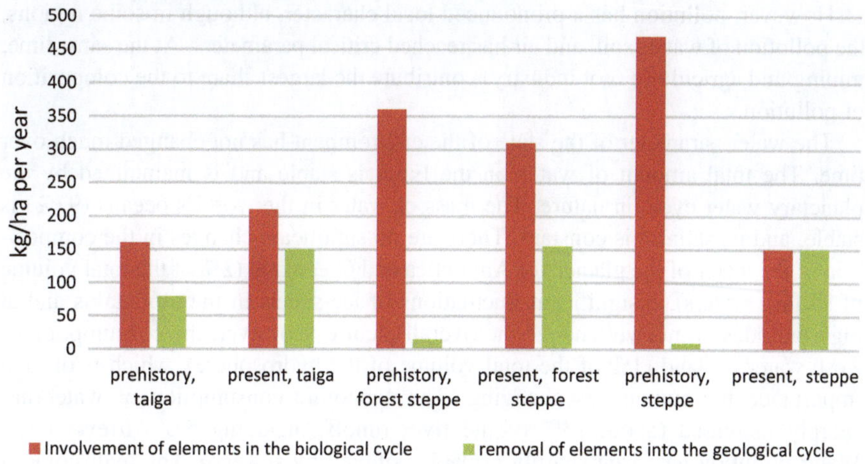

Fig. 4 Involvement of elements (total N, P, K, Ca, Mg, S) in the biological cycle and removal into the geological cycle, kg/ha per year. Source: Rozanov (1984)

The application of mineral and organic fertilizers can only partially improve the situation since, of the total amount of nitrogen introduced into the soil, agricultural plants use only 40–50%. The rest is washed out in the form of nitrates polluting waters or volatilizes into the atmosphere, polluting it with nitrogen oxides. A significant part of phosphate fertilizers is also washed into fresh water, causing the eutrophication of water (Bykovskaya et al., 1999). The introduction of large amounts of nitrogen fertilizers with a lack of phosphorus leads to the poisoning of the products produced, the transformation of hard wheat varieties into soft ones, and, generally, the replacement of a large variety of plant and animal species with a small number of cultivated plant species and animals adapted to agrocenoses.

The total content of all chemical elements on the planet is almost unchanged, but their cycles have undergone transformations as a result of anthropogenic activity and the synthesis of new compounds and substances. The biological cycle is disturbed on half of the land surface (48%): anthropogenic deserts comprise 5%, badlands 3%, residential areas 2%, arable land, gardens, sown meadows 13%, secondary low-productive forests 15%, and pastures 10% (FAO, 2022). If we exclude glaciers, rocks, and deserts, then this is 60% of the productive land surface. The geological cycle of elements is also subject to deformation due to: (1) agricultural soil erosion and an increase in the amount of solid runoff, (2) the annual movement of thousands of cubic meters of soil, (3) extraction of fossil fuels and their conversion into carbon dioxide, (4) extraction of ores, (5) redistribution of salts as a result of irrigation and the construction of reservoirs, (6) use of fertilizers and pesticides, (7) pollution of the environment by industrial, municipal, and agricultural waste, (8) synthesis of substances that did not exist earlier in nature, (9) dispersion of geochemical accumulations in the technogenesis.

However, pollution has a pronounced local character, although in some regions, the pollution of water, soil, and air has reached critical parameters. At the same time, mining and agriculture (not industry) contribute the largest share to the composition of pollution.

The water parameter of the state of the environment has not changed much over time. The total amount of water on the Earth is stable and is maintained by the planetary water cycle in nature. The mass of water in the world's oceans (97%) is stable, and its salinity is constant. There are no significant changes in the composition of the water of the glaciers of Antarctica and Greenland (2% of the total volume of surface waters). Insignificant fluctuations of ice sheets in the mountains and at high latitudes do not yet change the overall picture. However, the consumption of fresh water on land (1% of the total volume of the hydrosphere), which is of vital importance for humans, is changing. The economic consumption of water has sharply increased (about 13% of the river runoff, including 5.6% irreversibly). Water pollution has risen sharply (16% is polluted by sewage). The hydrology of the land has changed: small rivers disappear, large rivers are regulated by reservoirs, and groundwater flow changes due to the drainage of swamps and deforestation. The drying up of floodplains and the development of soil erosion are intensified by active deforestation (GWP Technical Committee, 2004).

The main feature of modern changes in the water regime of Russian rivers is a significant increase in water content during low periods, especially in the winter months. Another feature is the increase in the interannual runoff variability, especially the seasonal one. Against the general background of a trend toward an increase in river flow, both abnormally high-water and abnormally low-water years and seasons are possible. Features of modern changes in maximum water discharges are determined by the conditions of their formation. So, it has significantly decreased in the last decades for a large area of the European part of Russia, where the maximum runoff is formed during spring flood. On the other hand, in regions where maximum water flow is formed by rain floods (the Black Sea coast of the Caucasus, the Kuban and Amur basins), catastrophic floods occurred at the end of the last and beginning of this century, which were not observed before. Estimates of possible changes in river runoff based on the analysis of the reproduction of the components of the water balance of watersheds carried out with models of the general circulation of the atmosphere and ocean, show that in the coming decades, the forecasted water regime will be close to the conditions observed in the last 30–35 years in terms of its main parameters.

Forests cover almost a third of the land surface. The total forest area in the world is 4.06 billion hectares (ha), or 31% of the total land area. This means that there are 0.52 hectares per capita, but forests are unevenly distributed among the regions of the world. The largest part of the forests (45%) is located in the tropical zone, followed by boreal (27%), temperate (16%), and subtropical (11%) belts. More than a half (54%) of the world's forests are found in just five countries: the Russian Federation (20%), Brazil (12%), Canada (9%), the United States of America (8%), and China (5%) (FAO, 2020).

Table 5 Projected loss of forest area

Areas of deforestation	Projected reduction (million ha) by 2030
The Amazon River	23–48
Choco Darien	3
Cerrado	11
Gran Chaco Atlantic Forest	10
The Congo River basin	12
Coastal forests of East Africa	12
Borneo	22
Sumatra	5
New Guinea	7
Greater Mekong	15–30
Australia	6
Total	137–177

Data source: Lanly (1982)

During the existence of civilization, the area of forests has decreased by half; meanwhile, the area of the evaporating surface of the forest canopy is close to the area of the world ocean, and the global production is 40 GtC/year (Gorshkov et al., 2000; Martin, 2015) (Table 5).

One hectare of forest absorbs about 10 tons of CO_2 per year. To absorb 1 billion tons of CO_2 annually emitted into the atmosphere for 50 years, we need 1 million km^2 of forest. In this case, excessive emissions that are not adsorbed by the existing flora and the world ocean could be reduced by 10%.

Deforestation began in prehistoric times, accelerated during European colonial expansion (Gadgil & Guha, 1992), and continues to this day. In general, according to a review of satellite and ground-based data, the observed climate change has a stimulating effect on forest productivity. The increase in productivity of northern ecosystems is associated both directly with an increase in air temperature and indirectly with changes in soil temperature and the position and dynamics of the roof of permafrost due to its landscape-forming role. Changes in climatic characteristics lead to a change in the ranges of species, mainly of two types: (1) a shift in the ranges upwards in mountainous areas and (2) a shift in the ranges to the north. In the most temperate parts, forest area is now expanding from a historic low (Dudley et al., 2006), but this is more than offset by losses in the tropics (Pan et al., 2011). Many tropical forests that were partly deforested several decades ago (Lanly, 1982) have now practically disappeared.

Since 1990, the area of forests in the world has decreased by 178 million hectares, which is approximately equal to the territory of Libya. The rate of net loss of forest area between 1990 and 2020 has slowed noticeably as deforestation has declined in some countries, and forest area has increased in others through afforestation and natural forest expansion. The rate of net loss of forest area decreased from 7.8 million hectares per year in 1990–2000 to 5.2 million hectares per year in 2000–2010 and 4.7 million hectares per year in 2010–2020. However, in the last decade, the net loss of forest area has occurred at a slower rate. The annual rate of deforestation in the last 5 years (2015–2020) is estimated at 10 million ha, up from 12 million ha in 2010–2015 (FAO, 2020).

Despite the fact that the rate of overall deforestation is slowing down, the area of tropical forests in 2010–2015 decreased by 5.5 million ha every year (Keenan et al., 2015); other types of forests have been degraded (Sloan & Sayer, 2015) or overgrazed, converted to shrubland, or plantations. Up to 70% of the world's forests are at risk of further degradation (Haddad et al., 2015). The net loss of forest area is expected to continue for several decades. Deforestation has a serious impact on the soil, especially if forests grow on peat, where deforestation threatens to emit large amounts of carbon, or in dry areas, where the disappearance of trees leads to rapid soil erosion.

Urbanization has a serious impact on global ecology. In many parts of the world, urban land is growing faster than the urban population. While the urban population is projected to reach 5 billion in 2030 and 6.3 billion in 2050, the urban area is expected to triple from its 2000 baseline over the same period, an increase of 1.2 million km^2. Nearly 90% of this growth is likely to occur in Asia and Africa, where urban populations are projected to rise to 56% and 64%, respectively. According to current estimates, Africa's urban population will increase by more than 300 million between 2000 and 2030 (UN Population Division, 2018).

Thus, by 2050, the world's urban population is expected to grow to approximately 6.3 billion people. This growth often leads to the expansion of urban areas, some of them being built on top of fertile soils and agricultural land, resulting in the irreversible loss of arable land. Globally, about 2–3% of the land surface area is now urbanized; by 2050, this indicator is expected to increase to 4–5% (UN Population Division, 2018; Ellis et al., 2010; Ellis, 2011; Hooke et al., 2012). At the same time, urban areas are expected to triple in size by 2030 in some countries. Estimated losses from urbanization between 2000 and 2030 range from 1.6 to 3.3 million hectares of highly fertile agricultural land per year (Derpsch et al., 2010). In addition to the direct use ("withdrawal") of land, the "ecological footprint" of urban settlements extends far beyond their borders. For example, a positive correlation has been established between tropical deforestation and urban population growth, as well as agricultural exports.

The increase in urban areas is likely to lead to significant reductions in biodiversity, such as:

- Extensive urbanization in eastern Afromontana, the Guinean forests of West Africa and the Western Ghats, and the biodiversity hubs of Sri Lanka will result in urban expansion of approximately 1900%, 920%, and 900%, respectively, by 2030 (more than 2000 times), resulting in a serious reduction in biodiversity.
- In already reduced and highly fragmented habitats, such as the Mediterranean and South American Atlantic forests, relatively small habitat loss can lead to a disproportionate increase in species extinction rates.

- The five centers of biodiversity with the largest areas expected to become urban are predominantly coastal areas or islands that are particularly important for endemic species.

Among other factors that cause severe damage to nature and which require special attention are armed conflicts (see Appendix).

2 Prospects for Further Change in the Situation

The response to these challenges is quite simple: use less polluting energy sources [e.g., biogas generators [Budzianowski, 2016)], photovoltaic motors,[3] more efficient energy-saving solutions,[4] and apply land use and management practices that prioritize the conservation of carbon in the soil.

Soil health can be improved by limiting soil damage and increasing soil organic matter, for example, through the use of improved crop varieties [including varieties of deep-rooted plants (Kell, 2012) and cover crops (Poeplau & Don, 2015)], crop rotation (Burney et al., 2010) and, in some cases, no-tillage (West & Post, 2002). Soil erosion minimization interventions cover a wide spectrum from engineering interventions (such as terracing, lagoon construction ([Pansak et al., 2008]) and waterway improvement) to vegetative interventions [such as agroforestry, cover crops (Agus & Widianto, 2004)]. Tillage farming can dramatically improve the physical properties of topsoil (Alvarez et al., 2009). Measures to reduce wind erosion include the use of drought-tolerant species, pasture rotation, and windbreak planting, combined with no-tillage and stubble conservation tillage practices (Fryrear & Skidmore, 1985). Halting soil degradation and the accumulation of soil organic matter also helps to mitigate climate change through soil carbon sequestration and, at the same time, helps to increase the resilience of agricultural systems (Amanullah et al., 2016). Increased content of organic carbon in the soil leads to an increase in crop yields, especially in areas with low and variable rainfall (Branca et al., 2013). Soil salinization is best prevented by using high-quality irrigation water and ensuring proper drainage through the use of drainage pipes and/or drainage channels; periodic gypsum of soils may also be required. Preventing soil compaction requires specific management, as recovery can take many decades. Long-term limited or anti-erosion tillage is considered an effective approach in many regions of the world (Derpsch et al., 2010). While crucial to long-term soil health, these measures often do not provide immediate, tangible benefits for farmers and land users. For this reason, there are no direct stimuli for farmers to take conservation action, especially when they do not have land ownership, and stronger incentives are needed for that to happen (Hammad & Børresen, 2006).

[3] See, e.g., http://www.agrophotovoltaik.de/english/agrophotovoltaics/

[4] See chapter "Climate and Energy. Energy Transition Scenarios and Global Temperature Changes Based on Current Technologies and Trends" (Akaev & Davydova, 2023, this volume) for detail.

Example of South Korea Sixty years ago, South Korea's GDP did not differ much from that of Kenya or Tanzania. Currently, the average wage in the country is approximately equal to the Australian one. Within a couple of generations, South Korea reached the ranks among the richest countries in the world. One of the reasons for this success was massive efforts to restore the environment. During World War II and the ensuing civil war, the country experienced devastating environmental degradation characterized by the ecological crisis; most of the forests disappeared as a result of the conflict and fuelwood gathering. Since then, the Korean government has been implementing one of the most impressive reforestation programs in history, planting 2.8 million hectares of forests and increasing forest growth 12 times. As a result, much of the country is now covered with mature forests. In addition, Korea has established a system of protected areas covering an area of 16,000 km^2, which are extremely popular among the highly urbanized population of this country; in 2007, 38 million people visited national parks, 99% of which were domestic tourists.

In some parts of the world, organic agriculture is no longer the prerogative of individual fields and is becoming the main or only model of production.

Example of India In January 2016, Sikkim became the first state in India to go completely organic. It took 10 years to convert 75,000 hectares of agricultural land in the state of Sikkim to the status of certified organic farms (Chamling, 2010). The state now produces 800,000 tons of produce, nearly 65% of India's total organic production of 1.24 million tons. Sikkim is a state that is a model for the world because, while protecting ecosystem services, this model demonstrates that going organic does not harm productivity or development.

Example of Bhutan In 2011, Bhutan set itself an ambitious goal of converting the country's entire agricultural system to organic production in the 2020s. If successful, Bhutan will become the world's first country with completely organic food production. There are 700,000 people in this country, most of them farmers, so the only problem is to show that the benefits of such a system outweigh the costs and that the use of natural fertilizers alone does not affect yields. Bhutan's organic strategy is to move from stage to stage, region to region, and product to product, demonstrating that such innovations are necessary to find ways to naturally eradicate diseases and increase crop yields (Neuhoff et al., 2014).

All existing projection scenarios up to 2050 show that the strongest regional changes in land use are expected in Sub-Saharan Africa[5]; however, the best land is already being exploited, and expansion will increasingly take place on less productive land, resulting in lower yields (by 20%) (Rasmussen et al., 2016; Herrmann et al., 2014; Fensholt et al., 2012). In several regions, such as South Asia, there is

[5]See chapter "Africa – the Continent of the Future. Challenges and Opportunities" (Grinin & Korotayev, 2023a, this volume) and chapter "Modeling Social Self-Organization and Historical Dynamics. The Future W-Society" (Akaev & Davydova, 2023, this volume) for detail.

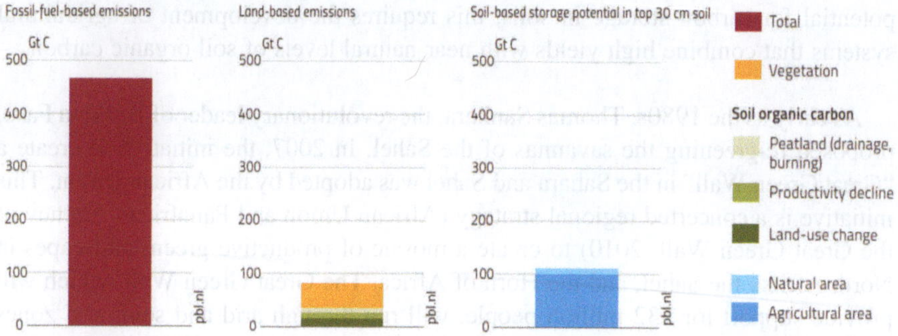

Fig. 5 Potential of soils for carbon sequestration. Source: UNCCD (2017)

either little or only marginal land left for agrarian expansion. As a result of ongoing changes in land use and the deterioration of soils, vegetation cover, and biological diversity, extensive changes in the state of land resources are also forecasted in the future.

In terms of the average relative abundance of species, the projected loss of biodiversity will continue at 4–12% until 2050 and will also proceed for much of the second half of the twenty-first century. In addition, changes in land cover and soil quality affect the likelihood of floods and droughts. These factors are even more intense in drylands, where more than average population growth is also expected.

Global soil organic carbon has been reduced by 176 Gt compared to its natural, undisturbed state. Losses of organic carbon in the humusosphere due to dehumification amounted to 313×10^{15} g over the last 10,000 years, including 90×10^{15} over the last 300 years and 38×10^{15} over the last 50 years (760 million tons per year). Since 1883, the humus content in the black soils of Russia has decreased from 10–12% to 4–7% (Kovalev & Kovaleva, 2016). The annual intake of CO_2 from soils into the atmosphere due to "breathing" is 60×10^{15} g of carbon. That is, the total loss of carbon by soils is two times greater than its entry into the atmosphere due to anthropogenic activities (Orlov et al., 2002). Therefore, the soil cover on which vegetation is restored is a carbon sink area, while the soils of the developed territories are sources of carbon dioxide (Fig. 5).

If current trends continue, anthropogenic emissions of carbon into the atmosphere from soil and vegetation on land for the period 2020–2050 will amount to an additional 80 Gt, which is equivalent to about 8 years at the current level of global carbon emissions from fossil fuels. Reducing these projected inland emissions would allow a larger global carbon budget of 170 to 320 Gt (i.e., the amounts of CO_2 emissions that are still tolerable) to be left untouched without jeopardizing the goal of keeping global temperature rise below 2 °C.[6] Despite the planet's significant

[6] See chapter "Climate and Energy. Energy Transition Scenarios and Global Temperature Changes Based on Current Technologies and Trends" (Akaev & Davydova, 2023, this volume) for detail; see also Akaev and Davydova (2021).

potential for carbon storage in soils, this requires the development of agricultural systems that combine high yields with near-natural levels of soil organic carbon.

As early as the 1980s, Thomas Sankara, the revolutionary leader of Burkina Faso, proposed re-greening the savannas of the Sahel. In 2007, the initiative to create a "Great Green Wall" in the Sahara and Sahel was adopted by the African Union. This initiative is a concerted regional strategy (African Union and Panafrican Agency of the Great Green Wall, 2010) to create a mosaic of productive green landscapes in North Africa, the Sahel, and the Horn of Africa. The Great Green Wall, which will provide support for 232 million people, will run through arid and semi-arid zones north and south of the Sahara and will be a 15 km wide and 7775 km long belt stretching from Dakar to Djibouti; the area of the main part of this belt will be 780 million hectares. About 10 million hectares will need to be restored annually (UNCCD, 2016). The goal of this wall is to reverse the effects of land degradation by 2025 and regional land transformation by 2050. Many changes have been already implemented[7]:

- Ethiopia—15 million hectares of degraded land restored, watersheds improved, tenure security improved; incentives are applied to encourage community involvement;
- Nigeria—five million hectares restored, including 319 km of windbreaks; 20,000 jobs created. In northern Nigeria, 5000 farmers were trained in land restoration; more than 500 young people got jobs as forest guards;
- Senegal—11.4 million trees planted; 1500 km of fire barriers; promotion of natural revegetation applied on 10,000 hectares of land; in general, 24,600 hectares of degraded land have been restored;
- Sudan—2000 hectares of restored land.

Deserts occupy almost one-fifth of the entire territory of China, while most of the territory, especially the dry regions of western China (also among the poorest regions), is at risk of desertification. The livelihoods of 400 million people are threatened or affected by land degradation and expanding deserts. Rapid industrialization and urbanization have swallowed up agricultural land, exacerbating the already serious problem. In addition, wood logging has made the land vulnerable to sand invasion. A prolonged drought in northwestern China has made the situation even more serious, intensifying dust storms and sandstorms. Since 1978, a "Great Green Wall" of trees, shrubs, and grasses has been erected in the Kubuqi Desert to protect northern cities; it cost US$6.3 million and slowed down the desertification from about 3400 km^2 per year in the 1990s to 2000 km^2 per year since 2001. By 2010, 12,452 km^2 of land affected by desertification had been restored, according to

[7] See, e.g., African Union and Panafrican Agency of the Great Green Wall (2010) and Ivie Ihejirika (2016).

a government study, although desertification still increased in some areas (State Forestry Administration, 2011).

However, the reality is much more complex than any model. The soil cover absorbs excess carbon dioxide, and the work of carbon landfills and carbon farms is based on this feature. At the same time, the initial amount of carbon in vegetation and humus is approximately the same; however, plants absorb 64–83% (101–115 GtC) of all absorbed CO_2, while humus absorbs 6–7% (7–10 Gt C), and litter absorbs 10% (12–14 GtC). However, the more intensive the carbon reserves of humus grow, the stronger the dependence of annual production on assimilable nitrogen.

All global models indicate that the secular trend of turning forest land into a fund of agricultural land will continue until at least 2050. Future needs for land use for agriculture will affect forests, savannahs, and grasslands. Consequently, the loss of natural habitats and associated impacts on biodiversity can be expected to worsen.

So, humankind has found itself in the conditions of survival in the "unstable biosphere". The state of bifurcation can last for several centuries. Moderate optimism is based on the information that the reducing impact of modern civilization on the biological cycle has not reached the level of the Wurm maximum. This means that the elasticity of the global system is greater than we think.

Gradual change in the population's social structure and mentality under the influence of a growing economy has already begun to lead to a decrease in population growth and limitation in resource consumption. Widespread implementation of eco-policy measures in a number of high-income countries, preservation of quite vast undeveloped territories as the basis for maintaining biodiversity within the north of Eurasia and America, the ability of the biosphere to function geologically for a long time in conditions of homeostasis disturbances, and the inertia of many planetary systems and processes (the world ocean, for example)—all this allows us to forecast the possibility of long-term sustainable development, but not simultaneous in different states. To develop a science-based strategy for sustainable development, the following priorities must be recognized:

– ecological and economic optimization of natural-anthropogenic and anthropogenic systems by greening technologies and greening production[8] (Fig. 6) [this

[8]Success stories: In 2014, India became the first country in the world to adopt a national agroforestry policy that encourages practices to grow trees, crops, and livestock at the same time on the same plot of land. Farmers have been growing trees on their plots for generations to help maintain the condition of the soil, as well as supply food, timber, and fuel. Intensive goat farming has led to the degradation of more than 1.5 million hectares of subtropical scrubland in the Eastern Cape of South Africa, resulting in the emergence of open desert-like landscapes in these areas with surface temperatures reaching 70 ° C. The reduction or loss of almost all ecosystem services provided by these bushes have led to a decrease in farmers' incomes and a decline in the local economy. The question was how to restore the functionality of this ecosystem in order to extract the maximum environmental and economic benefits. In the early 1970s, a livestock farm owner living in the vicinity of Uitenhage took one small but important step to solve this problem. He built a barn at the foot of a degraded slope that flooded every time it rained. The farmer decided to try to restore the surface of the slope by turning it back into dense scrub to increase rainwater capture and prevent his

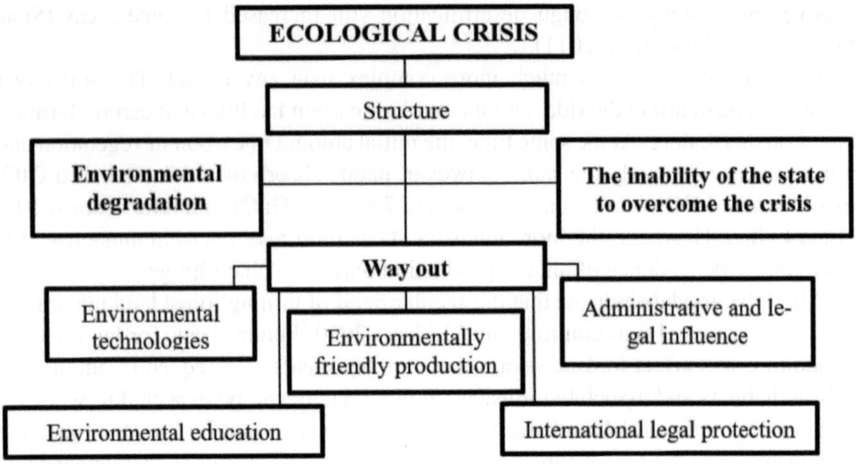

Fig. 6 Ways out of the global ecological crisis

topic is also a subject of one of the reports to The Club of Rome (Wijkman & Skånberg, 2017)][9];

- a multifunctional approach to land resources; land use planning, taking into account land degradation neutrality (land degradation neutralization is a new approach to control land degradation, designed to encourage actions aimed at preventing or reducing degradation, as well as restoring degraded land in order to achieve the goal of zero balance loss of healthy, productive land at the national level);

barn from flooding. Together with other farmers, he set about restoring the shrub cover using cuttings from a local succulent tree, the elephant bush (*Portulacaria afra*); this resulted in improved soil quality and increased carbon stocks, and the area's animal feed productivity and income increased ten times. Based on the results of these advanced farmers and pastoralists, the South African government decided to invest in large-scale restoration of degraded bushland. The Subtropical Thicket Restoration Program was established, with approximately $8 million spent between 2004 and 2016. To date, over 10,000 hectares of nature reserves, private lands, and areas of the Addo Elephant National Park have been covered with elephant bush seedlings.

[9] On optimum scenarios that could contribute to solving the climatic and environmental problems in technological, economic, social, demographic, and political dimensions see chapter "Technology. Limitless Possibilities, Effective Control" (Grinin & Grinin, 2023, this volume), chapter "Economics. Optimizing Growth" (Grinin et al., 2023b, this volume), chapter "Demography. Toward Optimization of Demographic Processes" (Korotayev et al., 2023a), chapter "Socio-Political Transformations. A Difficult Path to Cybernetic Society" (Grinin et al., 2023c, this volume), chapter "High Income and Low Income Countries. Toward a Common Goal at Different Speeds" (Grinin & Korotayev, 2023b, this volume), chapter "Future Political Change. Toward a More Efficient World Order" (Grinin et al., 2023a, in this volume), chapter "Africa – the continent of the future: challenges and opportunities" (Grinin & Korotayev, 2023a, this volume), chapter "Modeling Social Self-Organization and Historical Dynamics. Modern Society and a Look into the Global Future: Cybernetic W-Society" (Akaev et al., 2023b, this volume), and chapter "Modeling Social Self-Organization and Historical Dynamics. Africa's Futures" (Korotayev et al., 2023b, this volume).

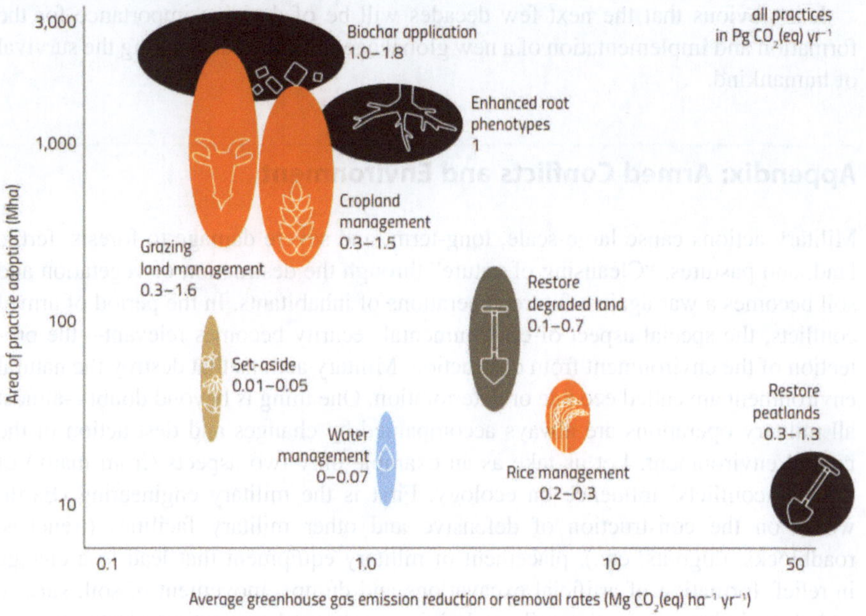

Fig. 7 Reducing greenhouse gas emissions from agriculture (UNCCD, 2017). Global potential for agricultural-based GHG mitigation practices, where 1Pg (Pentagram) equals 1 billion metric tons and Mg (Megagram) equals 1 metric ton. Source: Paustian et al. (2016)

- protection of natural systems and biodiversity through legal influence at administrative and international levels[10];
- formation of a new approach to territorial planning aimed at minimizing the consequences of uncontrolled urban sprawl and infrastructure development;
- prevention of total losses: the creation of incentives for environmentally sound consumption of natural resources and extraction of natural raw materials[11];
- formation of the resilience of communities and ecosystems to external influences, increasing the adaptive capacity of regions through a combination of environmental measures, sustainable management, and restoration of land resources (Fig. 7);
- environmental education and enlightenment of the population in order to create a new worldview ("third consciousness" after the agrarian and industrial-technological waves).[12]

[10]These issues are covered in several reports to the Club of Rome (Perissi & Bardi, 2021; Borgese, 1986, 1998).

[11]Issues of uncontrolled spending of natural resources of the planet are covered also in A Report to the Russian Association of the Club of Rome by Ugo Bardi (2014).

[12]In their report to the Club of Rome, Ilaria Perissi and Ugo Bardi (2021: 181) propose using simulation games as part of the education of the population, which make it possible to more easily explain what is not intuitive knowledge for people. By the latter, they understand the issues of

It is obvious that the next few decades will be of decisive importance for the formation and implementation of a new global agenda aimed at ensuring the survival of humankind.

Appendix: Armed Conflicts and Environment

Military actions cause large-scale, long-term, and severe damage to forests, fertile land, and pastures. "Cleansing of nature" through the destruction of vegetation and soil becomes a war against future generations of inhabitants. In the period of armed conflicts, the special aspect of environmental security becomes relevant—the protection of the environment from destruction. Military actions that destroy the natural environment are called ecocide or deterioration. One thing is beyond doubt—almost all military operations are always accompanied by changes and destruction of the natural environment. Let us take as an example only two aspects (from many) of military conflicts' influence on ecology. First is the military engineering (Earth) works on the construction of defensive and other military facilities (trenches, roadblocks, dugouts, etc.), placement of military equipment that lead to a change in relief, formation of artificial excavations and dumps, movement of soil, surface and deep influence on the soil, underlying rocks and vegetation, destruction of vegetation, wind and water erosion, change in the water-air regime of soils, disturbance of natural soil process, growth of buried soils.

Second, is the direct effect associated with changes in the surface due to explosions, and the indirect effect is caused by shock waves and disturbance of the stability of the soil cover. On the plains, the indirect effect is relatively small, but in the mountains, it is significant and depends on the steepness of the slope and the mass of soils moving due to the activation of erosion processes. Landslides create huge masses of friable gravelly deposits at the foot of the slopes.

It is already clear that the destruction of nature by military action will have profound long-term consequences. However, in any environmental impact, the greatest danger is not immediately occurring effects. More worrying is the prospect of those significant, slowly accumulating changes that can occur due to poorly understood chemical reactions.

saving natural resources, since intuitively people strive to achieve prosperity through economic growth, increasing profits, and the negative consequences of this are not intuitively felt, this requires additional reflection, to which the majority of the world's population is not too inclined. Therefore, the authors propose to use more familiar and easy-to-learn formats of education and enlightenment on the issues of saving the biosphere from anthropogenic impact. An example of such a game is described in detail in Section "Moby Dick: The Game of Fishing" of this report. Also, there are several other reports to the Club of Rome, devoted to the issues of education in terms of educational program on global problems and processes (Andersen, 2020; Bozesan, 2020; Berg, 2019; Korten, 2015; Giarini & Malitza, 2003; Vester, 2002; Peter, 1998; Botkin et al., 1979). On some interesting points with respect to interconnections between ecological situation and globalization see Kiss, 2020.

The following measures to improve the environment and rehabilitate contaminated territories will be needed in the territories affected by military actions:

- identification of sources of pollution, localization, and elimination of oil pollution of the territory at the locations of military equipment, wells, and oil refineries;
- examination and assessment of the degree of pollution of surface and groundwater used as drinking water by cities and towns;
- carrying out works to prevent pollution of surface waters;
- work on localization and extraction of oil products from long-term man-made deposits;
- inspection of radioactive waste disposal sites;
- restoration of the state service for monitoring the state of the environment.

References

African Union and Panafrican Agency of the Great Green Wall. (2010). *Harmonised regional strategy for implementation of the "Great Green Wall Initiative of the Sahara and the Sahel"*. African Union.

Agus, F., & Widianto, D. (2004). *Practical guidelines for upland soil conservation*. World Agroforestry Centre (ICRAF).

Akaev, A., & Davydova, O. (2021). A mathematical description of selected energy transition scenarios in the 21st century, intended to realize the main goals of the Paris Climate Agreement. *Energies, 14*(9), 1–28. https://doi.org/10.3390/en14092558

Akaev, A., & Davydova, O. (2023). Modeling social self-organization and historical dynamics. The future W-Society. In V. Sadovnichy et al. (Eds.), *Reconsidering the limits to growth. A report to the Russian Association of the Club of Rome* (pp. 53–70). Springer. https://doi.org/10.1007/978-3-031-34999-7_4

Akaev, A., Malkov, S., Davydova, O., Kovaleva, N., Malkov, A., Grinin, L., & Korotayev, A. (2023a). Modeling social self-organization and historical dynamics. Industrial society. In V. Sadovnichy et al. (Eds.), *Reconsidering the limits to growth. A report to the Russian Association of the Club of Rome* (pp. 387–417). Springer. https://doi.org/10.1007/978-3-031-34999-7_18

Akaev, A., Malkov, S., Grinin, L., Bilyuga, S., Davydova, O., Grinin, A., Kovaleva, N., Malkov, A., Musieva, J., & Korotayev, A. (2023b). Modeling social self-organization and historical dynamics. Modern society and a look into the global futures: Cybernetic W-Society. In V. Sadovnichy et al. (Eds.), *Reconsidering the limits to growth. A report to the Russian Association of the Club of Rome* (pp. 461–490). Springer. https://doi.org/10.1007/978-3-031-34999-7_20

Alvarez, C. R., Taboada, M. A., Gutierrez Boem, F. H., Bono, A., Fernandez, P. L., et al. (2009). Topsoil properties as affected by tillage systems in the Rolling Pampa region of Argentina. *Soil Science Society of America Journal, 73*, 1242–1250.

Amanullah, A., Khan, S. U., Iqbal, A., & Fahad, S. (2016). Growth and productivity response of hybrid rice to application of animal manures, plant residues and phosphorus. *Frontiers in Plant Sciences, 7*, 1440–1447.

Andersen, L. R. (2020). *Bildung – Keep growing. A report to the Club of Rome*. Nordic Bildung.

Bardi, U. (2014). *Extracted: How the quest for mineral wealth is plundering the planet. A report to The Club of Rome*. Chelsea Green Publishing.

Berg, C. (2019). *Sustainable action: Overcoming the barriers. A report to the Club of Rome*. Routledge.

Borgese, E. M. (1986). *The future of the Oceans. A report to the Club of Rome*. The Club of Rome.

Borgese, E. M. (1998). *The Oceanic circle. A report to the Club of Rome*. United Nations University Press.

Botkin, J., Elmandjra, M., & Malitza, M. (1979). *No limits to learning. Bridging the human gap. A report to The Club of Rome*. Pergamon Press.

Bozesan, M. (2020). *Integral investing. From profit to prosperity. A report to the Club of Rome*. Springer.

Branca, G., Lipper, L., McCarthy, N., & Jolejole, M. C. (2013). Food security, climate change, and sustainable land management. A review. *Agronomy for Sustainable Development, 33*, 635–650.

Budzianowski, W. (2016). Renewable and sustainable. *Energy Reviews, 54*, 1148–1171.

Burney, J. A., Davis, S. J., & Lobell, D. B. (2010). Greenhouse gas mitigation by agricultural intensification. *Proceedings of the National Academy of Sciences, 107*, 12052–12057.

Bykovskaya, T. K., Kovaleva, N. O., & Paramonova, T. A. (1999). *Ekoogicheskiye problemy sel'skokhozyaystvennogo proizvodstva i sostoyaniye pochv Rossii*. Agrarnaya nauka.

Chamling, P. (2010). *Sikkim Organic Mission 2015*. Food Security and Agriculture Development Department, Government of Sikkim.

Chesnokov, V. (2013). V.I.Vernadskiy – velikiy nauchnyy novator i gumanist (k 150-letiyu so dnya rozhdeniya V.I.Vernadskogo). *Vek globalizatsii, 1*(11), 1–18.

De Graaf, I. E. M., Sutanudjaja, E. H., van Beek, L. P. H., & Bierkens, M. F. P. (2014). A high resolution global scale groundwater model. *Hydrology and Earth System Sciences Discussions, 11*(5), 5217–5250.

Dellink, R., Chateau, J., Lanzi, E., & Magné, B. (2015). Long-term economic growth projections in the shared socioeconomic pathways. *Global Environmental Change, 42*, 200–214.

Derpsch, R., Friedrich, T., Kassam, A., & Li, H. W. (2010). Current status of adoption of notill farming in the world and some of its main benefits. *International Journal of Agricultural and Biological Engineering, 3*, 1–25.

Dobrovolsky, G. (1997). Tikhiy krizis planety. *Vestnik Rossiyskoy Akademii Nauk, 67*(4), 313–320.

Dobrovolsky, G., & Nikitin, Y. (1990). *Funktsii pochv v biosfere i ekosistemakh: ekologicheskoye znacheniye pochv*. Nauka.

Dokuchayev, V. (1899). *Pochvennyye zony severnogo polushariya*. MP.

Dudley, N., Schlaepfer, R., Jackson, W., Jeanrenaud, J. P., & Stolton, S. (2006). *Forest quality: Assessing forests at a landscape scale*. Earthscan.

Ellis, E. C. (2011). Anthropogenic transformation of the terrestrial biosphere. *Philosophical Transactions of the Royal Society of London A. Mathematical, Physical and Engineering Sciences, 369*, 1010–1035.

Ellis, E. C., Klein Goldewijk, K., Siebert, S., Lightman, D., & Ramankutty, N. (2010). Anthropogenic transformation of the biomes, 1700 to 2000. *Global Ecology and Biogeography, 19*, 589–606.

FAO. (2020). *Global'naya otsenka lesnykh resursov 2020 goda. Osnovnyye vyvody*. FAO.

FAO. (2022). *FAOSTAT database*. FAO. https://www.fao.org/faostat/en/#home

Fensholt, R., Langanke, T., Rasmussen, K., Reenberg, A., Prince, S. D., et al. (2012). Greenness in semi-arid areas across the globe 1981–2007—an Earth observing satellite based analysis of trends and drivers. *Remote Sensing of Environment, 121*, 144–158.

Fryrear, D. W., & Skidmore, E. L. (1985). Methods for controlling wind erosion. In R. F. Follett & B. A. Stewart (Eds.), *Soil erosion and crop productivity* (pp. 43–59). CABI.

Gadgil, M., & Guha, R. (1992). *This fissured land: An ecological history of India*. Oxford University Press.

Giarini, O., & Malitza, M. (2003). *The double helix of learning and work. A report to The Club of Rome*. UNESCO.

Golubev, G. N. (1999). *Geoekologiya*. GEOS.

Gorshkov, V. G., Gorshkov, V. V., & Makarieva, A. M. (2000). *Biotic regulation of the environment: Key issue of global change*. Springer.

Grinin, L., & Grinin, A. (2023). Technology. Limitless possibilities and effective control. In V. Sadovnichy et al. (Eds.), *Reconsidering the limits to growth. A report to the Russian Association of the Club of Rome* (pp. 139–154). Springer. https://doi.org/10.1007/978-3-031-34999-7_8

Grinin, L., & Korotayev, A. (2023a). Africa – the continent of the future. Challenges and opportunities. In V. Sadovnichy et al. (Eds.), *Reconsidering the limits to growth. A report to the Russian Association of the Club of Rome* (pp. 225–238). Springer. https://doi.org/10.1007/978-3-031-34999-7_13

Grinin, L., & Korotayev, A. (2023b). High income and low income countries. Toward a common goal at different speeds. In V. Sadovnichy et al. (Eds.), *Reconsidering the limits to growth. A report to the Russian Association of the Club of Rome* (pp. 207–224). Springer. https://doi.org/10.1007/978-3-031-34999-7_12

Grinin, L., Grinin, A., & Korotayev, A. (2023a). Future political change. Toward a more efficient world order. In V. Sadovnichy et al. (Eds.), *Reconsidering the limits to growth. A report to the Russian Association of the Club of Rome* (pp. 191–206). Springer. https://doi.org/10.1007/978-3-031-34999-7_11

Grinin, L., Grinin A., & Malkov, S. (2023b). Economics. Optimizing growth. In V. Sadovnichy et al. (Eds.), *Reconsidering the limits to growth. A report to the Russian Association of the Club of Rome* (pp. 155–168). Springer. https://doi.org/10.1007/978-3-031-34999-7_9

Grinin, L., Grinin, A., & Malkov, S. (2023c). Socio-political transformations. A difficult path to cybernetic society. In V. Sadovnichy et al. (Eds.), *Reconsidering the limits to growth. A report to the Russian Association of the Club of Rome* (pp. 169–189). Springer. https://doi.org/10.1007/978-3-031-34999-7_10

GWP Technical Committee. (2004). *Catalyzing change: A handbook for developing integrated water resources management (IWRM) and water efficiency strategies*. Global Water Partnership.

Haddad, N. M., Brudvig, L. A., Clobert, J., Davies, K. F., Gonzalez, A., et al. (2015). Habitat fragmentation and its lasting impact on Earth's ecosystems. *Science Advances, 1*(2), 1–9. https://doi.org/10.1126/sciadv.1500052

Hammad, A., & Børresen, T. (2006). Socioeconomic factors affecting farmers' perceptions of land degradation and stonewall terraces in central Palestine. *Environmental Management, 37*, 380–394.

Hannah, L., Lohse, D., Hatchinson, C., Carr, J. L., & Lankerani, A. (1994). A preliminary inventory of human disturbance of world ecosystems. *Ambio, 23*(4–5), 246–250.

Herrmann, S. M., Sall, I., & Oumar, S. (2014). People and pixels in the Sahel: A study linking coarse-resolution remote sensing observations to land users' perceptions of their changing environment in Senegal. *Ecology and Society, 19*(3), 1–18. https://doi.org/10.5751/ES-06710-190329

Hooke, R. L., Martín-Duque, J. F., & Pedraza, J. (2012). Land transformation by humans: A review. *GSA Today, 22*, 4–10.

Ivie Ihejirika, P. (2016). Desertification: Ensuring sustainable future for communities through GGW. *Leadership* 03.02.2016. http://leadership.ng/features/505887/desertificationensuring-sustainable-future-communities-ggw

Jiang, L., & O'Neill, B. C. (2015). Global urbanization projections for the Shared Socioeconomic Pathways. *Global Environmental Change, 42*, 192–199.

Kazanin, A. (2020). Globalizing Petro-Arctic: What Do Producing and Non-Producing Arctic States Have to Say ? *Journal of Globalization Studies, 11*(2), 76–87. https://doi.org/10.30884/jogs/2020.02.06

Keenan, R. J., Reams, G. A., Achard, F., de Freitas, J. V., Grainger, A., et al. (2015). Dynamics of global forest area: Results from the FAO Global Forest Resources Assessment 2015. *Forest Ecology and Management, 352*, 9–20.

Kell, D. (2012). Large-scale sequestration of atmospheric carbon via plant roots in natural and agricultural ecosystems: Why and how. *Philosophical Transactions of the Royal Society, B, 367*, 1589–1597.

Kiss, E. (2020). The Present Dialogue between Nature and Society or the Mutuality of Nature Preservation and Society Protection. *Journal of Globalization Studies, 11*(1), 86–101. https://doi.org/10.30884/jogs/2020.01.0010.30884/jogs/2020.01.06

Klige, R. (2006). *Sovremennyye global'nyye izmeneniya prirodnoy sredy*. Nauchnyy mir.

Korotayev, A., Malkov, S., & Musieva J. (2023a). Demography. Toward optimization of demographic processes. In V. Sadovnichy et al. (Eds.), *Reconsidering the limits to growth. A report to the Russian Association of the Club of Rome* (pp. 97–116). Springer. https://doi.org/10.1007/978-3-031-34999-7_6

Korotayev, A., Shulgin, S., Ustyuzhanin, V., Zinkina, J., & Grinin, L. (2023b). Modeling social self-organization and historical dynamics. Africa's futures. In V. Sadovnichy et al. (Eds.), *Reconsidering the limits to growth. A report to the Russian Association of the Club of Rome* (pp. 491–519). Springer. https://doi.org/10.1007/978-3-031-34999-7_21

Korten, D. C. (2015). *Change the story, change the future. A report to The Club of Rome*. Berrett-Koehler Publishers.

Kovalev, I., & Kovaleva, N. (2016). Otrazheniye sotsial'no-ekologicheskogo krizisa XIX veka v rabotakh V.V. Dokuchayeva (na primere antropogennoy evolyutsii chernozemov Tambo-skoy ravniny v istoricheskoye vremya). *Zhivyye i biokosnyye sistemy, 16*, 90–119.

Kovalev, I., & Kovaleva, N. (2020). Ekologicheskiye funktsii pochv i vyzovy so-vremennosti. *Ekologicheskiy vestnik Severnogo Kavkaza. KGFU, 16*(2), 4–17.

Lanly, J. P. (1982). *Tropical forest resource*. FAO.

Lutz, W., Butz, W. P., & Samir, K. E. (Eds.). (2014). *World population and human capital in the twenty-first century*. Oxford University Press.

Malkov, S., Kovaleva, N., Grinin, L., & Korotayev, A. (2023). Modeling social self-organization and historical dynamics. Agrarian society. In V. Sadovnichy et al. (Eds.), *Reconsidering the limits to growth. A report to the Russian Association of the Club of Rome* (pp. 337–385). Springer. https://doi.org/10.1007/978-3-031-34999-7_17

Martin, C. (2015). *On the edge. A report to The Club of Rome*. Ingram.

Neuhoff, D., Tashi, S., Rahmann, G., & Denich, M. (2014). Organic agriculture in Bhutan: Potential and challenges. *Organic Agriculture, 4*, 209–221.

Orlov, D., Sadovnikova, L., & Lozanovskaya, I. (2002). *Ekologiya i okhrana biosfery pri khimicheskom zagryaznenii*. Vysshaya shkola.

Pan, Y., Birdsey, R. A., Fang, J., Houghton, R., Kauppi, P. E., et al. (2011). A large and persistent carbon sink in the world's forests. *Science, 333*, 988–993.

Pansak, W., Hilger, T. H., Dercon, G., Kongkaew, T., & Cadisch, G. (2008). Changes in the relationship between soil erosion and N loss pathways after establishing soil conservation systems in uplands of Northeast Thailand. *Agriculture, Ecosystems and Environment, 128*(3), 167–176.

Paustian, K., Lehmann, J., Ogle, S., Reay, D., Robertson, G. P., & Smith, P. (2016). Climate-smart soils. *Nature, 532*, 49–57.

Perissi, I., & Bardi, U. (2021). *The empty sea. The future of the blue economy. A report to the Club of Rome*. Springer. https://doi.org/10.1007/978-3-030-51898-1

Peter, L. (1998). *The limits to social cohesion. A report to The Club of Rome*. Routledge.

Poeplau, C., & Don, A. (2015). Carbon sequestration in agricultural soils via cultivation of cover crops – A meta-analysis. *Agriculture, Ecosystems and Environment, 200*, 33–41.

Randers, J. (2012). *2052: A global forecast for the next forty years*. Chelsea Green Publishing.

Rasmussen, K., D'haen, S., Fensholt, R., Fog, B., Horion, S., et al. (2016). Environmental change in the Sahel: Reconciling contrasting evidence and interpretations. *Regional Environmental Change, 16*(3), 673–680.

Reymers, N. (1994). Ekologiya (teorii, zakony, pravila printsipy i gipotezy). *Rossiya molodaya, 4*, 27–44.

Rozanov, B. (1984). *Osnovy ucheniya ob okruzhayushchey srede: ucheb. posobiye.* MGU.

Rozanov, A., & Rozanov, B. (1990). Ekologicheskiye posledstviya antropogennykh izmeneniy pochv. Itogi nauki i tekhniki. *Pochvovedeniye i agrokhimiya, 7*, 35–52.

Sloan, S., & Sayer, J. A. (2015). Forest resource assessment of 2015 shows positive trends but forest loss and degradation persist in poor tropical countries. *Forest Ecology and Management, 352*, 134–145.

State Forestry Administration. (2011). *A bulletin of status quo of desertification and sandification in China.* State Forestry Administration.

Stoorvogel, J. J., Bakkenes, M., Temme, A. J., Batjes, N. H., & Brink, B. J. E. (2017a). S World: A global soil map for environmental modelling. *Land Degradation and Development, 28*, 22–33.

Stoorvogel, J. J., Bakkenes, M., Brink, B. J. E., & Temme, A. J. (2017b). To what extent did we change our soils? A global comparison of natural and current conditions. *Land Degradation and Development, 28*, 1982–1991. https://doi.org/10.1002/ldr.2721

Sutanudjaja, E. H., van Beek, L. P., Wada, Y., Wisser, D., de Graaf, I. E., et al. (2014). Development and validation of PCR-GLOBWB 2.0: A 5 arc min resolution global hydrology and water resources model. *Geophysical Research Abstracts, 16*. https://ui.adsabs.harvard.edu/abs/2014EGUGA..16.9993S/abstract

Trumper, K., Bertzky, M., Dickson, B., van der Heijden, G., Jenkins, M., & Manning, P. (2009). *The natural fix? The role of ecosystems in climate mitigation. A UNEP rapid response assessment.* United Nations Environment Programme, UNEP-WCMC.

UN Convention to Combat Desertification. (2017). *The Global Land Outlook* (1st ed.). United Nations.

UN Population Division. (2018). *World Urbanization Prospects: The 2018 Revision.* United Nations.

UNCCD. (2016). The Great Green Wall: Hope for the Sahara and Sahel. .

Van Vuuren, D. P., Kriegler, E., O'Neill, B. C., Ebi, K. L., Riahi, K., et al. (2014). A new scenario framework for climate change research: Scenario matrix architecture. *Climatic Change, 122*(3), 373–386.

Vester, F. (2002). *The art of interconnected thinking. A report to The Club of Rome.* Mcb Verlag.

von Weizsäcker, E. U., & Wijkman, A. (2018). *Come on! Capitalism, short-termism, population and the destruction of the planet – A report to the Club of Rome.* Springer.

West, T. O., & Post, W. M. (2002). Soil organic carbon sequestration rates by tillage and crop rotation. *Soil Science Society of America Journal, 66*, 1930–1940.

Wijkman, A., & Skånberg, K. (2017). *The circular economy and benefits for society. Jobs and climate clear winners in an economy based on renewable energy and resource efficiency. A report to The Club of Rome.* Club of Rome.

Natalia Kovaleva heads the Laboratory of Ecological Soil Science of the Faculty of Soil Science of Lomonosov Moscow State University. She is also a visiting professor at Kazan Federal (Volga Region) University. Natalia Kovaleva has authored and co-authored over 700 scientific publications including such monographs as *Recreational territories and ecological well-being of the population* (1996), *Soil science: history, sociology, methodology* (2005), *Soil in the modern world* (2008), *Soils in the biosphere and human life (2012), Protected areas—the basis of environmental safety* (2018), *Soil biomarkers* (2020).

Demography: Toward Optimization of Demographic Processes

Andrey Korotayev, Sergey Malkov, and Jameelah Musieva

Abstract

The present chapter analyzes one of the most important factors in world development—the dynamics of global demographic growth, which demonstrates that the fears of uncontrolled population growth expressed in the framework of the report to the Club of Rome "Limits to Growth" were partially justified and were typical for the period before the 1980s. However, statistical evidence after more than half a century demonstrates that the situation has changed, and in the 1960s and early 1970s, there was a peak of global demographic growth, after which a slowdown began. According to the UN forecasts, by the end of this century, there will be a stabilization of the population of the Earth. The chapter provides an explanation for the change in the dynamics of the global demographic transition—most of economically high-income states and a significant part of developing countries have moved to the second phase of the demographic transition, in which the birth rate falls to a level corresponding to a simple

A. Korotayev (✉)
HSE University, Moscow, Russia

Russian Presidential Academy of National Economy and Public Administration (RANEPA), Moscow, Russia

Lomonosov Moscow State University, Institute for African Studies, Russian Academy of Sciences, Moscow, Russia

S. Malkov
Lomonosov Moscow State University, Moscow, Russia

HSE University, Moscow, Russia

Institute of Economics, Russian Academy of Sciences, Moscow, Russia

J. Musieva
HSE University, Moscow, Russia

Faculty of Global Studies, Lomonosov Moscow State University, Moscow, Russia

V. Sadovnichy et al. (eds.), *Reconsidering the Limits to Growth*, World-Systems Evolution and Global Futures, https://doi.org/10.1007/978-3-031-34999-7_6

replacement of generations or below that level. At the same time, a new problem has arisen associated with a decline in fertility to the "lowest-low" level—a tendency formed to the negative natural population change in many countries, which is sometimes compensated by migration processes. Along with that, the process of population aging is developing in many countries of the world, with an increase in life expectancy (LE) alongside low fertility. The indicated trends in the stabilization of the world population occur unevenly, with a fairly significant number of countries (mainly the countries of Tropical Africa) in which the second phase started not long ago, and fertility rates are still very high. At the same time, there is an acceleration of urbanization of the population in many developing countries. The chapter also notes the mutual influence of demographic processes and developments in various spheres of society and provides scenarios for their possible subsequent evolution, highlighting as a possible optimum scenario the one in which the stabilization of the Earth's population will reduce the degree of negative anthropogenic impact on the environment, but a significant global depopulation will also be avoided. Given the unevenness of demographic processes, different approaches to stabilization have been noted: stimulating birth rate in countries with the lowest-low fertility and accelerating fertility transition in the countries with very high birth rates.

1 Current Situation

Malthusian fears reflected in the first report to the Club of Rome "The Limits to Growth" (Meadows et al., 1972), at that time were fully justified, since they were based on the analysis of statistical data since 1900. To better understand the situation and solid reasons for such apprehensions, it appears appropriate to consider the dynamics of world population growth over a longer period, from the beginning of our era to 1970 (see Fig. 1):

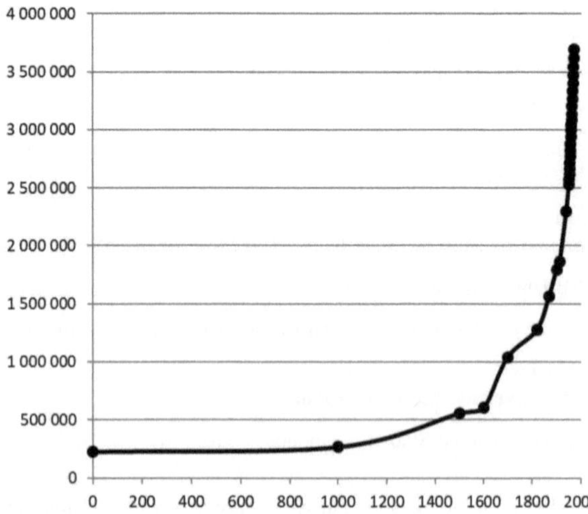

Fig. 1 Dynamics of the world population from the beginning of our era to the middle of the twentieth century (thousands). Data source: Maddison Project (2020)

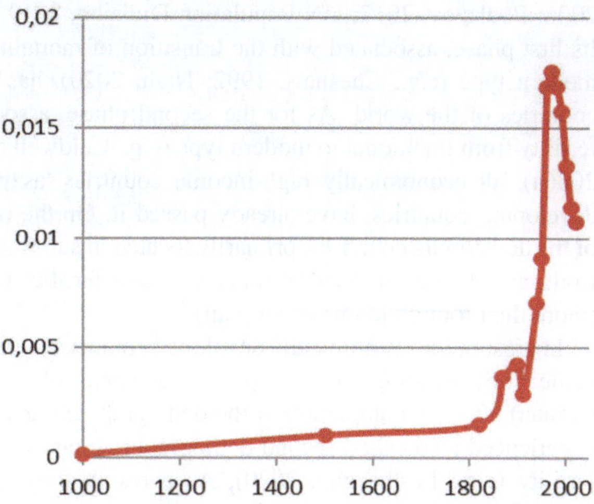

Fig. 2 Dynamics of averaged rates of the relative annual increase in the population of the Earth over the past 1000 years. Data source: Maddison Project (2020)

We can infer from Fig. 1 that the demographic dynamics in the period from the beginning of our era to the middle of the twentieth century are not exponential but hyperbolic. This was first noted by Heinz von Foerster and his colleagues in 1960 in their article "Doomsday: Friday, 13 November, AD 2026" (von Foerster et al., 1960), where the title of the article indicates the point of singularity obtained through a hyperbolic approximation of statistical data on the dynamics of the population of the Earth. Since the World3 model was based on the statistical data from the first half of the twentieth century, which reflected rapid and ever-accelerating demographic and economic growth, it is natural that the inertial forecast according to the model inevitably led to the depletion of resources at the beginning of the twenty-first century and to a demographic collapse. At the same time, the authors of the report believed that the development of technologies was unable to rectify the situation, and the most realistic way to prevent a collapse was a rapid and radical slowdown in demographic growth (decrease in the birth rate).

Sixty years have passed. What do the statistics show now? Figure 2 presents a graph describing the dynamics of the relative annual growth rate of the Earth's population:

We can see that the maximum rate of demographic growth was observed in the 1960s and the early 1970s (i.e., 50–60 years ago). Since then, a steady decline has begun. Instead of the expected singularity, UN forecasts project a rapid deceleration of population growth and stabilization of the Earth's population by the end of the twenty-first century, or even the start of global depopulation (UN Population Division, 2022a, 2022b; see also Alkema et al., 2011; Raftery et al., 2012; Grinin & Korotayev, 2015; Kaneda et al., 2016; Korotayev et al., 2015; Lutz et al., 2018; Randers et al., 2018; Wittgenstein Centre, 2022; see also Chapter "Modeling Social Self-Organization and Historical Dynamics. Modern Society and a Look into the Global Future: Cybernetic W-Society" (Akaev et al., 2023, this volume)).

What are the causes for sharp fluctuations in the rates of demographic dynamics observed over the past 200 years? The fact is that during this period, global demographic transition has been taking place and is now nearing its end (Kapitza,

2006; Podlazov, 2017; UN Population Division, 2019, 2022b; Korotayev, 2020b). Its first phase, associated with the transition of mortality from the traditional to the modern type (e.g., Chesnais, 1992; Nath, 2020), has already been passed by all countries of the world. As for the second phase, associated with the transition of fertility from traditional to modern type (e.g., Caldwell et al., 2006; Korotayev et al., 2006a), all economically high-income countries, as well as a significant part of developing countries, have already passed it. On the other hand, a significant part of the developing countries, primarily located in sub-Saharan Africa, are still very far from completing the fertility transition, and fertility rates there remain very high (more than four children per woman).[1]

In most of the economically developed countries, fertility rates did not settle at the value corresponding to a simple replacement of generations (2.1 children per woman) after the completion of the demographic transition. Rather, these countries experienced a so-called second demographic transition from low to the lowest-low fertility (e.g., Lesthaeghe, 2020). As a result, they show a steady trend toward depopulation, which in some first-world countries is still offset by the migration increase (e.g., Vollset et al., 2020).

Another important consequence of the second demographic transition is the acceleration of global aging processes against the background of decreasing birth rate combined with an increase in life expectancy (LE).[2] The process of global aging is currently more pronounced in most high-income countries, but it actually covers the entire world, which is evident from the increase in median age in all countries of the world.[3]

As a result of the processes described above, the overall growth rate of the world population is currently slowing down (e.g., Korotayev, 2020b). However, this process is very uneven. A large number of high-income and low-income countries (including Russia) are already experiencing absolute population declines (see Table 1, and, e.g., UN Population Division, 2019, 2022a, 2022b; Vollset et al., 2020). At the same time, exponential population growth continues in most countries of Tropical Africa because demographic inertia continues to compensate for some decline in the total fertility rate.[4]

---

[1] See chapter "Africa – the Continent of the Future. Challenges and Opportunities" (Grinin & Korotayev, 2023, this volume) and chapter "Modeling Social Self-Organization and Historical Dynamics. Africa's Futures" (Korotayev et al., 2023, this volume), as well as (Zinkina & Korotayev, 2014b; Korotayev & Zinkina, 2014, 2015; Nzimande & Mugwendere, 2018; Schoumaker, 2019; May & Rotenberg, 2020).

[2] See chapter "Global Ageing – an Integral Problem of the Future. How to Turn a Problem into a Development Driver?" (Grinin et al., 2023d, this volume).

[3] See chapter "Global Ageing – an Integral Problem of the Future. How to Turn a Problem into a Development Driver?" (Grinin et al., 2023d, this volume); see also (Goldstone, 2015; Goldstone et al. 2015; Zimmer, 2016; Bengtson, 2018; Fichtner, 2018; Mitchell & Walker, 2020).

[4] Zinkina and Korotayev (2014a), Korotayev et al. (2016), Nzimande and Mugwendere (2018), Schoumaker (2019), May and Rotenberg (2020); see also chapter "Africa – the Continent of the Future. Challenges and Opportunities" (Grinin & Korotayev, 2023, this volume) and chapter "Modeling Social Self-Organization and Historical Dynamics. Africa's Futures" (Korotayev

Table 1 Shares of population of the regions in the total population of the Earth

Region	North America (%)	Latin America (%)	Western Europe (%)	Eastern Europe (%)	Russia and CIS countries (%)	Japan and South Korea (%)	China (%)	India (%)	Southeast Asia (%)	Middle East and North Africa (%)	Sub-Saharan Africa (%)	Australia, New Zealand, Oceania (%)
2020	4.76	8.37	5.62	1.44	2.87	2.66	19.15	17.74	5.27	17.83	13.72	0.52
2100 upper scenario	2.49	6.48	2.26	0.50	1.15	0.75	6.76	14.01	4.22	22.55	38.43	0.43
2100 medium scenario	2.83	7.24	2.70	0.60	1.40	1.02	9.61	15.53	4.68	22.33	31.58	0.47
2100 lower scenario	3.29	7.92	3.24	0.73	1.59	1.27	11.33	16.34	4.97	22.09	26.75	0.49

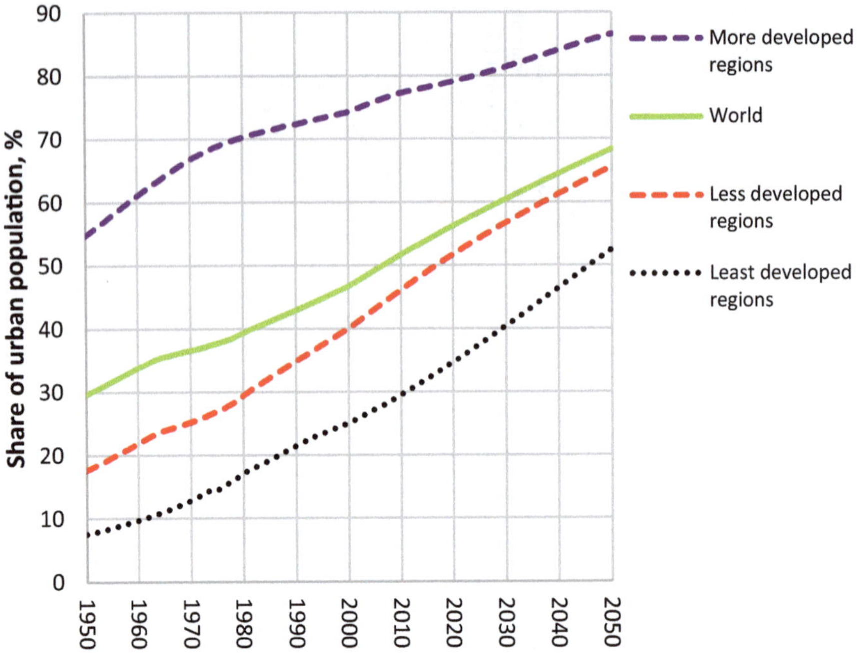

Fig. 3 Dynamics of the urban population share in the total population of the world and specific World System zones (empirical estimates for 1950–2020 with a forecast up to 2050). Data sources: UN Population Division (2018, 2022a)

The processes of urbanization of the World System are unfolding rapidly. While in 1950, noticeably less than one-third of the world population lived in cities, already in the mid-2000s, the share of city dwellers exceeded one-half of all the people on the Earth, and by the middle of this century, according to the forecast of the UN Population Division (2018), it will be noticeably more than two-thirds. At the same time, even in the countries of the World System core [according to the classification of Immanuel Wallerstein (1987)], only slightly more than one-half of the population lived in cities in 1950. By 2050, the proportion of city dwellers there will approach 90%. However, the growth rate of this share in the core countries will noticeably slow down as it approaches saturation, which apparently corresponds to this very level. In semi-peripheral countries, as early as 1950, only less than one-fifth of the population lived in cities, while by the middle of this century, this proportion will be about two-thirds. Finally, even in the countries of the World System periphery, more than one-half of the population will live in cities, while in the middle of the last century, this proportion was noticeably below 10% there (see Fig. 3):

et al., 2023, this volume) for a discussion of demographic problems in Tropical Africa. In addition, see chapter "High Income and Low Income Countries. Towards a Common Goal at Different Speeds" (Grinin et al., 2023a, this volume).

2 Demography as a Factor in World Development

Demography is among the most important factors in world development, as it exerts a decisive influence on the main global socio-natural processes. In essence, population size is an "order parameter" that determines the most important features of world dynamics at all phases of global human history. Let us note some important interrelations of the demographic factor with other factors of historical development.

2.1 The Impact of Demography on Climate

The impact of population growth on climate is due to the growing anthropogenic impact on the environment, primarily due to an increase in CO_2 emissions in the course of economic activity.[5] In this regard, the projected slowdown in the growth of the Earth's population should have a stabilizing effect on the climate, which should intensify with a gradual transition to low-carbon (carbon-free) energy.

Reverse effect:

- global warming can have a significant impact on demographic dynamics, rather indirectly, through the growing problems of providing food to the population of Tropical Africa and South Asia, since these regions have not yet fully escaped the Malthusian trap (e.g., Korotayev & Zinkina, 2015).

2.2 The Impact of Demography on the Environment

Population growth tends to increase CO_2 emissions and the amount of waste (recyclable, nonrecyclable, and toxic), impact natural biocenoses in a negative way, and cause additional deterioration in the quality of the environment (air, soil, and water bodies),[6] respectively, the projected slowdown in the growth of the Earth's population should reduce the severity of environmental problems. However, it must be borne in mind that in the scenarios implying the most rapid decline in fertility, the scale of depopulation is rampant (see Table 1 and Fig. 4), so it is necessary to look for a middle ground that would reduce the risks of environmental destabilization as much as possible, on the one hand, but, on the other hand, would prevent the risks of excessive depopulation.

---

[5] See chapter "Climate and Energy. Energy Transition Scenarios and Global Temperature Changes Based on Current Technologies and Trends" (Akaev & Davydova, 2023, this volume) and chapter "Ecology. Life in the "Unstable Biosphere"" (Kovaleva, 2023, this volume).

[6] See chapter "Climate and Energy. Energy Transition Scenarios and Global Temperature Changes Based on Current Technologies and Trends" (Akaev & Davydova, 2023, this volume) and chapter "Ecology. Life in the "Unstable Biosphere" (Kovaleva, 2023, this volume).

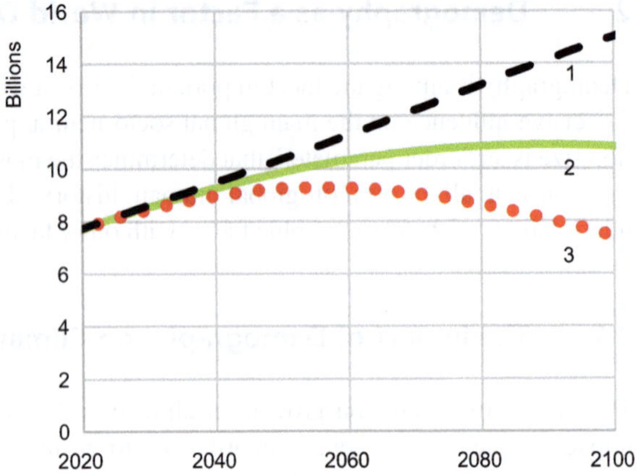

Fig. 4 Possible scenarios for world demographic dynamics up to 2100, billion people (1—upper scenario, 2—middle scenario, 3—lower scenario)

Reverse effect:

- environmental degradation can have an impact on demography through a decrease in the quality of life and an increase in food problems, which are most critical in the countries of Tropical Africa and South Asia, since these regions have not yet fully escaped the Malthusian trap (e.g., Korotayev & Zinkina, 2015).

2.3 The Impact of Demographics on Technology

For tens of thousands of years of human existence, demographic growth has had a powerful stimulating effect on the development of technology (Boserup, 1965; Taagepera, 1976, 1979; Kremer, 1993; Tsirel, 2004; Korotayev, 2005, 2007, 2020a, b; Korotayev et al., 2006a, b; Korotayev & Malkov, 2016; Podlazov, 2017; Grinin et al. 2020, 2022). In the long term, the global slowdown in population growth can be seen as one of the factors behind the slowdown in technological growth;

- there is a viewpoint suggesting that population aging can also contribute to a slowdown in the pace of technological growth, but this issue requires further study, including taking into account the dynamics of perspective ages; population aging can also lead to an acceleration in the development of medical technologies, as their costs will rise along with an increasing demand for such technologies[7];
- in case of depopulation, a decrease in the number of working-age population can stimulate the development of labor-saving technologies.

[7] See chapter "Technology. Limitless Possibilities, Effective Control" (Grinin & Grinin, 2023, this volume), chapter "Global Ageing – an Integral Problem of the Future. How to Turn a Problem into a Development Driver?" (Grinin et al., 2023d, this volume), and Grinin et al. (2017, 2020, 2022).

Reverse effect:

- impact on fertility: expansion of contraceptive use leads to a decrease in unwanted births; the development of artificial insemination technologies contributes to an increase in the birth rate among older women;
- impact on mortality: the development of medical technologies leads to a decrease in mortality (including infant mortality) in developing countries and to a general increase in life expectancy [see chapter "Global Ageing – an Integral Problem of the Future. How to Turn a Problem into a Development Driver?" (Grinin et al., 2023d, this volume)];
- impact on migration: on the one hand, the development of transport, information, communication, and financial technologies increases opportunities for migration; on the other hand, the development of remote work options contributes to the reduction of migration [see chapter "Future Political Change. Toward a More Efficient World Order" (Grinin et al., 2023b, this volume) and Grinin et al. (2022)].

2.4 The Impact of Demographics on Economy

Changes in the labor force size (as a result of changes in the population age structure and migration processes)—in most countries, it is expected to significantly decrease by the end of the century;

- changes in the population age structure cause changes in demand, as well as in the structure of budget expenditures [see chapter "Global Ageing – an Integral Problem of the Future. How to Turn a Problem into a Development Driver?" (Grinin et al., 2023d, this volume)];
- changes in the age structure affect economic growth in different ways at different stages of the demographic transition. For many countries of the world, the demographic bonus is relevant, as economic growth rates can increase when dependency ratios decline (Bloom & Williamson, 1998; Bloom & Canning, 2008; Bloom et al., 2007; Hawksworth & Cookson, 2008: 7–10; Lee & Mason, 2006, 2011; Barsukov, 2019; Groth et al., 2019; Kotschy et al., 2020; Korotayev et al., 2022). The most modernized countries, however, are facing a demographic onus due to population aging.[8]

[8] Ogawa et al. (2005), Komine and Kabe (2009), Park and Shin (2015), Goldstone (2015), Barsukov (2019), Hsu and Lo (2019), Warsito (2019); see also chapter "Economics. Optimizing Growth" (Grinin et al., 2023c, this volume); chapter "Global Ageing – an Integral Problem of the Future. How to Turn a Problem into a Development Driver?" (Grinin et al., 2023d, this volume) and chapter "Socio-Political Transformations. A Difficult Path to Cybernetic Society" (Grinin et al., 2023e, this volume).

Reverse effect:

- improvement in the material living conditions of the population (GDP per capita) due to economic growth affects the reduction in mortality and an increase in life expectancy, and vice versa. Conversely, an economic slowdown in countries that have completed the demographic transition may contribute to a further decline in fertility.

2.5 The Impact of Demography on the Social Sphere

Population density and urbanization that are still growing globally require transformations of social institutions;

- the growing proportion of older people affects the value system of society. "Global ageing" often provokes negative connotations. However, some studies show that with regard to value system, global aging can also lead to some positive effects due to increased support for prosocial values[9];
- Migration can lead to social disproportions and differentiation.

Reverse effect:

- an increase in the level of education in most cases leads to a decrease in fertility and is also correlated with a decrease in mortality (Singh & Casterline, 1985; Soares, 2005; Korotayev et al., 2006a; Kebede et al., 2019; Vollset et al., 2020);
- social stratification affects the differentiation of mortality rates;
- changes in values (e.g., the spread of the "childfree" ideology) affect birth rates.

2.6 The Impact of Demographics on Politics

Population aging can potentially reduce the intensity of violent destabilization processes due to disappearing youth bulges (e.g., Cincotta & Weber, 2021; Zinkina & Andreev, 2020);

- population aging can lead to increased conservative and right-wing orientations in society and provide greater support for conservative and right-wing political parties.[10]

[9] See, for example, chapter "Global Ageing – an Integral Problem of the Future. How to Turn a Problem into a Development Driver?" (Grinin et al., 2023d, this volume), and also Korotayev et al. (2018), Shulgin et al. (2019), Korotayev et al. (2019), Mayr and Freund (2020), Korotayev et al. (2021).

[10] See chapter "Global Ageing – an Integral Problem of the Future. How to Turn a Problem into a Development Driver?" (Grinin et al., 2023d, this volume); see also Van Hiel and Brebels (2011), Tilley and Evans (2014), Korotayev et al. (2018).

Reverse effect:

– the influence of politics on demographic processes is channeled primarily through changes in legislation and the implementation of state programs (targeted to support the birth rates, reduce mortality, regulate migration, etc.).[11]

3 Retrospective Analysis of Changes in the Demographic Sphere in the Historical Perspective (for the Last 8000 years)

World population growth accelerated significantly after the Neolithic revolution, which increased the anthropological carrying capacity of the Earth (e.g., Livi-Bacci, 2017; Korotayev, 2020a). Population growth occurred to a large extent due to increased birth rates and a smoothing out of the peaks of catastrophic mortality, while the overall mortality of early farmers was, as a rule, higher than that of hunter-gatherers. The general downward trend in life expectancy (LE) after the Neolithic revolution reached its lowest among intensive farmers (Cohen, 1989, 1998, 2009; Cohen & Armelagos, 1984; Storey, 1985; Cohen & Crane-Kramer, 2007; Ember et al., 2017; Algaze, 2018; Fagan & Durrani, 2018; Marklein et al., 2019). But the trend toward population growth took place due to a certain increase in fertility after the Neolithic (Agrarian) Revolution, the growth of the anthropological carrying capacity of the Earth, and the smoothing out of catastrophic fluctuations in mortality. As a result, a steady, systematic increase in life expectancy began to be observed only as part of the demographic transition that began in the most modernized countries in the early nineteenth century (Chesnais, 1992; Reher, 2011; Dyson, 2010; Livi-Bacci, 2017).

The demographic transition is a transition from the traditional type of reproduction, characterized by high mortality and high fertility, to its modern type, characterized by low mortality and low fertility. In the first phase of the transition, there is a drastic decrease in mortality due to a radical change in the structure of causes of death; thus, the transition from the traditional to the modern type of mortality proceeds. In modern societies, improved food security, development of water supply and sewerage systems, dramatic progress of health care technologies and expansion of access to them for the wide masses of the population, as well as the dissemination of modern medical knowledge achieved in the course of modernization, have made it possible to establish effective control over many types of mortality (Chesnais, 1992; Caldwell et al., 2006; Dyson, 2010). At the second stage of the demographic transition, changes affect fertility. In traditional societies, "the

[11]For more detail see chapter "Socio-Political Transformations. A Difficult Path to Cybernetic Society" (Grinin et al., 2023e, this volume), chapter "Developed and Developing Countries. Towards a Common Goal at Different Speeds" (Grinin et al., 2023a, this volume); and chapter "Africa – the Continent of the Future. Challenges and Opportunities" (Grinin & Korotayev, 2023, this volume).

objective goal [of demographic regulation], reflected in cultural norms … has always been a high birth rate" that was necessary in conditions of high mortality in order to guarantee that the population would not die out (Vishnevsky, 2005: 111). After a significant decrease in mortality, a decrease in birth rate becomes a necessary condition for maintaining demographic equilibrium. "Low fertility, … which, in combination with low mortality, reliably ensures the continuity of the process of renewal of generations, is now turning into a means of achieving a more general demographic goal (simple or slightly expanded reproduction, and thereby the goal of birth control)" (Vishnevsky, 2005: 114). The current situation is described in Sect. 1.

4 Demographic Development Scenarios

As shown above, the most important feature of modern demographic dynamics is that in all regions of the world, except for Tropical Africa, a second demographic transition has proceeded quite far (or has already been completed), which is accompanied by a decrease in birth rates and a corresponding slowdown in population growth, or even population decline. The only uncertain parameter is the speed at which birth rates will decrease worldwide (at the same time, the demographic situation will develop differently in different countries and regions). This uncertainty determines a set of probable demographic scenarios that were calculated based on mathematical models, taking into account current trends and hypotheses about further dynamics of demographic processes:

- the upper scenario implies continued growth of the world population in the twenty-first century with the stabilization of its numbers in the twenty-second century;
- the medium scenario implies stabilization of the world population by the end of the twenty-first century;
- the lower scenario implies stabilization of the world population with a subsequent decline in the second half of the twenty-first century.

The upper scenario is derived from the assumption of continued demographic divergence, implying that fertility in high- and middle-income countries remains at its current low levels or even declines further. Another assumption is that birth rates in Tropical African countries cease to decline at values greater than required for population replacement, approximately at 2.3 children per woman, while birth rates of the Islamic countries of North Africa and the Middle East stop declining at a level just above simple replacement of generations (about 2.1 children per woman).

The medium scenario is based on the assumption of demographic convergence, which implies, on the one hand, that the countries that have not yet completed the demographic transition will see their fertility rates decline at the fastest possible speed to the level of simple replacement of generations. On the other hand, in countries with ultra-low fertility, rates will increase and return to the level of simple replacement of generations. This scenario seems to be more favorable in comparison

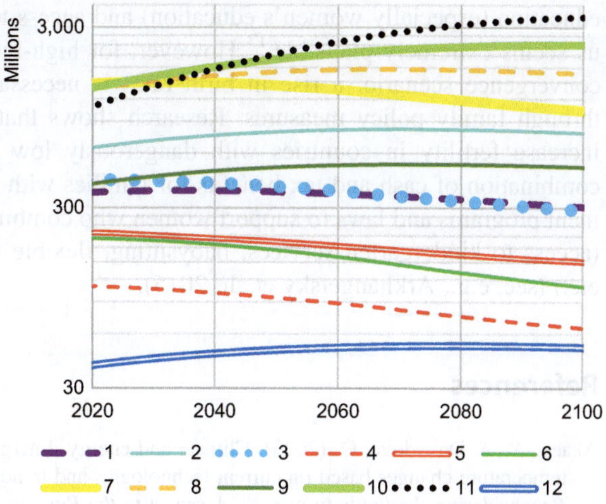

Fig. 5 Population dynamics in various regions of the world according to the medium scenario, millions, logarithmic scale (1—North America, 2—Latin America, 3—Western Europe, 4—Eastern Europe, 5—Russia and CIS countries, 6—Japan and Korea, 7—China, 8—India, 9—Southeast Asia, 10—Middle East and North Africa, 11—Sub-Saharan Africa, 12—Australia, New Zealand, Oceania)

with the upper scenario, as it will contribute to smoothing the tension between the center and the periphery of the World System, optimizing migration flows, preventing sociodemographic collapses in the least developed countries,[12] etc.

The lower scenario assumes a decrease in birth rates below the replacement level in all countries of the world without exception, which will cause global depopulation.

The calculations for these scenarios are shown in Fig. 4. These calculations correlated quite well with the UN projections presented in the recent issues of the *World Population Prospects* (UN Population Division, 2019, 2022b).

Calculations demonstrate the inevitability of a significant change in population numbers in different regions and civilizations, which will significantly affect world processes in the course of the twenty-first century (see Fig. 5).

We can see that, by the end of the twenty-first century, Tropical Africa will definitely come out on top in population, overtaking China and India. Moreover, in 2100, every second inhabitant of the Earth will be from Africa or the Middle East. These changes are typical for all scenarios and get more pronounced from lower to upper scenarios (see Table 1).

We believe that the most favorable scenario for solving existing global problems is the medium demographic development scenario (the convergence scenario). Stabilization of the Earth's population will slow down the negative anthropogenic impact on the natural environment and ensure a systematic reduction of this pressure.

According to IHME calculations (Vollset et al., 2020), in order to achieve a favorable demographic development scenario for the most demographically lagging countries (that is, to complete the demographic transition as quickly as possible) it would be sufficient to achieve sustainable development goals in the field of

[12]See chapter "Modeling Social Self-Organization and Historical Dynamics. Africa's Futures" (Korotayev et al., 2023, this volume) for detail.

education (especially women's education) and access to family planning, which to us seems extremely plausible.[13] However, for high-income countries to enter the convergence scenario, a rise in birth rates is necessary, which can be achieved through family policy measures. Research shows that the most effective way to increase fertility in countries with dangerously low birth rates is to provide a combination of cash and tax benefits for families with children, as well as government programs and laws to support women who combine work and raising children (access to kindergarten services, babysitting, flexible working hours for mothers, etc.) (see, e.g., Arkhangelsky et al., 2015).

References

Akaev, A., & Davydova, O. (2023). Climate and energy. Energy transition scenarios and global temperature changes based on current technologies and trends. In V. Sadovnichy et al. (Eds.), *Reconsidering the limits to growth. A report to the Russian Association of the Club of Rome* (pp. 53–70). Springer. https://doi.org/10.1007/978-3-031-34999-7_4

Akaev, A., Malkov, S., Grinin, L., Bilyuga, S., Davydova, O., Grinin, A., Kovaleva, N., Malkov, A., Musieva, J., & Korotayev, A. (2023). Modeling social self-organization and historical dynamics. Modern society and a look into the global futures: Cybernetic W-society. In V. Sadovnichy et al. (Eds.), *Reconsidering the limits to growth. A report to the Russian Association of the Club of Rome* (pp. 419–460). Springer. https://doi.org/10.1007/978-3-031-34999-7_19

Algaze, G. (2018). Entropic cities: The paradox of urbanism in ancient Mesopotamia. *Current Anthropology, 59*(1), 23–54. https://doi.org/10.1086/695983

Alkema, L., Raftery, A. E., Gerland, P., Clark, S. J., Pelletier, F., Buettner, T., & Heilig, G. K. (2011). Probabilistic projections of the total fertility rate for all countries. *Demography, 48*(3), 815–839. https://doi.org/10.1007/s13524-011-0040-5

Arkhangelsky, V., Bogevolnov, J., Goldstone, J., Khaltourina, D., Korotayev, A., Malkov, A., Novikov, K., Ryazantsev, S., Rybalchenko, S., Shulgin, S., Steck, P., Yuriev, Y., Zinkina, J., & Zvereva, N. (2015). *Critical 10 years. Demographic policies of The Russian Federation: Successes and challenges.* Russian Presidential Academy of National Economy and Public Administration (RANEPA).

Barsukov, V. N. (2019). From the demographic dividend to population ageing: World trends in the system-wide transition. *Economic and Social Changes: Facts, Trends, Forecast, 12*(4), 167–182. https://doi.org/10.15838/esc.2019.4.64.11

Bengtson, V. (Ed.). (2018). *Global ageing and challenges to families.* Routledge.

Bloom, D. E., & Canning, D. (2008). Global demographic change: Dimensions and economic significance. *Population and Development Review, 34*, 17–51.

Bloom, D. E., & Williamson, J. G. (1998). Demographic transitions and economic miracles in emerging Asia. *World Bank Economic Review, 12*(3), 419–455.

Bloom, D. E., Canning, D., Fink, G., & Finlay, J. (2007). Does age structure forecast economic growth? *International Journal of Forecasting, 23*(4), 569–585. https://doi.org/10.1016/j.ijforecast.2007.07.001

Boserup, E. (1965). The conditions of agricultural growth: The economics of agrarian change under population pressure. .

[13]Cf., e.g., Soares (2005), Zinkina and Korotayev (2014a), Kebede et al. (2019); see also chapter "Modeling Social Self-Organization and Historical Dynamics. Africa's Futures" (Korotayev et al., 2023, this volume).

Caldwell, J. C., Caldwell, B. K., Caldwell, P., McDonald, P. F., & Schindlmayr, T. (2006). *Demographic transition theory*. Springer. https://doi.org/10.1007/978-1-4020-4498-4
Chesnais, J. C. (1992). *The demographic transition: Stages, patterns, and economic implications*. Clarendon Press.
Cincotta, R., & Weber, H. (2021). Youthful age structures and the risks of revolutionary and separatist conflicts. In A. Goerres & P. Vanhuysse (Eds.), *Global political demography: Comparative analyzes of the politics of population change in all world regions* (pp. 57–92). Palgrave. https://doi.org/10.1007/978-3-030-73065-9_3
Cohen, M. N. (1989). *Health and the rise of civilization*. Yale University Press.
Cohen, M. N. (1998). Were early agriculturalists less healthy than food collectors? In C. R. Ember, M. Ember, & P. N. Peregrine (Eds.), *Research frontiers in anthropology* (pp. 61–83). Prentice Hall.
Cohen, M. N. (2009). Introduction: Rethinking the origins of agriculture. *Current Anthropology, 50*(5), 591–595. https://doi.org/10.1086/603548
Cohen, M. N., & Armelagos, G. J. (Eds.). (1984). *Paleopathology at the origins of agriculture*. Academic Press.
Cohen, M. N., & Crane-Kramer, G. (Eds.). (2007). *Bioarchaeological interpretations of the human past: Local, regional, and global perspectives series*. University Press of Florida.
Dyson, T. (2010). *Population and development. The demographic transition*. Zed Books.
Ember, C. R., Ember, M., Peregrine, P. N., Hoppa, R. D., & Fowler, K. (2017). *Physical anthropology and archaeology*. Pearson.
Fagan, B. M., & Durrani, N. (2018). *People of the earth: An introduction to world prehistory*. Routledge.
Fichtner, J. J. (2018). Global ageing and public finance. *Business Economics, 53*(2), 72–78.
Goldstone, J. A. (2015). Population ageing and global economic growth. In J. A. Goldstone, L. Grinin, & A. Korotayev (Eds.), *History & mathematics: Demography & ageing* (pp. 147–155). Uchitel.
Goldstone, J. A., Grinin, L., & Korotayev, A. (2015). Research into global ageing and its consequences. In J. A. Goldstone, L. Grinin, & A. Korotayev (Eds.), *History & mathematics: Political demography & global ageing* (pp. 5–9). Uchitel.
Grinin, L., & Grinin, A. (2023) Technology. Limitless possibilities, effective control. In V. Sadovnichy et al. (Eds.), *Reconsidering the limits to growth. A report to the Russian Association of the Club of Rome* (pp. 139–154). Springer. https://doi.org/10.1007/978-3-031-34999-7_8
Grinin, L., & Korotayev, A. (2015). *Great divergence and great convergence. A global perspective*. Springer. https://doi.org/10.1007/978-3-319-17780-9
Grinin, L., & Korotayev, A. (2023). Africa – The continent of the future. Challenges and opportunities. In V. Sadovnichy et al. (Eds.), *Reconsidering the limits to growth. A report to the Russian Association of the Club of Rome* (pp. 225–238). Springer. https://doi.org/10.1007/978-3-031-34999-7_13
Grinin, L. E., Grinin, A. L., & Korotayev, A. (2017). Forthcoming Kondratieff wave, cybernetic revolution, and global ageing. *Technological Forecasting and Social Change, 115*, 52–68. https://doi.org/10.1016/j.techfore.2016.09.017
Grinin, L., Grinin, A., & Korotayev, A. (2020). Dynamics of technological growth rate and the forthcoming singularity. In A. Korotayev & D. LePoire (Eds.), *The 21st century singularity and global futures. A big history perspective* (pp. 287–344). Springer. https://doi.org/10.1007/978-3-030-33730-8_14
Grinin, L., Grinin, A., & Korotayev, A. (2022). COVID-19 pandemic as a trigger for the acceleration of the cybernetic revolution, transition from e-government to e-state, and change in social relations. *Technological Forecasting and Social Change, 175*, 1–16. https://doi.org/10.1016/j.techfore.2021.121348
Grinin, L., Malkov, S., & Korotayev, A. (2023a). High income and low income countries. Towards a common goal at different speeds. In V. Sadovnichy et al. (Eds.), *Reconsidering the limits to*

growth. A report to the Russian Association of the Club of Rome (pp. 207–224). Springer. https://doi.org/10.1007/978-3-031-34999-7_12

Grinin, L., Grinin, A., & Korotayev, A. (2023b). Future political change. Toward a more efficient world order. In V. Sadovnichy et al. (Eds.), *Reconsidering the limits to growth. A report to the Russian Association of the Club of Rome* (pp. 191–206). Springer. https://doi.org/10.1007/978-3-031-34999-7_11

Grinin, L., Grinin A., & Malkov, S. (2023c). Economics. Optimizing growth. In V. Sadovnichy et al. (Eds.), *Reconsidering the limits to growth. A report to the Russian Association of the Club of Rome* (pp. 155–168). Springer. https://doi.org/10.1007/978-3-031-34999-7_9

Grinin, L., Grinin, A., & Korotayev, A. (2023d). Global ageing – An integral problem of the future. How to turn a problem into a development driver? In V. Sadovnichy et al. (Eds.), *Reconsidering the limits to growth. A report to the Russian Association of the Club of Rome* (pp. 117–135). Springer. https://doi.org/10.1007/978-3-031-34999-7_7

Grinin, L., Grinin, A., & Malkov, S. (2023e). Socio-political transformations. A difficult path to cybernetic society. In V. Sadovnichy et al. (Eds.), *Reconsidering the limits to growth. A report to the Russian Association of the Club of Rome* (pp. 169–189). Springer. https://doi.org/10.1007/978-3-031-34999-7_10

Groth, H., May, J. F., & Turbat, V. (2019). Policies needed to capture a demographic dividend in sub-Saharan Africa. *Canadian Studies in Population, 46*(1), 61–72. https://doi.org/10.1007/s42650-019-00005-8

Hawksworth, J., & Cookson, G. (2008). *The world in 2050. Beyond the BRICs: A Broader Look at Emerging Market Growth Prospects*. PricewaterhouseCoopers.

Hsu, Y. H., & Lo, H. C. (2019). The impacts of population ageing on saving, capital formation, and economic growth. *American Journal of Industrial and Business Management, 9*(12), 2231–2246. https://doi.org/10.4236/ajibm.2019.912148

Kaneda, T., Greenbaum, C., & Patierno, K. (2016). *World Population data sheet*. Population Reference Bureau. https://www.prb.org/2016-world-population-data-sheet/

Kapitza, S. (2006). *Global population blow-up and after. Report to the Club of Rome*. Global Marshall Plan Initiative.

Kebede, E., Goujon, A., & Lutz, W. (2019). Stalls in Africa's decline partly result from disruptions in female education. *Proceedings of the National Academy of Sciences, 116*(8), 2891–2896. https://doi.org/10.1073/pnas.1717288116

Komine, T., & Kabe, S. (2009). Long-term forecast of the demographic transition in Japan and Asia. *Asian Economic Policy Review, 4*(1), 19–38. https://doi.org/10.1111/j.1748-3131.2009.01103.x

Korotayev, A. (2005). A compact macromodel of world system evolution. *Journal of World-Systems Research, 11*(1), 79–93. https://doi.org/10.5195/jwsr.2005.401

Korotayev, A. (2007). Compact mathematical models of world system development, and how they can help us to clarify our understanding of globalization processes. In G. Modelski, T. Devezas, & W. Thompson (Eds.), *Globalization as evolutionary process: Modeling global change* (pp. 133–160). Routledge.

Korotayev, A. (2020a). How singular is the 21st century singularity? In A. Korotayev & D. LePoire (Eds.), *The 21st century singularity and global futures. A big history perspective* (pp. 571–595). Springer. https://doi.org/10.1007/978-3-030-33730-8_26

Korotayev, A. (2020b). The 21st century singularity in the big history perspective. A re-analysis. In A. Korotayev & D. LePoire (Eds.), *The 21st century singularity and global futures. A big history perspective* (pp. 19–75). Springer. https://doi.org/10.1007/978-3-030-33730-8_2

Korotayev, A., & Malkov, A. (2016). A compact mathematical model of the world system economic and demographic growth, 1 CE – 1973 CE. *International Journal of Mathematical Models and Methods in Applied Sciences, 10*, 200–209.

Korotayev, A., & Zinkina, J. (2014). How to optimize fertility and prevent humanitarian catastrophes in tropical Africa. *African Studies in Russia, 6*, 94–107. https://elibrary.ru/item.asp?id=22501708

Korotayev, A., & Zinkina, J. (2015). East Africa in the Malthusian trap? *Journal of Developing Societies, 31*(3), 1–36. https://doi.org/10.1177/0169796X15590322

Korotayev, A., Malkov, A., & Khaltourina, D. (2006a). *Introduction to social macrodynamics: Compact macromodels of the world system growth*. KomKniga/URSS.

Korotayev, A., Malkov, A., & Khaltourina, D. (2006b). *Introduction to social macrodynamics: Secular cycles and millennial trends*. KomKniga/URSS.

Korotayev, A., Goldstone, J., & Zinkina, J. (2015). Phases of global demographic transition correlate with phases of the great divergence and great convergence. *Technological Forecasting and Social Change, 95*, 163–169. https://doi.org/10.1016/j.techfore.2015.01.017

Korotayev, A., Zinkina, J., Goldstone, J., & Shulgin, S. (2016). Explaining current fertility dynamics in tropical Africa from an anthropological perspective: A cross-cultural investigation. *Cross-Cultural Research, 50*(3), 251–280. https://doi.org/10.1177/1069397116644158

Korotayev, A., Shulgin, S., Zinkina, J., & Novikov, K. (2018). *The impact of population ageing on the global value system and political dynamics*. Russian Presidential Academy of National Economy and Public Administration.

Korotayev, A., Novikov, K., & Shulgin, S. (2019). Values of the elderly people in an ageing world. *Sociology of Power, 31*(1), 114–142. https://doi.org/10.22394/2074-0492-2019-1-114-142

Korotayev, A., Butovskaya, M., Shulgin, S., & Zinkina, J. (2021). Impact of population ageing on the global value system. *Vek Globalizatsii, 4*, 69–80. https://doi.org/10.30884/vglob/2021.04.05

Korotayev, A., Shulgin, S., Zinkina, J., & Slav, M. (2022). Estimates of the possible economic effect of the demographic dividend for sub-Saharan Africa for the period up to 2036. *Vostok (Oriens), 2*, 108–123. https://doi.org/10.31857/S086919080019128-7

Korotayev, A., Shulgin, S., Ustyuzhanin, V., Zinkina, J., & Grinin, L. (2023). Modeling social self-organization and historical dynamics: Africa's futures. In V. Sadovnichy et al. (Eds.), *Reconsidering the limits to growth. A report to the Russian Association of the Club of Rome* (pp. 461–490). Springer. https://doi.org/10.1007/978-3-031-34999-7_20

Kotschy, R., Urtaza, P. S., & Sunde, U. (2020). The demographic dividend is more than an education dividend. *Proceedings of the National Academy of Sciences, 117*(42), 25982–25984. https://doi.org/10.1073/pnas.201228611

Kovaleva, N. (2023). Ecology. Life in the "unstable biosphere". In V. Sadovnichy et al. (Eds.), *Reconsidering the limits to growth. A report to the Russian Association of the Club of Rome* (pp. 71–95). Springer. https://doi.org/10.1007/978-3-031-34999-7_5

Kremer, M. (1993). Population growth and technological change: One million B.C. to 1990. *The Quarterly Journal of Economics, 108*, 681–716. https://doi.org/10.2307/2118405

Lee, R., & Mason, A. (2006). What is the demographic dividend? *Finance and Development, 43*(3), 16–17.

Lee, R., & Mason, A. (2011). *Population ageing and the generational economy: A global perspective*. Edward Elgar.

Lesthaeghe, R. (2020). The second demographic transition, 1986–2020: Sub-replacement fertility and rising cohabitation—A global update. *Genus, 76*(1), 1–38. https://doi.org/10.1186/s41118-020-00077-4

Livi-Bacci, M. (2017). *A concise history of world population*. Wiley-Blackwell.

Lutz, W., Goujon, A., Kc, S., Stonawski, M., & Stilianakis, N. (2018). *Demographic and human capital scenarios for the 21st century: 2018 assessment for 201 countries*. Publications Office of the European Union.

Maddison Project (2020). Maddison Project database 2020. Groningen Growth and Development Centre. https://www.rug.nl/ggdc/historicaldevelopment/maddison/releases/maddison-project-database-2020

Marklein, K. E., Torres-Rouff, C., King, L. M., & Hubbe, M. (2019). The precarious state of subsistence: Reevaluating dental pathological lesions associated with agricultural and hunter-gatherer lifeways. *Current Anthropology, 60*(3), 341–368. https://doi.org/10.1086/703376

May, J. F., & Rotenberg, S. (2020). A call for better integrated policies to accelerate the fertility decline in sub-Saharan Africa. *Studies in Family Planning, 51*(2), 193–204. https://doi.org/10.1111/sifp.12118

Mayr, U., & Freund, A. M. (2020). Do we become more prosocial as we age, and if so, why? *Current Directions in Psychological Science, 29*(3), 248–254. https://doi.org/10.1177/0963721420910811

Meadows, D. H., Meadows, D. L., Randers, J., & Behrens, W. W. (1972). *The limits to growth: A report for the Club of Rome's project on the predicament of mankind*. Universe Books.

Mitchell, E., & Walker, R. (2020). Global ageing: Successes, challenges and opportunities. *British Journal of Hospital Medicine, 81*(2), 1–9. https://doi.org/10.12968/hmed.2019.0377

Nath, S. K. (2020). Demographic transition and economic growth. *Solid State Technology, 63*(5), 3142–3148.

Nzimande, N., & Mugwendere, T. (2018). Stalls in Zimbabwe fertility: Exploring determinants of recent fertility transition. *Southern African Journal of Demography, 18*(1), 59–110.

Ogawa, N., Kondo, M., & Matsukura, R. (2005). Japan's transition from the demo-graphic bonus to the demographic onus. *Asian Population Studies, 1*(2), 207–226.

Park, D., & Shin, K. (2015). Impact of population ageing on Asia's future growth. In J. A. Goldstone, L. E. Grinin, & A. V. Korotayev (Eds.), *History & mathematics: Political demography & global ageing* (pp. 107–132). Uchitel.

Podlazov, A. (2017). A theory of the global demographic process. *Herald of the Russian Academy of Sciences, 87*(3), 256–266. https://doi.org/10.1134/S1019331617030054

Raftery, A. E., Li, N., Ševčíková, H., Gerland, P., & Heilig, G. K. (2012). Bayesian probabilistic population projections for all countries. *Proceedings of the National Academy of Sciences, 109*(35), 13915–13921. https://doi.org/10.1073/pnas.1211452109

Randers J., Rockström J., Per Espen Stoknes P. E., Golüke U., Collste D., & Cornell S. (2018). *How to achieve the sustainable development goals within planetary boundaries. A report to the Club of Rome from Stockholm Resilience Centre and BI Norwegian Business School.* Stockholm Resilience Centre.

Reher, D. (2011). Economic and social implications of the demographic transition. *Population and Development Review, 37*, 11–33. https://doi.org/10.1111/j.1728-4457.2011.00376.x

Schoumaker, B. (2019). Stalls in fertility transitions in sub-Saharan Africa: Revisiting the evidence. *Studies in Family Planning, 50*(3), 257–278. https://doi.org/10.1111/sifp.12098

Shulgin, S., Zinkina, J., & Korotayev, A. (2019). Religiosity and ageing: Age and cohort effects and their implications for the future of religious values in high-income OECD countries. *Journal for the Scientific Study of Religion, 58*(3), 591–603. https://doi.org/10.1111/jssr.12613

Singh, S., & Casterline, J. (1985). The socio-economic determinants of fertility. In J. Cleland & J. Hobcraft (Eds.), *Reproductive change in developing countries. Insights from world fertility survey* (pp. 199–222). Oxford University Press.

Soares, R. (2005). Mortality reductions, educational attainment, and fertility choice. *American Economic Review, 95*(3), 580–601. https://doi.org/10.1257/0002828054201486

Storey, R. (1985). An estimate of mortality in a pre-Columbian urban Population. *American Anthropologist, 87*, 515–535.

Taagepera, R. (1976). Crisis around 2005 AD? A technology-population interaction model. *General Systems, 21*, 137–138.

Taagepera, R. (1979). People, skills, and resources: An interaction model for world Population growth. *Technological Forecasting and Social Change, 13*, 13–30. https://doi.org/10.1016/0040-1625(79)90003-9

Tilley, J., & Evans, G. (2014). Ageing and generational effects on vote choice: Combining cross-sectional and panel data to estimate APC effects. *Electoral Studies, 33*, 19–27. https://doi.org/10.1016/j.electstud.2013.06.007

Tsirel, S. (2004). On the possible reasons for the hyperexponential growth of the earth Population. In M. G. Dmitriev & A. P. Petrov (Eds.), *Mathematical modeling of social and economic dynamics* (pp. 367–369). Russian State Social University.

UN Population Division. (2018). *World urbanization prospects 2018*. United Nations.
UN Population Division. (2019). *World population prospects 2019*. United Nations.
UN Population Division. (2022a). *United Nations population division database*. United Nations. http://www.un.org/esa/population
UN Population Division. (2022b). *World population prospects 2022*. United Nations.
Van Hiel, A., & Brebels, L. (2011). Conservatism is good for you: Cultural conservatism protects self-esteem in older adults. *Personality and Individual Differences, 50*(1), 120–123. https://doi.org/10.1016/j.paid.2010.09.002
Vishnevsky, A. (2005). *Izbrannyye demograficheskiye trudy. Vol. 1. Demograficheskaya teoriya i demograficheskaya istoriya*. Nauka.
Vollset, S. E., Goren, E., Yuan, C.-W., Cao, J., Smith, A. E., Hsiao, T., Murray, C. J. L., et al. (2020). Fertility, mortality, migration, and population scenarios for 195 countries and territories from 2017 to 2100: A forecasting analysis for the global burden of disease study. *The Lancet, 396*(10258), 17–23. https://doi.org/10.1016/S0140-6736(20)30677-2
von Foerster, H., Mora, P. M., & Amiot, L. W. (1960). Doomsday: Friday, 13 November, A.D. 2026: At this date human population will approach infinity if it grows as it has grown in the last two millennia. *Science, 132*(3436), 1291–1295. https://doi.org/10.1126/science.132.3436.1291
Wallerstein, I. (1987). World-systems analysis. In A. Giddens & J. H. Turner (Eds.), *Social theory today* (pp. 309–324). Cambridge University Press.
Warsito, T. (2019). Attaining the demographic bonus in Indonesia. *Jurnal Pajak Dan Keuangan Negara (PKN), 1*(1), 6–16.
Wittgenstein Centre. (2022). *Wittgenstein Centre Human Capital Data Explorer*. http://dataexplorer.wittgensteincentre.org/wcde-v2/
Zimmer, Z. (2016). *Global ageing in the twenty-first century: Challenges, opportunities and implications*. Routledge.
Zinkina, J. V., & Andreev, A. I. (2020). The demographic history and current age structure in Latin America: The youth bulge and implications for sociopolitical stability. *Journal of Globalization Studies, 11*(2), 23–34. https://doi.org/10.30884/jogs/2020.02.02
Zinkina, J., & Korotayev, A. (2014a). Explosive population growth in tropical Africa: Crucial omission in development forecasts (emerging risks and way out). *World Futures, 70*(4), 271–305. https://doi.org/10.1080/02604027.2014.894868
Zinkina, J., & Korotayev, A. (2014b). Projecting Mozambique's demographic futures. *Journal of Futures Studies, 19*(2), 21–40.

Andrey Korotayev heads the Centre for Stability and Risk Analysis at HSE University in Moscow, Russia. He is also a Senior Research Professor at the Institute of Oriental Studies and Institute for African Studies, Russian Academy of Sciences, as well as at the Faculty of Global Studies of Lomonosov Moscow State University and the Russian Presidential Academy of National Economy and Public Administration. He has authored or co-authored over 650 scholarly publications, including such monographs as *Ancient Yemen* (Oxford University Press, 1995), *World Religions and Social Evolution of the Old World Oikumene Civilizations: A Cross-Cultural Perspective* (The Edwin Mellen Press, 2004), *Introduction to Social Macrodynamics: Compact Macromodels of the World System Growth* (URSS, 2006; with Daria Khaltorina and Artemy Malkov), *Introduction to Social Macrodynamics: Secular Cycles and Millennial Trends* (URSS, 2006; with Daria Khaltorina and Artemy Malkov), *Great Divergence and Great Convergence. A Global Perspective* (Springer, 2015; with Leonid Grinin), *Economic Cycles, Crises, and the Global Periphery* (Springer, 2016; with Leonid Grinin), *Islamism, Arab Spring, and the Future of Democracy. World System and World Values Perspectives* (Springer, 2019; with Leonid Grinin and Arno Tausch). He is a laureate of a Russian Science Support Foundation in "The Best Economists of the Russian Academy of Sciences" Nomination (2006); in 2012 he was awarded the Gold Kondratieff Medal by the International N. D. Kondratieff Foundation.

Sergey Malkov heads the Center for Long-Term Forecasting and Strategic Planning at Lomonosov Moscow State University; he also works as a Senior Research Professor at the Institute of Economics of the Russian Academy of Sciences as well as at the Centre for Stability and Risk Analysis, HSE University, Moscow. He is the author of over 300 scholarly publications, including such monographs as *Social self-organization and historical process: possibilities of mathematical modeling* (URSS, 2009), *Modeling and forecasting of world dynamics* (ISPI RAS, 2012, together with Victor Sadovnichy and others), *A new theory of the reproduction of capital: development and practical application* (Nestor-Istoriya, 2016, together with Vladimir Mayevsky and others), *The twenty-first Century Singularity and Global Futures. A Big History Perspective* (Springer, 2020, together with David J. LePoire and others). In 2022, he was awarded the Kondratieff Award by the Russian Academy of Sciences.

Jameelah Musieva is a research assistant at Lomonosov Moscow State University, Faculty of Global Studies as well as the Centre for Stability and Risk Analysis at HSE University, Moscow. She is the author and co-author of more than 10 scholarly publications, including such articles as "The Concept of "Quality of Life" (2019), "Methodology for Cross-Country Assessment of the Quality of Life Based on the LQI Index" (2019), "A Cross-Country Analysis of the Quality of Life" (2020), and "Quality of Life in the BRICS Countries Based on the LQI Index" (2021).

Global Aging: An Integral Problem of the Future. How to Turn a Problem into a Development Driver?

Leonid Grinin, Anton Grinin, and Andrey Korotayev

Abstract

This chapter analyzes the process of global population aging, which has a significant impact on all areas of public life in the twenty-first century. Already today, in almost two dozen countries of the world, 20% of the population is over 65 years old. According to the WHO, the 60+ age group in the world already exceeds the number of children under 5 years of age. At the same time, by 2050 the number of 65+ will exceed the number of adolescents and young people aged 15–24. The influence of the population aging process on various areas of social activity will be determined by its scale and depth, i.e. by the possible demographic structure of the future society. At the same time, the ability of modern society to cope with increasing risks associated with this process is problematic. Among the risks and problems associated with population aging, one finds economic slowdowns, pension crisis, problems of care for the growing numbers of disabled and frail people, maintaining the financial savings of the elderly, the issue of ageism, and the vulnerability of the elderly during periods of crisis and pandemics and consequently maintaining life expectancy as well as a stable democracy and even world order. These risks and problems are of particular

L. Grinin
HSE University, Moscow, Russia

Institute of Oriental Studies, Russian Academy of Sciences, Moscow, Russia

A. Grinin
HSE University, Moscow, Russia

Faculty of Global Studies, Lomonosov Moscow State University, Moscow, Russia

A. Korotayev (✉)
HSE University, Moscow, Russia

Institute of Oriental Studies, Russian Academy of Sciences, Moscow, Russia

Faculty of Global Studies, Lomonosov Moscow State University, Moscow, Russia

V. Sadovnichy et al. (eds.), *Reconsidering the Limits to Growth*, World-Systems Evolution and Global Futures, https://doi.org/10.1007/978-3-031-34999-7_7

concern given that by the mid-twenty-first century, two-thirds of people over 65 will live in middle and low-income countries. Moreover, aging will affect even the currently poorest countries in sub-Saharan Africa with young populations. In view of this, the problem of social security and health care for the elderly is becoming increasingly acute and requires the adoption of proactive measures. The problem of population aging is given insufficient attention by such international organizations as the UN and WHO; in general, studies of this issue are limited in the public sphere as well as in scientific discourse. However, even now the, global aging is an extremely important problem and, we suppose, it will become the most crucial problem in the future. So, insufficient attention to it is threatening societal tensions, crisis, and intergenerational conflicts as well as political and social instability. In the present chapter, we approach this process from the point of view of the transition from perceiving it as a problem to considering it as providing new opportunities associated with the presence of worldly wisdom and experience in a more mature and older generation. Today, concrete measures must be taken in different spheres, and special long-term programs must be adopted to create a higher health-related quality of life (HRQoL) for the older generation and to combat the widespread stereotype of the end of life and the feeling of doom. Grinin et al. point out the importance of technologies, the development of which can be stimulated precisely by the process of global aging and the need to increase life expectancy. As a background for the forthcoming technological wave (which we denote as the Cybernetic Revolution), global aging may create an acute demand for labor-saving technologies, as well as give a powerful impulse in the field of medicine. Progress in the latter will help to extend the working age and improve health-related (biological) quality of life.

1 Statement of the Problem

Global aging is an extremely important process that has a big effect today and will increase its influence in the future. However, global aging and especially its wide influence on all spheres of society is given unjustifiably little attention even in academic journals and publications.[1]

[1] Although, of course, there is a lot of research on various aspects of the aging process and its impact on society, especially on the economy (about this phenomenon see, e.g., Clark & Spengler, 1980; Lee et al., 1988; Bös & Weizsäcker, 1989; Peterson, 1999; Fukuyama, 2002; Jackson et al., 2008, 2013; Lee et al., 2010; Lee & Mason, 2011; UNFPA, 2012; UN Population Division, 2013, 2020; Popova & Yanik, 2014; Goldstone, 2015; Goldstone et al., 2015; Orlická, 2015; Bloom & Luca, 2016; Maestas et al., 2016; Zimmer, 2016; van Dullemen, 2017; Bengston, 2018; Fichtner, 2018; Randers et al., 2018; Yaşar & Yaşar, 2018; Barsukov, 2019; Hsu & Lo, 2019; Kapelyushnikov, 2019a, 2019b; Shcherbakova, 2019; Mitchell & Walker, 2020; Jackson, 2021; Medici, 2021). Here we try to explore global aging as an integral problem of the future (Grinin & Korotayev, 2015, 2016; Grinin et al., 2016; Grinin et al., 2017, 2020); in fact, there is a huge horizon of work here, since the phenomenon permeates the entire society and affects all processes in it.

At the same time, the UN and related organizations, especially the WHO, pay quite a lot of attention to the process of global aging, which occurs at different pace in different types of countries. This work has been actively carried out since the late 1970s (Popova & Yanik, 2014; Kamiya & Hertog, 2020; Medici, 2021; UN Population Division, 2013, 2015a, 2017, 2019a, 2020; WHO, 2017, 2021a, 2021b). Many different programs, decisions, declarations, etc., have been adopted. In general, a set of well-defined problems, tasks, and attitudes contribute to a more active and humane inclusion of the elderly people in social processes and also help to ensure that aging is perceived not only as a growing burden on society but also as an opportunity to point the societal development in a certain direction, using assets of people of mature and old age—their unique experience and life wisdom. In general, human rights approaches are being developed, along with ways to combat the bullying of the elderly people and setting them free from a sense of doom, the end of life, and the feeling that they are a burden for everyone and themselves (e.g., Kamiya & Hertog, 2020; UN Population Division, 2019a, 2020; WHO, 2017, 2021b). Much has been done to bring together data on national aging and adapt strategic documents at the national level, etc. Nevertheless, one should agree that major approaches of international organizations contain several delayed risks associated with changing the hierarchy of goals and focus shifting to the solution of an ever-increasing number of current and local tasks at the expense of strategic ones (Popova & Yanik, 2014). But, of course, international organizations are unable to solve all problems, let alone work out a common strategy, since they are tied to complex consensus of opinions.

In addition, even international organizations and the European Union hardly pay attention to all aspects of this problem. The analysis of the impact of global aging processes mainly covers several aspects: pension, budget, legal, gender, and medical in terms of aging-associated diseases. However, aging has many more important aspects. In other words, *the amount of research on this problem—namely, the problem of growing influence of aging on all aspects of societal life from year to year and in the future—is clearly disproportionate to its significance*. And this situation of strange silence has been preserved for several decades. Thus, in 1999 Peter Peterson reasonably argued that the aging of the higher-income countries would change our collective future more than anything else (the graying of the higher-income countries' population—that may actually do more to reshape our collective future than anything else) and at the same time, he wondered why so little is said about the future manifestations of the global aging process (Peterson, 1999: 42). And 10 years before him, Dieter Bös and Robert K. Von Weizsäcker wrote about ignoring the problem of aging (Bös & Weizsäcker, 1989: 345).

We believe that in the final phase of the Cybernetic revolution[2] (Grinin, 2006; Grinin & Grinin, 2015a, 2015b; Grinin & Grinin, 2016; Grinin et al., 2017, 2020, 2021) and after it, the most important task will be to adapt the global society to the

[2] About it see chapter "Technology. Limitless Possibilities and Effective Control" (Grinin & Grinin, 2023b).

process of global aging. At the same time, every society at each of its levels must actively integrate this problem into everyday life in order to accelerate the processes of adaptation to aging and the institutionalization of changed age proportions in society. We will turn to these aspects in more detail later.[3] While this chapter focuses more on the current aspects of the global aging process and the near future, as well as on the projections of the number of older people in the twenty-first century, the above-mentioned chapters are its direct continuation and present our vision of changes in the society under the influence of the global aging process throughout the twenty-first century: during and after the Cybernetic Revolution in the last third of the twenty-first century.

So, aging is already and will further remain the most important process and factor of the twenty-first (and at least of the first decades of the twenty-second) century, that is, for the next hundred or more years, unless there happens some considerable setback in terms of life expectancy. In terms of healthcare needs, the elderly are beginning to replace children in higher-income countries. But elderly care requires more money and effort from society, despite the fact that the main concerns so far fall on the family and relatives (see below).

It is necessary to explain how and why global aging will affect almost all spheres of life and society in the future. First of all, of course, the most important thing is the limits of the process and, above all, what the proportion of the elderly in the demographic structure of society will be. We present some estimated forecasts below. The aging process actually strongly affects (a) economic vectors (employment, consumption, and its structure, growth rates, investment, etc.); (b) political vectors (e.g., whether older people will be the leading electoral stratum; what part of the budget will be provided to them); (c) social vectors (to what extent social programs will be oriented toward them; how much the professional structure will be adapted for them); and of course, (d) the medical system. We have already mentioned the pension system above.

It is also necessary to mention here such important points as (1) preservation of savings and property of the elderly people (and not only theirs, of course); (2) social lifts, which may be hindered by ageist tendencies in society.[4] Conservatism, as inherent to the aging society, is also worth mentioning in particular, since this factor will have extensive influence (see, e.g., Grinin et al., 2020). By the way, one should note that both the media and their ideological orientation can change significantly.

At present, it is becoming obvious that while in technological terms, the developmental vector has already been fully designated toward the extension of life expectancy, from the political, financial, and economic perspectives, it is not at all clear if modern society is ready to solve the problems associated with increasing

[3] See chapter "Socio-political transformations. A difficult path to cybernetic society" (Grinin et al., 2023c), chapter "Modeling social self-organization and historical dynamics. Global phase transitions" (Malkov et al., 2023).

[4] About ageism and its possible evolution see chapter "Socio-political transformations. A difficult path to cybernetic society" (Grinin et al., 2023c, this volume) and chapter "Modeling social self-organization and historical dynamics. Global phase transitions" (Malkov et al., 2023, this volume).

lifespan. To what extent, for example, is it ready to simply preserve the savings and property of older people, and not only theirs, of course? How safe will they be in the situation of unprecedented growth of public debt and increasing power of bankers and financiers? Will they burn out in financial crises, defaults and inflation, or in the emergent (with the help of technology) manipulations and prohibitions (e.g., there are attempts to ban cash); zero and negative rates (Grinin, 2021), as a result of which pension money depreciates? It is sad that this topic is not mainstream at all, and yet it is fundamental for understanding the future of aging generations.

With the threat of possible permanent pandemics, the increasing influence of the media, AI-technology and many other things, what will be the position of older people? Will society ensure a real increase in life expectancy and in the quality of biological (health-related) life? Or will we face a serious social crisis, a backslide [the rehearsal of which we have witnessed recently with the pandemic when the elderly turned out to be the most vulnerable group (e.g., UN Population Division, 2020: 17–26)], which we will have to overcome in order to advance the process of increasing life expectancy and to improve the quality of biological life?

We will discuss various developmental scenarios below. It is very important to take into account that in any case, social structure, coupled with the political one, will become increasingly age-related and the role of age gradations will be more and more important in sociopolitical structures. We will also discuss the impact of aging on democracy below.

Already today, the share of the population aged 65+ is 20% or more in 22 countries and territories, mostly in Southern and Eastern Europe, but Japan is the leader with 28.4% (see Fig. 1; see also Shcherbakova, 2019; Kumaya & Hertog, 2020; Medici, 2021).

In several countries, the share of the population aged 80+ has already exceeded 6%, again Japan is the leader with 10.2%[5] (see Fig. 2):

The number of people aged 60 and over already exceeds the number of children under 5 years of age (WHO, 2021b). There are many aspects associated with global aging, and there will be even more of them, while the problems are likely to become more acute, at least for a couple of decades.

The population aging projections, even the inertial ones, are quite impressive. The growth rate of the proportion of people over 60 is several times higher than the overall growth rate of the Earth's population (UN Population Division, 2013, 2022a, 2022b). In the early 1980s, there were no "aging countries" in the world where older people would consume more than young people. In 2010, there were already 23 "aging countries," and by 2040, there will be 89 (UNCTAD, 2012). According to the most recent UN medium projection, by 2100, 2.47 billion people, more than a quarter of the world's population, will be over 65 and only 2.30 billion will be under 20 (UN Population Division, 2013, 2022a, 2022b). The number of those over 80 will increase sixfold, from about 140 million in 2020 to over 860 million by the end of

[5] An even higher percentage (16.6%) is found in Monaco (UN Population Division, 2022a), but this is, of course, a very special country.

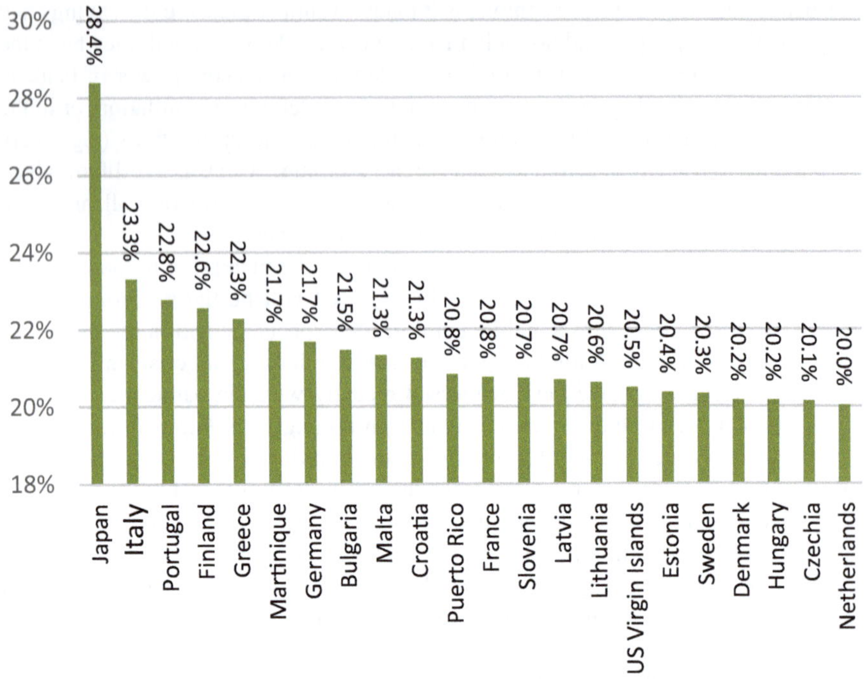

Fig. 1 Countries and territories of the world with the share of the population aged 65+ >20% in 2020. *Source*: Authors' calculations on the basis of data provided by the UN Population Division database (2022a)

the twenty-first century (Gallagher, 2020); thus, compared to 1950, the number of the 80-year-olds should increase by almost two orders of magnitude by 2100! (see Fig. 3):

Respectively, the global ratio of people over 80 for every child aged under 5 is projected to rise from 0.23 in 2021 to 1.73 in 2100.

In general, all population projections are unanimous in predicting a dramatic aging of the world population in the coming decades (Alkema et al., 2011; Kaneda et al., 2016; Lutz et al., 2018; Raftery et al., 2012; UN Population Division, 2019a, 2019b, 2022a, 2022b; Vollset et al., 2020; Wittgenstein Center, 2022) (see Fig. 4).

The year 2100 seems very far away, but it must be clearly understood that the elderly population growth rate increases with increasing median age and the pace of aging continues to climb (Zimmer, 2016; Shcherbakova, 2019; UN Population Division, 2019a, 2020, 2022a, 2022b; Kumaya & Hertog, 2020; Medici, 2021). From 2020 to 2030, the proportion of people over 60 in the global population will increase by 34% (WHO, 2021b; UN Population Division, 2022a, 2022b). Drastic changes in the path of global and country aging will occur as early as in the 2050s. Thus, the population of 90 and over year olds will increase by a factor of 3.6 by 2050

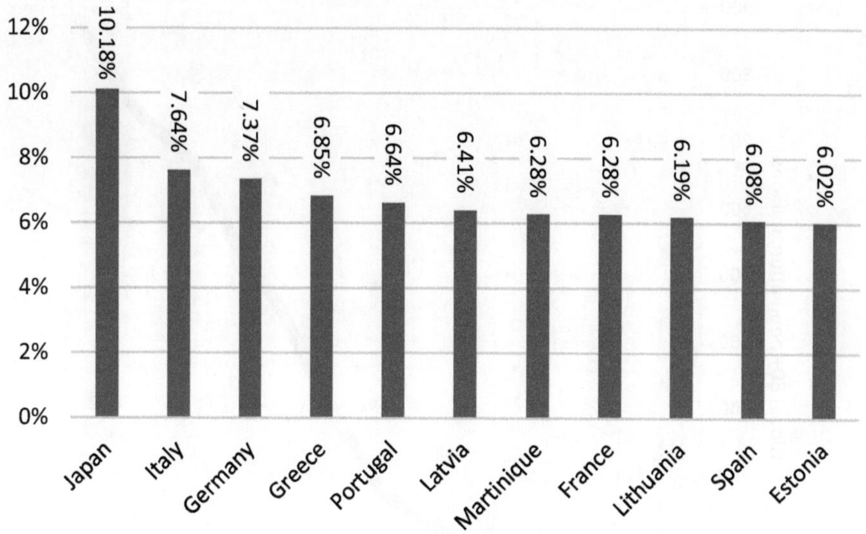

Fig. 2 Countries and territories of the world with the share of the population aged 80+ >6% in 2021. *Source*: Authors' calculations on the basis of data provided by the UN Population Division database (2022a)

(up to 82.9 from 22.8 million people in 2021 (UN Population Division, 2022a, 2022b; cp. Shcherbakova, 2019).

Between 2000 and 2050, the proportion of the world population aged over 60 will more than double from about 9.9% to 22%. The absolute number of people aged 60 and over is projected to increase from 1.1 billion in 2021 to 1.4 billion by 2030 and 2.1 billion by 2050, and it may reach 3.1 billion in 2100. The number of people aged 80 years and over is expected to triple to 458 million between 2021 and 2050 (UN Population Division, 2022a, 2022b; cp. WHO, 2021b).

By 2050, about 34% of the population over the age of 60 will live in Europe, and 25% will reside in Latin America and Asia. Although Africa has the youngest population of all major regions, it is also projected to rapidly age, with the percentage of people aged 60 years or older rising from 5% today (in 2022) to 9% by 2050. Keep in mind that older people will be increasingly concentrated in economically less developed regions (according to the UN classification).[6] UN Population Division, 2015a, 2015b, 2022a, 2022b; WHO, 2017: 11). In 2050, approximately 65% of older people will live in low- and middle-income countries (WHO, 2021b, 2022a, 2022b). Therefore, the problem of pension and social security for the elderly can become extremely acute. As has already been mentioned, the number of people aged 60 years and over (1,109 million) already now, in 2022, is almost double the number of children under 5 years of age (663 million) (UN Population Division, 2022a, 2022b; cp. WHO, 2021b). It is a somewhat scary perspective that by 2050 there will

[6]UN Population Division (2022a, 2022b); cp. WHO (2017: 11, 2021b).

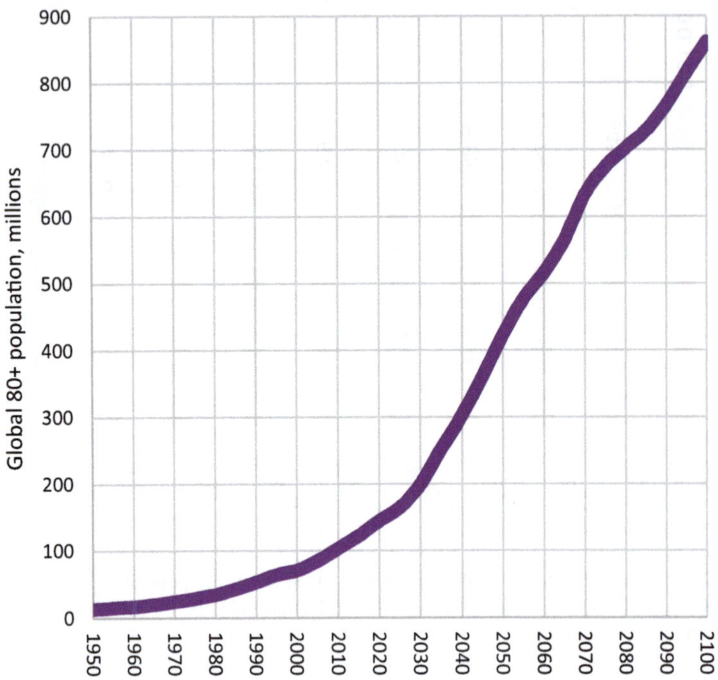

Fig. 3 Global population aged 80+, millions; UN Population Division empirical estimates for 1950–2021 and UN Population Division medium forecast scenario for 2022–2100. *Data source*: Authors' calculations on the basis of data provided by the UN Population Division database (2022a)

be almost twice more people aged 60 and over (2,132 million) than teenagers and young people in the 15–24 age group (1,299 million) (UN Population Division, 2022a, 2022b; cp. WHO, 2021b).

Already by 2030, fundamental transformations will occur. In 2030, the proportion of the population aged 65 and over will be 20% or more already in 59 countries/territories of the world (UN Population Division, 2022a, 2022b; cp. Shcherbakova, 2019).[7] In 2030, people aged 60 and over will make up one-sixth (16.6%) of the world's inhabitants (UN Population Division, 2022a, 2022b; cp. WHO, 2021a).

Such forecasts should make politicians think twice and start changing the course. But many people think this is not that serious and do not understand the urgency of

[7]The proportion of the population over the age of 60 has already reached 36% in Japan (UN Population Division, 2022a, 2022b; cp. WHO, 2021b). It is significant that this was forecasted quite a long time ago, at least as early as in 2008 (Jackson et al., 2008). But the trend has not changed, the recent forecasts confirm it. We are afraid that the prediction of the same authors that by 2050 the proportion of older people in Japanese society will reach 40% may also come true (Jackson et al., 2008); in fact, according to the most recent UN medium projection, in Japan in 2050 this share will be 44%, whereas, according to their lowest fertility projection, it will be even 47% (UN Population Division, 2022a, 2022b). And, according to one of IHME's forecasts (see above), this share may already reach 40% by 2090 even at the global level (Vollset et al., 2020).

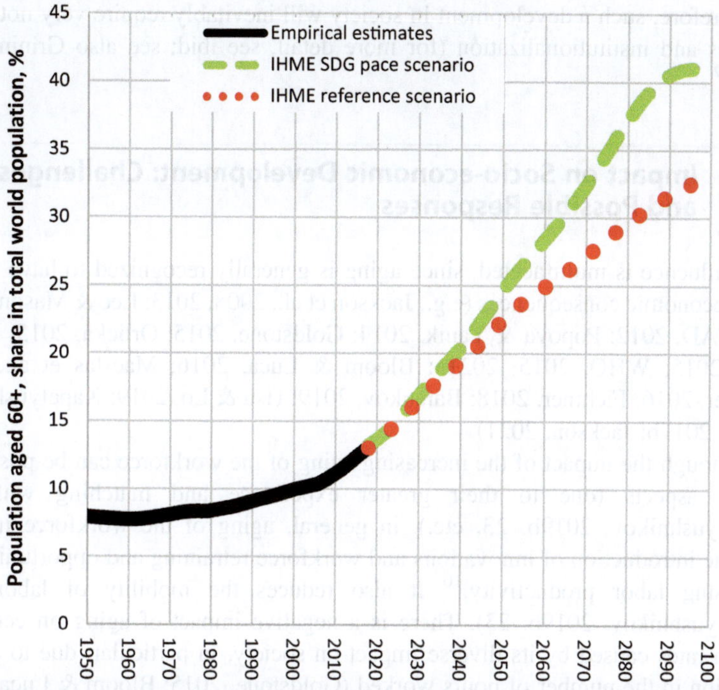

Fig. 4 Population aged 60+, share in total population of the world (%); Institute for Health Metrics and Evaluation at the University of Washington (IHME) empirical estimates for 1950–2020; IHME reference scenario for 2021–2100; IHME Sustainable Development Goals (SDG) pace scenario for 2021–2100. The IHME SDG pace scenario "shows what we can expect population trends to look like if every country and territory meets the SDGs for education and contraceptive met need by 2030" (Vollset et al., 2020: 1295)

decisions and the complexity of measures that could if not reverse, then at least slow down the trend (about this see, e.g., Gallagher, 2020).

We believe that the process of global aging and increasing life expectancy, if it continues, will entail very serious changes in sociopolitical life.[8]

However, *modern society is largely unprepared for such changes and adaptations, hence the development of these trends can cause serious tensions and conflicts.*

[8]For some of these changes see chapter "Socio-Political Transformations. A Difficult Path to Cybernetic Society" (Grinin et al., 2023c, this volume) and chapter "Modeling Social Self-Organization and Historical Dynamics. Global Phase Transitions" (Malkov et al., 2023, this volume). See also chapter "Modeling Social Self-Organization and Historical Dynamics. Modern Society and a Look into the Global Futures: Cybernetic W-Society" (Akaev et al., 2023, this volume).

Therefore, such a development in society will inevitably require very noticeable changes and institutionalization (for more detail, see ibid; see also Grinin et al., 2020).[9]

2 Impact on Socio-economic Development: Challenges and Possible Responses

This influence is multifaceted, since aging is generally recognized to have serious macroeconomic consequences (e.g., Jackson et al., 2008, 2013; Lee & Mason, 2011; UNCTAD, 2012; Popova & Yanik, 2014; Goldstone, 2015; Orlická, 2015; Park & Shin, 2015; WHO, 2015, 2021a; Bloom & Luca, 2016; Maestas et al., 2016; Zimmer, 2016; Fichtner, 2018; Barsukov, 2019; Hsu & Lo, 2019; Kapelyushnikov, 2019a, 2019b; Jackson, 2021).

Although the impact of the increasing aging of the workforce can be positive in certain aspects (due to their greater experience and matching with jobs [Kapelyushnikov, 2019b: 23, etc.), in general, aging of the workforce impedes both the introduction of innovations and workforce retraining and opportunities for increasing labor productivity.[10] It also reduces the mobility of labor force (Kapelyushnikov, 2019b: 23). There is a negative impact of aging on economic growth rates caused by its diverse impact on society, in particular, due to a sharp reduction in the number of hours worked (Goldstone, 2015; Bloom & Luca, 2016; Kapelyushnikov, 2019b: 42). The strongest impact, of course, is associated with a sharp reduction in labor supply. We can agree that currently and in the near future, this is the main challenge that economies will face in the context of population aging (Kapelyushnikov, 2019a, 2019b).

Strangely enough, the topic of the looming pension crisis as the main challenge to the welfare state is not widely discussed. However, the crisis is approaching, and in some cases it is already detected. It will become especially acute in case of major collapses in the securities markets where pension funds are invested, as well as possible state defaults. Those funds that are invested in public debts of high-income countries, in fact, may be already considered as spent funds. And since economic

[9]In this regard, the report of Giarini & Leidtke (1996) is quite noteworthy, as it analyzes employment options for both those who cannot find work for one reason or another, and for people over 60 years of age. Both categories are offered jobs that match their capabilities, i.e. maximally simplified, but captivating and entertaining, while giving them the opportunity to earn money to meet their basic needs.

[10]According to some reports, labor productivity is declining due to the population aging. Maestas et al. (2016) found out that with an increase in the share of older workers in the number of employed by 10 percentage points, labor productivity declines in all age groups by an average of 3–5%. The reasons for this are not entirely clear. Maestas et al. (2016) believe that this may be due to the earlier exit of the most productive workers from the labor market (Kapelyushnikov, 2019b: 29). But, perhaps, the general increase in conservatism affects younger groups as well.

growth is weak, and the impending debt crisis can be very acute, the pension system becomes a hostage of the general economic situation.

A serious problem is providing home care for vulnerable people and its economic component (e.g., Kamya & Herzog, 2020; UN Population Division, 2019a, 2020). Most people who need care stay with families. This applies to all countries, including the high-income ones. For example, in OECD countries, about 10% of adults are involved in home care (OECD, 2011). Those are huge numbers! Specifically, in the United States, at least 43.5 million people aged 18 and over provide unpaid care for an older person who is a family member or a friend (this is equal to 19% of adults in the USA). The majority of caregivers are women (67%) with an average age of 50 and almost 20 h of care per week (NAC & AARP, 2009, 2020). In Japan in 2012, women aged 50–65 represented the majority of people leaving or changing jobs to care for their parents (WHO, 2017: 14).

For the most part, of course, care is provided by women. Thus, the burden of caring for children, previously fell on women, but now it has been partly replaced by the burden of caring for the elderly and relatives in need. It is mostly women of a young and relatively young age who are likely to care for children, while the care for the vulnerable is carried out by women of a mature and elderly age. We can agree with the World Health Organization that, in addition to the general losses associated with nonparticipation of a huge number of women in labor force, there are also significant social, political, environmental and economic inequalities that are a separate burden on women in these countries, and need to be thought over (WHO, 2017: 14).

There is also a number of other problems. In particular, the problem of ageism, that is various types of age discrimination. Not so long ago, the danger of the increasing number of old people in nursing homes also became clear, since pandemics (as COVID-19 showed) gather their victims especially densely there (see, e.g., UN Population Division, 2020: 17–24; about other COVID influences, that also negatively affect elderly see Grinin, 2020; Irshad, 2020). There are threats of confrontation between young and old states in the future, which could blow up the World System order. Democracy is also in danger due to population aging, and in addition, there is a risk of society's bias toward gerontocracy. It is especially dangerous if gerontocracy encompasses political and other elites, which is observed today, for example, in the United States, and is emerging in China and the Russian Federation. There are a number of important social aspects associated with aging that can have a strong impact on future social relations in society.[11]

In general, the problem of aging requires a special report to the Club of Rome. The main prerequisites for our further forecasts about its impact on modern and future society can be summarized in two theses:

[11]We will touch these aspects in chapter "Socio-Political Transformations. A Difficult Path to Cybernetic Society" (Grinin et al., 2023c), chapter "Modeling Social Self-Organization and Historical Dynamics. Global Phase Transitions" (Malkov et al., 2023), and chapter "Modeling Social Self-Organization and Historical Dynamics. Modern Society and a Look into the Global Futures: Cybernetic W-Society" (Akaev et al., 2023, this volume).

1. The deeper is the process of global aging in society resulting from extending life expectancy and low birth rates, the more this process depends on the development of medical, biotechnological, and many other technological trends and on the development of technologies in general.
2. The process of global aging will be unable to develop in the future, not only without the development of innovative technologies in medicine, biotechnologies etc., including self-regulating systems[12] and others but also without profound changes in society, including social innovations deeply integrated into societal life.[13]

See chapter "Socio-Political Transformations. A Difficult Path to Cybernetic Society" (Grinin et al., 2023d, this volume), chapter "Technology. Limitless Possibilities and Effective Control" (Grinin & Grinin, 2023b), and chapter "Analyzing Social Self-Organization and Historical Dynamics. Future Cybernetic W-Society: Socio-Political Aspects" (Grinin & Grinin, 2023a, this volume) for more details on the scenarios that may be associated with the implementation of these prerequisites.[14]

3 Is It Possible to Use the Problem of Aging as a Development Driver?

Yes, it is possible. There is good news here as well. Official documents of the WHO and some other organizations rightly point out that aging can have positive aspects for our benefit. Firstly, it gives an opportunity to use the unique life, professional and other experiences of older people, who now live much longer than before. Secondly, in the field of value orientations, global aging may have some positive effects due to increasing support for prosocial values at the global level (Korotayev et al., 2021, 2023). Thirdly, self-organization of these people is to be expected, with the healthier and more energetic of them helping the less healthy and less energetic in their

[12] See chapter "Technology. Limitless Possibilities and Effective Control" (Grinin & Grinin, 2023b) for details.

[13] The global aging is associated with increasing life expectancy. And in order to extend the lifespan, the development of medical and other technologies is necessary: the better the medicine (that is, technologies) is developed, the longer people live. If advancement in this field stops, then the life expectancy will stop growing as well or will even decrease so that the global aging will slow down, and may reverse in older societies, as it was in Russia in the 1990s, when life expectancy in this country significantly decreased (see, e.g., Korotayev & Khaltourina, 2008).

[14] On optimum scenarios for this in economic, social, demographic, and political dimensions see also chapter "Economics. Optimizing Growth" (Grinin et al., 2023c, this volume), chapter "Demography. Toward Optimization of Demographic Processes" (Korotayev et al., 2023), chapter "High Income and Low Income Countries. Toward a Common Goal at Different Speeds" (Grinin et al., 2023b, this volume), chapter "Future Political Change. Toward a More Efficient World Order" (Grinin et al., 2023a, this volume), and chapter "Africa—the Continent of the Future. Challenges and Opportunities" (Grinin & Korotayev, 2023, this volume).

communities. Fourth, with the development of medicine, older people can work for a long time, thereby bringing economic benefits to society without being a burden to it. Fifth, older people can play an important role in social life. However, not enough attention has been paid to the role of the global aging process as an engine of technology. We would like to pay a special attention to this.

We believe that global aging can become a driver of technological growth. It is worth noting that the impact of global aging on the speed and direction of scientific-technological progress is understudied (Galor & Weil, 2000; de Grey & Rae, 2008; Prettner, 2013; Tsirel, 2020), and global aging affects technological, economic, political, social and other spheres in various ways (Fukuyama, 2002; Goldstone et al., 2015; Harper, 2006; Powell & Khan, 2013; Goldstone, 2015; Coleman & Rowthorn, 2015; Park & Shin, 2015; Haas, 2015; Zimmer, 2016; see also chapters in this volume mentioned in the footnote above). Our study shows that in the course of the Cybernetic revolution (in the 2030s–2060s), aging will spur technological progress[15] (Grinin & Grinin, 2015a, 2015b; Grinin et al., 2020, 2021), since labor-saving technologies will be required (due to the lack of labor resources) along with strong technological development in medicine to extend the working age and improve the biological quality of life of the elderly. Adaptation to aging, which was mentioned above, is impossible without the powerful development of technologies allowing to fight the most dangerous diseases (oncology, dementia, etc.), and: (a) making easier the lives of the elderly and the disabled, and here achievements are extremely important in a variety of areas: medicine, biotechnology and pharmaceuticals, artificial intelligence, cognitive technologies, etc.; (b) facilitating and reducing the cost of caring for the disabled and the elderly with the help of robots, artificial intelligence and, in general, a special techno-biological observation environment; (c) facilitating participation in the labor process for the elderly with the help of the latest technologies. All this will lead to a huge breakthrough, a new technological wave, one of the most important drivers of which will be global aging, which will pose (and is already posing) very serious and fundamentally new problems for society.

We think that, as a result of the largely symbiotic development of the process of global aging and the adaptation of society to it, on the one hand, and the powerful development of self-regulating systems, on the other hand, a new type of society will arise—the cybernetic society. It is a society that will be formed as a result of the completion of the Cybernetic Revolution, and will be (a) super-technologized; (b) socio-technologically regulated at all levels; and (c) a society where, as a result of the aging process, the division into age categories will become socially much more significant than today (that is, age will become a much more important social marker than today). Its main features are described in chapter "Socio-Political Transformations. A Difficult Path to Cybernetic Society" (Grinin et al., 2023d, this volume), and it is discussed in more detail in chapter "Technology. Limitless

[15] See chapter "Technology. Limitless Possibilities and Effective Control" (Grinin & Grinin, 2023b).

Possibilities and Effective Control" (Grinin & Grinin, 2023b, this volume) and chapter "Analyzing Social Self-Organization and Historical Dynamics. Future Cybernetic W-Society: Socio-Political Aspects" (Grinin & Grinin, 2023a, this volume).

However, as we note in chapter "Technology. Limitless Possibilities and Effective Control" (Grinin & Grinin, 2023b, this volume),[16] technological development not only brings relief in a number of aspects, but also creates threats to the individual freedom, dignity, and privacy. This is especially true for older people, as they are especially psychologically vulnerable. Therefore, adaptation to aging requires the adaptation of technological innovations to the principles of a free society. And this is a serious problem to solve, which is already felt very clearly today and already causes social protests.

Of course, technological progress cannot always advance at a rapid pace, especially if the population of the Earth strives for stabilization. A successful adaptation to aging will lead to a certain stabilization, to what can be considered a sustainable development. The fact is that due to the increasing proportion of the elderly population and the growth of general conservatism in society, the tendency to accelerate technological progress will slow down until its pace subsides by the end of the twenty-first and the beginning of the twenty-second centuries. At the same time, depending on the scenario of population growth in the twenty-first century, the beginning of a slowdown in the advancement of technological progress and the pace of its slowdown will change in some aspects.

Thus, the process of global aging turns out to be truly integral and it is necessary to pay as much attention to it as possible. And the sooner the better for the future of the world.

References

Akaev, A., Malkov, S., Grinin, L., Bilyuga, S., Davydova, O., Grinin, A., Kovaleva, N., Malkov, A., Musieva, J., & Korotayev, A. (2023). Modeling social self-organization and historical dynamics. Modern Society and a look into the global futures: Cybernetic W-society. In V. Sadovnichy et al. (Eds.), *Reconsidering the limits to growth. A report to the Russian Association of the Club of Rome* (pp. 461–490). Springer. https://doi.org/10.1007/978-3-031-34999-7_20

Alkema, L., Raftery, A. E., Gerland, P., Clark, S. J., Pelletier, F., Buettner, T., & Heilig, G. K. (2011). Probabilistic projections of the total fertility rate for all countries. *Demography, 48*(3), 815–839. https://doi.org/10.1007/s13524-011-0040-5

Barsukov, V. N. (2019). From the demographic dividend to population aging: World trends in the system-wide transition. *Economic and Social Changes: Facts, Trends, Forecast, 12*(4), 167–182. https://doi.org/10.15838/esc.2019.4.64.11

Bengtson, V. (Ed.). (2018). *Global aging and challenges to families*. Routledge.

Bloom, D. E., & Luca, D. L. (2016). *The global demography of aging: Facts, explanation, future*. Institute for the Study of Labor.

[16]Chapter "Modeling Social Self-Organization and Historical Dynamics. Global Phase Transitions" (Malkov et al., 2023).

Bös, D., & Von Weizsäcker, R. K. (1989). Economic consequences of an aging population. *European Economic Review, 33*(2–3), 345–354.

Clark, R. J., & Spengler, J. J. (1980). *The economics of individual population aging*. Cambridge University Press.

Coleman, D., & Rowthorn, R. (2015). Population decline – Making the best of inevitable destiny? In J. A. Goldstone, L. Grinin, & A. Korotayev (Eds.), *History & mathematics: Demography & aging* (pp. 26–41). Uchitel.

Fichtner, J. J. (2018). Global aging and public finance. *Business Economics, 53*(2), 72–78.

Fukuyama, F. (2002). *Our post-human future: Consequences of the bio-technology revolution*. Farrar, Straus, and Giroux.

Gallagher, J. (2020, July 15). Fertility rate: 'Jaw-dropping' global crash in children being born. *BBC*. https://www.bbc.com/news/health-53409521

Galor, O., & Weil, D. N. (2000). Population, technology, and growth: From Malthusian stagnation to the demographic transition and beyond. *American Economic Review, 90*(4), 806–828.

Giarini, O., & Leidtke, P. M. (1996). *The employment dilemma. A report to the Club of Rome*. Wilhelm Heyne Verlag.

Goldstone, J. A. (2015). Population aging and global economic growth. In J. A. Goldstone, L. Grinin, & A. Korotayev (Eds.), *History & mathematics: Demography & aging* (pp. 147–155). Uchitel.

Goldstone, J. A., Grinin, L., & Korotayev, A. (2015). Research into global aging and its consequences. In J. A. Goldstone, L. Grinin, & A. Korotayev (Eds.), *History & mathematics: Political demography & global aging* (pp. 5–9). Uchitel.

Grey, A. B. d., & Rae, M. (2008). *Ending aging: The rejuvenation breakthroughs that could reverse human aging in our lifetime*. St. Martin's Press.

Grinin, L. (2006). *Proizvoditel'nyye sily i istoricheskiy protsess*. URSS.

Grinin, L. E. (2020). How Can COVID-19 Change Geopolitics and Economy? *Journal of Globalization Studies, 11*(2), 121–134. https://doi.org/10.30884/jogs/2020.02.09

Grinin, L. (2021). Otritsatel'nyye stavki i drugiye noveyshiye finansovyye tekhnologii. *Obshchestvo i ekonomika, 2*, 18–30.

Grinin, L., & Grinin, A. (2015a). Kiberneticheskaya revolyutsiya i shestoy tekhno-logicheskiy uklad. *Istoricheskaya psikhologiya i sotsiologiya istorii, 8*(1), 172–197.

Grinin, L., & Grinin, A. (2015b). *Ot rubil do nanorobotov. Mir na puti k epokhe samoupravlyayemykh sistem (istoriya tekhnologiy i opisaniye ikh budushchego)*. Uchitel.

Grinin, L., & Grinin, A. (2016). *The cybernetic revolution and the forthcoming epoch of self-regulating systems*. Uchitel.

Grinin, L., & Grinin, A. (2023a). Analyzing social self-organization and historical dynamics. Future Cybernetic W-society: Sociopolitical aspects. In V. Sadovnichy et al. (Eds.), *Reconsidering the limits to growth. A report to the Russian Association of the Club of Rome* (pp. 521–530). Springer. https://doi.org/10.1007/978-3-031-34999-7_22

Grinin, L., & Grinin, A. (2023b). Technology. Limitless possibilities and effective control. In V. Sadovnichy et al. (Eds.), *Reconsidering the limits to growth. A report to the Russian Association of the Club of Rome* (pp. 139–154). Springer. https://doi.org/10.1007/978-3-031-34999-7_8

Grinin, L., Grinin, A., & Korotayev, A. (2017). The MANBRIC-technologies in the forthcoming technological revolution. In T. Devezas, J. Leitão, & A. Sarygulov (Eds.), *Industry 4.0 – Entrepreneurship and structural change in the new digital landscape: What is coming on along with the fourth industrial revolution* (pp. 243–261). Springer. https://doi.org/10.1007/978-3-319-49604-7_13

Grinin, L., Grinin, A., & Korotayev, A. (2020). A quantitative analysis of worldwide long-term technology growth: From 40,000 BCE to the early 22nd century. *Technological Forecasting and Social Change, 155*, 1–19. https://doi.org/10.1016/j.techfore.2020.119955

Grinin, L., Grinin, A., & Korotayev, A. (2021). COVID-19 pandemic as a trigger for the acceleration of the cybernetic revolution, transition from E-government to E-state, and change in social relations. *Technological Forecasting and Social Change, 175*, 1–15. https://doi.org/10.1016/j.techfore.2021.121348

Grinin, L., Grinin, A., & Korotayev, A. (2023a). Future political change. Toward a more efficient world order. In V. Sadovnichy et al. (Eds.), *Reconsidering the limits to growth. A report to the Russian Association of the Club of Rome* (pp. 191–206). Springer. https://doi.org/10.1007/978-3-031-34999-7_11

Grinin, L., Malkov, S., & Korotayev, A. (2023b). High income and low income countries. Toward a common goal at different speeds. In V. Sadovnichy et al. (Eds.), *Reconsidering the limits to growth. A report to the Russian Association of the Club of Rome* (pp. 207–224). Springer. https://doi.org/10.1007/978-3-031-34999-7_12

Grinin, L., Grinin, A., & Malkov, S. (2023c). Economics. Optimizing growth. In V. Sadovnichy et al. (Eds.), *Reconsidering the limits to growth. A report to the Russian Association of the Club of Rome* (pp. 155–168). Springer. https://doi.org/10.1007/978-3-031-34999-7_9

Grinin, L., Grinin, A., & Malkov, S. (2023d). Socio-political transformations. A difficult path to cybernetic society. In V. Sadovnichy et al. (Eds.), *Reconsidering the limits to growth. A report to the Russian Association of the Club of Rome* (pp. 169–189). Springer. https://doi.org/10.1007/978-3-031-34999-7_10

Grinin, L., & Korotayev, A. (2015). Global'noye stareniye naseleniya, shestoy tekhnologicheskiy uklad i mirovaya finansovaya Sistema. In L. Grinin & A. Korotayev (Eds.), *Istoriya i matematika: futurologicheskiye i metodologicheskiye aspekty* (pp. 31–56). Uchitel.

Grinin, L., & Korotayev, A. (2016). Global population aging, the sixth Kondratieff wave, and the global financial system. *Journal of Globalization Studies, 7*(2), 11–31.

Grinin, L., & Korotayev, A. (2023). Africa – The continent of the future. Challenges and opportunities. In V. Sadovnichy et al. (Eds.), *Reconsidering the limits to growth. A report to the Russian Association of the Club of Rome* (pp. 225–238). Springer. https://doi.org/10.1007/978-3-031-34999-7_13

Grinin, L., Korotayev, A., & Grinin, A. (2016). Global'noye stareniye naseleniya, shestoy kondrat'yevskiy tsikl i kiberneticheskaya revolyutsiya. *Problemy teorii i praktiki upravleniya, 7*, 54–59.

Haas, M. L. (2015). Population aging and the future of the great powers. In J. A. Goldstone, L. Grinin, & A. Korotayev (Eds.), *History & mathematics: Demography & aging* (pp. 133–146). Uchitel.

Harper, S. (2006). Addressing the implications of global aging. *Journal of Population Research, 23*(2), 205–223.

Hsu, Y. H., & Lo, H. C. (2019). The impacts of population aging on saving, capital formation, and economic growth. *American Journal of Industrial and Business Management, 9*(12), 2231–2246. https://doi.org/10.4236/ajibm.2019.912148

Irshad, M. S. (2020). Covid-19 and Global 'At-Risk Community': From Benefit-Sharing to Risk-Sharing of Economic Crisis. *Journal of Globalization Studies, 11*(2), 135–145. https://doi.org/10.30884/jogs/2020.02.10

Jackson, R., Howe, N., Strauss, R., & Nakasima, K. (2008). *The greying of the great powers. Demography and geopolitics in the 21st century.* Center for Strategic & International Studies.

Jackson, R., Macaranas, R., & Peter, T. (2013). *U.S. development policy in an aging world. New challenges and new priorities for a new demographic era. A research report from project on U.S. leadership.* Center for Strategic and International Studies.

Jackson, R. (2021). *The shape of things to come. The macro challenges of population aging.* Concord Coalition & the Global Aging Institute.

Kamiya, Y., & Hertog, S. (2020). *Measuring household and living arrangements of older persons around the world.* United Nations, Department of Economics and Social Affairs, Population Division.

Kaneda, T., Greenbaum, C., & Patierno, K. (2016). *World population data sheet.* Population Reference Bureau. https://www.prb.org/2016-world-population-data-sheet/

Kapelyushnikov, R. (2019a). Fenomen stareniya naseleniya: ekonomicheskiye effekty. *Ekonomicheskaya politika, 14*(2), 8–63.

Kapelyushnikov, R. (2019b). *Fenomen stareniya naseleniya: ekonomicheskiye effekty*. Institut ekonomiki RAN.

Korotayev, A., & Khaltourina, D. (2008). The Russian demographic crisis in cross-national perspective. In D. W. Blum (Ed.), *Russia and globalization: Identity, security, and society in an era of change* (pp. 37–78). Johns Hopkins University Press.

Korotayev, A., Butovskaya, M., Shulgin, S., & Zinkina, J. (2021). Vliyaniye global'nogo stareniya na global'nuyu sistemu tsennostey. *Vek Globalizatsii, 4*, 69–80. https://doi.org/10.30884/vglob/2021.04.05

Korotayev, A., Malkov, S., & Musieva, J. (2023). Demography. Toward optimization of demographic processes. In V. Sadovnichy et al. (Eds.), *Reconsidering the limits to growth. A report to the Russian Association of the Club of Rome* (pp. 97–116). Springer. https://doi.org/10.1007/978-3-031-34999-7_6

Lee, R. D., Arthur, W. B., & Rodgers, G. (1988). *Economics of changing age distributions in high income countries*. Oxford University Press.

Lee, R., & Mason, A. (2011). *Population aging and the generational economy: A global perspective*. Edward Elgar.

Lee, R. D., Mason, A., & Cotlear, D. (2010). *Some economic consequences of global aging. A discussion note for the World Bank*. World Bank.

Lutz, W., Goujon, A., Kc, S., Stonawski, M., & Stilianakis, N. (2018). *Demographic and human capital scenarios for the 21st century: 2018 assessment for 201 countries*. Publications Office of the European Union.

Maestas, N., Mullen, K. J., & Powell, D. (2016). *The effect of population aging on economic growth, the labor force and productivity*. NBER.

Malkov, S., Grinin, L., Grinin, A., Musieva, J., & Korotayev, A. (2023). Modeling social self-organization and historical dynamics. Global phase transitions. In V. Sadovnichy et al. (Eds.), *Reconsidering the limits to growth. A report to the Russian Association of the Club of Rome* (pp. 419–460). Springer. https://doi.org/10.1007/978-3-031-34999-7_19

Medici, A. C. (2021). *Health sector challenges and policies in the context of aging populations*. United Nations, Department of Economics and Social Affairs, Population Division.

Mitchell, E., & Walker, R. (2020). Global aging: Successes, challenges and opportunities. *British Journal of Hospital Medicine, 81*(2), 1–9. https://doi.org/10.12968/hmed.2019.0377

NAC & AARP. (2009). *Care giving in the U.S.: A focused look at those caring for someone the 50 or older*. https://assets.aarp.org/rgcenter/il/caregiving_09.pdf

NAC & AARP. (2020). *Caregiving in the U.S. 2020: A focused look at family caregivers of adults age 50*. https://www.caregiving.org/wp-content/uploads/2021/05/AARP1340_RR_Caregiving50Plus_508.pdf

OECD. (2011). *Help wanted? Providing and paying for long-term care*. https://www.oecd.org/els/health-systems/help-wanted-9789264097759-en.htm

Orlická, E. (2015). Impact of population aging and elderly poverty on macroeconomic aggregates. *Procedia Economics and Finance, 30*, 598–605.

Park, D., & Shin, K. (2015). Impact of population aging on Asia's future growth. In J. A. Goldstone, L. Grinin, & A. Korotayev (Eds.), *History & mathematics: Political demography & global aging* (pp. 107–132). Uchitel.

Peterson, P. G. (1999). Gray down: The global aging crisis. *Foreign Affairs, 78*(1), 42–55.

Popova, S., & Yanik, A. (2014). Problemy global'nogo stareniya naseleniya: Analiz dokumentov i strategii OON. *Mezhdunarodnoye pravo i mezhdunarodnyye organizatsii, 3*, 429–443.

Powell, J. L., & Khan, H. T. (2013). Aging and globalization: A global analysis. *Journal of Globalization Studies, 4*(1), 137–146.

Prettner, K. (2013). Population aging and endogenous economic growth. *Journal of Population Economics, 26*(2), 811–834.

Raftery, A. E., Li, N., Ševčíková, H., Gerland, P., & Heilig, G. K. (2012). Bayesian probabilistic population projections for all countries. *Proceedings of the National Academy of Sciences, 109*(35), 13915–13921. https://doi.org/10.1073/pnas.1211452109

Randers, J., Rockström, J., Stoknes, P. E., Golüke, U., Collste, D., & Cornell, S. (2018). *Transformation is feasible. A report to The Club of Rome*. Stockholm Resilience Centre.
Shcherbakova, Y. (2019). Stareniye naseleniya mira po otsenkam OON 2019 goda. *Demoskop Weekly* (pp. 837–838). http://demoscope.ru/weekly/2019/0837/barom01.php
Tsirel, S. (2020). Future technological achievements as a challenge for post-singularity human society. In A. V. Korotayev & D. J. LePoire (Eds.), *The 21st century singularity and global futures. A big history perspective* (pp. 419–437). Springer. https://doi.org/10.1007/978-3-030-33730-8_19
UN Population Division. (2013). *World population aging 2013*. United Nations, Department of Economics and Social Affairs, Population Division.
UN Population Division. (2015a). *World population aging 2015*. United Nations, Department of Economics and Social Affairs, Population Division.
UN Population Division. (2015b). *World population prospects 2015*. United Nations, Department of Economics and Social Affairs, Population Division.
UN Population Division. (2017). *World population aging 2017*. United Nations, Department of Economics and Social Affairs, Population Division.
UN Population Division. (2019a). *World population aging 2019*. United Nations, Department of Economics and Social Affairs, Population Division.
UN Population Division. (2019b). *World population prospects 2019*. United Nations, Department of Economics and Social Affairs, Population Division.
UN Population Division. (2020). *World population aging 2020*. United Nations, Department of Economics and Social Affairs, Population Division.
UN Population Division. (2022a). *United Nations population division database*. United Nations, Department of Economics and Social Affairs, Population Division. http://www.un.org/esa/population
UN Population Division. (2022b). *World population prospects 2022*. United Nations, Department of Economics and Social Affairs, Population Division.
UNCTAD. (2012). *Doklad o mirovykh investitsiyakh, 2012 god. Konferentsiya Organizatsii ob"yedinennykh natsiy po torgovle i razvitiyu*. https://unctad.org/system/files/official-document/wir2012overview_ru.pdf
UNFPA. (2012). *Aging in the twenty-first century: A celebration and a challenge*. UNFPA and HelpAge International. https://www.unfpa.org/publications/aging-twenty-first-century
van Dullemen, C. E. (2017). The politics of aging, global challenges and national solutions. *Journal of Aging and Neuropsychology., 01*(2017), 1–4. https://doi.org/10.29011/JANP-104.100004
Vollset, S. E., Goren, E., Yuan, C.-W., Cao, J., Smith, A. E., Hsiao, T., et al. (2020). Fertility, mortality, migration, and population scenarios for 195 countries and territories from 2017 to 2100: A forecasting analysis for the Global Burden of Disease Study. *The Lancet, 396*(10258), 17–23. https://doi.org/10.1016/S0140-6736(20)30677-2
WHO – World Health Organization. (2017). *Global strategy and action plan on aging and health*. World Health Organization.
WHO – World Health Organization. (2021a). *Aging and health*. https://www.who.int/news-room/fact-sheets/detail/aging-and-health
WHO – World Health Organization. (2021b). *Ageism is a global challenge*. https://www.who.int/news/item/18-03-2021-ageism-is-a-global-challenge-un
Wittgenstein Centre. (2022). *Wittgenstein Centre Human Capital Data Explorer*. http://dataexplorer.wittgensteincentre.org/wcde-v2/
Yaşar, E., & Yaşar, M. (2018). Global wealth and aging population. In R. Yilmaz & G. Löschnigg (Eds.), *Studies on Balkan and near eastern social sciences* (Vol. 2, pp. 47–59). Peter Lang.
Zimmer, Z. (2016). *Global aging in the twenty-first century: Challenge, opportunities and implicationss*. Routledge.

Leonid Grinin is a Senior Research Professor at Centre for Stability and Risk Analysis at HSE University in Moscow and a Senior Research Professor at the Institute of Oriental Studies, Russian Academy of Sciences. He is co-editor of the international journals *Social Evolution & History* and the *Journal of Globalization Studies*, as well as co-editor of the international yearbooks *Evolution, History & Mathematics*, and *Kondratieff Waves*. He is author of more than 600 publications, including 33 monographs in Russian, English, Spanish, German, and Chinese. His current research focuses on comparative political studies, theory of revolution, political anthropology, global economy, global history, historical sociology, and futurology. Among his monographs are *Great Divergence and Great Convergence. A Global Perspective* (Springer, 2015; with Andrey Korotayev), *The Cybernetic Revolution and the Forthcoming Epoch of Self-Regulating Systems* (Uchitel, 2016, with Anton Grinin), *Economic Cycles, Crises, and the Global Periphery* (Springer, 2016; with Andrey Korotayev and Arno Tausch), *Big History of Globalization. The Emergence of a Global World System* (Springer 2019, in co-authorship with Julia Zinkina, David Christian, and others), *Islamism, Arab Spring, and the Future of Democracy. World System and World Values Perspectives* (Springer, 2019; with Andrey Korotayev and Arno Tausch). In 2004 he was awarded as a winner of the best publication of *Socis* journal during the year for the article "People of Publicity—a New Social Layer?". In 2012 he was awarded with the Gold Kondratieff Medal by the International N. D. Kondratieff Foundation. In 2016 he received the Belyaev award for the book *From Bifaces to Nanorobots* (the book published in Russian in co-authorship with Anton Grinin).

Anton Grinin is a Russian scientist specializing in the study of modern technological trends and futures studies, as well as evolutionist and philosopher. He has a PhD in Biology and works as a Research Fellow for Lomonosov Moscow State University, Russia, and at the Centre for Stability and Risk Analysis at HSE University in Moscow. He is the author of more than 100 publications in Russian and English, including 3 monographs. Awards: Gold Kondratieff Young Scholars Medal and Alexander Belyaev Literature Award. Anton Grinin focuses on the issues of the technological revolution (the Cybernetic revolution) evolving in the current decades of the twenty-first century and the resultant profound transformations in economy and society as well as related economic, political, ethic-legal and other risks, including political revolutions. He also investigates the correlation between the Cybernetic revolution and Kondratieff waves as well as between the Cybernetic revolution and global aging.

Andrey Korotayev heads the Centre for Stability and Risk Analysis at HSE University in Moscow, Russia. He is also a Senior Research Professor at the Institute of Oriental Studies and Institute for African Studies, Russian Academy of Sciences, as well as at the Faculty of Global Studies of Lomonosov Moscow State University and at the Russian Presidential Academy of National Economy and Public Administration in Moscow. He has authored or co-authored over 650 scholarly publications, including such monographs as *Ancient Yemen* (Oxford University Press, 1995), *World Religions and Social Evolution of the Old World Oikumene Civilizations: A Cross-Cultural Perspective* (The Edwin Mellen Press, 2004), *Introduction to Social Macrodynamics: Compact Macromodels of the World System Growth* (URSS, 2006; with Daria Khaltourina and Artemy Malkov), *Introduction to Social Macrodynamics: Secular Cycles and Millennial Trends* (URSS, 2006; with Daria Khaltourina and Artemy Malkov), *Great Divergence and Great Convergence. A Global Perspective* (Springer, 2015; with Leonid Grinin), *Economic Cycles, Crises, and the Global Periphery* (Springer, 2016; with Leonid Grinin and Arno Tausch), *Islamism, Arab Spring, and the Future of Democracy. World System and World Values Perspectives* (Springer, 2019; with Leonid Grinin and Arno Tausch). He is a laureate of the Russian Science Support Foundation Award in 'The Best Economists of the Russian Academy of Sciences' Nomination (2006); in 2012 he was awarded with the Gold Kondratieff Medal by the International N. D. Kondratieff Foundation.

Part III

Problems, Forecasts, Solutions (Technology, Economics, Socio-Political Development)

Technologies: Limitless Possibilities and Effective Control

Leonid Grinin and Anton Grinin

Abstract

This chapter examines technologies' current and future development in the framework of the Cybernetic Revolution—the third of the largest production (or technological) revolutions after the Agrarian and Industrial ones. The Cybernetic Revolution is a fundamental transition from industrial production to the production of services and goods based on the widespread implementation of self-regulating systems, that is, systems that can function in the absence or with minimum involvement of people and independently make complex decisions. This transition has already started and will continue up to the 2070s. The Cybernetic Revolution began its active development in the 1950s and has now finished its modernization phase. At the moment, the key technologies are information and communication technology and artificial intelligence, whose role in society is gradually increasing, and they come with benefits and potential risks. However, Grinin & Grinin assume that from the 2030s, the new—final—phase of the Cybernetic Revolution will start. Its major technological breakthroughs will lead to self-regulating systems' formation and widespread implementation. So, Grinin & Grinin assume that new technologies will emerge. They forecast that it will be a set of technological spheres, and the MANBRIC complex/convergence is taking shape and will actively develop in the final phase of the Cybernetic Revolution (in the 2030s–2070s). The MANBRIC is an abbreviation formed from the initial letters of the seven breakthrough areas:

L. Grinin (✉)
HSE University, Moscow, Russia

Institute of Oriental Studies, Russian Academy of Sciences, Moscow, Russia

A. Grinin
HSE University, Moscow, Russia

Faculty of Global Studies, Lomonosov Moscow State University, Moscow, Russia

V. Sadovnichy et al. (eds.), *Reconsidering the Limits to Growth*, World-Systems Evolution and Global Futures, https://doi.org/10.1007/978-3-031-34999-7_8

Medicine-Additive-Nano-Bio-Robotics-Info-Cognitive technologies. These technological fields closely interact and corroborate each other and will continue to do so increasingly in the future. Due to its specific characteristics, medicine will be an integral part of the MANBRIC complex. Grinin & Grinin also offer some scenarios for further technological development. They significantly depend on the areas where technological breakthroughs will start. The main developmental scenario is presented as a breakthrough that will occur in the 2030s in the field of medicine, especially at the nexus of its new directions and some areas of the MANBRIC. There will be the introduction of innovations based on self-regulating systems in various fields of social activity (economy, medicine, biology, and socio-administrative structures). Grinin & Grinin describe the most favorable scenario and recommend how to move toward this scenario.

Let us start by saying that, in our opinion, there are not enough systematic studies on the future development of technologies and their impact on the society.[1] However, some researchers present interesting scenarios, and there are also some (albeit too few) works that provide consistent forecasts of technological development based on identified developmental trends (Modis, 1999; Fukuyama, 2002; Martino, 2003; Randers, 2012; Rifkin, 2011; Fücks, 2013; Farmer & Lafond, 2015; Schwab, 2016; von Weizsäcker & Wijkman, 2018; Schwab & Malleret, 2020). Unfortunately, however, these scenarios and forecasts are often very contradictory.

One may divide the authors of these works into several categories.

The first one may be called the technological hyperoptimists (or "machine superoptimizers") (Eden et al., 2012: 127; see also Linstone, 2014). They believe that the development of AI will soon result in a complete and radical change of humans as a biological species and of humanity in general (e.g., Yampolskiy & Fox, 2012; Kurzweil, 2004, 2005).

The second category is formed by technological pessimists—see, for example, Maddison (2007), or Teulings and Baldwin (2014) who believe that at present we deal with a slowdown of scientific progress [see also LePoire, 2005; Phillips, 2011; the Huebner–Modis debate on the possible declining trend in worldwide innovation (Huebner, 2005; Modis, 2005)].

The third category is the pessimist-alarmists who claim that technological progress proceeds too fast, and this threatens humanity with disasters. Among them are many authors of reports to the Club of Rome (e.g., Meadows et al., 1972, 2004; Wijkman & Rockström, 2012; Bardi, 2014).

The fourth category is the balanced optimists and pessimists who realize that technological progress brings not only achievements but also serious problems that

[1] Most of the researches are related to forecasts of the development of some large areas, such as the development of bio-, nano- and information technologies (e.g., Venkatesh et al., 2003), or new fields such as nanomedicine (Moghimi, 2005), cloning (Gurdon & Colman, 1999), or nanorobots (Mallouk & Sen, 2009).

humanity has not faced before (e.g., Gabor et al., 1978; Scliaff & Friedrichs, 1982; Pestel, 1989; Rifkin, 2011; Fücks, 2013; Schwab, 2016; von Weizsäcker & Wijkman, 2018; Lovins et al., 2018; Randers, 2012; Randers et al., 2018; Berg, 2019; Komlosy 2021, 2023). And different researchers have different proportions of optimism and pessimism.

In the latter category, we may distinguish careful optimists who urge to think in advance of the negative consequences without disregarding the progress (e.g., Fukuyama, 2002; Schwab & Malleret, 2020). The authors of the present chapter also belong to this category and think that we should not be afraid of the future but be careful and try to prepare for very serious problems and consequences.

1 Current Situation

At the moment, technological development is determined by the continuing Cybernetic Revolution. *The Cybernetic Revolution is a fundamental transition from the industrial production principle to the production of services and goods based on the widespread implementation of self-regulating systems, that is, systems that can not only function in the absence (or with minimal involvement) of humans, but can also make complex decisions independently* (see Grinin, 2006; Grinin & Grinin, 2015a, 2015b, 2016; Grinin et al., 2017a, 2017b, 2020, 2021).[2] This revolution is called cybernetic after the scientific discipline of Cybernetics. This revolution will result in profound changes not only in production management but also in the management of all other spheres of life on the basis of self-regulating systems, since Cybernetics is the science concerned with such regulatory and targeted systems. We base our analysis of these systems on some ideas from the field which studies the control of various complex regulated systems (see, e.g., Wiener, 1948; Ashby, 1956; Beer, 1959, 1994; Foerster & Zopf, 1962; Umpleby & Dent, 1999; Tesler, 2004; McClelland & Fararo, 2006; Dupuy, 2009; Wieser & Slunecko, 2014).[3]

[2] We should point out that automatization which was considered as the major characteristics of the start of the Cybernetic Revolution (see, e.g., Lilley, 1966, 1976; Philipson, 1962; Bernal, 1965; Benson & Lloyd, 1983) appears to be only the initial level of future self-regulating systems.

[3] There are a number of works and books on the coming revolution which different scholars denote by different terms; however, we denote it as the Cybernetic Revolution. Of course, any term is always conventional, yet, it emphasizes the leading qualities of a phenomenon or a process. At the same time, it is not useful to simply enumerate the third, fourth or any other in succession industrial revolution with respect to a fundamentally new phenomenon, which are the current and future waves of technological innovation. In this context, we may mention "Third Industrial Revolution" (Rifkin, 2011) and "The Fourth Industrial Revolution" (Schwab, 2016). These books present interesting predictions, but their authors simply rely on empirical observations of current trends or reproduce someone else's forecasts and do not rely on a scientifically derived paradigm. Meanwhile, forecasts based mainly on contemporary data or on profound and long-term trends and regularities are completely different things. From the titles it can be deduced that these authors actually believe that all three or four revolutions have similar nature, which, as the reader will see, is not the case at all since the nature of every technological revolution is unique.

The Cybernetic Revolution has already passed its initial phase (in the 1950s–1980s) when information technologies emerged,[4] and it is currently completing its intermediate phase of modernization (the 1990s–2020s). So, the most important part of the development of technologies currently includes information and digital technologies, which act as an integral part of most technological areas. This diffusion of information, digital, and communication technologies has recently acquired a much larger scale due to various forms of artificial intelligence (neural networks, remote contact, linguistic systems, face recognition systems, location monitoring, determining preferences, and many others). Artificial intelligence (AI) is and will remain an indispensable part of many self-regulating systems, just as the electric motor is an indispensable part of many machines. Most recently, AI has been increasingly affecting social relations, control, and even administrative relations. At present, information and digital technologies provide the most striking examples of the development of self-regulating systems. The prospects for the development of information-digital and communication technologies are great; however, these technologies alone seem to be unable to provide a future technological breakthrough, but only as an integral part of other technologies (for the analysis of different aspects of AI see e.g., Abi Younes et al., 2020; Eden et al., 2012; Graglia & Von Huelsen, 2020; Gregg, 2018; Hengstler et al., 2016; Liu et al., 2020; Michalski et al., 2013; Montes & Goertzel, 2019; Plebe & Perconti, 2020; Russell et al., 2003).[5]

Among the most promising technologies that may become leading in the future are medical and biotechnologies which are dynamically developing, with the latter evolving in several directions: "red" biotechnologies (pharmaceuticals and medicine), "green" biotechnologies (related to agriculture and environmental improvement), and "white" biotechnologies. The latter is used to produce a wide range of products obtained via biocatalysis and fermentation. Good future perspectives seem also to await robotics, including drones and self-driving cars. Green energy (solar, wind) and related industries are also actively developing. However, despite huge investments, there are rather considerable technological obstacles here. Carbon energy is drifting from the more polluting fuels (coal and oil) to the less polluting gas; liquefied gas technologies and means of its delivery are developing

[4]About the initial phase of the Cybernetic Revolution, which was called the Scientific-Technical Revolution (see, e.g., Bernal, 1965; Philipson, 1962; Benson & Lloyd, 1983; Sylvester & Klotz, 1983).

[5]It is important to point out, that due to this rapid development, artificial intelligence technology poses a major challenge for the humanity (Eden et al., 2012). In the recent years, communication technologies, data analysis and surveillance technology have considerably advanced. As a result, the problem has become even more acute. A number of works have already been devoted to the analysis of various aspects of this problem in the present and future (e.g. Westin, 1966; Ashman et al., 2014; Cecere et al., 2015; Moustaka et al., 2019; Schwartz, 1999; Solove, 2008; Brammer et al., 2020; Alharbi, 2020).

dynamically. However, a rollback has recently started due to the energy crisis, so coal consumption has started to grow again.[6]

Despite the active development in these directions, the current phase of the Cybernetic Revolution is inferior to its initial phase (especially the period of the 1950s–1970s) in terms of the pace of technological development, so we can consider it a period of a slowing down technological process (see, e.g., Polterovich, 2009; Grinin et al., 2020; Grinin & Korotayev, 2015). We see that technological process in many areas, including engineering and transport, energy, chemicals, and other industries, as well as many, if not most, types of services, is not accelerating but rather slowing down. This situation is fully explained (and forecasted) by the theory of production/technological revolutions.[7]

2 Technology as the Most Important Factor of Historical Development

In general, the entire human history, especially during the last few centuries, may be regarded (albeit with significant qualifications) as a history of scientific and technological achievements (Kremer, 1993; Galor & Tsiddon, 1997; Kayal, 1999; Galor & Weil, 2000; Carree, 2003; Phillips, 2011; Grinin & Grinin, 2015a, 2016; Grinin et al., 2017a, 2017b, 2020, 2021; Widdowson, 2021). According to the theory of production revolutions, the Cybernetic Revolution is the third largest technological revolution in history after the Agrarian Revolution (12,000–5,000 years ago) and the industrial revolution (the 1430s–1830s).[8] These production revolutions radically changed production processes, formed new principles of production (craft-agrarian and trade-industrial ones, respectively), and provided unprecedented scope for the development of new productive forces, which ultimately and completely changed human society and the very way of life. Production revolutions are rare but recurring phenomena with typical similarities (Grinin, 2007; Grinin & Grinin, 2016; Grinin et al., 2017a). In particular, they have a similar structure and consist of three phases. The initial phase lays the foundations for a new production principle and new production systems and their complete cycle. The modernization phase enormously expands the scope of the new production principle and improves its technologies. The final phase of the production revolution develops a new production principle

[6] For more information on modern energy problems see chapter "Climate and Energy. Energy Transition Scenarios and Global Temperature Changes Based on Current Technologies and Trends" (Akaev & Davydova, 2023, this volume).

[7] But there is also positive news regarding energy, see chapter "Climate and Energy. Energy Transition Scenarios and Global Temperature Changes Based on Current Technologies and Trends" (Akaev & Davydova, 2023, this volume).

[8] About the Agrarian or Neolithic Revolution (see, e.g., Childe, 1934, 1944, 1948; Reed, 1977; Harris & Hillman, 1989); about the Industrial Revolution (see, e.g., Cipolla, 1976; Allen, 2009, 2011; Grinin & Korotayev, 2015; about evolution of Marxist theories on Industrial Revolution see Grinin and Grinin 2023b).

until its full potential is revealed, after which the production principle advances to its mature forms, completely transforming all sectors of the economy and spheres of life.

Technological development is the most important factor in human history. It is largely due to the development of technologies that expanded the resource base and improved living conditions that the population of the Earth has grown 10,000 times since the Neolithic. Throughout the history of humankind, technologies have directly influenced all spheres of life, including the economy, social sphere, and politics.[9] At present, this influence has increased even more.

3 Scenarios of Technological Development in the Twenty-First Century

The ongoing Cybernetic Revolution determines the main scenario for developing technologies in the twenty-first century. As a result, there will be revolutionary changes in production management along with other spheres of life based on the mass introduction of self-regulating systems. At the same time, such systems will evolve within technology and biology, medicine (including robots, nano- and biochips, artificial organs, etc., and also in control of health and aging[10]), in genetic engineering influencing the human body, in the functioning of waste-free production and self-cleaning (including self-cleaning with the help of biotechnologies and nantechnologies). These systems will also spread in socio-administrative areas, regulating some relations in society, especially between administrative, controlling, and governing bodies of power, on the one hand, and people, on the other [we call such self-regulating systems socio-technical (Grinin et al., 2021)].[11]

In this respect, we should emphasize the following two points: (1) the final phase of the Cybernetic Revolution will inevitably accelerate the pace of technological progress (Ibid.); (2) regardless of the areas in which major technological breakthroughs begin, they will eventually lead to the formation and widespread implementation of self-regulating systems.

However, future developmental scenarios significantly depend on the areas where these technological breakthroughs will start, and the first cycles of activity based on self-regulating systems will be formed. The main developmental scenario is described as a breakthrough in the 2030s in the field of medicine, especially at the nexus of its new directions and some areas of MANBRIC-technologies (see below)

---

[9]For the relationship between historical and technological development see chapter "Modeling Social Self-Organization and Historical Dynamics. Global Phase Transitions" (Malkov et al., 2023, this volume; see also Grinin & Grinin, 2023a, this book and Grinin & Grinin, 2023b).

[10]See chapter "Global Aging–an Integral Problem of the Future. How to Turn a Problem into a Development Driver?" (Grinin et al., 2023a, this volume) for details.

[11]For more about these systems see chapter "Analyzing Social Self-Organization and Historical Dynamics. Future Cybernetic W-Society: Socio-Political Aspects" (Grinin & Grinin, 2023, this volume).

especially such as bio- and nanotechnologies, and perhaps in collaboration with information and digital technologies, robotics, as well as additive and cognitive technologies. All these technologies are used very actively in modern medicine, including additive technologies (i.e., 3D printers) for creating artificial organs and tissues, nanotechnologies for creating vaccines and drugs, and robots for performing operations. In the future, the relationship of all these areas with medicine will strengthen noticeably, and in some areas, they will fundamentally increase. So, we may assume that the MANBRIC complex or MANBRIC convergence is taking shape and will actively develop during the final phase of the Cybernetic Revolution (in the 2030s–2070s). MANBRIC is an abbreviation consisting of the first letters of the seven leading areas: Medicine-Additive-Nano-Bio-Robotics-Info-Cognitive technologies.[12] These technological fields are and will closely interact and support each other.[13]

Of course, there are also differences in defining the leading directions of development of the coming revolution. Some researchers (e.g., Rifkin, 2011; Fücks, 2013) assume it is the new green energy, while Schwab (2016) believes this will be a complex of leading spheres, but primarily the AI (he is not original here, see, e.g., Kurzweil, 2004, 2005). Others put focus on biotechnological development (Sylvester & Klotz, 1983; Fukuyama, 2002; see also Nefiodow, 1996, 2017). With respect to the sixth technological paradigm, there is a widely used idea connected with the notion of NBIC[14]-technology (or NBIC-convergence) (see Lynch, 2004; Bainbridge & Roco, 2005; Dator, 2006; Kovalchuk, 2011; Akaev, 2012). There are also some researchers (e.g., Jotterand, 2008) who give this role to another set of technological directions—GRAIN (Genomics, Robotics, Artificial Intelligence, and Nano-technology).

Our arguments make us believe that the system will be wider and have a different structure.

Medicine will be the central integrative link in this complex. The leading role of medicine in the final phase of the Cybernetic Revolution (at least in 2030–2040/2050s) is accounted for by the following factors:

1. population aging and its consequences: (a) the need for additional labor force in the high income and some low-income countries; (b) an increasing pension and

---

[12]The order of the letters in the acronym does not reflect our understanding of the relative importance of areas of the complex. For example, biotechnologies will be more important than nanotechnologies, let alone additive manufacturing. The order is determined simply by the convenience of pronunciation.

[13]This became clear during the period of COVID-19 pandemic. For examples of the increasing interdependence between medicine, biotechnology, information technology, and additive technologies as well as nanotechnologies (see Choong et al., 2020; Chauhan et al., 2020; Palestino et al., 2020; Tang et al., 2021; Weiss et al., 2020).

[14]Nano-Bio-Info-Cogno.

other social burdens on budgets and the need to raise the retirement age, which requires elderly people to have sufficiently good health to continue their work;[15]
2. growing incomes and expanding middle class in developing countries, which should lead to an increase in culture and health costs;
3. the ever-increasing accumulation of assets in pension and social funds, which to a large extent should be used specifically to improve the quality of life of the elderly, especially medicine.

All these factors worked until 2020, and they will continue to work. But after 2020, the pandemic (and the danger of new pandemics) has been added to them, and thus, dramatically increased the interest in health and medicine issues in society. Thus, the coronavirus pandemic has greatly accelerated the development of medicine and biotechnologies, bringing them to the forefront of the development of technology and science (see Grinin et al., 2021). The coronavirus pandemic is most likely bringing the start of the final phase of the Cybernetic Revolution closer, and with it, perhaps, the beginning of the upward phase of the sixth long (Kondratieff) wave and, accordingly, the sixth technological paradigm (Grinin et al., 2017a).

As a result, medical technologies will rapidly develop under the influence of the aging population (Phillips, 2011), and this will expand the search for opportunities to create "smart," self-regulating systems, including robots, which can largely replace human labor, especially in the service sector (Frey & Osborne, 2017), including complex services (e.g., in the fields of elder care, education, medicine, etc.) (DeCanio, 2016).

The indicated above acceleration of technological development will be observed (with possible fluctuations) during the final phase of the Cybernetic Revolution and for some time after it (that is, in the last third of the twenty-first century), since the new scientific cybernetic production principle, after acquiring its mature features, will actively expand and introduce self-regulating systems in more and more areas. However, by the end of the twenty-first century and the beginning of the twenty-second century, we can assume a slowdown in the pace of technological development, which will be associated with large-scale population aging (see Grinin et al., 2020).[16]

Additional scenarios of technological development in the twenty-first century may depend on many factors. In particular, a sharp aggravation of the climate problem can bring to the fore the task of switching to new energy sources, which will significantly change many trends (in particular, perhaps, the rapid development of self-driving electric vehicles). Currently, the focus of European countries (as well

[15] For more details see chapter "Global Aging–an Integral Problem of the Future. How to Turn a Problem into a Development Driver?" (Grinin et al., 2023a, this volume).

[16] See chapter "Socio-Political Transformations. A Difficult Path to Cybernetic Society" (Grinin et al., 2023b, this volume) and chapter "Modeling Social Self-Organization and Historical Dynamics. Global Phase Transitions" (Malkov et al., 2023, this volume) for more details.

as the United States) is put on the development of "green" energy and a sharp increase in the ability to store electricity. This can also be viewed as an alternative scenario for future technological development (with the reduction of carbon energy). Our baseline carbon-free energy scenario is outlined in chapter "Climate and Energy. Energy Transition Scenarios and Global Temperature Changes Based on Current Technologies and Trends" (Akaev & Davydova, 2023, this volume). At the same time, we believe that the transition to mass implementation of self-regulating systems will largely solve the problems of energy efficiency and energy saving, as well as other important technological problems in the "green" energy industry. This transition will create a system for the utilization of CO_2, making significant progress toward reducing carbon emissions. But it is also possible to consider switching to fundamentally new energy sources, such as hydrogen energy and thermonuclear fusion. The latter can provide large amounts of cheap energy, but the prospects of these sources are still unclear.

Another alternative scenario is the one in which information and communication technology (ICT and AI) will continue to lead. This idea is well expressed in the theory of the fourth industrial revolution (Schwab, 2016), but the potential of new areas, including medicine and biotechnology, is clearly underestimated there.

Let us mention another point that could be potentially important in the future for providing food for the growing population of the Earth, which, according to the forecasts of the UN and others, will continue to grow for a number of decades, and possibly until the end of the twenty-first century and later (Alkema et al., 2011; Raftery et al., 2012; Podlazov, 2017; Lutz et al., 2018; UN Population Division, 2019, 2022; Korotayev, 2020; Vollset et al., 2020; Wittgenstein Centre, 2022).[17] Here, biotechnologies (the marine ones, let us call them "blue") could play a colossal role. We are talking about the widespread artificial cultivation of seafood (the so-called aquaculture). It is a growing area but has so far played a limited role. Due to the depletion of fish and seafood in the world's oceans, a transition is possibly akin to the one that took place thousands of years ago, from an appropriating economy to a producing one (this transition is called the Agrarian Revolution, see above). Today, the volume of aquaculture production is already approaching 50% of the world's production of fish and seafood (and amounts to tens of millions of tons). But potentially, it is possible to produce several times more food (moreover, protein) at sea farms (see, e.g., Weizsäcker & Wijkman, 2018). With vast undeveloped water areas and new technologies, the transition to artificial farming can solve the problem of a growing population.

In general, we can draw an important conclusion that the Cybernetic Revolution, its completion, and the transition of the cybernetic production principle into mature phases will introduce a very large number of smart and self-regulating systems into

[17] See chapter "Demography. Toward Optimization of Demographic Processes" (Korotayev et al., 2023b, this volume), chapter "Africa–the Continent of the Future. Challenges and Opportunities" (Grinin & Korotayev, 2023, this volume), and chapter "Modeling Social Self-Organization and Historical Dynamics. Africa's Futures" (Korotayev et al., 2023a, this volume).

our lives, that can significantly advance our economy toward reducing the exploitation of resources, growing energy efficiency, solving the problem of labor shortage, etc. *In fact, huge opportunities for saving skilled labor and resources are opening up. However, with the spread of self-regulating systems developed on the basis of highly advanced artificial intelligence technologies, there sharply increases the danger that human basic rights will be violated while a total control will be established over their behavior and consumption etc. And here the society must find the ways to effectively control the faceless and anonymous technological power that has hugely increased and can suppress freedom, initiative, and human capabilities.*

In this regard, the question arises of how to switch to a favorable scenario in the development of technologies. A clear understanding is needed that the increasing pace of technological progress leads to inevitable changes in other areas of development. Therefore, humankind needs: (a) forecasting and possibly taking preliminary actions against negative changes; (b) political and legal mechanisms, in addition to the development of a scientifically-based and practice-oriented perspective; (c) a comprehensive, systematic view of the problems for more profitable channeling of technological development. The latter actually involves the international sphere. Attempts to channel technological progress have been made for a long time; this is especially evident in the military and green energy in recent decades. However, much needs to be institutionalized here (including expertocracy, the combination of private and public interest, optimization of growth, and different developmental pace of different countries).[18] We can provide many examples here. In general, there is a need to channel the development of productive forces toward greater friendliness to nature; to consider nature protection as the most important part of technology development. For example, sea farms, which are mentioned in the "Come on!" report (Weizsäcker & Wijkman, 2018) can become a means to preserve biodiversity in the ocean.

Thus, technology is the most important component of the social system, which largely determines the nature and direction of changes in all spheres of society.

References

Abi Younes, G., Ayoubi, C., Ballester, O., Cristelli, G., de Rassenfosse, G., Foray, D., et al. (2020). COVID-19: Insights from innovation economists. *Science and Public Policy, paaa036*. https://doi.org/10.1093/scipol/scaa028

Akaev, A. (2012). Matematicheskiye osnovy innovatsionno-tsiklicheskoy teorii ekonomicheskogo razvitiya Shumpetera – Kondrat′yeva. In A. Akaev et al. (Eds.), *Kondrat′yevskiye volny: aspekty i perspektivy* (pp. 110–135). Uchitel.

Akaev, A., & Davydova, O. (2023). Climate and energy. Energy transition scenarios and global temperature changes based on current technologies and trends In: V. Sadovnichy et al. (Eds.),

[18]See chapter "High Income and Low Income Countries. Toward a Common Goal at Different Speeds" (Grinin et al., 2023c, this volume).

Reconsidering the limits to growth. A report to the Russian Association of the Club of Rome (pp. 53–70). Springer. https://doi.org/10.1007/978-3-031-34999-7_4

Alharbi, F. S. (2020). Dealing with data breaches amidst changes in technology. *International Journal of Computer Science and Security (IJCSS), 14*(3), 108–115.

Alkema, L., Raftery, A. E., Gerland, P., Clark, S. J., Pelletier, F., Buettner, T., & Heilig, G. K. (2011). Probabilistic projections of the total fertility rate for all countries. *Demography, 48*(3), 815–839. https://doi.org/10.1007/s13524-011-0040-5

Allen, R. C. (2009). *The British industrial revolution in global perspective*. Cambridge University Press.

Allen, R. C. (2011). *Global economic history: A very short introduction*. Oxford University Press.

Ashby, R. (1956). *An introduction to cybernetics*. Chapman and Hall.

Ashman, H., Brailsford, T., Cristea, A. I., Sheng, Q. Z., Stewart, C., Toms, E. G., & Wade, V. (2014). The ethical and social implications of personalization technologies for e-learning. *Information & Management, 51*(6), 819–832. https://doi.org/10.1016/j.im.2014.04.003

Bainbridge, W. S., & Roco, M. (2005). *Managing Nano-Bio-Info-Cogno innovations: Converging technologies in society*. Springer.

Bardi, U. (2014). *Extracted: How the quest for mineral wealth is plundering the planet. A report to the Club of Rome*. Chelsea Green Publishing.

Beer, S. (1959). *Cybernetics and management*. English Universities Press.

Beer, S. (1994). *Decision and control: The meaning of operational research and management cybernetics*. Wiley.

Benson, I., & Lloyd, J. (1983). *New technology and industrial change: The impact of the scientific-technical revolution on labour and industry*. Kogan Page.

Berg, C. (2019). *Sustainable action: Overcoming the barriers. A report to the Club of Rome*. Routlege.

Bernal, J. D. (1965). *Science in history* (3rd ed.). Hawthorn Books.

Brammer, S., Branicki, L., & Linnenluecke, M. (2020). COVID-19, societalization and the future of business in society. *Science of the Total Environment, 34*(4), 2–7. https://doi.org/10.5465/amp.2019.0053

Carree, M. A. (2003). Technological progress, structural change and productivity growth: A comment. *Structural Change and Economic Dynamics, 14*(1), 109–115.

Cecere, G., Le Guel, F., & Soulié, N. (2015). Perceived internet privacy concerns on social networks in Europe. *Technological Forecasting and Social Change, 96*, 277–287. https://doi.org/10.1016/j.techfore.2015.01.021

Chauhan, G., Madou, M. J., Kalra, S., Chopra, V., Ghosh, D., & Martinez-Chapa, S. O. (2020). Nanotechnology for COVID-19: Therapeutics and vaccine research. *ACS Nano, 14*(7), 7760–7782. https://doi.org/10.1021/acsnano.0c04006

Childe, V. G. (1934). *New light on the most ancient east*. Routledge & Kegan.

Childe, V. G. (1944). *Progress and archaeology*. Watts.

Childe, V. G. (1948). *What happened in history*. Penguin Books.

Choong, Y. Y. C., Tan, H. W., Patel, D. C., Choong, W. T. N., Chen, C.-H., Low, H. Y., et al. (2020). The global rise of 3D printing during the COVID-19 pandemic. *Nature Reviews Materials, 5*(9), 637–639. https://doi.org/10.1038/s41578-020-00234-3

Cipolla, C. M. (Ed.). (1976). *The industrial revolution, 1700–1914*. Harvester Press.

Dator, J. (2006). Alternative futures for K-waves. In T. C. Devezas (Ed.), *Kondratieff waves, warfare and world security* (pp. 311–317). IOS Press.

DeCanio, S. J. (2016). Robots and humans – complements or substitutes? *Journal of Macroeconomics, 49*, 280–291.

Dupuy, J. (2009). *On the origins of cognitive science. The mechanization of the mind*. MIT Press.

Eden, A. H., Moor, J. H., Soraker, J. H., Steinhart, E., & (Eds.). (2012). *Singularity hypotheses: A scientific and philosophical assessment*. Springer. https://doi.org/10.1007/978-3-642-32560-1

Farmer, J. D., & Lafond, F. (2015). *How predictable is technological progress?* http://arxiv.org/abs/1502.05274

Frey, C. B., & Osborne, M. A. (2017). The Future of Employment: How Susceptible are Jobs to Computerisation? *Technological Forecasting and Social Change, 114*, 254–280.
Fücks, R. (2013). *Intelligent Wachsen. Die Grune Revolution*. Carl Hanser Verlag.
Fukuyama, F. (2002). *Our post-human future: Consequences of the bio-technology revolution*. Farrar, Straus, and Giroux.
Gabor, D., Colombo, U., King, A., & Galli, R. (1978). *Beyond the age of waste*. Pergamon Press.
Galor, O., & Tsiddon, D. (1997). Technological progress, mobility and economic growth. *The American Economic Review, 87*(3), 363–382.
Galor, O., & Weil, D. N. (2000). Population, technology, and growth: From Malthusian stagnation to the demographic transition and beyond. *American Economic Review, 90*(4), 806–828.
Graglia, M. A. V., & Von Huelsen, P. G. (2020). The sixth wave of innovation: Artificial intelligence and the impacts on employment. *Journal on Innovation and Sustainability RISUS, 11*(1), 3–17. https://doi.org/10.23925/2179-3565.2020v11i1p3-17
Gregg, B. (2018). The coming political: Challenges of artificial intelligence. *Digital Culture & Society, 4*(1), 157–180. https://doi.org/10.25969/mediarep/13530
Grinin, L. (2006). *Proizvoditel'nyye sily i istoricheskiy protsess*. URSS.
Grinin, L. (2007). Production revolutions and periodization of history: A comparative and theoretic-mathematical approach. *Social Evolution & History, 6*(2), 11–55.
Grinin, L., & Grinin, A. (2015a). Kiberneticheskaya revolyutsiya i shestoy tekhno-logicheskiy uklad. *Istoricheskaya psikhologiya i sotsiologiya istorii, 8*(1), 172–197.
Grinin, L., & Grinin, A. (2015b). *Ot rubil do nanorobotov. Mir na puti k epokhe samoupravlyayemykh sistem (istoriya tekhnologiy i opisaniye ikh budushchego)*. Uchitel.
Grinin, L., & Grinin, A. (2016). *The cybernetic revolution and the forthcoming epoch of self-regulating systems*. Uchitel.
Grinin, L., & Grinin, A. (2023a). Analyzing social self-organization and historical dynamics. Future cybernetic W-society: Socio-political aspects. In: V. Sadovnichy et al. (Eds.), *Reconsidering the limits to growth. A report to the Russian Association of the Club of Rome* (pp. 491–519). Springer. https://doi.org/10.1007/978-3-031-34999-7_21
Grinin, L., & Grinin, A. (2023b). Historical Materialism: Does the Concept Have a Future? *Social Evolution & History, 22*(1), 143–178. https://doi.org/10.30884/seh/2023.01.07
Grinin, L., Grinin, A., & Korotayev, A. (2017a). Forthcoming Kondratieff wave, cybernetic revolution, and global aging. *Technological Forecasting and Social Change, 115*, 52–68. https://doi.org/10.1016/j.techfore.2016.09.017
Grinin, L., Grinin, A., & Korotayev, A. (2017b). The MANBRIC-technologies in the forthcoming technological revolution. In T. Devezas, J. Leitão, & A. Sarygulov (Eds.), *Industry 4.0 – Entrepreneurship and structural change in the new digital landscape: What is coming on along with the fourth industrial revolution* (pp. 243–261). Springer. https://doi.org/10.1007/978-3-319-49604-7_13
Grinin, L., Grinin, A., & Korotayev, A. (2020). A quantitative analysis of worldwide long-term technology growth: From 40,000 BCE to the early 22nd century. *Technological Forecasting and Social Change, 155*, 1–19. https://doi.org/10.1016/j.techfore.2020.119955
Grinin, L., Grinin, A., & Korotayev, A. (2021). COVID-19 pandemic as a trigger for the acceleration of the cybernetic revolution, transition from E-government to E-state, and change in social relations. *Technological Forecasting & Social Change, 175*, 1–16. https://doi.org/10.1016/j.techfore.2021.121348
Grinin, L., Grinin, A., & Korotayev, A. (2023a). Global aging – An integral problem of the future. How to turn a problem into a development driver? In: V. Sadovnichy et al. (Eds.), *Reconsidering the limits to growth. A report to the Russian Association of the Club of Rome* (pp. 117–135). Springer. https://doi.org/10.1007/978-3-031-34999-7_7
Grinin, L., Grinin, A., & Malkov, S. (2023b). Socio-political transformations. A difficult path to cybernetic society. In: V. Sadovnichy et al. (Eds.), *Reconsidering the limits to growth. A report to the Russian Association of the Club of Rome* (pp. 169–189). Springer. https://doi.org/10.1007/978-3-031-34999-7_10

Grinin, L., & Korotayev, A. (2015). *Great divergence and great convergence. A global perspective*. Springer.

Grinin, L., & Korotayev, A. (2023). Africa – The continent of the future. Challenges and opportunities. In: V. Sadovnichy et al. (Eds.), *Reconsidering the limits to growth. A report to the Russian Association of the Club of Rome* (pp. 225–238). Springer. https://doi.org/10.1007/978-3-031-34999-7_13

Grinin, L., Malkov, S., & Korotayev, A. (2023c). High income and low income countries. Toward a common goal at different speeds. In: V. Sadovnichy et al. (Eds.), *Reconsidering the limits to growth. A report to the Russian Association of the Club of Rome* (pp. 207–224). Springer. https://doi.org/10.1007/978-3-031-34999-7_12

Gurdon, J. B., & Colman, A. (1999). The future of cloning. *Nature, 402*(6763), 743–746.

Harris, D., & Hillman, G. (Eds.). (1989). *An evolutionary continuum of people – Plant interaction. Foraging and farming. The evolution of plant exploitation*. Unwin Hyman.

Hengstler, M., Enkel, E., & Duelli, S. (2016). Applied artificial intelligence and trust—The case of autonomous vehicles and medical assistance devices. *Technological Forecasting and Social Change, 105*, 105–120. https://doi.org/10.1016/j.techfore.2015.12.014

Huebner, J. (2005). A possible declining trend for worldwide innovation. *Technological Forecasting and Social Change, 72*, 980–986.

Jotterand, F. (2008). *Emerging conceptual, ethical and policy issues in bionanotechnology*. Springer.

Kayal, A. (1999). Measuring the pace of technological progress: Implications for technological forecasting. *Technological Forecasting and Social Change, 60*(3), 237–245.

Komlosy, A. (2021). Entanglements of Catching-Up: Rethinking 'Industrial Revolution' from a Global Perspective. *Journal of Globalization Studies, 12*(1), 79–100. https://doi.org/10.30884/jogs/2021.01.05

Komlosy, A. (2023). Turn of an Era. Corona, Big Data and the Cybernetic Future. *Journal of Globalization Studies, 14*(1), 190–192. https://doi.org/10.30884/jogs/2023.01.12

Korotayev, A. (2020). The 21st century singularity in the big history perspective. A re-analysis. In A. Korotayev & D. LePoire (Eds.), *The 21st century Singularity and global futures. A big history perspective* (pp. 19–75). Springer. https://doi.org/10.1007/978-3-030-33730-8_2

Korotayev, A., Shulgin, S., Ustyuzhanin, V., Zinkina, J., & Grinin, L. (2023a). Modeling social self-organization and historical dynamics. Africa's futures. In: V. Sadovnichy et al. (Eds.), *Reconsidering the limits to growth. A report to the Russian Association of the Club of Rome* (pp. 461–490). Springer. https://doi.org/10.1007/978-3-031-34999-7_20

Korotayev, A., Malkov, S., & Musieva, J. (2023b). Demography. Toward optimization of demographic processes. In: V. Sadovnichy et al. (Eds.), *Reconsidering the limits to growth. A report to the Russian Association of the Club of Rome* (pp. 97–116). Springer. https://doi.org/10.1007/978-3-031-34999-7_6

Kovalchuk, M. V. (2011). Konvergentsiya nauk i tekhnologiy – proryv v budushcheye. *Rossiiskie nanotekhnologii, 3*, 134–158.

Kremer, M. (1993). Population growth and technological change: One million B.C. to 1990. *The Quarterly Journal of Economics, 108*(3), 681–716. https://doi.org/10.2307/2118405

Kurzweil, R. (2004). The law of accelerating returns. In C. Teuscher (Ed.), *Alan Turing: Life and legacy of a great thinker* (pp. 381–416). Springer. https://doi.org/10.1007/978-3-662-05642-4_16

Kurzweil, R. (2005). *The singularity is near: When humans transcend biology*. Penguin.

LePoire, D. J. (2005). Application of logistic analysis to the history of physics. *Technological Forecasting and Social Change, 72*(4), 471–479. https://doi.org/10.1016/S0040-1625(03)00044-1

Lilley, S. (1966). *Men, machines and history: The story of tools and machines in relation to social progress*. International Publishers.

Lilley, S. (1976). Technological progress and the industrial revolution, 1700–1914. In C. M. Cipolla (Ed.), *The industrial revolution, 1700–1914. The Fontana economic history of Europe* (pp. 187–254). The Harvest Press.
Linstone, H. (2014). Review of: Singularity hypotheses: A scientific and philosophical assessment, Amnon H. Eden, James H. Moor, Johnny H. Søraker, Eric Steinhart (Eds.), Springer Verlag (2012). *Technological Forecasting and Social Change, 82*, 226–227. https://doi.org/10.1016/j.techfore.2013.06.011
Liu, J., Chang, H., Forrest, J. Y.-L., & Yang, B. (2020). Influence of artificial intelligence on technological innovation: Evidence from the panel data of china's manufacturing sectors. *Technological Forecasting and Social Change, 158*, 1–12. https://doi.org/10.1016/j.techfore.2020.120142
Lovins, L. H., Wallis, S., Wijkman, A., & Fullerton, J. (2018). *A finer future. Creating an economy in service to life*. New Society Publishers.
Lutz, W., Goujon, A., Kc, S., Stonawski, M., & Stilianakis, N. (2018). *Demographic and human capital scenarios for the 21st century: 2018 assessment for 201 countries*. Publications Office of the European Union.
Lynch, Z. (2004). Neurotechnology and society 2010–2060. *Annals of the New York Academy of Sciences, 1031*, 229–233.
Maddison, A. (2007). *Contours of the world economy, 1–2030*. Oxford University Press.
Malkov, S., Grinin, L., Grinin, A., Musieva, J., & Korotayev, A. (2023). Modeling social self-organization and historical dynamics. Global phase transitions. In: V. Sadovnichy et al. (Eds.), *Reconsidering the limits to growth. A report to the Russian Association of the Club of Rome* (pp. 387–417). Springer. https://doi.org/10.1007/978-3-031-34999-7_18
Mallouk, T. E., & Sen, A. (2009). Powering nanorobots. *Scientific American, 300*(5), 72–77.
Martino, J. P. (2003). A review of selected recent advances in technological forecasting. *Technological Forecasting and Social Change, 70*(8), 719–733.
McClelland, K. A., & Fararo, T. J. (Eds.). (2006). *Purpose, meaning, and action: Control systems theories in sociology*. Palgrave Macmillan.
Meadows, D. H., Meadows, D. L., Randers, J., & Behrens, W. W. (1972). *The limits to growth: A report for the Club of Rome's project on the predicament of mankind*. Universe Books.
Meadows, D., Randers, J., & Meadows, D. (2004). *Limits to growth: The 30-year update*. Earthscan.
Michalski, R. S., Carbonell, J. G., & Mitchell, T. M. (2013). *Machine learning. An artificial intelligence approach*. Springer.
Modis, T. (1999). Technological forecasting at the stock market. *Technological Forecasting and Social Change, 62*(3), 173–202.
Modis, T. (2005). Discussion of Huebner article. *Technological Forecasting and Social Change, 72*, 987–1000.
Moghimi, S. M. (2005). Nanomedicine: Current status and future prospects. *The FASEB Journal, 19*(3), 311–330. https://doi.org/10.1096/fj.04-2747rev
Montes, G. A., & Goertzel, B. (2019). Distributed, decentralized, and democratized artificial intelligence. *Technological Forecasting and Social Change, 141*, 354–358. https://doi.org/10.1016/j.techfore.2018.11.010
Moustaka, V., Theodosiou, Z., Vakali, A., Kounoudes, A., & Anthopoulos, L. G. (2019). Enhancing social networking in smart cities: Privacy and security borderlines. *Technological Forecasting and Social Change, 142*, 285–300. https://doi.org/10.1016/j.techfore.2018.10.026
Nefiodow, L. (1996). *Der sechste Kondratieff. Wege zur Produktivität und Vollbeschäftigung im Zeitalter der Information*. Rhein-Sieg-Verlag.
Nefiodow, L. (2017). *The Sixth Kondratieff: A new long wave in the global economy*. CreateSpace.
Palestino, G., García-Silva, I., González-Ortega, O., & Rosales-Mendoza, S. (2020). Can nanotechnology help in the fight against COVID-19? *Expert Review of Anti-infective Therapy, 18*(9), 849–864. https://doi.org/10.1080/14787210.2020.1776115
Pestel, E. (1989). *Beyond the limit to growth*. The Club of Rome.

Philipson, M. (Ed.). (1962). *Automation: Implications for the future*. Vintage.
Phillips, F. (2011). The state of technological and social change: Impressions. *Technological Forecasting and Social Change, 78*(6), 1072–1078. https://doi.org/10.1016/j.techfore.2011.03.020
Plebe, A., & Perconti, P. (2020). Plurality: The end of singularity? In A. V. Korotayev & D. J. LePoire (Eds.), *The 21st century singularity and global futures* (pp. 163–184). Springer. https://doi.org/10.1007/978-3-030-33730-8_8
Podlazov, A. (2017). A theory of the global demographic process. *Herald of the Russian Academy of Sciences, 87*(3), 256–266. https://doi.org/10.1134/S1019331617030054
Polterovich, V. (2009). Gipoteza ob innovatsionnoy pauze i strategiya modernizatsii. *Voprosy ekonomiki, 6*, 4–23.
Raftery, A. E., Li, N., Ševčíková, H., Gerland, P., & Heilig, G. K. (2012). Bayesian probabilistic population projections for all countries. *Proceedings of the National Academy of Sciences, 109*(35), 13915–13921. https://doi.org/10.1073/pnas.1211452109
Randers, J. (2012). *2052. A report to the Club of Rome*. Chelsea Green Publishing.
Randers, J., Rockström, J., Stoknes, P. E., Golüke, U., Collste, D., & Cornell, S. (2018). *Transformation is feasible. A report to the Club of Rome*. Stockholm Resilience Centre.
Reed, C. A. (Ed.). (1977). *The origins of agriculture*. Mouton.
Rifkin, J. (2011). *The third industrial revolution. How lateral power is transforming energy, the economy, and the world*. Palgrave Macmillan.
Russell, S. J., Norvig, P., Canny, J. F., Malik, J. M., & Edwards, D. D. (2003). *Artificial intelligence: A modern approach*. Prentice Hall Upper.
Schwab, K., & Malleret, T. (2020). *COVID-19: The great reset*. World Economic Forum.
Schwab, K. (2016). *The fourth industrial revolution*. World Economic Forum.
Schwartz, P. M. (1999). Internet privacy and the state. *Connecticut Law Review, 32*, 815–829.
Scliaff, A., & Friedrichs, G. (1982). *Microelectronics and society: For Better or for Worse. A report to The Club of Rome*. Pergamon Press.
Solove, D. J. (2008). *Understanding privacy*. Harvard University Press.
Sylvester, E., & Klotz, L. C. (1983). *The gene age: Genetic engineering and the next industrial revolution*. Scribner.
Tang, Z., Zhang, X., Shu, Y., Guo, M., Zhang, H., & Tao, W. (2021). Insights from nanotechnology in COVID-19 treatment. *Nano Today, 36*, 1–9. https://doi.org/10.1016/j.nantod.2020.101019
Tesler, G. S. (2004). *Novaya kibernetika*. Logos.
Teulings, C., & Baldwin, R. (Eds.). (2014). *Secular stagnation: Facts, causes, and cures*. CEPR.
Umpleby, S. A., & Dent, E. B. (1999). The origins and purposes of several traditions in systems theory and cybernetics. *Cybernetics and Systems, 30*, 79–103.
UN Population Division. (2019). *World population prospects 2019*. UN Population Division.
UN Population Division. (2022). *United Nations Population Division Database*. United Nations. http://www.un.org/esa/population
Venkatesh, V., Morris, M. G., Davis, G. B., & Davis, F. D. (2003). User acceptance of information technology: Toward a unified view. *MIS Quarterly, 27*(3), 425–478.
Vollset, S. E., Goren, E., Yuan, C.-W., Cao, J., Smith, A. E., Hsiao, T., et al. (2020). Fertility, mortality, migration, and population scenarios for 195 countries and territories from 2017 to 2100: A forecasting analysis for the Global Burden of Disease Study. *The Lancet, 396*(10258), 1285–1306. https://doi.org/10.1016/S0140-6736(20)30677-2
von Foerster, H., & Zopf, G. W. (1962). Principles of self-organization. In M. C. Yovits & S. Cameron (Eds.), *Self-organization* (pp. 31–50). Pergamon Press.
Weiss, C., Carriere, M., Fusco, L., Capua, I., Regla-Nava, J. A., Pasquali, M., et al. (2020). Toward nano-technology-Enabled approaches against the COVID-19 pandemic. *ACS Nano, 14*(6), 6383–6406. https://doi.org/10.1021/acsnano.0c03697
Weizsäcker, E. U., & Wijkman, A. (2018). *Come on! Capitalism, short-termism, population and the destruction of the planet. A report to the Club of Rome*. Springer. https://doi.org/10.1007/978-1-4939-7419-1

Westin, A. F. (1966). Science, privacy, and freedom: Issues and proposals for the 1970's. Part I. The current impact of surveillance on privacy. *Columbia Law Review, 66*(6), 1003–1050.

Widdowson, M. (2021). From Covid-19 to Zero-Gravity: Complex Crises and Production Revolutions. *Journal of Globalization Studies, 12*(1), 119–146. https://doi.org/10.30884/jogs/2021.01.07

Wiener, N. (1948). *Cybernetics, or control and communication in the animal and the machine*. MIT Press.

Wieser, M., & Slunecko, T. (2014). Cybernetics, radical behaviorism, cultural psychology, cultural-historical psychology. In T. Teo (Ed.), *Encyclopedia of critical psychology*. Springer.

Wijkman, A., & Rockström, J. (2012). *Bankrupting nature*. Routledge.

Wittgenstein Centre. (2022). *Wittgenstein Centre Human Capital Data Explorer*. Wittgenstein Centre. http://dataexplorer.wittgensteincentre.org/wcde-v2/

Yampolskiy, R. V., & Fox, J. (2012). Artificial general intelligence and the human mental model. In A. H. Eden, J. H. Moor, J. H. Søraker, & E. Steinhart (Eds.), *Singularity hypotheses: A scientific and philosophical assessment* (pp. 129–145). Springer. https://doi.org/10.1007/978-3-642-32560-1_7

Leonid Grinin is a Senior Research Professor at the Centre for Stability and Risk Analysis at HSE University in Moscow, as well as a Senior Research Professor at the Institute of Oriental Studies, Russian Academy of Sciences. He is co-editor of the international journals *Social Evolution & History* and the *Journal of Globalization Studies*, as well as co-editor of the international yearbooks *Evolution, History & Mathematics*, and *Kondratieff Waves*. He is the author of more than 600 publications, including 33 monographs in Russian, English, Spanish, German, and Chinese. His current research focuses on comparative political studies, theory of revolution, political anthropology, global economy, global history, historical sociology, and futurology. Among his monographs are *Great Divergence and Great Convergence. A Global Perspective* (Springer, 2015; with Andrey Korotayev), *The Cybernetic Revolution and the Forthcoming Epoch of Self-Regulating Systems* (Uchitel, 2016, with Anton Grinin), Economic Cycles, Crises, and the Global Periphery (Springer, 2016; with Andrey Korotayev and Arno Tausch), *Big History of Globalization. The Emergence of a Global World System* (Springer 2019, in co-authorship with Julia Zinkina, David Christian, and others), *Islamism, Arab Spring, and the Future of Democracy. World System and World Values Perspectives* (Springer, 2019; with Andrey Korotayev and Arno Tausch). In 2004 he was awarded as a winner of the best publication of *Socis* journal during the year for the article "People of Publicity—a New Social Layer?". In 2012 he was awarded the Gold Kondratieff Medal by the International N. D. Kondratieff Foundation. In 2016 he received the Belyaev Award for the book *From Bifaces to Nanorobots* (the book published in Russian in co-authorship with Anton Grinin).

Anton Grinin is a Russian scientist specializing in the study of modern technological trends and futures studies, as well as an evolutionist and philosopher. He has a Ph.D. in Biology and works as a Research Fellow for Lomonosov Moscow State University, Russia. As well as Centre for Stability and Risk Analysis at HSE University, Moscow. He is the author of more than 100 publications in Russian and English, including 3 monographs. Awards: Gold Kondratieff Young Scholars Medal and Alexander Belyaev Literature Award. Anton Grinin focuses on the issues of the technological revolutions (the Cybernetic revolution) evolving in the current decades of the twenty-first century and the resultant profound transformations in economy and society as well as related economic, political, ethic-legal, and other risks, including political revolutions. He also investigates the correlation between the Cybernetic revolution and Kondratieff waves and between the Cybernetic revolution and global aging.

Economics: Optimizing Growth

Leonid Grinin, Anton Grinin, and Sergey Malkov

Abstract

The chapter examines the issues of mutual influence of technological and economic growth, the pace of which, after unprecedentedly high rates in the middle of the twentieth century, stabilized and began to slow down. Limited opportunities for extensive growth, a decrease in demand due to the stabilization of demographic growth, a worsening problem of inequality, the transfer of most industries to low-income countries, as well as an aggravation of the environmental problem, are identified as causes of this slowdown. These circumstances lead to the question of whether the current system of capitalist relations ceases to be relevant. This chapter discusses the problems of future economic growth and available resources, its rates, as well as possibilities to transform capitalism, to support tendencies toward lessening of consumerism, and to move to a less stratified society. Grinin et al. emphasize that the problems of inequality, injustice, and overconsumption cannot be solved only through increasing regulation and redistribution, as is sometimes suggested. This is a way to create more contradictions and to split society (as we could see during the pandemic). Grinin et al. suggest that there are more complex but more effective ways through

L. Grinin (✉)
HSE University, Moscow, Russia

Institute of Oriental Studies, Russian Academy of Sciences, Moscow, Russia

A. Grinin
HSE University, Moscow, Russia

Faculty of Global Studies, Lomonosov Moscow State University, Moscow, Russia

S. Malkov
Lomonosov Moscow State University, Moscow, Russia

HSE University, Moscow, Russia

Institute of Oriental Studies, Russian Academy of Sciences, Moscow, Russia

V. Sadovnichy et al. (eds.), *Reconsidering the Limits to Growth*, World-Systems Evolution and Global Futures, https://doi.org/10.1007/978-3-031-34999-7_9

simultaneous use of regulations, development of innovative technologies, and keeping optimal rates of economic growth. Reducing social tension currently requires reducing wealth inequality and strengthening economic regulation, but these measures can suppress economic activity and extinguish economic growth. Without the personal initiative and personal interest of millions of people, the internal economic engine of society stalls; the common good cannot be created by redistribution alone. The development of the economy in conjunction with a correct social policy (redistribution) is the only way to raise the standard of living. And regarding inequality, it should not be about trying to eliminate inequality in principle but only about how to smooth it out without destroying the driving forces of development and, at the same time, reducing social tensions. The task of society is to find ways to channel development in the direction of a course that is beneficial for it, to find the optimal combination between private initiative and social programs, and between the growth of production as an important characteristic of economic development (without increasing the burden on the environment) and consumption optimization. In general, the authors' vision of economic development is that it is necessary to optimize it both in its pace and indicators and also take into account the need to more quickly pull growing economies closer to high-income ones. There are several factors, including the new technological wave—the Cybernetic Revolution—that are accelerating the development of third-world countries and providing conditions for world economic growth. At the same time, of course, the directions, sources, and centers of economic growth will change. So, the modern world economy has faced serious problems and limitations in growth opportunities, which has begun to slow down the growth of high-income and low-income countries. However, Grinin et al. do not believe that the opportunities for economic growth have been exhausted. On the contrary, the new technologies of the Cybernetic Revolution in 2030–2040 can give an impetus to an increase in economic growth rates compared to current rates. The authors also assume that in connection with the global aging of the population, the model of consumption in society will also change, and consequently, the model of the economy itself will begin to transform.

1 Current Situation

Technological changes throughout human history have brought with them changes in the economic sphere, causing economic growth. At the same time, economic growth led to demographic growth and increased the demand for innovation, thereby stimulating further technological changes (e.g., Berger, 1986; Kremer, 1993; Maddison, 2007; Allen, 2011; Grinin et al., 2013, 2020). Thus, a positive feedback loop appeared, which caused the extreme unevenness of economic dynamics in the history of humankind: periods of slow economic development were followed by periods of rapid economic and technological growth, then, again, an epoch of slowdown and stagnation, then again rapid growth, etc. During those growth epochs, labor productivity grew, market relations expanded, interregional trade

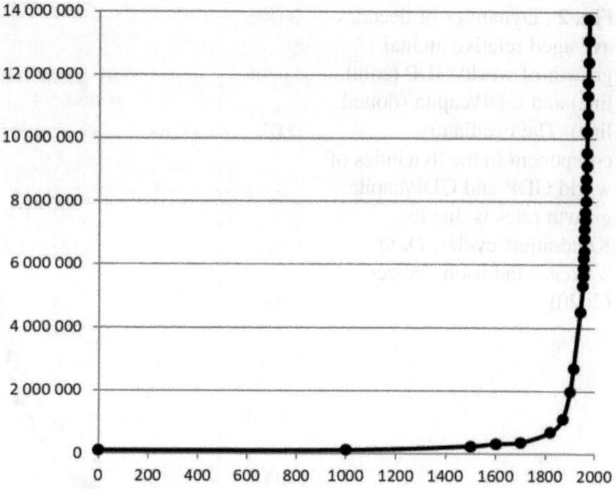

Fig. 1 Dynamics of world GDP from the beginning of common era to the mid-twentieth century (US $1990 million at PPP). *Data source*: Maddison Project (2020)

actively developed, and competition for sales markets intensified. During the periods of stagnation, the share of subsistence farming grew, and competitive mechanisms were replaced by distributive ones (Maddison, 2007; Maddison Project, 2020; Polterovich, 2009; Grinin, 2006, 2007, 2022; Grinin & Grinin, 2015, 2016). Changes in economic life inevitably influenced political life: the conditions for economic growth stimulated domestic political competition, and contributed to the emergence of republican forms of government; in the epochs of stagnation, political structures of a hierarchical type (for example, hereditary monarchies) prevailed (Grinin, 2008, 2011, 2012; Grinin & Korotayev, 2010, 2014, 2015, 2017; Grinin et al., 2010, 2016b, 2019, 2022).

The twentieth century was characterized by unprecedentedly high rates of inventive activity and economic growth (see Fig. 1):

However, since the late twentieth century, some slowdown has begun to be felt both in the technological and especially in the economic spheres (see Fig. 2[1]).

At the same time, the processes of inhibition have the strongest effect on the core of the World System (Grinin & Korotayev, 2015; Grinin et al., 2016b). The reasons for the slowdown include: (1) a decrease in the opportunities for extensive expansion [since economic globalization has ended (Grinin & Korotayev, 2020; Grinin, 2020; Grinin et al., 2021)], (2) a decrease in domestic demand (depopulation in many high-income countries, aging,[2] and demand saturation), (3) rising inequality

[1] See also Wallerstein et al. (2013), Maddison (2007), Teulings and Baldwin (2014), Grinin et al. (2020, 2022), Grinin & Grinin (2021a, 2021b), as well as chapter "Modeling Social Self-Organization and Historical Dynamics. Global Phase Transitions" (Malkov et al., 2023, this volume) and chapter "Technology. Limitless Possibilities and Effective Control" (Grinin & Grinin, 2023, this volume).

[2] See Grinin and Korotayev (2016), Grinin et al. (2017), Grinin and Grinin (2019), as well as chapter "Demography. Toward Optimization of Demographic Processes" (Korotayev et al., 2023) and chapter "Global Aging—an Integral Problem of the Future. How to Turn a Problem into a Development Driver?" (Grinin et al. 2023, this volume) for details.

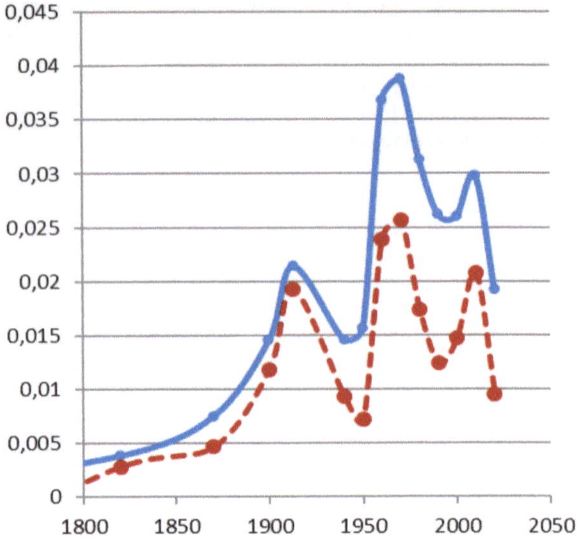

Fig. 2 Dynamics of decades-averaged relative annual growth of world GDP (solid line) and GDP/capita (dotted line). The oscillatory component in the dynamics of world GDP and GDP/capita growth rates is due to Kondratieff cycles. *Data source*: Maddison Project (2020)

(Randers, 2012; Calhoun, 2013; Schwab & Malleret, 2020; Schwab & Vanham, 2021), the disintegration of the middle class (Collins, 2013) and, as a result, a drop in demand among a large part of the population (Wallerstein, 2013; Mann, 2013); (4) transfer of many industries to low-income countries (Grinin & Korotayev, 2015); (5) increase in environmental costs;[3] (6) domination of financial capital (Calhoun, 2013; Grinin, 2021; Efimovich 2020; see also Bulatov et al. 2021) and others, including disaters like the COVID-19 pandemic (see Grinin 2020; Irshad 2020; Chumakov & Yurchenko, 2021; Hussain, 2021; Widdowson, 2021; Komlosy, 2023); revolutions (see, e.g. Lawson 2022; Grinin and Grinin 2022); and wars. All these reduce profits (Wallerstein, 2013), and this, in turn, reduces the efficiency of the functioning of capitalism as a system (which requires a "positive sum game," the absence of which is perceived as depression or crisis). On the other hand, there is an economic catch up of the world-system periphery to the core (Grinin & Korotayev, 2015; Grinin et al., 2016b) and, accordingly, a rather high economic growth there due to: (1) technology transfer and the development of scientific and technological revolution, (2) maintaining opportunities for extensive growth in the Periphery in the context of continuing population growth and undersaturated demand; (3) rapid diffusion of technological and other innovations.

[3] See chapter "Climate and Energy. Energy Transition Scenarios and Global Temperature Changes Based on Current Technologies and Trends" (Akaev & Davydova, 2023, this volume) and chapter "Ecology. Life in the "Unstable Biosphere" (Kovaleva, 2023, this volume) for details.

2 Prospects for Further Economic Development

The modern world economy has faced serious problems and limitations in growth opportunities, which have begun to slow down the growth of not only high-income but also low-income countries, including China. However, we do not believe that the opportunities for economic growth have been exhausted. On the contrary, as mentioned above, the new technologies of the Cybernetic Revolution in 2030–2040 can give an impetus to an increase in economic growth rates compared to current rates. A powerful development of AI and robotics will greatly reduce employment in the traditional service sector (which is offset by the withdrawal of the aging population from the labor process), and the development of additive technology may lead to a transition from the production of standardized goods to small wholesale or individual production.[4] We assume that in connection with global aging of the population, the model of consumption in society will also change, and consequently, the model of the economy itself will begin to transform (Grinin et al., 2020).

With the growing importance of medicine and biotechnology, the share of consumption of health-related goods and services will increase. In addition, the spread of self-regulating and self-managing systems will further smooth out the boundaries between the daily life and the economy. With the development of information and communication technologies, remote work and remote services will become more widespread.

At the same time, a new technological wave will lead to significant changes in the structure of employment—both negative and positive ones. On the one hand, the reduction in the proportion of people of working age will increase the labor shortage. But on the other hand, there will be a very serious transformation of professional skills; professions will change (and some will disappear), and remote education and remote work will flourish.[5] This will require closer economic ties between the high-income and the low-income countries, as discussed in chapter "High Income and Low Income Countries. Towards a Common Goal at Different Speeds" (Grinin et al., 2023, this volume; see also Grinin et al. 2022).

In our opinion, in the high-income and medium-income countries, as well as in many low-income countries, there is a real or potential shortage of labor resources. However, we believe that, although the future technological wave will bring major changes in the structure of employment with inevitable costs and negative

[4] See chapter "Technology. Limitless Possibilities and Effective Control" (Grinin & Grinin, 2023, this volume) and chapter "Global Aging—an Integral Problem of the Future. How to Turn a Problem into a Development Driver?" (Grinin et al. 2023, this volume) for details.

[5] Ambivalent consequences of technological progress were discussed in the report to The Club of Rome *Microelectronics and society: for better or for worse* (Scliaff & Friedrichs, 1982). The authors drew attention to the risks associated with the dismissal of workers employed in traditional industries, the use of electronic dossiers to exercise control over people or put pressure on them, limiting the opportunities for the free development of personality and human robotization. Works of such scholars as Phelps (2013), or Banerjee and Duflo (2019a, 2019b) are also dedicated to this topic.

consequences, including increased instability in some countries (e.g., Weizsäcker & Wijkman, 2018), it will also be able to alleviate the problems of the limited labor force (associated with a decrease in the proportion of young people and the aging of the population). This will happen not so much due to demographic growth but primarily due to an increase in the quality of life—first of all, thanks to breakthroughs in medicine that increase the working age and improve the quality of biological (or health-related) life. Artificial intelligence contributes to expanding opportunities in remote work, which, accordingly, can reduce unemployment, making the labor market largely global. Consequently, both the problem of over-population and the problem of labor shortages act as regional-global ones.

However, the World System cannot ignore the limited resources or climate issues for too long.[6] Therefore, discussions have now intensified about whether economic growth is needed at all or whether it is time to abandon it for the sake of climate. Here we agree with Ralf Fücks that zero growth is an unrealistic prospect, that "it is no longer a question of whether the growth of the world economy will continue, the question is rather how it will continue" (Fücks, 2013: 18).

For some time now, social thinkers have been saying more frequently that neoliberalism, which puts private initiative and growth in the first place, should be abandoned, and other principles should be put in place of its principles, since it is wrong that neoliberalism favors competition over solidarity and economic growth over social welfare, that the main goal of companies can no longer be simply the unbridled pursuit of financial profit; now they are obligated to serve all their stakeholders, and not just those who own shares.[7]

[6] As part of the issue of limited resources for further development and for high quality of life, the report to the Club of Rome *Factor Five. Scan of the Earth* (von Weizsacker, 1998) is quite interesting as it dwells on a revolutionary increase in productivity in the use of energy; from hypercars to low-calorie beef; materials, from subsoil drip irrigation to e-books, transportation, video conferencing to CyberTran, and demonstrating how much more today can be done with much less. It explains how markets can be organized and taxes restructured to remove perverse incentives and make rewards more efficient so that wealth can grow without increase in consumption. The author is convinced that this way profits will increase, pollution and waste will decrease, and the quality of life will improve. Moreover, the benefits will be universal: progress will no longer depend on increasing the productivity of fewer and fewer people. Instead, more people and fewer resources can be used. For many low income countries, the efficiency revolution may offer the only real chance to prosper in a reasonable amount of time. This issue has been covered in several other reports to the Club of Rome (Wijkman & Skånberg, 2017; Bardi, 2014, 2017; Maxton & Randers, 2016; Martin, 2015; Randers, 2012). This topic is also widely studied by different scholars across the world, among which it is worth highlighting the monograph *How much have global problems cost the world?* by Lomborg (2013) who carried out an impressive quantitative research on measuring the damage caused by global problems.

[7] The need to change the economic rationality of market capitalism, which has become the imperative of the of modern society, in favour of the concept of "dignity" and "culture" (which is more important than "cost/price") is mentioned in several reports to the Club of Rome (Andersen, 2020; Bozesan, 2020; Kuenkel, 2019; Berg, 2019; Lovins et al., 2018; Korten, 2015). Same ideas are being expressed by some other analysts, like Collier and Kay (2020) and Collier (2018). Among the possible solutions the following measures are suggested: to adapt the market system to reduce the negative impact of consumerism. The environmental and social costs of consumerism must be

Caring for a more equitable distribution, reducing the level of inequality is necessary. Otherwise, economic growth will not save society from increasing social tensions. It happens, on the contrary, that rapid gross domestic product (GDP) growth without a proper social policy leads to revolutionary moods as expectations from economic development begin to greatly exceed the possibilities (Davies, 1962, 1969; Gurr, 1968, 1974, 2015; Grinin, 2012, 2013; Grinin & Korotayev, 2012). Recent cases of such a paradox are Tunisia and Egypt in 2011. In the event of unexpected crisis, a powerful movement of discontent rises against the backdrop of previous growth. This was exactly what happened in Egypt and Tunisia in 2011. Both countries developed quite actively in the previous period, but the outbreak of the world crisis with rising food prices caused a socio-economic crisis, which was intensified by an excess of young, educated people (the result of the rapid development of higher education) who could not find a job that satisfies them (see: Grinin et al., 2016a, 2019).

Thus, economic growth must be combined with a reasonable and far-sighted social policy. However, to divide something, it is necessary to produce it; and in order to live better, it is necessary to produce more and eliminate the most severe distortions in distribution. But production is still primary here. Reducing or stabilizing the size of the social "pie" (especially taking into account the aging of the society) inevitably leads to the fact that it is redistribution, strict regulation, etc., that become the main thing. And this, in turn, inevitably leads to a decrease in entrepreneurial activity and other negative consequences, which in the long term are often extremely difficult or impossible to change.

For many centuries, the desire for profit has been an incentive for individual entrepreneurial activity; and competitive market relations, despite periodically occurring distortions, eventually led to economic growth. The problem is that the positive feedback that is inherent in market relations and market competition inevitably led to an increase in wealth stratification, which gave rise to social tension and instability in society. Reducing social tension currently requires reducing wealth inequality and strengthening economic regulation, but these measures can suppress economic activity and extinguish economic growth. Without the personal initiative and personal interest of millions of people, the internal economic engine of society stalls; the common good cannot be created by redistribution alone. The development of the economy in conjunction with a consistent social policy (redistribution) is the only way to raise the standard of living. And regarding inequality, it should not be about trying to eliminate inequality in principle but only about how to smooth it out without destroying the driving forces of development and, at the same time, reducing social tensions.

---

reflected in market prices, the development of less resource-intensive business models, production systems must be adapted accordingly, and products must be designed for a longer period of use, their renewal and recycling; facilitating the transition to a post-material society; reducing inequality that drives consumption; changing the perception of consumption as something that makes a used product unusable and breaks down; prosperity and a more modest lifestyle; struggle against selfishness and lack of meaning (Berg, 2019: 198–204).

Therefore, although we are on the verge of truly major changes, because of which, most likely, the regulation of the economy and technological development will intensify; however, the opportunities and reserves for normal economic growth will remain for at least the next half a century or even until the end of the century. There are several factors, including the anticipated new technological wave [see chapter "Technology. Limitless Possibilities and Effective Control" (Grinin & Grinin, 2023, this volume)], the accelerating development of third world countries and a number of others, which will provide conditions for world economic growth. At the same time, of course, the directions, sources, and centers of economic growth will change. The task of society is to find ways to channel development in the direction of a course that is beneficial for it, to find the optimal combination between private initiative and social programs, and between the growth of production as an important characteristic of economic development (without increasing the burden on the environment[8]) and consumption optimization.

In this regard, it makes sense to touch upon the problem of measuring the level of economic development and using the gross domestic product (GDP) indicator for this purpose. In the twentieth century, GDP and GDP per capita became the main macroeconomic indicators that assessed the economic success of a country (on the problem of economic growth and GDP see Grinin & Grinin, 2021a, 2021b). At the same time, in essence, GDP is an extensive indicator that reflects only the quantitative side, but not the qualitative aspect of economic development.[9] In the twentieth century, this made a certain sense since the main task for most countries of the world was the fight against poverty and escaping the Malthusian trap (Korotayev et al., 2014). But since the end of the twentieth century, in high-income countries, where a relatively high level of well-being has been achieved, other tasks have come to the fore. These tasks are not associated with an increase in material consumption but rather with an increase in the quality of life, a decrease in energy consumption, and the solution of environmental problems.[10] Under these conditions, the GDP indicator largely ceases to function as a universal measure of the level of economic development and should be used with certain reservations. There are two horizons here:

[8]See chapter "Climate and Energy. Energy Transition Scenarios and Global Temperature. Changes Based on Current Technologies and Trends" (Akaev & Davydova, 2023, this volume) and chapter "Ecology. Life in the "Unstable Biosphere"" (Kovaleva, 2023, this volume).

[9]The report to the Club of Rome *Taking Nature into Account* (Van Dieren, 1995) was also devoted to this topic. The author, together with WWF experts, questions the relevance of using GDP, as well as the need to find a new measure of the "fiscal" health of states, including ethical, historical, economic and environmental evidence in favor of including environmental factors. Similarly, the call to stop using GDP as a key development goal is made in the report to the Club of Rome by Wijkman and Rockström (2012).

[10]See chapter "Climate and Energy. Energy Transition Scenarios and Global Temperature Changes Based on Current Technologies and Trends" (Akaev & Davydova, 2023, this volume) and chapter "Ecology. Life in the "Unstable Biosphere"" (Kovaleva, 2023, this volume).

- near horizon. As long as there is nothing to really replace the "GDP" indicator, it can be used (especially for low-income countries that need to get out of poverty and backwardness and are primarily faced with the task of increasing the level of material well-being of their citizens), but its inherent limitations must be taken into account;
- distant horizon. For cybernetic society, in the context of accelerating global aging processes, new integral indicators are needed that reflect not only quantitative but also qualitative aspects of economic and social development. Work on the formation of such indicators are underway [see, for example, Ahmad et al., 1991; ul Haq, 1995; Banerjee and Duflo, 2019a, b; Masood, 2021; Colman, 2022, as well as chapter "Modeling Social Self-Organization and Historical Dynamics. Global Phase Transitions" (Malkov et al., 2023, this volume)].[11] It is necessary to gradually change the target socio-economic indicators, transferring, in particular, the needs from personal consumption to public consumption, introducing environmental and health-saving indicators, etc. But this must be done gradually and carefully, each time assessing the practical expediency of innovations.

In general, our vision of economic development is that it is necessary to optimize both its pace and indicators, and also take into account the need to pull growing economies closer to high income ones more quickly.

Regarding the question of the fate of capitalism [which is very much related to the problems of the future economy (Grinin et al., 2022)], we will return to it in chapter "High Income and Low Income Countries. Towards a Common Goal at Different Speeds" (Grinin et al., 2023, this volume), where we will talk about the differences in the development of the World System zones. One of our conclusions is that, in any case, capitalism will evolve, and the role of regulation and the social functions of business will increase (especially in developed countries). However, at the same time, while in some areas (particularly in medicine), a very significant regulation of economic activity is inevitable, in other areas (particularly in robotics and the production of self-driving cars), the economic mechanisms of the market-corporate type may grow stronger. In those low-income countries that continue the processes of economic and socio-political modernization, capitalism may mature, which would result in the strengthening of market institutions.

[11] Report to The Club of Rome in memory of Aurelio Peccei "Beyond The Limit To Growth" (Pestel, 1989) discussed topical problems of our time and promising opportunities for their solution in a global context, taking into account both the achievements of science and technology, including microelectronics, biotechnology, nuclear and other types of energy, and the international situation in the world, which determines the socio-economic and political contradictions between different countries. Eduard Pestel, as it were, summed up the 15-year debate about the limits of growth and emphasized that the issue is not growth as such, but the quality of growth. Repeating the considerations expressed in the second report to the Club of Rome, he proposed moving to a strategy of differentiated development, characterized not by static, but by dynamic equilibrium. Similar, but more updated in comparison to 1989, ideas are presented in a few other reports to the Club of Rome (Lovins et al., 2018; Wijkman & Skånberg, 2017; von Weizsäcker et al., 2010).

References

Ahmad, E., Drèze, J., Hills, J., & Sen, A. (1991). *Social security in developing countries*. Oxford Scholarship.

Akaev, A., & Davydova, O. (2023). Climate and energy. Energy transition scenarios and global temperature. Changes based on current technologies and trends. In V. Sadovnichy et al. (Eds.), *Reconsidering the limits to growth. A report to the Russian Association of the Club of Rome (pp. 53–70)*. Springer. https://doi.org/10.1007/978-3-031-34999-7_4

Allen, R. C. (2011). *Global economic history: A very short introduction*. Oxford University Press.

Andersen, L. R. (2020). *Bildung – Keep growing. A report to the Club of Rome*. Nordic Bildung.

Banerjee, A. V., & Duflo, E. (2019a). *Good economics for hard times*. Allen Lane.

Banerjee, A., & Duflo, E. (2019b). *Good economics for hard times*. Public Affairs.

Bardi, U. (2017). *The Seneca effect. Why growth is slow but collapse is rapid. A report to The Club of Rome*. Springer. https://doi.org/10.1007/978-3-319-57207-9

Bardi, U. (2014). *Extracted: how the quest for mineral wealth is plundering the planet. A report to The Club of Rome*. Chelsea Green Publishing.

Berg, C. (2019). *Sustainable action: Overcoming the barriers. A report to the Club of Rome*. Routledge.

Berger, P. L. (1986). *The capitalist revolution*. Basic Books.

Bozesan, M. (2020). *Integral investing. From profit to prosperity. A report to the Club of Rome*. Springer. https://doi.org/10.1007/978-3-030-54016-6

Bulatov, A., Habarta, A., & Sergeev, E. (2021). International financial centers as channels for foreign investment inflow to European Cities. *Journal of Globalization Studies, 12*(1), 102–118. https://doi.org/10.30884/jogs/2021.01.06

Calhoun, C. (2013). What threatens capitalism now? In I. Wallerstein et al. (Eds.), *Does capitalism have a future?* (pp. 216–265). Oxford University Press.

Chumakov, A. N., & Yurchenko, P. S. (2021). Global pandemic in terms of objective causes and subjective speculations. *Journal of Globalization Studies, 12*(2), 157–168. https://doi.org/10.30884/jogs/2021.02.08

Collier, P. (2018). *The future of capitalism: Facing the new anxieties.* .

Collier, P., & Kay, J. (2020). *Greed is dead: Politics after individualism*. Penguin Books.

Collins, R. (2013). The end of middle-class work: No more escapes. In I. Wallerstein et al. (Eds.), *Does capitalism have a future?* (pp. 23–60). Oxford University Press.

Colman, R. (2022). *What really counts: The case for a sustainable and equitable economy*. Columbia University Press.

Davies, J. C. (1962). Toward a theory of revolution. *American Sociological Review, 27*(1), 5–19.

Davies, J. C. (1969). Toward a theory of revolution. In B. McLaughlin (Ed.), *Studies in social movements. A social psychological perspective* (pp. 85–108). Free Press.

Efimovich, A. L. (2020). Negative rates and other financial technologies in modern economic reality on the world-system scale. *Journal of Globalization Studies, 11*(2), 63–76. https://doi.org/10.30884/jogs/2020.02.05

Fücks, R. (2013). *Intelligent Wachsen. Die Grüne Revolution*. Carl Hanser Verlag.

Grinin, L. (2006). Periodization of history: A theoretic-mathematical analysis. In L. Grinin, V. de Munck, & A. Korotayev (Eds.), *History & mathematics: Analyzing and modeling global development* (pp. 10–38). KomKniga/URSS.

Grinin, L. (2007). Production revolutions and periodization of history: A comparative and theoretic-mathematical approach. *Social Evolution & History, 6*(2), 11–55.

Grinin, L. (2008). Early state, developed state, mature state: Statehood evolutionary sequence. *Social Evolution & History, 7*(1), 67–81.

Grinin, L. (2011). *The evolution of statehood. From early state to global society*. Lambert Academic Publishing.

Grinin, L. (2012). *Macrohistory and globalization*. Uchitel.

Grinin, L. (2013). State and socio-political crises in the process of modernization. *Social Evolution & History, 12*(2), 35–76.

Grinin, L. (2020). How can COVID-19 change geopolitics and economy? *Journal of Globalization Studies, 11*(2), 121–134. https://doi.org/10.30884/jogs/2020.02.09

Grinin, L. (2022). Revolutions and modernization traps. In J. A. Goldstone, L. Grinin, & A. Korotayev (Eds.), *New waves of revolutions in the 21st century – Understanding the causes and effects of disruptive political changes (pp. 219–238).* Springer. https://doi.org/10.1007/978-3-030-86468-2_8

Grinin, L., & Grinin, A. (2015). *Ot rubil do nanorobotov. Mir na puti k epokhe samoupravlyayemykh sistem (istoriya tekhnologiy i opisaniye ikh budushchego).* Uchitel.

Grinin, L., & Grinin, A. (2016). *The cybernetic revolution and the forthcoming epoch of self-regulating systems.* Uchitel.

Grinin, L., & Grinin, A. (2019). Global'noye stareniye, tempy nauchno-tekhnicheskogo progressa i izmeneniye sovremennoy modeli potrebleniya v XXI – nachale XXII v. *Istoricheskaya psikhologiya i sotsiologiya istorii, 2*, 36–56.

Grinin, L. E., & Grinin, A. L. (2022). The current wave of revolutions in the world-system and its zones. *Journal of Globalization Studies, 13*(2), 171–189. https://doi.org/10.30884/jogs/2022.02.12

Grinin, L., & Grinin, A. (2021a). Razmyshleniya ob ekonomicheskom roste i Budushchem. Part 1. Globalizm vs rost VVP i«zakat Zapada». *Filosofiya i obshchestvo, 3*, 5–34. https://doi.org/10.30884/jfio/2021.03.01

Grinin, L., & Grinin, A. (2021b). Razmyshleniya ob ekonomicheskom roste i budushchem. Part 2. Globalizm, neosotsializm, spaseniye klimata i ekonomicheskiy rost. *Filosofiya i obshchestvo, 4*, 5–38. https://doi.org/10.30884/jfio/2021.04.01

Grinin, L., & Grinin, A. (2023). Technology. Limitless possibilities, effective control. In V. Sadovnichy et al. (Eds.), *Reconsidering the limits to growth. A report to the Russian Association of the Club of Rome (pp. 139–154).* Springer. https://doi.org/10.1007/978-3-031-34999-7_8

Grinin, L., Grinin, A., & Korotayev, A. (2017). The MANBRIC-technologies in the forthcoming technological revolution. In T. Devezas, J. Leitão, & A. Sarygulov (Eds.), *Industry 4.0 – Entrepreneurship and structural change in the new digital landscape: What is coming on along with the fourth industrial revolution* (pp. 243–261). Springer. https://doi.org/10.1007/978-3-319-49604-7_13

Grinin, L., Grinin, A., & Korotayev, A. (2020). A quantitative analysis of worldwide long-term technology growth: From 40,000 BCE to the early 22nd century. *Technological Forecasting and Social Change, 155*, 1–19. https://doi.org/10.1016/j.techfore.2020.119955

Grinin, L., Grinin, A., & Korotayev, A. (2021). Global trends and forecasts of the 21st century. *World Futures, 77*(5), 335–370. https://doi.org/10.1080/02604027.2021.1949939

Grinin, L., Grinin, A., & Korotayev, A. (2023). Global aging – An integral problem of the future. How to turn a problem into a development driver? In: V. Sadovnichy et al. (Eds.), Reconsidering the limits to growth. A report to the Russian Association of the Club of Rome (pp. 117–135). Springer. https://doi.org/10.1007/978-3-031-34999-7_7

Grinin, L., Issaev, L., & Korotayev, A. (2016a). *Revolyutsii i nestabilnost na Blizhnem Vostoke* (2nd ed.). Moskovskaya redaktsiya izdatelstva Uchitel.

Grinin, L., & Korotayev, A. (2010). *Globalnyy krizis v retrospektive. Kratkaya istoriya podemov i krizisov: ot Likurga do Alana Grinspena.* LIBROKOM.

Grinin, L., & Korotayev, A. (2012). Does "Arab Spring" mean the beginning of World System reconfiguration? *World Futures, 68*(7), 471–505. https://doi.org/10.1080/02604027.2012.697836

Grinin, L., & Korotayev, A. (2014). Interaction between Kondratieff waves and Juglar cycles. In L. Grinin, T. Devezas, & A. Korotayev (Eds.), *Kondratieff Waves. Juglar – Kuznets – Kondratieff* (pp. 25–95). Uchitel.

Grinin, L., & Korotayev, A. (2015). *Great divergence and great convergence. A global perspective*. Springer. https://doi.org/10.1007/978-3-319-17780-9
Grinin, L., & Korotayev, A. (2016). Global population aging, the sixth Kondratieff wave, and the global financial system. *Journal of Globalization Studies, 7*(2), 11–31.
Grinin, L., & Korotayev, A. (2017). The future of the global economy in the light of inflationary and deflationary trends and long cycles theory. *World Futures, 74*(2), 84–103. https://doi.org/10.1080/02604027.2017.1357934
Grinin, L., & Korotayev, A. (2020). Seven weaknesses of the U.S., Donald Trump, and the future of American hegemony. *World Futures, 77*(1), 23–54. https://doi.org/10.1080/02604027.2020.1801309
Grinin, L., Korotayev, A., & Malkov, S. (2010). A mathematical model of Juglar cycles and the current global crisis. In L. Grinin, A. Korotayev, & P. Herrmann (Eds.), *History & mathematics: Processes and models of global dynamics* (pp. 138–187). Uchitel.
Grinin, L., Korotayev, A., & Tausch, A. (2016b). *Economic cycles, crises, and the global periphery*. Springer. https://doi.org/10.1007/978-3-319-41262-7
Grinin, L., Korotayev, A., & Tausch, A. (2019). *Islamism, Arab Spring, and the future of democracy. World system and world values perspectives*. Springer. https://doi.org/10.1007/978-3-319-91077-2
Grinin, L., Malkov, S., Grinin, A., & Korotayev, A. (2022). Will capitalism die? Reflections on the capitalism of the past, present and future. *Sotsiologicheskiy Zhurnal = Sociological Journal, 28*(2), 100–130. https://doi.org/10.19181/socjour.2022.28.2.8988
Grinin, L., Malkov, S., & Korotayev, A. (2023). High income and low income countries. Towards a common goal at different speeds. In V. Sadovnichy et al. (Eds.), *Reconsidering the limits to growth. A report to the Russian Association of the Club of Rome (pp. 207–224)*. Springer. https://doi.org/10.1007/978-3-031-34999-7_12
Grinin, L., Markov, A., & Korotayev, A. (2013). On similarities between biological and social evolutionary mechanisms: Mathematical modeling. *Cliodynamics, 4*(2), 185–228. https://doi.org/10.21237/C7clio4221334
Gurr, T. (1968). A causal model of civil strife: A comparative analysis using new indices. *American Political Science Review, 62*(04), 1104–1124.
Gurr, T. (1974). Persistence and change in political systems, 1800–1971. *American Political Science Review, 68*(4), 1482–1504. https://doi.org/10.2307/1959937
Gurr, T. (2015). *Why men rebel*. Routledge.
Hussain, M. (2021). The changing dynamics of liberal international order: Covid-19 and its global implications. *Journal of Globalization Studies, 12*(2), 143–156. https://doi.org/10.30884/jogs/2021.02.07
Irshad, M. S. (2020). Covid-19 and global 'at-risk community': From benefit-sharing to risk-sharing of economic crisis. *Journal of Globalization Studies, 11*(2), 135–145. https://doi.org/10.30884/jogs/2020.02.10
Komlosy, A. (2023). Turn of an Era. Corona, Big Data and the Cybernetic Future. *Journal of Globalization Studies, 14*(1), 190–192. https://doi.org/10.30884/jogs/2023.01.12
Korotayev, A., Malkov, S., & Grinin, L. (2014). A trap at the escape from the trap? Some demographic structural factors of political instability in modernizing social systems. *History & Mathematics, 4*, 201–267. https://www.elibrary.ru/item.asp?id=22661111
Korotayev, A., Malkov, S., & Musieva, J. (2023). Demography. Toward optimization of demographic processes. In V. Sadovnichy et al. (Eds.), *Reconsidering the limits to growth. A report to the Russian Association of the Club of Rome (pp. 97–116)*. Springer. https://doi.org/10.1007/978-3-031-34999-7_6
Korten, D. C. (2015). *Change the story, change the future. A report to The Club of Rome*. Berrett-Koehler Publishers.
Kovaleva, N. (2023). Ecology. Life in the "unstable biosphere". In V. Sadovnichy et al. (Eds.), *Reconsidering the limits to growth. A report to the Russian Association of the Club of Rome* (pp. 71–95). Springer. https://doi.org/10.1007/978-3-031-34999-7_5

Kremer, M. (1993). Population growth and technological change: One million B.C. to 1990. *The Quarterly Journal of Economics, 108*, 681–716. https://doi.org/10.2307/2118405

Kuenkel, P. (2019). *Stewarding sustainability transformations. A report to the Club of Rome.* Springer. https://doi.org/10.1007/978-3-030-03691-1

Lawson, G. (2022). Revolutions in the Twenty-First Century. *Journal of Globalization Studies, 13*(2), 190–192. https://doi.org/10.30884/jogs/2022.02.13

Lomborg, B. (2013). *How much have global problems cost the world? A scorecard from 1900 to 2050.* Cambridge University Press.

Lovins, L. H., Wallis, S., Wijkman, A., & Fullerton, J. (2018). *A finer future. Creating an economy in service to life. A report to The Club of Rome.* New Society Publishers.

Maddison, A. (2007). *Contours of the world economy, 1–2030.* Oxford University Press.

Maddison Project. (2020). *Maddison Project Database 2020.* Groningen Growth and Development Centre.. https://www.rug.nl/ggdc/historicaldevelopment/maddison/releases/maddison-project-database-2020

Malkov, S., Grinin, L., Grinin, A., Musieva, J., & Korotayev, A. (2023). Modeling social self-organization and historical dynamics. Global phase transitions. In V. Sadovnichy et al. (Eds.), *Reconsidering the limits to growth. A report to the Russian Association of the Club of Rome (pp. 387–417).* Springer. https://doi.org/10.1007/978-3-031-34999-7_18

Mann, M. (2013). The end may be nigh, but for whom? In I. Wallerstein et al. (Eds.), *Does capitalism have a future?* (pp. 71–97). Oxford University Press.

Martin, C. (2015). *On the edge. A report to The Club of Rome.* Ingram.

Masood, E. (2021). *GDP: The world's most powerful formula and why it must now change.* Icon Books.

Maxton, G., & Randers, J. (2016). *Reinventing prosperity. A report to The Club of Rome.* Greystone Books.

Pestel, E. (1989). *Beyond the limit to growth. A report to The Club of Rome.* The Club of Rome.

Phelps, E. (2013). *Mass flourishing: How grassroots innovation created jobs, challenge, and change.* Princeton University Press. https://doi.org/10.2307/j.ctt32bbrz

Polterovich, V. (2009). Gipoteza ob innovatsionnoy pauze i strategiya modernizatsii. *Voprosy ekonomiki, 6*, 4–23.

Randers, J. (2012). *2052. A global forecast for the next forty years. A report to The Club of Rome.* Chelsea Green Publishing.

Schwab, K., & Malleret, T. (2020). *COVID-19: The great reset.* Forum Publishing.

Schwab, K., & Vanham, P. (2021). *Stakeholder capitalism.* Wiley.

Scliaff, A., & Friedrichs, G. (1982). *Microelectronics and society: For better or for worse. A report to The Club of Rome.* Pergamon Press.

Teulings, C., & Baldwin, R. (Eds.). (2014). *Secular stagnation: Facts, causes, and cures.* CEPR.

ul Haq, M. (1995). *Reflections on human development.* Oxford University Press.

Van Dieren, W. (1995). *Taking nature into account. A report to the Club of Rome.* 3Island Press.

von Weizsäcker, E., Hargroves, K., Smith, M. H., Desha, C., & Stasinopoulos, P. (2010). *Factor five. A report to The Club of Rome.* Earthscan.

von Weizsäcker, E. U., & Wijkman, A. (2018). *Come on! Capitalism, short-termism, population and the destruction of the planet. A report to the Club of Rome.* Springer. https://doi.org/10.1007/978-1-4939-7419-1

von Weizsacker, E. U. (1998). *Factor four. A report to The Club of Rome.* Taylor & Francis.

Wallerstein, I. (2013). Structural crisis, or why capitalists may no longer find capitalism rewarding. In I. Wallerstein et al. (Eds.), *Does capitalism have a future? (pp. 9–35).* Oxford University Press.

Wallerstein, I., Collins, R., Mann, M., Derluguian, G., & Calhoun, C. (2013). *Does capitalism have a future?* Oxford University Press.

Widdowson, M. (2021). From Covid-19 to zero-gravity: complex crises and production revolutions. *Journal of Globalization Studies, 12*(1), 119–146. https://doi.org/10.30884/jogs/2021.01.07

Wijkman, A., & Rockström, J. (2012). *Bankrupting nature. A report to the Club of Rome*. Routledge.

Wijkman, A., & Skånberg, K. (2017). *The circular economy and benefits for society. Jobs and climate clear winners in an economy based on renewable energy and resource efficiency. A report to The Club of Rome*.

Leonid Grinin is a Senior Research Professor at the Centre for Stability and Risk Analysis at HSE University in Moscow, as well as a Senior Research Professor at the Institute of Oriental Studies, Russian Academy of Sciences. He is co-editor of the international journals *Social Evolution & History* and the *Journal of Globalization Studies*, as well as co-editor of the international yearbooks *Evolution, History & Mathematics*, and *Kondratieff Waves*. He is the author of more than 600 publications, including 33 monographs in Russian, English, Spanish, German and Chinese. His current research focuses on comparative political studies, theory of revolution, political anthropology, global economy, global history, historical sociology, and futurology. Among his monographs are *Great Divergence and Great Convergence. A Global Perspective* (Springer, 2015; with Andrey Korotayev), *The Cybernetic Revolution and the Forthcoming Epoch of Self-Regulating Systems* (Uchitel, 2016, with Anton Grinin), *Economic Cycles, Crises, and the Global Periphery* (Springer, 2016; with Andrey Korotayev and Arno Tausch), *Big History of Globalization. The Emergence of a Global World System* (Springer 2019, in co-authorship with Julia Zinkina, David Christian, and others), *Islamism, Arab Spring, and the Future of Democracy. World System and World Values Perspectives* (Springer, 2019; with Andrey Korotayev and Arno Tausch). In 2004 he was awarded as a winner of the best publication of *Socis* journal during the year for the article "People of Publicity—a New Social Layer?". In 2012 he was awarded with the Gold Kondratieff Medal by the International N. D. Kondratieff Foundation. In 2016 he received the Belyaev Award for the book *From Bifaces to Nanorobots* (the book published in Russian in co-authorship with Anton Grinin).

Anton Grinin is a Russian scholar of modern technological trends and future studies, as well as evolutionist and philosopher. He has a Ph.D. in Biology and works as a Research Fellow for Moscow State University, Russia as well as at the Centre for Stability and Risk Analysis at HSE University. He is the author of more than 100 publications in Russian and English, including 3 monographs. Awards: Gold Kondratieff Young Scholars Medal and Alexander Belyaev Literature Award. Anton Grinin focuses on the issues of the technological revolutions (the Cybernetic revolution) evolving in the current decades of the twenty-first century and the resultant profound transformations in economy and society as well as related economic, political, ethic-legal, and other risks, including political revolutions. He also investigates the correlation between the Cybernetic revolution and Kondratieff waves as well as between the Cybernetic revolution and global aging.

Sergey Malkov heads the Center for Long-Term Forecasting and Strategic Planning at Lomonosov Moscow State University; he also works as a Senior Research Professor at the Institute of Economics of the Russian Academy of Sciences as well as at the Centre for Stability and Risk Analysis at HSE University, Moscow. He is the author of over 300 scholarly publications, including such monographs as *Social self-organization and historical process: possibilities of mathematical modeling* (URSS, 2009), *Modeling and forecasting of world dynamics* (ISPI RAS, 2012, together with Victor Sadovnichy and others), *A new theory of the reproduction of capital: development and practical application* (Nestor-Istoriya, 2016, together with Vladimir Mayevsky and others), *The 21st Century Singularity and Global Futures. A Big History Perspective* (Springer, 2020, together with David J. LePoire and others). In 2022 he was awarded with the Kondratieff Award by the Russian Academy of Sciences.

Sociopolitical Transformations: A Difficult Path to Cybernetic Society

Leonid Grinin, Anton Grinin, and Sergey Malkov

Abstract

Since the second half of the twentieth century, there have been radical changes in all spheres of life, which qualitatively distinguish the new phase of development from the previous ones. The chapter is devoted to the ongoing transformations in the social and political sphere, which include: the increasing digitalization of all spheres of life together with their positive and negative consequences, the change in the demographic structure of the population (including global aging), the development of global interactions, the increasing interdependence of the countries of the world, etc. The most important factor in these transformations is technological development, the Cybernetic Revolution unfolding before our eyes. These transformations lead to a serious restructuring of the social structure and sociopolitical changes. Grinin et al. consider the shape of future society, which they call "Cybernetic W-society" in different areas: technological, demographic, economic, political and administrative, international relations, social, and environmental. They also consider different scenarios of future Cybernetic W-society. According to them, as a result of the completion of the Cybernetic

L. Grinin (✉)
HSE University, Moscow, Russia

Institute of Oriental Studies, Russian Academy of Sciences, Moscow, Russia

A. Grinin
HSE University, Moscow, Russia

Faculty of Global Studies, Lomonosov Moscow State University, Moscow, Russia

S. Malkov
Lomonosov Moscow State University, Moscow, Russia

HSE University, Moscow, Russia

Institute of Economics, Russian Academy of Sciences, Moscow, Russia
e-mail: s@malkov.org

V. Sadovnichy et al. (eds.), *Reconsidering the Limits to Growth*, World-Systems Evolution and Global Futures, https://doi.org/10.1007/978-3-031-34999-7_10

Revolution and the achievement by the global aging of an advanced phase, a new society will be formed–Cybernetic W-society. It will be an elderly society, with the institutionalization of age differences, relying almost entirely on smart (cybernetic) technologies and AI-based self-regulating systems. Grinin et al. also consider other scenarios and aspects of this society. They analyze two extreme alternatives which are not likely to be implemented in their pure forms but rather indicate the limits within which the actual trajectory of global evolution will proceed: (1) "new feudalism," (2) "World-Organism." The former alternative essentially involves the reincarnation of a totalitarian-type system in the cybernetic era, whereas the latter alternative is a society where, in the presence of a deep professional specialization of the population, there is no discrimination, management is carried out on the basis of network technologies (network technologies are used to "synchronize" collective action and to exercise direct democracy). The second variant of development—"World-Organism"—is considered as favorable because it assumes cooperation, instead of competition, as a fundamental principle in development. Grinin et al. also note that ideology, which is of a universal nature, is of great importance in an effective transition to the second developmental option, which allows overcoming the confrontation between representatives of different faiths, nationalities, and civilizations. Grinin et al. point out that it would be preferable if we could move to a future society that is based on cooperation rather than competition and, at the same time, preserves the privacy, individuality, and initiative of its citizens.

1 Current Situation

The modern world is rapidly changing in all its aspects. It is clear that the era of rapid demographic and economic growth initiated by the Industrial Revolution 200 years ago is ending (see Chapter "Economics. Optimizing Growth" [Grinin et al., 2023c, this volume] above). The upward hyperbolic trends of global demographic and economic characteristics were bound to be interrupted, which is currently happening.[1] The era of qualitative changes is coming.

The rate of demographic growth has repeatedly experienced sharp changes in the history of humankind, which marked changes in the phases of historical development.[2] Now we see a similar case. Since the second half of the twentieth century,

[1] See Chapter "Demography. Toward Optimization of Demographic Processes" (Korotayev et al., 2023, this volume), Chapter "Economics. Optimizing Growth" (Grinin et al., 2023c this volume), and Chapter "Modeling Social Self-Organization and Historical Dynamics. Global Phase Transitions" (Malkov et al., 2023a, this volume); see also Pereira et al., 2022; Perissi & Bardi, 2021; Andersen, 2020; Bozesan, 2020; Berg, 2019; Randers et al., 2018, von Weizsäcker & Wijkman, 2018; Lovins et al., 2018; Wijkman & Skånberg, 2017.

[2] See Chapter "Modeling Social Self-Organization and Historical Dynamics. Global Phase Transitions" (Malkov et al., 2023a this volume). See also Mokyr & Voth, 2010; Vishnevskiy, 2005; Kapitza, 2006, 2009; Korotayev et al., 2006.

there have been radical changes in all spheres of life, signifying the global transition of humankind to a new phase of historical development.[3] The following changes qualitatively distinguish the new phase of development from the previous ones.

1. There is a rapid deceleration of world population growth due to the second phase of the demographic transition, which has engulfed all the regions of the planet. At the same time, the world population will age, and life expectancy will increase. Under these conditions, the proportion of older ages will increase, while the proportion of the young will fall. The age pyramid changes from triangular to rectangular.[4]
2. Digitalization and total computerization of all spheres of life will continue and accelerate. The entire population of the world will be covered by information networks, everyone will be under control, and human autonomy will practically disappear. People will increasingly transfer their decision-making function to self-regulating and self-managing systems, artificial intelligence and, accordingly, will increasingly depend on the algorithms embedded in them (see Chapter "Technology. Limitless Possibilities and Effective Control" [Grinin & Grinin, 2023b, this volume]).
3. The world will be more and more technogenic. However, if earlier, nature was the object of technological transformations, now people per se (biotechnologies, genetic engineering, artificial organs, cloning, etc.) and their consciousness (cognitive technologies, neural interfaces, neuropsychological effects, psychotropic substances, etc.) are the objects of changes.
4. The world has become unified and interdependent, global resource constraints and environmental problems have arisen,[5] and the irreversible impact of human activity on nature and climate has been realized (see Chapter "Climate and Energy. Scenarios of Energy Transition and Global Temperature Changes Based on Current Technologies and Trends" [Akaev & Davydova, 2023, this volume]) along with the need to take into account the arising limitations in practical activities.
5. The process of globalization has gradually proceeded over the course of many centuries (Zinkina et al., 2019; Grinin & Korotayev, 2014; Sheffield et al., 2013; Grinin, 2021; see also Pakhomov, 2021), and the world has become unified,

[3] See Chapter "Macrohistorical Approach" (Sadovnichy et al., 2023, this volume), Chapter "Modeling Social Self-Organization and Historical Dynamics. Global Phase Transitions" (Malkov et al., 2023a, this volume), and Chapter "Modeling Social Self-Organization and Historical Dynamics. Modern Society and Problems of Global Transition" (Akaev et al., 2023b, this volume).

[4] See Chapter "Demography. Toward Optimization of Demographic Processes" (Korotayev et al., 2023, this volume) and Chapter "Global Aging—an Integral Problem of the Future. How to Turn a Problem into a Development Driver?" (Grinin et al., 2023d, this volume).

[5] See Chapter "Economics. Optimizing Growth" (Grinin et al., 2023c, this volume) and Chapter "Ecology. Life in the "Unstable Biosphere"" (Kovaleva, 2023, this volume).

entangled in information, economic, and communication ties.[6] While in earlier interstate relations, it was typical to divide into "us" and "them" (that is, those who were not subject to moral norms adopted when interacting with "ours"—see Chapter "Modeling Social Self-Organization and Historical Dynamics. Agrarian Society" [Malkov et al., 2023b, this volume]), now, despite the temporary aggravation of contradictions between blocs of countries, the task of forming common ideological principles of co-existence comes to the fore.

These are fundamental changes; they are imperative. These changes will exert a principal influence on the formation of future society. The most important factor determining the features of global processes is the fundamental technological transformation. It was the fundamental technological transformation throughout human history that led to cardinal changes in the social structure and the specific features of the societies' functioning (see Chapter "Technology. Limitless Possibilities and Effective Control" [Grinin & Grinin, 2023b, this volume]; see also Grinin, 2006; Grinin & Grinin, 2015a, 2015b, 2015c, 2016; Grinin et al., 2017a, 2017b).

- The Neolithic Revolution comprised the transition of humankind from foraging (hunting and gathering) to food production (agriculture and animal husbandry), which led to the formation of agrarian society with its estate structure, the important role of religion, and cyclical dynamics (see Chapter "Modeling Social Self-Organization and Historical Dynamics. Agrarian Society" [Malkov et al., 2023b, this volume]).
- The Industrial Revolution was based on the replacement of manual labor with machine labor, which led to the formation of industrial society with its internal dynamism, instability, focus on constant change and continuous growth (see Chapter "Modeling Social Self-Organization and Historical Dynamics. Industrial Society" [Akaev et al., 2023a, this volume]).
- The Cybernetic Revolution; currently, the invention of high-speed computing devices, their introduction into all spheres of life and, on their basis, the gradual replacement of intellectual labor by machine labor, and in the final phase, the widespread use of self-regulating systems will lead (and is already leading) to the formation of a society, the functioning of which decisively depends on smart and super smart systems, their algorithms and capabilities. In this regard, it is logical to call this emerging society "cybernetic" (see Chapter "Technology. Limitless Possibilities and Effective Control" [Grinin & Grinin, 2023b, this volume],

[6]The process of deglobalization which has begun recently is not likely to stop the globalization process completely. Deglobalization, in our understanding, is a sort of ebb which is bound to be replaced with a new and stronger tide of globalization (see Grinin & Korotayev, 2021; Grinin et al., 2021b). In fact, Chase-Dunn et al. explain the possibility of deglobalization through the structural character of globalization that has both cycle and upward trend components, as periods of greater global integration have been followed by periods of deglobalization on a long-term stair-step toward the greater connectedness of humanity (Chase-Dunn et al., 2020).

Grinin, 2006; Grinin & Grinin, 2015a, 2015b, 2015c, 2016; Grinin et al., 2017a, 2017b).[7] What are its features?

2 Cybernetic Society: Super-Technogenic Society of the Elderly

Below we will consider how the first three of the cardinal changes listed above affect the shape of future society. The most important integral processes that can be associated with the formation of a cybernetic society will include a paradigm shift in technology and the process of global aging based on the increase in life expectancy.

The paradigm shift in technology is associated with the ongoing Cybernetic Revolution, a fundamental transition from the industrial production principle to the production of services and goods based on the widespread introduction of self-regulating and self-managing systems, that is, systems that can not only function in the absence (or with minimal participation) of people but also make complex decisions independently. Accordingly, cybernetic society will rely almost everything on smart (cybernetic) technologies and self-regulating systems. The Cybernetic Revolution passed its initial phase in the second half of the twentieth century; in the 2030s, it will complete its middle (modernization) phase and begin its final phase, which may end in the 2060s–2070s. Its main scenario is a breakthrough in medicine, especially coupled with bio- and nanotechnologies, as well as with digital technologies and robotics. All of these areas are very actively used in modern medicine, but in the future, their relationship with medicine will fundamentally intensify. MANBRIC convergence is already taking shape and will be actively developing in the final phase of the Cybernetic Revolution (see Chapter "Technology. Limitless Possibilities and Effective Control" [Grinin & Grinin, 2023b, this volume]; see also Grinin & Grinin, 2015a, 2015b, 2015c, 2016; Grinin et al., 2017a, 2017b, 2020, 2021a).

The second important factor influencing the formation of future society is global aging. Global aging as an integral problem is discussed in detail in Chapter "Global Aging—an Integral Problem of the Future. How to Turn a Problem into a Development Driver?" (Grinin et al., 2023b, this volume) and in our previous research

[7] About all the three production revolutions see Chapter "Technology. Limitless Possibilities and Effective Control" (Grinin & Grinin, 2023b, this volume); see also Grinin, 2006; Grinin & Grinin, 2015a, 2015b, 2015c, 2016; Grinin et al., 2017a, 2017b. About the Agrarian or Neolithic Revolution see, e.g., Childe, 1948; Reed, 1977; Harris & Hillman, 1989; Bellwood, 2004; about the Industrial Revolution see, e.g., Cipolla, 1976; Allen, 2009, 2011; Goldstone, 2015; Mokyr, 1999; Mokyr & Voth, 2010; Grinin & Korotayev, 2015). About the initial phase of the Cybernetic Revolution, which was called the Scientific-Technical Revolution, see e.g., Bernal, 1965; Philipson, 1962; Lilley, 1976; Benson & Lloyd, 1983; Sylvester & Klotz, 1983; about the future technologies in this revolution see Fukuyama, 2002; Rifkin, 2011; Fücks, 2013; Schwab, 2016.

(Grinin & Korotayev, 2016; Grinin et al. 2020, 2021b).[8] Our analysis allows us to draw the following conclusions:

- the deeper the process of global aging goes, the more it depends on the development of medicine, biotechnology, and other technologies;
- the process of global aging will not be able to develop in the future without profound changes in society. In other words, the processes of global aging and the growth of life expectancy (LE) will entail very serious changes in sociopolitical life;
- although the phenomenon of global aging permeates the entire society, affects all processes in it, and its influence will constantly be increasing, modern society, in many respects, is not yet ready for such changes. Accordingly, developing these trends can cause serious tensions and internal conflicts. However, sooner or later, such development will inevitably require very noticeable sociopolitical changes and, as a result, their institutionalization.

The main directions of institutionalization will be related to:

1. de jure or de facto different rights for different age groups of people. Of course, such a difference in rights exists today in relation to minors and the rights of people reaching a certain age to receive a pension. Other than this, legally, adult people are formally equal in rights. However, we believe that in the future, the number of rights or restrictions associated with age will increase greatly;
2. self-organization of age groups in various forms, from clubs and mutual aid societies to world organizations; from mutual assistance to lobbying interests;
3. the structuring of the electorate by age categories and the formation of appropriate political and legal programs;
4. the strengthening of the ideological orientation when discussing the rights and obligations of age groups (including all kinds of campaigns like what we see in the feminist movement);
5. the inclusion of age in all types of planning and activities (e.g., in the planning of buildings, in legal acts and laws, in requirements for employers, or landlords). Here, to a certain extent, the institutionalization of disabled people today can serve as a guideline for the development of the institutionalization of age tomorrow;

[8] On various aspects of the aging process and its impact on society see, e.g., Clark & Spengler, 1980; Lee et al., 1988; Bös & Von Weizsäcker, 1989; Peterson, 1999; Fukuyama, 2002; Jackson et al., 2008, 2013; Lee et al., 1988; Lee & Mason, 2011; UNFPA, 2012; UN Population Division, 2013, 2020; Popova & Yanik, 2014; Goldstone, 2015; Goldstone et al., 2015; Orlická, 2015; Bloom & Luca, 2016; Maestas et al., 2016; McDaniel & Zimmer, 2016; van Dullemen, 2017; Bengston, 2018; Fichtner, 2018; Randers et al., 2018; Yaşar & Yaşar, 2018; Barsukov, 2019; Hsu & Lo, 2019; Kapelyushnikov, 2019; Shcherbakova, 2019; Mitchell & Walker, 2020; Jackson, 2021; Medici, 2021). Here we try to explore global aging as an integral problem of the future (Grinin & Korotayev, 2015, 2016; Grinin et al., 2016; Grinin et al., 2017a, 2020).

6. the expansion of judicial and legal activity to protect the age rights of people.

In general, one can assume a rather difficult and conflicting path of adaptation to aging. Accordingly, the problem of ageism, which is already acute today, will manifest itself in ever-new aspects and directions, greatly changing society.[9] We assume that ageism will go through several stages in its development, including quite conflicting ones. But the laws, rules, and institutions gradually worked out by experience will lead to a situation where a system will arise for taking into account the interests of different age groups, fixing their rights, duties, and restrictions, up to quotas in political representation and holding posts (as is observed today in relation to quotas for women and people of color).

Thus, cybernetic society is a society that will take shape (and is already taking shape) as a result of the completion of the Cybernetic Revolution. It will be closely connected, on the one hand, with very deep technological changes in the management and regulation of our life activities at all levels. And on the other hand, it is inextricably linked with the process of global aging because aging has become an integral part of society. As it develops and the institutions of society adapt to it, it will change all spheres of society: technological, economic, consumption, social, ideological, etc.

Thus, cybernetic society is a society that will emerge as a result of the completion of the Cybernetic Revolution, which will be (a) supertechnological; (b) sociotechnologically regulated at all levels; (c) a society where, as a result of the aging process, the division into age categories will become socially much more significant than today (that is, age will become a much more important social marker than today).

In short, *cybernetic society is an elderly society, with the institutionalization of age differences, almost entirely based on smart (cybernetic) technologies and self-regulating systems.*

Let us consider the expected shape of cybernetic society in its individual areas (see Fig. 1):

1. in the technological sphere, active development of self-regulating and self-managing systems is expected, which will allow: (a) ensuring an increase in life expectancy and an improvement in the quality of biological life, which will intensify the process of global aging; (b) changing political, administrative, managerial, social and economic relations;
2. in the demographic sphere, we expect: (a) stabilization of the population (or movement toward it) on a global scale[10]; (b) an increase in the proportion

[9] On the problem of ageism today see Chapter "Global Aging—an Integral Problem of the Future. How to Turn a Problem into a Development Driver?" (Grinin et al., 2023b, this volume).

[10] See Chapter "Demography. Toward Optimization of Demographic Processes" (Korotayev et al., 2023, this volume).

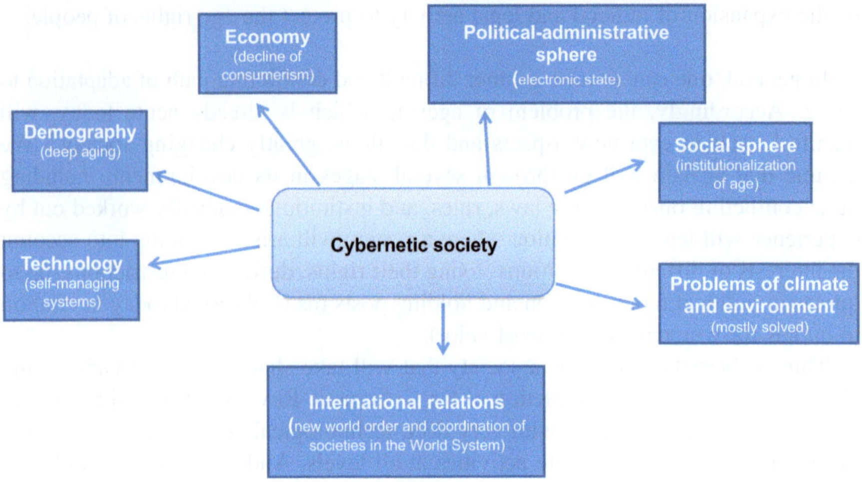

Fig. 1 Subsystems of cybernetic society

of the elderly population in some societies, which will lead to a noticeable increase in the social and political significance of these age cohorts;

3. in the economic sphere, we expect that: (a) the aging of the population will lead to certain conservatism in needs and demand (Grinin et al., 2020). This will noticeably reduce consumerism and rebuild the economic model of supply, demand, and consumption. In particular, the movement from the acquisition of goods toward the improvement of the biological quality of life and healthy lifestyle will intensify; (b) capitalism as an economic structure will persist,[11] while market relations will change toward greater stability and predictability (that is, market volatility will decrease) due to the fact that self-managing systems will be leading technologies, and they essentially require certain stability;
4. in politics and administration, the development of self-regulating systems will entail very significant, one might say revolutionary, changes in the field of governance because many social and administrative relations will be regulated by these systems. This will take place both at the level of individual administrative units and cities (so-called smart cities), and at the level of the state as a whole. As a result, what we have called the "electronic state" will emerge. This will lead to a sharp reduction in the number of bureaucrats, as well as to the "cheapening" of the state, and to a reduction in the vices associated with management (corruption, red tape, etc. [see Grinin et al., 2021a]). But this will also create a number of problems. On the one hand, with the development of social self-regulating

[11] For the evolution of capitalism, see Chapter "High Income and Low Income Countries. Toward a Common Goal at Different Speeds" (Grinin et al., 2023e, this volume) and Chapter "Modeling Social Self-Organization and Historical Dynamics. Global Phase Transitions" (Malkov et al., 2023a, this volume).

systems, a transition to direct democracy through permanent electronic voting is very likely. But on the other hand, using self-regulating systems in governance will lead to increased technological and political control. We anticipate this will result in a special kind of political regime, like democratic authoritarianism. Finally, the role of the elderly in politics will increase dramatically and, accordingly, age will become an important line of political agitation (for more detail see Chapter "Analyzing Social Self-Organization and Historical Dynamics. Future Cybernetic W-Society: Socio-Political Aspects" [Grinin & Grinin, 2023a, this volume]; see also Grinin et al., 2021a). We also suppose that most of revolutionary events and revolutionary waves will have finished by this time; however, the first half of the 21st century and, perhaps some later will be abound of revolutions (see Lawson, 2022; Grinin & Grinin, 2022).

5. in international relations, as a result of changes in individual societies, as well as the emergence of a new world order, the coordination of societies will increase. It is also very likely that countries with older populations will become more peaceful, with less ambition to wage war[12];
6. in the social sphere, it is expected that: (a) age will become a very important social marker; (b) regulation using the self-regulating systems will cover society from top to bottom, which can cause social tensions; (c) social tensions between different age cohorts are very likely, as a result of which the institutionalization of age is expected to occur; (d) the importance of the relationship between the individual and technology will increase, and a kind of sociomedical -technological sphere will arise, in which people will be forced to stay almost constantly. This will affect, first of all, the elderly and the disabled, whose lives will be significantly regulated. In our opinion, only in such a sociotechnological environment is it possible to move toward increasing life expectancy. It can be assumed that society as a whole will become more solidary; however, ageism, the contradictions between age strata, between conservatism and the need for technological progress, between the rights to personal freedom and sociotechnological restrictions will be felt quite strongly (for more details see Chapter "Analyzing Social Self-Organization and Historical Dynamics. Future Cybernetic W-Society: Socio-Political Aspects" [Grinin & Grinin, 2023a, this volume]);
7. in the field of ecology and its impact on the natural environment, an important role will be played by the understanding that an increase in life expectancy and its biological quality is possible only under certain environmental standards. Otherwise, the harm to health from bad ecology will undermine all the successes of medicine. Thus, we assume that a good ecological situation should already be a prerequisite for cybernetic society to reach its maturity. Of course, one cannot rule out a scenario of worsening climate and environmental conditions. However, this will definitely delay the transition to a cybernetic society.

[12] Chapter "Future Political Change. Toward a More Efficient World Order" (Grinin et al., 2023a, this volume).

Thus, we believe that the role of age groups in the social division of cybernetic society will expand, and the positions of different generations will be explicitly taken into account in various relations and laws. But the path to this will be uneasy, often conflicting, and will affect the interests of almost every person. However, we think that as a result of the completion of the Cybernetic Revolution and complex sociodemographic transformations, in the last third or quarter of the twenty-first century, a cybernetic society will develop in many countries, that is, a society that will be: (a) super technological; (b) sociotechnologically regulated at all levels; (c) a society where, as a result of the aging process, the division into age categories will become a socially much more important social marker than today (for more on cybernetic society, see Chapter "Analyzing Social Self-Organization and Historical Dynamics. Future Cybernetic W-Society: Socio-Political Aspects" [Grinin & Grinin, 2023a, this volume] and Chapter "Modeling Social Self-Organization and Historical Dynamics. Global Phase Transitions" [Malkov et al., 2023a, this volume]).

3 Cybernetic Society: Options for Sociopolitical Development

Let us now consider the impact on the shape of future society exerted by the last (in order, but not in importance) of the cardinal changes listed above, which reflect the macroeconomic and macrosocial transformation of the World System as a whole. These changes are related to the economic globalization, with the moving to world unification and its inner interdependence, though this moving is temporaly stopping.

These changes are important because modern society inherited its main features from industrial society, which was formed in the conditions of continuous economic growth and economic expansion, continuous expansion into the World System periphery, which determined the features of its internal organization and functioning. Now, however, industrial society is facing some limits to growth (or precisely, it needs growth optimization) [see Chapter "Economics. Optimizing Growth" (Grinin et al., 2023c, this volume)]. The former type of development is becoming less possible, the transition to a new model of society, more adequate to the new historical era, is inevitable. There have already been great transitions in the history of humankind (see Fig. 2).[13]

[13] On the uneven nature of historical development see Chapter "Modeling Social Self-Organization and Historical Dynamics. Global Phase Transitions" (Malkov et al., 2023a, this volume). One of transitions of this type is the transition from the era of antiquity to the era of the Middle Ages (it included the decline and collapse of the Western Roman Empire; some researchers point to a certain fundamental similarity between this empire and the modern USA). Demographic growth in the "Axial Age" (Jaspers, 1953) was associated with a powerful technological transformation of the Iron Age (with the massive introduction of iron tools in all spheres of life) and the resultant economic growth.

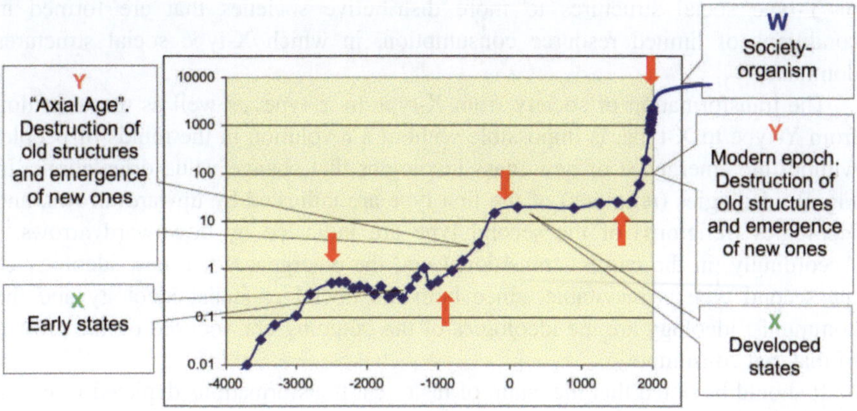

Fig. 2 Correlation between demographic, social, and spiritual development in world history (solid line represents the dynamics of the world urban population, millions, logarithmic scale [*data source* Korotayev, 2006: 52]). Arrows mark turning points in the formation of ideological currents (see commentary in the text)

For example, demographic and economic growth in modern times was associated with the Industrial Revolution (with the replacement of manual labor by machines).[14]

Global technological transformations that resulted in a significant increase in labor productivity, revolutions in military affairs, development of transport, etc., triggered the processes of a sharp increase in the scale of social systems:

- in the "Axial Age" these processes led to the formation of vast empires that absorbed smaller states;
- in the modern era, these processes have led to globalization in the full sense of the word in the economic, information, and communication spheres.

Sooner or later, the epochs of growth result in transformations accompanied by changes and optimization in resource consumption.[15] It is important that, at the same time, cardinal changes take place not only in the economy. The ideological sphere transits from competitive, continuous growth-oriented societies with the dominance

[14]See Cipolla, 1976; Mokyr, 1999; Mokyr & Voth, 2010; Allen, 2009, 2011; Goldstone, 2009; Malkov, 2009; Grinin & Grinin, 2015a, 2016; Grinin & Korotayev, 2015; see also Chapter "Modeling Social Self-Organization and Historical Dynamics. Industrial Society" (Akaev et al., 2023a, this volume).

[15]On the ways to optimize growth rates and new resources see Chapter "Economics. Optimizing Growth" (Grinin et al., 2023c, this volume). In this chapter, we also point out that the task of the society is to find ways to channel development in the direction of a course that is beneficial for it, to find the optimal combination between private initiative and social programs, between the growth of production as an important characteristic of economic development (without increasing the burden on the environment) and consumption optimization.

of Y-type social structures to more distributive societies that are formed in conditions of limited resource consumption, in which X-type social structures dominate.[16]

The transformation of society from X-type to Y-type, as well as the transition from Y-type to X-type, is impossible without a revolution in the minds of people, without the emergence of new mass ideologies that change value orientations. In Fig. 2, ideologies (religions) of the first type are indicated by upward arrows, and ideologies (religions) of the second type are indicated by downward arrows.[17] Accordingly, in the current transitional era, the emergence of a new ideology of the second type is inevitable, since both the liberal capitalist ideology and the communist ideology are the ideologies of the outgoing era, focused on the growth of material consumption.

It should be noted that the chain of historical transformations depicted in Fig. 2 will not end with a return to social X-structures. After the completion of the Cybernetic Revolution, a society (called "cybernetic" above) with a fundamentally new social structure (let us call it a W-type structure), shaped by the unique changes in human history outlined at the beginning of this chapter, may emerge. These changes make the following social transformations inevitable:

1. the transition from a liberal market economy to a more distributive one[18] (see Chapter "Economics. Optimizing Growth" [Grinin et al., 2023c, this volume]);
2. change in the demographic structure of society (decrease in the proportion of young people), its aging, the gradual transition from a "growth society" to a "society of stability" (see Chapter "Global Aging—an Integral Problem of the Future. How to Turn a Problem into a Development Driver?" [Grinin et al., 2023b, this volume]);
3. transition to self-regulating systems and decreasing of human factor in production processes (due to the Cybernetic Revolution) leading to a decrease in the importance of personal experience, the dependence of a person and society as a whole on information technologies[19];

[16]For more details on X- and Y-type social structures, see Chapter "Modeling Social Self-Organization and Historical Dynamics. Global Phase Transitions" (Malkov et al., 2023a, this volume), as well as, for example, Malkov, 2009; Kirdina, 2001, 2004, 2014.

[17]So, for example, the religion that helped in the transition from European medieval X-society to Y-society of modern times was Protestantism with its spirit of industry stimulation and the idea that successful people were God's chosen ones (Weber, 2002 [1905]). About ideologies that stimulated the transition to capitalism see Korotayev, 2003, 2004; Korotayev et al., 2023.

[18]This is a consequence of the decline in the overall rate of profit and the end of an era when the "positive sum game" was the rule, not the exception to the rule.

[19]See Chapter "Technology. Limitless Possibilities and Effective Control" (Grinin & Grinin, 2023b, this volume); about the socio-medical-technological sphere see above and Chapter "Analyzing Social Self-Organization and Historical Dynamics. Future Cybernetic W-Society: Socio-Political Aspects" (Grinin & Grinin, 2023a, this volume).

4. total control over human behavior through the systems with artificial intelligence and cognitive technologies (about its negative consequences, see Chapter "Technology. Limitless Possibilities and Effective Control" [Grinin & Grinin, 2023b, this volume]);
5. virtualization of communication, the formation of a network society and, probably, a decrease in empathy in human relations due to a decrease in the intensity of personal contacts (due to predominant online communication);
6. deep changes in professional specialization (see Grinin & Grinin, 2015b), increase in the social role of age divisions.[20] One cannot exclude the possibility of emergence of "new estates";
7. meeting the basic material needs of the population. In the context of technological progress, material goods become available to the vast majority of people, while at the same time the value of intangible goods, especially information, increases;
8. increasing role of ideology and influence on consciousness as a means of controlling people's behavior.

The question is what social shape these transformations will take, what form the social structure of future cybernetic society will acquire (above we called it the W-structure). There are options here. Let us consider this question in more detail.

In future Cybernetic W-society, due to the increasing role of age markers and what we have called the economic type of alienation,[21] the former social divisions will become less significant. What will be the new social structure? There are also various options available. One such option, as we see it, is an increase in the fragmentation and multiplicity of social groups, which will result in a decrease in the factors of social cohesion.[22] This raises the question: how will the "synchronization" of the actions of these social actors be ensured? Another feature of future society is the strengthening of distributive principles (in contrast to the dominance of liberal market principles in outgoing industrial society).[23]

---

[20] See Chapter "Analyzing Social Self-Organization and Historical Dynamics. Future Cybernetic W-Society: Socio-Political Aspects" (Grinin & Grinin, 2023a, this volume), Chapter "Global Aging—an Integral Problem of the Future. How to Turn a Problem into a Development Driver?" (Grinin et al., 2023b, this volume), and Chapter "Socio-Political Transformations. A Difficult Path to Cybernetic Society" (Grinin et al., 2023d, this volume) for more detail.

[21] See Chapter "Analyzing Social Self-Organization and Historical Dynamics. Future Cybernetic W-Society: Socio-Political Aspects" (Grinin & Grinin, 2023a, this volume).

[22] But it is worth noting that for the social structure of feudal and industrial, as well as modern societies, the fragmentation of social division is also characteristic. The identification of large groups—classes (to a lesser extent, estates)—is largely a result of scientific classification.

[23] Malkov (2021), using the basic model of economic competition, shows that when moving from a "positive-sum game" to a "zero-sum game", competitive interactions inevitably lead to the victory of one of the parties and to economic monopoly. Under these conditions, liberal market mechanisms are transformed into distributive ones dependent on the monopolist. The question is who the distributor will be, and on what principles the distribution will be carried out. In the modern world, such monopolists can be the state, transnational companies and financial capital (in the economic sphere), Internet giants (in the information sphere).

There are a number of varieties of the future ways; more precisely, there is a continuum of variants of the future social structure and order in different societies. Below we discuss only two polar versions of the future societal order.

From our point of view, option No. 1 implies the formation of a hierarchical totalitarian society controlled by an explicit or implicit, formal or informal "global authority" (a global elite that includes representatives of financial capital, transnational companies, the internet giants that control resource and information flows). The basic material needs of the population are provided for with an unconditional basic income, but the population is deprived of property and any real opportunity to influence the fundamentally important decisions made by the "global (or regional) authority." At the same time, people's mentality is controlled, which makes it possible to ensure the necessary "synchronization" of society. Digital technologies are used for total control over the behavior of the population. This is option #1 of future W-society.

In essence, this is a sort of new feudalism, the reincarnation of the X-system of a totalitarian type in the cybernetic era. But there is an essential difference from the feudalism of the Middle Ages. In medieval states, there was internal competition of elites for power and external competition of states with each other, which, despite all the costs, served as an incentive for further development (including technological development). In option No. 1 of future W-society, in the context of globalization and monopoly in the field of governance and distribution of material goods, competition is suppressed, which will inevitably lead to stagnation and subsequent degradation.

An alternative to option No. 1 can be option No. 2 of future W-society under the conditional name "World-Organism." This is a society where, in the presence of a deep professional specialization of the population, there is no discrimination, and governance is carried out on the basis of network technologies (network technologies are used to "synchronize" collective activities and implement direct democracy). An important role is played by ideology (in the form of ethical norms and self-restrictions that suppress antagonistic competition), which establishes a system of rules for developing coordinated decisions. At the same time, nonantagonistic competition is not suppressed; on the contrary, it is encouraged but channeled into the sphere of sports, science, and technology, creative professions, etc. The activities of bureaucrats are controlled by society (this is ensured by information transparency and network platforms), which makes it possible to form a governing apparatus based on meritocratic principles. Of course, it is only a metaphor, but such a society is certain sense acts as a biological organism, where all organs act in concert without suppressing each other to achieve a common goal.

These options are two extreme alternatives. In their pure form, they may not be implemented.

4 How to Switch to a Favorable Development Scenario

This raises the question of what social mechanisms will ensure the fulfillment of moral imperatives and curb the "free rider effect" (that is, the desire of certain members of society to evade self-restrictions)? In Y-society, the fight against violators of the legally established rules of conduct is based on a rigid system of punishments (administrative law, criminal law, etc.), but the rule "unless caught stealing, one is not a thief" really works. In religious X-society, the rules of conduct are fixed in the sacred books, and the threat in case of their violation is the Lord's punishment or karma, which in the minds of believers is a serious deterrent (but this threat does not affect the minds of nonreligious people). In Cybernetic W-society, the information transparency of a person and the network organization of society (among other things, on the basis of the above-mentioned electronic state and a possible variant of direct democracy) becomes the means of identifying and suppressing the "free rider effect." At the same time, information transparency can be used both to create a "digital concentration camp" (in option No. 1 of W-society) and to establish effective control of society over the work of officials, to suppress corruption, and to form a governance apparatus based on meritocratic principles (in option No. 2 of W-society). The way how the opportunities of universal computerization and information transparency will be used in W-society depends on society itself. And it is necessary to think in advance about the checks and balances of the possibility of total electronic control.

Thus, the transition to a favorable path of development will not happen by itself, and it requires purposeful and coordinated actions of the entire world community (Bondarenko, 2021). This work on forming a new world order based not on competition (as it has been so far in human history) but on cooperative principles is the most important task of our time or common future, depending on how this task is accomplished. It appears appropriate to mention here that, in some sense, the term "world order" means the conventional view of an asymmetric relationship between countries in which some of them impose a certain "order" to others. This does not fully correspond to what will take place when the formation of the cybernetic society is finished when more equal connections between countries can be expected. So, the term "new world order" has some defects in this sense. Perhaps, it will be better to use a new term, which has been absent until now. Perhaps, "world consensual system" would be suitable, or something like that.

In conclusion, we want to say that it would be preferable if we could move to a future society that is based on cooperation rather than competition and, at the same time, preserves privacy, individuality, and initiative of its citizens.

References

Akaev, A., & Davydova, O. (2023). Climate and energy. Scenarios of energy transition and global temperature changes based on current technologies and trends. In V. Sadovnichy et al. (Eds.),

Reconsidering the limits to growths. A report to the Russian Association of the Club of Rome (pp. 53–70). Springer. https://doi.org/10.1007/978-3-031-34999-7_4

Akaev, A., Malkov, S., Davydova, O., Kovaleva, N., Malkov, A., Grinin, L., & Korotayev, A. (2023a). Modeling social self-organization and historical dynamics. Industrial society. In V. Sadovnichy et al. (Eds.), *Reconsidering the limits to growths. A report to the Russian Association of the Club of Rome* (pp. 337–385). Springer. https://doi.org/10.1007/978-3-031-34999-7_17

Akaev, A., Malkov, S., Grinin, L., Bilyuga, S., Davydova, O., Grinin, A., Kovaleva, N., Malkov, A., Musieva, J., & Korotayev, A. (2023b). Modeling social self-organization and historical dynamics. Modern society and problems of global transition. In V. Sadovnichy et al. (Eds.), *Reconsidering the limits to growths. A report to the Russian Association of the Club of Rome* (pp. 419–460). Springer. https://doi.org/10.1007/978-3-031-34999-7_19

Allen, R. C. (2009). *The British industrial revolution in global perspective*. Cambridge University Press.

Allen, R. C. (2011). *Global economic history: A very short introduction*. Oxford University Press.

Andersen, L. R. (2020). *Bildung–Keep growing. A report to the Club of Rome. A report to the Club of Rome*. Nordic Bildung.

Barsukov, V. N. (2019). From the demographic dividend to Population aging: World trends in the system-wide transition. *Economic and Social Changes: Facts, Trends, Forecast, 12*(4), 167–182. https://doi.org/10.15838/esc.2019.4.64.11

Bellwood, P. (2004). *First farmers: The origins of agricultural societies*. Wiley-Blackwell.

Bengtson, V. (Ed.). (2018). *Global aging and challenges to families*. Routledge.

Benson, I., & Lloyd, J. (1983). *New technology and industrial change: The impact of the scientific-technical revolution on labour and industry*. Kogan Page.

Berg, C. (2019). *Sustainable action: Overcoming the barriers. A report to the Club of Rome*. Routledge.

Bernal, J. D. (1965). *Science in history* (3rd ed.). Hawthorn Books.

Bloom, D. E., & Luca, D. L. (2016). *The global demography of aging: Facts, explanation, Future*. Elsevier.

Bondarenko, V. (2021). A unified developmental strategy for the global world: A necessity and opportunity. *Journal of Globalization Studies, 12*(2), 169–180. https://doi.org/10.30884/jogs/2021.02.09

Bös, D., & Von Weizsäcker, R. K. (1989). Economic consequences of an aging Population. *European Economic Review, 33*(2–3), 345–354.

Bozesan, M. (2020). *Integral investing. From Profit to Prosperity. A report to the Club of Rome*. Springer. https://doi.org/10.1007/978-3-030-54016-6

Cipolla, C. M. (Ed.). (1976). *The industrial revolution, 1700–1914*. Harvester Press.

Chase-Dunn, C., Kim, J., & Alvarez, A. (2020). *Comparing waves of structural deglobalization: A world-systems perspective*. Institute for Research on World-Systems, University of California-Riverside.

Childe, V. G. (1948). *What happened in history*. Penguin Books.

Clark, R. J., & Spengler, J. J. (1980). *The economics of individual Population aging*. Cambridge University Press.

Fichtner, J. J. (2018). Global aging and public finance. *Business Economics, 53*(2), 72–78.

Fücks, R. (2013). *Intelligent Wachsen*. Carl Hanser Verlag.

Fukuyama, F. (2002). *Our post-human future: Consequences of the bio-technology revolution*. Farrar, Straus, and Giroux.

Goldstone, J. A. (2009). *Why Europe? The Rise of the West in World History, 1500–1850*. McGraw-Hill.

Goldstone, J. A. (2015). Population aging and global economic growth. In J. A. Goldstone, L. Grinin, & A. Korotayev (Eds.), *History & mathematics: Demography & Aging* (pp. 147–155). Uchitel.

Goldstone, J. A., Grinin, L., & Korotayev, A. (2015). Research into global aging and its consequences. In J. A. Goldstone, L. Grinin, & A. Korotayev (Eds.), *History & mathematics: Political demography & global aging* (pp. 5–9). Uchitel.

Grinin, L. (2006). *Proizvoditel'nyye sily i istoricheskiy protsess*. URSS.

Grinin, L. (2021). How globalization shifts the global economic and political balance. *Journal of Chinese Philosophy, 48*(2), 222–234.

Grinin, L., & Grinin, A. (2015a). Kiberneticheskaya revolyutsiya i shestoy tekh-nologicheskiy uklad. *Istoricheskaya psikhologiya i sotsiologiya istorii, 8*(1), 172–197.

Grinin, L., & Grinin, A. (2015b). *Ot rubil do nanorobotov. Mir na puti k epokhe samoupravlyayemykh sistem (istoriya tekhnologiy i opisaniye ikh budushchego)*. Uchitel.

Grinin, A., & Grinin, L. (2015c). The cybernetic revolution and historical process. *Social Evolution and History, 14*(1), 125–184.

Grinin, L., & Grinin, A. (2016). *The cybernetic revolution and the forthcoming epoch of self-regulating systems*. Uchitel.

Grinin, L. E., & Grinin, A. L. (2022). The current wave of revolutions in the world-system and its zones. *Journal of Globalization Studies, 13*(2), 171–189. https://doi.org/10.30884/jogs/2022.02.12

Grinin, L., & Grinin, A. (2023a). Analyzing social self-organization and historical dynamics. Future cybernetic W-society: Sociopolitical aspects. In V. Sadovnichy et al. (Eds.), *Reconsidering the limits to growths. A report to the Russian association of the Club of Rome* (pp. 491–519). Springer. https://doi.org/10.1007/978-3-031-34999-7_21

Grinin, L., & Grinin, A. (2023b). Technology. Limitless possibilities, effective control. In V. Sadovnichy et al. (Eds.), *Reconsidering the limits to growths. A report to the Russian association of the Club of Rome* (pp. 139–154). Springer. https://doi.org/10.1007/978-3-031-34999-7_8

Grinin, L., Grinin, A., & Korotayev, A. (2017a). Forthcoming Kondratieff wave, cybernetic revolution, and global aging. *Technological Forecasting and Social Change, 115*, 52–68. https://doi.org/10.1016/j.techfore.2016.09.017

Grinin, L., Grinin, A., & Korotayev, A. (2017b). The MANBRIC-technologies in the forthcoming technological revolution. In T. Devezas, J. Leitão, & A. Sarygulov (Eds.), *Industry 4.0–Entrepreneurship and structural change in the new digital landscape: What is coming on along with the fourth industrial revolution* (pp. 243–261). Springer. https://doi.org/10.1007/978-3-319-49604-7_13

Grinin, L., Grinin, A., & Korotayev, A. (2020). A quantitative analysis of worldwide long-term technology growth: From 40,000 BCE to the early 22nd century. *Technological Forecasting and Social Change, 155*, 1–16. https://doi.org/10.1016/j.techfore.2020.119955

Grinin, L., Grinin, A., & Korotayev, A. (2021a). COVID-19 pandemic as a trigger for the acceleration of the cybernetic revolution, transition from E-government to E-state, and change in social relations. *Technological Forecasting and Social Change, 175*, 1–16. https://doi.org/10.1016/j.techfore.2021.121348

Grinin, L., Grinin, A., & Korotayev, A. (2021b). Global trends and forecasts of the 21st century. *World Futures, 77*(5), 335–370. https://doi.org/10.1080/02604027.2021.1949939

Grinin, L., Grinin, A., & Korotayev, A. (2023a). Future political change. Toward a more efficient world order. In V. Sadovnichy et al. (Eds.), *Reconsidering the limits to growths. A report to the Russian Association of the Club of Rome* (pp. 191–206). Springer. https://doi.org/10.1007/978-3-031-34999-7_11

Grinin, L., Grinin, A., & Korotayev, A. (2023b). Global aging—An integral problem of the future. How to turn a problem into a development driver? In V. Sadovnichy et al. (Eds.), *Reconsidering the limits to growths. A report to the Russian association of the Club of Rome* (pp. 117–135). Springer. https://doi.org/10.1007/978-3-031-34999-1_7

Grinin, L., Grinin, A., & Malkov, S. (2023c). Economics. Optimizing growth. In V. Sadovnichy et al. (Eds.), *Reconsidering the limits to growths. A report to the Russian association of the Club of Rome* (pp. 155–168). Springer. https://doi.org/10.1007/978-3-031-34999-7_9

Grinin, L., Grinin, A., & Malkov, S. (2023d). Socio-political transformations. A difficult path to cybernetic society. In V. Sadovnichy et al. (Eds.), *Reconsidering the limits to growths. A report to the Russian association of the Club of Rome* (pp. 169–189). Springer. https://doi.org/10.1007/978-3-031-34999-7_10

Grinin, L., Malkov, S., & Korotayev, A. (2023e). High income and low income countries. Toward a common goal at different speeds. In V. Sadovnichy et al. (Eds.), *Reconsidering the limits to growths. A report to the Russian association of the Club of Rome* (pp. 207–224). Springer. https://doi.org/10.1007/978-3-031-34999-7_12

Grinin, L., & Korotayev, A. (2014). Globalization shuffles cards of the world pack: In which direction is the global economic-political balance shifting? *World Futures, 70*(8), 515–545. https://doi.org/10.1080/02604027.2014.982474

Grinin, L., & Korotayev, A. (2015). *Great divergence and great convergence. A global perspective.* Springer. https://doi.org/10.1007/978-3-319-17780-9

Grinin, L., & Korotayev, A. (2016). Global Population aging, the sixth Kondratieff wave, and the global financial system. *Journal of Globalization Studies, 7*(2), 11–31.

Grinin, L., & Korotayev, A. (2021). Seven weaknesses of the U.S., Donald Trump, and the future of American hegemony. *World Futures, 77*(1), 23–54. https://doi.org/10.1080/02604027.2020.1801309

Grinin, L., Korotayev, A., & Tausch, A. (2016). *Economic cycles, crises, and the global periphery.* Springer. https://doi.org/10.1007/978-3-319-41262-7

Harris, D., & Hillman, G. (Eds.). (1989). *An evolutionary continuum of people–plant interaction. Foraging and farming. The evolution of plant exploitation.* Unwin Hyman.

Hsu, Y. H., & Lo, H. C. (2019). The impacts of Population aging on saving, capital formation, and economic growth. *American Journal of Industrial and Business Management, 9*(12), 2231–2246. https://doi.org/10.4236/ajibm.2019.912148

Jackson, R. (2021). *The shape of things to come. The macro challenges of population aging.* Concord Coalition & the Global Aging Institute.

Jackson, R., Howe, N., Strauss, R., & Nakasima, K. (2008). *The graying of the great powers. Demogarphy and Geopololitics in the 21st century.* Center for Strategic & International Studies.

Jackson, R., Macaranas, R., & Peter, T. (2013). *U.S. development policy in an aging world. New Challeges and new priorities for a new demographic era. A research report from project on U.S. leadership.* Center for Strategic & International Studies.

Jaspers, K. (1953). *The origin and goal of history.* Yale University Press.

Kapelyushnikov, R. I. (2019). *Fenomen stareniya naseleniya: Ekonomicheskiye effekty.* Institut ekonomiki RAN.

Kapitza, S. (2006). *Global population blow-up and after. Report to the Club of Rome.* Global Marshall Plan Initiative.

Kapitza, S. (2009). *Obshchaya teoriya rosta chelovechestva. Kak ros i kuda idet mir cheloveka.* Nikitsky Klub.

Kirdina, S. (2001). *Institutsional'nyye matritsy i razvitiye Rossii.* IEiOPP SO RAN.

Kirdina, S. (2004). *X- i Y-ekonomiki: Institutsional'nyy analiz.* Nauka.

Kirdina, S. (2014). Institutional matrices theory, or X&Y theory: The main provisions and applications. *Journal of Institutional Studies, 6*(3), 13–33.

Korotayev, A. (2003). Unilineal descent groups and deep Christianization: A cross-cultural comparison. *Cross-Cultural Research, 37*(1), 132–156. https://doi.org/10.1177/1069397102238925

Korotayev, A. (2004). *World religions and social evolution of the old world Oikumene civilizations: A cross-cultural perspective.* Edwin Mellen Press.

Korotayev, A. (2006). The world system urbanization dynamics: A quantitative analysis. In P. Turchin, L. Grinin, V. C. de Munck, & A. Korotayev (Eds.), *History & mathematics: Historical dynamics and development of complex societies* (pp. 44–62). KomKniga/URSS.

Korotayev, A., Butovskaya, M., Shulgin, S., & Zinkina, J. (2021). Impact of Population aging on the global value system. *Vek Globalizatsii, 4*, 69–80. https://doi.org/10.30884/vglob/2021.04.05

Korotayev, A., Malkov, A., & Khaltourina, D. (2006). *Introduction to social macrodynamics: Secular cycles and millennial trends*. KomKniga/URSS.

Korotayev, A., Malkov, S., & Musieva, J. (2023). Demography. Toward optimization of demographic processes. In V. Sadovnichy et al. (Eds.), *Reconsidering the limits to growths. A report to the Russian association of the Club of Rome* (pp. 97–116). Springer. https://doi.org/10.1007/978-3-031-349997_6

Kovaleva, N. (2023). Ecology. Life in the "unstable biosphere". In V. Sadovnichy et al. (Eds.), *Reconsidering the limits to growths. A report to the Russian association of the Club of Rome* (pp. 71–95). Springer. https://doi.org/10.1007/978-3-031-34999-7_5

Lawson, G. (2022). Revolutions in the twenty-first century. *Journal of Globalization Studies, 13*(2), 190–192. https://doi.org/10.30884/jogs/2022.02.13

Lee, R. D., Arthur, W. B., & Rodgers, G. (1988). *Economics of changing age distributions in developed countries*. Oxford University Press.

Lee, R., & Mason, A. (2011). *Population aging and the generational economy: A global perspective*. Edward Elgar.

Lilley, S. (1976). Technological progress and the industrial revolution, 1700–1914. In C. M. Cipolla (Ed.), *The industrial revolution, 1700–1914. The Fontana economic history of Europe* (pp. 187–254). The Harvest Press.

Lovins, L. H., Wallis, S., Wijkman, A., & Fullerton, J. (2018). *A finer future. Creating an economy in service to life*. New Society Publishers.

Maestas, N., Mullen, K. J., & Powell, D. (2016). *The effect of population aging jn economic growth, The laboe Force and Prodactivity*. NBER.

Malkov, S. (2009). *Sotsial'naya samoorganizatsiya i istoricheskiy protsess: voz-mozhnosti matematicheskogo modelirovaniya*. Librokom.

Malkov, S. (2021). Razmyshleniya na temu: yest' li budushcheye u kapitalizma? *Informatsionnyye voyny, 4*(60), 41–49.

Malkov, S., Grinin, L., Grinin, A., Musieva, J., & Korotayev, A. (2023a). Modeling social self-organization and historical dynamics. Global phase transitions. In V. Sadovnichy et al. (Eds.), *Reconsidering the limits to growths. A report to the Russian association of the Club of Rome* (pp. 387–417). Springer. https://doi.org/10.1007/978-3-031-34999-7_18

Malkov, S., Kovaleva, N., Grinin, L., & Korotayev, A. (2023b). Modeling social self-organization and historical dynamics. Agrarian society. In V. Sadovnichy et al. (Eds.), *Reconsidering the limits to growths. A report to the Russian association of the Club of Rome* (pp. 309–335. Springer. https://doi.org/10.1007/978-3-031-34999-7_16

McDaniel, S. A., & Zimmer, Z. (2016). *Global aging in the twenty-first century: Challenges, opportunities and implications*. Routledge.

Medici, A. C. (2021). *Health sector challenges and policies in the context of aging populations*. United Nations, Department of Economics and Social Affairs, Population Division.

Mitchell, E., & Walker, R. (2020). Global aging: Successes, challenges and opportunities. *British Journal of Hospital Medicine, 81*(2), 1–9. https://doi.org/10.12968/hmed.2019.0377

Mokyr, J. (1999). *The British industrial revolution: An economic perspective*. Westview Press.

Mokyr, J., & Voth, H.-J. (2010). Understanding growth in Europe, 1700–1870: Theory and evidence. In S. Broadberry & K. H. O'Rourke (Eds.), *The Cambridge economic history of modern Europe 1700–1870* (Vol. 1, pp. 7–42). Cambridge University Press.

Orlická, E. (2015). Impact of population aging and elderly poverty on macroeconomic aggregates. *Procedia Economics and Finance, 30*, 598–605.

Pakhomov, O. (2021). The cold war and countercultural origins of globalization. *Journal of Globalization Studies, 12*(2), 103–118. https://doi.org/10.30884/jogs/2021.02.05

Pereira, C. A., Bardi, U., Bateson, N., Bologna, G., Cheng, Y. H., van Dieren, W., Dixson-Declève, S., et al. (2022). *Limits and beyond: 50 years on from the limits to growth what did we learn and what's next?* Exapt Press.

Perissi, I., & Bardi, U. (2021). *The empty sea. The future of the blue economy*. Springer. https://doi.org/10.1007/978-3-030-51898-1

Peterson, P. G. (1999). Gray down: The global aging crisis. *Foreign Affairs, 78*(1), 42–55.
Philipson, M. (Ed.). (1962). *Automation: Implications for the future*. Vintage.
Popova, S., & Yanik, A. (2014). Problemy global'nogo stareniya naseleniya: analiz dokumentov i strategii OON. *Mezhdunarodnoye pravo i mezhdunarodnyye organizatsii, 3*, 429–443.
Randers, J., Rockström, J., Stoknes, P. E., Golüke, U., Collste, D., & Cornell, S. (2018). *Transformation is feasible. A report to the Club of Rome*. Stockholm Resilience Centre.
Reed, C. A. (1977). *The origins of agriculture*. Mouton.
Rifkin, J. (2011). *The third industrial revolution. How lateral power is transforming energy, the economy, and the world*. Palgrave Macmillan.
Sadovnichy, V., Akaev, A., Ilyin, I., Malkov, S., Grinin, L., Aleshkovski, I., and Korotayev, A. (2023). Machrohistorical approach. In: V. Sadovnichy et al. (eds.), Reconsidreing the limits to growth. A report to the Russian Association of the Club of Rome (pp. 17–33). Springer. http://doi.org/10.1007/978-3-031-34999-7_2
Schwab, K. (2016). *The fourth industrial revolution*. World Economic Forum.
Shcherbakova, Y. (2019). Stareniye naseleniya mira po otsenkam OON goda. *Demoskop Weekly*, 837–838.
Sheffield, J., Grinin, L., & Korotayev, A. (2013). *Globalization: Yesterday, today, and tomorrow*. Emergent Publications.
Sylvester, E., & Klotz, L. C. (1983). *The gene age: Genetic engineering and the next industrial revolution*. Scribner.
UN Population Division. (2013). *World Population aging 2013*. United Nations, Department of Economics and Social Affairs, Population Division.
UN Population Division. (2020). *World Population aging 2020*. United Nations, Department of Economics and Social Affairs, Population Division.
van Dullemen, C. E. (2017). The politics of aging, global challenges and National Solutions. *Journal of Aging and Neuropsychology., 01*(2017), 1–4. https://doi.org/10.29011/JANP-104.100004
Vishnevskiy, A. (2005). *Izbrannyye demograficheskiye trudy*. Nauka.
von Weizsäcker, E. U., & Wijkman, A. (2018). *Come on!* Springer. https://doi.org/10.1007/978-1-4939-7419-1
Weber, M. (2002 [1905]). *The protestant ethic and the spirit of capitalism*. Penguin Books.
Wijkman, A., & Skånberg, K. (2017). *The circular economy and benefits for society. Jobs and climate clear winners in an economy based on renewable energy and resource efficiency. A report to the Club of Rome*. Club of Rome.
Yaşar, E., & Yaşar, M. (2018). Global wealth and aging Population. In R. Yilmaz & G. Löschnigg (Eds.), *Studies on Balkan and near eastern social sciences* (Vol. 2, pp. 47–59). Peter Lang.
UNFPA (2012). *Aging in the twenty-first century: A Celebration and A challenge*. UNFPA and HelpAge International. https://www.unfpa.org/publications/aging-twenty-first-century
Zinkina, J., Christian, D., Grinin, L., Ilyin, I., Andreev, A., Aleshkovski, I., Shulgin, S., & Korotayev, A. (2019). *A big history of globalization: The emergence of a global world system*. Springer. https://doi.org/10.1007/978-3-030-05707-7

Leonid Grinin is a Senior Research Professor at Centre for Stability and Risk Analysis at HSE University in Moscow, as well as a Senior Research Professor at the Institute of Oriental Studies, Russian Academy of Sciences. He is co-editor of the international journals *Social Evolution & History* and the *Journal of Globalization Studies*, as well as co-editor of the international yearbooks *Evolution, History & Mathematics*, and *Kondratieff Waves*. He is author of more than 600 publications, including 33 monographs in Russian, English, Spanish, German and Chinese. His current research focuses on comparative political studies, theory of revolution, political anthropology, global economy, global history, historical sociology, and futurology. Among his monographs are *Great Divergence and Great Convergence. A Global Perspective* (Springer, 2015; with Andrey Korotayev), *The Cybernetic Revolution and the Forthcoming Epoch of Self-*

Regulating Systems (Uchitel, 2016, with Anton Grinin), *Economic Cycles, Crises, and the Global Periphery* (Springer, 2016; with Andrey Korotayev and Arno Tausch), *Big History of Globalization. The Emergence of a Global World System* (Springer 2019, in co-authorship with Julia Zinkina, David Christian, and others), *Islamism, Arab Spring, and the Future of Democracy. World System and World Values Perspectives* (Springer, 2019; with Andrey Korotayev and Arno Tausch). In 2004 he was awarded as a winner of the best publication of *Socis* journal during the year for the article "People of Publicity—a New Social Layer?". In 2012 he was awarded with the Gold Kondratieff Medal by the International N. D. Kondratieff Foundation. In 2016 he received the Belyaev Award for the book *From Bifaces to Nanorobots* (the book published in Russian in co-authorship with Anton Grinin).

Anton Grinin is a Russian scientist specializing in the study of modern technological trends and futures studies, as well as evolutionist and philosopher. He has a Ph.D. in Biology and works as a Research Fellow for Lomonosov Moscow State University, Russia as well as at the Centre for Stability and Risk Analysis, HSE University, Moscow. He is the author of more than 100 publications in Russian and English, including 3 monographs. Awards: Gold Kondratieff Young Scholars Medal and Alexander Belyaev Literature Award.

Sergey Malkov heads the Center for Long-Term Forecasting and Strategic Planning at Lomonosov Moscow State University; he also works as a Senior Research Professor at the Institute of Economics of the Russian Academy of Sciences and at the Centre for Stability and Risk Analysis of HSE University, Moscow. He is the author of over 300 scholarly publications, including such monographs as *Social self-organization and historical process: possibilities of mathematical modeling* (URSS, 2009), *Modeling and forecasting of world dynamics* (ISPI RAS, 2012, together with Victor Sadovnichy and others), *A new theory of the reproduction of capital: development and practical application* (Nestor-Istoriya, 2016, together with Vladimir Mayevsky and others), *The twenty-first Century Singularity and Global Futures. A Big History Perspective* (Springer, 2020, together with David J. LePoire and others). In 2022 he was awarded with the Kondratieff Award by the Russian Academy of Sciences.

Future Political Change. Toward a more Efficient World Order

Leonid Grinin, Anton Grinin, and Andrey Korotayev

Abstract

In the chapter, Grinin et al. give the outline of a new world order whose emergence will be caused by a new technological wave (the final phase of the Cybernetic Revolution), aggravation of global problems (shortage of resources, demographic disproportions, climate change, and so on), and constant transformation of international relations. The chapter deals with issues of changing world order in the current period of global transformations. This period is called the reconfiguration of the World System. Grinin et al. offer short-term, medium-term, and long-term forecasts for the development of the world order, which are based on the assumption that the change in the world order is an inevitable process that happens when former leaders lose their positions. For a new world order to emerge, the balance of power must change significantly. The process of weakening the U.S. hegemony has been continuing for more than two decades. The U.S. position will be further weakening. So, the modern world order will undergo transformations too, until in the middle of the twenty-first century a new world order emerges in its main features. The World System directions of development, the global problems, necessity to avoid military conflicts, as well as the aspiration

L. Grinin
HSE University, Moscow, Russia

Institute of Oriental Studies, Russian Academy of Sciences, Moscow, Russia

A. Grinin
HSE University, Moscow, Russia

Faculty of Global Studies, Lomonosov Moscow State University, Moscow, Russia

A. Korotayev (✉)
HSE University, Moscow, Russia

Institute of Oriental Studies, Russian Academy of Sciences, Moscow, Russia

Faculty of Global Studies, Lomonosov Moscow State University, Moscow, Russia

V. Sadovnichy et al. (eds.), *Reconsidering the Limits to Growth*, World-Systems Evolution and Global Futures, https://doi.org/10.1007/978-3-031-34999-7_11

for progress require a stable balance of forces and interests, as well as a coordination of the activities of international actors, i.e., new principles of the world order. Grinin et al. formulate such principles. However, the authors note that the transition to a new more efficient world order will not be smooth, it will be accompanied by crises, tensions, conflicts, and disturbances, which implies a high degree of uncertainty. Along with this, Grinin et al. emphasize the importance of changing the balance of power for the formation of a more just order based on cooperation, which is facilitated by the active growth of low-income countries (especially in Africa), and the necessity to direct this development to the right track. Finally, Grinin et al. analyze various scenarios of political changes and the establishment of a new world.

In the previous chapter (Chapter "Socio-Political Transformations. A Difficult Path to Cybernetic Society" [Grinin, Grinin, & Malkov, 2023c, this volume]), we concentrated mostly on rather distant horizons of the formation of sociopolitical organization of future Cybernetic W-society (approximately from the end of the twenty-first century to the first decades of the twenty-second century). So, we did not pay sufficient attention to future international political relations. In this chapter, we will give an outline of a new world order and will consider closer forecasting horizons—from the next decade to the middle of the century, and, in less detail, until the end of the twenty-first century.

First of all, let us briefly formulate the main ideas from which we proceed:

- the transition to a new world order within three to four decades is inevitable;
- the new world order will require a fairly stable balance of forces and interests, new models of supranational governance and coordination of global processes;
- for the formation of a new world order, the balance of power must change significantly, since a more or less stable new order always relies on a new balance of power;
- we need a new world order with new principles that would reduce conflict and stimulate cooperation. We will formulate them below;
- the formation of a new order can stimulate the development of low-income countries (especially in Africa) and direct this development to the right track;
- the transition to the new system will be turbulent. In addition, in the course of the phase transition that has begun, various variants of transition systems will arise;
- different scenarios of the new world order are possible. However, the main scenario, in our opinion, will not be unipolar or bipolar; perhaps it will represent a certain group of major players (countries and unions). We can expect the emergence of a kind of Gx, i.e., a political forum, or a club of forces, or even an organization of three, five, seven, or more major powers and blocs capable of somehow organizing a new order (see Grinin et al., 2021).

Our main forecasts in this chapter will be presented at three levels:

1. Short-term forecasts (10–15 years).
2. Medium-term forecasts (until the 2050–2060s).
3. Long-term forecasts (until 2100).

1 Short-Term Forecasts (10–15 Years)

Politically (as well as economically), it will be a turbulent period. This means an increase in contradictions and conflicts, not excluding armed ones (for example, National Intelligence Council, 2017: 6), similar to what we could observe between Armenia and Azerbaijan, Russia, and Ukraine. This period is sometimes called the reconfiguration of the World System (Grinin & Korotayev, 2012, 2015; Grinin, 2022; Grinin et al., 2019). Within the theory of production revolutions,[1] the reconfiguration period can be explained by a vigorous diffusion and improvement of leading innovations in the middle (modernization) phase of the Cybernetic Revolution, which we are going through right now. However, for a new wave of innovations, for the transition to the final phase of the Cybernetic Revolution,[2] various parts of the World System should catch up with each other in their development (as much as possible). In other words, it is necessary to involve in the innovation process not only the core of the World System but also its semi-periphery (and part of the periphery). The gradual convergence of the periphery and the core of the World System in the level of economic development, defined as the "Great Convergence" (Grinin & Korotayev, 2015), has been unfolding since the end of the last century and will continue further.[3]

However, in the process of innovation diffusion and the Great Convergence, the political component lags behind the economic one. At the current level of globalization, we see that economic globalization is far ahead of political. Thus, to start a new technological breakthrough, a catch-up of the political component is required. This means some slowdown or even a temporary setback of economic globalization, as well as the emergence of various political and international tensions and conflicts since the process of reconfiguration of political institutions is painful. This explains the expected turbulence in the next 10–15 years. While we cannot agree with Wallerstein's (2013) claim that capitalism is more than ever in its final crisis

[1] On it see Chapter "Technology. Limitless Possibilities and Effective Control" (Grinin & Grinin, 2023, this volume); see also Grinin, 2007; Grinin & Grinin, 2015, 2016.

[2] See Chapter "Technology. Limitless Possibilities and Effective Control" (Grinin & Grinin, 2023, this volume); see also Grinin & Grinin, 2015, 2016; Grinin et al., 2017, 2020).

[3] On such economic trends see Chapter "High Income and Low Income Countries. Toward a Common Goal at Different Speeds" (Grinin & Korotayev, 2023b, this volume), Chapter "Economics. Optimizing Growth" (Grinin et al., 2023a, this volume), and Chapter "Africa—the Continent of the Future. Challenges and Opportunities" (Grinin & Korotayev, 2023a, this volume) for more detail.

phase,[4] we agree with him that the coming years will be quite turbulent or even very turbulent. At the same time, it is possible that sudden (and at the first glance inexplicable) political crises will break out in the most unexpected places, including in the core of the World System (salient examples are provided by the recent events in France, Great Britain, the USA, or Canada [see, for example, Kipfer, 2019; Cutts et al., 2020; Updegrove et al., 2020]). Due to the particularly rapid economic development of the countries of the periphery, as well as due to the mosaic nature and insufficient development of their political systems, serious political upheavals are possible there too.

We have witnessed many revolutionary and protest events in low-income countries in the previous decade,[5] and we should expect them to continue, if not increase, in the next decade.[6] It can also be a manifestation of the fact that the political component is lagging behind the economic one.

Thus, due to reconfiguration, economic globalization is rolling back (at least in terms of the introduction of import tariffs, the crisis of the WTO, massive sanctions, and the WHO crisis),[7] and political globalization is clearing the space for future changes. It is important that the formation of a new world order includes the destruction of the old one. Naturally, this is a painful and dangerous process, the acute phase of which we are currently observing. There are many pessimistic views that point to signs of crises in various areas (Taylor, 2020; Manheim, 2020).

In general, as a result of the weakening of the United States and the unipolar world order, in recent years, there has been an increase in "disorder" in the world (Grinin, 2016; Grinin et al., 2016).[8] With Donald Trump and his desire to change

[4] See Chapter "Modeling Social Self-Organization and Historical Dynamics. Modern Society and Problems of Global Transition" (Akaev et al., 2023, this volume) and Chapter "Economics. Optimizing Growth" (Grinin et al., 2023a, this volume), as well as Grinin et al., 2022.

[5] Campante & Chor, 2012; Dabashi, 2012; Malik & Awadallah, 2013; Davis, 2016; Acemoglu et al., 2017; Haas & Lesch, 2017; Steinert-Threlkeld, 2017; Korotayev et al., 2018; Sadovskaya et al., 2019; Baskan, 2019; Grinin et al., 2019; Goldstone et al., 2022a, 2022b, 2022c; Msellemu & Kessy 2022; Grinin & Grinin 2022; Issaev et al., 2021, 2022; Lawson 2022.

[6] See Grinin, 2020, 2022; Goldstone et al., 2022a, 2022b, 2022c; Grinin & Korotayev, 2022; Sadovskaya et al. 2021; see also Chapter "High Income and Low Income Countries. Towards a Common Goal at Different Speeds" (Grinin & Korotayev, 2023b, this volume). Some researchers even speak about the current ongoing world revolution. They think that this world revolution will continue and believe that there are some chances that a United Front of the New Global Left will have an opportunity to fundamentally transform the capitalist World System into a democratic and collectively rational global commonwealth (see, e.g., Chase-Dunn & Nagy, 2022; Chase-Dunn et al., 2022). Unfortunately, on the other hand, the influence of radicals, especially radical Islamists, is also likely to increase among revolutionary movements (see Grinin 2020, 2022; Grinin & Korotayev 2022; Vasiliev, 2022; Korotayev & Khokhlova, 2023).

[7] See Grinin & Korotayev, 2020; Grinin et al., 2021; see also Chapter "Economics. Optimizing Growth" (Grinin et al., 2023a, this volume) and Chapter "Socio-Political Transformations. A Difficult Path to Cybernetic Society" (Grinin et al., 2023c, this volume).

[8] During several decades many researchers have been repeatedly pointing out that the United States is increasingly losing its leadership role and will lose it further. Some of them regard this process inevitable, some recommend the U.S. to change radically its foreign policy to avoid losing its

many established relationships, the "disorder" only grew worse (Grinin & Korotayev, 2020; see also Irkhin and Moskalenko, 2022). We suppose that the U.S. position will be weakening further, and this will inevitably lead to the formation of a new world order during the next two-three decades.

So, in the foreign policy of many countries, unexpected zigzags and turns are possible in the near future, regardless of which administration of which party rules in the country. The coronavirus pandemic, which has created a crisis situation in many areas of society (Mitroff, 2020) and provoked significant changes in society (Hofbauer & Komlosy, 2020), can also become an important milestone in the aggravation of internal political confrontation and the struggle between world elites (including representatives of the U.S. business and finance oriented to the world market) and business oriented to the domestic consumer market of the USA and its manufacturing sector (Green et al., 2020).

Thus, in the short term, there is still no solid basis for establishing a new world order. However, it is possible that the process of weakening old alliances will proceed further, and the formation of new alliances will intensify (although these alliances can often be very opportunistic and short-lived). Therefore, we called this (and subsequent) period the "epoch of new coalitions" (see Grinin, 2009; Grinin & Korotayev, 2011).

The search for new forms and principles of organization is a complex evolutionary process in which the most promising forms of new relationships and alliances gradually crystallize. Over time, there will be more and more noticeable competition among the forces claiming leadership in the world. They will have to act under the slogans of a more just world order, and, of course, for such a policy they will need new alliances and allies. Thus, a regrouping of forces on the world and regional arenas will be inevitable. Since many of these alliances will be unstable, there will be a certain natural selection of principles and forms, as well as ideologies and other symbolic resources of these associations. Therefore, the selection process will not be a short-term phenomenon, but rather a medium-term one.

2 Medium-Term Forecasts (until the 2050–2060s)

Building a new system of world order will require a period of trial and error; therefore, we have difficult and turbulent years ahead of us, years of radical transformation of the balance of power between different countries and alliances. This process will also cover part of the medium term and will take place under favorable conditions in the 2020s–2040s.

Therefore, the period of turbulence may continue for some time (until the late 2030s or mid-2040s), but, in our opinion, the confrontation will be less serious. Tensions will continue due to the weakening of the United States because of its

positions (see, for example, Wallerstein, 2003; Kaphen, 2004; Todd, 2003; Zakaria, 2008; National Intelligence Council, 2008, 2012, 2017; Kissinger, 2001, 2014; see also Komlosy, 2019).

desire to prevent the rise of new leaders and also because of the attempts of other powers to defend their interests in all possible ways. At the same time, the currently existing blocs (for example, the EU, USMCA, OPEC, etc.) may become weaker. Sociopolitical crises in large low-income countries, in particular in China and India (for example, Vague, 2015; Bishop, 2018), cannot be ruled out as well, since there are many reasons for this. Thus, the weakening of political ties or unsuccessful reforms can worsen the situation.

However, we believe that these countries are able to cope with large shocks without catastrophic consequences. Nevertheless, in the 2050s–2060s, the contours and principles of a new world order should already be formed (otherwise, the tension and disorder in the World System will increase significantly). We disagree with the still popular idea (e.g., von Weizsaecker & Wijkman, 2018; Kurtz, 2019) that world development is moving toward something that could be called a world government. The new order is more likely to be based on other grounds.[9]

It makes sense to make a few important remarks here.

First of all, for the formation of a new world order, the balance of power must change significantly, since a more or less stable order always relies on a certain balance of power (Kissinger, 1994, 2001, 2014). Of course, a lot depends on the military power of countries. However, it will gradually change. In any case, this may be true for China, India, and other states whose military power is growing (Holslag, 2017; Amin, 2019; Bandeira, 2019; Marwah, 2019). In general, the volume of production and the combined economic and demographic power of countries will increasingly determine the balance of forces in the world and their influence on the formation of the world order. In the context of a stagnant and aging population in their countries, Western businesses (primarily TNCs) will inevitably rush to growing markets in low-income countries.[10] The transfer of various forms of economic activity to the low-income countries will also continue. Together with an increase in the effectiveness of economic and financial strategies in the low-income countries, this will inevitably lead to a change in the balance of power and the rules of the world order. It should also be borne in mind that the proportion of migrants from the low-income countries and their descendants in the population of high-income countries is growing. By the 2040s and 2050s, their numbers will increase greatly, along with their influence on the culture of Europe and the United States, politics, and relations with their countries of origin, which can become another big trump card for the low-income countries.

[9] However, we consider a similar scenario (global ruling elite) as a negative one in Chapter "Socio-Political Transformations. A Difficult Path to Cybernetic Society" (Grinin et al., 2023c, this volume) and Chapter "Modeling Social Self-Organization and Historical Dynamics. Global Futures and Cybernetic W-Society" [Grinin et al., 2023b, this volume]). See also below in this chapter.

[10] See Chapter "Global Aging—an Integral Problem of the Future. How to Turn a Problem into a Development Driver?" (Grinin et al., 2023d, this volume), Chapter "High Income and Low Income Countries. Toward a Common Goal at Different Speeds" (Grinin & Korotayev, 2023b, this volume), and Chapter "Demography. Toward Optimization of Demographic Processes" (Korotayev et al., 2023a, this volume).

In short, the balance of world power in the next three to four decades will change significantly in favor of those states (or their alliances) that are now considered low-income countries. In the end, the movement toward making common decisions will begin. Thus, it will be difficult to ignore the tenfold superiority of the population of the low-income countries over the developed ones (which will be established by about the 2050s).

Second, it is important to understand that for the formation of a new world order, new models of development and pluralism of regimes, ensuring the necessary diversity is of great evolutionary importance. Only competition between different models will allow choosing the best ones for the future. This can be an important part of creating common interests and rules. One should understand that democracy is far from universal (especially in multiethnic and multiconfessional regions without stable nations and states, often with a strong influence of religion).

There will also be signs of a possible crisis in democracy in its modern form aggravated by population aging, which will require its modernization, especially at supranational level. Democracy, having certain advantages on a national scale (but not everywhere and not always), can lose them at the supranational level.[11]

For example, the European Union, which, it would seem, was a model of supranational formations, no longer looks like it because of several serious problems. In particular, super-democratic procedures and the demand for consensus create a governance that is contradictory de jure and de facto (in the latter case, stronger states "twist the arms" of weaker ones, but at the same time, outward forces, by manipulating these weak actors, can exert their influence on the whole union). Therefore, the principle of democracy is likely to work only partially for such unions (see Grinin, 2016; Grinin et al., 2016, 2019 for details).

The rejection of "super-democracy" as a system of supranational governance needs new forms. One of our assumptions is the growing role of experts and international expert bodies. Even now, their real role is sometimes quite significant, as, for example, the role of rating agencies. This is why the possibility of development turning into something like an expertocracy (similar to the increasing role of corporate managers in the 1920s and 1930s [see Burnham, 1941]) cannot be ruled out. Accordingly, there may be expert bodies that can play a political role. For example, the nomination of political experts on the basis of their professional qualities may become the prerogative of national academies (scientific or other institutions) or the result of public elections of specialists chosen by professional communities. The purpose of such expert meetings is to conduct an examination (directly or with the help of relevant expert organizations) and scientific analysis of various acts and decisions for their compliance with all scientific achievements. Today, many decisions (including medical regulation, family law, parenting,

[11] On the problems of democracy, see, for example, MacEoin, 2018; Tsirel, 2020; on its connection with aging, see Peterson, 1999; Fukuyama, 2002; Jackson et al., 2008, 2013; Grinin et al., 2020, as well as Chapter "Socio-Political Transformations. A Difficult Path to Cybernetic Society" (Grinin et al., 2023c, this volume) and Chapter "Modeling Social Self-Organization and Historical Dynamics. Global Futures and Cybernetic W-Society" (Grinin et al., 2023b, this volume).

technological regulations, and even climate legislation, etc.) are made on an emotional basis, without calculations and forecasts of possible consequences. It is important to take into account the constant changes in the development of science and technology in order to regulate these processes. Given the inevitable limitations (and possible bias) of expert decisions, they will nevertheless be more balanced and more appropriate to the needs of modern society. Thus, experts could become a kind of branch of power, their role might increase, and the positions of expert organizations will in a certain sense resemble the guilds of the Middle Ages.

Unlike politicians, experts have powers of a different nature, so they can act disregarding voters and society, and, moreover, their own citizenship. Specialists belonging to different nationalities (but united by a single professional ethos) are often closer to each other than to representatives of their countries with a lower level of education (Grinin, 2016; Grinin et al., 2016; Grinin et al., 2019).

Global population aging may significantly re-divide the World System, in particular, into the aging North and the younger South.[12] This, on the one hand, will strengthen ties in the world (because the South and the North need each other), but, on the other hand, it can complicate their relations (see Grinin & Korotayev, 2015, 2016 for more details).

3 Long-Term Forecasts (until 2100)

As has been mentioned above, a new world order is likely to emerge in the medium term. New blocks and alliances will appear. Thus, in our opinion, the new world order will be neither unipolar nor bipolar. As for the prospects of the possible global hegemony of China, which are pointed out by some researchers (for example, Komlosy, 2019), it has been demonstrated (Grinin, 2013; Grinin et al., 2015; Grinin et al., 2021) that, on the one hand, its economic and political positions will strengthen in the coming decades, but, on the other hand, the country, even assuming all possible future successes, will not be able to take the US position in the World System for a number of reasons.

As we have pointed out above, we can expect the emergence of some new Gx system, i.e., the unification of three to five to seven (up to 15) major powers and blocs (including Asian and African ones), which will create another "concert of great powers" to influence the order in the World System. It is possible that any significant international action will require, in addition to real interest, a certain ideological justification. Considering the world stage as a "Great Chessboard" (Brzezinski, 1997), where the strongest wins, and minor pieces can be exchanged and sacrificed, may no longer be relevant.

[12] See Fukuyama, 2002; Jackson et al., 2008, 2013; Powell & Khan, 2013; Goldstone, 2015; Goldstone et al., 2015; Haas, 2015; Zimmer, 2016; Bengtson, 2018; Fichtner, 2018; see also Chapter "Global Aging—an Integral Problem of the Future. How to Turn a Problem into a Development Driver?" (Grinin et al., 2023d, this volume) and Chapter "Demography. Toward Optimization of Demographic Processes" (Korotayev et al., 2023a, this volume).

The world stage will be seen as a common area of interest in which it is necessary to establish rules that are beneficial to all participants and to follow them. A kind of oligarchy of major players (countries and unions) is likely to be created, which would also act as the supreme arbiter in interstate disputes. The nature of national interests, the causes of international rivalry, the forms of conflicts and litigation will gradually take on a different form than in the past and present (see Grinin, 2016; Grinin et al., 2016; Grinin & Korotayev, 2015). Within the oligarchy, rivalry will certainly continue, but consensus and engagement with the oligarchy can do more than deepen the rivalry. Just like the EU countries, which, despite all the contradictions within the union, believe that maintaining unity is better than a split (once again, this was proved by the position of the continental EU members on Brexit).

Slogans about the common benefit will gradually increase in foreign policy, although behind the statement about the need to "better represent global interests" someone's selfish goals may, as always, be hidden. But this transformation will somehow lead to very significant (and rather positive) changes.

In this case, the problem of unification and clash of national and supranational, group interests, and global interests will remain the most important for a long time to come. This process, which began in the 2010s, will finally end only in the 2070s–2080s. At the same time, global ideological motives can be strengthened, as well as participation in certain common programs. It is possible that the climate problem, which now occupies an important place in world political agenda, will become aggravated.[13] It is possible that new international regulators will be created (perhaps where the value of the vote of representatives will depend on the population of their country or bloc). Undoubtedly, one of the stumbling blocks will be the problem of equality in the global "concert," especially (which is quite likely) in the case of the creation of any new collective bodies to replace the UN or other organizations. Also, in the 2060s–2080s, African powers will rise and may claim their place in the new order (see Chapter "Africa—the Continent of the Future. Challenges and Opportunities" [Grinin & Korotayev, 2023a, this volume] and Chapter "Modeling Social Self-Organization and Historical Dynamics. Africa's Futures" [Korotayev, Shulgin et al., 2023a, b, this volume] for more details). At the same time, it will be difficult to prevent countries from deviating from generally accepted rules and choosing an independent path.

Alternative scenario: catastrophic changes occur as a result of global warming, leading to a collapse of the economy, a global migration crisis, a big war, etc. The following sub-scenarios are possible here:

[13] See, for example, von Weizsaecker & Wijkman, 2018; Gills & Morgan, 2022, as well as Chapter "Climate and Energy. Scenarios of Energy Transition and Global Temperature Changes Based on Current Technologies and Trends" (Akaev & Davydova, 2023, this volume).

(a) the world is rolling back; countries try to save themselves alone. Due to the complexity of the current situation, returning to the pre-crisis level requires a lot of effort and time;
(b) the foundations of the new world order are formed in an accelerated manner, the leading countries form something like a single body that begins to play the role of a world government to overcome the crisis.

The new world order will require new principles to reduce confrontation and encourage cooperation. These principles will begin to take shape in the first half of the medium term and will be visibly established in its second half and strengthened in the initial phase of the long term. Let us formulate a short list of these principles:

1. pluralism of political regimes. This principle implies (a) the legitimacy of all regimes; (b) the idea that there is no better regime (including a democratic one), each one has their pluses and minuses; and (c) pluralism itself with respect to the differences between regimes can be very useful for the international system (similar to the pluralism of religions for society);
2. rejection of the forcible imposition of institutions (including democracy);
3. recognition of the value of stability, order, legitimacy, and refusal to inspire internal revolutions. This means the following: a) social order, guaranteed security, and economic growth must be ensured even under authoritarian regimes and be recognized as more important than formal democratic rights and b) support for antigovernment protests, threats, anarchy, and overthrow of power (and even more so incitement to overthrow the regime) can be carried out only in exceptional cases, legalized by generally recognized international institutions;
4. a ban on the use of global economic, financial, rating organizations, mass media, etc. as instruments of foreign policy and the overthrow of regimes;
5. guarantees of state borders, restriction of all kinds of separatist movements, under whatever slogans they may be carried out. This would greatly reduce international friction;
6. refusal from forcible introduction of new ethical and legal norms (especially in sexual and family relations) and from discrimination under this pretext;
7. demonopolization of the media throughout the world and the prohibition of attempts by any political or other forces to monopolize these assets. As the media have become super-powerful and global in scope, extremely influencing the formation of public opinion and ideology, checks and balances are needed in this area both at the national and international levels.

Obviously, the recognition of the principles does not in itself guarantee their implementation, but it is important to avoid today's practice, when the ideals of democracy, the fight against corruption, or for human rights are used as a battering ram in the geopolitical struggle.

References

Acemoglu, D., Hassan, T. A., & Tahoun, A. (2017). The power of the street: Evidence from Egypt's Arab spring. *The Review of Financial Studies, 31*(1), 1–42.

Akaev, A., & Davydova, O. (2023). Climate and energy. Scenarios of energy transition and global temperature changes based on current technologies and trends. In V. Sadovnichy et al. (Eds.), *Reconsidering the limits to growth. A report to the Russian Association of the Club of Rome* (pp. 53–70). Springer. https://doi.org/10.1007/978-3-031-34999-7_4

Akaev, A., Malkov, S., Grinin, L., Bilyuga, S., Davydova, O., Grinin, A., Kovaleva, N., Malkov, A., Musieva, D., & Korotayev, A. (2023). Modeling social self-organization and historical dynamics: Modern society and a look into the global future: Cybernetic W-society. In V. Sadovnichy et al. (Eds.), *Reconsidering the limits to growth. A report to the Russian Association of the Club of Rome* (pp. 419–460). Springer. https://doi.org/10.1007/978-3-031-34999-7_19

Amin, K. (2019). The implications of China's regional military power projections: Challenges for India's military strategy. Global local interactions. *Journal of International Relations, 1*(1), 14–26.

Bandeira, L. A. M. (2019). A new US strategy to confront China's and Russia's growing military power. In *The world disorder* (pp. 61–76). Springer. https://doi.org/10.1007/978-3-030-03204-3_6

Baskan, B. (2019). Turkey's slow revolution. *Journal of Globalization Studies, 10*(2), 91–100.

Bengtson, V. (2018). *Global aging and challenges to families*. Routledge.

Bishop, M. (2018). China crisis? In *The coming crisis* (pp. 103–111). Palgrave Macmillan.

Brzezinski, Z. (1997). *The grand chessboard: American primacy and its geostrategic imperatives*. Basic Books.

Burnham, J. (1941). *The managerial revolution*. What is Happening in the World.

Campante, F. R., & Chor, D. (2012). Why was the Arab world poised for revolution? Schooling, economic opportunities, and the Arab spring. *The Journal of Economic Perspectives, 26*(2), 167–187.

Chase-Dunn, C., & Nagy, S. (2022). Global inequality and world revolutions: Past, present and future. In J. A. Goldstone, L. Grinin, & A. Korotayev (Eds.), *Handbook of revolutions in the 21st century: The new waves of revolutions, and the causes and effects of disruptive political change* (pp. 999–1022). Springer. https://doi.org/10.1007/978-3-030-86468-2_39

Chase-Dunn, C., Stäbler, R., Breckenridge-Jackson, I., & Herrera, J. (2022). Articulating the web of transnational social movements. In J. A. Goldstone, L. Grinin, & A. Korotayev (Eds.), *Handbook of revolutions in the 21st century: The new waves of revolutions, and the causes and effects of disruptive political change* (pp. 939–969). Springer. https://doi.org/10.1007/978-3-030-86468-2_37

Cutts, D., Goodwin, M., Heath, O., & Surridge, P. (2020). Brexit, the 2019 general election and the realignment of British politics. *The Political Quarterly, 91*(1), 7–23.

Dabashi, H. (2012). *The Arab spring: The end of postcolonialism*. Zed Books.

Davis, J. (2016). *The Arab spring and Arab thaw: Unfinished revolutions and the quest for democracy*. Routledge.

Fichtner, J. J. (2018). Global aging and public finance. *Business Economics, 53*(2), 72–78.

Fukuyama, F. (2002). *Our post-human future: Consequences of the bio-technology revolution*. Farrar, Straus, and Giroux.

Gills, B., & Morgan, J. (Eds.). (2022). *Economics and climate emergency*. Routledge.

Goldstone, J. A. (2015). Population ageing and global economic growth. In J. A. Goldstone, L. Grinin, & A. Korotayev (Eds.), *History & mathematics: Demography & Ageing* (pp. 147–155). Uchitel.

Goldstone, J. A., Grinin, L., & Korotayev, A. (2015). Research into global ageing and its consequences. In J. A. Goldstone, L. Grinin, & A. Korotayev (Eds.), *History & mathematics: Political Demography & Global Ageing* (pp. 5–9). Uchitel.

Goldstone, J. A., Grinin, L., & Korotayev, A. (2022a). Conclusion. How many revolutions will we see in the 21st century? In J. A. Goldstone, L. Grinin, & A. Korotayev (Eds.), *Handbook of revolutions in the 21st century: The new waves of revolutions, and the causes and effects of disruptive political change* (pp. 1037–1061). Springer. https://doi.org/10.1007/978-3-030-86468-2_41

Goldstone, J. A., Grinin, L., & Korotayev, A. (2022b). The Phenomenon and theories of revolutions. In J. A. Goldstone, L. Grinin, & A. Korotayev (Eds.), *Handbook of Revolutions in the 21st Century. The New Waves of Revolutions, and the Causes and Effects of Disruptive Political Change* (pp. 37–68). Springer. https://doi.org/10.1007/978-3-030-86468-2_2

Goldstone, J. A., Grinin, L., & Korotayev, A. (2022c). Introduction. Changing yet persistent: Revolutions and revolutionary events. In J. A. Goldstone, L. Grinin, & A. Korotayev (Eds.), *Handbook of revolutions in the 21st century: The new waves of revolutions, and the causes and effects of disruptive political change* (pp. 1–34). Springer. https://doi.org/10.1007/978-3-030-86468-2_1

Green, J., Edgerton, J., Naftel, D., Shoub, K., & Cranmer, S. J. (2020). Elusive consensus: Polarization in elite communication on the COVID-19 pandemic. *Science Advances, 6*(28), 2717.

Grinin, L. (2007). Production revolutions and periodization of history: A comparative and theoretic-mathematical approach. *Social Evolution & History, 6*(2), 11–55.

Grinin, L. (2009). The state in the past and in the future. *Herald of the Russian Academy of Sciences, 79*(5), 480–486.

Grinin, L. (2013). The Tiger and the dragon. Development models and perspectives of India and China. *Journal of Globalization Studies, 4*(1), 5–31.

Grinin, L. (2016). Evolution of world order. In L. Grinin & A. Korotayev (Eds.), *Evolution and big history: Dimensions, trends, and forecasts* (pp. 75–95). Uchitel.

Grinin, L. (2020). Revolutions as a factor of destabilization and reconfiguration of the world system in the 21st century and the Afroasian macrozone of instability. In L. Grinin, A. Korotayev, & D. Bykanova (Eds.), *System monitoring of global and regional risks* (pp. 146–180). Uchitel.

Grinin, L. (2022). Revolutions of the 21st century as a factor of the world system reconfiguration. In J. A. Goldstone, L. Grinin, & A. Korotayev (Eds.), *Handbook of revolutions in the 21st century: The new waves of revolutions, and the causes and effects of disruptive political change* (pp. 975–999). Springer. https://doi.org/10.1007/978-3-030-86468-2_38

Grinin, L., & Grinin, A. (2015). *Ot rubil do nanorobotov. Mir na puti k epokhe samoupravlyayemykh sistem (istoriya tekhnologiy i opisaniye ikh budushchego)*. Uchitel.

Grinin, L., & Grinin, A. (2016). *The cybernetic revolution and the forthcoming epoch of self-regulating systems*. Uchitel.

Grinin, L. E., & Grinin, A. L. (2022). The current wave of revolutions in the world-system and its zones. *Journal of Globalization Studies, 13*(2), 171–189. https://doi.org/10.30884/jogs/2022.02.12

Grinin, L., & Grinin, A. (2023). Technologies: Limitless possibilities and effective control. In V. Sadovnichy et al. (Eds.), *Reconsidering the limits to growth. A report to the Russian Association of the Club of Rome* (pp. 139–154). Springer. https://doi.org/10.1007/978-3-031-34999-7_8

Grinin, L., Grinin, A., & Korotayev, A. (2017). The MANBRIC-Technologies in the Forthcoming Technological Revolution. In T. Devezas, J. Leitão, & A. Sarygulov (Eds.), *Industry 4.0–Entrepreneurship and structural change in the new digital landscape: What is coming on along with the fourth industrial revolution* (pp. 243–261). Springer. https://doi.org/10.1007/978-3-319-49604-7_13

Grinin, L., Grinin, A., & Korotayev, A. (2020). A quantitative analysis of worldwide long-term technology growth: From 40,000 BCE to the early 22nd century. *Technological Forecasting and Social Change, 155*, 1–19. https://doi.org/10.1016/j.techfore.2020.119955

Grinin, L., Grinin, A., & Korotayev, A. (2021). Global trends and forecasts of the 21st century. *World Futures, 77*(5), 335–370. https://doi.org/10.1080/02604027.2021.1949939

Grinin, L., Grinin, A., & Malkov, S. (2023a). Economics. Optimizing growth. In V. Sadovnichy et al. (Eds.), *Reconsidering the limits to growth. A report to the Russian Association of the Club of Rome* (pp. 155–168). Springer. https://doi.org/10.1007/978-3-031-34999-7_9

Grinin, L., Grinin, A., & Malkov, S. (2023b). Modeling social self-organization and historical dynamics. Global futures and cybernetic W-society. In V. Sadovnichy et al. (Eds.), *Reconsidering the limits to growth. A report to the Russian Association of the Club of Rome (pp. 521–530).* Springer. https://doi.org/10.1007/978-3-030-*****-*_22

Grinin, L., Grinin, A., & Malkov, S. (2023c). Socio-political transformations: A difficult path to cybernetic society. In V. Sadovnichy et al. (Eds.), *Reconsidering the limits to growth. A report to the Russian Association of the Club of Rome* (pp. 169–189). Springer. https://doi.org/10.1007/978-3-031-34999-7_10

Grinin, L., Grinin, A., & Korotayev, A. (2023d). Global aging–An integral problem of the future. How to turn a problem into a development driver? In V. Sadovnichy et al. (Eds.), *Reconsidering the limits to growth. A report to the Russian Association of the Club of Rome* (pp. 117–135). Springer. https://doi.org/10.1007/978-3-031-34999-7_7

Grinin, L., Ilyin, I., & Andreev, A. (2016). World order in the past, present, and future. *Social Evolution and History, 15*(1), 58–84.

Grinin, L., & Korotayev, A. (2011). The coming epoch of new coalitions: Possible scenarios of the near future. *World Futures, 67*(8), 531–563. https://doi.org/10.1080/02604027.2011.625749

Grinin, L., & Korotayev, A. (2012). Does "Arab spring" mean the beginning of world system reconfiguration? *World Futures, 68*(7), 471–505. https://doi.org/10.1080/02604027.2012.697836

Grinin, L., & Korotayev, A. (2015). *Great divergence and great convergence*. Springer. https://doi.org/10.1007/978-3-319-17780-9

Grinin, L., & Korotayev, A. (2016). Global population ageing, the sixth Kondratieff wave, and the global financial system. *Journal of Globalization Studies, 7*(2), 11–31.

Grinin, L., & Korotayev, A. (2020). Seven weaknesses of the U.S., Donald Trump, and the future of American hegemony. *World Futures, 77*(1), 23–54. https://doi.org/10.1080/02604027.2020.1801309

Grinin, L., & Korotayev, A. (2023a). Africa: The continent of the future. Challenges and opportunities. In V. Sadovnichy et al. (Eds.), *Reconsidering the limits to growth. A report to the Russian Association of the Club of Rome* (pp. 225–238). Springer. https://doi.org/10.1007/978-3-031-34999-7_13

Grinin, L., & Korotayev, A. (2023b). High-income and low-income countries. Toward a common goal at different speeds. In V. Sadovnichy et al. (Eds.), *Reconsidering the limits to growth. A report to the Russian Association of the Club of Rome* (pp. 207–224). Springer. https://doi.org/10.1007/978-3-031-34999-7_12

Grinin, L., & Korotayev, A. (2022). The future of revolutions in the 21st century and the world system reconfiguration. *World Futures, 1–22*, 69. https://doi.org/10.1080/02604027.2022.2050342

Grinin, L., Korotayev, A., & Tausch, A. (2019). *Islamism, Arab spring, and the future of democracy*. Springer. https://doi.org/10.1007/978-3-319-91077-2

Grinin, L., Malkov, S., Grinin, A., & Korotayev, A. (2022). Will capitalism die? Reflections on the capitalism of the past, present and future. *Sotsiologicheskiy Zhurnal = Sociological Journal, 28*(2), 100–130. https://doi.org/10.19181/socjour.2022.28.2.8988

Grinin, L., Tsirel, S., & Korotayev, A. (2015). Will the explosive growth of China continue? *Technological Forecasting and Social Change, 95*, 294–308. https://doi.org/10.1016/j.techfore.2014.06.023

Haas, M. L. (2015). Population ageing and the future of the great powers. In J. A. Goldstone, L. Grinin, & A. Korotayev (Eds.), *History & mathematics: Demography & ageing* (pp. 133–146). Uchitel.

Haas, M. L., & Lesch, D. W. (Eds.). (2017). *The Arab spring: The hope and reality of the uprisings*. Westview Press.

Hofbauer, H., & Komlosy, A. (2020). Corona-Krise: Anschub für eine kybernetische Wende. *Telepolis, 20*(05), 2020. https://www.heise.de/tp/features/Corona-Krise-Anschub-fuer-eine-kybernetische-Wende-4723168.html

Holslag, J. (2017). *Trapped giant: China's military rise*. Routledge.

Irkhin, A., & Moskalenko, O. (2022). The global crisis and the USA search for its new identity: Trump's attempt of transition from globalism to isolationism. *Journal of Globalization Studies, 13*(2), 21–33. https://doi.org/10.30884/jogs/2022.02.02

Issaev, L., Fain, E., & Korotayev, A. (2021). Impact of the Arab Spring on terrorist activity in the Sahel. *Ideology and Politics Journal, 3*(19), 34–49. https://doi.org/10.36169/2227-6068.2021.03.00003

Issaev, L. M., Korotayev, A. V., & Bobarykina, D. A. (2022). The global terrorist threat in the Sahel and the origins of terrorism in Burkina Faso. *Vestnik RUDN. International Relations, 22*(2), 411–421. https://doi.org/10.22363/2313-0660-2022-22-2-411-421

Jackson, R., Howe, N., Strauss, R., & Nakasima, K. (2008). *The graying of the great powers. Demogarphy and Geopololitics in the 21st century*. Center for Strategic & International Studies.

Jackson, R., Macaranas, R., & Peter, T. (2013). U.S. development policy in an aging world. New Challeges and new priorities for a new demographic era. In *A research report from project on U.S. leadership*. Center for Strategic & International Studies.

Kaphen, C. (2004). *Zakat Ameriki. Uzhe skoro*. AST.

Kipfer, S. (2019). What colour is your vest? Reflections on the yellow vest movement in France. *Studies in Political Economy, 100*(3), 209–231.

Kissinger, H. (1994). *Diplomacy*. Simon & Schuster.

Kissinger, H. (2001). *Does America need a foreign policy? Toward a diplomacy for the 21st century*. Simon & Schuster.

Kissinger, H. (2014). *World order*. Penguin Press.

Komlosy, A. (2019). Crises, long waves, and world system analysis. *Journal of Globalization Studies, 10*(2), 55–76.

Korotayev, A., Meshcherina, K., & Shishkina, A. (2018). A wave of global sociopolitical destabilization of the 2010s: A quantitative analysis. *Democracy and Security, 14*(4), 331–357. https://doi.org/10.1080/17419166.2018.1517337

Korotayev, A., & Khokhlova, A. (2023). MENA sociopolitical destabilization dimensions in comparative global perspective. *Journal of Globalization Studies, 14*(1), 67–94. https://doi.org/10.30884/jogs/2023.01.05

Korotayev, A., Malkov, S., & Musieva, J. (2023a). Demography: Toward optimization of demographic processes. In V. Sadovnichy et al. (Eds.), *Reconsidering the limits to growth. A report to the Russian Association of the Club of Rome* (pp. 97–116). Springer. https://doi.org/10.1007/978-3-031-34999-7_6

Korotayev, A., Shulgin, S., Ustyuzhanin, V., Zinkina, J., & Grinin, L. (2023b). Modeling social self-organization and historical dynamics: Africa's futures. In V. Sadovnichy et al. (Eds.), *Reconsidering the limits to growth. A report to the Russian Association of the Club of Rome* (pp. 461–490). Springer. https://doi.org/10.1007/978-3-031-34999-7_20

Kurtz, D. V. (2019). Politics, culture, rhetoric, global warming: From local to global realities. *Journal of Globalization Studies, 10*(1), 3–18.

Lawson, G. (2022). Revolutions in the twenty-first century. *Journal of Globalization Studies, 13*(2), 190–192. https://doi.org/10.30884/jogs/2022.02.13

MacEoin, D. (2018). Death of democracy. *Gatestone Institute International Policy Council, 07*(02), 2018. https://www.gatestoneinstitute.org/11807/democracy-death

Malik, A., & Awadallah, B. (2013). The economics of the Arab spring. *World Development, 45*, 296–313.

Manheim, D. (2020). The fragile world hypothesis: Complexity, fragility, and systemic. *Futures, 122*, 1–16. https://doi.org/10.1016/j.futures.2020.102570

Marwah, O. (2019). *Military power and policy in Asian states: China, India*. Routledge.

Mitroff, I. I. (2020). Corona virus: A prime example of a wicked mess. *Technological Forecasting and Social Change, 157*, 1–12. https://doi.org/10.1016/j.techfore.2020.120071

Msellemu, S. A., & Kessy, E. M. (2022). 'Arab Spring' and its destabilizing effects to the world. *Journal of Globalization Studies, 13*(1), 119–132. https://doi.org/10.30884/jogs/2022.01.08

National Intelligence Council. (2008). *Global trends 2025: A transformed world*. National Intelligence Council.

National Intelligence Council. (2012). *Global trends 2030: Alternative worlds*. National Intelligence Council.

National Intelligence Council. (2017). *Global trends: Paradox of progress*. National Intelligence Council.

Peterson, P. G. (1999). Gray down: The global aging crisis. *Foreign Affairs, 78*(1), 42–55.

Powell, J. L., & Khan, H. T. (2013). Ageing and globalization: A global analysis. *Journal of Globalization Studies, 4*(1), 137–146.

Sadovskaya, L., Fakhrutdinova, N., & Kochanova, T. (2019). Global political destabilization and modern civic protest movements in sub-Saharan Africa. *Journal of Globalization Studies, 10*(2), 77–90.

Sadovskaya, L., Fakhrutdinova, N., & Kochanova, T. (2021). Evolution of civic protest movements in sub-Saharan Africa: From independence to the present day. *Social Evolution & History, 20*(2), 157–173. https://doi.org/10.30884/seh/2021.02.07

Steinert-Threlkeld, Z. C. (2017). Spontaneous collective action: Peripheral mobilization during the Arab spring. *American Political Science Review, 111*(2), 379–403.

Taylor, K. B. (2020). The passing of western civilization. *Futures, 122*, 1–14. https://doi.org/10.1016/j.futures.2020.102582

Todd, E. (2003). *After the empire: The breakdown of American order*. Columbia University Press.

Tsirel, S. (2020). Future technological achievements as a challenge for post-singularity human society. In A. Korotayev, D. LePoire, & D. (Eds.), *The 21st century singularity and global futures. A big history perspective* (pp. 419–437). Springer. https://doi.org/10.1007/978-3-030-33730-8_19

Updegrove, A. H., Cooper, M. N., Orrick, E. A., & Piquero, A. R. (2020). Red states and black lives: Applying the racial threat hypothesis to the black lives matter movement. *Justice Quarterly, 37*(1), 85–108.

Vague, R. (2015). The coming China crisis. *Democracy, 36*, 16–28.

Vasiliev, A. (2022). Contradictions between Islamists in the regional and global context: Wahhabis, Muslim Brothers, al-Qaeda. *Journal of Globalization Studies, 13*(1), 133–147. https://doi.org/10.30884/jogs/2022.01.09

Wallerstein, I. (2003). *The decline of American power. The U.S. in a chaotic world*. New Press.

Wallerstein, I. (2013). Structural crisis, or why capitalists can no longer find capitalism rewarding. In I. Wallerstein, R. Collins, M. Mann, G. Derluguian, & C. Calhoun (Eds.), *Does Capitalism Have a Future?* (pp. 9–36). Oxford University Press.

von Weizsaecker, E., & Wijkman, A. (Eds.). (2018). *Come on! capitalism, short-termism, population and the destruction of the planet. A Report to the Club of Rome*. Springer.

Zakaria, F. (2008). *The post-American world*. W.W. Norton.

Zimmer, Z. (2016). *Global ageing in the twenty-first century: Challenges, opportunities and implications*. Routledge.

Leonid Grinin is Senior Research Professor at Centre for Stability and Risk at HSE University in Moscow, as well as Senior Research Professor of the Institute of Oriental Studies, Russian Academy of Sciences. He is co-editor of the international journals *Social Evolution & History* and the *Journal of Globalization Studies*, as well as co-editor of the international yearbooks *Evolution, History & Mathematics*, and *Kondratieff Waves*. He is author of more than 600 publications, including 33 monographs in Russian, English, Spanish, German, and Chinese. His current research focuses on comparative political studies, theory of revolution, political anthropology, global economy, global history, historical sociology, and futurology. Among his monographs are *Great Divergence and Great Convergence. A Global Perspective* (Springer, 2015; with Andrey Korotayev), *The Cybernetic Revolution and the Forthcoming Epoch of Self-Regulating Systems* (Uchitel, 2016, with Anton Grinin), Economic Cycles, Crises, and the Global Periphery (Springer, 2016; with Andrey Korotayev and Arno Tausch), *Big History of Globalization. The Emergence of a Global World System* (Springer 2019, in co-authorship with Julia Zinkina, David Christian and others), *Islamism, Arab Spring, and the Future of Democracy. World System and World Values Perspectives* (Springer, 2019; with Andrey Korotayev and Arno Tausch). In 2004, he was awarded as a winner of the best publication of *Socis* journal during the year for the article "People of Publicity–a New Social Layer?" In 2012, he was awarded with the Gold Kondratieff Medal by the International N. D. Kondratieff Foundation. In 2016, he received the Belyaev award for the book *From Bifaces to Nanorobots* (the book published in Russian in co-authorship with Anton Grinin).

Anton Grinin is a Russian scientist specializing in the study of modern technological trends and futures studies, as well as evolutionist and philosopher. He has a PhD in Biology and works as a Research Fellow for Lomonosov Moscow State University, Russia, as well as at the Centre for Stability and Risk at HSE University in Moscow. He is the author of more than 100 publications in Russian and English, including 3 monographs. Awards: Gold Kondratieff Young Scholars Medal, Alexander Belyaev Literature Award. Anton Grinin focuses on the issues of the technological revolution (the Cybernetic revolution) evolving in the current decades of the twenty-first century and the resultant profound transformations in economy and society as well as related economic, political, ethic-legal, and other risks, including political revolutions. He also investigates the correlation between the Cybernetic revolution and Kondratieff waves as well as between the Cybernetic revolution and global aging.

Andrey Korotayev heads the Centre for Stability and Risk at HSE University in Moscow, Russia. He is also Senior Research Professor at the Institute of Oriental Studies and Institute for African Studies, Russian Academy of Sciences, as well as at the Faculty of Global Studies of Lomonosov Moscow State University and the Russian Presidential Academy of National Economy and Public Administration. He has authored or co-authored over 650 scholarly publications, including such monographs as *Ancient Yemen* (Oxford University Press, 1995), *World Religions and Social Evolution of the Old World Oikumene Civilizations: A Cross-Cultural Perspective* (The Edwin Mellen Press, 2004), *Introduction to Social Macrodynamics: Compact Macromodels of the World System Growth* (URSS, 2006; with Daria Khaltorina and Artemy Malkov), *Introduction to Social Macrodynamics: Secular Cycles and Millennial Trends* (URSS, 2006; with Daria Khaltorina and Artemy Malkov), *Great Divergence and Great Convergence. A Global Perspective* (Springer, 2015; with Leonid Grinin), *Economic Cycles, Crises, and the Global Periphery* (Springer, 2016; with Leonid Grinin), *Islamism, Arab Spring, and the Future of Democracy. World System and World Values Perspectives* (Springer, 2019; with Leonid Grinin and Arno Tausch). He is a laureate of Russian Science Support Foundation Award in "The Best Economists of the Russian Academy of Sciences" Nomination (2006); in 2012, he was awarded with the Gold Kondratieff Medal by the International N. D. Kondratieff Foundation.

High-Income and Low-Income Countries. Toward a Common Goal at Different Speeds

Leonid Grinin, Sergey Malkov, and Andrey Korotayev

Abstract

The chapter discusses the necessity of choosing different strategies for different groups of countries for a successful movement to the recognized worldwide goals. In this respect, strategies as well as policies depend on different causes and conditions, including demographics, natural resources, historical traditions, etc., but especially on levels of development of countries which are very uneven. The most important and deep gap is the one between the income levels of high-income and low-income countries. However, it happens quite often when high-income countries as well as their (or world) organizations impose on low-income countries strict and almost single-type programs for economic and social reforming, transforming of agricultural or energetic spheres, and so on for achievement of approved world goals (as democracy, carbon neutrality, sustainable development, decrease of inequality and injustice, increase in state efficiency, rates of economic growth, consumptions optimizing, etc.). These programs are worked out on the basis of high-income countries' experience and

L. Grinin
HSE University, Moscow, Russia

Institute of Oriental Studies, Russian Academy of Sciences, Moscow, Russia

S. Malkov
Lomonosov Moscow State University, Moscow, Russia

HSE University, Moscow, Russia

Institute of Economics, Russian Academy of Sciences, Moscow, Russia
e-mail: s@malkov.org

A. Korotayev (✉)
HSE University, Moscow, Russia

Institute of Oriental Studies, Russian Academy of Sciences, Moscow, Russia

Faculty of Global Studies of Lomonosov Moscow State University, Moscow, Russia

V. Sadovnichy et al. (eds.), *Reconsidering the Limits to Growth*, World-Systems Evolution and Global Futures, https://doi.org/10.1007/978-3-031-34999-7_12

they are often not suitable for many low-income countries. Grinin et al. maintain that there should not be single-type recommendations, programs, etc. for all countries for their taking part in the world-wide movement toward common goals. There are common world goals, but for their achievement different countries would need strategies, policies, programs, and measures of different types, which would take into consideration specifics of different societies. Though there still is a great gap between high-income and low-income countries, Grinin & Korotayev note that the process of "Great Divergence" (when the abovementioned gap had been increasing since the early nineteenth century) in the recent decades was replaced by the "Great Convergence," when this difference in development is reduced by increasing the level of development in less prosperous states. This convergence is characterized by heterogeneity—different rates of development of certain states, on the basis of which fast-growing states and less high-income states (mainly the countries of Tropical Africa) are distinguished. This feature requires different approaches in solving global problems common to all countries of the world. In view of this, Grinin et al. draw attention to the need to form global strategies for moving toward sustainable development, taking into account differences in development and available opportunities. In some countries, the level of development of capitalism has reached a level at which it is already exhausting itself and transformations are required, while there are countries in which full-fledged market relations have not yet been formed. Grinin et al. maintain that in the coming decades, the world will experience a complex continuum of types, transitions, combinations, and forms of economic structures. The authors analyze two main types. The high-income countries will experience a shift toward the strengthening of distributive institutions. The low-income countries that continue the processes of economic and socio-political modernization, in their turn, will proceed toward the strengthening of capitalist market institutions though with some important restrictions preventing exploitation and environmental degradation. All this will take place during the final phase of the Cybernetic Revolution, with the development of the process of internationalization of the labor force (through the development of remote employment) along with the internationalization of capital. Thus, the most important task in the coming decades is to increase the flexibility of the socioeconomic development of all countries of the world. Grinin & Korotayev believe that it is necessary to find ways to combine all the best features of capitalism in order to provide the necessary dynamism and innovativeness of development, at the same time softening all the distortions that have formed as a result of the development of capitalism over the past twenty years, bearing in mind that not so much industrial capitalism as financial and global financial capitalism is responsible for it.

1 Variation in Levels of Countries' Development Demands and Different Types of Strategies and Programs for Achieving Common World Goals

As a result of the industrial technological transition, the world has become connected and unified. The "Great Divergence" (the initial separation of the industrialized Western countries from the countries of the periphery of the World System) was replaced by the "Great Convergence," that is, the rapid catch-up of the economies of the low-income countries to the high-income ones (for detail see Grinin & Korotayev, 2015). Convergence issues will play an increasingly important role. However, the low-income countries are catching up at very different speeds. Some of them are rapidly growing (especially China) and have achieved notable success in raising their living standards (for example, the countries of the Persian Gulf, or Mexico). Others, the so-called "less developed" countries, are not proceeding fast enough—many of these are in Tropical Africa. Therefore, the most serious problem in this plan is the integration of Africa into world development (on this see Chapter "Africa—the Continent of the Future. Challenges and Opportunities" [Grinin & Korotayev, 2023, this volume]).

Since different regions of the world have different rates of development, approaches to solving the most important world problems are significantly transformed for different groups of countries.

We note in some chapters of this report to the Russian Association of the Club of Rome[1] that the development of technology in the high-income countries has made it possible for the first time in the history of humankind to satisfy the material needs of the vast majority of the population. However, this is not true for many low-income countries. Although the number of people below the UN poverty line has been decreasing significantly, in general, the standard of living in many countries, especially in Tropical Africa and some parts of Asia, leaves much to be desired.[2] Therefore, it is impossible to preserve the standard of living of these countries at the present level. Hence, the question of economic growth rates, which we raised in chapter "Economics. Optimizing Growth" [Grinin et al., 2023b, this volume],

[1] For example, Chapter "Macrohistorical approach" (Sadovnichy et al., 2023a, this volume), Chapter "Future Society and Transition to It" (Sadovnichy et al., 2023b, this volume), Chapter "Conclusion. Reconsidering The Limits—suggestions (Come On!)" (Sadovnichy et al., 2023c, this volume), Chapter "Global aging–An integral problem of the future. How to turn a problem into a development driver?" (Grinin et al., 2023c, this volume), and Chapter "Modeling Social Self-Organization and Historical Dynamics. Modern Society and a Look into the Global Futures: Cybernetic W-Society" (Akaev et al., 2023b, this volume).

[2] Bertrand Schneider in his report to the Club of Rome (Schneider, 1995) dwelled on wasted and misplaced aid, aid diversion scandals, development profiteers. The author proposed a new development concept: "The well-being of people and societies everywhere" (WISE) and a number of strategic applications of this concept, for example, empowerment through property rights, a radical new approach to development financing, measures to combat corruption, education, population control, strengthening the role of civil society and NGOs. These measures require tough decisions involving not only international institutions, but also member states.

sounds differently for the high-income and the low-income countries. For the former, the slowdown in economic growth might be painful, but not fatal; for the latter, it is critical, since it can lead to a serious humanitarian crisis.

As Ralf Fücks rightly notes, "the debate around society beyond growth does not take into account the prospect of global growth dynamics in the coming decades. *Whether the world economy will grow or not is not decided in Europe*" (Fücks, 2013: 18). He emphasizes that it is no longer a question of whether the growth of the world economy will continue, but rather how it will continue. Zero growth is unrealistic and, given the scale of poverty, highly undesirable. It is also impossible to continue in the same vein (Fücks, 2013). Fücks's book rightly proposes embarking on a third path, i.e., environmentally sustainable and socially acceptable growth (ibid.).

Therefore, the general task of changing the consumption model of society and individuals means focusing on the optimization of consumption and the reduction of excessive consumption for high-income countries (although this must be done very carefully and accurately, since rather large population groups in high-income countries underconsume). For the low-income countries, the optimization of consumption (including the aspect of moving toward the reduction of carbon emissions) should proceed with an emphasis on taking into account the strong underconsumption of their populations.

Therefore, *the general movement toward a more climate- and ecofriendly economy (as well as toward a more equitable distribution) should proceed at different speeds and allow bringing the levels of development of different groups of countries closer to each other.*[3]

[3]This issue has been considered in a number of reports to the Club of Rome, though not recently. In Guernier's (1980) report, an attempt was made to comprehend the historical forms of economic development of the countries of Asia, Africa and Latin America. It contained a call to choose an alternative path of development, which would involve the orientation of low income countries not so much on outside assistance as on self-sufficiency. The report emphasized the need to reorganize agriculture on a priority basis of greater centralization of the financial and organizational efforts of low income countries. Report of Lenoir (1984) analyzed the problems of food shortages in low income countries and considered scenarios for the further development of these countries, taking into account the possibilities of importing food products from high income countries or their orientation towards self-sufficiency. In this study, preference was given to such ways of development of the third world, in which the emphasis was on self-sufficiency. At the same time, this report considered specific development strategies, according to which priority was given to independent agricultural communities and their self-government. Schneider in his report *The Barefoot Revolution* (see Schneider, 1988) is giving an overview of many small projects carried out by various teams of researchers in Asia, Africa and Latin America. Based on this review, it was emphasized that the activities of non-governmental organizations that took part in the development of small projects turned out to be more effective than the work of state institutions. However, according to Schneider, it is necessary to move from micro-projects to integrated development projects based on a global vision of the prospects of humankind. This is a new approach to development based on the understanding that the initiative of action is moving from the North to the South, the priorities of political activity are given to meeting the needs of the rural population, more attention is paid not to the export and import of food, but to its local production and consumption. What is happening today in Asia, Africa and Latin America can be called a "barefoot revolution". It is characterized by a

With regard to demographic problems, we can say that they vary greatly even within low-income countries. Some of them, like China, have practically the same problems as the high-income ones, that is, very low birth rates, an aging population, and the exhaustion of the demographic dividend (see Chapter "Demography. Toward the Optimization of Demographic Processes" [Korotayev et al., 2023a, this volume] and Chapter "Global Aging—an Integral Problem of the Future. How to Turn a Problem into a Development Driver?" [Grinin et al., 2023c, this volume]). Others, like India, have entered the second phase of the demographic transition, their birth rate has fallen and continues to fall, but the population is still young, and the demographic dividend is huge. Accordingly, they are faced with the task of providing this young population with jobs, so their economy should grow. The third group (Tropical Africa and some Asian countries) has recently entered the second phase of the demographic transition, these countries still have to solve the issue of reducing the birth rate (see Chapter "Africa—the Continent of the Future. Challenges and Opportunities" [Grinin & Korotayev, 2023, this volume] and Chapter "Modeling Social Self-Organization and Historical Dynamics. Africa's Futures" [Korotayev et al., 2023b, this volume]).

To conclude this paragraph, we would like to say about the necessity of choosing different strategies for different groups of countries for a successful movement to the recognized worldwide goals. In this respect, strategies as well as policies depend on different causes and conditions, including demographics, natural resources, historical traditions, etc., but especially on levels of development of countries that are very uneven. The most important and deep gap is the one between the income levels of high-income and low-income countries. However, it happens quite often when high-income countries as well as their (or world) organizations impose on low-income countries strict and almost single-type programs for economic and social reforming, transforming of agricultural or energetic spheres and so on for achieving of approved world goals (such as democracy, carbon neutrality, sustainable development, decrease of inequality, injustice, and state efficiency). These programs are worked out on the basis of high-income countries' experience, and they are often not suitable for many low-income countries.

Thus, we maintain that there should not be a single type of recommendations, programs, etc. for all countries for their participation in the worldwide movement to common goals. There are common world goals, but, for their achievement, different countries would need different types of strategies, policies, programs, and measures, which would take into consideration specifics of different societies.

peaceful character and is carried out within the framework of respect for the laws. This revolution is not yet correlated with political parties and ideological structures. However, according to Schneider, over time it can develop into a political movement or become a destructive force in the hands of extremist leaders.

2 The Problem of Economic Growth and the Future of Capitalism

The question of the future of capitalism, including whether it has a future at all, is not new. But the crisis of 2008 and subsequent major changes gave this problem a new dimension. We noted above (see Chapter "Economics. Optimizing Growth" [Grinin et al., 2023b; see also Grinin et al. 2022] that the era of capitalism, based on liberal market principles and focused on maximizing consumption and profits, has approached a milestone beyond which capitalism will continue its transformation toward even more regulation than today, including the regulation on the international scale. The question of how quickly and in what directions such a transformation will go and how it will manifest itself in different groups of countries (the high-income ones, the low-income ones, and the former socialist ones) and regions are very important and is one of the main intrigues of the world development in the twenty-first century. *While in the high-income countries with a long history of capitalist development capitalism has come to the point of serious transformation, many low-income countries suffer from the dominance of archaic, pre-capitalist relations.*

After the crisis of 2008, *a new reason for increasing criticism of capitalism has clearly manifested itself (and has become the most important one)*, namely the reduction in profitability within the United States and in profitable investments in the world in general. In the collective monograph *Does Capitalism Have a Future?* (Wallerstein et al., 2013) Immanuel Wallerstein even titled his chapter "Why Capitalists May No Longer Find Capitalism Rewarding" (Wallerstein, 2013). In his opinion, this can happen because capitalism no longer leads to such an increase in capital as before.

In fact, it has been argued that neoliberalism should be replaced by some semblance of socialism, and some macrosociologists openly use the term socialism when describing the evolution of the contemporary societies in the future,[4] while others prefer euphemisms.

Recipes and forecasts here vary greatly, but the mainstream is increasingly inclined toward something that is more or less similar to socialist (or state capitalist) ideas. Therefore, it is not at all surprising that in the past few years, there has been more and more talk about the fact that capitalism is bad, that the desire for profit, as well as for economic growth, harms the climate, the environment, humanity as a whole, that it is necessary to replace it with something else.[5] However, even some critics of capitalism doubt, due to various reasons, that it might be really replaced.

---

[4]Thus, Randal Collins (2013) predicts that capitalism will most likely be replaced by socialism. It lies in state ownership or tight control over the sources of profit, central planning and redistribution. The pursuit of profit should be replaced by concern for the broad masses of the population.

[5]To this one may add the criticism of such a basic modern indicator of economic growth as GDP. The GDP indicator is often subjected to generally fair criticism (see Chapter "Analyzing Social Self-Organization and Historical Dynamics. Life Quality Index" [Malkov et al., 2023a, this volume]; see also, e.g., Easterlin, 2010; Costanza et al., 2014; McElwee & Daly, 2014; Goldsmith, 2019; Kapoor & Debroy, 2019), but often criticism is mainly intended to hide the fact that the potential of high income states has weakened. Hence the desire to question the general need for

For example, Jorgen Randers (2012) maintains that, though some changes are bound to happen, the result is likely not to be so strong to call it "the end of capitalism." We also find that, though the idea to replace capitalism with a better system sounds attractive, there is no such an alternative system yet, and it is not likely to appear very soon. The capitalist system will transform gradually. The main question is—into what it will evolve in the long run. We discuss this question further.

We note that expanded reproduction, that is, economic growth, is a feature of the capitalist economy, which has been noted by many of its researchers, in particular Karl Marx, as well as Simon Kuznets (1966). Without the pursuit of profit, there is no capitalism. This raises a very important question: in the light of the observed trend toward a slowdown in global demographic and economic indicators, will there not be a situation in the future when there will be neither growth nor the opportunity to increase profit? And, if that happens, what is the future of capitalism?

In our opinion, when discussing the fate of capitalism, it is necessary to clearly separate the concepts of "*capitalism as a socio-economic system*" and "*capitalism as a sector of economy,*" which are often mixed, giving rise to confusion in reasoning (for more details, see Chapter "Modeling Social Self-Organization and Historical Dynamics. Global Phase Transitions" [Malkov et al., 2023b, this volume], as well as Grinin et al., 2022).

Capitalism as an economic sector is a way of doing business, a sector of the economy with its characteristic form of ownership, values, and way of life. Capitalism as an economic sector can be a part of a socioeconomic system that includes several sectors/modes, occupying certain niches within it. But here it is extremely important whether this sector is the leading one in the general system of sectors, or just an additional one (see below).

The socioeconomic system (SES) has a higher level of organization than the economic sector. SES is understood as an integral set of interrelated and interacting social and economic institutions and relations. Accordingly, capitalism *as a socio-economic system* is an SES in which the capitalist economic sector occupies the dominant position, determining the nature of social relations and institutions and subjugating or strongly influencing other sectors. Since the most important characteristics of capitalism are private ownership of the means of production and the desire for profit, then for the sustainable functioning of "capitalism as a socio-economic system" it is necessary to provide conditions under which, in the presence of competitive market relations, the vast majority of economic agents in society would have positive profit (excess of income over expenses). This is possible only under conditions of continuous and, moreover, substantial (considering the significant variation of profitability in different areas of business) economic growth, so that profit (even if small) goes not only to the leaders but also to those who are lagging behind in the economic race, in order to give them a chance in the next production

economic growth, but the rejection of growth as a goal objectively leads to the stagnation of society (see Grinin & Grinin, 2021a, 2021b, 2021c; Grinin, 2020 for details).

cycle to increase their efforts and try to get ahead. That is, the "positive-sum game[6]" (a game theory term) must be constantly maintained, which is possible only with continuous growth.

When growth slows down or stops as a result of depression or stagnation, then "capitalism as a socio-economic system" loses stability and must be supported by special measures on the part of the state (see Keynes, 2016 [1936]).

"Capitalism as an economic sector" arose a very long time ago, but in the era of agrarian societies, it occupied only a relatively small economic niche (mainly in the area related to trade).[7]

During the Industrial Revolution, especially since the late 18th—early nineteenth century, the countries of the world system core experienced a rapid increase in labor productivity and created conditions for a "positive sum game" supported by the influx of raw materials from the countries of the world-system periphery (Grinin & Grinin, 2015, 2016; Grinin & Korotayev, 2015; Korotayev & Grinin, 2017). This allowed capitalism to strengthen and widen capitalist sector and then to transform it into a socioeconomic system with the corresponding socio-political institutions of the Y-type (see Chapter "Modeling Social Self-Organization and Historical Dynamics. Industrial Society" (Akaev et al., 2023a, this volume]), while the socio-political institutions of the X-type remained dominant in the countries of the periphery.[8]

At the same time, at the beginning of the development of industrial capitalism, the growth of profits and the expansion of production occurred both due to the growth of productivity and due to the increased exploitation of wage workers. In mature capitalism, with its limitations of the working hours and minimum wage requirements,[9] profits are increasingly dependent on the growth of labor productivity. Since the second half of the twentieth century, economic growth in high-income countries has increasingly been driven by increased investment in human capital. And now this trend has begun to intensify in low-income countries.

Is it right to consider capitalism in the late 20th–early 21st centuries and capitalism in the 18th–19th centuries as the same thing? A possible answer is that although the main features of capitalism were preserved, they were significantly changed due to the growth of general wealth, the shifts in the redistribution system, and a

[6] A "positive-sum game" is a situation where, as a result of the interaction of players, the gain of one player does not mean the loss of the other, and the total result of the game is greater than zero.

[7] See Chapter "Modeling Social Self-Organization and Historical Dynamics. Agrarian Society" (Malkov et al., 2023c, this volume) and Chapter "Modeling Social Self-Organization and Historical Dynamics. Global Phase Transitions" (Malkov et al., 2023b, this volume).

[8] For more details on X-type and Y-type societies see Chapter "Economics. Optimizing Growth" (Grinin et al., 2023b, this volume), Chapter "Socio-Political Transformations. A Difficult Path to Cybernetic Society" (Grinin et al., 2023d, this volume), and Chapter "Analyzing Social Self-Organization and Historical Dynamics. Future Cybernetic W-Society: Socio-Political Aspects" (Grinin & Grinin, 2023a, this volume); see also Kirdina, 2014.

[9] The limitation of the exploitation of wage workers was largely the result of the intensified struggle of workers for their rights and the emergence of the USSR and the socialist countries as an alternative socio-economic system to capitalism.

significantly increased role of the state. This allows smoothing out the social contradictions generated by liberal market relations.[10]

In the light of the above said, the problem of the fate of capitalism can be seen as follows. The current transition to a unified world, the emergence of serious climate and environmental problems,[11] and the slowdown in economic and demographic growth indicate that the era of extensive economic growth is coming to an end. In the face of diminishing opportunities for the "positive-sum game," the stability and effectiveness of socio-political institutions of the Y-type are declining, and with them the "capitalism as a socio-economic system" is falling as well. Under these conditions, the introduction of additional methods of regulation is inevitable; their task will be to harmonize the interests of various socioeconomic and socio-political actors in order to increase the stability of society. In essence, we are talking about a combination of Y- and X-type institutions, about creating such a balance of them that, using redistributive mechanisms, would ensure social stability, but at the same time would not suppress economic activity. This is not a trivial task, because, as shown in Chapter "Economics. Optimizing Growth" (Grinin et al., 2023b, this volume) and Chapter "Modeling Social Self-Organization and Historical Dynamics. Global Phase Transitions" (Malkov et al., 2023b, this volume), those requirements are not easily combined with each other. In essence, we are talking about the formation of a new system of institutions, a new image of society, adequate to the new era, which we called "Cybernetic W-society" above in Chapter "Socio-Political Transformations. A Difficult Path to Cybernetic Society" (Grinin et al., 2023d, this volume). So, in the coming decades, the world will experience a complex continuum of types, transitions, combinations, and forms of economic structures; various combinations of X- and Y-structures and institutions as well as their evolution into new W-structures and institutions. Anyhow among this great number of combinations, we can single out two variants of further transformation of capitalism.

The first way This is the way of the low-income countries, in which the processes of socioeconomic modernization and catching up with the economically high-income countries of the modern world-system core[12] will continue. In those

[10]The greatest success in the combination of Y- and X-institutions of socio-economic development was achieved by the countries of Northern Europe.

[11]See Chapter "Climate and Energy. Energy Transition Scenarios and Global Temperature Changes Based on Current Technologies and Trends" (Akaev & Davydova, 2023, this volume) and Chapter "Ecology. Life in the "Unstable Biosphere"" (Kovaleva, 2023, this volume).

[12]Note that ongoing processes of the Great Convergence and World System reconfiguration imply the growing decrease of differences between the world-system core and periphery, as well as the movement of some former peripheral countries to the world-system core and vice versa. For example, China can be hardly characterised now as "peripheral" country, though this was quite appropriate a few decades ago. In the present world, it is rather a part of the world-system core, whereas Australia (that was often characterised as a core state [e.g., Wallerstein, 2004]), should now be more appropriately characterized as a peripheral country with respect to China (note, for example, that the main source of foreign currency for Australia now is the exportation of raw iron ore to China).

countries, market Y-institutions are underdeveloped. Their leading sectors in the coming decades will be *industrial and post-industrial production of the first type* (that is, simple, unskilled, or insufficiently skilled types of services).

Despite the fact that in these countries capitalism will be significantly limited by the regulation of the state and the modern level of the rights of workers (including women, children, etc.), the level of exploitation in them (as it was in China) will remain high (low wages, long working hours, weak labor protection and social guarantees, etc.). In a sense, this is the level of capitalism of the first half to the middle of the twentieth century in high-income countries. Given the gigantic mass of under-qualified young people and the middle-aged population in low-income countries, there can be no other option; otherwise, it is impossible to provide everyone with work. Gradually, the limitations of exploitation will reach the level of the second half to the end of the twentieth century in high-income countries.

It is possible that these countries (mainly African) will become not just suppliers of raw materials and semi-finished products but also to a noticeable extent (like the countries of Asia today) a world workshop and (like today's India) a powerful supplier of complex services, while they will combine classical and postclassical capitalism along with state socialism and state capitalism (that is, Y- and X-type institutions will combine and interact). Capital from the high-income countries will be transferred to these growing economies (as the conditions for extensive growth will still be present there). However, the socio-political situation in the low-income countries will be unstable due to the presence of "youth bulges" (especially in the countries of Tropical Africa[13]), imperfections and immaturity of political institutions, the presence of so-called "development traps."[14]

In general, these countries are facing a difficult path to find their own modernization model adapted to local conditions. This path will be difficult because of socioeconomic instability and a significant number of socio-political risks. It will be possible to pass it successfully only with the support of high-income countries, if *option No. 2* of the movement toward a future global society is implemented.[15] If the transition to the future society follows *option No. 1*, then the instability of the modernization process in low-income countries is highly likely to be used by the elites of the high-income countries to maintain "development traps" and

[13] See Chapter "Modeling Social Self-Organization and Historical Dynamics. Modern Society and a Look into the Global Futures: Cybernetic W-Society" (Akaev et al., 2023b, this volume).

[14] See Chapter "Modeling Social Self-Organization and Historical Dynamics. Industrial Society" (Akaev et al., 2023a, this volume); see also, Korotayev et al., 2011; Grinin, 2022; Grinin and Grinin 2022.

[15] It may be denoted as a "world-organism", see Chapter "Socio-Political Transformations. A Difficult Path to Cybernetic Society" (Grinin et al., 2023d, this volume) and Chapter "Analyzing Social Self-Organization and Historical Dynamics. Future Cybernetic W-Society: Socio-Political Aspects" (Grinin & Grinin, 2023a, this volume).

socio-political instability in their low-income counterparts in order to slow down (or reverse) the "Great Convergence" that began in the late twentieth century.[16]

The second way This is the path of modern high-income countries. Their resources for growth so far (before the new technological wave) have been seriously exhausted. As already mentioned, since the second half of the last century, regulatory measures in the economic and social sphere have been strengthened in the high-income countries in order to smooth out the social contradictions generated by liberal market relations. At the same time, the type of economy itself is changing:

(a) there is a gradual transition from the predominant manufacturing of goods (which is now moving to low-income countries) to the post-industrial production of the second type, that is, to the production of complex, highly skilled types of services. These are financial, education, legal, consulting, rating, cultural, medical, etc. services (remote services in the field of medicine and education should be singled out separately). This trend appeared quite a long time ago and is clearly visible today, but it will intensify as the share of these services in the global division of labor will increase;
(b) accordingly, corporations in the high-income countries will increasingly focus on the export of complex highly sophisticated types of services, including financial ones, while, in order to minimize costs, they will begin to actively attract employees from the low-income countries through remote hiring. This, by the way, can increase the flow of financial resources to low-income countries (see Grinin, 2021);
(c) the use of remote work will expand and attract huge numbers of highly qualified workers from the developing periphery[17] (taking into account the fact that English and French are official languages in most of African countries). This trend is also visible today but will become even more noticeable in the future. It should be noted that this is one of the important reserves for future economic development, which will compensate for the lack of demographic resources in the high-income countries. Thus, remote work will begin to partially replace emigration.

Thus, it can be stated that demographic stagnation and the aging of society have significantly reduced resources in the high-income countries, but this is not yet the case for the world as a whole (on global aging see Chapter "Global Aging—an Integral Problem of the Future. How to Turn a Problem into a Development Driver?"

---

[16] Grinin & Korotayev, 2014, 2015; Grinin, 2013; Grinin et al., 2016; Korotayev et al., 2011a, 2011b, 2015, 2020; Korotayev & de Munck, 2013, 2014; Korotayev & Zinkina, 2014; see also Chapter "Future Political Change. Toward a More Efficient World Order" (Grinin et al., 2023a, this volume) and Chapter "Modeling Social Self-Organization and Historical Dynamics. Global Phase Transitions" (Malkov et al., 2023b, this volume).

[17] See note 12 above.

[Grinin et al., 2023c, this volume]). The new resources include the growing young population and telework opportunities in many low-income countries, as well as the great opportunities there for expansion of education and urbanization.[18] The main thing, however, is the coming new technological wave, the final phase of the Cybernetic Revolution (see Chapter "Technology. Limitless Possibilities and Effective Control" [Grinin & Grinin, 2023b, this volume]; Grinin & Grinin, 2015, 2016; Grinin et al., 2017a, 2017b, 2020, 2021a). As for the end of the twenty-first century and the beginning of the twenty-second century, we expect a slowdown in the pace of technological growth, as well as a reduction in the role of consumerism and, in general, the modern consumption model due to global aging (Grinin et al., 2021b; Grinin et al., 2020). But even then, the development will not stop, but transform into a new type. How it will relate to modern capitalism is a question for a separate study.

From the foregoing, we can make important assumptions about the possible transformations of the World System:

1. it appears appropriate to mention once again that in the forthcoming decades, the world will experience a complex continuum of types, transitions, combinations, and forms of economic structures (with an increase in the proportion of post-industrial forms). At the same time, the high-income countries will experience a shift toward the strengthening of distributive institutions. The low-income countries that continue the processes of economic and socio-political modernization, in their turn, will proceed toward the strengthening of capitalist market institutions though with some important restrictions preventing exploitation and environmental degradation;
2. the process of internationalization of the labor force will intensify (due to the development of remote hiring), along with the internationalization of capital;
3. a new technological wave (the 2030s–2070s)—the final phase of the Cybernetic Revolution, which will be an era of wide diffusion of AI-based self-regulating and self-managing systems—will bring new resources and expand the scope for production and services, which will support economic growth in both high-income and low-income countries (see Chapter "Technology. Limitless Possibilities and Effective Control" [Grinin & Grinin, 2023b, this volume);
4. at the same time, in some areas (particularly in medicine), a very significant regulation of economic activity is inevitable, while in other areas (particularly in robotics and the production of self-controlled machines [cars, etc.])—the economic mechanisms of the market-corporate type may get stronger[19];

[18] See Chapter "Modeling Social Self-Organization and Historical Dynamics. Africa's Futures" (Korotayev et al., 2023b, this volume); see also Medvedev & Korotayev, 2020; Korotayev et al., 2021, 2022; Sawyer et al., 2022.

[19] However, it is possible that the development of industrial robots may be delayed by Africa, which has a chance to turn into a new workshop of the world, to some extent repeating the path of China (see Chapter "Africa—the Continent of the Future. Challenges and Opportunities" (Grinin & Korotayev, 2023, this volume)). But this is a very complex issue that requires a serious study,

5. thus, the most important task in the coming decades is to increase the flexibility of the socioeconomic development of all countries of the world, namely, to gain the ability to regulate the ratio of Y-type market mechanisms and X-type distribution mechanisms, depending on the areas of activity and tasks to be solved, without the pressure of ideological dogmas.

With regard to the above discussion about the fate of capitalism, we can say the following. "Capitalism as a socio-economic system" is a product of the outgoing industrial age. What the institutional shape of the W-society of the future cybernetic era will be is not yet clear, the options mentioned above are possible: it all depends on which way the world community will eventually follow. It is only clear that "capitalism as a socio-economic system" will no longer exist, but "capitalism as a sector of economy" will remain and most likely it will occupy an important economic niche. In this case, the question becomes fundamental: will capitalism be the leading sector or will the leading role be taken by other ways of organizing production and distribution? Or will some kind of symbiotic way of life with capitalism still remaining a leader? In any case, it is obvious that, undoubtedly, it will evolve, the role of regulation and the social functions of business will increase. *We believe that it is necessary to find ways to combine all the best features of capitalism in order to provide the necessary dynamism and innovativeness of development, at the same time softening all the distortions that have formed as a result of the development of capitalism over the past twenty years, bearing in mind that not so much industrial capitalism as financial and global financial capitalism is responsible for it.*

References

Akaev, A., & Davydova, O. (2023). Climate and energy. Scenarios of energy transition and global temperature changes based on current technologies and trends. In V. Sadovnichy et al. (Eds.), *Reconsidering the limits to growth. A report to the Russian Association of the Club of Rome (pp. 53–70)*. Springer. https://doi.org/10.1007/978-3-031-34999-7_4

Akaev, A., Malkov, S., Davydova, O., Kovaleva, N., Malkov, A., Grinin, L., & Korotayev, A. (2023a). Modeling social self-organization and historical dynamics. Industrial society. In V. Sadovnichy et al. (Eds.), *Reconsidering the limits to growth. A report to the Russian Association of the Club of Rome (pp. 387–417)*. Springer. https://doi.org/10.1007/978-3-031-34999-7_18

Akaev, A., Malkov, S., Grinin, L., Bilyuga, S., Davydova, O., Grinin, A., Kovaleva, N., Malkov, A., Musieva, J., & Korotayev, A. (2023b). Modeling social self-organization and historical dynamics. Modern society and a look into the global futures: Cybernetic W-society. In V. Sadovnichy et al. (Eds.), *Reconsidering the limits to growth. A report to the Russian Association of the Club of Rome* (pp. 419–460). Springer. https://doi.org/10.1007/978-3-031-34999-7_19

given the large civilizational differences between China and Africa (see Grinin & Korotayev, 2016; Grinin et al., 2019: 27–57).

Collins, R. (2013). The end of middle-class work: No more escapes. In I. Wallerstein et al. (Eds.), *Does capitalism have a future?* (pp. 37–70). Oxford University Press.
Costanza, R., Kubiszewski, I., Giovannini, E., Lovins, H., McGlade, J., Pickett, K. E., Ragnarsdóttir, K. V., Roberts, D., De Vogli, R., & Wilkinson, R. (2014). Development: Time to leave GDP behind. *Nature News, 505*(7483), 283–285. https://doi.org/10.1038/505283a
Easterlin, R. A. (2010). Well-being, front and center: A note on the Sarkozy report. *Population and Development Review, 36*(1), 119–124. https://doi.org/10.1111/j.1728-4457.2010.00320.x
Fücks, R. (2013). *Intelligent Wachsen. Die Grune Revolution*. Carl Hanser Verlag.
Goldsmith, C. (2019). Why GDP is no longer the most effective measure of economic success. *World Finance, 25*(07), 2019. https://www.worldfinance.com/strategy/why-gdp-is-no-longer-the-most-effective-measure-of-economic-success
Grinin, A. (2020). Analiziruya global'nyye problemy XXI v. Obzor i prognoz na osnove doklada Rimskomu klubu «Come On!». *Vek globalizatsii, 4*, 47–64.
Grinin, L. (2013). Globalizatsiya tasuyet mirovuyu kolodu (Kuda sdvigayetsya global'nyy ekonomiko-politicheskiy balans). *Vek globalizatsii, 2*, 63–78.
Grinin, L., Ilyin, I., & Andreev, A. (2016). World order in the past, present, and future. *Social Evolution and History, 15*(1), 58–84.
Grinin, L. (2021). Otritsatel'nyye stavki i drugiye noveyshiye finansovyye tekhnologii. *Obshchestvo i ekonomika, 2*, 18–30.
Grinin, L. (2022). Revolutions of the 21st century as a factor of the world system reconfiguration. In J. A. Goldstone, L. Grinin, & A. Korotayev (Eds.), *Handbook of revolutions in the 21st century: The new waves of revolutions, and the causes and effects of disruptive political change* (pp. 975–999). Springer. https://doi.org/10.1007/978-3-030-86468-2_38
Grinin, L., & Grinin, A. (2015). *Ot rubil do nanorobotov. Mir na puti k epokhe samoupravlyayemykh sistem (istoriya tekhnologiy i opisaniye ikh budushchego)*. Uchitel.
Grinin, L., & Grinin, A. (2016). *The cybernetic revolution and the forthcoming epoch of self-regulating systems*. Uchitel.
Grinin, L., & Grinin, A. (2021a). Analyzing the global problems of the twenty-first century. A review and forecast based on the report to the Club of Rome 'come on!'. *Journal of Globalization Studies, 12*(2), 181–195. https://doi.org/10.30884/jogs/2021.02.10
Grinin, L., & Grinin, A. (2021b). Razmyshleniya ob ekonomicheskom roste i budushchem. Part 1. Globalizm vs rost VVP i «zakat Zapada». *Filosofiya i obshchestvo, 3*, 5–34. https://doi.org/10.30884/jfio/2021.03.01
Grinin, L., & Grinin, A. (2021c). Razmyshleniya ob ekonomicheskom roste i budushchem. Part 2. Globalizatsiya i ekonomicheskiy rost. *Filosofiya i obshchestvo, 4*, 5–38. https://doi.org/10.30884/jfio/2021.04.01
Grinin, L. E., & Grinin, A. L. (2022). The current wave of revolutions in the world-system and its zones. *Journal of Globalization Studies, 13*(2), 171–189. https://doi.org/10.30884/jogs/2022.02.12
Grinin, L., & Grinin, A. (2023a). Analyzing social self-organization and historical dynamics. Future cybernetic W-society: Sociopolitical aspects. In V. Sadovnichy et al. (Eds.), *Reconsidering the limits to growth. A report to the Russian Association of the Club of Rome* (pp. 491–519). Springer. https://doi.org/10.1007/978-3-031-34999-7_21
Grinin, L., & Grinin, A. (2023b). Technology. Limitless possibilities, effective control. In V. Sadovnichy et al. (Eds.), *Reconsidering the limits to growth. A report to the Russian Association of the Club of Rome* (pp. 139–154). Springer. https://doi.org/10.1007/978-3-031-34999-7_8
Grinin, L., Grinin, A., & Korotayev, A. (2017a). Forthcoming Kondratieff wave, cybernetic revolution, and global aging. *Technological Forecasting and Social Change, 115*, 52–68. https://doi.org/10.1016/j.techfore.2016.09.017
Grinin, L., Grinin, A., & Korotayev, A. (2017b). The MANBRIC-Technologies in the Forthcoming Technological Revolution. In T. Devezas, J. Leitão, & A. Sarygulov (Eds.), *Industry 4.0–Entrepreneurship and structural change in the new digital landscape: What is coming on*

along with the fourth industrial revolution (pp. 243–261). Springer. https://doi.org/10.1007/978-3-319-49604-7_13

Grinin, L., Grinin, A., & Korotayev, A. (2020). A quantitative analysis of worldwide long-term technology growth: From 40,000 BCE to the early 22nd century. *Technological Forecasting and Social Change, 155*, 1–19. https://doi.org/10.1016/j.techfore.2020.119955

Grinin, L., Grinin, A., & Korotayev, A. (2021a). Global trends and forecasts of the 21st century. *World Futures, 77*(5), 335–370. https://doi.org/10.1080/02604027.2021.1949939

Grinin, L., Grinin, A., & Korotayev, A. (2021b). COVID-19 pandemic as a trigger for the acceleration of the cybernetic revolution, transition from E-government to E-state, and change in social relations. *Technological Forecasting and Social Change, 175*, 1–16. https://doi.org/10.1016/j.techfore.2021.121348

Grinin, L., Grinin, A., & Korotayev, A. (2023a). Future political change. Toward a more efficient world order. In V. Sadovnichy et al. (Eds.), *Reconsidering the limits to growth. A report to the Russian Association of the Club of Rome* (pp. 191–206). Springer. https://doi.org/10.1007/978-3-031-34999-7_11

Grinin, L., Grinin, A., & Malkov, S. (2023b). Economics. Optimizing growth. In V. Sadovnichy et al. (Eds.), *Reconsidering the limits to growth. A report to the Russian Association of the Club of Rome* (pp. 155–168). Springer. doi:https://doi.org/10.1007/978-3-031-34999-7_9

Grinin, L., Grinin, A., & Korotayev, A. (2023c). Global aging–An integral problem of the future. How to turn a problem into a development driver? In V. Sadovnichy et al. (Eds.), *Reconsidering the limits to growth. A report to the Russian Association of the Club of Rome* (pp. 117–135). Springer. https://doi.org/10.1007/978-3-031-34999-7_7

Grinin, L., Grinin, A., & Malkov, S. (2023d). Socio-political transformtions. A difficult path to cybernetic society. In V. Sadovnichy et al. (Eds.), *Recosidering the limits to growth. A report to the Russian Association of the Club of Rome* (pp. 117–135). Springer. https://doi.org/10.1007/978-3-031-34999-7.

Grinin, L., & Korotayev, A. (2014). Globalization shuffles cards of the world pack: In which direction is the global economic-political balance shifting? *World Futures, 70*(8), 515–545. https://doi.org/10.1080/02604027.2014.982474

Grinin, L., & Korotayev, A. (2015). *Great divergence and great convergence: A global perspective*. Springer. https://doi.org/10.1007/978-3-319-17780-9

Grinin, L., & Korotayev, A. (2016). *Blizhniy Vostok. Indiya i Kitay v globalizatsionnykh protsessakh*. Uchitel.

Grinin, L., & Korotayev, A. (2023). Africa–The continent of the future. Challenges and opportunities. In V. Sadovnichy et al. (Eds.), *Reconsidering the limits to growth. A report to the Russian Association of the Club of Rome* (pp. 225–238). Springer. https://doi.org/10.1007/978-3-031-34999-7_13

Grinin, L., Korotayev, A., & Tausch, A. (2019). *Islamism, Arab spring, and the future of democracy. World system and world values perspectives*. Springer. https://doi.org/10.1007/978-3-319-91077-2

Grinin, L., Malkov, S., Grinin, A., & Korotayev, A. (2022). Will capitalism die? Reflections on the capitalism of the past, present and future. *Sotsiologicheskiy Zhurnal = Sociological Journal, 28*(2), 100–130. https://doi.org/10.19181/socjour.2022.28.2.8988

Guernier, M. (1980). *Tiers-Monde: Trois Quarts Du Monde. A report to The Club of Rome*. Dunod.

Kapoor, A., & Debroy, B. (2019). GDP is not a measure of human Well-being. *Harvard Business Review, 10*. https://hbr.org/2019/10/gdp-is-not-a-measure-of-human-well-being

Keynes, J. M. (2016) [1936]. *The general theory of employment, interest, and money*. Stellar Classics.

Kirdina, S. (2014). Institutional matrices theory, or X & Y theory: The main provisions and applications. *Journal of Institutional Studies, 6*(3), 13–33.

Korotayev, A., Bilyuga, S., & Shishkina, A. (2020). Which countries generate Kondratieff waves in global GDP growth rate dynamics in the contemporary world? *Journal of Globalization Studies, 11*(1), 33–63. https://doi.org/10.30884/jogs/2020.01.03

Korotayev, A., & de Munck, V. (2013). Advances in development reverse inequality trends. *Journal of Globalization Studies, 4*(1), 105–124.

Korotayev, A., & de Munck, V. (2014). Advances in development reverse global inequality trends. *Globalistics and Globalization Studies, 3*, 164–183.

Korotayev, A., Goldstone, J., & Zinkina, J. (2015). Phases of global demographic transition correlate with phases of the great divergence and great convergence. *Technological Forecasting and Social Change, 95*, 163–169. https://doi.org/10.1016/j.techfore.2015.01.017

Korotayev, A., & Grinin, L. (2017). The technological activity and competition in the middle ages and modern history: A quantitative analysis. In L. Grinin & A. Korotayev (Eds.), *History & mathematics: Economy, demography, culture, and cosmic civilizations* (pp. 78–102). Uchitel.

Korotayev, A., Malkov, S., & Musieva, J. (2023a). Demography. Toward optimization of demographic processes. In V. Sadovnichy et al. (Eds.), *Reconsidering the limits to growth. A report to the Russian Association of the Club of Rome* (pp. 97–116). Springer. https://doi.org/10.1007/978-3-031-34999-7_6

Korotayev, A., Shulgin, S., Ustyuzhanin, V., Zinkina, J., & Grinin, L. (2023b). Modeling social self-organization and historical dynamics. Africa's futures. In V. Sadovnichy et al. (Eds.), *Reconsidering the limits to growth. A report to the Russian Association of the Club of Rome* (pp. 461–490). Springer. https://doi.org/10.1007/978-3-031-34999-7_20

Korotayev, A., Romanov, D., Zinkina, J., & Slav, M. (2022). Urban youth and terrorism: A quantitative analysis (are youth bulges relevant anymore?). *Political Studies Review.*, 147892992210759. https://doi.org/10.1177/14789299221075908

Korotayev, A., Sawyer, P., Gladyshev, M., Romanov, D., & Shishkina, A. (2021). Some sociodemographic factors of the intensity of anti-government demonstrations: Youth bulges, urbanization, and protests. *Sociologiceskoe Obozrenie = Russian Sociological Review, 20*(3), 98–128. https://doi.org/10.17323/1728-192x-2021-3-98-128

Korotayev, A., & Zinkina, J. (2014). On the structure of the present-day convergence. *Campus-Wide Information Systems, 31*(2–3), 139–152. https://doi.org/10.1108/CWIS-11-2013-0064

Korotayev, A., Zinkina, J., Bogevolnov, J., & Malkov, A. (2011a). Global unconditional convergence among larger economies after 1998? *Journal of Globalization Studies, 2*(2), 25–62.

Korotayev, A., Zinkina, J., Kobzeva, S., Bogevolnov, J., Khaltourina, D., Malkov, A., & Malkov, S. (2011b). A trap at the escape from the trap? Demographic-structural factors of Po-litical instability in modern Africa and West Asia. *Cliodynamics: The Journal of Quantitative History and Cultural Evolution, 2*(2), 276–303. https://doi.org/10.21237/C7clio22217

Kovaleva, N. (2023). Ecology. Life in the "unstable biosphere". In V. Sadovnichy et al. (Eds.), *Reconsidering the limits to growth. A report to the Russian Association of the Club of Rome* (pp. 71–95). Springer. https://doi.org/10.1007/978-3-031-34999-7_5

Kuznets, S. (1966). *Modern economic growth*. Yale University Press.

Lenoir, R. (1984). *Le Tiers Monde Peut Se Nourrir. A report to The Club of Rome*. Fayard.

McElwee, S., & Daly, L. (2014). Why we should abolish the GDP. *Washington Post, 05*(06), 2014. https://www.washingtonpost.com/posteverything/wp/2014/06/05/why-we-should-abolish-the-gdp/

Malkov, S., Bilyuga, S., & Musieva, J. (2023a). Analyzing social self-organization and historical dynamics. Life quality index. In V. Sadovnichy et al. (Eds.), *Reconsidering the limits to growth. A report to the Russian Association of the Club of Rome* (pp. 521–530). Springer. https://doi.org/10.1007/978-3-031-34999-7_22

Malkov, S., Grinin, L., Grinin, A., Musieva, J., & Korotayev, A. (2023b). Modeling social self-organization and historical dynamics. Global phase transitions. In V. Sadovnichy et al. (Eds.), *Reconsidering the limits to growth. A report to the Russian Association of the Club of Rome* (pp. 387–417). Springer. https://doi.org/10.1007/978-3-031-34999-7_18

Malkov, S., Kovaleva, N., Grinin, L., & Korotayev, A. (2023c). Modeling social self-organization and historical dynamics. Agrarian society. In V. Sadovnichy et al. (Eds.), *Reconsidering the limits to growth. A report to the Russian Association of the Club of Rome* (pp. 309–335). Springer. https://doi.org/10.1007/978-3-031-34999-7_16

Medvedev, I., & Korotayev, A. (2020). Perspectives of Urbanization Transition in Africa, 2020–2050: Some Observations. In D. Bondarenko (Ed.), *Africa: History and Results of Decolonization* (pp. 75–78). Institute for African Studies of the Russian Academy of Sciences. https://elibrary.ru/item.asp?id=48242642

Randers, J. (2012). *2052–A global forecast for the next forty years*. Chelsea Green Publishing.

Sadovnichy, V., Akaev, A., Ilyin, I., Malkov, S., Grinin, L., Aleshkovski, I., & Korotayev, A. (2023a). Macrohistorical approach. In V. Sadovnichy et al. (Eds.), *Reconsidering the limits to growth. A report to the Russian Association of the Club of Rome* (pp. 17–33). Springer. https://doi.org/10.1007/978-3-031-34999-7_2

Sadovnichy, V., Akaev, A., Ilyin, I., Malkov, S., Grinin, L., Andreev, A., & Korotayev, A. (2023b). Future society and transition to it. In V. Sadovnichy et al. (Eds.), *Reconsidering the limits to growth. A report to the Russian Association of the Club of Rome* (pp. 35–50). Springer. https://doi.org/10.1007/978-3-031-34999-7_3

Sadovnichy, V., Akaev, A., Ilyin, I., Malkov, S., Grinin, L., Ivanov, D., & Korotayev, A. (2023c). Conclusion. Reconsidering the limits–Suggestions (come on!). In V. Sadovnichy et al. (Eds.), *Reconsidering the limits to growth. A report to the Russian Association of the Club of Rome* (pp. 225–238). Springer. https://doi.org/10.1007/978-3-031-34999-7_13

Sawyer, P., Romanov, D., Slav, M., & Korotayev, A. (2022). Urbanization, the youth, and protest: A cross-National Analysis. *Cross-Cultural Research, 56*(2–3), 125–149. https://doi.org/10.1177/10693971211059762

Schneider, B. (1988). *The barefoot revolution. A report to The Club of Rome*. The Club of Rome.

Schneider, B. (1995). *The scandal and the shame. A report to The Club of Rome*. The Club of Rome.

Wallerstein, I. (2004). *World-systems analysis: An introduction*. Duke University Press.

Wallerstein, I. (2013). Structural crisis, or why capitalists may no longer find capitalism rewarding. In I. Wallerstein et al. (Eds.), *Does capitalism have a future?* (pp. 9–35). Oxford University Press.

Wallerstein, I., Collins, R., Mann, M., Derluguian, G., & Calhoun, C. (2013). *Does capitalism have a future?* Oxford University Press.

Leonid Grinin is Senior Research Professor at the Centre for Stability and Risk Analysis at HSE University in Moscow, as well as Senior Research Professor of the Institute of Oriental Studies, Russian Academy of Sciences. He is co-editor of the international journals *Social Evolution & History* and the *Journal of Globalization Studies*, as well as co-editor of the international yearbooks *Evolution, History & Mathematics*, and *Kondratieff Waves*. He is author of more than 600 publications, including 33 monographs in Russian, English, Spanish, German, and Chinese. His current research focuses on comparative political studies, theory of revolution, political anthropology, global economy, global history, historical sociology, and futurology. Among his monographs are *Great Divergence and Great Convergence. A Global Perspective* (Springer, 2015; with Andrey Korotayev), *The Cybernetic Revolution and the Forthcoming Epoch of Self-Regulating Systems* (Uchitel, 2016, with Anton Grinin), *Economic Cycles, Crises, and the Global Periphery* (Springer, 2016; with Andrey Korotayev and Arno Tausch), *Big History of Globalization. The Emergence of a Global World System* (Springer 2019, in co-authorship with Julia Zinkina, David Christian and others), *Islamism, Arab Spring, and the Future of Democracy. World System and World Values Perspectives* (Springer, 2019; with Andrey Korotayev and Arno Tausch). In 2004 he was awarded as a winner of the best publication of *Socis* journal during the year for the article "People of Publicity—a New Social Layer?". In 2012 he was awarded with the Gold Kondratieff Medal by the International N. D. Kondratieff Foundation. In 2016 he received the Belyaev award for the book *From Bifaces to Nanorobots* (the book published in Russian in co-authorship with Anton Grinin).

Sergey Malkov heads the Center for Long-Term Forecasting and Strategic Planning at Lomonosov Moscow State University; he also works as a Senior Research Professor at the Institute of Economics of the Russian Academy of Sciences, as well as at the Centre for Stability and Risk Analysis, HSE University, Moscow. He is the author of over 300 scholarly publications, including such monographs as *Social self-organization and historical process: possibilities of mathematical modeling* (URSS, 2009), *Modeling and forecasting of world dynamics* (ISPI RAS, 2012, together with Victor Sadovnichy and others), *A new theory of the reproduction of capital: development and practical application* (Nestor-Istoriya, 2016, together with Vladimir Mayevsky and others), *The twenty-first Century Singularity and Global Futures. A Big History Perspective* (Springer, 2020, together with David J. LePoire and others). In 2022 he was awarded with the Kondratieff Award by the Russian Academy of Sciences.

Andrey Korotayev heads the Centre for Stability and Risk Analysis at HSE University in Moscow, Russia. He is also Senior Research Professor at the Institute of Oriental Studies and Institute for African Studies, Russian Academy of Sciences, as well as at the Faculty of Global Studies of Lomonosov Moscow State University and the Russian Presidential Academy of National Economy and Public Administration. He has authored or co-authored over 650 scholarly publications, including such monographs as *Ancient* Yemen (Oxford University Press, 1995), *World Religions and Social Evolution of the Old World Oikumene Civilizations: A Cross-Cultural Perspective* (The Edwin Mellen Press, 2004), *Introduction to Social Macrodynamics: Compact Macromodels of the World System Growth* (URSS, 2006; with Daria Khaltorina and Artemy Malkov), *Introduction to Social Macrodynamics: Secular Cycles and Millennial Trends* (URSS, 2006; with Daria Khaltorina and Artemy Malkov), *Great Divergence and Great Convergence. A Global Perspective* (Springer, 2015; with Leonid Grinin), *Economic Cycles, Crises, and the Global Periphery* (Springer, 2016; with Leonid Grinin and Arno Tausch), *Islamism, Arab Spring, and the Future of Democracy. World System and World Values Perspectives* (Springer, 2019; with Leonid Grinin and Arno Tausch). He is a laureate of Russian Science Support Foundation Award in 'The Best Economists of the Russian Academy of Sciences' Nomination (2006); in 2012 he was awarded with the Gold Kondratieff Medal by the International N. D. Kondratieff Foundation.

Africa: The Continent of the Future. Challenges and Opportunities

Leonid Grinin and Andrey Korotayev

Abstract

Grinin and Korotayev's main approach in this chapter is formulated in its title "Africa—the Continent of the Future. Challenges and Opportunities." And it is really so. Africa is the most rapidly growing continent, and its growth (demographic, economic, cultural, etc.) will inevitably influence the World System (and the further, the stronger). The growth of African societies will bring at the same time both challenges and opportunities. Grinin and Korotayev discuss both of them in detail. Grinin and Korotayev consider Africa as the main intrigue of demographic transformations in the twenty-first century. But they also think that this continent can constitute a very important intrigue in other respects (economic, political, ethnical etc.). No doubt, in the future, African countries will produce many surprises both positive and negative. Grinin and Korotayev suppose that, with respect to the humankind's future development, Africa will be a great source of both problems and their solutions. For example, the problem of surplus labor force will continue to be acute in Africa. But on the other hand, in the context of declining birth rates in most of regions of the planet, these are African labor resources that can partially help solve the problem of labor shortages in high income countries. The growth of the African economy should be inevitable already in connection with the growth of the population, especially urban one, and its escape from extreme poverty. On the one hand, African growth

L. Grinin
HSE University, Moscow, Russia

Institute of Oriental Studies, Russian Academy of Sciences, Moscow, Russia

A. Korotayev (✉)
HSE University, Moscow, Russia

Faculty of Global Studies, Lomonosov Moscow State University, Moscow, Russia

Institute for African Studies, Russian Academy of Sciences, Moscow, Russia

V. Sadovnichy et al. (eds.), *Reconsidering the Limits to Growth*, World-Systems Evolution and Global Futures, https://doi.org/10.1007/978-3-031-34999-7_13

could greatly influence ecology (especially in agrarian sector). But on the other hand, the future African growth can become an increasingly important source of global growth, which is close to its exhaustion. Besides, this growth is able to partially solve the problem of the shortage of many resources, at least for some significant time. Much will also depend on how deeply, systematically and quickly the green energy transition will develop. For its implementation, in addition to a host of technological and other restraining problems, an important limiting factor is the shortage of many metals and components (like silicon) necessary for green energy, electric vehicles and other equipment. Some countries of Africa are rich in resources that are in great demand in the modern electronics industry and green energy; in addition, one cannot but mention the reserves of rare earth metals. The need for these minerals can activate foreign capital and accelerate the industrial development of Africa, which will undoubtedly affect the decline in fertility, as well as the growth of interest in Africa among the leading geopolitical and economic players with all the pros and cons of their active penetration to this continent. Therefore, the world community faces an important task of trying to use the African resources as efficiently as possible to solve global problems, with the maximum benefit for the Africans themselves. Grinin and Korotayev also discuss problems of current and future instability in many African countries, connected with rapid growth, modernization, weakness of statehood and national historical experience, traditions, and so on. So in some sense Grinin and Korotayev expect that the twenty-first century will be the African century.

The main approach of this chapter is formulated in its title "Africa—the Continent of the Future. Challenges and Opportunities." And it is really so. Africa is the most rapidly growing continent, and its growth (demographic, economic, cultural, etc.) will inevitably influence the World System (and the further, the stronger). The growth of African societies will bring at the same time both challenges and opportunities. Below we will discuss both of them in detail. Africa is the main intrigue of demographic transformations in the twenty-first century. We also think that this continent can constitute a very important intrigue in other respects (economic, political, ethnical, etc.). No doubt, in the future, African countries will produce many surprises both positive and negative.

We suppose that, with respect to the humankind's future development, Africa will be a great source of both problems and their solutions. For example, as we will see below the problem of surplus labor force will continue to be acute in Africa. But on the other hand, in the context of declining birth rates in most of regions of the planet, these are African labor resources that can partially help solve the problem of labor shortages in high-income countries. The growth of the African economy should be inevitable already in connection with the growth of the population, especially urban one, and its escape from extreme poverty. On the one hand, African growth could greatly influence ecology (especially in agrarian sector). But on the other hand, the future African growth can become an increasingly important source of global

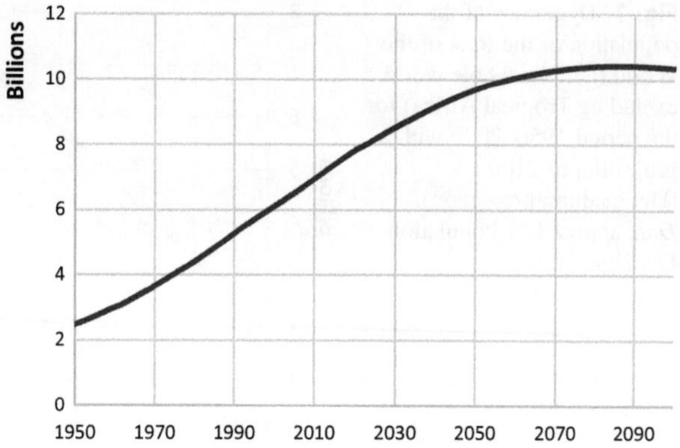

Fig. 1 Dynamics of the world population—empirical estimates (until 2022) and the medium forecast of the UN Population Division until 2100. *Data source* UN Population Division, 2022

growth, which is close to its exhaustion. Besides, this growth is able to partially solve the problem of the shortage of many resources for the green energy transition.

1 Africa as the Main Intrigue of Demographic Transformations in the Twenty-First Century

Despite the slowdown in population growth (see Chapter "Demography. Toward the Optimization of Demographic Processes" [Korotayev et al., 2023a, this volume]), the demographic growth in the World System will continue for quite a long time, stabilizing under the medium scenario only by the end of the twenty-first century (see Fig. 1), and under the high scenario even much later.

And the timing of the stabilization of the world's population will depend to a large extent on the demographic development of the African countries.[1] Of course, the growth of the population of our planet in the second half—the last third of the twenty-first century will not be a "population explosion" or "population bomb"[2]; but still, even from the current level, the population of the Earth can increase by 2–3 billion (see Chapter "Demography. Toward the Optimization of Demographic Processes" [Korotayev et al., 2023a, this volume]). Here, Tropical Africa and several

[1] African population growth and problems connected with it have been already touched upon in some reports to the Club of Rome though not separately, but rather in connection with the issues of the world population growth. However, it is worth singling out Aklilu Lemma's report to The Club of Rome "Africa Beyond Famine" (Lemma, 1989) that considered the problems of population, urbanization, the debt of African states to world banks, desertification of the African continent and famine, affecting millions of people. The need for the development of local self-government in various communities is noted. Great hope was placed on the combination of the use of new technologies and traditional forms of agricultural production, which should contribute to resolving the most acute crisis situations in African countries.

[2] To use Ehrlich's (1975) expression.

Fig. 2 Dynamics of the population of the Rest of the World (i.e., the whole world excluding Tropical Africa) for the period 1950–2020 with a projection to 2100 (UN medium projection). *Data source* UN Population Division, 2022

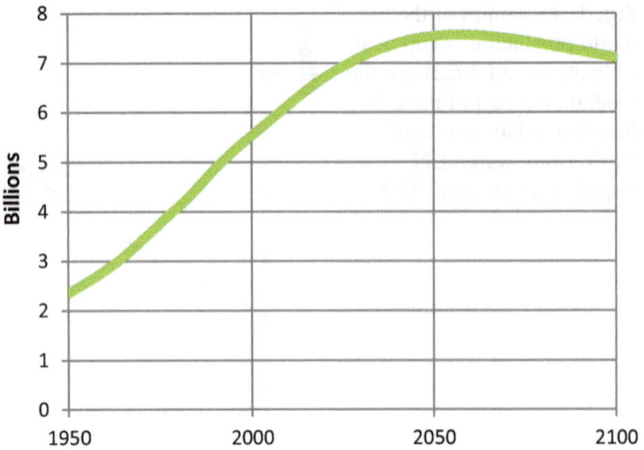

Fig. 3 Dynamics of the population of Tropical Africa for the period 1950–2020 with a projection to 2100 (UN medium projection). *Data source* UN Population Division, 2022

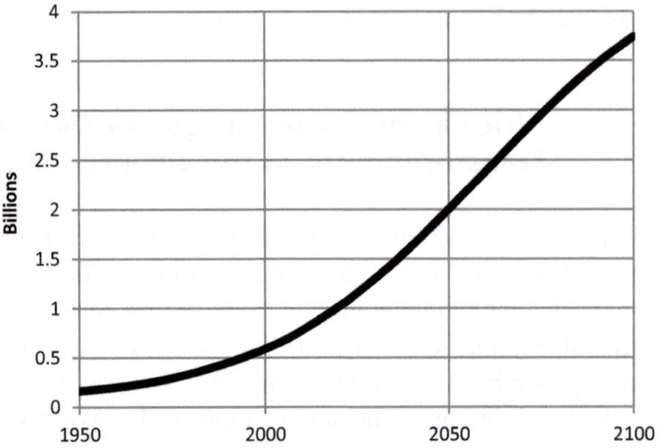

countries from Asia (like Yemen, Afghanistan) will be the main part of the world providing the bulk of the growth (see Figs. 2, 3 and 4[3]):

One can point to traditional and religious (including Islamic) ideology, and a number of other factors preventing active entry into the second phase of the demographic transition in Tropical Africa (e.g., Korotayev et al., 2016a; Grinin, 2020a). In most countries of this part of the continent, fertility is declining rather

[3]See also Chapter "Demography. Toward Optimization of Demographic Processes" (Korotayev et al., 2023a, this volume), Chapter "High Income and Low Income Countries. Toward a Common Goal at Different Speeds" (Grinin et al., 2023, this volume), the section on the possible scenarios of the world dynamics and the section on the mathematical modeling of the African demographic futures in Chapter "Modeling Social Self-Organization and Historical Dynamics. Africa's Futures" (Korotayev et al., 2023b, this volume), as well as Korotayev & Zinkina, 2014; Korotayev et al., 2016b; Zinkina & Korotayev, 2014a, 2017.

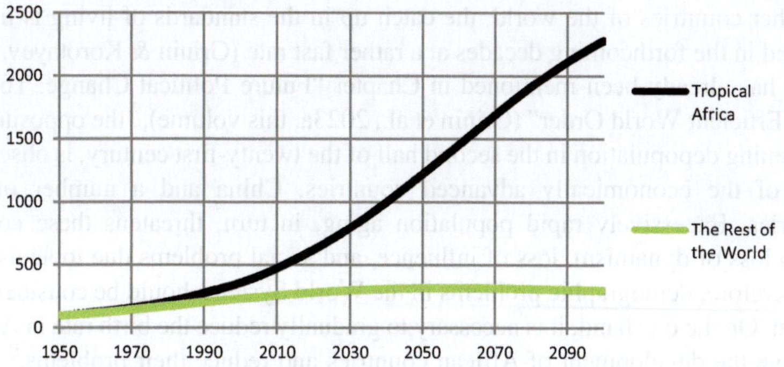

Fig. 4 Relative dynamics of the population of Tropical Africa and the Rest of the World for 1950–2020 projected to 2100 (UN medium projection), 100 = 1950 level. *Data source* UN Population Division, 2022

slowly[4] (see, for example, Zinkina & Korotayev, 2014b; Nzimande & Mugwendere, 2018; Kebede et al., 2019; Schoumaker, 2019; May & Rotenberg, 2020); however, the overall trend—the less developed the country, the lower the level of prevalence of modern education and urbanization in it, the higher the birth rate—is quite visible here as well. Hence, as Tropical African countries develop, a reduction in the birth rate is inevitable, but it is not clear at what speed this will go (and the overall growth of the Earth's population at any given moment in the forthcoming decades will depend on this). There is also no doubt that the type of agriculture (still mostly extensive hoe agriculture in combination with millions of people engaged in semi-nomadic animal husbandry) greatly contributes to a high birth rate (see Korotayev et al., 2016a).

On the one hand, the modernization of agriculture in Tropical Africa should develop (in some places this is happening), on the other hand, this situation conflicts with the environmental situation in the world, since the increase in environmental pressure caused by modernization can become destructive.[5] *This, along with demographics that have not yet been sufficiently controlled, can also become a factor of pressure on the demographic policy of African countries.*

At the same time, the risks of certain types of destabilization may grow even with the growth in the quality of life/standard of living[6] and, given the still very large gap between the GDP per capita between most Tropical African countries and almost all

[4]In contrast to Northern and Southern Africa where fertility is much lower. However, in almost all countries of Tropical Africa, it is still declining. For example, in Niger in 2005–2010 the total fertility rate was 7.55 children per woman, and is currently 6.95 (UN Population Division, 2022).

[5]See Chapter "Economics. Optimizing Growth" (Grinin et al., 2023c, this volume) for more details; see also Grinin, 2020d.

[6]E.g., Bilyuga et al., 2016; Korotayev et al., 2016a, 2017a, 2017b, 2018a; Vaskin et al., 2018; Korotayev et al., 2017, 2018b, 2020, 2021c; Grinin, 2012, 2013, 2022; Korotayev et al. 2021b.

the other countries of the world, the catch up in the standards of living is likely to proceed in the forthcoming decades at a rather fast rate (Grinin & Korotayev, 2015).

As has already been mentioned in Chapter "Future Political Change. Toward a More Efficient World Order" (Grinin et al., 2023a, this volume),[7] the opposite trend, threatening depopulation in the second half of the twenty-first century, is observed in most of the economically advanced countries, China and a number of other countries. Excessively rapid population aging, in turn, threatens these countries with a loss of dynamism, loss of influence, and social problems due to ageism.[8]

Therefore, demographic problems in the World System should be considered in a system. On the one hand, it is necessary to gradually reduce the birth rate in Africa to improve the development of African countries and reduce their problems,[9] and on the other hand, it is very desirable to overcome the limit of depopulation, otherwise we will lose many nations and create a new kind of problem. According to UN projections (see UN Population Division, 2022), between 2020 and 2100, 90 countries will experience population decline. In other words, we need to think about the problem of optimizing demographic processes in terms of quantitative development and keeping in mind the growth of the biological and material quality of life, which will (perhaps with noticeable fluctuations) continue all the time until the end of the century. At the moment, the world crude birth rate is 18 per 1000, whereas the average total fertility rate is around 2.3–2.4 births per woman. Accordingly, for the normal reproduction of the population, the values of the crude fertility rate of the order of 11 births per 1000 people are needed (or 2.1 children per woman in terms of total fertility rate).

[7] See also Chapter "Demography. Toward Optimization of Demographic Processes" (Korotayev et al., 2023a, this volume), Chapter "High Income and Low Income Countries. Toward a Common Goal at Different Speeds" (Grinin et al., 2023, this volume), and the section on the possible scenarios of the world dynamics in Chapter "Modeling Social Self-Organization and Historical Dynamics. Modern Society and a Look into the Global Futures: Cybernetic W-Society" (Akaev et al., 2023, this volume).

[8] On aging see Chapter "Global Aging—an Integral Problem of the Future. How to Turn a Problem into a Development Driver?" (Grinin et al., 2023b, this volume); on ageism see Chapter "Socio-Political Transformations. A Difficult Path to Cybernetic Society" (Grinin et al., 2023d, this volume) and Chapter "Analyzing Social Self-Organization and Historical Dynamics. Future Cybernetic W-Society: Socio-Political Aspects" (Grinin & Grinin, 2023a, this volume) for more details.

[9] See Chapter "Modeling Social Self-Organization and Historical Dynamics. Africa's Futures" (Korotayev et al., 2023b, this volume), as well as Korotayev & Zinkina, 2014; Korotayev et al., 2016b; Zinkina & Korotayev, 2014a, 2017.

2 Africa is the Main Reserve of People of Working Age in the World System

On the one hand, the problem of surplus labor force is and will continue to be even more acute in Africa.[10]

Rapid population growth in Africa against the background of declining fertility is creating and will continue to create a huge demographic dividend in many African countries (see, e.g., Bloom et al., 2013; Groth & May, 2017; Jimenez & Pate, 2017; Hasan et al., 2019; Korotayev et al., 2022). At present, of course, a poorly literate working age population (especially females) does not represent a very great value for the world economy. However, as the level and quality of education increase, these cohorts can become a very powerful resource for the deficient working-age population of aging high-income countries[11] as well as a source of youth for the education systems of these countries. Already today, a large number of African students study in European, American, and Asian universities, a large number of specialists from there work in one way or another for these economies. In the context of the strong development of remote work (see Chapter "Global Aging—an Integral Problem of the Future. How to Turn a Problem into a Development Driver?" [Grinin et al., 2023b, this volume] for more details), more Africans will be connected with economically high income countries. In this regard, high-income countries face a very important task in providing more assistance than is being done today in the development of the education system and the possible recruitment of specialists for remote work. This is a project for decades. And this is one of the ways to reduce unemployment in Africa, raise living standards in these countries, and bring European and African cultures closer together. There are also possible ways to bring European pension systems closer to the governments of African countries (for more details, see Grinin & Korotayev, 2010, 2016).

In a word, we see that, on the one hand, Africa is a big problem for the modern World System in terms of poorly controlled population growth, which can lead to serious environmental, climatic, social, food, and other problems[12] that in one way or another will concern the entire World System. Therefore, the leading powers are interested in seeing the demographic development of Africa proceed in a more manageable way. But, on the other hand, in the context of a declining birth rate, it is African labor resources that can partially help solve the problem of labor shortages in the World System. Thus, high-income countries need to develop a common program of influence on African countries in a direction more acceptable to the

[10]See Chapter "Modeling Social Self-Organization and Historical Dynamics. Africa's Futures" (Korotayev et al., 2023b, this volume), as well as, for example, Zinkina & Korotayev, 2014a; Korotayev & Zinkina, 2015; Zinkina & Korotayev, 2017; Korotayev et al., 2022.

[11]See Chapter "Future Political Change. Toward a More Efficient World Order" (Grinin et al., 2023a, this volume) and Chapter "Global Ageing—an Integral Problem of the Future. How to Turn a Problem into a Development Driver?" (Grinin et al., 2023b, this volume).

[12]See Chapter "Modeling Social Self-Organization and Historical Dynamics. Africa's Futures" (Korotayev et al., 2023b, this volume) for more details.

humankind so that this program does not infringe on the interests of African states. Meanwhile, not enough attention has been paid to Africa so far.

What has been said is all the more important because the multidirectional vectors of demographic development will inevitably lead to the formation of the World System model: the old North—the young South (see Chapter "Future Political Change. Toward a More Efficient World Order" [Grinin et al., 2023a, this volume], as well as Fukuyama, 2002; Goldstone et al., 2015; Haas, 2015). And it is worth considering in advance that such a division would not bring serious conflicts but would become a source of cooperation between different groups of countries.

3 Problems of Sociopolitical Destabilization in Africa

Fast population growth, an excess of young people in conditions of immature states and national-ethnic tensions, rapid urbanization, and other factors threaten Africa with very serious upheavals and destabilization processes.[13] Even a cursory analysis of a number of factors characteristic of the current state and development of many African countries suggests that Africa will be the most restless continent in the twenty-first century. These factors include the following: (a) insufficient maturity of political relations and public administration (see, e.g., Bondarenko, 2021), leading to the fact that many countries of Tropical Africa are weak or failed states, unable to resist criminal and terrorist groups, protect their own population; (b) archaism of the economy and social relations; (c) ongoing processes of modernization and urbanization; (d) Africa, including the Sahel and North Africa, is becoming a territory of increasing geopolitical rivalry between major powers.

All this leads to the conclusion that the period of elimination of these shortcomings and the maturation of societies will also be a period of extremely high risk of destabilization processes. In particular, we believe that in the twenty-first century it is Africa that will be the continent where the largest number of political, social, and national revolutions will occur due to the fact that African countries are still in the process (often in the initial phases) of modernization, urbanization, the formation of ethno-political nations, and the development of modern statehood (see Grinin, 2020a, 2020b, 2020c, 2020d, 2022; Ustyuzhanin et al., 2021; Goldstone et al., 2022; Grinin & Korotayev, 2022). The world is interested in the least possible conflict development of African societies; therefore, it is necessary to think about how to help Africa in order to reduce the number, depth, and destructive power of conflicts as much as possible, although the further development there is very unlikely to proceed entirely without such conflicts.

[13] See, for example, Moller, 1968; Mesquida & Weiner, 1999; Goldstone, 2002; Urdal, 2004, 2006, 2008, 2012; Korotayev et al., 2011; Yair & Miodownik, 2016; Weber, 2019; Cincotta & Weber, 2021; Grinin, 2020b, 2020c, 2020d; Korotayev et al., 2021a; Romanov et al., 2021; Korotayev et al., 2023c; Korotayev & Zinkina, 2022; Sawyer et al., 2022, as well as Chapter "Modeling Social Self-Organization and Historical Dynamics. Africa's Futures" (Korotayev et al., 2023b, this volume).

4 Growth of the African Economy

The growth of the African economy, which should be inevitable already in connection with the growth of the population and its extreme poverty, can become an increasingly important source of global growth. And this growth is able to partially solve the problem of the shortage of many resources, at least for some significant time. Much will also depend on how deeply, systematically, and quickly the green energy transition will develop.[14] For its implementation, in addition to a host of technological and other restraining problems, an important limiting factor is the shortage of many metals and components (like silicon) necessary for green energy, electric vehicles, and other equipment. Some countries of Africa (for example, Chad, DRC, Zambia, Namibia, and Zimbabwe) are rich in ores of manganese, chromites, bauxites, gold, platinoids, cobalt, vanadium, copper, uranium, antimony, beryllium, graphite, titanium, nickel, bismuth, lithium, tantalum, niobium, tin, and tungsten—much of this is in great demand in the modern electronics industry and green energy; in addition, one cannot but mention the reserves of rare earth metals. The need for these minerals can activate foreign capital and accelerate the industrial development of Africa, which will undoubtedly affect the decline in fertility, as well as the growth of interest in Africa among the leading geopolitical and economic players with all the pros and cons of their active penetration to this continent.

Therefore, the world community faces an important task of trying to use the African resource as efficiently as possible to solve global problems, with the maximum benefit for the Africans themselves.

Concluding this chapter, we would like to emphasize once again that Africa is the most rapidly developing continent, and its growth (demographic, economic, cultural, etc.) will inevitably influence (in different ways, both positive and negative) the World System (and the further, the stronger). So, in some sense, we expect that the twenty-first century will be the African century.

References

Akaev, A., Malkov, S., Grinin, L., Bilyuga, S., Davydova, O., Grinin, A., Kovaleva, N., Malkov, A., Musieva, J., & Korotayev, A. (2023). Modeling social self-organization and historical dynamics. Modern society and a look into the global futures: Cybernetic W-society. In V. Sadovnichy et al. (Eds.), *Reconsidering the limits to growth. A report to the Russian Association of the Club of Rome* (pp. 419–460). Springer. https://doi.org/10.1007/978-3-031-34999-1_19

Bilyuga, S., Shishkina, A., Zinkina, J., & Korotayev, A. (2016). Global correlation between GDP per capita and the level of sociopolitical destabilization between 1960 and 2014: A preliminary quantitative analysis. *Globalistics and Globalization Studies, 5*, 213–219. https://www.elibrary.ru/item.asp?id=28935474&pff=1

[14] On possibilities to decrease carbon emission through future innovations see Chapter "Technology. Limitless Possibilities and Effective Control" (Grinin & Grinin, 2023b, this volume).

Bloom, D. E., Humair, S., Rosenberg, L., Sevilla, J. P., & Trussell, J. (2013). *A demographic dividend for sub-Saharan Africa: Source, magnitude, and realization*. IZA Discussion Papers 7855, Institute of Labor Economics (IZA).

Bondarenko, D. M. (2021). Nation-building in post-colonial states: Historical past and present-day realities. *Journal of Globalization Studies, 12*(1), 18–37. https://doi.org/10.30884/jogs/2021.01.02

Cincotta, R., & Weber, H. (2021). Youthful age structures and the risks of revolutionary and separatist conflicts. In A. Goerres & P. Vanhuysse (Eds.), *Global political demography: Comparative analyzes of the politics of Population change in all world regions* (pp. 57–92). Palgrave. https://doi.org/10.1007/978-3-030-73065-9_3

Ehrlich, P. R. (1975). *The Population bomb*. Rivercity Press.

Fukuyama, F. (2002). *Our post-human future: Consequences of the bio-technology revolution*. Farrar, Straus, and Giroux.

Goldstone, J. A. (2002). Population and security: How demographic change can Lead to violent conflict. *Journal of International Affairs, 56*(1), 3–21.

Goldstone, J. A., Grinin, L., & Korotayev, A. (2015). Research into global ageing and its consequences. In J. A. Goldstone, L. Grinin, & A. Korotayev (Eds.), *History & mathematics: Political Demography & Global Ageing* (pp. 5–9). Uchitel.

Goldstone, J. A., Grinin, L., & Korotayev, A. (2022). Conclusion. How many revolutions will we see in the 21st century? In J. A. Goldstone, L. Grinin, & A. Korotayev (Eds.), *Handbook of revolutions in the 21st century: The new waves of revolutions, and the causes and effects of disruptive political change* (pp. 1037–1061). Springer. https://doi.org/10.1007/978-3-030-86468-2_41

Grinin, L. E. (2012). State and socio-political crises in the process of modernization. *Cliodynamics: The Journal of Quantitative History and Cultural Evolution, 3*(1), 124–157.

Grinin, L. E. (2013). State and socio-political crises in the process of modernization. *Social Evolution & History, 12*(2), 35–76.

Grinin, L. (2020a). Demograficheskiye protsessy kak bazovyy i dlitel'nyy faktor vozmozhnoy destabilizatsii v stranakh afraziyskoy makrozony nestabil'nosti. In L. Grinin, A. Korotayev, & D. Bykanova (Eds.), *Sistemnyy monitoring global'nykh I regional'nykh riskov* (pp. 181–199). Uchitel.

Grinin, L. (2020b). Dvizheniye k ustoychivym demokraticheskim otnosheniyam i natsional'nomu gosudarstvu kak dlitel'nyy faktor vozmozhnoy destabilizatsii v stranakh afraziyskoy makrozony nestabil'nosti. In L. Grinin, A. Korotayev, & D. Bykanova (Eds.), *Sistemnyy monitoring global'nykh i regional'nykh riskov* (pp. 694–724). Uchitel.

Grinin, L. (2020c). Dvizheniye obshchestv k zrelym sotsial'no-politicheskim i ekonomicheskim otnosheniyam kak dlitel'nyy faktor vozmozhnoy destabilizatsii v stranakh afraziyskoy makrozony nestabil'nosti. In L. Grinin, A. Korotayev, & D. Bykanova (Eds.), *Sistemnyy monitoring global'nykh i regional'nykh riskov* (pp. 666–693). Uchitel.

Grinin, L. (2020d). Radikal'nyy islamizm, religioznyy faktor i destabilizatsiya v stranakh afraziyskoy makrozony nestabil'nosti. In L. Grinin, A. Korotayev, & D. Bykanova (Eds.), *Sistemnyy monitoring global'nykh i regional'nykh riskov* (pp. 810–828). Uchitel.

Grinin, L. (2022). Revolutions of the 21st century as a factor of the world system reconfiguration. In J. A. Goldstone, L. Grinin, & A. Korotayev (Eds.), *Handbook of revolutions in the 21st century: The new waves of revolutions, and the causes and effects of disruptive political change* (pp. 975–999). Springer. https://doi.org/10.1007/978-3-030-86468-2_38

Grinin, L., & Grinin, A. (2023a). Analyzing social self-organization and historical dynamics. Future cybernetic W-society: Sociopolitical aspects. In V. Sadovnichy et al. (Eds.), *Reconsidering the limits to growth. A report to the Russian Association of the Club of Rome* (pp. 491–519). Springer. https://doi.org/10.1007/978-3-031-34999-7_21

Grinin, L., & Grinin, A. (2023b). Technology. Limitless possibilities, effective control. In V. Sadovnichy et al. (Eds.), *Reconsidering the limits to growth. A report to the Russian*

Association of the Club of Rome (pp. 139–154). Springer. https://doi.org/10.1007/978-3-031-34999-7_8

Grinin, L., Grinin, A., & Korotayev, A. (2023a). Future political change. Toward a more efficient world order. In V. Sadovnichy et al. (Eds.), *Reconsidering the limits to growth. A report to the Russian Association of the Club of Rome* (pp. 191–206). Springer. https://doi.org/10.1007/978-3-031-34999-7_11

Grinin, L., Grinin, A., & Korotayev, A. (2023b). Global ageing–An integral problem of the future. How to turn a problem into a development driver? In V. Sadovnichy et al. (Eds.), *Reconsidering the limits to growth. A report to the Russian Association of the Club of Rome* (pp. 117–135). Springer. https://doi.org/10.1007/978-3-031-34999-7_7

Grinin, L., Grinin, A., & Malkov, S. (2023c). Economics. Optimizing growth. In V. Sadovnichy et al. (Eds.), *Reconsidering the limits to growth. A report to the Russian Association of the Club of Rome* (pp. 155–168). Springer. https://doi.org/10.1007/978-3-031-34999-7_9

Grinin, L., Grinin, A., & Malkov, S. (2023d). Socio-political transformations. A difficult path to cybernetic society. In V. Sadovnichy et al. (Eds.), *Reconsidering the limits to growth. A report to the Russian Association of the Club of Rome* (pp. 169–189). Springer. https://doi.org/10.1007/978-3-031-34999-7_10

Grinin, L., & Korotayev, A. (2010). Will the global crisis Lead to global transformations? The coming epoch of new coalitions. *Journal of Globalization Studies, 1*(2), 166–183.

Grinin, L., & Korotayev, A. (2015). Global'noye stareniye naseleniya, shestoy tekhnologicheskiy uklad i mirovaya finansovaya sistema. In L. Grinin, A. Korotayev, & V. Bondarenko (Eds.), *Kondrat'yevskiye volny: naslediye i sovremennost* (pp. 107–132). Uchitel.

Grinin, L., & Korotayev, A. (2016). Global Population ageing, the sixth Kondratieff wave, and the global financial system. *Journal of Globalization Studies, 7*(2), 11–31.

Grinin, L., & Korotayev, A. (2022). The future of revolutions in the 21st century and the world system reconfiguration. *World Futures., 79*, 69. https://doi.org/10.1080/02604027.2022.2050342

Grinin, L., Malkov, S., & Korotayev, A. (2023). High income and low income countries. Towards a common goal at different speeds. In V. Sadovnichy et al. (Eds.), *Reconsidering the limits to growth. A report to the Russian Association of the Club of Rome* (pp. 207–224). Springer. https://doi.org/10.1007/978-3-031-34999-7_12

Groth, H., & May, J. F. (Eds.). (2017). *Africa's population: In search of a demographic dividend*. Springer. https://doi.org/10.1007/978-3-319-46889-1

Haas, M. L. (2015). Population ageing and the future of the great powers. In J. A. Goldstone, L. Grinin, & A. Korotayev (Eds.), *History & mathematics: Demography & ageing* (pp. 133–146). Uchitel.

Hasan, R., Loevinsohn, B., Moucheraud, C., Ahmed, S. A., Osorio-Rodarte, I., Suzuki, E., Pradhan, E., Madhavan, S., Troiano, S., Sexton, M., Mustapha, F. A., Odutolu, O., & Okunola, O. (2019). Nigeria's demographic dividend? In *Policy note in support of Nigeria's ERGP 2017–2020*. World Bank.

Jimenez, E., & Pate, M. A. (2017). Reaping a demographic dividend in Africa's largest country: Nigeria. In H. Groth & J. F. May (Eds.), *Africa's population: In search of a demographic dividend* (pp. 33–52). Springer.

Kebede, E., Goujon, A., & Lutz, W. (2019). Stalls in Africa's fertility decline partly result from disruptions in female education. *Proceedings of the National Academy of Sciences, 116*(8), 2891–2896. https://doi.org/10.1073/pnas.171728811

Korotayev, A., Bilyuga, S., & Shishkina, A. (2016a). GDP per capita, protest intensity and regime type: A quantitative analysis. *Sravnitelnaya Politika-Comparative Politics, 7*(4), 72–94. https://doi.org/10.18611/2221-3279-2016-7-4(25)-72-94

Korotayev, A., Bilyuga, S., & Shishkina, A. (2017a). Correlation between GDP per capita and protest intensity: A quantitative analysis. *Polis-Politicheskiye Issledovaniya, 2*, 155–169. https://doi.org/10.17976/jpps/2017.02.11

Korotayev, A., Bilyuga, S., & Shishkina, A. (2017b). GDP per capita, intensity of anti-government demonstrations and level of education. Cross-national analysis. *Politeia-Journal of Political Theory, Political Philosophy and Sociology of Politics, 84*(1), 127–143. https://doi.org/10.30570/2078-5089-2017-84-1-127-143

Korotayev, A., Bilyuga, S., & Shishkina, A. (2018a). GDP per capita and protest activity: A quantitative reanalysis. *Cross-Cultural Research, 52*(4), 406–440. https://doi.org/10.1177/1069397117732328

Korotayev, A., Malkov, S., & Musieva, J. (2023a). Demography. Toward the optimization of demographic processes. In V. Sadovnichy et al. (Eds.), *Reconsidering the limits to growth. A report to the Russian association of the club of Rome* (pp. 97–116). Springer. https://doi.org/10.1007/978-3-031-34999-7_6

Korotayev, A., Romanov, D., Zinkina, J., & Slav, M. (2023). Urban youth and terrorism: A quantitative analysis (are youth bulges relevant anymore?). *Political Studies Review, 21*(3), 550–574. https://doi.org/10.1177/14789299221075908

Korotayev, A., Sawyer, P., Gladyshev, M., Romanov, D., & Shishkina, A. (2021a). Some sociodemographic factors of the intensity of anti-government demonstrations: Youth bulges, urbanization, and protests. *Sociologiceskoe Obozrenie = Russian Sociological Review, 20*(3), 98–128. https://doi.org/10.17323/1728-192x-2021-3-98-128

Korotayev, A., Sawyer, P., Grinin, L., Romanov, D., & Shishkina, A. (2020). Socio-economic development and anti-government protests in light of a new quantitative analysis of global databases. *Sotsiologicheskiy Zhurnal, 26*(4), 61–78. https://doi.org/10.19181/socjour.2020.26.4.7642

Korotayev, A., Sawyer, P., & Romanov, D. (2021b). Socio-economic development and protests. A quantitative reanalysis. *Comparative Sociology, 20*(2), 195–222. https://doi.org/10.1163/15691330-bja10030

Korotayev, A., Shulgin, S., Ustyuzhanin, V., Zinkina, J., & Grinin, L. (2023b). Modeling social self-organization and historical dynamics. Africa's futures. In V. Sadovnichy et al. (Eds.), *Reconsidering the limits to growth. A report to the Russian association of the club of Rome* (pp. 461–490). Springer. https://doi.org/10.1007/978-3-031-34999-7_20

Korotayev, A., Shulgin, S., Zinkina, J., & Slav, M. (2022). Estimates of the possible economic effect of the demographic dividend for sub-saharan Africa for the period up to 2036. *Vostok (Oriens), 2*, 108–123. https://doi.org/10.31857/S086919080019128-7

Korotayev, A., Vaskin, I., & Bilyuga, S. (2017). Olson-Huntington hypothesis on a bell-shaped relationship between the level of economic development and sociopolitical destabilization: A quantitative analysis. *Sociologiceskoe Obozrenie = Russian Socioloogical Review, 16*(1), 9–49. https://doi.org/10.17323/1728-192X2017-1-9-49

Korotayev, A., Vaskin, I., Bilyuga, S., & Ilyin, I. (2018b). Economic development and sociopolitical destabilization: A re-analysis. *Cliodynamics: The Journal of Quantitative History and Cultural Evolution, 9*(1), 59–118. https://doi.org/10.21237/c7clio9137314

Korotayev, A., Vaskin, I., & Tsirel, S. (2021c). Economic growth, education, and terrorism: A re-analysis. *Terrorism and Political Violence, 33*(3), 572–595. https://doi.org/10.1080/09546553.2018.1559835

Korotayev, A., & Zinkina, J. (2014). How to optimize fertility and prevent humanitarian catastrophes in tropical Africa. *African Studies in Russia, 6*, 94–107. https://elibrary.ru/item.asp?id=22501708

Korotayev, A., & Zinkina, J. (2015). East Africa in the Malthusian trap? *Journal of Developing Societies, 31*(3), 1–36. https://doi.org/10.1177/0169796X15590322

Korotayev, A., & Zinkina, J. (2022). Egypt's 2011 revolution. A demographic structural analysis. In J. A. Goldstone, L. Grinin, & A. Korotayev (Eds.), *Handbook of revolutions in the 21st century: The new waves of revolutions, and the causes and effects of disruptive political change* (pp. 651–683). Springer. https://doi.org/10.1007/978-3-030-86468-2_25

Korotayev, A., Zinkina, J., Goldstone, J., & Shulgin, S. (2016b). Explaining current fertility dynamics in tropical Africa from an anthropological perspective: A cross-cultural investigation. *Cross-Cultural Research, 50*(3), 251–280. https://doi.org/10.1177/1069397116644158

Korotayev, A., Zinkina, J., Kobzeva, S., Bogevolnov, J., Khaltourina, D., Malkov, A., & Malkov, S. (2011). A trap at the escape from the trap? Demographic-structural factors of political instability in modern Africa and West Asia. *Cliodynamics: The Journal of Quantitative History and Cultural Evolution, 2*(2), 276–303. https://doi.org/10.21237/C7clio22217

Lemma, A. (1989). *Africa beyond famine. A report to The Club of Rome*. The Club of Rome.

May, J. F., & Rotenberg, S. (2020). A call for better integrated policies to accelerate the fertility decline in sub-Saharan Africa. *Studies in Family Planning, 51*(2), 193–204.

Mesquida, C. G., & Weiner, N. I. (1999). Male age composition and severity of conflicts. *Politics and the Life Sciences, 18*, 113–117. https://doi.org/10.1017/S0730938400021158

Moller, H. (1968). Youth as a force in the modern world. *Comparative Studies in Society and History, 10*(3), 237–260. https://doi.org/10.2307/177801

Nzimande, N., & Mugwendere, T. (2018). Stalls in Zimbabwe fertility: Exploring determinants of recent fertility transition. *Southern African Journal of Demography, 18*(1), 59–110.

Romanov, D., Meshcherina, K., & Korotayev, A. (2021). The share of youth in the Total Population as a factor of intensity of non-violent protests: A quantitative analysis. *Polis-Politicheskie Issledovaniya, 3*, 166–181. https://doi.org/10.17976/jpps/2021.03.11

Sawyer, P., Romanov, D., Slav, M., & Korotayev, A. (2022). Urbanization, the youth, and protest: A cross-National Analysis. *Cross-Cultural Research, 56*(2–3), 125–149. https://doi.org/10.1177/10693971211059762

Schoumaker, B. (2019). Stalls in fertility transitions in sub-Saharan Africa: Revisiting the evidence. *Studies in Family Planning, 50*(3), 257–278. https://doi.org/10.1111/sifp.12098

UN Population Division. (2022). *United Nations population division database*. United Nations. http://www.un.org/esa/population

Urdal, H. (2004). *The devil in the demographics: The effect of youth bulges on domestic armed conflict, 1950–2000*. World Bank.

Urdal, H. (2006). A clash of generations? Youth bulges and political violence. *International Studies Quarterly, 50*(3), 607–629. https://doi.org/10.1111/j.1468-2478.2006.00416.x

Urdal, H. (2008). Population, resources, and political violence: A subnational study of India, 1956–2002. *Journal of Conflict Resolution, 52*(4), 590–617. https://doi.org/10.1177/0022002708316741

Urdal, H. (2012). Youth bulges and violence. In J. A. Goldstone, E. P. Kaufmann, & M. D. Toft (Eds.), *Political demography: How population changes are reshaping international security and national politics* (pp. 117–132). Oxford University Press.

Ustyuzhanin, V., Grinin, L., & Korotayev, A. (2021). Revolyutsionnyye sobytiya XXI veka v afraziyskoy makrozone nestabil'nosti i nekotorykh drugikh mir-sistemnykh zonakh: predvaritel'nyy kolichestvennyy analiz. *Sistemnyy monitoring global'nykh i regional'nykh riskov, 12*, 106–114.

Vaskin, I., Tsirel, S., & Korotayev, A. (2018). Economic growth, education and terrorism: Experience in quantitative analysis. *Sotsiologicheskiy zhurnal, 24*(2), 28–65. https://doi.org/10.19181/socjour.2018.24.2.5844

Weber, H. (2019). Age structure and political violence: A re-assessment of the "youth bulge" hypothesis. *International Interactions, 45*(1), 80–112. https://doi.org/10.1080/03050629.2019.1522310

Yair, O., & Miodownik, D. (2016). Youth bulge and civil war: Why a country's share of young adults explain only non-ethnic wars. *Conflict Management and Peace Science, 33*(1), 25–44. https://doi.org/10.1177/0738894214544613

Zinkina, J., & Korotayev, A. (2014a). Explosive population growth in tropical Africa: Crucial omission in development forecasts (emerging risks and way out). *World Futures, 70*(4), 271–305. https://doi.org/10.1080/02604027.2014.894868

Zinkina, J., & Korotayev, A. (2014b). Projecting Mozambique's demographic futures. *Journal of Futures Studies, 19*(2), 21–40.
Zinkina, J., & Korotayev, A. (2017). *Sotsial'no-demograficheskoye razvitiye stran Tropicheskoy Afriki: Klyuchevyye faktory riska, modifitsiruyemyye upravlyayushchiye parametry, rekomendatsii*. Lenand/URSS.

Leonid Grinin is Senior Research Professor at the Centre for Stability and Risk Analysis at HSE University in Moscow, as well as Senior Research Professor of the Institute of Oriental Studies, Russian Academy of Sciences. He is co-editor of the international journals *Social Evolution & History* and the *Journal of Globalization Studies*, as well as co-editor of the international yearbooks *Evolution, History & Mathematics*, and *Kondratieff Waves*. He is author of more than 600 publications, including 33 monographs in Russian, English, Spanish, German, and Chinese. His current research focuses on comparative political studies, theory of revolution, political anthropology, global economy, global history, historical sociology, and futurology. Among his monographs are *Great Divergence and Great Convergence. A Global Perspective* (Springer, 2015; with Andrey Korotayev), *The Cybernetic Revolution and the Forthcoming Epoch of Self-Regulating Systems* (Uchitel, 2016, with Anton Grinin), *Economic Cycles, Crises, and the Global Periphery* (Springer, 2016; with Andrey Korotayev and Arno Tausch), *Big History of Globalization. The Emergence of a Global World System* (Springer 2019, in co-authorship with Julia Zinkina, David Christian, and others), *Islamism, Arab Spring, and the Future of Democracy. World System and World Values Perspectives* (Springer, 2019; with Andrey Korotayev and Arno Tausch). In 2004, he was awarded as a winner of the best publication of *Socis* journal during the year for the article "People of Publicity—a New Social Layer?." In 2012, he was awarded with the Gold Kondratieff Medal by the International N. D. Kondratieff Foundation. In 2016, he received the Belyaev award for the book *From Bifaces to Nanorobots* (the book published in Russian in co-authorship with Anton Grinin).

Andrey Korotayev heads the Centre for Stability and Risk Analysis at HSE University in Moscow, Russia. He is also Senior Research Professor at the Institute of Oriental Studies and Institute for African Studies, Russian Academy of Sciences, as well as at the Faculty of Global Studies of Lomonosov Moscow State University and the Russian Presidential Academy of National Economy and Public Administration. He has authored or co-authored over 650 scholarly publications, including such monographs as *Ancient Yemen* (Oxford University Press, 1995), *World Religions and Social Evolution of the Old World Oikumene Civilizations: A Cross-Cultural Perspective* (The Edwin Mellen Press, 2004), *Introduction to Social Macrodynamics: Compact Macromodels of the World System Growth* (URSS, 2006; with Daria Khaltorina and Artemy Malkov), *Introduction to Social Macrodynamics: Secular Cycles and Millennial Trends* (URSS, 2006; with Daria Khaltorina and Artemy Malkov), *Great Divergence and Great Convergence. A Global Perspective* (Springer, 2015; with Leonid Grinin and Arno Tausch), *Economic Cycles, Crises, and the Global Periphery* (Springer, 2016; with Leonid Grinin), *Islamism, Arab Spring, and the Future of Democracy. World System and World Values Perspectives* (Springer, 2019; with Leonid Grinin and Arno Tausch). He is a laureate of Russian Science Support Foundation Award in "The Best Economists of the Russian Academy of Sciences" Nomination (2006); in 2012 he was awarded with the Gold Kondratieff Medal by the International N. D. Kondratieff Foundation.

Part IV

Modeling Social Self-Organization and Historical Dynamics – from Agrarian to Cybernetic W-Society

Modeling Social Self-Organization and Historical Dynamics. An Overview

Victor Sadovnichy, Askar Akaev, Ilya Ilyin, Sergey Malkov, Leonid Grinin, Vasily Popov, and Andrey Korotayev

Abstract

Sadovnichy et al. present this chapter as an introductory one to the Mathematical Part which consists of nine chapters, including the present one. In the present chapter, Sadovnichy et al. explain the design of the Mathematical Part, which mathematical apparatus is used, how the chapters and conclusions of this part are

V. Sadovnichy · I. Ilyin
Lomonosov Moscow State University, Moscow, Russia
e-mail: info@rector.msu.ru; dekanat@fgp.msu.ru

A. Akaev
Lomonosov Moscow State University, Moscow, Russia

HSE University, Moscow, Russia

S. Malkov
Lomonosov Moscow State University, Moscow, Russia

HSE University, Moscow, Russia

Institute of Economics, Russian Academy of Sciences, Moscow, Russia
e-mail: s@malkov.org

L. Grinin
HSE University, Moscow, Russia

Institute of Oriental Studies, Russian Academy of Sciences, Moscow, Russia

V. Popov
Voronezh State University of Engineering Technologies, Voronezh, Russia
e-mail: pvn@vsuet.ru

A. Korotayev (✉)
Lomonosov Moscow State University, Moscow, Russia

HSE University, Moscow, Russia

Institute of Oriental Studies, Russian Academy of Sciences, Moscow, Russia

V. Sadovnichy et al. (eds.), *Reconsidering the Limits to Growth*, World-Systems Evolution and Global Futures, https://doi.org/10.1007/978-3-031-34999-7_14

connected with those in the previous chapters of the present report, etc. Sadovnichy et al. also show what this Mathematical Part is about, why it is needed, how it is organized, and so on. The Mathematical Part of the present Report to the Russian Association of the Club of Rome mostly presents results of mathematical modeling of various aspects of historical, present, and future reality and forecasts, but it also includes important verbal qualitative sections. Its necessity is dictated by the fact that, although in the previous parts of the present report, Sadovnichy et al. have realized most of its goals, however, in order for its conclusions to become more convincing and scientific, they must also be supported by logical and mathematical modeling. This Mathematical Part is called "Modeling Social Self-Organization and Historical Dynamics—from Agrarian to Cybernetic W-Society." This name reflects its structure, since the chapters of this part consider social systems of the past, present, and future. On the other hand, it implements important methodological assumptions. In accordance with these assumptions, the objects of research and modeling in the Mathematical Part are the basic processes (including social and political ones) that determine features of interaction between various spheres of life at respective phases of historical development. The logic of the Mathematical Part leads us from ancient societies to modern and future ones, through understanding how phase transitions occur from one type of society to another, and allows us to make forecasts about the future society. The chapters of this part demonstrate how we see the combining of the world systems, historical and evolutionary approaches, a systematic view of society, and mathematical modeling within a single research program.

As has been mentioned in the Introductory Chapter[1] to the present Report to the Club of Rome, the main goals of the book are (1) to give an analysis of changes through which the world system has come to its present state, based on an integrated approach (that incorporates the world systems, historical, and evolutionary approaches), on mathematical modeling, as well as on a systematic view of society, in which changes in one subsystem cause transformations in others; (2) to define the main vectors of transformations of the World System; (3) to make a detailed forecast of the development of all the main subsystems of society and the world system, while presenting three or four horizons of changes (from short-term to ultra-long-term up to 100 years); and (4) to present different development scenarios and make recommendations on how to switch to the most favorable development scenario.

In the previous parts of our report, we have realized most of these goals. However, in order for our conclusions to become more convincing and scientific, we consider it necessary to accompany the report with a mathematical part. It is also important to demonstrate how we see the combining of the world systems, historical and evolutionary approaches, a systematic view of society, and mathematical

[1] See Chapter "Introduction. Hoping for the Future" (Sadovnichy et al., 2023, this volume).

modeling within a single research program. This part mainly presents results of mathematical modeling for various aspects of historical, present and future reality, and our forecasts, but it also includes important verbal qualitative sections.

The present chapter is an introductory one to this part. In the present chapter, we explain the design of the Mathematical Part, which mathematical apparatus we use, how the chapters and conclusions of this part are connected with those in the previous chapters of the present report, etc.

This Mathematical Part consists of nine chapters, including the present one. It is titled "Modeling Social Self-Organization and Historical Dynamics." This name reflects both its structure, since, in its specific chapters, the previous types of society (agrarian, industrial), as well as the future society and its types, are considered, and it implements the specified methodological assumption: the changes taking place in the world are fundamental and long-term, therefore, they should be considered in a broad historical context (and not just as a continuation of the trends of the twentieth century). Accordingly, when modeling world dynamics, historical approach should be implemented, which implies viewing the current situation as a certain phase of macrohistorical development, as a certain stage of the historical transformation of the world system. It also stresses an uneven character of the historical process which proceeds sometimes in gradual evolutional way and sometimes proceeds in uneven jerks (taking the form of revolutions) occurring from time to time. The Chapter "Modeling Social Self-Organization and Historical Dynamics. Global Phase Transitions" is devoted almost entirely to this aspect.

The chapters of this part are very important with respect to the possibility to set out in more detail some topics that are briefly touched upon in the main text of the report, but which, in the opinion of the authors, require more detailed coverage and explanation (links to the relevant chapters of the Mathematical Part are indicated in the main text of this report). They also give an opportunity for more scrupulous insight into forecasts about the future society and ways to it. Thus, they help to build a scientific foundation for our forecasts. In addition, the chapters contain visual materials (graphs, drawings), mathematical models, and calculation results that illustrate and substantiate the conclusions of the report.

So, the chapters of the Mathematical Part are interconnected with the chapters of the main parts of our report. At the same time, in some respects, the materials of the Mathematical Part chapters are independent, as they consistently set out the main elements of the system of mathematical modeling of world dynamics, which was developed at Lomonosov Moscow State University under the guidance of Victor Sadovnichy and Askar Akaev (see Sadovnichy et al., 2012, 2014, 2017) and which has been used for the preparation of this report.

In accordance with this and other methodological assumptions, mentioned in the Introductory Chapter to this report,[2] the objects of research and modeling in these chapters are the basic processes (including social and political ones) that determine features of interaction of various spheres of life at the respective phases of historical

[2] See Chapter "Introduction. Hoping for the Future" (Sadovnichy et al., 2023, this volume).

development. Thus, we try to apply systems modeling in many areas, some of them where modelers rarely engage at all, such as socio-political transformations. This could be considered as a bold move, or even stretching too far the possibilities of modeling. But we believe that it is necessary to try moving in this direction. Accordingly, the modeling is carried out at a high level of aggregation. The purpose of modeling is not so much to determine the specific values of variables as to reveal the logic of their long-term dynamics. It is reasonable to present the results of modeling not in the form of graphs that reflect the change in the values of certain variables over time within specific scenario conditions, but rather with the help of phase portraits. Analysis of the phase portraits makes it possible to identify "order parameters" (i.e., key factors on which the type of phase portrait depends) and characteristics of the sustainability of economic and social development at the respective phases of historical development (based on an analysis of the conditions under which a radical transformation of the phase portrait occurs).

The main problem that had to be faced in the course of research was the extreme multiparameter nature of the subject area under consideration. The desire to take into account various details inevitably required the introduction of additional parameters and relationships into the system of equations, which made it opaque and difficult to analyze. Therefore, the most important task of the study was the task of reducing the dimensions of the system of basic equations as much as possible by highlighting the most significant processes that determine long-term dynamics. The model equations were constructed in order to capture the most important regularities of these processes. Also, when constructing and analyzing phase portraits, Tikhonov's theorem was used (Tikhonov & Samarskii, 2011), which, by dividing the model variables into "fast" and "slow," makes this analysis more transparent and reveals the most important points of the phases of historical development (based on an analysis of the conditions under which a radical transformation of the phase portrait occurs).

The reference points in the creation of this modeling system were the works of Forrester (1971, 1975, 1984), Dennis Meadows (Meadows et al., 1972, 1992, 2004; Meadows & Booth Sweeney, 2010, 2016), Randers (2012; Randers et al., 2018), and other authors of relevant reports to the Club of Rome, as well as the works of Russian scientists—Vernadsky (1998, 2012, 2013, 2015, 2018), Prigogine (1997, 2017, Prigogine & Stengers, 1984), Moiseev (1982, 1990, 1996, 2022; Moiseev et al., 1985), Gvishiani (1977, 1980, 1984, 1986, 2010; Gvishiani et al., 1991), Kapitza (1992, 1996, 1999a, 1999b, 2003, 2006, 2010), Gorshkov (1987, 1995, 2012; Gorshkov et al., 2000, 2004), Yury Romanovsky (Romanovsky et al., 1975); Chernavsky (2001), Dubovsky (1995, 2004, 2010), Victor Gelovani (Gelovani et al., 1975, 1979, 2005, 2009; Gelovani & Dubovsky, 1990) and many others who contributed to the development of mathematical models of world dynamics. Like all of them, we faced an extremely difficult task: to describe with minimal mathematical means (avoiding the complexity of the mathematical apparatus) the main patterns of world dynamics as clearly and convincingly as possible and, on the basis of this, to make a forecast of possible scenarios for further world development. The task was complicated by the fact that, from the very beginning, we sought to describe world development on the basis of a unified approach over a long period of time (hundreds and thousands of years) in order to avoid "historical myopia" and subjectivity in the interpretation of current events. The task was even more complicated due to the fact that we initially sought to model not only interactions in the "society-nature" system but also interactions within society, leading to social and political changes (Grinin et al., 2021; Kurdyumov et al., 1999; Makarieva et al., 2008; Prigogine & Stengers, 2018). This task turned out to be the most difficult due to the extreme diversity of social systems of the past and the present, their internal structure, types of political regimes, etc. Therefore, we took the path of identifying and modeling the basic mechanisms of social self-organization in various external conditions, as a result of which the institutional shape of social systems is ultimately formed. Based on this, it was possible to identify the basic institutional structures that were characteristic of social systems in the historical periods under consideration, which made it possible to consider social dynamics as a process of historical transformation of these structures. The results of the analysis of world dynamics based on this approach are presented in the chapters of the Mathematical Part.

As has been mentioned above, Mathematical Part "Modeling Social Self-Organization and Historical Dynamics" consists of nine chapters: (1) An Overview[3]; (2) A General Approach[4]; (3) Agrarian Society[5]; (4) Industrial Society[6]; (5) Global Phase Transitions[7]; (6) Modern Society and Problems of Global Transition[8]; (7) Africa's Futures[9]; (8) Global Futures and Cybernetic W-Society[10]; and (9) Life Quality Index.[11] Thus, the logic of the chapters of this part leads us from ancient societies to modern ones, through understanding how phase transitions occur from one type of society to another, which allows us to make predictions about the future society.

Chapter "Modeling Social Self-Organization and Historical Dynamics. A General Approach" (Akaev et al., 2023a this volume) outlines the methodological approach

[3] The present chapter.

[4] See Chapter "Modeling Social Self-Organization and Historical Dynamics. A General Approach" (Akaev et al., 2023, this volume).

[5] See Chapter "Modeling Social Self-Organization and Historical Dynamics. Agrarian Society" (Malkov et al., 2023, this volume)

[6] See Chapter "Modeling Social Self-Organization and Historical Dynamics. Industrial Society" (Akaev et al., 2023, this volume).

[7] See Chapter "Modeling Social Self-Organization and Historical Dynamics. Global Phase Transitions" (Malkov et al., 2023, this volume).

[8] See Chapter "Modeling Social Self-Organization and Historical Dynamics. Modern Society and Problems of Global Transition" (Akaev et al., 2023, this volume).

[9] See Chapter "Modeling Social Self-Organization and Historical Dynamics. Africa's Futures" (Korotayev et al., 2023, this volume).

[10] See Chapter "Analyzing Social Self-Organization and Historical Dynamics. Future Cybernetic W-Society: Socio-Political Aspects" (Grinin & Grinin, 2023a, this volume).

[11] See Chapter "Modeling Social Self-Organization and Historical Dynamics. Life Quality Index" (Malkov et al., 2023, this volume).

to the mathematical modeling of social self-organization and historical dynamics, which has been used for the preparation of the present report. Chapter "Modeling Social Self-Organization and Historical Dynamics. Agrarian Society" (Malkov et al., 2023c, this volume) presents the results of modeling the dynamics of an agrarian society. It is shown that in agrarian societies, the mechanisms of social self-organization mainly lead to the formation of social structures of the so-called X-type, which are characterized by the institutional combination "distributive economy—directive (hierarchical) management system—the primacy of collectivism in the public consciousness." Chapter "Modeling Social Self-Organization and Historical Dynamics. Industrial Society" (Akaev et al., 2023b, this volume) presents the results of modeling the dynamics of an industrial society. It is shown that in such societies, the mechanisms of social self-organization mainly lead to the formation of social structures of the so-called Y-type, which are characterized by the institutional combination of "a market economy of a liberal type—an adaptive (democratic) control system—the primacy of individualism in the public mind." It is shown that a necessary condition for the effective functioning of Y-type social structures is economic growth, in the absence of which one observes their crisis and transformation.

Chapter "Modeling Social Self-Organization and Historical Dynamics. Global Phase Transitions" (Malkov et al., 2023b, this volume) presents the results of modeling transitional eras, during which cardinal changes take place in the "society—nature" interactions and in social relations. Chapter "Modeling Social Self-Organization and Historical Dynamics. Modern Society and Problems of Global Transition" (Akaev et al., 2023c, this volume) presents the results of modeling the modern world dynamics. It is shown that the modern historical period is a period of transition from the "epoch of growth," which followed the industrial revolution of the early nineteenth century, to the "era of deceleration," one of the clearest indicators of which is a rapid decrease in the growth rate of the Earth's population and its aging. Alternative options for further development are considered.

Chapter "Modeling Social Self-Organization and Historical Dynamics. Africa's Futures" (Korotayev et al., 2023, this volume) presents forecasts for the emergence of large-scale political and demographic collapses and for the economic growth of some countries in Sub-Saharan Africa where the likelihood of armed civil conflicts and population impoverishment is the highest in the coming decades. It is shown that the main way of mitigating the risks of sociodemographic collapses is rapid progress toward achieving the Sustainable Development Goals (SDGs) in the very near future, which seems impossible without the support of the world community. This chapter is tightly interconnected with Chapter "Africa—the Continent of the Future. Challenges and Opportunities" (Grinin et al., 2023a, this volume).

Chapter "Analyzing Social Self-Organization and Historical Dynamics. Future Cybernetic W-Society: Socio-Political Aspects" (Grinin & Grinin, 2023a, this volume) discusses the future cybernetic W-society, which will emerge as a result of completion of the Cybernetic Revolution and an apogee of the global aging processes in the late twenty-first century.[12] It provides details on this subject preliminarily described in Chapter "Socio-Political Transformations. A Difficult Path to Cybernetic Society" (Grinin et al., 2023b, this volume).

Finally, Chapter "Modeling Social Self-Organization and Historical Dynamics. Life Quality Index" (Malkov, Bilyuga, & Musieva, 2023a, this volume) proposes a quantitative measurement of the development of countries and the well-being of the population that could serve as an alternative to both GDP per capita and the Human Development Index.

Acknowledgements This work was done with the support of MSU Program of Development, Project No 23A-SCH05-03.

References

Akaev, A., Malkov, S., Bilyuga, S., Malkov, A., Musieva, J., & Korotayev, A. (2023a). Modeling social self-organization and historical dynamics. A general approach. In V. Sadovnichy et al. (Eds.), *Reconsidering the limits to growth. A report to the Russian association of the Club of Rome* (pp. 253–307). Springer. https://doi.org/10.1007/978-3-031-34999-7_15

Akaev, A., Malkov, S., Davydova, O., Kovaleva, N., Malkov, A., Grinin, L., & Korotayev, A. (2023b). Modeling social self-organization and historical dynamics. industrial society. In V. Sadovnichy et al. (Eds.), *Reconsidering the limits to growth. A report to the Russian association of the Club of Rome* (pp. 337–385). Springer. https://doi.org/10.1007/978-3-031-34999_17

Akaev, A., Malkov, S., Grinin, L., Bilyuga, S., Davydova, O., Grinin, A., Kovaleva, N., Malkov, A., Musieva, J., & Korotayev, A. (2023c). Modeling social self-organization and historical dynamics. Modern society and a look into the global futures: Cybernetic W-society. In V. Sadovnichy et al. (Eds.), *Reconsidering the limits to growth. A report to the Russian Association of the Club of Rome* (pp. 419–460). Springer. https://doi.org/10.1007/978-3-031-34999-7_19

Chernavsky, D. (2001). *Sinergetika i informatsiya: dinamicheskaya teoriya informatsii*. Nauka.

Dubovsky, S. (1995). The Kondratiev cycle as an object of modelling. *Matematicheskoe modelirovanie, 7*(6), 65–74.

Dubovsky, S. (2004). *Raspredeleniye energii i dokhodov v ekonomicheskom razvitii. matematicheskiye modeli*. ROKHOS.

Dubovsky, S. (2010). Systems analysis and global sustainable development. In *Global Security and International Political Economy–Volume II* (p. 235).

Forrester, J. W. (1971). *World dynamics*. Wright-Allen Press.

Forrester, J. W. (1975). *Collected papers of Jay W. Forrester*. Pegasus Communications.

Forrester, J. W. (1984). *Principes des systèmes (Science des systèmes. Série Dynamique des systèmes)*. Presses universitaires de Lyon.

Gelovani, V., & Dubovsky, S. (1990). Global modeling of the potential world system. *International Political Science Review, 11*, 207–218. https://doi.org/10.1177/019251219001100205

[12] See Chapter "Technology. Limitless Possibilities, Effective Control" (Grinin & Grinin, 2023b, this volume) and Chapter "Global Aging—an Integral Problem of the Future. How to Turn a Problem into a Development Driver?" (Grinin et al., 2023, this volume).

Gelovani, V., Britkov, V., & Dubovsky, S. (2009). *SSSR i Rossiya v global'noy sisteme (1985-2030 gg.). Rezul'taty global'nogo modelirovaniya*. Librokom.
Gelovani, V., Dubovsky, S., & Yurchenko, V. (1979). Methodological problems in global models. *SIMULATION, 33*, 19–23.
Gelovani, V., Egorov, V., Mitrofanov, V., & Piontkovskii, A. (1975). A certain control problem in the Forrester global dynamic model. *Soviet Physics Doklady, 20*, 19–25.
Gelovani, V., Kurakin, P., Malinetskii, G., & Makhov, S. (2005). Steady-state solutions in a modified Forrester model. *Doklady Mathematics, 71*(2), 186–188.
Gorshkov, V. (1987). *Predely ustoychivosti biosfery i okruzhayushchey sredy*. LIYaF.
Gorshkov, V. (1995). *Fizicheskiye i biologicheskiye osnovy ustoychivosti zhizni*. VINITI.
Gorshkov, V. (2012). *Physical and biological bases of life stability: Man, biota, Environment*. Springer.
Gorshkov, V., Makarieva, A., & Gorshkov, V. V. (2000). *Biotic regulation of the environment: Key issues of global Change*. Springer.
Gorshkov, V., Makarieva, A., & Gorshkov, V. V. (2004). Revising the fundamentals of ecological knowledge: The biota–environment interaction. *Ecological Complexity, 1*(1), 17–36.
Grinin, L., & Grinin, A. (2023a). Analyzing social self-organization and historical dynamics. Future cybernetic W-society: Sociopolitical aspects. In V. Sadovnichy et al. (Eds.), *Reconsidering the limits to growth. A report to the Russian Association of the Club of Rome* (pp. 491–519). Springer. https://doi.org/10.1007/978-3-031-34999-7_21
Grinin, L., & Grinin, A. (2023b). Technology. Limitless possibilities, effective control. In V. Sadovnichy et al. (Eds.), *Reconsidering the limits to growth. A report to the Russian Association of the Club of Rome* (pp. 139–154). Springer. https://doi.org/10.1007/978-3-031-34999-7_8
Grinin, L., Grinin, A., & Korotayev, A. (2021). COVID-19 pandemic as a trigger for the acceleration of the cybernetic revolution, transition from E-government to E-state, and change in social relations. *Technological Forecasting and Social Change, 175*, 1–17. https://doi.org/10.1016/j.techfore.2021.121348
Grinin, L., Grinin, A., & Korotayev, A. (2023a). Global aging–An integral problem of the future. How to turn a problem into a development driver? In V. Sadovnichy et al. (Eds.), *Reconsidering the limits to growth. A report to the Russian Association of the Club of Rome* (pp. 117–135). Springer. https://doi.org/10.1007/978-3-031-34999-7_7
Grinin, L., Grinin, A., & Malkov, S. (2023b). Socio-political transformations. A difficult path to cybernetic society. In V. Sadovnichy et al. (Eds.), *Reconsidering the limits to growth. A report to the Russian association of the Club of Rome* (pp. 169–189). Springer. https://doi.org/10.1007/978-3-031-34999-7_10
Gvishiani, D. (1977). *Metodologicheskiye problemy modelirovaniya global'nogo razvitiya*. VNIISI.
Gvishiani, D. (1980). Methodological problems of global development modelling. In E. Velikhov et al. (Eds.), *Science, technology and the future: Soviet scientists' analysis of the problems of and prospects for the development of science and technology and their role in society* (pp. 21–35). Pergamon.
Gvishiani, D. (Ed.). (1984). *Systems research: Methodological problems*. Pergamon Press.
Gvishiani, D. (1986). *Priroda modeley i modeley prirody*. Mysl.
Gvishiani, D. (2010). *Mosty v budushchee. Rimskiy klub, IFIAS, Venskiy sovet*. URSS.
Gvishiani, D., Velikhov, E., Leybin, V., et al. (1991). *Modelirovaniye protsessov mirovogo razvitiya i sotrudnichestva*. Nauka.
Kapitza, S. (1992). A mathematical model for global population growth. *Matematicheskoe modelirovanie, 4*(6), 65–79.
Kapitza, S. (1996). Phenomenological theory of world population growth. *Physics Uspekhi, 39*(1), 57–71. https://doi.org/10.1070/PU1996v039n01ABEH000127
Kapitza, S. (1999a). *Obshchaya teoriya rosta chelovechestva. Skol'ko lyudey zhilo, zhivet i budet zhit' na Zemle*. Nauka.

Kapitza, S. (1999b). *An essay on the demographic imperative: Non-linear theory of growth of humankind*. The Edwin Mellen Press.

Kapitza, S. (2003). The statistical theory of global population growth. In *Formal descriptions of developing systems* (pp. 11–35). Springer. https://doi.org/10.1007/978-94-010-0064-2_2

Kapitza, S. (2006). *Global population blow-up and after. A report to the Club of Rome*. Global Marshall Plan Initiative.

Kapitza, S. (2010). On the theory of global population growth. *Physics Uspekhi, 53*(12), 1287–1296. https://doi.org/10.3367/UFNe.0180.201012g.1337

Korotayev, A., Shulgin, S., Ustyuzhanin, V., Zinkina, J., & Grinin, L. (2023). Modeling social self-organization and historical dynamics. Africa's futures. In V. Sadovnichy et al. (Eds.), *Reconsidering the limits to growth. A report to the Russian Association of the Club of Rome* (pp. 461–490). Springer. https://doi.org/10.1007/978-3-031-34999-7_20

Kurdyumov, S., Kapitza, S., & Malinetskii, G. (1999). *Synergetics and the study of the future (Russian geographical, geopolitical and futurological studies)*. The Edwin Mellen Press Ltd.

Makarieva, A., Gorshkov, V., & Li, B. (2008). Energy budget of the biosphere and civilization: Rethinking environmental security of global renewable and non-renewable resources. *Ecological Complexity, 5*(4), 281–288.

Malkov, S., Bilyuga, S., & Musieva, J. (2023a). Modeling social self-organization and historical dynamics. Life quality index. In V. Sadovnichy et al. (Eds.), *Reconsidering the limits to growth. A report to the Russian Association of the Club of Rome* (pp. 521–530). Springer. https://doi.org/10.1007/978-3-031-34999-7_22

Malkov, S., Grinin, L., Grinin, A., Musieva, J., & Korotayev, A. (2023b). Modeling social self-organization and historical dynamics. Global phase transitions. In V. Sadovnichy et al. (Eds.), *Reconsidering the limits to growth. A report to the Russian association of the Club of Rome* (pp. 387–417). Springer. https://doi.org/10.1007/978-3-031-34999-7_18

Malkov, S., Kovaleva, N., Grinin, L., & Korotayev, A. (2023c). Modeling social self-organization and historical dynamics. Agrarian society. In V. Sadovnichy et al. (Eds.), *Reconsidering the limits to growth. A report to the Russian Association of the Club of Rome* (pp. 309–335). Springer. https://doi.org/10.1007/978-3-031-34999-7_16

Meadows, D., & Booth Sweeney, L. (2010). *The systems thinking playbook: Exercises to stretch and build learning and systems thinking capabilities*. Chelsea Green Publishing.

Meadows, D., & Booth Sweeney, L. (2016). *The climate change playbook: 22 systems thinking games for more effective communication about climate change*. Chelsea Green Publishing.

Meadows, D., Meadows, D., & Randers, J. (1992). *Beyond the limits: Confronting global collapse, envisioning a sustainable future*. Post Mills.

Meadows, D., Meadows, D., Randers, J., & Behrens, W. W. (1972). *The limits to growth. A report to The Club of Rome*. Potomac Associates.

Meadows, D., Randers, J., & Meadows, D. (2004). *Limits to growth. The 30-year update*. Chelsea Green Publisher.

Moiseev, N. (1982). Chelovek, okruzhayushchaya sreda, obshchestvo. Problema formalizatsii opisaniya. Nauka.

Moiseev, N. (1990). *Chelovek i noosfera*. Molodaya gvardiya.

Moiseev, N. (1996). Limits of predictability for Biospheric processes. In Y. A. Kravtsov & J. B. Kadtke (Eds.), *Predictability of complex dynamical systems* (pp. 169–185). Springer. https://doi.org/10.1007/978-3-642-80254-6_10

Moiseev, N. (2022). *How far it is to tomorrow…*. Birkhäuser. https://doi.org/10.1007/978-3-030-96651-5

Moiseev, N., Alexandrov, V., & Tarko, A. (1985). *Chelovek i biosfera: opyt sistemnogo analiza i eksperimenty s modelyami*. Nauka.

Prigogine, I. (1997). *The end of certainty*. Free Press.

Prigogine, I. (2017). *Non-equilibrium statistical mechanics (Dover books on physics)*. Dover Publications.

Prigogine, I., & Stengers, I. (1984). *Order out of chaos: Man's new dialogue with nature*. Bantam.

Prigogine, I., & Stengers, I. (2018). *Order out of chaos: Man's new dialogue with nature* (2nd ed.). Verso Books.
Randers, J. (2012). *2052. A global forecast for the next forty years. A report to the club of Rome*. Chelsea Green Publishing.
Randers, J., Rockström, J., Stoknes, P. E., Golüke, U., Collste, D., & Cornell, S. (2018). *Transformation is feasible. A report to the Club of Rome*. Stockholm Resilience Centre.
Romanovsky, Y., Stepanova, N., & Chernavsky, D. (1975). *Matematicheskoye modelirovaniye v biofizike*. Nauka.
Sadovnichy, V., Akaev, A., Ilyin, I., Malkov, S., Grinin, L., Sayamov, Y., & Korotayev, A. (2023). Introduction. Hoping for the future. In V. Sadovnichy et al. (Eds.), *Reconsidering the limits to growth. A report to the Russian association of the Club of Rome* (pp. 1–14). Springer. https://doi.org/10.1007/978-3-031-34999-7_1
Sadovnichy, V., Akaev, A., Korotayev, A., & Malkov, S. (2012). *Modelirovaniye i prognozirovaniye mirovoy dinamiki*. ISPI RAN.
Sadovnichy, V., Akaev, A., Korotayev, A., & Malkov, S. (2014). *Kompleksnoye modelirovaniye i prognozirovaniye razvitiya stran BRIKS v kontekste mirovoy dinamiki*. Nauka.
Sadovnichy, V., Akaev, A., Korotayev, A., Malkov, S., & Sokolov, V. (2017). *Analiz i modelirovaniye mirovoy i stranovoy dinamiki*. LENAND.
Tikhonov, A., & Samarskii, A. (2011). *Equations of mathematical physics*. Dover Publications.
Vernadsky, V. (1998). *The biosphere*. Copernicus.
Vernadsky, V. (2012). *Biosphere and noosphere*. Iris-Press.
Vernadsky, V. (2013). *Collected works: In 24 volumes*. Nauka.
Vernadsky, V. (2015). *The study of life and the new physics*. 21st Century Science & Technology.
Vernadsky, V. (2018). *Geochemistry and the biosphere: Essays*. Synergetic Press.

Viktor Sadovnichy is Doctor of Physics and Mathematics, Professor, Full Member of the Russian Academy of Sciences, Rector of Lomonosov Moscow State University since 1992. Fields of scientific research: mathematics, mechanics and computer science. Sadovnichy's research deals with the theory of differential operators and functional analysis, as well as mathematical modeling of long term social and economic dynamics. He has obtained important theoretical and applied results related to the issues of dynamic simulation of controlled flights, in particular, of spacecraft and aircraft. This work formed the basis for creation of the first simulator of zero-gravity in the Earth conditions in the world.

Askar Akaev is the First President of the Kyrgyz Republic and Foreign Full Member of the Russian Academy of Sciences. He works as Chief Research Professor in the Prigogine Institute for Mathematical Investigations of Complex Systems at Lomonosov Moscow State University. He is also Academic Supervisor of the Centre for Stability and Risk Analysis at HSE University, Moscow, Russia. Askar Akaev has authored and co-authored over 300 scientific publications.

Ilya Ilyin is a well-known Russian researcher, Head of the Department of Globalistics, Dean of the Faculty of Global Studies at Lomonosov Moscow State, the author of more than 350 academic papers. He is the Chairman of the Presidium of the Society for Global Studies and President of the International Public Organization for the Promotion of Public Diplomacy.

Sergey Malkov heads the Center for Long-Term Forecasting and Strategic Planning at Lomonosov Moscow State University; he also works as a Senior Research Professor at the Centre for Stability and Risk Analysis at HSE University in Moscow. He is the author of over 300 scholarly publications.

Leonid Grinin is Senior Research Professor at the Centre for Stability and Risk Analysis at HSE University in Moscow, as well as Senior Research Professor of the Institute of Oriental Studies, Russian Academy of Sciences. He is the author of more than 600 publications, including more than 33 monographs in Russian, English, and Chinese.

Vasily Popov is Doctor of Biological Sciences, Professor, Rector of Voronezh State University of Engineering Technologies, laureate of the State Prize of the Russian Federation in the field of science and technology for young scientists in 1998, Honorary Diploma of the President of the Russian Federation, Professor of the Russian Academy of Sciences.

Andrey Korotayev heads the Centre for Stability and Risk Analysis at HSE University in Moscow, Russia. He is also Senior Research Professor at the Institute of Oriental Studies and Institute for African Studies, Russian Academy of Sciences, as well as at the Faculty of Global Studies of Lomonosov Moscow State University and the Russian Presidential Academy of National Economy and Public Administration. He has authored or co-authored over 650 scholarly publications.

Modeling Social Self-Organization and Historical Dynamics. A General Approach

Askar Akaev, Sergey Malkov, Stanislav Bilyuga, Artemy Malkov, Jameelah Musieva, and Andrey Korotayev

Abstract

Akaev et al. outline their methodological approach to mathematical modeling of social self-organization and historical dynamics, which is used in the study of processes occurring in the world and in forecasting further world development. They describe basic equations (in the demographic, ecological, economic, technological, social, and political spheres) and the order of their use in modeling world dynamics. Based on these equations, a set of basic models has been developed that describes interactions in the "society—nature" system, as well

A. Akaev
Lomonosov Moscow State University and HSE University, Moscow, Russia

S. Malkov
Lomonosov Moscow State University, Moscow, Russia

HSE University, Moscow, Russia

Institute of Economics, Russian Academy of Sciences, Moscow, Russia

S. Bilyuga
Faculty of Global Studies, Lomonosov Moscow State University, Moscow, Russia
e-mail: bilyuga@fgp.msu.ru

A. Malkov
Moscow Institute of Physics and Technology, Moscow, Russia
e-mail: as@malkov.org

J. Musieva
HSE University, Moscow, Russia

Faculty of Global Studies, Lomonosov Moscow State University, Moscow, Russia

A. Korotayev (✉)
HSE University, Moscow, Russia

Lomonosov Moscow State University, Moscow, Russia

Institute of Oriental Studies, Russian Academy of Sciences, Moscow, Russia

V. Sadovnichy et al. (eds.), *Reconsidering the Limits to Growth*, World-Systems Evolution and Global Futures, https://doi.org/10.1007/978-3-031-34999-7_15

as interactions within society between the main social groups (including economic and political interactions). Finally, Akaev et al. propose a technique for analyzing modeling results based on constructing and studying features of phase portraits of respective equation systems.

1 General Provisions

1.1 Principles of Modeling of Social Self-Organization and Historical Dynamics

I. The recent decades have seen a significant progress in the development of historical dynamics models. The models currently available can be divided into three groups:
   1. concept models based on identifying and analyzing the general patterns and presenting them in the form of cognitive schemes that describe logical relationships between various factors influencing historical processes.[1] Such models are characterized by a high degree of generalization. These are not mathematical models, but purely logical, conceptual ones;
   2. mathematical simulation models describing specific economic processes, historical events and phenomena (Forrester, 1971, 1975, 1984; Meadows et al., 1972, 1992, 2004; Randers, 2012; Randers et al., 2018; Pavlovskiy, 2000; Pavlovskiy et al., 2005; Moiseev et al., 1984), economic CGE models (e.g., Bergman, 1991, 2005; Decaluwé & Martens, 1988; Dellink et al., 2020) etc. Such models carefully consider and describe the factors and processes that influence the phenomena under consideration. The disadvantage of such models is that they are "tied" to a specific socioeconomic process or historical event; their applicability is limited by imposed conditions and a specified space-time interval. Predictive models are based mainly on the extrapolation of existing trends, which limits their ability to account for cyclical dynamics and development traps;
   3. mathematical models that are intermediate between the two types mentioned above. Their task is to identify the basic patterns that characterize the course of the processes under consideration, to analyze the processes of self-organization, to take into account nonlinear relationships that can result in abrupt changes in the state of systems, and in socioeconomic and political phase transitions.[2] Parametric analysis of such models makes it possible to

[1]E.g., Goldstein, 1988; Claessen & Skalník, 1978; Gellner, 1984, 1988; Wallerstein, 1974, 2004; Modelski & Thompson, 1996; Arrighi & Silver, 1999; Denemark et al., 2000; Wallerstein et al., 2013; North et al., 2009; Frank & Gills, 1993; Diakonoff, 1999; Modelski et al., 2008; Allen, 2011; McNeill, 2013; Goldstone, 2009, 2014, 2016; Grinin, 2011; Grinin et al., 2016; Chase-Dunn & Hall, 1997; Chase-Dunn & Anderson, 2005; Chase-Dunn et al., 2010; Gills & Thompson, 2012; Chase-Dunn & Lerro, 2016; Rozov, 2019; Zinkina et al., 2019; Pomeranz, 2021.

[2]E.g., Akaev et al., 2012; Akaev & Sadovnichy, 2018, 2019; Sadovnichy et al., 2012, 2014, 2017; Komlos & Nefedov, 2002; Chernavsky, 1990, 2004a, 2004b; Nefedov, 2004, 2013, 2014a, 2014b;

determine the areas of stability of various states, the possibilities and conditions for the transition from one state to another, and the occurrence of development traps.

For our topic, models of the latter type are of greatest interest, since they have the ability to generalize and, at the same time, allow taking into account specific historical conditions and features. The basis for creating such models is a mathematical description of the mechanisms of social self-organization and evolution, considering the specific conditions of the functioning of social systems. When creating models, the achievements of complexity science/synergetics and the theory of non-linear dynamic systems (e.g., Haken, 1978, 1980, 1983, 1984, 2006a, 2006b; Jantsch, 1980; Chernavsky, 2004a, 2004b; Prigogine, 1997, 2017; Prigogine & Stengers, 1984, 2018; Allen, 2011) are used.

Since the end of the last century, under the influence of complexity science/synergetics, researchers have begun to view social systems as dynamic developing systems with their inherent internal instability. One could observe the emergence of new relevant research directions, such as "synergistic economics," cliodynamics, and sociophysics.

To date, a lot of studies have been carried out that have made it possible to understand the general properties of social systems and, despite partial differences, to analyze the features of their behavior in various conditions in different historical epochs. These features include the following (Malkov, 2009):

1. Specific features of social systems:
   (i) the multielement nature, the vulnerability of elements to external destabilizing influences. In order to survive, the elements need to be active, use their energy, which requires constant replenishment of their resource base;
   (ii) elements strive to maximize their own "utility function" in order to create reserves of resources, thus ensuring survival;
   (iii) the resource base available for use is limited;
   (iv) antagonism (contradiction) of the interests of the elements of the systems results from this due to the need to fight for limited resources; creating coalitions in the struggle for a resource; emergence of "friend-friend," "friend-foe," "foe-foe" relations, and institutions that consolidate those interactions;
   (v) nonlinear interactions of the elements, characterized by positive feedback (PFB); the strongest (having more advantages) wins in a competitive

Grinin & Korotayev, 2015: 186–198; Turchin, 2006, 2018; Turchin & Korotayev, 2006, 2020; Turchin & Nefedov, 2009; Turchin et al., 2021, 2022; Korotayev, 2005, 2007, 2020; Markov & Korotayev, 2007; Markov et al., 2010; Korotayev & Khaltourina, 2006; Korotayev & Malkov, 2012; Korotayev & Malkov, 2016; Korotayev & Zinkina, 2017; Korotayev et al., 2006a, 2006b, 2007, 2011, 2015a, 2016b, 2021c, 2022; Malkov, 1998, 2007, 2009, 2020; Malkov & Bilyuga, 2015; Aruka & Kirman, 2017; Malkov et al., 2020.

struggle, as a result of which it becomes even stronger (power gives rise to power, money makes money, etc.). The nonlinear process of increasing the advantages is limited either by the total amount of the resource, or by the individual capabilities of the subject to master and process the acquired resource, as well as the ability to maintain their advantages in the face of competition;

(vi) the threshold nature of the processes: the real return on the efforts made to achieve any goal only becomes visible when the efforts exceed a certain threshold value. Prior to this, the return on the applied efforts is insignificant;

(vii) the inertia of the behavior of subjects, the presence of some delay $\Delta\tau$ of the reaction of subjects to changes in external conditions;

(viii) the ability of the elements of a social system to consciously set goals and to reflect on their own actions and actions of other actors.

2. The nonlinear nature of the interaction of actors and limited resources (properties (v) and (iii)) determine the specifics of dynamic processes in social systems: the dialectic of development is described not as "thesis—antithesis—synthesis", but rather as "thesis—antithesis—temporary compromise." Contradictions are permanent. They are not "removed" in the process of social system development, but always exist, generating chaos at the micro level. At the macro level, a temporary compromise is achieved due to external conditions and existing resource constraints. When they change, the dynamic balance is disturbed, a new struggle of actors begins, leading to a new compromise level. Competition is immanent. There will be no development in social systems if there are no contradictions.
3. The general principle of social dynamics is the combination of chaos at the micro level and order at the macro level; everything (within the limits of objective regularities) is tried, but only something resistant to external and internal destabilizing processes and influences is fixed in the form of social structures.
4. Ensuring the stability of the social systems (necessary for their "survival") with permanent contradictions between their elements (property (iv)) is a challenging task. Through trial and error, social systems develop mechanisms of checks and balances and ways of compensation for internal stresses, allowing to achieve the stability of the system as a whole. These mechanisms include:
   - ideological and sociopsychological means of restraining aggressive intentions: religious imperatives, norms of morality, etc. (containment of chaos);
   - limiting the arbitrariness of actions of the elements of the system with the help of laws and regulations aimed at smoothing out the existing contradictions (limitation of chaos);
   - direct power regulation of relations between elements, carried out by the social system control bodies (chaos control).

   Social systems that manage to develop an effective system for compensating internal stresses survive. Social systems that fail to do so eventually disintegrate or perish in competition with other systems.
5. At different phases of evolution, social systems have different sensitivities and vulnerabilities to external influences. The most effective influences are intentional external impacts during periods of crisis when the system is largely chaotic. In

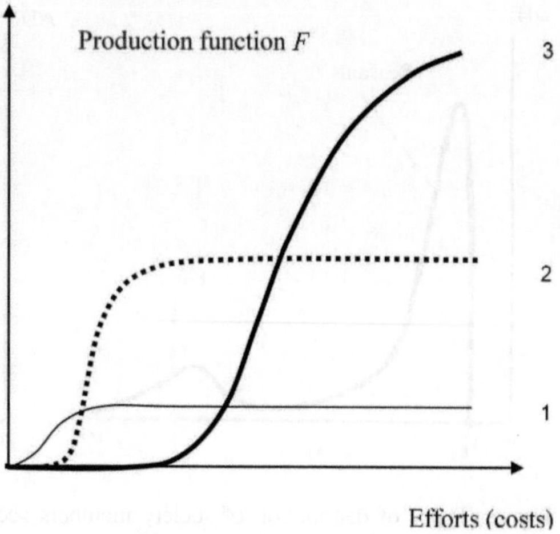

Fig. 1 Typical dependence of the output on the effort expended (investment in production) for hunter-gatherers (1) and for agricultural, (2) industrial, and (3) production

this case, even a low-intensity external influence can set the direction for the development of the system and influence the nature of its further self-organization.

6. Social systems are characterized by the possibility of multivariate development, due to the fact that they usually have not one, but several quasi-stable states (attractors) with their own basins of attraction. After some time, the system will end up in one of the attractors (this property is called equifinality). However, which attractor it will be exactly, depends on many factors and, to a large extent, on the purposeful activity of the social system control body.

II. The basic economic process in any social system is the production, redistribution, and consumption of resources necessary for its functioning. Projecting the above features of social systems onto economic processes allows us to formulate the following:
   1. in socioeconomic systems, there are always restrictions on the economic resources available for use (which include raw materials, energy, production resources, labor resources, effective demand, etc.);
   2. production is a nonlinear process. The production function (that is, the dependence of the amount of output on the effort and/or investment expended) has the following properties: (a) it becomes saturated when costs are increasing against the background of unchanged production technologies (due to the inevitably limited resources); (b) threshold nature (which is associated with the presence of the so-called fixed production costs). The type of typical production functions F is shown in Fig. 1 (the form F is

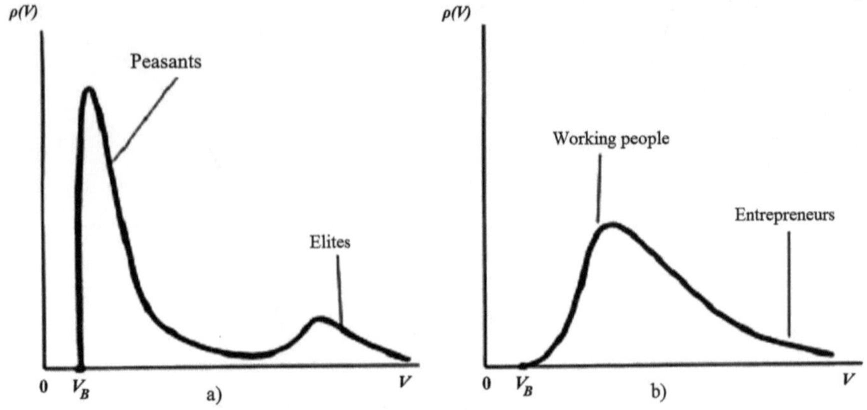

Fig. 2 Types of distribution of society members according to their savings $\rho(V)$: (**a**) bimodal structure (agrarian society); (**b**) unimodal structure (developed industrial society) (V is savings; V_B is the level of physical survival)

determined by the features of production, the technologies used, the specifics of production relations).[3]

At the same time, the production sector of a socioeconomic system consists of heterogeneous subsystems that differ in terms of specific qualities, age of fixed capital, and other characteristics. Relations between subsystems can be both cooperative and competitive;

3. consumption is a nonlinear process characterized by (a) saturation of demand for goods and services designed to meet everyday physiological needs and (b) the inability of demand for elite (prestigious) goods and durable goods intended to meet social needs to become fully saturated;
4. redistribution is a nonlinear process with positive feedback: the greater the economic or political (military) power, the greater the opportunities for non-equivalent redistribution of the product through taxes, forced seizures, monopoly prices, etc.

III. The maximization of individual utility functions under these conditions faces the following problems:
   1. a permanent shortage of resources for everyone, which leads to a constant struggle over resources (within society and between societies);
   2. the uneven distribution of resources among individuals and social groups, which leads to social tensions (an indicator of unevenness is the economic structure of society (ESS), that is, the density of distribution of members of society according to their savings; see Fig. 2).

[3]The type of production function is relatively stable for a society of hunter-gatherers and for an agrarian society. For an industrial society, the type of production function is constantly changing due to changes in technology and increases in labor productivity.

Due to the existence of positive feedbacks, social systems are characterized by internal instability. How, nevertheless, to achieve the social system resilience to external and internal destabilizing influences? For this, three main problems must be solved:

A. to ensure the possibility of economic survival (providing for the material needs of the society members);
B. to ensure sociopsychological stability in the presence of antagonistic interests of the elements of the social system (reduction of conflict within the social system);
C. to ensure effective governance. Since social system is a distributed system, the task of governance is to ensure the coordinated functioning of all its subsystems. Efficiency conditions:

on the part of bureaucrats, the ability to manage and the availability of funds necessary for governance;

on the part of the ruled, their consent to be ruled in conditions of social stratification.

Under the conditions of internal instability of the social system, the main function of management is to coordinate the interactions of socioeconomic subsystems and to duly remove the emerging contradictions and disproportions.

As a result of solving these problems, a structure of society is formed that is stable and has a chance of survival. The key feature of the processes of social structuring is the nature of positive feedbacks (PFB) (initiating changes) and the nature of negative feedbacks (NF) (determined by existing restrictions on changes of an objective and subjective nature). If the problems of ensuring stability cannot be solved, society perishes (disintegrates).

Methods for solving stability problems essentially depend on given external conditions (the most important of which are characteristics of available resources) and on characteristics of resource use (on forms and parameters of production function, as well as on principles and methods of managing socio-economic processes). Society seeks to maintain the stability of social structures; however, as the external environment (natural and technogenic) changes, conditions for the functioning of society change, and the stability of the old structures decreases, they disintegrate and form new structures under the influence of new positive feedbacks.

IV. The above approaches to the analysis of social interactions form the basis for modeling the functioning and evolution of social systems:

- Sect. 3.1.1 below provides a basic model of socioeconomic interactions;
- a fundamental feature of social systems is an important role of competition both in interactions between individual social actors within society and in interactions of societies with each other. Section 3.1.2 provides a basic model of economic competition;
- Section 3.2.1 provides a basic model of the sociopolitical stability of a society in the presence of competitive relations;
- Section 3.2.2 provides a basic model of political competition;

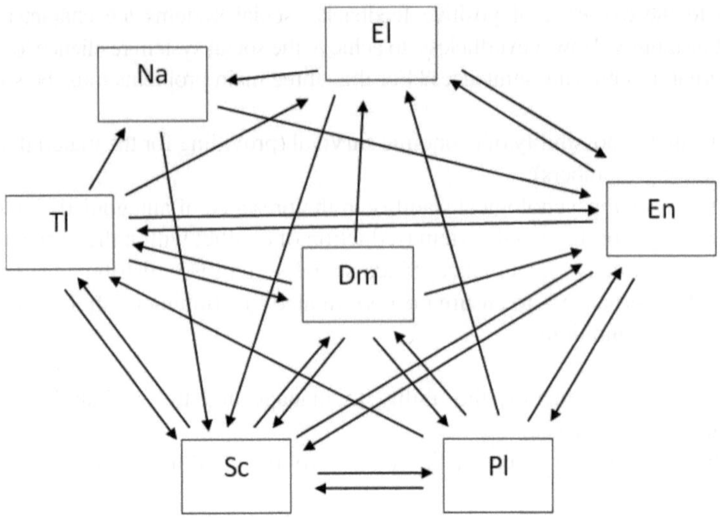

Fig. 3 Scheme of interaction of particular models in modeling historical dynamics

- Section 3.2.3 provides a basic model for the interaction of social systems in the presence of competitive relations.

Processes of social self-organization, the formation, and transformation of institutional structures are analyzed below in Chapter "Modeling Social Self-Organization and Historical Dynamics. Global Phase Transitions" (Malkov et al., 2023a, this volume).

Further, these basic mathematical models are used to analyze characteristics of an agrarian society (Chapter "Modeling Social Self-Organization and Historical Dynamics. Agrarian Society" [Malkov et al., 2023b, this volume]), an industrial society (Chapter "Modeling Social Self-Organization and Historical Dynamics. Industrial Society" [Akaev et al., 2023a, this volume]), as well as to forecast future changes and analyze options for further world development (Chapter "Modeling Social Self-Organization and Historical Dynamics. Modern Society and a Look into the Global Futures: Cybernetic W-Society" [Akaev et al., 2023b, this volume]).

1.2 Cognitive Scheme for Modeling Historical Dynamics

An important requirement for the modeling system of historical dynamics is the need for an integrated approach, that is, taking into account in their relationship both the interaction of society with the environment, and social interactions.

The scheme of interactions between various fields of activity, which is the basis for the historical dynamics modeling system, is shown in Fig. 3.

Designations

The vertices of the digraph are models for individual areas: Na—climate, natural environment, El—ecology, Tl—technology, Dm—demographics, En—economy (production, economic relations), Sc – sociosphere (social interactions), Pl—politics (public administration, political interactions).

Digraph edges (arrows) represent the influence of one sphere on another:

(A → B) is the influence of sphere A on sphere B (for example, (El → Na) is the influence of ecology on climate). The nature of the influence of various spheres on each other is determined by features of a historical epoch under consideration. The equations of the system of models represent a mathematical description of the interactions between the spheres indicated in Fig. 3 in relation to a respective epoch.

Consideration levels

- world dynamics (global processes);
- regional dynamics (for example, the Eurasian space, Tropical Africa, etc.);
- country dynamics (for example, Russia, USA, China, etc.).

1.3 Basic Equations (by Spheres) and their Use in Modeling the World Dynamics

In this section, we present equations that reflect basic processes in relation to each of the seven areas indicated in Fig. 3. In Chapter "Modeling Social Self-Organization and Historical Dynamics. Agrarian Society" (Malkov et al., 2023b, this volume), Chapter "Modeling Social Self-Organization and Historical Dynamics. Industrial Society" (Akaev et al., 2023a, this volume) and Chapter "Modeling Social Self-Organization and Historical Dynamics. Modern Society and a Look into the Global Futures: Cybernetic W-Society" (Akaev et al., 2023b, this volume), those equations are specified in relation to historical epochs considered in respective chapters.

Basic Equations of the "Demography" Block (Dm)

The basic equation of *demographic* dynamics characterizes the change in the size of society population as a result of the dynamics of death rates and birth rates (Malkov, 2009).[4]

$$dN/dt = (\text{birth rate}) - (\text{death rate}) \approx r \cdot N \cdot G(x_0, x) \tag{1}$$

The expression $G(x_0,x)$ on the right side of the Eq. (1) characterizes the dependence of population growth (that is, the difference between births and deaths) on the level of well-being (x is the material product per capita; x_0 is the level of the

[4]For simplicity, it is assumed that migration processes can be neglected. If necessary, terms that take into account migration processes can be added to the right side of Eq. (1).

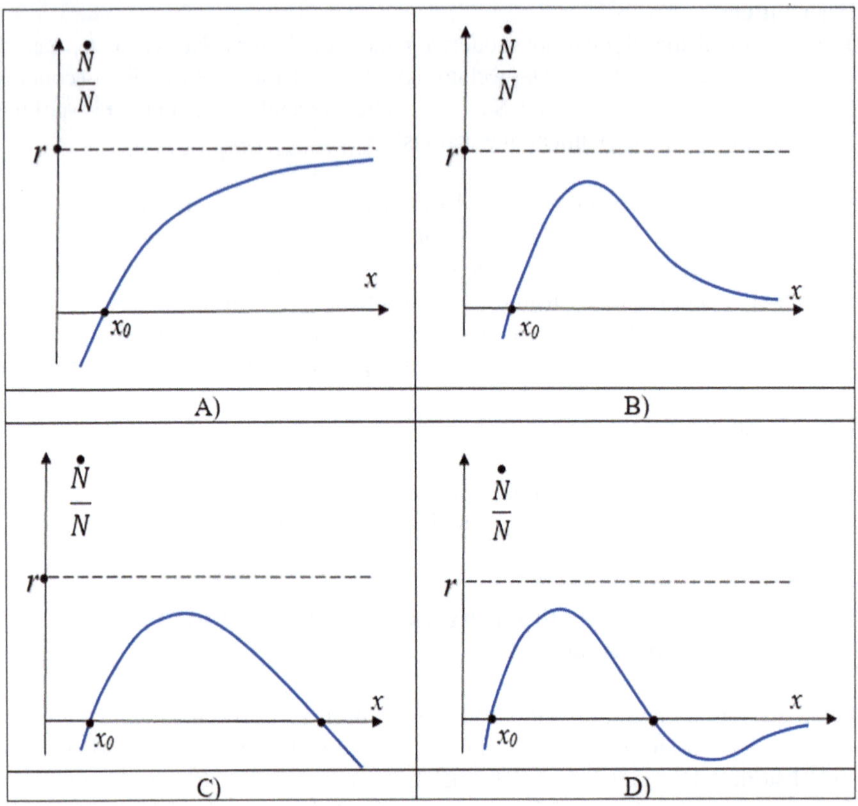

Fig. 4 (**a, b, c, d**) Variants of the dependence of population growth on the level of well-being

"subsistence minimum" (physical survival level): when the level of material support decreases below x_0, death rates start to exceed the birth rates. The graph of the derivative *dN/dt* as a function of *x* depending on the nature of the expression $G(x_0, x)$, can have different forms (see Fig. 4).

The dependence of relative population growth on *x* shown in Fig. 4a is typical for a preindustrial society. For the developed industrial societies with a high level of urbanization, the dependences shown in Fig. 4b and c are typical, characterizing the second phase of the demographic transition when, simultaneously with an increase in the level of well-being, there is a decrease in the birth rates (transition from a large family model to a small family one). The dependence shown in Fig. 4d reflects the situation typical, for example, for modern Scandinavian countries, when, after a period of decline in the birth rate, it is restored, and the demographic dynamics enters the mode close to the one of simple reproduction.

Basic Equations of the "Economy" Block (En)

The basic equation of economic dynamics characterizes dynamics of the level of material well-being of a society's members (Malkov, 2009)[5]:

$$dx/dt = \text{(consumer goods production per capita)} - \text{(consumption per capita)} \quad (2)$$

If in the long run, the rate of consumer goods production outpaces the rate of consumption, then welfare increases; otherwise, welfare falls.

Within the basic model, the right side of expression (2) is specified according to the following algorithm:

- first, the value of the gross domestic product is determined (the total volume of production in society, including nonconsumer products);
- then nonconsumer products (that is, what households do not consume directly) are subtracted from this value and the result is divided by population. Thus, we obtain the value of the production of consumer goods per capita;
- then the average per capita consumption of consumer products is determined, taking into account the depreciation of durable products.

In general, expression (2) can be represented as follows:

$$dx/dt = (F(R^{-}, I_1, S) - I_1 - I_2)/N - Q, \quad (3)$$

where $F(R^{-}, I_1)$ is the production function (the total amount of final products produced per unit of time); R^{-} is the natural resources used in the production process per unit of time; S is the level of technological development; I_1 is the part of the manufactured products used for investment purposes (to ensure the next production cycle); I_2 is the fraction of manufactured products used for purposes not directly related to future production or consumption (for example, to improve environmental conditions, transport infrastructure, life support systems, etc.); and Q is the consumption of consumer products.

The amount of products produced in society per unit of time, F (production function) depends on the number of producers, on the labor means and technologies used, as well as on the availability of necessary natural resources. Qualitatively different situations emerge when (1) there are no restrictions on resources (resources are available in excess) and (2) when resources are in short supply.

In the first case, the quantity of products produced is determined by the number of producers, means of labor, and technologies (resources are not a constraint). Then the production function can be written as.

[5]Consumer products include both everyday products (food, clothing, shoes, etc.) and durable products (housing, vehicles, etc.).

$$F = A_R(S) \cdot V, \tag{4}$$

where V is a number of production units (one production unit consists of producers and labor instruments required to process one unit of resource into the final product); $A_R(S)$ is a coefficient that depends on the level of technology (the larger the value of $A_R(S)$, the greater the amount of the final product produced from one unit of resources). Thus, production growth will take place if the number of production units V increases and/or labor productivity increases.

In the second case, the availability of resources is decisive, while the number of production units V is in excess and they are not fully utilized. So, in this case, in the basic model, the production function F can be written as.

$$F = A_R(S) \cdot R^-, \tag{5}$$

where R^- is a number of units of natural resources used in the production process per unit of time; $A_R(S)$ is a coefficient depending on the level of technology. Thus, in the case under consideration, the growth of production will take place if the amount of resources used in production increases and/or the efficiency of its use increases.

In view of the foregoing, the resulting expression for the production function F, which combines both cases, can be written as:

$$F = min\ \{A_R(S) \cdot V; A_R(S) \cdot R^-\}, \tag{6}$$

that is, the minimum of expressions (4) and (5) is chosen as F.

The products produced in the current time period are further used:

(a) to ensure production in the following periods (I_1 is investment costs accounting for the depreciation of the old means of production and the creation of new ones, as well as preservation /renewal/ increase in the resource base for production, creation/improvement of production technologies, vocational training);
(b) to transform the external environment in order to create comfortable/acceptable conditions for life (I_2 is the costs of life-supporting infrastructure such as communal services, housing construction, transport, and transport infrastructure) and to ensure acceptable properties of the external environment (ecology, waste disposal, retention of environmental characteristics within acceptable limits);
(c) for consumption in the broad sense of the word, that is, to ensure the life cycle of particular members of society (Q is the cost of food, clothing, health care, education, upbringing).

The $A_R(S)$ function in Eqs. (4) to (6) is determined based on statistical data.

Basic Equations of the "Technologies" Block (Tl)

Michael Kremer (1993) proposes a formula for the dynamics of the technology development level S[6] in the following form:

$$dS/dt \approx c \cdot S \cdot N \tag{7}$$

where c (in our interpretation) is a coefficient that could be different in different historical epochs (depending on various social factors). The meaning of formula (7) is as follows: the relative increase in the number of technological innovations is proportional to the number of inventors, who make up a certain percentage of the total population N.

Andrey Korotayev and his colleagues (Korotayev et al., 2006a, 2007) propose to consider the value of GDP per capita production as a quantitative indicator reflecting the level of technology development S, and also to introduce a variable.

$$S' = S - m,$$

where m is a constant coefficient that plays the role of a "subsistence minimum" (the share of the produced product that is strictly directed to maintain the achieved number of population). Thus, the value of S' has the meaning of resource per capita that can be spent on additional purposes, such as expanded reproduction of the population, science, art, entertainment, etc. Accordingly, Eq. (7) is proposed to be written as:

$$dS'/dt \approx c \cdot S' \cdot N = c \cdot N \cdot (S - m). \tag{8}$$

Formula (8) makes technological progress dependent on the presence of resources excessing the "subsistence minimum" m in the society, that is, it is believed that the poorer the society, the less inventive activity in it.[7]

Taking into account (7) and (8), the basic equation for technological progress can be written as:

$$dS/dt \approx c \cdot N \cdot (S - S_O). \tag{9}$$

where the value of GDP production per capita is used as a quantitative indicator of the level of technology development S, and S_O is the part of GDP per capita necessary to ensure the "subsistence minimum". Unlike the value of m in formula (8), the value of S_O in formula (9) is not constant, it changes from era to era and

[6] Note that symbols that we use in our models are different from the ones used by Kremer.

[7] This can also be seen with respect to the increase in the speed of technological progress by dozens of times in industrial societies compared to agrarian ones (e.g., Grinin et al., 2020).

consistently grows, because the more complex a society is, the more efforts and means are required to ensure basic conditions for life support.[8]

Basic Equations of the "Nature" Block (Na)

This block includes basic equations that characterize the dynamics of the most important parameters of the natural environment that affect people's lives. First of all, these are climate changes (a generalized indicator is the average temperature of the atmosphere in the surface layer) and resource provision (a generalized indicator is the amount of natural resources available for use).

Dynamics of Atmospheric Temperature Increase ΔT Due to Anthropogenic Factors Is

$$\Delta T = f(W_{CO2}) = f(W_{CO2}(E(N(t)), S)), \tag{10}$$

where ΔT depends to a large extent on CO_2 emissions (W_{CO2}), which, in turn, depend on energy consumption (E) that depends on the population ($N(t)$), and technologies (S).

The Basic Equation for the Expenditure of Natural Resources *R* Is

$$dR/dt = R^{+} - R^{-}, \tag{11}$$

where R is the total amount of the available resource; R^{-} is resource reduction per unit of time; and R^{+} is resource increase per unit of time. It is advisable to divide resources into the following categories:

(a) nonreproducible resources (example: coal, oil, and gas);
(b) reproducible resources (example: forest, fish, and animals);
(c) reusable resources (example: land, water, and air).

Equation for the non-reproducible resources (subsoil resources):

$$dR/dt = R^{+} - R^{-} = k^{+}{}_{R} \cdot q_R \cdot I_1 - F/A_R(S), \tag{12}$$

where $k^{+}{}_{R} \cdot q_R \cdot I_1$ is anthropogenic restoration/increase in the amount of resource per unit of time (for example, discovery of new mineral deposits, expansion of the resource base due to new types of resources, etc.); q_R is the share of the total investment costs I_1 directed to the restoration/increase of the amount of the resource; and $k^{+}{}_{R}$ is the efficiency of restoring/increasing the amount of a resource based on the current technological level.

[8] For example, living in a city requires much more basic costs than living in a rural area (including the cost of centralized water supply, sewerage, roads, transport, etc., without which life in a large city is impossible).

Equation for reproducible single use resources (bioresources reproduced in nature):

$$dR/dt = R^{+} - R^{-} = R^{+}{}_{na} + k^{+}{}_{R} \cdot q_{R} \cdot I_{1} - R^{-}{}_{na} - k^{-}{}_{R} \cdot F, \tag{13}$$

where $R^{+}{}_{na}$ is natural recovery of the amount of resource per unit of time; $R^{-}{}_{na}$ is natural loss of resource per unit of time; $k^{+}{}_{R} \cdot q_{R} \cdot I_{1}$ is anthropogenic restoration/increase in the amount of a resource per unit of time (for example: planting forests, breeding fish and animals, etc.); q_R is the share of the total investment costs I_1 directed to the restoration/increase of the amount of the resource; $k^{+}{}_{R}$ is the efficiency of resource recovery/increase based on the current technological level; $k^{-}{}_{R} \cdot F$ is anthropogenic resource consumption; $k^{-}{}_{R}$ is the coefficient of proportionality.

Equation for reusable resources (land, soils):

$$dR/dt = R^{+} - R^{-} = k^{+}{}_{R} \cdot q_{R} \cdot I_{1} - f^{-}{}_{R}(t), \tag{14}$$

where $k^{+}{}_{R} \cdot q_{R} \cdot I_{1}$ is anthropogenic restoration/increase in the amount of resource per unit of time (for example, plowing of virgin lands, soil recultivation); $f^{-}{}_{R}(t)$ is anthropogenic decrease in the amount of resource per unit of time (for example, withdrawal of land from agricultural circulation, desertification).

For reusable resources, changing their A_R quality over time is important:

$$A_{R}(t) = A^{+}{}_{R}(S) \cdot A^{-}{}_{R}, \tag{15}$$

where $A^{+}{}_{R}(S)$ is the anthropogenic improvement of the quality of the resource through technology; $A^{-}{}_{R}$ is anthropogenic deterioration of the quality of the resource due to its repeated exploitation.

Basic Equations of the "Ecology" Block (EI)

This block includes basic equations that reflect the impact of human activities on the natural environment.

Dynamics of the Accumulation of Waste Products W_i

$$dW_{i}/dt = W_{i}{}^{+} - W_{i} - = k^{+}{}_{Wi} \cdot F - W_{i}{}^{-}{}_{na} - k^{-}{}_{Wi} \cdot q_{Wi} \cdot I_{2}, \tag{16}$$

where W_i is the total amount of waste products of the *ith* type; W_i^{+} is the increase in waste products per unit of time; $k^{+}{}_{Wi}$ is the coefficient of proportionality, the value of which depends on the development of technologies; $W_i{}^{-}{}_{na}$ is natural processing (utilization) of waste products of the *ith* type per unit of time; $k^{-}{}_{Wi} \cdot q_{Wi} \cdot I_2$ is anthropogenic processing (utilization) of waste products per unit of time (for example, disposal, recycling, etc.); q_{Wi} is the share of the total costs I_2 allocated for the processing (utilization) of waste products of the *ith* type; $k^{-}{}_{Wi}$ is the efficiency of processing (utilization) of waste products of the *ith* type based on the current technological level.

Dynamics of Anthropogenic Change in CO_2 Balance in the Atmosphere[9]

$$dW_{CO2}/dt = W^{+}{}_{CO2}(E(N(t)), S)) - W^{-}{}_{CO2}(S), \qquad (17)$$

where $W^{+}{}_{CO2}$ is the anthropogenic generation of CO_2, and $W^{-}{}_{CO2}(S)$ is CO_2 utilization.

Dynamics of Changes in the Characteristics of the External Environment U_j as a Result of Anthropogenic Impact

$$U_j = f_j(W_i), \qquad (18)$$

where the form and characteristics of function $f_j(W_i)$ are determined on the basis of statistical and other data in each particular case. The dynamics of change in U_j is compared with the permissible limits (boundaries) determined in each specific case on the basis of additional studies.

Basic Equations of the "Sociosphere" Block (Sc)

Social structures vary greatly from country to country, as they are determined by local conditions, traditions, peculiarities of social interactions, etc. What is important for us are not particular features, but the fundamental conditions for the stable functioning of social systems and the reproducibility of their main characteristics over long historical periods, despite the constant change of generations.

The most important aspects of ensuring the stability of the social system functioning include an effective economy (material aspect), stability and efficiency of public administration (organizational aspect), spiritual, and ideological unity of population (sociopsychological aspect). The first aspect characterizes the economic and military independence of the state, and the ability of people to satisfy their material needs. The second aspect characterizes the extent to which it is possible to ensure the coherence and effectiveness of the actions of all economic, political, and social structures in the state. The third aspect characterizes the desire of people for concerted action. Destabilization of any of these spheres makes the state unable to maintain its sovereignty and leads to the actual disintegration of state structures or their submission to external forces.[10] At the same time, all these spheres influence each other, which leads to nonlinear synergistic effects that accelerate destabilization processes. Features of this mutual influence depend on structural and functional features of the social system. Analysis of the social system stability in each particular case can be carried out according to the following scheme:

1. analysis and formalization of basic economic processes and economic interactions between key social groups and actors in the form of a logical and

[9] This factor is singled out among other factors of changing environmental characteristics due to the importance of its impact on climate change.

[10] About the role of the state in society and its social life see, e.g., Grinin, 2008, 2011.

mathematical model. The model should have a dynamic character and should make it possible to identify stable states of the economic system (dynamic attractors), as well as the ratio of social system parameters under which the existence of these states is possible;
2. identification and analysis of the main contradictions (multidirectional interests) between the key actors of the social system in the economic sphere. This analysis should make it possible to understand in which direction each of the actors wants to change the parameters of the system;
3. determining the current equilibrium state of the system parameters and the means by which they are achieved in organizational and socio-psychological spheres; finding explanation of how the current values of the parameters are maintained in the presence of existing contradictions (through laws, traditions, private agreements, etc.);
4. determining how and why the current equilibrium can be disturbed (both as a result of changes in external conditions, and as a result of the desire of actors to improve their position);
5. determining whether means used in organizational and socio-psychological spheres are sufficient to maintain the stability of the system. Critical values of key parameters should also be determined, upon reaching which the system destabilizes. The risk of destabilization should be assessed.

In the general case, if the number of social groups in the analyzed social system is *N*, then the basic processes that reflect the economic interaction of these groups and the socioeconomic stability of the system as a whole can be written as follows.

The equation for the dynamics of material resources of the *ith* social group $X_i(t)$ $(i = 1...N)$ has the form:

$$dX_i(t)/dt = \text{(income)} - \text{(consumption)} - \text{(investment)} - \text{(net taxes)} \qquad (19)$$

where "net taxes" are taxes minus subsidies, allowances, and other transfers received by the relevant social group from the state budget; "income" includes, among other things, payments from the state budget.

The equation for the dynamics of material resources of the state $X_\Gamma(t)$ (balance of government revenues and expenditures) has the form:

$$dX_\Gamma(t)/dt = \text{(net taxes)} - \text{(consumption)} \qquad (20)$$

where "consumption" is expenditures on the state apparatus, as well as on public procurement.

The resulting system of equations has the form:

$$\begin{cases} \dfrac{dX_i}{dt} = F_i - Q_i - I_i - T_i \\ \dfrac{dX_\Gamma}{dt} = \sum_{i=1}^{N} T_i - Q_\Gamma \end{cases} \qquad (21)$$

where F_i is the production function of the *ith* social group (including income from government contracts); Q_i is consumer spending of the *ith* social group; Q_Γ is expenditures on the state apparatus and other public expenditures; I_i is investments of the *ith* social group, and T_i is net taxes of the *ith* social group.

In the course of socioeconomic interactions, each social group seeks to increase their income and reduce their tax burden. Naturally, these aspirations conflict with each other due to the limited overall resource base. Multidirectional interests lead to social and political contradictions and conflicts. As a result of social interaction (confrontation), a temporary equilibrium is established, reflecting the compromise reached. The mechanism for reaching a compromise is a political one based on the results of interaction between parties and leaders defending the interests of the relevant social groups. The result of the political struggle (and the corresponding temporary equilibrium state of the social system) ultimately depends on how significant the economic and political support of social groups is to their leaders (the level of support may be reflected, for example, in election results). Thus, in the *long run*, the compromise state reflects the fundamental relationship between the economic and political influence of main social groups, and, in this sense, it is natural. However, in the *short term*, political decisions are made by political actors on the basis of opportunistic considerations, as well as on the basis of their expectations about future changes in the situation (rather than current realities). Therefore, current decisions of political actors can be very volatile, and the political situation can change in a bizarre way (instability of this kind often occurs during political crises in times of political uncertainty). But only those states of the social system that reflect the real role of key social groups in economic and political processes can be stable and long term.

Equations (21), reflecting the processes of interaction between social strata, if necessary, should be supplemented by equations that reflect the features of interaction *within* social strata (for example, reflecting competitive interactions between producers of goods and services that affect economic dynamics and technological development).

Depending on the ratios of the parameters, the system of Eqs. (21) may have either one stable state (attractor), or a few of them, or no stable state (attractor) at all. The characteristics of attractors make it possible to judge the properties of a social system and analyze ways to ensure its stability (including ways to coordinate the interaction of socioeconomic subsystems of the social system).

Despite the fact that the described algorithm is mainly focused on the study of economic interactions, it actually allows us to study the issues of socio-psychological stability of society and the effectiveness of governance systems, since these factors significantly affect parameters of system (21) and its resistance to destabilization.

Basic Equations of the "Politics" Block (Pl)

Political interaction is a type of relationship between political actors in which they exert mutual influence on each other and seek to influence the decision-making on issues of mutual interest.

The purpose of political actions of particular political actors, as a rule, is to promote their interests, stimulate (coerce) the other side to make decisions that are beneficial to these political actors, or decisions that political actors consider appropriate in a certain situation.[11]

Since we are primarily interested in the sustainability of the functioning of social systems in various conditions, the main topics of analysis and modeling include.

- ensuring the internal sociopolitical stability of society despite the difference in interests of individual social groups and political actors;
- the formation of the political structure of society and of relations such as "state power—society," "elite—counter-elite," taking into account the peculiarities of political competition within a society;
- interstate political interactions (interactions that go beyond individual social system).

The basic fotmulas for modeling the above topics are equations that describe the competition of political actors, as well as equations that describe the possibility of coordination of actions (synchronization) of actors despite the presence of differences in their interests. The equations and corresponding models are considered in more detail in Sections 2 and 3.2 of Chapter "Modeling Social Self-Organization and Historical Dynamics. Agrarian Society" (Malkov et al., 2023, this volume), Sect. 2 of Chapter "Modeling Social Self-Organization and Historical Dynamics. Industrial Society" (Akaev et al., 2023a, this volume), and Sect. 2 of Chapter "Modeling Social Self-Organization and Historical Dynamics. Global Phase Transitions" (Malkov et al., 2023a, this volume).

Using Basic Equations as a Tool for Modeling World Dynamics

When using mathematical methods for analyzing and modeling the world dynamics, we continue the tradition of system dynamics by Forrester—Meadows (Forrester, 1971; Meadows et al., 1972), but there are a number of important differences:

- while various modifications of World3 (Meadows et al., 1992, 2004; Randers et al., 2018) model mostly concentrated on modeling the material flows (resources, pollution, population, industrial, and agricultural production), for us modeling the *social* processes and social interactions is no less important[12];
- while the World models (Forrester, 1971; Meadows et al., 1972; Randers et al., 2018) extrapolate the existing trends into the future, our main emphasis is on the analysis of the *mechanisms* of global processes, their changes in the past and possible changes in the future;

[11] About theories of political agency, see, e.g., Karasev, 2022.

[12] For our analysis of advantages and disadvantages of World3 see Chapter "Introduction. Hoping for the Future" (Sadovnichy et al., 2023, this volume).

- the World models are aimed at the solution of the Cauchy problem (Hartman, 1964a, 1964b)[13] (that is, initial values of variables are set and their further dynamics is calculated for various options of changing model parameters); in contrast, our main method is the construction of *phase portraits*, as well as the analysis of social self-organization and the features of the formation of stable (reproducible) social structures in the prevailing conditions.

The problem of using mathematical modeling for long-term forecasting is as follows. Usually, many equations reflecting various processes are written and then solved on a computer with various combinations of parameters included in these equations. This procedure produces a lot of graphs, but it is difficult to analyze them (due to the complexity of the models themselves), the results are opaque. How can one enrich the content and at the same time simplify the procedure for obtaining results and make these results more transparent and understandable?

For long-term modeling and forecasting, the main goal is to determine the conditions under which the society functions stably during its successive historical transformations and, most importantly, to analyze the possible options for future development in terms of their stability, both in relation to the "society-nature" interaction and in relation to social interactions within society. Due to the multivariance of the trajectories of specific societies, it is not these trajectories themselves that are important, but the general patterns of the dynamics of societies of the type under consideration and the analysis of the presence of stable states that are reproduced over long periods of time. Therefore, it is advisable to consider not individual trajectories, which can be numerous, but phase portraits of the basic equations characterizing the types of societies under consideration, since it is the phase portrait that allows us to see the overall picture and understand the logic of a social system evolution.

The construction of a phase portrait is possible if the dynamics of the indicators of the analyzed social system can be described in the language of ordinary differential equations:

$$dX/dt = f(X, a, t) = X^{+} - X^{-} \tag{22}$$

where $\boldsymbol{X} = (X_1, \ldots, X_n)$ is the vector of the dependent variables characterizing the state of the social system; dX/dt is the rate of change of variables $\boldsymbol{X}$; t is time; $f(X, a, t)$ is a vector function (nonlinear in the general case) that reflects the change in these variables over time by comparing the processes leading to an increase in the value of the corresponding variable (function X^{+}) and the processes leading to a decrease in this variable (function X^{-}); $\boldsymbol{a} = (a_1, \ldots, a_m)$ is the vector of system parameters that generally depend on time.

[13] The Cauchy problem for a first-order differential equation is to find a solution that, for a given value of the argument, takes on a given value, i.e. satisfies the initial condition.

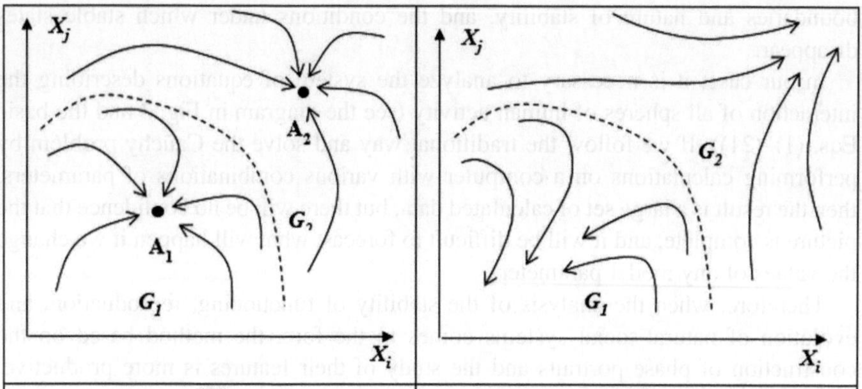

(A)	(B)
The situation when the system has two stable states A_1 and A_2, to which the system tends over time (*G_1* and *G_2* are the areas of attraction of attractors A_1 and A_2)	**The situation when, with a change in the parameter a_k, the stable states A_1 and A_2 disappear and the system becomes unstable: in the *G_2* area, the variables X_i and X_j grow without limit, while in the *G_1* area they fall**

Fig. 5 (**a, b**) Phase portrait of a system with two attractors $\mathbf{A_1}$ and $\mathbf{A_2}$ and their corresponding areas of attraction G_1 and G_2 (left graph) and transformation of the phase portrait with a change in the value of the parameter a_k (right graph)

Solutions of Eqs. (22) can be represented as trajectories $X(a, t)$ in the phase space of the system (see Fig. 5):

In Fig. 5, points $\mathbf{A_1}$ and $\mathbf{A_2}$ are stable states of the system (attractors) of the radial fixed point type, to which the system tends as a result of its evolution; areas G_1 and G_2 are areas of attraction of these points (if the system is located at any point of the phase space belonging to these areas, then over time it will end up, respectively, at point $\mathbf{A_1}$ or $\mathbf{A_2}$). Phase trajectory analysis makes it possible to draw conclusions about the nature of the evolution of the system, to determine the areas of its deterministic behavior and the areas of *bifurcations* (that is, the areas of parameters at which instability occurs and the number and/or type of system solutions (22) change). As a rule, the transition from a stable to an unstable state and vice versa occurs when any of parameters a_i of system (22) changes (this situation is shown in Fig. 5a and b). In this case, this parameter is called the order parameter. By decreasing (or increasing) the values of the order parameters, it is possible to influence the behavior of the system, to change its state. Identification of the order parameters in model (22) and their critical values (that is, the values at which bifurcations occur) helps to choose the best ways to control real systems.

Thus, the description of the processes of social system functioning in the language of differential Eqs. (22), which describe the mechanisms for key characteristic change and the subsequent construction of phase portraits (see Fig. 5) makes it possible to understand the patterns of evolution of these systems, as well as the features of social self-organization, the parameters of stable states (if any), the boundaries and nature of stability, and the conditions under which stable states disappear.

In our case, it is necessary to analyze the system of equations describing the interaction of all spheres of human activity (see the diagram in Fig. 3 and the basic Eqs. (1)–(21)). If we follow the traditional way and solve the Cauchy problem by performing calculations on a computer with various combinations of parameters, then the result is a large set of calculated data, but there will be no confidence that the picture is complete, and it will be difficult to forecast what will happen if we change the values of any model parameters.

Therefore, when the analysis of the stability of functioning, reproduction, and evolution of natural-social systems comes to the fore, the method based on the construction of phase portraits and the study of their features is more productive. The problem is that when there are many variables (and hence equations), the phase portrait turns out to be multidimensional, which reduces visibility and complicates the analysis.

In our case, when it is required to reveal the fundamental laws of world dynamics without drowning in detail, it is advisable to use the basic equations that reflect the most important processes. It should be taken into account that these processes proceed at different rates. This facilitates the task of analysis and macro modeling, since it makes it possible to use the Tikhonov's theorem (Tikhonov, 1952). Since the characteristic times of the processes described by the system of basic Eqs. (1)–(21) are quite different from each other, in accordance with this theorem, when considering specific situations, it is possible to replace some of the differential equations with algebraic equations and to consider some slowly changing variables to be determined outside the analyzed system of equations. Thus, it becomes possible to consider the interactions shown in Fig. 3 stage-by-stage, with a reduction in the dimension of the phase portraits considered at each of the stages.

The characteristic times of the processes reflected in Fig. 3 are as follows: natural processes (Na)—hundreds of years; environmental changes (El)—tens and hundreds of years; technological changes (Tl)—hundreds of years[14]; demographic changes (Dm)—the duration of a generation (20–25 years); economic processes (En)—a year (in agriculture) or less (in industry) if production cycles are considered, decades or more if the dynamics of economic structures is considered; social processes (Sc)—the duration of a generation and more; political processes (Pl)—the duration of a generation and more if evolutionary processes are considered, years if situations of instability (reforms, revolutions, and wars) are considered.

Taking into account the above differences in the characteristic times of the processes, it becomes possible to single out the following groups of interactions into specialized blocks: "society-nature" interactions, with specific subgroups of interactions Dm—Tl—En (demographic-economic processes, characteristic time is decades) and En—Tl—El—Na (human influence on nature and the reverse influence of the natural environment on human society, characteristic time is decades to

[14] In times of technological revolutions, the characteristic time of change can be reduced to decades.

hundreds of years), as well as social interactions, with specific subgroups of interactions Sc—En—Dm (socioeconomic processes, characteristic time is decades) and Sc—Pl—Dm (sociopolitical processes, characteristic time is decades). In the history of human society, demographic and economic processes (Dm—Tl—En) were decisive. They took place in interaction with slower natural processes (En—Tl—El—Na) and determined the nature of faster socioeconomic (Sc—En—Dm) and sociopolitical (Sc—Pl—Dm) processes. Such a hierarchy of interactions allows us to consider the variety of processes shown in Fig. 3, step by step, moving from one topic to another, each time concentrating on the most important processes for the considered block of issues, while taking into account all the others as additional ones. With this approach, the dimension of the phase space at each stage can be reduced, the phase portrait can be made clear (as shown in Fig. 5), and slower variables can be considered as factors affecting the shift of equilibrium states (points $\mathbf{A_1}$ and $\mathbf{A_2}$ in Fig. 5) over longer time intervals.

So, the approach we use to analyze the stability of interactions in the "society-nature" system, as well as in the system of social interactions within society at various stages of historical development, is described below.

2 Modeling the Dynamics and Stability of Social Systems: Interactions in the "Society-Nature" System

Interactions of the "society-nature" type are determined primarily by the interactions Na—El—Tl—Dm—En (see Fig. 6). Analyzing and modeling them with the use of the basic equations from Sect. 1.3 specified for a particular epoch makes it possible to identify the stable states of the system (if any), the boundaries and nature of stability, as well as critical conditions under which stable states disappear.

As has been mentioned, technological, demographic, and economic processes (Dm—Tl—En interactions) that took place when interacting with slower natural processes (En—Tl—El—Na interactions) played a decisive role in the history of human society. Accordingly, when modeling the "society-nature" interaction, it is advisable, on the basis of Tikhonov's theorem (Tikhonov, 1952), to separate the description of faster demographic and economic processes (Dm—Tl—En interactions) and the description of slower natural processes (En—Tl—El—Na interactions) into two separate models.

2.1 Basic Model Demography—Technology—Economy (Dm—Tl—En)

Equations (1), (3), (9) adapted to the given historical epoch are the key ones for the mathematical description of demographic and economic processes. They characterize the demographic dynamics determined by the opportunities for economic development (the available resource base), which, in turn, depend on the level of technological development:

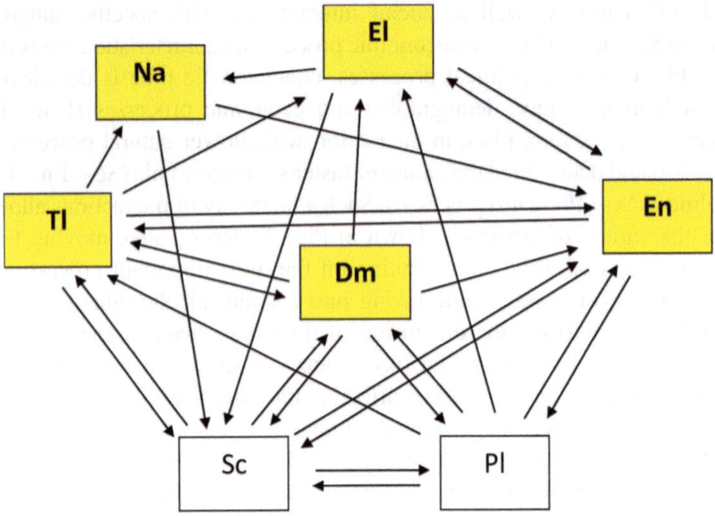

Fig. 6 The subsystem of interactions that determines stability in the "society-nature" system. *Designations:* the vertices of the digraph are models for individual areas. *Na* climate, natural environment, *El* ecology, *Tl* technology, *Dm* demographics, *En* economy (production, economic relations), *Sc* sociosphere (social interactions), *Pl* politics (public administration, political interactions)

$$dN/dt = (\text{birth rate}) - (\text{death rate}) \approx r \cdot N \cdot G(x_0, x) \tag{23}$$

dx/dt = (production of consumer products[15] per capita) – (consumption per capita) = $(F(R^{-}, I_1, S) - I_1 - I_2)/N - Q$, (24)

$$dS/dt \approx c \cdot N \cdot (S - S_0). \tag{25}$$

(See Sect. 1.3 for an explanation of the equations).

2.2 Base Model En-El-Na

The main factor of human influence on nature is economic activity, determined by the development of technology. This influence leads to a change in the natural environment, biochemical and biological cycles, landscapes, climate, which, in turn, affects life and can lead to a loss of stability in the interaction of human society and nature. The key equations for the mathematical description of the impact of human activity on nature and the reverse impact of the natural environment on

[15]Consumer products include both everyday products (food, clothing, shoes, etc.) and durable products (housing, vehicles, etc.).

human society are the basic Eqs. (10), (12), (13), (14), (16), (17), (18) from Sect. 1.3 adapted to a respective historical epoch:

$$dW_i/dt = k^+{}_{Wi} \cdot F - W_i{}^-{}_{na} - k^-{}_{Wi} \cdot q_{Wi} \cdot I_2 \quad (26)$$

$$dW_{CO2}/dt = W^+{}_{CO2}(E(N(t)), S)) - W^-{}_{CO2}(S), \quad (27)$$

$$dR/dt = k^+{}_R \cdot q_R \cdot I_1 - F/A_R(S), \quad (28)$$

$$dR/dt = R^+{}_{na} + k^+{}_R \cdot q_R \cdot I_1 - R^-{}_{na} - k^-{}_R \cdot F, \quad (29)$$

$$dR/dt = k^+{}_R \cdot q_R \cdot I_1 - f^-{}_R(t), \quad (30)$$

$$U_j = f_j(W_i), \quad (31)$$

$$\Delta T = f(W_{CO2}) = f(W_{CO2}(E(N(t)), S)). \quad (32)$$

(See Sect. 1.3 for an explanation of the equations).

3 Modeling the Dynamics and Stability of Social Systems: Social Interactions

The key spheres for social interactions are En—Dm—Tl—Sc—Pl (see Fig. 7). At the same time, it is reasonable to single out socioeconomic processes (Sc—En—Dm) leading to the stratification of society and to the appearance of social groups with their own interests, as well as sociopolitical processes (Sc—Dm—Pl) ensuring

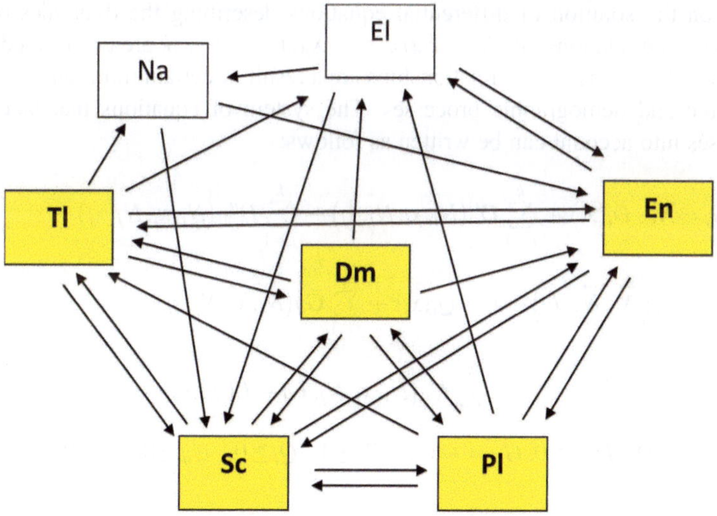

Fig. 7 Subsystem of interactions that determines socioeconomic and sociopolitical stability

the consistency of the interaction of these social groups (despite the difference in interests) and determining the interaction of the society with the outside world. The purpose of the analysis and modeling of social interactions using the basic equations from Sect. 1.3 (specified for a respective epoch) is to identify stable states of the system (if any), the boundaries and nature of stability, as well as critical conditions under which stable states disappear.

3.1 Socioeconomic Stability

Equations (1), (21) from Sect. 1.3, adapted to a respective historical epoch and subject areas, are the key ones for the mathematical description of socioeconomic processes. They reflect economic relations, dynamics of social groups' sizes, patterns of formation of social structure, basic socioeconomic relations, and interactions in a respective historical period and also allow us to determine conditions for the socioeconomic stability of society (that is, the conditions that ensure the stability of the social structure and social interactions).

Basic Model of Socioeconomic Interactions

Possibilities of achieving the socioeconomic stability of social systems are studied with the help of the analysis and modeling of interactions between various social groups in the society under consideration. The following algorithm can be used to describe those interactions at the macroeconomic level. In a social system, the main subsystems (social groups or institutions) are distinguished, which determine the features of its functioning. The following are used as the main characteristics for describing each of the subsystems: size N_i of a respective social group (i.e., the number of group members) and the total material savings X_i of the group members. Based on the solution of differential equations describing the dynamics of these quantities, the relationships $N_i = N_i(t)$, $X_i = X_i(t)$, $i = 1,\ldots,k$ are determined for the selected subsystems. These relationships are a result of certain intra- and intergroup economic and demographic processes. The system of equations that takes these processes into account can be written as follows:

$$\begin{cases} \dot{N}_i = N_i \cdot D_i(x_i) + \sum\limits_{j=1}^{k} D'_{ij}(N_i, x_i, N_j, x_j) - \sum\limits_{j=1}^{k} D''_{ij}(N_i, x_i, N_j, x_j) \\ \dot{X}_i = F_i\left(\vec{X}, \vec{N}, \vec{I}\right) - N_i \cdot Q_i(x_i) + \sum\limits_{j=1}^{k} G_{ij}(N_i, x_i, N_j, x_j) - \\ \qquad - \sum\limits_{i=1}^{k} C_{ij}(N_i, x_i, N_j, x_j) - I_i(x_i, N_i) \end{cases}$$
$$D'_{ij} = -D'_{ji}, D''_{ij} \geq 0, G_{ij} = -G_{ji}, \quad F_i \geq 0, \quad Q_i \geq 0, \quad C_{ij} \geq 0, \quad I_i \geq 0; i, j = \overline{1, k}, \tag{33}$$

where the dot above the symbol means time differentiation; $x_i = X_i/N_i$ is average specific savings of the representatives of the *ith* group; D_i is the demographic coefficient representing the difference between birth rate and death rate (quantitative characteristics of which depend on social and economic conditions); D'_{ij} is the rate of transition from group j to group i; D''_{ij} is mortality in group i due to antagonistic interaction with group j; F_i is a function of production understood as the amount of output produced by members of a social group per unit of time;

Q_i is the function of individual consumption;

G_{ij} is the amount of material goods withdrawn in one way or another by group i from group j or transferred from group i to group j (for example, rents, taxes, subsidies, etc.);

C_{ij} is the costs of group i for managing group j;

I_i is the costs for the production of material goods (investments in production).

Model (33) uses average values. Averaging is carried out within each social group for all its members. If necessary, the model can take into account the diversity of individual characteristics within social groups. The rules for dividing society into groups depend on the tasks being solved in a respective study.

Depending on the ratios of the parameters, the system of Eqs. (33) may have either one, or several, or no stable states (attractors). Characteristics of attractors make it possible to understand the properties of a social system and analyze the ways to ensure its stability (including ways to coordinate the interaction of socioeconomic subsystems).

Despite the fact that the described algorithm is focused mainly on the study of economic and demographic processes, it actually allows us to study the issues of socio-psychological stability of society and the effectiveness of governance systems as well, since these factors significantly affect the parameters of the system (33) and its resistance to destabilization.

Basic Model of Economic Competition

Economic interactions play an important role not only in relations between different social groups but also in relations within groups. In this sense, intragroup economic competition (for example, competition between entrepreneurs or firms) is an important factor in economic development. A large number of works use mathematical modeling of economic competition considering specific situations at the micro- and macroeconomic levels. The meaning of economic competition lies in the fact that several economic actors work in the same economic niche, offering potential buyers certain products similar in their consumer properties, thereby competing for consumer demand, which is limited. For the case when N economic actors compete, the model of economic competition in a generalized form can be represented as follows:

$$
\begin{aligned}
dM_i/dt &= (income\, from\ the\ sale\ of\ products) \\
&- (production, transport\ and\ other\ costs\ associated\ with\ economic\ activity) \\
&= k_i \cdot M_H - h_i \cdot F_i
\end{aligned}
\tag{34}
$$

where M_i is funds of the ith economic actor; M_H is the total volume of consumer demand; k_i is the share of total consumer demand which goes to the ith economic actor per unit of time from the sale of its products; $h_i{\cdot}F_i$ is the costs of the ith actor associated with economic activity, where F_i is the volume of products produced by the ith economic actor, h_i is the compliance coefficient (in the generalized model (34) it is assumed that the costs of the ith actor are proportional to the output). The specific values of k_i, M_H, h_i, and F_i are determined by the specifics of the situation under consideration and, in the general case, they depend both on external conditions and on actions of economic actors themselves. The results of concretization of the model (34) in relation to different historical epochs are given in Sect. 2.1 of Chapter "Modeling Social Self-Organization and Historical Dynamics. Industrial Society" (Akaev et al., 2023a, this volume).

3.2 Sociopolitical Stability

Modeling the dynamics of sociopolitical systems is a complex task, since sociopolitical interactions are characterized by the presence of *positive feedbacks* (see Sect. 1.1), which makes such systems internally unstable. The fundamental requirement for mathematical models that claim to adequately describe sociopolitical processes is the requirement that they reflect properties (i) - (viii) of social systems listed in Sect. 1.1. Below we discuss approaches to modeling sociopolitical systems that make it possible to take into account those properties.

Each society consists of social groups that have different interests and are often in competitive (and even conflict) relations with each other. Under these conditions, sociopolitical stability is only possible if there are mechanisms in society that ensure the coherence of actions of social groups despite existing internal contradictions. The mechanisms of influence of some actors on others can be very different: through a system of rules and norms, through bilateral and multilateral agreements, through coercion, etc. As a result of interaction, decisions made can converge and synchronize (while interests remain different); in this case, actors will act in a coordinated manner, and the social system will function as a single organism. However, if there is no synchronization of decisions, there will be a divergence of actors and destabilization of the system.

The quantitative assessment of the difference in interests of actors can be carried out within the "information field" theory (Bukharin et al., 2009), according to which each social actor at a given point in time can be positioned in a multidimensional information space of values (beliefs). Techniques for understanding the value space based on the results of sociological surveys have now been worked out in various

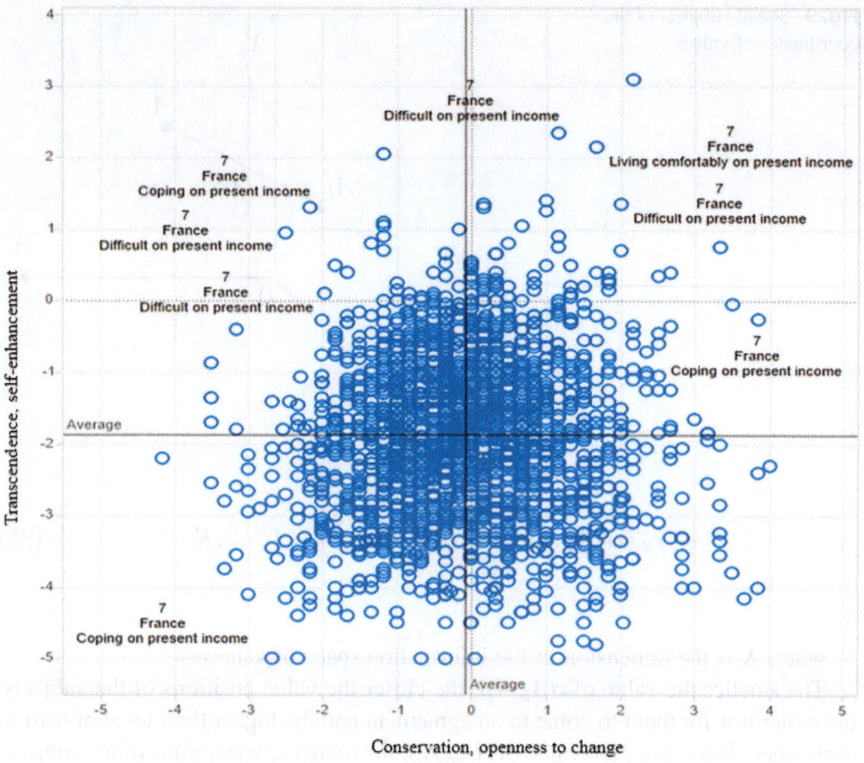

Fig. 8 Distribution of respondents in France in the value space according to the results of the sixth wave of the ESS survey (2012)

scientific projects[16] (as an example, Fig. 8 shows the distribution of French respondents in the value space based on the results of the 2012 ESS European Social Survey[17]).

The positioning of social subjects in the value space is displayed by points. If these points are determined, it becomes possible to identify the distance between subjects in the value space (that is, to determine the degree of difference between their value systems). If the coordinates of the *i*th and *j*th subjects in the space of values are displayed by points A_i and A_j (see Fig. 9), the distance r_{ij} between these points can be set as follows:

[16]See, e.g., Grinin et al., 2019; Korotayev et al., 2019; Korotayev 2019; Korotayev et al., 2019; Korotayev et al., 2020; Shulgin et al., 2017.

[17]European Social Survey. URL: http://www.europeansocialsurvey.org/data/country_index.html

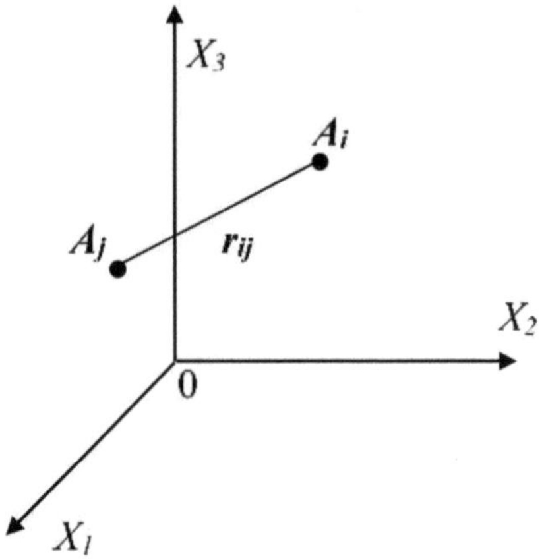

Fig. 9 Social subjects in the coordinates of values

$$r_{ij} = r(A_i, A_j)\sqrt{\sum_{k=1}^{k} (X_{ik} - X_{jk})^2, k = 1, \ldots, K} \tag{35}$$

where K is the dimension of the information space of values.

The smaller the value of $\boldsymbol{r(A_i, A_j)}$, the closer the value positions of the subjects, the easier it is for them to come to an agreement and the higher their level of trust in each other. The greater the value of $\boldsymbol{r(A_i, A_j)}$, the more the viewpoints of the subjects diverge, the more difficult it is for them to agree and coordinate their positions. Thus, the value of $\boldsymbol{r_{ij}}$ can be a quantitative measure of the difference in the interests of social subjects.

Despite the fact that the value of $\boldsymbol{r_{ij}}$ is almost always greater than zero, the subjects $\boldsymbol{A_i}$ and $\boldsymbol{A_j}$ under certain circumstances can act in a coordinated manner, while under other circumstances they will act inconsistently. To assess the degree of consistency/mismatch of their actions in the considered time interval Δt, the *mismatch index* S_{ij} is introduced in (Malkov & Bilyuga, 2015), which is defined as follows:

if the subjects $\boldsymbol{A_i}$ and $\boldsymbol{A_j}$ always acted in a coordinated manner (took the same actions) within the time interval Δt, then the value of S_{ij} in this period is equal to zero;

if the subjects $\boldsymbol{A_i}$ and $\boldsymbol{A_j}$ always acted in an uncoordinated way (that is, differently) within the time interval Δt, then the value of S_{ij} in this period is equal to one;

if the subjects $\boldsymbol{A_i}$ and $\boldsymbol{A_j}$ sometimes acted in a coordinated manner, and at other times acted in an uncoordinated way within the time interval Δt, and the share of mismatched actions in their total number was s, then the value of S_{ij} in this period is equal to s.

Fig. 10 Dynamics of the mismatch index $S_{ij}(t)$ of the positions of Brazil and the United States in their voting in the UN General Assembly, 1946–2014

Thus, the index S_{ij} takes a value between 0 and 1, while the value 0 means complete consistency, and the value 1 means complete mismatch between the actions of subjects A_i and A_j.

By recording values of S_{ij} at successive times t, we obtain function $S_{ij}(t)$. The form of this function in the general case will be irregular (of the type of chaotic oscillations), while the time-averaged oscillation amplitude S_{ij} will be the smaller when:

(a) the value of $r(A_i, A_j)$ is smaller (that is, the smaller the difference in the value attitudes of subjects A_i and A_j);
(b) the desire to coordinate their positions under the influence of external or internal circumstances is greater.

Similar situations arise with voting in international organizations. As an example, Fig. 10 shows the dynamics of the discordance index between Brazil and the United States in their voting in the UN General Assembly from 1946 to 2014 (when constructing the index, the data of Bailey et al. (2015) were used; the mismatch index was calculated using Lijphart's (1963) method).

The dynamics of the index shown in Fig. 10 is irregular and resembles chaotic oscillations. At the same time, it is clear that over the 70 years under consideration, the nature of the consistency of the positions of Brazil and the United States has changed significantly. Before 1960, Brazil mostly voted in agreement with the position of the United States (the values of the $S_{ij}(t)$ index are low). However, starting from the second half of the 1970s, the positions of Brazil and the United States began to diverge significantly (the values of the $S_{ij}(t)$ index are high). This means that during the period under review, the regime of coordination between the positions of Brazil and the United States changed to a regime of discordance.

It is necessary to distinguish between the inconsistency of *opinions, decisions, and actions*, since it is not infrequent for the members of the same team to have different opinions, but act in concert.

Discordance of *opinions* occurs when people express their independent opinion on some issue (without being influenced by other information subjects). In this case, the *index of opinion discordance* $S^M{}_{ij}$ of actors $\boldsymbol{A}_i$ and $\boldsymbol{A}_j$ depends primarily on the value of $\boldsymbol{r}_{ij}$, which characterizes the distance between them in the information space: the larger the value of $\boldsymbol{r}_{ij}$, the greater the expected value of the index of opinion discordance (IOD).

Discordance of *decisions* occurs when people learn opinions of other people on the issue under consideration and make a decision taking into account these opinions. Accordingly, the *index of decision discordance* $S^P{}_{ij}$ (IDD) of actors $\boldsymbol{A}_i$ and $\boldsymbol{A}_j$ depends not only on the value of $\boldsymbol{r}_{ij}$ but also on how ready these actors are to take into account each other's opinions, as well as on the ability of some actors to purposefully influence decisions of others (on the basis of authority as well as through coercion). As a rule, such an influence on decision-making occurs if actors are somehow dependent on each other, when there are some significant connections between them that they do not want or cannot eliminate. Therefore, even if actors have different opinions, they make agreed decisions (the tool for finding agreed decisions is, for example, parliament, where various parties, in the course of discussing the existing problems, seek a mutually acceptable compromise and adopt laws that become binding for all). Accordingly, in the same situations, the IOD value is usually less than the IDD value.

Discordance of *actions* occurs when people do not coordinate their actions unknowingly or intentionally with actions of other people. Accordingly, the *index of action discordance* $S^\partial{}_{ij}$ (IAD) of actors $\boldsymbol{A}_i$ and $\boldsymbol{A}_j$ depends both on the value of $\boldsymbol{r}_{ij}$ and on how ready these subjects are to take into account each other's decisions in their actions. As a rule, such readiness arises when the actors are interested in the results of common activities and understand that inconsistency can lead to an undesirable result for them. For this reason, the value of the IAD is usually less than the value of the IOD and IDD (for example, in developed democracies, even if the opposition party in parliament opposes some law, but it is adopted by the majority, then the members of this party will comply with it).

Thus, in a normally functioning society, for the time-averaged values of the discordance indices (the averaged values of the indices are further denoted by underscores), the following relation must be fulfilled:

$$S^M{}_{ij} > S^P{}_{ij} > S^\partial{}_{ij} \rightarrow 0. \tag{36}$$

For the stability of society, it is important that the condition $S^\partial{}_{ij} \rightarrow 0$ be fulfilled, i.e., that there should be consistency in the actions of individuals and social groups (while their opinions may differ significantly, as happens in pluralistic societies). If $S^\partial{}_{ij}$ increases, the threat of society destabilization increases as well: its normal functioning is disrupted, social tension grows, conflict situations become more frequent, as a result of which life-support systems begin to fail, which leads to an even greater disorganization and causes discontent and protest of the population. It is important that the process of destabilization, as a rule, has a threshold character (Malkov & Bilyuga, 2015): the reduction of connections that unite members of society down to a certain level (and the corresponding decrease in their interest in concerted actions) does not lead to catastrophic consequences. However, their further decrease (below this level) starts the processes of disorganization, which are growing like a snowball, and which often can no longer be stopped even with the help of extraordinary measures.

Thus, considering the dynamics of the discordance indices and their averaged values, it is possible to analyze the change in the degree of coherence/discordance of the actions of the society members and assess the risk of the transition from a stable to an unstable state of the society. This analysis should be carried out substantively, in relation to a specific situation, taking into account actual circumstances. At the same time, of particular importance per se is to identify general patterns and conditions of strengthening/decreasing the coherence of society members. In this case, it is advisable to consider those processes in a generalized way with the use of mathematical models of a general kind.

Basic Model of Political Stabilization/Destabilization

Mathematical Formalization of the Assessment of Sociopolitical Stability/Destabilization in the Social System Functioning

A methodical approach to the analysis of sociopolitical stability/destabilization in the functioning of socioeconomic systems has been outlined above on the basis of modeling the coherence/discordance of the actions of society members and various social groups. The mathematical analog of this process is the process of synchronization/desynchronization of interconnected dynamic systems. Processes of this type have been considered in a large number of studies.[18] Of the greatest interest are studies that consider the synchronization/desynchronization of chaotic (having positive Lyapunov exponents λ) systems, since we are primarily interested in situations of interaction between subjects that are initially localized at different points in the information space and often have different opinions on the same issues. The mathematical formulation of the problem of synchronization of chaotic systems is as follows. There are m dynamical systems with positive Lyapunov exponents, each of which has its own phase trajectory $x_i(t)$, where $i = 1,2,\ldots,m$. In the absence of links between the systems, the trajectories $x_i(t)$ are in discordance, and the systems are not synchronized (systems are considered to be synchronized if their trajectories coincide, i.e., $x_i(t) = x_j(t)$). It is necessary to answer the question: under what conditions are chaotic dynamical systems synchronized?

Studies show that synchronization of chaotic systems is possible only if there is a certain structure of connections and their intensity exceeds a certain threshold (Malkov, 1998, 2007). Therefore, the problem of effective control of systems with

[18] See, e.g., literature review in: Moon, 2008.

chaotic dynamics is to create an optimal structure (in terms of topology and intensity) of connections among these systems, transferring them to a synchronized (coordinated) state.

Mathematical tools for analyzing the synchronization conditions of chaotic dynamical systems are presented in (Dmitriyev et al., 1996). This toolkit can be used to analyze the conditions for the coordination/discordance of the actions of social actors if:

discordance index $S_{ij}(t)$ of social subjects $\boldsymbol{A_i}$ and $\boldsymbol{A_j}$ at the moment t matches the value of the difference $|x_i(t) - x_j(t)|$;

the degree of difference in the interests of actors (characterized by the value $\boldsymbol{r(A_i, A_j)}$) is identified with the degree of randomness of function $|x_i(t) - x_j(t)|$ (characterized by the value of the Lyapunov exponents λ);

the intensity of relations between social actors $\boldsymbol{A_i}$ and $\boldsymbol{A_j}$ in the mathematical model can be characterized by value α_{ij}, which for different tasks can be described by different indicators (for example, if the interaction of states is considered, then in some cases α_{ij} can be used as a value proportional to the volume of trade between them; if the interaction of people is considered, then in some cases it is possible to use a value proportional to the strength of the information impact on each other as α_{ij}, etc.).

A mathematical model that uses these assumptions and a calculation procedure that can be used to study the influence of various types of interactions on the coordination (or vice versa, discordance) of the actions of social actors is described in (Malkov & Bilyuga, 2015).

In real socioeconomic and political systems, as a rule, it is possible to change $\boldsymbol{r(A_i, A_j)}$, as well as the structure and strength of connections α_{ij} between social actors. Thus, it is possible to intentionally (or unconsciously) transfer the system from a stable state to an unstable one and vice versa. Analyzing the real dynamics of changes in distances $\boldsymbol{r(A_i, A_j)}$ and connections α_{ij} between actors within a social system, it is possible to assess risks of sociopolitical destabilization of society.

Main Features and Patterns of Sociopolitical Stability/Destabilization of the Social System Functioning (According to the Results of Modeling)

The above theoretical model and its calculation version make it possible to identify important features of ensuring the stability of social system functioning and the conditions under which destabilization occurs. Some important results of the analysis are shown below.

Consider (at a generalized level) the impact of the interaction between the central government $\boldsymbol{A_1}$, the local government $\boldsymbol{A_2}$ and population $\boldsymbol{A_3}$ on the social system stability. We will consider them as independent social actors with a certain localization in the information space and divergent interests. In the language of the mathematical model, we can say that we have a social system consisting of three elements $\boldsymbol{A_1}$, $\boldsymbol{A_2}$, and $\boldsymbol{A_3}$, each of which tends to move along its own (different from the others) phase trajectory $x_1(t)$, $x_2(t)$, and $x_3(t)$. When these elements interact, six effects α_{12}, α_{13}, α_{21}, α_{23}, α_{31}, α_{32} can be expected (see Fig. 11). However, in terms of real cases, only some of them play a decisive role.

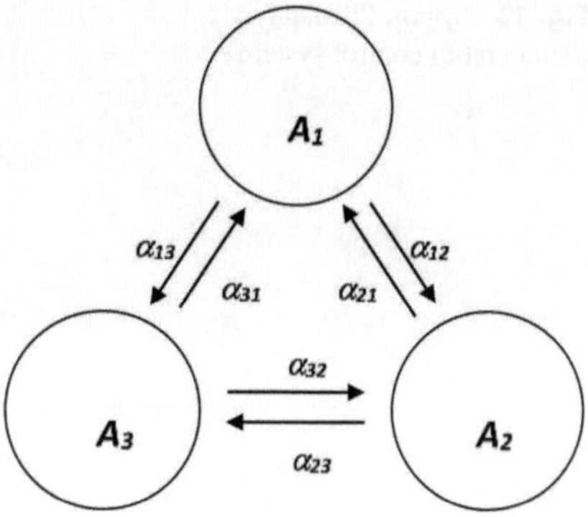

Fig. 11. A complete scheme of connections in a system of three elements

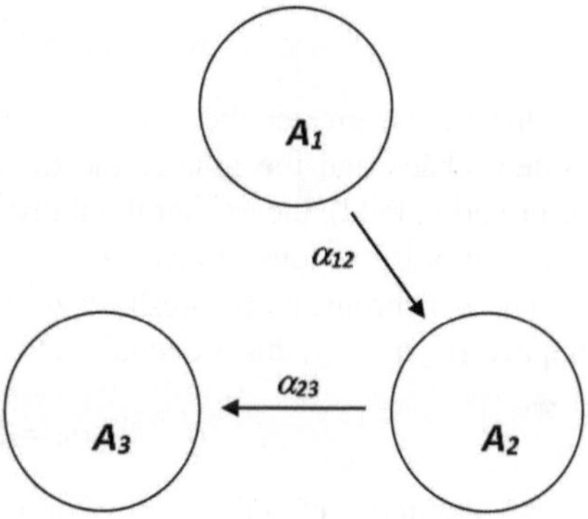

Fig. 12 Option 1: rigid directive control system

The task is to build the control system of the social system in such a way (that is, in terms of the model, to provide such a structure and intensity of connections α_{ij}) that makes it possible to synchronize the phase trajectories $x_1(t)$, $x_2(t)$, and $x_3(t)$ (to ensure that the index of discordance in the actions of social actors is zero).

The results of mathematical analysis show that there are two most effective and least expensive options for managing the social system under consideration, which ensure the synchronization of its elements.

Option 1: rigid directive control system. In this case, the links α_{12} and α_{23} are decisive, and feedback from objects to subjects of control is practically absent or insignificant (see Fig. 12).

Model calculations show that an effective control of such a system (that is, synchronization of the dynamics of all subsystems) is possible only if the intensity of control efforts exceeds a certain threshold value:

$$min(\alpha_{12}, \alpha_{23}) > \alpha_0. \tag{37}$$

Within the framework of the model (Malkov & Bilyuga, 2015), the value α_0 is related to the Lyapunov exponent λ by the relation.

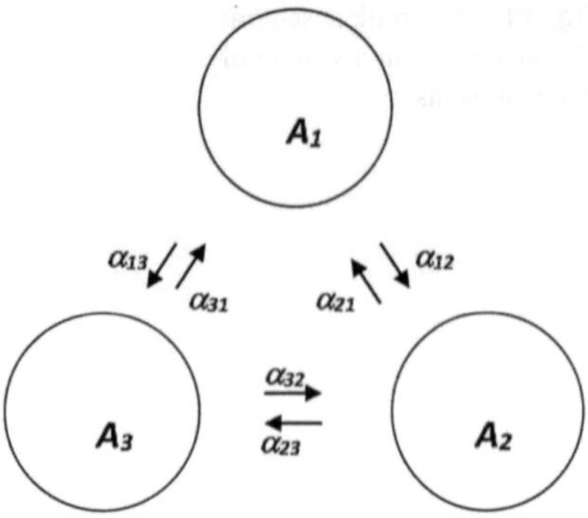

Fig. 13 *Option 2*: adaptive (democratic) control system

$$\alpha_0 = 1 - exp(-\lambda), \tag{38}$$

that is, the greater the value of λ (and, accordingly, the higher the degree of system chaos and the greater the distance $\boldsymbol{r(A_i, A_j)}$ between social actors in the information field), the greater the threshold value α_0 and the more efforts are needed to synchronize actions of actors.

The synchronization threshold α' (that is, the minimum total intensity of the impacts α_{12} and α_{23} that are required to ensure the synchronization mode) is equal to:

$$\alpha' = 2 \cdot \alpha_0 = 2 \cdot (1 - \exp(-\lambda)). \tag{39}$$

An alternative effective control option is an adaptive (democratic) system (*option* 2), in which all direct and feedback connections α_{ij} are involved (see Fig. 13). At the same time, α_{13} characterizes the direct impact of the central authorities on the objects of control, for example, through media, official ideology, etc.; α_{31} characterizes the reverse impact of the population on the central government, for example, through elections, referendums, mass actions (strikes, demonstrations), etc.

Mathematical calculations show (Malkov & Bilyuga, 2015) that the synchronization threshold for *option 2* is minimal when all connections have the same intensity: $\alpha_{ij} = \nu$, while:

$$\alpha' = \alpha_{12} + \alpha_{13} + \alpha_{21} + \alpha_{23} + \alpha_{31} + \alpha_{32} = 6 \cdot \nu = 2 \cdot \alpha_0, \tag{40}$$

that is, the system synchronization threshold is the same as for *option 1* (see (39)). However, if we compare the efforts of the controlling bodies (that is, values $\alpha'' = \alpha_{12} + \alpha_{13} + \alpha_{21} + \alpha_{23}$), then with an adaptive control system the synchronization threshold is lower ($\alpha''=1{,}5{\cdot}\alpha_0$) than with a directive one ($\alpha''=2{\cdot}\alpha_0$). In addition, even with a high degree of chaos (for large λ), there is a safety margin for relations, since $\nu \approx 1/3$, which is much lower than the maximum allowable value for the intensity of relations, which is equal to one within the model. So, the adaptive system copes better with the task of governance. This option corresponds to market relations in the economy and a democratic regime of government.

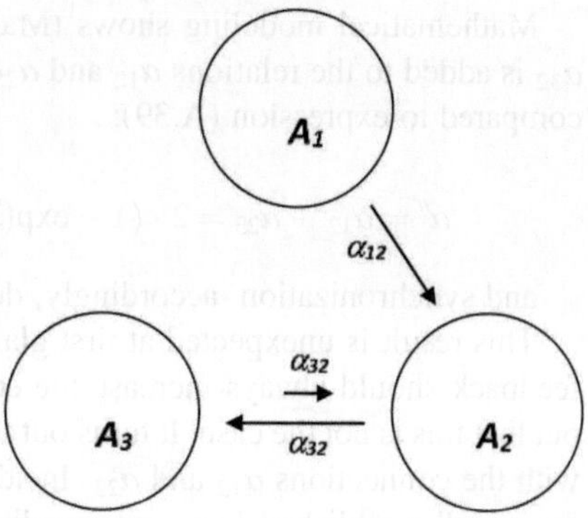

Fig. 14 Option 1A: directive management system with additional feedback α_{32} (example: democratic reforms in an authoritarian country)

Options 1 and 2 are two extreme cases (minimum and maximum connections in the system). Our analysis of intermediate situations, when, in addition to α_{12} and α_{23}, not all the remaining relations but only some of them are involved, shows that in this case, the necessary costs for ensuring controllability tend to *increase*. Therefore, intermediate situations are usually unstable and gravitate toward *options* 1 and 2 discussed above. Even though the directive system (*option* 1) requires more costs for its maintenance compared to the adaptive system (*option* 2), it is stable in the sense that sporadic (fluctuating) additional connections do not lower, but usually raise the synchronization threshold. Therefore, the system tends to return to its original state again. This is the reason for the historical vitality of centralized directive systems, which, in particular, include monarchy and various forms of authoritarian rule. The transition to another stable state—an adaptive system (option 2) characteristic of democratic regimes—requires the simultaneous actualization of all links α_{ij}, which is possible only as a result of active purposeful activity on the part of the "center" (reform) or on the part of controlled subsystems (revolution). In this case, the intensity of the added relations should not be arbitrary, but quite definite, otherwise the effect will turn out to be the opposite and will lead not to stabilization, but to even greater destabilization of the system.

This feature, paradoxical at first glance, is the reason for the complexity of the transition from a directive to an adaptive control system. As an example, consider a situation where an authoritarian government decides to grant more economic and political freedoms to the population. In the language of the model, this is ***option 1A***, which implies the introduction of an additional relation α_{32} into the control system of the *option 1*, which ensures the influence of population A_3 on the local authorities A_2 (this situation is shown in Fig. 14).

Relation α_{32} realizes the feedback of the objects of governance on their immediate supervisors, for example, through meetings of labor collectives or through councils of servicemen in the army (as was the case after the February 1917 revolution in Russia).

Mathematical modeling shows (Malkov & Bilyuga, 2015) that when feedback α_{32} is added to the relations α_{12} and α_{23}, the synchronization threshold α'"increases compared to expression (A.39):

$$\alpha'' = \alpha_{12} + \alpha_{23} = 2 \cdot (1 - \exp(-\lambda)) + 2 \cdot \sqrt{(1 - \exp(-\lambda)) \times \alpha_{32}} \qquad (41)$$

and synchronization, accordingly, decreases.

This result is unexpected at first glance: intuitively, it seems that the addition of feedback should always increase the consistency of subsystems. However, it turns out that this is not the case. It turns out that in this case the connection α_{32} "conflicts" with the connections α_{12} and α_{23}. Incidentally, this provides additional explanation for a well-established fact—intermediate political regimes (partial autocracies and partial democracies) turn out to be much more liable to sociopolitical destabilization than both full autocracies and full (consolidated) democracies.[19]

Option 1A illustrates the difficulties of the transition from an authoritarian system of governance (*option* 1) to an adaptive democratic one (*option* 2) at the initial stage of this process. No less difficulties arise at the later stages of this process, when formally all connections α_{ij} in the system have already been actualized, but their intensity is still relatively weak and close to the synchronization threshold (40). Therefore, the weakening of any relation can be critical and lead to the destabilization of the system. This effect is the basis for the implementation of "color revolutions" technologies.[20] Analysis shows that "color revolutions" are most successful in countries that have taken the path of expanding democratic institutions, but these institutions are still relatively weak and their "safety margins" are small (Mitchell, 2012, 2022). In this situation, the relation α_{13} is the most vulnerable, which characterizes the influence of the central government $\boldsymbol{A_1}$ on population $\boldsymbol{A_3}$. Usually, in order to implement "color revolutions," the opposition finds a reason to accuse the central government of some "sins" (for example, corruption) and begins to organize actions of civil disobedience (antigovernment demonstrations, manifestations, political strikes, pickets, refusals to obey the demands of the authorities, sit-ins, blocking highways, etc.). "Civil disobedience" means a decrease in the effectiveness of the relation α_{13}. At the same time, since the government declares adherence to democratic principles, it is limited in the use of force to restore order and strengthen the relation α_{13} (for example, by banning demonstrations, arresting protest leaders, strengthening measures to ensure law and order, etc.).

[19] See, e.g., Gurr, 1974; Muller & Weede, 1990; Mansfield & Snyder, 1995; Ellingsen & Gleditsch, 1997; Gates et al., 2006; Ulfelder & Lustik, 2007; Vreeland, 2008; Goldstone et al., 2010; Korotayev et al., 2014, 2015b, 2015c, 2016a, 2016b, 2022; Gaibulloev et al., 2017; Slinko et al., 2017; Vaskin et al., 2018; Korotayev et al., 2019, 2021; Korotayev et al., 2021; Khokhlov & Korotayev, 2022; Korotayev et al., 2016b, 2021; Zhdanov & Korotayev, 2022; Beissinger, 2022; Walter, 2022.

[20] About color revolutions, their distinctive features, and revolutionary technologies see Mitchell, 2012, 2022; Filin et al., 2022; Goldstone et al., 2022a; Grinin, 2022; Grinin & Korotayev, 2022; see also Lawson, 2019; Selbin, 2022.

Fig. 15 *Option 2A*: adaptive governance without communication α_{13} (example: "color revolutions" in countries with weak democracies)

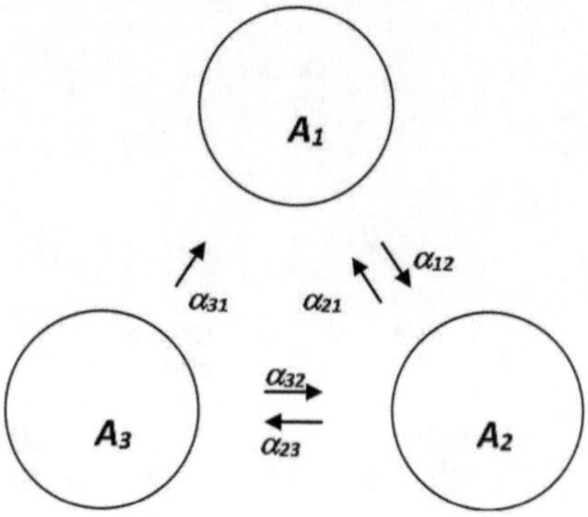

Accordingly, *option* 2 of the control system (see Fig. 13) is transformed into ***option 2A*** (see Fig. 15), in which the system connectivity decreases, and the synchronization threshold increases.

Reducing the connectivity of the system can bring it into a "subthreshold" state and lead to political destabilization and, sometimes, revolutionary change of power.[21]

Thus, the above examples show that the appearance of additional feedback in social systems does not in itself lead to an improvement in controllability. Moreover, in many cases, the system synchronization threshold does not decrease, but increases. Related to this is the possibility of destabilizing social systems in the course of democratic reforms and the transition from a directive (authoritarian) system of governance to an adaptive (democratic) one. Protest movements are also dangerous for young democracies, as they weaken the coherence of the system when the latter is close to the threshold of synchronization. This determines the effectiveness of "soft power," when certain world powers support protest movements in order to achieve their geopolitical interests.

Basic Model of the Interaction of Main Political Actors and Social Groups

The above method to model the sociopolitical stability of social systems is the basis for the development of a model of interaction of main political actors and social groups. The task of the model is to assess the current level of sociopolitical stability and forecast its expected level in a particular society, depending on its social structure, the main social "faults" (conflicts), the structure and intensity of relations and mutual influence of the main social actors.

The first stage of creating a model is cognitive modeling, which results in the preparation of a cognitive scheme that displays the main actors of the process under consideration and the most significant connections (mutual influences) between them. We will be primarily interested in the processes of change of power in the

[21] On nature and definitions of revolution see, e.g., Goldstone, 2014; Lawson, 2019; Goldstone et al., 2022b; Beissinger, 2022.

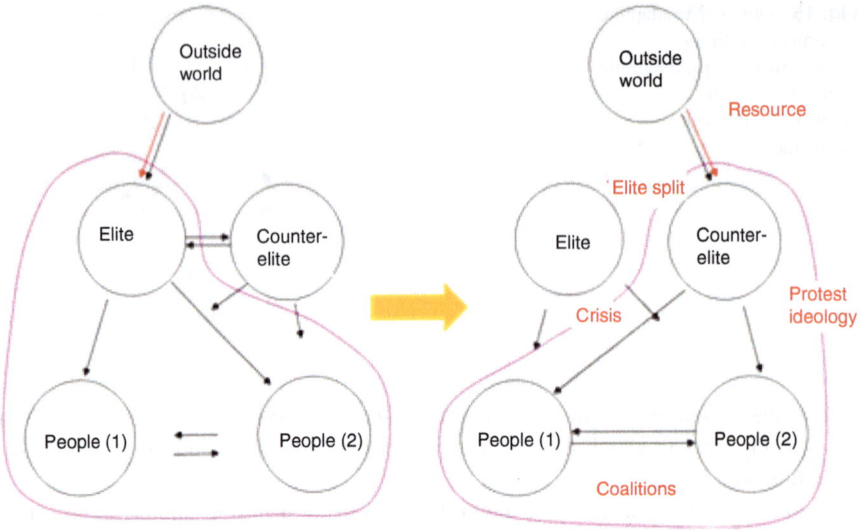

Fig. 16 A generalized scheme of the change of power in a country. The left part of the scheme shows the state of the sociopolitical system before the change of power. Its right part shows the state of the sociopolitical system after the change of power

country (violent and nonviolent types, including the revolutions). Accordingly, the basic cognitive scheme of the model can be represented as follows (see Fig. 16):

The left part of the scheme in Fig. 16 shows the state of the sociopolitical system before the change of power. Its right part shows the state of the sociopolitical system after the change of power. The main actors of the political process are presented in the scheme, including the ruling elite, the counter-elite claiming power (legally or illegally), certain social groups of the population (there are two of them in the figure, but in reality there are more of them), and the outside world (external forces that influence the political process in the country). The arrows show the strength of influence (power, economic, ideological, etc.) of the actors on each other. The essence of the process under consideration lies in the fact that, as a result of the political process, the counter-elite achieves a radical increase in its influence on the population and seizes control of society. In the scheme, the increase in influence is reflected by a change in the length of the arrows that characterize the strength of the influence of actors on each other. The solid line delineates the "zone of synchronization," which includes actors acting in concert, in accordance with the rules established by the ruling elite (before the change of power) or the counter-elite (after the change of power). "Synchronization" of actors occurs when mutual connections in the zone under consideration exceed a certain level of intensity. If the level of intensity of mutual ties in the zone under consideration is below the critical threshold level, then synchronization of social actors does not occur, they begin to act inconsistently, a situation of social chaos sets in. Thus, by decreasing or increasing the intensity of ties between actors, one can achieve both destabilization (chaotization) of the political situation and its stabilization, increasing the sociopolitical stability of the system.

With this approach, in order to determine conditions for sociopolitical stabilization/destabilization one should analyze the influence of various factors (economic, demographic, social, political, ideological, and psychological) on the ties that ensure the coherence of actors, as well as to identify conditions under which the intensity of these ties is already sufficient (or not sufficient) to ensure synchronization of these actors according to the rules established by the ruling elite. At the same time, this approach allows us to consider in dynamics the various scenarios of weakening/strengthening ties in the social system and, accordingly, to analyze various scenarios for the transition of the social system from a stable to unstable state and vice versa. Such a dynamic approach is convenient for the use in systems for monitoring the sociopolitical situation and supporting decision-making in the field of national security.

The operationalization of this generalized cognitive model to a specific situation in a specific country is carried out on the basis of expert analysis according to the following scheme:

- first, the main social groups of the population are identified, who are aware of their commonality on the basis of national, territorial, confessional, professional, or any other indicator and oppose themselves to other groups (on the principle of greater or lesser trust: "friend-friend" or "friend-foe"). It is determined how strong the contradictions between these groups are, whether they can cause an open conflict and destabilize the political situation (in the future, the main attention should be paid to groups with a high level of contradictions). These social groups determine the composition of the actors displayed in the cognitive scheme of Fig. 16 in the category "people" (where *i* is the number of the social group);
- the main actors in the ruling elite and in the opposition (counter-elite), who have real power or are capable of claiming power in the future, are identified. These actors are displayed in the cognitive scheme of Fig. 16 in the categories of "elite" and "counter-elite";
- the nature and strength of the influence of "elite" and "counter-elite" actors on various social groups of the "population/people" are determined, as well as the reverse influence of these population groups on the "elite" and "counter-elite";
- it is determined which splits (conflicts) in society are the most significant (dangerous) in terms of possible destabilization;
- a forecast is made as to which conflicts tend to decrease or increase in the forecast period (for example, until 2025–2030) if the situation develops in an inertial way;
- it is recorded what measures are taken by the ruling "elite" (if any) to reduce the splits;
- it is determined which social groups the competing groups of the elite rely on, which splits within the elite are the most dangerous from the point of view of possible destabilization.

On the basis of this information a cognitive scheme similar to the one shown in Fig. 16, characterizing the situation in the country in question, is formed. On the basis of this information, a cognitive scheme similar to the one shown in Fig. 16, characterizing the situation in the country in question, is formed.[22] This cognitive scheme serves as the basis for the formation of a simulation mathematical model using quantitative values for the links α_{ij} between the main actors. The model makes it possible, by varying the strength of the connections α_{ij}, to determine which of them are the most critical in terms of destabilization (de-synchronization) of the social system (the task of identifying the "bottlenecks"). The cognitive scheme is then used to form options for countering destabilization, and a simulation model based on it is used to assess the impact of the proposed measures to counter destabilization on the state of the social system.

At the same time, the analysis of measures to reduce the risks of destabilization must take into account the peculiarities of the mechanisms for ensuring sociopolitical stability in particular states. Measures that are effective in some countries may be ineffective or even have the opposite effect in other countries (see Malkov, 2020 and Chapter "Modeling Social Self-Organization and Historical Dynamics. Global Phase Transitions" [Malkov et al., 2023a, this volume]).

Basic Model of Political Competition

The difference between social actors in the space of values (see Fig. 9) predetermines the inevitability of political struggle and competition for political influence (an example of this is the competitive struggle between the elite and counter-elite, reflected in the cognitive scheme of Fig. 16).

The basic model of competition is set out in the works of Dmitri Chernavsky (Chernavsky, 2004a, 2004b). The model is formed on the basis of formalization of general patterns that characterize the interaction of political, socioeconomic, sociocultural, information, and biological systems. In a generalized form, it is a system of differential equations describing the change in the balance of forces of various actors as a result of competition:

$$du_i/dt = \boldsymbol{A}_i(u_i) - \boldsymbol{B}_i(u_i) + \sum\nolimits_{j \neq i} \boldsymbol{C}_{ij}(u_i, u_j); i, j = 1, 2, 3, \ldots, N. \quad (42)$$

Here t is the time; u_i is an indicator that characterizes the "strength" (degree of influence, dominance, economic power, etc.) of the ith political actor at time t. Term $\boldsymbol{A}_i(u_i)$ describes the reproduction (renewal) of the "strength" of the ith actor. Term $\boldsymbol{B}_i(u_i)$ describes the decrease in the "strength" of the ith actor due to natural processes not related to the competitive struggle (negative feedback in the social system, resource constraints, etc.) and due to intraspecific struggle (competition of subsystems with each other). Terms $\boldsymbol{C}_{ij}(u_i,u_j)$ describe the interaction between the actors. In the case when this interaction is competitive, the terms are negative, since

[22] For an example of the application of such a methodology to a specific case see, e.g., Malkov et al., 2020.

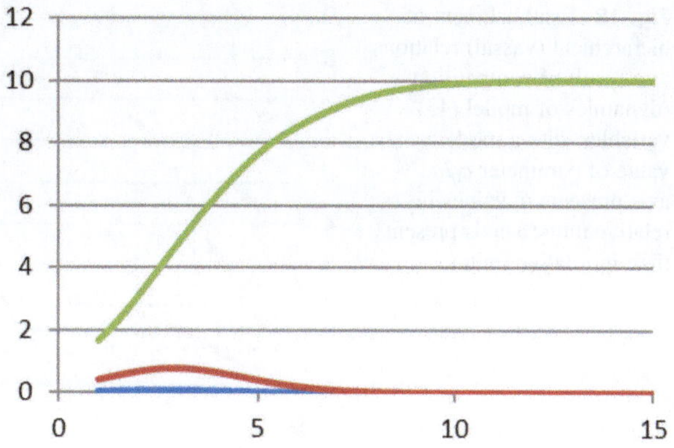

Fig. 17 Establishment of political monopoly as a result of competition (dynamics of the model (43) variables at a high value of the parameter c_{ij}/a_i; *y*-axis presents u_i values in relative units; *x*-axis presents time in relative units)

in the competitive struggle actors tend to suppress (weaken) each other. In the process of competitive struggle, coalitions (temporary alliances) can arise when several actors combine their efforts in order to jointly resist a common enemy. In this case, terms $\boldsymbol{C}_{ij}(u_i,u_j)$, describing the interaction of coalition members, take positive values, since they help each other in the fight against a common enemy.

In a simplified form (keeping only the linear and quadratic terms on the right-hand sides of Eqs. (42)), the basic competition model can be written as follows:

$$du_i/dt = a_i \cdot u_i - b_i \cdot u_i^2 + \sum\nolimits_{j \neq i} c_{ij} \cdot u_i \cdot u_j; i,j = 1, 2, 3, \ldots, N; a_i > 0; \\ b_i > 0. \quad (43)$$

Despite its simplicity, system (43) shows complex behavior. Consider it for a case of competitive interaction of three actors ($i, j = 1, 2, 3$). In this case, we will assume that initially all actors compete with each other ($c_{ij} < 0$); however, if one of the actors becomes significantly stronger in relation to the other two, they unite against it and begin to act cooperatively (their mutual coefficients c_{ij} become positive) in order to weaken it.[23] It can be shown that the "order parameter" (that is, the parameter on the change of which the behavior of the system depends critically) in this case is the ratio c_{ij}/a_i, which reflects the intensity of competition (the ratio of efforts expended on competition to the rate of reproduction of actors' power). Typical situations are the following:

(a) if the competitive struggle is fierce (high value of the parameter c_{ij}/a_i) and the forces are unequal, this leads to the complete suppression (destruction) of weaker actors by the strong ones (see Fig. 17), or hierarchical relations are established with the subordination of the weak to the dominant (see Fig. 18);
(b) if the forces are approximately equal and the value of c_{ij}/a_i parameter is not very high, then changeable dynamics can take place, when one or the other of the

[23] We will call such a situation "alliance of the weak against the strong".

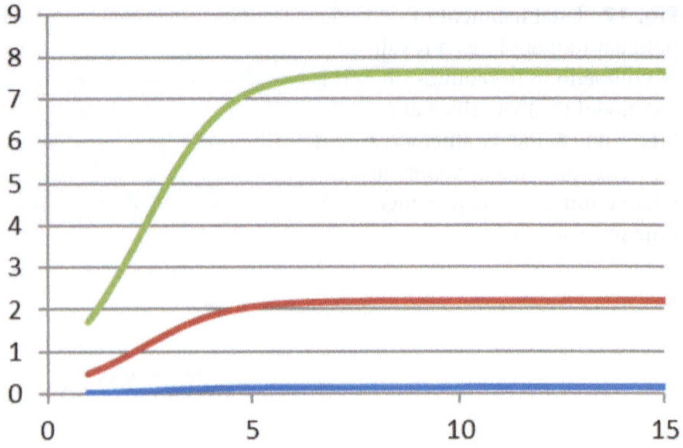

Fig. 18 Establishment of hierarchical (vassal) relations as a result of competition (dynamics of model (43) variables with a moderate value of parameter c_{ij}/a_i; *y*-axis presents u_i values in relative units; *x*-axis presents time in relative units)

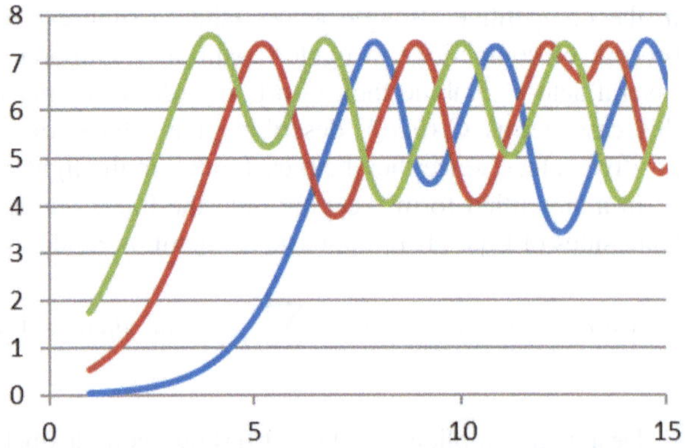

Fig. 19 Establishment of the regime of "checks and balances" ("alliance of the weak against the strong") as a result of competition (the dynamics of the model (43) variables with a low value of the parameter c_{ij}/a_i; *y*-axis presents u_i values in relative units; *x*-axis presents time in relative units)

actors pulls ahead, but then the rest of the actors unite against the leader and weaken it again by joint efforts.[24] As a result, a dynamic equilibrium is established in the system, in which leadership periodically passes from one actor to another, but none of them manages to maintain its leadership for a long time (see Fig. 19);

(c) there might be a situation when the actors refuse to compete with each other in favor of cooperation (this corresponds to the fact that the coefficients c_{ij} in model (A.43) become positive). In this case, equal relations are established between the actors and over time their political influence levels off (see Fig. 20).

Basic Model of Interstate Interactions

The study of the social system's resistance to external destabilizing influences (from other social systems) is carried out on the basis of the analysis and modeling of the competitive struggle of social systems with each other, which comprises both economic and political struggle (including the one involving military force). The basic model of social systems' competitive struggle is similar to model (42), but

[24] Such a regime is sometimes referred to as a "checks and balances" regime.

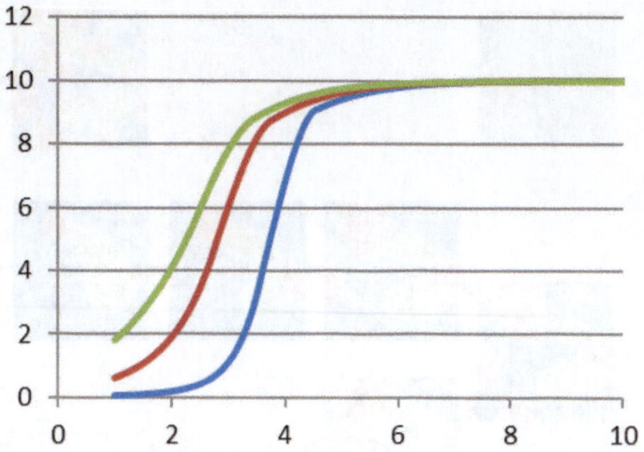

Fig. 20 Establishment of equal relations between political actors (dynamics of model (43) variables when antagonistic competition is eliminated; y-axis presents u_i values in relative units; x-axis presents time in relative units)

spatial variables can play a significant role in it since this struggle often unfolds in physical space (for example, classic "hot" wars with the conquest of territories). This basic model is described in the works of Dmitri Chernavsky and his colleagues (Chernavsky et al., 2005a, 2005b) and has the following form:

$$\partial u_i/\partial t = \boldsymbol{A}_i(u_i, x, y) - \boldsymbol{B}_i(u_i, x, y) + \sum_{j \neq i} \boldsymbol{C}_{ij}(u_i, u_j) + \boldsymbol{D}_i(u_i, x, y). \tag{44}$$

Here $i, j = 1, 2, 3, \ldots, N$; t is the time; x, y are spatial coordinates; u_i is an indicator that characterizes the "strength" (degree of influence, dominance, economic and military power, etc.) of the ith social system at time t at a point in space (x, y). Term $\boldsymbol{A}_i(u_i,x,y)$ describes the reproduction (renewal) of the "strength" of the ith social system. Term $\boldsymbol{B}_i(u_i,x,y)$ describes the decrease in the "strength" of the ith social system due to natural processes (negative feedback in social systems, resource constraints, etc.) and intraspecific struggle (internal competition of subsystems of the i-th social system). Terms $\boldsymbol{C}_{ij}(u_i,u_j)$ describe the competitive struggle between social systems. These terms are negative because, in a competitive struggle, actors tend to overwhelm each other. Term $\boldsymbol{D}_i(u_i,x,y)$ describes the distribution of the "force" of the i-th social system in space.

In a simplified form (keeping only linear and quadratic terms on the right-hand sides of Eq. (44)), the basic competition model can be written as follows:

$$\partial u_i/\partial t = a_i \cdot u_i - b_i \cdot u_i^2 + \sum_{j \neq i} c_{i,j} \cdot u_i \cdot u_j + D_i \cdot \Delta u_i, \tag{45}$$

where $i, j = 1, 2, 3, \ldots, N$; $a_i > 0$, $b_i > 0$, $c_{ij} < 0$; D_i is the coefficient characterizing the speed of propagation of the "force" of the ith SS in space; Δ is Laplace operator.[25]

[25] $\Delta = \frac{\partial^2}{\partial x^2} + \frac{\partial^2}{\partial y^2}$.

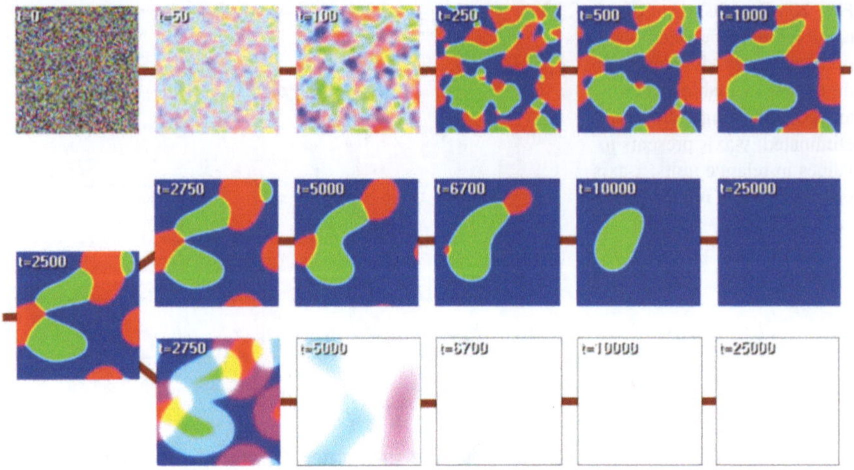

Fig. 21 Competitive interactions simulation results, in the form of successive frames for calculating the dynamics of the system described by Eq. (45) (numbers in the upper left corner of each frame corresponding to a serial number of a respective iteration cycle of the calculation)

Figure 21 shows the behavior of a system consisting of elements of three types (they are indicated in the figure in red, blue, and green), the interaction of which is described by Eq. (45).

The process of system development (45) is divided into a number of stages:

1. stage I (frames 1–4 from time $t = 0$ to $t = 250$): formation of "pure" areas (clusters) as a result of interspecific struggle, in which elements of a certain type predominate (different for different areas).
2. stage II (frames 5–7 from time $t = 500$ to $t = 2500$): struggle among clusters to expand the space they occupy. At this stage, the antagonistic interaction of clusters occurs only at interfaces between them, while the dynamics of changes slows down significantly compared to the previous stage. Ultimately, one cluster wins (that is, captures the entire space), but which one is not initially clear. It is important that the first two stages are stochastic in nature, the final result at these stages is unambiguously unpredictable.
3. stage III (starting from frame 7 from time $t = 2500$ to $t = 25{,}000$) is divided into two variants. The first option (upper branch of successive frames) corresponds to the continuation of the antagonistic struggle between clusters. Within this scenario, one of the clusters begins to dominate over the others and eventually completely suppresses them. At the end of this stage, a pure state is formed, which is stable and does not evolve further.

The second variant of stage III (the lower branch of successive frames) corresponds to the case when at the time $t = 2500$ the absolute values of the coefficients c_{ij} are reduced, while the values of the D_i coefficients increase, that is, the level of antagonism of cluster interaction is reduced with a simultaneous increase in the intensity of interaction. This reflects a situation where equal relations are gradually established between the actors without attempts at mutual dominance. As a result, the color of the frames becomes white, since the blue, red, and green colors that mark the actors in question, when mixed in equal proportions, make white color.

Depending on the objectives of the study, system (44) can model various aspects of the competitive struggle of social systems, such as economic, military-political, and ideological. On the other hand, the interaction of social systems may be noncompetitive (this applies, for example, to mutually beneficial trade relations, military alliances, etc.). In this case, the sign of $C_{ij}(u_i,u_j)$ is not negative, but positive.

A set of social systems of one type can constitute a megasystem (civilization) that opposes another megasystem (examples are the confrontation between agricultural and nomadic peoples, Christian and Muslim countries over long historical periods, etc.). At the same time, the social strata of one social system are subsystems that can be in competitive (up to antagonistic) relations with each other. Thus, social systems in modeling can be represented as complex structures, each level of which is described by basic dynamic equations of type (42) or (44). Examples of using models of this type are given below in Sect. 2.3 and Sect. 2.3 of Chapter "Modeling Social Self-Organization and Historical Dynamics. Industrial Society" (Akaev et al., 2023a, this volume) and Sect. 2.2 of Chapter "Modeling Social Self-Organization and Historical Dynamics. Modern Society and Problems of Global Transition" (Akaev et al., 2023b, this volume).

The basic mathematical models given in this Chapter are a basis for modeling various historical eras—the specification of the basic equations in relation to agrarian society is given in Chapter "Modeling Social Self-Organization and Historical Dynamics. Agrarian Society" (Malkov et al., 2023b, this volume), in relation to industrial society this is done in Chapter "Modeling Social Self-Organization and Historical Dynamics. Industrial Society" (Akaev et al., 2023a, this volume), in relation to the modern society this is done in Chapter "Modeling Social Self-Organization and Historical Dynamics. Modern Society and a Look into the Global Futures: Cybernetic W-Society" [Akaev et al., 2023b, this volume]). This allows identifying patterns of macrohistorical transformations and trying to look into the future on the basis of a general approach.

Acknowledgements This work was done with the support of MSU Program of Development, Project No 23A-SCH05-03.

References

Akaev, A., Malkov, S., Davydova, O., Kovaleva, N., Malkov, A., & Korotayev, A. (2023a). Modeling social self-organization and historical dynamics. Industrial society. In V. Sadovnichy et al. (Eds.), *Reconsidering the limits to growth. A report to the Russian association of the Club of Rome (pp. 337–385).* Springer. https://doi.org/10.1007/978-3-031-34999-7_17

Akaev, A., Malkov, S., Grinin, L., Bilyuga, S., Davydova, O., Grinin, A., Kovaleva, N., Malkov, A., Musieva, J., & Korotayev, A. (2023b). Modeling social self-organization and historical dynamics. Modern society and a look into the global futures: Cybernetic W-society. In V. Sadovnichy et al. (Eds.), *Reconsidering the limits to growth. A report to the Russian association of the Club of Rome (pp. 419–460)*. Springer. https://doi.org/10.1007/978-3-031-34999-7_19

Akaev, A., & Sadovnichy, V. (2018). Mathematical models for calculating the development dynamics in the era of digital economy. *Doklady Mathematics, 98*(2), 526–531.

Akaev, A., & Sadovnichy, V. (2019). On the choice of mathematical models for describing the dynamics of digital economy. *Differential Equations, 55*(5), 729–738.

Akaev, A., Sadovnichy, V., & Korotayev, A. (2012). On the dynamics of the world demographic transition and financial-economic crises forecasts. *The European Physical Journal Special Topics, 205*(1), 355–373. https://doi.org/10.1140/epjst/e2012-01578-2

Allen, R. C. (2011). *Global economic history: A very short introduction*. Oxford University Press.

Arrighi, G., & Silver, B. J. (1999). *Chaos and governance in the modern world system*. University of Minnesota Press.

Aruka, Y., & Kirman, A. (2017). *Economic foundations for social complexity science*. Springer.

Bailey, M., Strezhnev, A., & Voeten, E. (2015). Estimating dynamic state preferences from United Nations voting data. *Journal of Conflict Resolution, 61*(2), 430–456. https://doi.org/10.1177/0022002715595700

Beissinger, M. R. (2022). *The Revolutionary City: Urbanization and the global transformation of rebellion*. Princeton University Press.

Bergman, L. (1991). General equilibrium effects of environmental policy: A CGE-modeling approach. *Environmental and Resource Economics, 1*(1), 43–61.

Bergman, L. (2005). CGE modeling of environmental policy and resource management. *Handbook of Environmental Economics, 3*, 1273–1306.

Bukharin, S., Kovalev, V., & Malkov, S. (2009). O formalizatsii ponyatiya informatsionnogo polya. *Informatsionnyye voyny, 4*(12), 2–9.

Chase-Dunn, C., & Anderson, E. (Eds.). (2005). *The historical evolution of world-systems*. Springer. https://doi.org/10.1057/9781403980526

Chase-Dunn, C., & Hall, T. D. (1997). *Rise and demise: Comparing world-systems*. Westview Press.

Chase-Dunn, C., & Lerro, B. (2016). *Social change: Globalization from the stone age to the present*. Routledge.

Chase-Dunn, C., Niemeyer, R., Alvarez, A., Inoue, H., & Love, J. (2010). Cycles of rise and fall, upsweeps and collapses: Changes in the scale of settlements and polities since the bronze age. In L. Grinin, P. Herrmann, A. Korotayev, & A. Tausch (Eds.), *History and mathematics: Processes and models of global dynamics* (pp. 64–91). Uchitel.

Chernavsky, D. (1990). *Sinergetika i informatsiya*. Znaniye.

Chernavsky, D. (2004a). *Sinergetika i informatsiya (dinamicheskaya teoriya informatsii)*. URSS.

Chernavsky, D., Chernavskaya, N., Malkov, S., & Malkov, A. (2005a). Bor'Ba uslovnykh informatsiy. In S. Malkov & A. Korotayev (Eds.), *Istoriya i sinergetika: Matematicheskoye modelirovaniye sotsial'noy dinamiki* (pp. 88–102). KomKniga.

Chernavsky, D., Chernavskaya, N., Malkov, S., & Malkov, A. (2005b). Geopoliticheskiye protsessy kak ob″yekt matematicheskogo modelirovaniya. In S. Malkov & A. Korotayev (Eds.), *Istoriya i sinergetika: Matematicheskoye modelirovaniye sotsial'noy dinamiki* (pp. 103–116). KomKniga.

Chernavsky, D. (2004b) *Synergetics and information: Dynamic information theory*. 2nd ed.

Claessen, H. J. M., & Skalník, P. (Eds.). (1978). *The early state*. Mouton.

Decaluwé, B., & Martens, A. (1988). CGE modeling and developing economies: A concise empirical survey of 73 applications to 26 countries. *Journal of Policy Modeling, 10*(4), 529–568.

Dellink, R., Van der Mensbrugghe, D., & Saveyn, B. (2020). Shaping baseline scenarios of economic activity with CGE models: Introduction to the special issue. *Journal of Global Economic Analysis, 5*(1), 1–27.

Denemark, R. A., Friedman, J., Gills, B. K., & Modelski, G. (2000). *World system history. The social science of long-term change*. Routledge.
Diakonoff, I. (1999). *The paths of history*. Cambridge University Press.
Dmitriyev, A., Starkov, S., & Shirokov, M. (1996). Sinkhronizatsiya ansambley svyazannykh otobrazheniy. Izvestiya vuzov. *Prikladnaya nelineynaya dinamika, 4*(4–5), 40–58.
Gates, S., Hegre, H., Jones, M. P., & Strand, H. (2006). Institutional inconsistency and political instability: Polity duration, 1800–2000. *American Journal of Political Science, 50*(4), 893–908. https://doi.org/10.1111/j.1540-5907.2006.00222.x
Hartman, P. (1964a). *Ordinary differential equations*. John Wiley & Sons.
Ellingsen, T., & Gleditsch, N. P. (1997). Democracy and armed conflict in the third world. In K. D. Volden & D. Smith (Eds.), *Causes of conflict in third world* (pp. 69–81). North/-South Coalition & International Peace Research Institute.
Filin, N., Khodunov, A., & Koklikov, V. (2022). Serbian "Otpor" and the color revolutions' diffusion. In J. A. Goldstone, L. Grinin, & A. Korotayev (Eds.), *Handbook of revolutions in the 21st century: The new waves of revolutions, and the causes and effects of disruptive political change* (pp. 465–482). Springer. https://doi.org/10.1007/978-3-030-86468-2_17
Forrester, J. W. (1971). *World dynamics*. Wright-Allen Press.
Forrester, J. W. (1975). *Collected papers of jay W. Forrester*. Pegasus Communications.
Forrester, J. W. (1984). *Principes des systèmes (Science des systèmes. Série Dynamique des systèmes)*. Presses universitaires de Lyon.
Frank, A. G., & Gills, B. K. (Eds.). (1993). *The world system: Five hundred years, or five thousand*. Routledge.
Gaibulloev, K., Piazza, J. A., & Sandler, T. (2017). Regime types and terrorism. *International Organization, 71*(3), 491–522. https://doi.org/10.1017/S0020818317000169
Gellner, E. (1984). *Nations and nationalism*. Blackwell.
Gellner, E. (1988). *Plough, sword and book. The structure of human history*. The University of Chicago Press.
Gills, B. K., & Thompson, W. (2012). *Globalization and global history*. Routledge.
Goldstein, J. (1988). *Long cycles: Prosperity and war in the modern age*. Yale University Press.
Goldstone, J. A. (2009). *Why Europe? The rise of the west in world history. 1500–1850*. McGraw-Hill.
Goldstone, J. A. (2014). *Revolutions: A very short introduction*. Oxford University Press.
Goldstone, J. A. (2016). *Revolution and rebellion in the early modern world: Population change and state breakdown in England, France, Turkey, and China, 1600–1850*. Routledge.
Goldstone, J. A., Bates, R., Epstein, D., Gurr, T., Lustik, M., Marshall, M., Ulfelder, J., & Woodward, M. (2010). A global model for forecasting political instability. *The American Journal of Political Science, 54*(1), 190–208. https://doi.org/10.1111/j.1540-5907.2009.00426.x
Goldstone, J. A., Grinin, L., & Korotayev, A. (2022a). Introduction. Changing yet persistent: Revolutions and revolutionary events. In J. A. Goldstone, L. Grinin, & A. Korotayev (Eds.), *Handbook of revolutions in the 21st century: The new waves of revolutions, and the causes and effects of disruptive political change* (pp. 1–34). Springer. https://doi.org/10.1007/978-3-030-86468-2_1
Goldstone, J. A., Grinin, L., & Korotayev, A. (2022b). The phenomenon and theories of revolutions. In J. A. Goldstone, L. Grinin, & A. Korotayev (Eds.), *Handbook of revolutions in the 21st century: The new waves of revolutions, and the causes and effects of disruptive political change* (pp. 37–68). Springer. https://doi.org/10.1007/978-3-030-86468-2_2
Grinin, L. (2008). Early state, developed state, mature state: Statehood evolutionary sequence. *Social Evolution & History, 7*(1), 67–81.
Grinin, L. (2011). *The evolution of statehood. From early state to global society*. Lambert Academic Publishing.
Grinin, L. (2022). On revolutionary situations, stages of revolution, and some other aspects of the theory of revolution. In J. A. Goldstone, L. Grinin, & A. Korotayev (Eds.), *Handbook of*

revolutions in the 21st century: The new waves of revolutions, and the causes and effects of disruptive political change (pp. 69–104). Springer. https://doi.org/10.1007/978-3-030-86468-2_3

Grinin, L., Grinin, A., & Korotayev, A. (2020). A quantitative analysis of worldwide long-term technology growth: From 40,000 BCE to the early 22nd century. *Technological Forecasting and Social Change, 155*, 1–19. https://doi.org/10.1016/j.techfore.2020.119955

Grinin, L., & Korotayev, A. (2015). *Great divergence and great convergence. A global perspective*. Springer. https://doi.org/10.1007/978-3-319-17780-9

Grinin, L., & Korotayev, A. (2022). Revolutions, counterrevolutions, and democracy. In J. A. Goldstone, L. Grinin, & A. Korotayev (Eds.), *Handbook of revolutions in the 21st century: The new waves of revolutions, and the causes and effects of disruptive political change* (pp. 105–136). Springer. https://doi.org/10.1007/978-3-030-86468-2_4

Grinin, L., Korotayev, A., & Tausch, A. (2016). Economic cycles, crises, and the global periphery. *Springer*. https://doi.org/10.1007/978-3-319-41262-7

Grinin, L., Korotayev, A., & Tausch, A. (2019). *Islamism, Arab spring, and the future of democracy. World system and world values perspectives*. Springer. https://doi.org/10.1007/978-3-319-91077-2

Gurr, T. R. (1974). Persistence and change in political systems, 1800–1971. *American Political Science Review, 68*(4), 1482–1504.

Haken, H. (1978). *Synergetics: An introduction: Nonequilibrium phase transitions and self-organization in physics, chemistry, and biology*. Springer.

Haken, H. (1980). Synergetics. *Naturwissenschaften, 67*(3), 121–128.

Haken, H. (1983). *Advanced synergetics: Instability hierarchies of self-organizing systems and devices*. Springer.

Haken, H. (1984). *The science of structure: Synergetics*. Van Nostrand Reinhold.

Haken, H. (2006a). *Information and self-organization: A macroscopic approach to complex systems*. Springer.

Haken, H. (2006b). *Synergetik in der Psychologie Selbstorganisation verstehen und gestalten*. Hogrefe.

Hartman, P. (1964b). *Ordinary differential equations*. Wiley.

Jantsch, E. (1980). *The self-organizing universe: Scientific and human implications of the emerging paradigm of evolution*. Pergamon.

Karasev, D. (2022). The "problem of structure and agency" and contemporary sociology of revolution and social movements. In J. A. Goldstone, L. Grinin, & A. Korotayev (Eds.), *Handbook f revolutions in the 21st century: The new waves of revolutions, and the causes and effects of disruptive political change* (pp. 201–217). Springer. https://doi.org/10.1007/978-3-030-86468-2_7

Khokhlov, N., & Korotayev, A. (2022). Internet, political regime and terrorism: A quantitative analysis. *Cross-Cultural Research, 56*(4), 385–418. https://doi.org/10.1177/10693971221085343

Komlos, J., & Nefedov, S. (2002). A compact macromodel of pre-industrial population growth. *Historical Methods: A Journal of Quantitative and Interdisciplinary History, 35*(2), 92–94. https://doi.org/10.1080/01615440209604133

Korotayev, A. (2005). A compact macromodel of world system evolution. *Journal of World-Systems Research, 11*(1), 79–93. https://doi.org/10.5195/jwsr.2005.401

Korotayev, A. (2007). Compact mathematical models of world system development, and how they can help us to clarify our understanding of globalization processes. In G. Modelski, T. Devezas, & W. Thompson (Eds.), *Globalization as evolutionary process: Modeling global change* (pp. 133–160). Routledge.

Korotayev, A. (2020). The 21st century singularity in the big history perspective. A re-analysis. In A. Korotayev & D. LePoire (Eds.), *The 21st century singularity and global futures. A big history perspective* (pp. 19–75). Springer. https://doi.org/10.1007/978-3-030-33730-8_2

Korotayev, A., Bilyuga, S., & Shishkina, A. (2016a). GDP per capita, protest intensity and regime type: A quantitative analysis. *Sravnitelnaya Politika-Comparative Politics, 7*(4), 72–94. https://doi.org/10.18611/2221-3279-2016-7-4(25)-72-94

Korotayev, A., Goldstone, J. A., & Zinkina, J. (2015a). Phases of global demographic transition correlate with phases of the great divergence and great convergence. *Technological Forecasting and Social Change, 95*, 163–169. https://doi.org/10.1016/j.techfore.2015.01.017

Korotayev, A., Issaev, L., & Shishkina, A. (2014). The Arab spring. A quantitative analysis. *Arab Studies Quarterly, 36*(2), 149–169. https://doi.org/10.13169/arabstudquar.36.2.0149

Korotayev, A., Isaev, L., & Vasilev, A. (2015b). Quantitative analysis of 2013-2014 revolutionary wave. *Sotsiologicheskie Issledovaniya, 8*(376), 119–127.

Korotayev, A., Issaev, L., & Zinkina, J. (2015c). Center-periphery dissonance as a possible factor of the revolutionary wave of 2013–2014: A cross-national analysis. *Cross-Cultural Research, 49*(5), 461–488. https://doi.org/10.1177/1069397115595374

Korotayev, A., & Khaltourina, D. (2006). *Introduction to social macrodynamics: Secular cycles and millennial trends in Africa*. KomKniga/URSS

Korotayev, A., & Malkov, A. (2016). A compact mathematical model of the world system economic and demographic growth, 1 CE–1973 CE. *International Journal of Mathematical Models and Methods in Applied Sciences, 10*, 200–209.

Korotayev, A., Malkov, A., & Khaltourina, D. (2006a). *Introduction to social macrodynamics: Compact macromodels of the world system growth*. KomKniga/URSS.

Korotayev, A., Malkov, A., & Khaltourina, D. (2006b). *Introduction to social macrodynamics: Secular cycles and millennial trends*. KomKniga/URSS.

Korotayev, A., & Malkov, S. (2012). Mathematical models of the world-system development. In S. Babones & C. Chase-Dunn (Eds.), *Routledge handbook of world-systems analysis* (pp. 158–161). Routledge.

Korotayev, A., Grinin, L., & Grinin A. (2021c). Mathematical model of interaction between civilization center and tribal periphery: A description. *Social Evolution & History, 20*(2), 50–78. https://doi.org/10.30884/seh/2021.02.03

Korotayev, A., Grinin, L., & Grinin, A. (2022). Mathematical model of interaction between civilization center and tribal periphery: An analysis. *Social Evolution & History, 21*(1), 67–100. https://doi.org/10.30884/seh/2022.01.03

Korotayev, A., Malkov, A., & Khaltourina, D. (2007). *Zakony istorii: Matematicheskoye modelirovaniye razvitiya Mir-Sistemy. Demografiya, ekonomika, kul'tura*. KomKniga.

Korotayev, A., Meshcherina, K., Slinko, E., & Shishkina, A. (2019a). Value orientations of the Afrasian zone of instability: Gender dimensions. *Vostok (Oriens), 1*, 122–154. https://doi.org/10.31857/S086919080003963-6

Korotayev, A., Slinko, E., Shulgin, S., & Biluga, S. (2016b). Intermediate types of political regimes and sociopolitical instability (quantitative cross-national analysis). *Politeia-Journal of Political Theory, Political Philosophy and Sociology of Politics, 82*(3), 31–51. https://doi.org/10.30570/2078-5089-2016-82-3-31-51

Korotayev, A., Slinko, E., Meshcherina, K., & Zinkina, J. (2020). Variation of human values and modernization: Preliminary results. *Cross-Cultural Research, 54*(2–3), 238–272. https://doi.org/10.1177/1069397119874781

Korotayev, A., Tsirel, S., & Bilyuga, S. (2019b). Corruption, human values and attempts of violent changes of government in countries with various levels of per capita GDP: A comparative quantitative analysis. *Sravnitelnaya Politika-Comparative Politics, 10*(1), 98–123. https://doi.org/10.24411/2221-3279-2019-10007

Korotayev, A., Vaskin, I., & Romanov, D. (2019c). Democracy and terrorism: A re-analysis. *Russian Sociological Review, 18*(3), 9–48. https://doi.org/10.17323/1728-192x-2019-3-9-48

Korotayev, A., Vaskin, I., & Romanov, D. (2021a). Terrorism and democracy. A reconsideration. *Comparative Sociology, 20*(3), 344–379. https://doi.org/10.1163/15691330-bja10033

Korotayev, A., Vaskin, I., & Tsirel, S. (2021b). Economic growth, education, and terrorism: A re-analysis. *Terrorism and Political Violence, 33*(3), 572–595. https://doi.org/10.1080/09546553.2018.1559835

Korotayev, A., & Zinkina, J. (2017). Systemic boundary issues in the light of mathematical modeling of world-system evolution. *Journal of Globalization Studies, 8*(1), 78–96.

Korotayev, A., Zinkina, J., & Andreev, A. (2016c). Secular cycles and millennial trends. *Cliodynamics, 7*(2), 204–216. https://doi.org/10.21237/C7CLIO7230484

Korotayev, A., Zinkina, J., Slinko, E., & Meshcherina, K. (2019d). Human values and modernization: A global analysis. *Journal of Globalization Studies, 10*(1), 44–71. https://doi.org/10.30884/jogs/2019.01.04

Korotayev, A., Zinkina, J., Kobzeva, S., Bogevolnov, J., Khaltourina, D., Malkov, A., & Malkov, S. (2011). A trap at the escape from the trap? Demographic-structural factors of political instability in modern Africa and West Asia. *Cliodynamics: The Journal of Quantitative History and Cultural Evolution, 2*(2), 276–303. https://doi.org/10.21237/C7clio22217

Kremer, M. (1993). Population growth and technological change: One million B.C. to 1990. *The Quarterly Journal of Economics, 108*(3), 681–716. https://doi.org/10.2307/2118405

Lawson, G. (2019). *Anatomies of revolution*. Cambridge University Press.

Lijphart, A. (1963). The analysis of bloc voting in the general assembly: A critique and a proposal. *American Political Science Review, 57*(4), 902–917.

Malkov, S. (1998). Politika s tochki zreniya sinergetiki. *Strategicheskaya stabil'nost', 3*, 90–99.

Malkov, S. (2007). Dinamika politicheskikh sistem: modelirovaniye ustoychivosti i destabilizatsii. *Informatsionnyye voyny, 2*, 11–20.

Malkov, S. (2009). *Sotsial′naya samoorganizatsiya i istoricheskiy protsess: Vozmozhnosti matematicheskogo modelirovaniya*. LIBROKOM.

Malkov, S. (2020). Sotsial′no-politicheskaya nestabil'nost′: logiko-matematicheskiy analiz. *Istoriya i sovremennost′, 4*, 3–61.

Malkov, S., & Bilyuga, S. (2015). Model′ ustoychivosti/destabilizatsii politicheskikh system. *Informatsionnyye voyny, 1*(33), 7–18.

Malkov, S., Grinin, L., Grinin, A., & Korotayev, A. (2023a). Modeling social self-organization and historical dynamics. Global phase transitions. In V. Sadovnichy et al. (Eds.), *Reconsidering the limits to growth. A report to the Russian Association of the Club of Rome (pp. 387–417)*. Springer. https://doi.org/10.1007/978-3-031-34999-7_18

Malkov, S., Kovaleva, N., Grinin, L., & Korotayev, A. (2023b). Modeling social self-organization and historical dynamics. Agrarian society. In V. Sadovnichy et al. (Eds.), *Reconsidering the limits to growth. A report to the Russian Association of the Club of Rome (pp. 309–335)*. Springer. https://doi.org/10.1007/978-3-031-34999-7_16

Malkov, S., Meshcherina, K., & Korotayev, A. (2020). Evolution of stability of socioeconomic system functioning: Some approaches to modeling (with an application to the case of Egypt, 2011–2013). *Biosystems, 198*, 104231. https://doi.org/10.1016/j.biosystems.2020.104231

Mansfield, E. D., & Snyder, J. (1995). Democratization and the danger of war. *International Security, 20*(1), 5–38.

Markov, A., Anisimov, V., & Korotayev, A. (2010). Relationship between genome size and organismal complexity in the lineage leading from prokaryotes to mammals. *Paleontological Journal, 44*(4), 363–373. https://doi.org/10.1134/S0031030110040015

Markov, A., & Korotayev, A. (2007). Phanerozoic marine biodiversity follows a hyperbolic trend. *Palaeoworld, 16*(4), 311–318. https://doi.org/10.1016/j.palwor.2007.01.002

McNeill, W. H. (2013). *The pursuit of power: Technology, armed force, and society since AD 1000*. University of Chicago Press.

Meadows, D., Meadows, D., & Randers, J. (1992). *Beyond the limits: Confronting global collapse, Envisioning a Sustainable Future*. Post Mills.

Meadows, D., Meadows, D., Randers, J., & Behrens, W. W. (1972). *The limits to growth. A report to The Club of Rome*. Potomac Associates.

Meadows, D., Randers, J., & Meadows, D. (2004). *Limits to growth. The 30-year update*. Chelsea Green Publisher.
Mitchell, L. A. (2012). *The color revolutions*. Penn Press.
Mitchell, L. A. (2022). The "color" revolutions. Successes and limitations of non-violent protest. In J. A. Goldstone, L. Grinin, & A. Korotayev (Eds.), *Handbook of revolutions in the 21st century: The new waves of revolutions, and the causes and effects of disruptive political change* (pp. 435–445). Springer. https://doi.org/10.1007/978-3-030-86468-2_15
Modelski, G., & Thompson, W. R. (1996). *Leading sectors and world powers: The coevolution of global politics and economics*. University of South Carolina Press.
Modelski, G., Thompson, W. R., & Devezas, T. (2008). *Globalization as evolutionary process: Modeling global change*. Routledge.
Moiseev, N., Guseynov, A., Pavlovskiy, Y., & Ustinov, V. (1984). *Opyt imitatsionnogo modelirovaniya istoricheskogo protsessa*. Nauka.
Moon, F. C. (2008). *Chaotic and fractal dynamics: Introduction for applied scientists and engineers*. John Wiley & Sons.
Muller, E. N., & Weede, E. (1990). Cross-national variation in political violence: A rational action approach. *Journal of Conflict Resolution, 34*(4), 624–651.
Nefedov, S. A. (2004). A model of demographic cycles in a traditional society: The case of ancient China. *Social Evolution & History, 3*(1), 69–80.
Nefedov, S. A. (2013). Modeling Malthusian dynamics in pre-industrial societies. *Cliodynamics: The Journal of Quantitative History and Cultural Evolution, 4*(2), 229–240. https://doi.org/10.21237/C7clio4221335
Nefedov, S. A. (2014a). Modeling Malthusian dynamics in pre-industrial societies. *Social Evolution & History, 13*(1), 172–184.
Nefedov, S. A. (2014b). Modeling Malthusian dynamics in pre-industrial societies: Mathematical modeling. *History of Mathematics, 4*, 190–220.
North, D. C., Wallis, J. J., & Weingast, B. R. (2009). *Violence and social orders: A conceptual framework for interpreting recorded human history*. Cambridge University Press.
Pavlovskiy, Y. (2000). *Imitatsionnyye modeli i sistemy*. Fazis.
Pavlovskiy, Y., Belotelov, N., Brodskiy, Y., & Olenev, N. (2005). *Opyt imitatsionnogo modelirovaniya pri analize sotsial'no-ekonomicheskikh yavleniy*. Press.
Pomeranz, K. (2021). *The great divergence. China, Europe, and the making of the modern world economy* (2nd ed.). Princeton University Press. https://doi.org/10.1515/9780691217192
Prigogine, I. (1997). *The end of certainty*. Free Press.
Prigogine, I. (2017). *Non-equilibrium statistical mechanics (Dover books on physics)*. Dover Publications.
Prigogine, I., & Stengers, I. (1984). Order out of chaos: Man's new dialogue with nature. *Bantam*.
Prigogine, I., & Stengers, I. (2018). *Order out of chaos: Man's new dialogue with nature* (2nd ed.). Verso Books.
Randers, J. (2012). *2052. A global forecast for the next forty years. A report to the club of Rome*. Chelsea Green Publishing.
Randers, J., Rockström, J., Stoknes, P. E., Golüke, U., Collste, D., & Cornell, S. (2018). *Transformation is feasible. A report to the Club of Rome*. Stockholm Resilience Centre.
Rozov, N. (2019). *Filosofiya i teoriya istorii*. URSS.
Sadovnichy, V., Akaev, A., Ilyin, I., Malkov, S., Grinin, L., Sayamov, Y., & Korotayev, A. (2023). Introduction. Hoping for the future. In V. Sadovnichy et al. (Eds.), *Reconsidering the limits to growth. A report to the Russian Association of the Club of Rome (pp. 1–14)*. Springer. https://doi.org/10.1007/978-3-031-34999-7_1
Sadovnichy, V., Akaev, A., Korotayev, A., & Malkov, S. (2012). *Modelirovaniye i prognozirovaniye mirovoy dinamiki*. ISPI RAN.
Sadovnichy, V., Akaev, A., Korotayev, A., & Malkov, S. (2014). *Kompleksnoye modelirovaniye i prognozirovaniye razvitiya stran BRIKS v kontekste mirovoy dinamiki*. Nauka.

Sadovnichy, V., Akaev, A., Korotayev, A., Malkov, S., & Sokolov, V. (2017). *Analiz i modelirovaniye mirovoy i stranovoy dinamiki*. LENAND.

Selbin, E. (2022). All around the world: Revolutionary potential in the age of authoritarian revanchism. In J. A. Goldstone, L. Grinin, & A. Korotayev (Eds.), *Handbook of revolutions in the 21st century: The new waves of revolutions, and the causes and effects of disruptive political change* (pp. 415–433). Springer. https://doi.org/10.1007/978-3-030-86468-2_14

Shulgin, S., Zinkina, J., & Korotayev, A. (2017). "Neighbors in values": A new dataset of cultural distances between countries based on individuals' values, and its application to the study of global trade. *Research in International Business and Finance, 42*, 966–985. https://doi.org/10.1016/j.ribaf.2017.07.031

Slinko, E., Bilyuga, S., Zinkina, J., & Korotayev, A. (2017). Regime type and political destabilization in cross-national perspective: A re-analysis. *Cross-Cultural Research, 51*(1), 26–50. https://doi.org/10.1177/1069397116676485

Tikhonov, A. (1952). Sistemy differentsial′nykh uravneniy, soderzhashchiye malyye parametry pri proizvodnykh. *Matematicheskiy sbornik, 31*(3), 575–586.

Turchin, P. (2006). *War and peace and war: The life cycles of imperial nations*. Pi Press.

Turchin, P. (2018). *Historical dynamics: Why states rise and fall* (2nd ed.). Princeton University Press.

Turchin, P., Hoyer, D., Korotayev, A., Kradin, N., Nefedov, S., et al. (2021). Rise of the war machines: Charting the evolution of military technologies from the Neolithic to the industrial revolution. *PLoS One, 16*(10), e0258161. https://doi.org/10.1371/journal.pone.0258161

Turchin, P., & Korotayev, A. (2006). Population density and warfare: A reconsideration. *Social Evolution and History, 5*(2), 121–158.

Turchin, P., & Korotayev, A. (2020). The 2010 structural-demographic forecast for the 2010–2020 decade: A retrospective assessment. *PLoS One, 15*(8), e0237458. https://doi.org/10.1371/journal.pone.0237458

Turchin, P., & Nefedov, S. (2009). *Secular cycles*. Princeton University Press.

Turchin, P., Whitehouse, H., Gavrilets, S., Hoyer, D., François, P., Bennett, J. S., Feeney, K. C., Peregrine, P., Feinman, G., Korotayev, A., Kradin, N., Levine, J., Reddish, J., Cioni, E., Wacziarg, R., Mendel-Gleason, G., & Benam, M. (2022). Disentangling the evolutionary drivers of social complexity: A comprehensive test of hypotheses. *Science Advances, 8*(25), 1–18. https://doi.org/10.1126/sciadv.abn3517

Ulfelder, J., & Lustik, M. (2007). Modelling transitions to and from democracy. *Democratization, 14*(3), 351–387. https://doi.org/10.1080/13510340701303196

Vaskin, I., Tsirel, S., & Korotayev, A. (2018). Economic growth, education and terrorism: Experience in quantitative analysis. *Sotsiologicheskiy zhurnal, 24*(2), 28–65. https://doi.org/10.19181/socjour.2018.24.2.5844

Vreeland, J. R. (2008). The effect of political regime on civil war: Unpacking anocracy. *Journal of Conflict Resolution, 52*(3), 401–425. https://doi.org/10.1177/0022002708315594

Wallerstein, I. (1974). *The modern world-system*. Academic Press.

Wallerstein, I. (2004). *World-systems analysis: An introduction*. Duke University Press.

Wallerstein, I., Collins, R., Mann, M., Derluguian, G., & Calhoun, C. (2013). *Does capitalism have a future?* Oxford University Press.

Walter, B. (2022). *How civil wars start: And how to stop them*. Penguin Random House.

Zhdanov, A., & Korotayev, A. (2022). Factors of deconsolidation of a Liberal democratic regime: The case of the United States. *Cliodynamics: The Journal of Quantitative History and Cultural Evolution, 13*, 1–34. https://doi.org/10.21237/C7clio13054207

Zinkina, J., Christian, D., Grinin, L., Ilyin, I., Andreev, A., Aleshkovski, I., Shulgin, S., & Korotayev, A. (2019). *A big history of globalization: The emergence of a global world system. Spring*. https://doi.org/10.1007/978-3-030-05707-7

Askar Akaev is the First President of the Kyrgyz Republic and Foreign Full Member of the Russian Academy of Sciences. He works as Chief Research Professor in the Prigogine Institute for Mathematical Investigations of Complex Systems at Lomonosov Moscow State University. He is also Academic Supervisor of the Centre for Stability and Risk Analysis at HSE University, Moscow, Russia. Askar Akaev has authored and co-authored over 300 scientific publications, including such monographs as *Coherent Optical Computers* (1977), *Optical Methods of Information Processing* (1983), *Holographic Memory* (1996), *Selected Lectures on Optical Computers* (1996), *Economic Transition from the Point of View of a Physicist. A Mathematical Model of Emerging Economy* (2000), *Modeling and Forecasting the World Dynamics* (2012), *Quality of Education, Effectiveness of R&D, and Economic Growth. Quantitative Analysis and Mathematical Modeling* (2016), etc.

Sergey Malkov heads the Center for Long-Term Forecasting and Strategic Planning at Lomonosov Moscow State University; he also works as a Senior Research Professor at the Centre for Stability and Risk Analysis at HSE University in Moscow. He is the author of over 300 scholarly publications, including such monographs as *Social self-organization and historical process: possibilities of mathematical modeling* (URSS, 2009), *Modeling and forecasting of world dynamics* (ISPI RAS, 2012, together with V. A. Sadovnichy and others), and *The twenty-first Century Singularity and Global Futures. A Big History Perspective* (Springer, 2020, together with David J. LePoire and others).

Stanislav Bilyuga has a PhD degree in Political Sciences and is affiliated to the Faculty of Global Studies, Lomonosov Moscow State University. His research interests include the modeling of global trends and dynamics and forecasts of destabilization. He has authored and co-authored over 70 scientific publications concerning factors, trends, indices and methods of research on sociopolitical destabilization of various countries.

Artemy Malkov is a Russian mathematician, specialist in mathematical modeling of socio-historical processes, spatial historical dynamics, lecturer at the Moscow Institute of Physics and Technology. He is the author or co-author of more than 100 scientific publications, including such monographs as *Introduction to Social Macrodynamics: Compact Macromodels of the World System Growth* (KomKniga/URSS, 2006; with Andrey Korotaev and Daria Khaltourina), *Introduction to Social Macrodynamics: Secular Cycles and Millennial Trends* (KomKniga/URSS, 2006; with Andrey Korotayev and Daria Khaltourina), *Laws of History: Mathematical Modeling of the Development of the World System. Demography, Economics, Culture* (LENAND/URSS, 2019, in Russian; with Andrey Korotayev and Daria Khaltourina).

Jameelah Musieva is research assistant at Lomonosov Moscow State University, Faculty of Global Studies and Research Fellow at the Centre for Stability and Risk Analysis at HSE University in Moscow, Russia. She is the author/co-author of more than 10 scholarly publications, including such articles as "The Concept of "Quality of Life"" (2019), "Methodology for Cross-Country Assessment of The Quality of Life Based on The LQI Index" (2019), "A Cross-Country Analysis of The Quality of Life" (2020), "Quality of Life in The BRICS Countries Based on The LQI Index" (2021).

Andrey Korotayev heads the Centre for Stability and Risk Analysis at HSE University in Moscow, Russia. He is also Senior Research Professor at the Institute of Oriental Studies and Institute for African Studies, Russian Academy of Sciences, as well as at the Faculty of Global Studies of Lomonosov Moscow State University and the Russian Presidential Academy of National Economy and Public Administration. He has authored or co-authored over 650 scholarly publications.

Modeling Social Self-Organization and Historical Dynamics. Agrarian Society

Sergey Malkov, Natalia Kovaleva, Leonid Grinin, and Andrey Korotayev

Abstract

This chapter presents the results of applying the set of basic models described in the previous chapter to model the dynamics of an agrarian society. Both interactions in the "society–nature" system and social interactions within society between the main social groups (including economic and political interactions) are considered. It is shown that agrarian society is characterized by cyclical demographic and economic dynamics due to the limited resource base (primarily limited land resources) and the low level of technological development. Based on the modeling of social interactions, the main structural and functional features of a typical agrarian society are revealed. It is shown that in agrarian societies, the mechanisms of social self-organization mainly lead to the formation of social structures of the so-called X-type.

S. Malkov
Lomonosov Moscow State University, Moscow, Russia

HSE University, Moscow, Russia

Institute of Economics, Russian Academy of Sciences, Moscow, Russia
e-mail: s@malkov.org

N. Kovaleva
Soil Science Faculty, Lomonosov Moscow State University, Moscow, Russia

L. Grinin
HSE University, Moscow, Russia

Institute of Oriental Studies, Russian Academy of Sciences, Moscow, Russia

A. Korotayev (✉)
HSE University, Moscow, Russia

Institute for African Studies, Russian Academy of Sciences, Moscow, Russia

Faculty of Global Studies, Lomonosov Moscow State University, Moscow, Russia

V. Sadovnichy et al. (eds.), *Reconsidering the Limits to Growth*, World-Systems Evolution and Global Futures, https://doi.org/10.1007/978-3-031-34999-7_16

1 Modeling of Interactions in the "Society–Nature" System

1.1 Simulation of Dm-Tl-En Dynamics

A. There were numerous and rather diverse agrarian societies in human history (and even at present, there are a few countries in the world where the majority of population is employed in agriculture). What is important for us is the main typological features of agrarian societies in the pre-industrial era. These features include the following:
   - The main type of production activity in agrarian society is agricultural production (agriculture and/or animal husbandry).
   - The main resource R is land, while the area of fertile land is limited.
   - The main production unit is the household: a family that cultivates a plot of land of a certain acreage. Subsistence farming predominates; the main purpose of agricultural households is production for their own consumption.
   - Demographic dynamics is characterized by the relationship shown in Fig. 4a of Sect. 1.3 in Chapter "Modeling Social Self-Organization and Historical Dynamics. A General Approach" (Akaev et al., 2023a, this volume).

   Thus, basic Eqs. (23)–(25)[1] Dm-Tl-En model for a typical agrarian society takes the following form.

A1. Generalized equation of *demographic* dynamics.

Basic Eq. (23),[2] reflecting the nature of the demographic dynamics characteristic of an agrarian society and depicted in Fig. 4a[3] of the previous chapter, can be written as:

$$dN/dt \approx r \cdot N \cdot (1 - x_0/x) \tag{1}$$

In accordance with this equation, if the average level of well-being in society x falls below the critical value x_0, depopulation begins (due to an abnormal increase in mortality). Otherwise, when $x > x_0$, the population grows (primarily due to a decrease in child mortality against the background of high fertility).

A2. Generalized equation of *economic* dynamics.

Agrarian society mostly produces agricultural products intended for consumption (primarily food). Non-food products (I_2) represent a small share of the total output. Accordingly, in the first approximation, the value of I_2 can be neglected, and we can

[1]From the previous Chapter "Modeling Social Self-Organization and Historical Dynamics. A General Approach" (Akaev et al., 2023, this volume).

[2]From the previous chapter.

[3]From the previous Chapter "Modeling Social Self-Organization and Historical Dynamics. A General Approach" (Akaev et al., 2023, this volume).

assume that the main contribution to the value of I_1 is made by the sowing fund (i.e., the part of the crop reserved by the peasants for sowing next year).[4] So, Eq. (24)[5] takes the form:

$$dx/dt \approx (F(R^-, I_1) - I_1)/N - Q \approx (F(R^-, I_1) - I_1)/N - q \cdot x, \tag{2}$$

where $q{\cdot}x$ is consumption (it is assumed that consumption is proportional to savings: the more material goods quantity x, the greater their consumption; q is a proportionality coefficient).

When analyzing economic dynamics in an agrarian society, it is necessary to consider situations where (1) there are no resource constraints (resources are available in abundance) and (2) resources are in short supply.

In the first case, the population density is low, land[6] is not scarce. Total output (collected yield) depends on the number of production units (households), their labor input, the fertility of the land, and the average area of land cultivated by one household. Accordingly, in this case

$$F = A_R(S) \cdot V = \gamma \cdot R' \cdot N, \tag{3}$$

where R' is the average area of cultivated land per capita, and γ is the productivity of land.

In the second case, the population density is high, land is scarce. The total production (collected yield) depends only on the total area of cultivated land and the productivity of the land. Accordingly, in this case

$$F = A_R(S) \cdot V = \gamma \cdot R_O, \tag{4}$$

where R_O is the total area of cultivated land.

Accordingly, in the first case, Eq. (2) takes the form:

$$dx/dt \approx \gamma' \cdot R' - q \cdot x, \tag{5}$$

where γ' is the yield minus the seed fund. In this case, the welfare level exceeds x_O and is equal to $x_1 = \gamma' {\cdot} R'/q$.

In the second case, Eq. (2) takes the form:

$$dx/dt \approx \gamma' \cdot R_0/N - q \cdot x. \tag{6}$$

It can be seen that in the first case per capita incomes are constant, whereas in the second case, they decrease as the population grows. The critical parameter (order

[4] In this case, we neglect the volume of livestock production.

[5] From the previous chapter.

[6] We assume that in the absence of land shortage, the area of land that a household cultivates is determined by the needs and physical capabilities of the household.

parameter) of the transition from regime (5) to regime (6) and vice versa is the population in the society under consideration: if $N < R_0/R'$, then regime (5) is implemented; otherwise, we deal with regime (6). While in the first case, per capita incomes are constant; in the second case, they decrease as the population grows. The critical parameter (order parameter) of the transition from regime (5) to regime (6) and vice versa is the population in the society under consideration: if $N < R_0/R'$, regime (5) is implemented, and regime (6) is in the vice versa case.

A3. Generalized equation of *technological* dynamics.

In the basic model, it is assumed that production is carried out on the basis of traditional technologies, and the increase in labor productivity (if any) is very slow. This allows us to assume that[7]:

$$dS/dt \approx c \cdot N \cdot (S - S_0) \approx 0 \tag{7}$$

due to the low value of the coefficient $\boldsymbol{c}$ (see Sect. 2.2 about this). Since technologies in an agrarian society mainly affect the values of γ and R', these values in typical cases can be considered constant.

B. In accordance with (1)–(7), the phase portrait of a typical agrarian society has the following form (Fig. 1):

Region I in the phase portrait (when $N < R_0/R'$) corresponds to the situation where there are no restrictions on land resources. Region II in the phase portrait (when $N > R_0/R'$) corresponds to a situation where land resources are scarce. In area I, the equilibrium value of x_1 exceeds the critical level of survival x_0; therefore, in accordance with (1), the population is growing. However, in this case, the transition to area II is inevitable, where the system has a stable equilibrium state (attractor) with coordinates $x' = x_0$ (which means that this is a low-income society) and $N' = \gamma' \cdot R_0/(x_0 \cdot q)$ (N' *is* the demographic capacity of the territory[8]). In region II, an increase in N leads to a decrease in x and vice versa (that is, N and x are inversely proportional). This leads to a "Malthusian trap"[9] (which, among other things, means that a population exceeding the demographic capacity N' of the territory cannot live

[7] Slow technological growth is a characteristic feature of agrarian society; thus, we can assume that $dS/dt \approx 0$ in the basic model when considering typical cases. However, in some historical periods technological growth accelerated, which also entailed serious demographic and social changes. These situations are considered in Chapter "Modeling Social Self-Organization and Historical Dynamics. Global Phase Transitions" (Malkov et al., 2023, this volume).

[8] At the same time, it should be borne in mind that the demographic capacity of the territory can grow (and the attractor can shift towards higher values of N) if land productivity (γ) increases.

[9] On the issues connected with the "Malthusian trap" see, e.g., Malthus, 1798; Artzrouni & Komlos, 1985; Steinmann & Komlos, 1988; Kögel & Prskawetz, 2001; Komlos & Artzrouni, 1990; Steinmann et al., 1998; Wood, 1998; Korotayev et al., 2011, 2014; Zinkina & Korotayev, 2014a, 2014b; Korotayev et al., 2016; Korotayev & Zinkina, 2014, 2015, 2022.

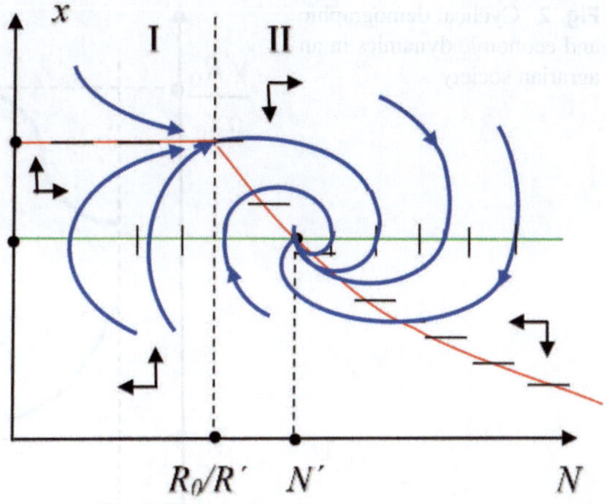

Fig. 1 Phase portrait of a typical agrarian society (blue arrows are phase trajectories, green and red lines are isoclines)

in this territory) and to sociodemographic cycles.[10] The main reason for sociodemographic cycles is that with unchanged resource base, demographic growth is accompanied by a decrease in the well-being x of most of the population up to the level of physical survival x_0, as a result of which social tension sharply increases, which ultimately leads to a social explosion,[11] civil war,[12] and state breakdown, which was usually accompanied by epidemics, catastrophic famines, barbarian invasions, and so on; and the final result was a full-scale sociodemographic collapse (e.g., Korotayev et al., 2006; Malkov, 2009; Turchin & Nefedov, 2009; Nefedov, 2014a, 2014b; Korotayev, 2017). The main result of sociodemographic collapses from the demographic point of view is the rapid decline in the population to a level below N', which in turn leads to an increase in x above the critical level x_0 and the beginning of a new sociodemographic cycle.

The sociodemographic cycles generated by the model for the case when civil wars (and what accompanies them) serve as a mechanism of rapid population decline are shown in Fig. 2 (arrows indicate a sharp decline in population as a result of sociodemographic collapses). The phase portrait of corresponding sociodemographic cycles is shown in Fig. 3 (arrow 1 shows a rapid decline in population as a result of the sociodemographic collapse, arrow 2 shows an increase in the welfare of the population in the post-collapse period as a result of an increase

[10] On the pre-modern and early modern sociodemographic cycles see, e.g., Usher, 1989; Chu & Lee, 1994; Komlos & Nefedov, 2002; Nefedov, 2004, 2013, 2014a, 2014b; Korotayev & Komarova, 2004; Korotayev & Khaltourina, 2006; Turchin & Korotayev, 2006; Turchin & Nefedov, 2009; Malkov, 2009; Korotayev et al., 2006, 2007, 2011; Lee & Zhang, 2010; Grinin & Korotayev, 2012; Korotayev et al., 2016; Goldstone, 2016; Korotayev & Zinkina, 2015; Korotayev, 2017; Turchin, 2018.

[11] On social explosion as a form of socio-political destabilization of the social systems, see Sect. 2.1.

[12] An alternative way to keep the value of x at a level above x_0 in an agrarian society was to continuously expand the area of the society R_0 (through migration or wars) or artificially limit population growth (for example, by infanticide, the practice of late marriages, etc.).

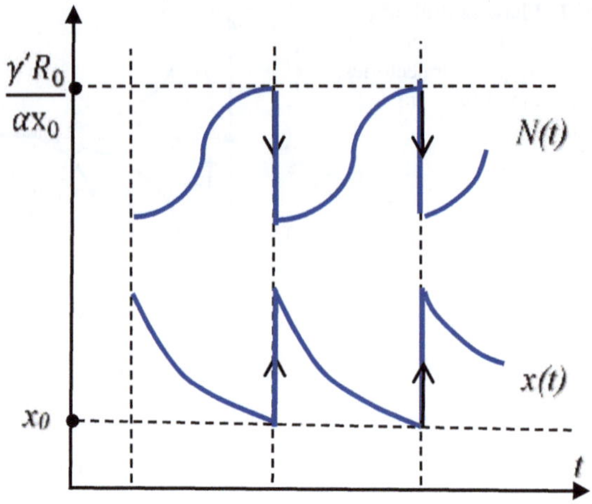

Fig. 2 Cyclical demographic and economic dynamics in an agrarian society

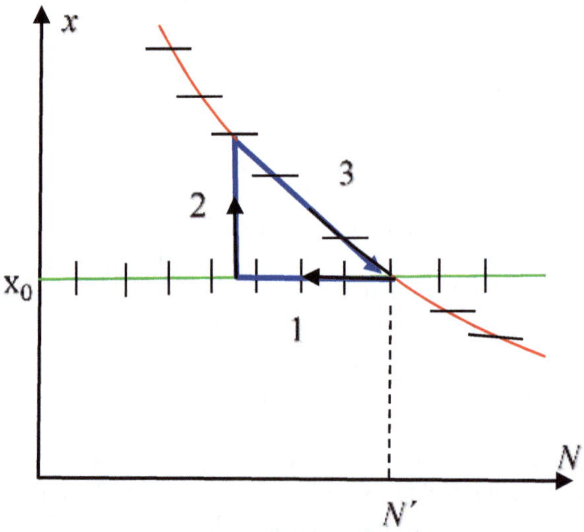

Fig. 3 Phase portrait of demographic cycles in an agrarian society

in per capita land resources, arrow 3 shows slow demographic growth accompanied by a decrease in land allotments per capita). Figure 4 shows historical data on the demographic dynamics of China, reflecting the presence of visible sociodemographic cycles (closely correlating in the Chinese case with dynastic cycles [e.g., Usher, 1989; Chu & Lee, 1994; Korotayev et al., 2006; Lee & Zhang, 2010; Turchin, 2018]).

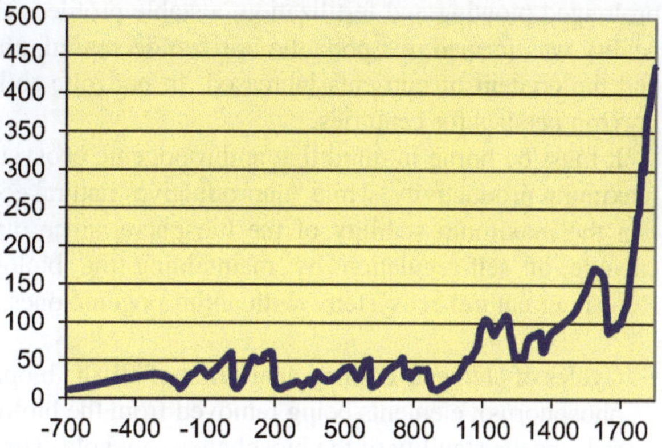

Fig. 4 Demographic dynamics in China (million people) since the seventh century BCE to the mid-nineteenth century CE. *Data source* Korotayev et al., 2006

1.2 Simulation of En-El-Na Dynamics

In the agrarian society era, the impact of human activity on the natural environment was rather slow and relatively weak, while the natural environment (and, in particular, the climate) changed according to its internal laws and also quite slowly throughout human history.[13] Therefore, in the basic model, the characteristics of the natural environment can be considered as slowly changing or constant parameters in relation to the demographic and economic variables.

In an agrarian society, land is the main resource. The change in the amount of this resource is given by Eq. (30).[14] In economic equations, the quality of this resource is reflected by the parameter γ (land productivity), which can change over time due to climate change and human activity. Eq. (15)[15] can be used to analyze changes induced by human activity.

The quality of the resource can be increased, for example, through the introduction of irrigation, agricultural improvements (e.g., a three-field system), technical innovations (e.g., wheeled plough) or fertilization (on such improvements, see, e.g., Korotayev & Grinin, 2013; Grinin & Grinin, 2015, 2016). With the help of agricultural technologies, barren and unproductive soils were transformed into highly productive land. For example, rice fields appeared on the site of floodplains, mangroves, marches; irrigated lands emerged in deserts; drained massifs took the place of swamps. The influence of human agricultural activity in the taiga zone over many centuries has led to the appearance of the humus-accumulative horizon, a new horizon in podzolic soils. With slash-and-burn farming systems, this horizon quickly disappeared after the arable land was abandoned and overgrown with forest. With

[13]See Chapter "Ecology. Life in the "Unstable Biosphere" (Kovaleva, 2023, this volume) and Chapter "Technology. Limitless Possibilities, Effective Control" (Grinin & Grinin, 2023, this volume).

[14]From the previous Chapter "Modeling Social Self-Organization and Historical Dynamics. A General Approach" (Akaev et al., 2023, this volume).

[15]From the previous chapter.

prolonged plowing and fertilization, a stable profile with a humus horizon and low acidity was formed, the podzolic horizon decreased, physical properties improved, and the content of nutrients increased. In podzolic soils of secondary forests, this horizon persists for centuries.

It must be borne in mind that anthropogenic ecosystems are aimed at achieving maximum productivity, while "unproductive" natural ecosystems are tuned to maintain the maximum stability of the biosphere, since, unlike agrocenoses, they are capable of self-regulation by maintaining the biological cycle of substances. Replacing natural ecosystems with anthropogenic ones leads to

- Cycles of elements coming apart, first of all, the biophilic ones (carbon, nitrogen, phosphorus); elements being removed from the biological cycles, which, in turn, reduces the stability of the biosphere as a whole. The introduction of mineral and organic fertilizers can only partially correct the situation, since out of the total amount of nitrogen introduced into the soil, agricultural plants use only 40–50%, the rest is washed out in the form of nitrates, polluting water bodies, or volatilizes into the atmosphere, polluting it with nitrogen oxides. A significant part of phosphate fertilizers is also washed into fresh water, causing eutrophication of water bodies. The introduction of large amounts of nitrogen fertilizers with a lack of phosphorus leads to poisoning of the products, the transformation of durum wheat varieties into soft ones, etc.
- A large variety of plant and animal species being replaced with a small number of cultivated plant species and animals adapted to agrocenoses.
- A sharp decrease in the resistance of plant communities to adverse environmental influences (outbreaks of infections, pests, droughts, frosts, etc.) as a result of the replacement of multi-species cenoses by mono-species ones;
- Changes in the hydrological and radiation regimes of large areas.
- Pollution of the atmosphere, pedosphere, hydrosphere by decomposition products of fertilizers, plant protection products, etc., eutrophication of water bodies.

The turning point in the evolution of the biosphere is associated with the emergence of agriculture in the Mesolithic and Neolithic (from 12,000 to 9000 years ago), which resulted in a significant transformation of the Earth's biosphere, leading to the replacement of primary natural ecosystems by agroecosystems. After the abandonment of depleted lands and the migration of the population, the vegetation cover was restored in some territories, but the soil (being a non-renewable resource) cannot return to its original state due to the irreversibility of many processes. Of the 2 billion hectares of land degraded during the existence of the humankind, 1.3 billion were made unsuitable for agriculture before 1700.

A major role in soil degradation was played by salinization processes due to high salinity of irrigation water on watersheds and/or due to rising groundwater levels in floodplains and deltas. Irrigation system of the Sumerians, built between the Tigris and Euphrates rivers in the second half of the fourth millennium BC, began to suffer from silting and salinization after 1000–1500 years of its use.[16] The irrigation systems of ancient Lebanon built by the Phoenicians in 1700 BCE were completely silted up. The irrigation systems of the pre-Inca culture of Peru built in 1000–500 BCE covered an area twice as large as modern ones, and fell into disrepair due to secondary salinization and silting of canals. The immediate causes of the decline of irrigation systems in the agrarian era included silting of distribution canals, silting of irrigated fields, rising groundwater, secondary soil salinization, wind and water erosion, destruction of terraces on slopes, and erosion of water supply channels (see, e.g., Korotayev, 1995a, 1995b, 2020a, 2021 with respect to South Arabia).

Thus, in an agrarian society, human activity influenced the natural environment; however, this influence proceeded rather slowly and was mainly of a localized nature (occurred in limited geographical areas of intensive agriculture and cattle breeding). More significant was the reverse effect of natural (primarily climatic) changes on human society. An example of this is data on a high degree of correlation between the dynamics of the Earth's population in the agrarian era and climate change. A large amount of data on the impact of climate change on human history is presented by Klimenko (2009) and other scholars (e.g., Korotayev et al., 1999; Lee, 2007; Zhang et al., 2007; Lee & Zhang, 2010; Parker, 2013; Thompson & Zakhirova, 2021; Fleitmann et al., 2022).

2 Stability of Social Interactions

2.1 Socio-Economic Stability

A. The model of interactions within and outside the economy of the main social groups in an agrarian society.

The key equations for the mathematical description of socio-economic processes in an agrarian society are Eqs. (21) from Sect. 1.3 of Chapter "Modeling Social Self-Organization and Historical Dynamics. A General Approach" (Akaev et al., 2023a, this volume) adapted to the historical era under consideration.

A.1) Formalization of basic economic processes. The main social groups in the type of agrarian society that we model are peasants and landowners.[17] Landowners are

[16] About the Agrarian Revolution and its turning points see Chapter "Technology. Limitless Possibilities, Effective Control" (Grinin & Grinin, 2023, this volume), as well as Grinin & Grinin, 2015, 2016.

[17] Actually, this chapter offers a model of a Western European feudal society, in which the role of the state and officials was weaker than in Asian, especially East Asian societies. This was one of the important reasons why the transition to a capitalist society happened in Europe (see Grinin & Korotayev, 2015). We have opted to concentrate on this type of agrarian societies that was typical for pre-modern Western Europe, because the first independent transition to industrial society took

owners of the land and have political power. Their income is constituted by rents (tire, corvée, etc.) from peasants dependent on them. These rents are spent on their own consumption and on administration (expenses on keeping peasants in obedience, on the means of coercion). Peasants are agricultural producers, dependent on landowners and forced to pay rent to them. Their income is the product of their labor, one part of which is spent on consumption, another part on paying rents, and the third part on ensuring production in the next time cycle (investment).

The basic economic processes that reflect the economic interaction of these social groups, in accordance with the methodological approach set out in Sect. 3 of Chapter "Modeling Social Self-Organization and Historical Dynamics. A General Approach" (Akaev et al., 2023a, this volume), can be described using a system of differential equations that reflect the dynamics of material (or cash[18]) flows in the system (see the basic model of socio-economic interactions (33) from Sect. 3 in Chapter "Modeling Social Self-Organization and Historical Dynamics. A General Approach" [Akaev et al., 2023a, this volume]). In a generalized form, these equations have the following form for a society consisting of landowners and peasants dependent on them.[19]

The equation for the dynamics of the material resources of peasants $X_1(t)$ has the following form:

$$dX_1(t)/dt = (produced\ products) - (consumption) - (investment) - (rents) \quad (8)$$

The equation for the dynamics of material resources of landowners $X_2(t)$ has the form:

$$dX_2(t)/dt = (rents\ collected) - (consumption) - (management\ costs) \quad (9)$$

or

$$\begin{cases} \frac{dX_1}{dt} = F - Q_1 - T - I \\ \frac{dX_2}{dt} = T - Q_2 - C \end{cases} \quad (10)$$

place in this part of the world, and we needed such a model for our further mathematical investigation of the spontaneous phase transition from the agrarian to industrial society. Thus, the specified mathematical model of agrarian society is not applicable to most other agrarian societies that were extremely diverse in their organization. For example, it is not applicable, to pre-modern agrarian societies of the Middle East, India, China, let alone Sub-Saharan Africa, New Guinea, or Pre-Columbian Americas. Of course, the Western European model is also very simplified.

[18] Accounting for the purchasing power of money, that is, taking into account their depreciation in the presence of inflation.

[19] To simplify the description, it is assumed that on time scale under consideration, the change in the population size can be neglected.

where F is the production function, Q_1 is the consumption expenditures of the peasants, Q_2 is the consumption expenditures of the landowners, T is rents, I is investment, and C is management costs. Agricultural production in an agrarian society depends on the land productivity and the size of the cultivated land. We will assume that these factors change little over time; therefore, in the first approximation, we can assume that F is a constant value. Also, in model (10), population is a constant, that is, $N_1 = const$, $N_2 = const$, where N_1 and N_2 are the populations of peasants and landowners, respectively.

Under fairly general assumptions about the nature of the dependence of the functions on the right sides of the equations on the variables X_1 and X_2, system (10) can be written as follows:

$$\begin{cases} \dfrac{dX_1}{dt} = F' - \alpha_1 \cdot X_1 - \beta \cdot F' \\ \dfrac{dX_2}{dt} = \beta \cdot F' - \alpha_2 \cdot X_2 - \dfrac{\gamma}{X_1 - X_{01}} \end{cases} \tag{11}$$

where F' is the modified production function (accounting for the investment costs, that is, $F' = F\text{-}I$); β is the rent coefficient (characterizing the share of rents in the value of F'); α_1, α_2, and γ are coefficients. The last term in the second equation accounts for the increase in management costs when social instability rises. We assume that social instability increases as the level of welfare of peasants $x_1 = X_1/N_1$ decreases and approaches the level of physical survival x_{01}. With an increase in X_1, social instability decreases).

A.2) Basic contradictions. The basic contradictions are economic (arising primarily over the distribution of the output) and social (over the rights to this distribution). In model (11), these are contradictions over the amount of rents, that is, the value of β, the share of rents in the output. Peasants want a decrease in the value of β, while landowners want it to increase. This divergence of interests leads to the fact that landowners make attempts to increase rents from time to time, and peasants fight for their reduction (in this case, the initiative in changing β, as a rule, belongs to landowners). As a result of social interaction (confrontation), a temporary equilibrium is established (and so is the value of β corresponding to it), fixing the achieved compromise for the period until the next change in the situation.

A.3) The current state of the social system, and the mechanisms for its maintenance and stabilization.

If (and when) a compromise on β is reached, the phase portrait of the system (11) has the form shown in Fig. 5, where $X_{01} = x_{01} \cdot N_1$.

We can infer from the figure that in an agrarian society with a fixed β, there is a state of stable equilibrium. The equilibrium values X_1 and X_2 (corresponding to the coordinates of attractor point **A** in Fig. 5) are

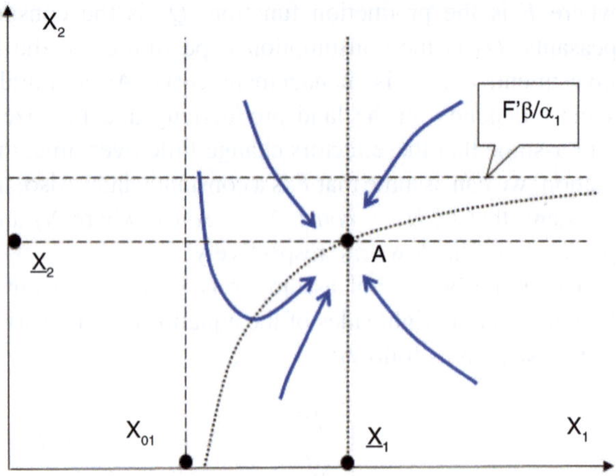

Fig. 5 Phase portrait of the system of Eqs. (11) describing the social system of an agrarian type (arrows are phase trajectories, dotted lines are isoclines, point **A** at the intersection of isoclines is a stable state of equilibrium)

$$\begin{cases} \underline{X}_1 = \dfrac{F' \cdot (1-\beta)}{\alpha_1} \\ \underline{X}_2 = \dfrac{1}{\alpha_2}\left(F' \cdot \beta - \dfrac{\gamma \cdot \alpha_1}{F' \cdot (1-\beta) - \alpha_1 \cdot X_{01}}\right) \end{cases} \quad (12)$$

It follows from (12) that at low values of β, the equilibrium shifts in favor of peasants, and at high values of β, it shifts in favor of landowners. At the same time, very high values of β are unfavorable for landowners, since in this case X_1 decreases to the level of physical survival of X_{01}, which entails a sharp increase in social instability due to the growing discontent of peasants. This, in turn, leads to an increase in the costs of landowners to maintain controllability, including the application of force, which devalues the increase in the value of β. In this regard, there is an optimal value of β for landowners (in terms of achieving the maximum value of X_2 in Eq. (12)), which is equal to:

$$\beta_{opt} = 1 - \frac{(\gamma \cdot \alpha_1)^{0,5} + X_{01} \cdot \alpha_1}{F'} \quad (13)$$

The maximum value of X_2 in this case is equal to:

$$\underline{X}_2 = \frac{1}{\alpha_2}\left(F' - 2 \cdot (\gamma \cdot \alpha_1)^{0,5} - X_{01} \cdot \alpha_1\right) \quad (14)$$

Thus, the current state of the social system is established as a compromise between peasants and landowners over the value of β (that is, over the amount of rents). At the same time, landowners are forced to bear the costs of ensuring manageability, such as forcing the peasants to pay rents (the last term of the second equation in model (11)). A reduction of these costs largely depends on the socio-psychological factors which are taken into account in model (11) through indicator γ. It can decrease if the peasants are ready to pay rents. This consent is provided:

- Either with a threat of severe punishment for disobedience (punitive measures that cause fear).[20] The key factor in the psychological impact of landowners on peasants in this case is utilitarian (based on rational analysis, pragmatism) component a^G (for the utilitarian component of psychological impact, see Ilyin et al., 2020).
- Through belief in the legitimacy of rents (based on tradition and/or sacredness of power, etc.). The key factor in the psychological impact of landowners on peasants in this case is deontological (based on ideas about morality, duty, and so on) component a^F (for the deontological component of the psychological impact, see Ilyin et al., 2020).

The second of these factors (the deontological one) is extremely important for agrarian societies. It ensures the reduction of social conflict and the voluntary payment of rents, although from the purely utilitarian viewpoint peasants are interested in their complete absence (their ideal is $\beta = 0$). The importance of the deontological factor for achieving social compromise determines the importance of ideology (religion, moral norms, traditions, etc.) in agrarian societies (Baskin, 2020). Ideology (religion, tradition) is the most important bond of an agrarian society.

A.4) Possible reasons for changes in the current state.

As already noted, the current values of β and γ are set on the basis of social compromise. As long as the compromise is maintained, there is a state of equilibrium in the society that is resistant to local perturbations (see the stable attractor in Fig. 5). This ensures the socio-political stability of agrarian societies.

The reasons for changing the current equilibrium (current values of β and γ) can be:

- Natural disasters leading to a decrease in F (in this case, the equilibrium value $X_1 = F' \cdot (1-\beta)/\alpha_1$ may decrease to the level of physical survival X_{01}, which forces the peasants to demand a reduction in rents, that is, the value of β);
- Demographic overproduction of the elites (landowners) (in this case, with constant rent revenues, the individual consumption of the representatives of the elites falls, and they tend to increase the rent burden on the peasants, that is, increase in β);
- Excessive decline of per capita productivity (as a result of an excessive increase in the number of peasants) (in this case, the individual consumption of peasants is reduced to the level of physical survival, which forces them to demand a reduction in rents, that is, a decrease in the value of β);

[20] This method is used by dictatorial regimes or invaders.

- a decrease in the legitimacy of power – for example, as a result of the interruption of the ruling dynasty or the violent usurpation of power (in this case, the deontological factor a^F decreases and, accordingly, the value of γ increases);
- the emergence of a common external threat – for example, the threat of a foreign invasion (in this case, in the face of a common enemy, the deontological factor a^F increases and, accordingly, the value of γ decreases).

A.5) Assessment of the social system destabilization risk.

If factors (causes) arise that violate the previously existing balance, the following development of events is possible:

- Either a transition to new values of β and γ (and to new equilibrium values of X_1 and X_2) on the basis of a new social compromise (this requires changes in the sphere of governance and/or socio-psychological sphere).
- Or social system destabilization, since a change in β and γ leads to X_1 or X_2 going beyond the range of permissible values,[21] as a result of which the stable attractor shown in Fig. 5 disappears and the system ceases to exist in its original form. In real life, this is expressed in the form of a civil (peasant) war/uprising/rebellion, leading to a change in the social and demographic composition of society (in which a new state of equilibrium is established after the end of war).

In this case, the transition from a stable state to destabilization occurs in a non-linear manner due to the presence of positive feedback in the destabilization processes. For example, in the case of the demographic overproduction of landowning elites, their desire to raise rents (that is, increase β in order not to reduce their per capita consumption) leads to a decrease in the deontological component a^F of the socio-psychological factor of stability (since the authorities violated traditions and thereby lost legitimacy), which leads to an increase in the value of γ (the necessary costs of ensuring controllability). And this, in turn, further reduces the per capita consumption of the elite since it takes away the already lacking resources. Due to this nonlinearity, the transition to destabilization has a threshold character, which must be taken into account in the analysis and modeling.

Mathematical models of these processes are presented in (Nefedov, 2004, 2013, 2014a, 2014b; Korotayev & Komarova, 2004; Korotayev et al., 2006; Gushchina & Malkov, 2014; Malkov, 2009; Grinin et al., 2009; Turchin, 2018). They show that the economic interaction of peasants and landowners, taking into account the influence of long-term demographic factors, ultimately leads to social destabilization and cyclical demographic and economic dynamics (Fig. 6).

[21] Permissible values mean the level of physical survival for peasants, and the level of "decent" existence for landowners, which tend to have different values in different societies.

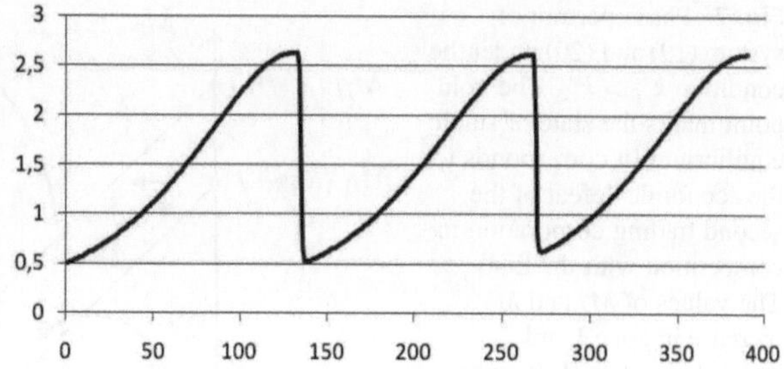

Fig. 6 Typical dynamics of demographic characteristics in accordance with the model (Gushchina & Malkov, 2014) that describes socio-economic interactions and demographic processes in agrarian societies (x-axis is years, y-axis is population, relative units)

B. Model of trade competition in an agrarian society.

In an agrarian society, subsistence farming dominates. Competitive market relations and the struggle for markets take place essentially only in the segment of international trade. At the same time, trade was mainly in elite goods, since only the elite possessed significant funds and could create demand for overseas goods. An example of trade competition in the Middle Ages is the rivalry between the Republics of Genoa and the Venice.

The basic model of economic competition is given in Chapter "Modeling Social Self-Organization and Historical Dynamics. A General Approach" (Akaev et al., 2023a, this volume) in Sect. 3.1.2. As applied to the case of competition between two competing trading corporations in a certain region (state), the equations of the basic model can be written as

$$dM_1/dt = k_1 \cdot M_H - h_1 \cdot F_1 = k'_1 \cdot (F_1/(F_1 + F_2)) \cdot M_H - h_1 \cdot F_1 \tag{15}$$

$$dM_2/dt = k_2 \cdot M_H - h_2 \cdot F_2 = k'_2 \cdot (F_2/(F_1 + F_2)) \cdot M_H - h_2 \cdot F_2 \tag{16}$$

where M_i is the funds of the *i*-th trading corporation; M_H is the total volume of consumer demand in the region under consideration; k_i is the share of total consumer demand which goes to the *i-th* trading corporation per unit of time from the sale of its goods in the given region; F_i is the volume of goods sold by the *i-th* trading corporation; h_i is the coefficient of proportionality (it is assumed that the expenses of the *i-th* actor are proportional to the volume of trade turnover).

Since sales revenues are proportional to the volumes of goods sold, then, assuming that the consumer properties of the goods of both corporations are the same, from (15) to (16), we obtain:

$$dM_1/dt = k'_1 \cdot (F_1/(F_1 + F_2)) \cdot M_H - h_1 \cdot F_1 \tag{17}$$

$$dM_2/dt = k'_2 \cdot (F_2/(F_1 + F_2)) \cdot M_H - h_2 \cdot F_2 \tag{18}$$

where k'_1 and k'_2 are coefficients reflecting the influence of non-economic factors on the amount of income.

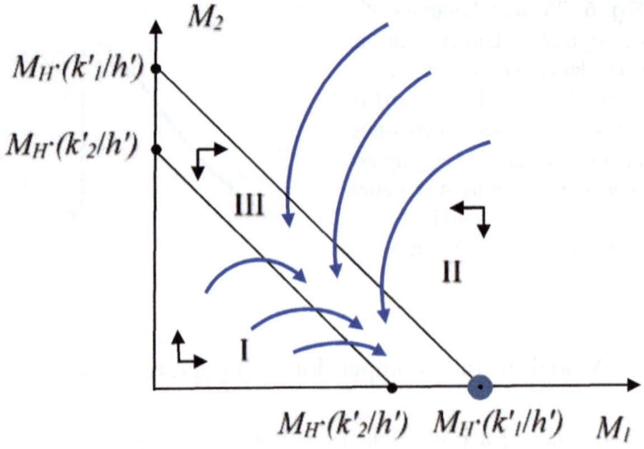

Fig. 7 Phase portrait of system (19) and (20) under the condition $k'_1 > k'_2$. The bold point marks the state of stable equilibrium (it corresponds to the economic defeat of the second trading corporation in competition with the first). The values of M_1 and M_2 increase in zone I and decrease in zone II. In zone III, the value of M_1 increases, and the value of M_2 decreases

In economic terms, let the specific (per unit of goods) costs of both corporations be the same ($h_1 = h_2 = h$), and the quantity of the offered goods F_i is proportional to M_i (that is, the volume of trade is proportional to the financial power of corporations: $F_i \sim M_i$). Then from (17) and (18) we get

$$dM_1/dt = k'_1 \cdot (\mathrm{M}_1/(\mathrm{M}_1 + \mathrm{M}_2)) \cdot M_H - h' \cdot \mathrm{M}_1 \quad (19)$$

$$dM_2/dt = k'_2 \cdot (\mathrm{M}_2/(\mathrm{M}_1 + \mathrm{M}_2)) \cdot M_H - h' \cdot \mathrm{M}_2 \quad (20)$$

System (19)–(20) does not have a stable state of equilibrium. The expressions for the isoclines of the system (19) and (20) are

$$M_2 = (k'_1/h') \cdot M_H - M_1 \quad (21)$$

$$M_2 = (k'_2/h') \cdot M_H - M_1 \quad (22)$$

For $k'_1 = k'_2$ the isoclines coincide, and for $k'_1 \neq k'_2$ they run parallel. The phase portrait of the system (19)–(20) for $k'_1 > k'_2$ is shown in Fig. 7.

It can be seen from the phase portrait that competition leads to the complete exclusion of one of the competitors from the region under consideration. Other things being equal, the non-economic factors of obtaining advantages (bribery, raider seizures, forceful methods, etc.) play a decisive role. Indeed, in the era of agrarian societies, trade competition often had a forceful character (an example is the repeated wars in the 13th–15th centuries between the Genoese and Venetian merchant republics) and essentially boiled down to the delimitation of zones of influence, within which the monopoly of one of the parties was established.

It makes sense to finish the section with Table 1 ("Demographic changes during the agrarian-handicraft principle of production[22]). Although the above demographic

[22]For agricultural and handicraft production and agricultural societies in general, see, e.g., Grinin & Korotayev, 2015; Grinin et al., 2020; see also Chapter "Technology. Limitless Possibilities, Effective Control" (Grinin & Grinin, 2023, this volume).

Table 1 Demographic Changes during the Agricultural-Craft Mode of Production

Stage/ Phase	Dates	Techno-social change	Demographic changes	Restrictions	The population of the Earth at the beginning of the period
The initial phase of the agrarian revolution	12–9 thousand years ago	The origin of agriculture and animal husbandry in some areas of the Middle East. Expansion of the ecological niche	The beginning of the formation of a new type of population reproduction with an increased birth rate	Very narrow territorial base of a new ecological niche	8–ten million[a]
Modernization phase of the agrarian revolution	9–5.5 thousand years ago	The development of agriculture and its diffusion	Formation of a new type of population reproduction with a high birth rate, but with a limited ecological niche	The limited ecological niche and the growth of conflict due to the complexity of societies	50 million[b]
The final phase of the agrarian revolution	5.5–3 thousand years ago	Transition to intensive farming and a pronounced expansion of the ecological niche	Development of a new type of population reproduction with a high birth rate already on the basis of a capacious ecological niche; rapid population growth in some places	Narrow territorial base of a new ecological niche in the valleys of large rivers	
Mature stages of the agrarian-handicraft principle of production	3 thousand years ago - 1430	Creating political conditions for the intensification of agriculture	Powerful development of a new type of population reproduction with a high birth rate based on a capacious ecological niche, greatly expanded geographically. Together with the influence of the state on the growth of production, internal peace and order, this leads to a strong population growth.	Malthusian trap	About 200–300 million at the beginning of our era; 400–500 million in the fifteenth century.[c]

[a] 10–15 million around 9–10 thousand ybp (Kapitza, 2006: 19; Kapitza, 2009: 38); see also Korotayev, 2020a, 2020b: 589–591.
[b] Valentey, 1974: 236; Vishnevsky, 1973: 31; Korotayev, 2020a, 2020b: 589–591.
[c] Estimates for 1 CE range from 170 (McEvedy & Jones, 1978) to 250 (Arab-Ogly, 1978: 88; Livi-Bacci, 2012: 25, table 1.3) and 300–330 million people (Durand, 1974; Shcherbakova, 2019; Akimov, 1999), and sometimes up to 400 million (see, e.g., Korotayev et al., 2007: 114); see also Korotayev, 2020a, 2020b: 589–591.

parameters in agrarian society were taken as constant, but in fact, of course, an already highly developed agrarian-handicraft society was modeled above, which in some ways was even more sophisticated in terms of development than the average, because its features became most prominent already in the period of late feudalism. Naturally, early farmers have not been taken into account in the modeling, although in terms of the duration of the periods, most of the period of agrarian society falls precisely on the societies of early farmers and pastoralists. The table below partially reflects the fact that it was not possible to show in the chapter, the fact of slow but still significant evolution of agrarian society in the long term. The table will also better prepare the readers for the next chapter "Modeling Social Self-Organization and Historical Dynamics. Industrial Society" (Akaev et al., 2023b, in this volume).

2.2 Socio-Political Stability

Since land is the main and most important resource in an agrarian society, the competitive struggle of political actors primarily aims at possessing more territory. At the same time, land is a limited resource; accordingly, the struggle takes on the form of a "zero-sum game".[23] In this case, as is shown by the basic model of political competition (see Sect. 3.2.2 of Chapter "Modeling Social Self-Organization and Historical Dynamics. A General Approach" [Akaev et al., 2023a, this volume]), a typical[24] situation is the establishment of political monopoly (see Fig. 17 in the previous chapter[25]) or hierarchical (vassal) relations within the political elite (see Fig. 18 in the previous chapter), which in real history often took the form of hereditary monarchies.

With a monarchical form of government and the power monopoly of the ruling elite, a directive governance system is established (see Fig. 12 from the previous chapter[26]). As is shown in Sect. 3.2.1 of Chapter "Modeling Social Self-Organization and Historical Dynamics. A General Approach" (Akaev et al., 2023a, this volume), this governance system is stable in the sense that deviations from it (for example, towards decentralization) tend to increase the synchronization

[23] "Zero-sum game" (antagonistic game) is a game theory term. Antagonistic game is a non-cooperative game in which two or more players participate, and the gain of one player is equal to the loss of the other.

[24] In some cases, when the strength of political competitors is approximately equal, an internally competitive regime of the "alliance of the weak against the strong" can be established (see Fig. 19 in Sect. 3.2.2 of Chapter "Modeling Social Self-Organization and Historical Dynamics. A General Approach" [Akaev et al., 2023, this volume]). Such a regime breeds fragmentation, and inability to unite against external enemies. Sooner or later, a strong external enemy appears, who conquers such a social system using the internal contradictions in it (see, for example, the conquest of Russian principalities by the Mongols, or the partitions of Poland in the eighteenth century)

[25] Section 3.2.2 of Chapter "Modeling Social Self-Organization and Historical Dynamics. A General Approach" (Akaev et al., 2023, this volume).

[26] Section 3.2.1 of Chapter "Modeling Social Self-Organization and Historical Dynamics. A General Approach" (Akaev et al., 2023, this volume).

threshold and reduce the controllability of the social system, so the ruling elite tends to restore a strong power vertical. This self-consistency is responsible for the historical persistence of centralized directive systems, which, in addition to monarchies, include various forms of authoritarian rule.

At the same time, political monopoly has serious disadvantages:

- The elite does not produce an internal demand for innovation, as it seeks to gain advantages not through innovative activity but through the imposition of duties on the population (taxes, rents, corvée, dues, etc.).
- The population also has no intention to innovate since it is concerned with the problems of survival and minimizing risks and damages, rather than maximizing income, as it is well aware that an increase in income will be noticed by the authorities and will only lead to higher taxes/rents.

Those points lead to a decrease in the value of coefficient c in the equation of technological dynamics (7) and to a slowdown in technological development in the agrarian era.

2.3 Interstate Interactions

In the era of agrarian societies, subsistence farming dominated, and economic ties between states were poorly developed. In the international stage, states acted not as partners but as competitors. From time to time, wars broke out between these states for territory, the main resource of that time (Young, 2022).

Accordingly, the basic model for describing interstate interactions in the agrarian era is the competitive struggle model (see Eqs. (44) from the previous chapter[27]). An agrarian society corresponds to stage I of the dynamics described by the basic model: the shaping, as a result of the interaction between individual ethnic groups, of territorial state entities (clusters) in competitive relations with each other. The simulation results of this process are presented in Fig. 8 (they correspond to frames 1–4 of Fig. 21 from Chapter "Modeling Social Self-Organization and Historical Dynamics. A General Approach" [Akaev et al., 2023a, this volume], Sect. 3.2.3).

In essence, the frames of calculation cycles depicted in Fig. 8 illustrate the process of medieval state formation in Europe in the era of feudal fragmentation. For comparison, Fig. 9 shows the political map of France during the period of feudal fragmentation (in the twelfth century CE).

[27] Section 3.2.3 of Chapter "Modeling Social Self-Organization and Historical Dynamics. A General Approach" (Akaev, et al., 2023, this volume); see also Korotayev et al., 2021, 2022 for a mathematical model of interaction between agrarian empires and their tribal peripheries.

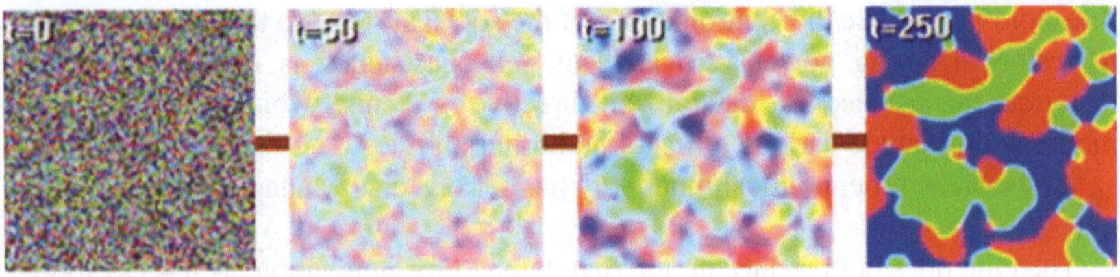

Fig. 8 Dynamics of cluster formation when modeling competitive interactions using model (Eq. (45) from the previous chapter (Chapter "Modeling Social Self-Organization and Historical Dynamics. A General Approach" (Akaev et al., 2023a, this volume).)) (numbers in the upper left corner of each frame corresponding to the serial number of the respective cycle of calculation)

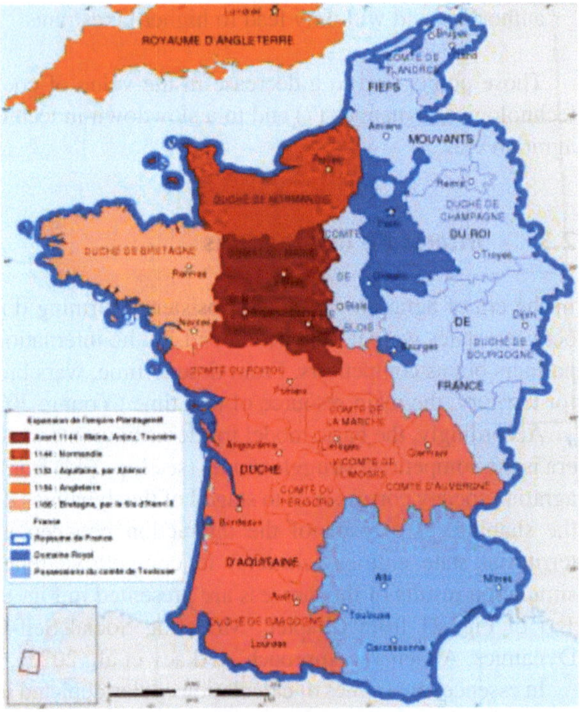

Fig. 9 Political map of France in the twelfth century CE. *Source* Wikimedia Commons (https://commons.wikimedia.org/wiki/File:France_1154-fr.svg)

2.4 Patterns of Social Self-Organization

The peculiarities of social interactions inherent in an agrarian society, which were discussed in Sections 2.1–2.3, allow us to discuss the mechanisms of self-organization of this society, as a result of which social structures of a certain type are formed and sustainably reproduced. Some researchers (Kirdina, 2004, 2014a, 2014b; Malkov, 2009) call them X-type social structures.

In Sect. 1.1 of Chapter "Modeling Social Self-Organization and Historical Dynamics. A General Approach" (Akaev et al., 2023a, this volume), it is shown that in order to ensure the stability of social systems facing external and internal destabilizing influences, three main problems must be solved:

(A) ensuring the possibility of *economic survival* (satisfying the material needs of members of society);
(B) ensuring the *socio-psychological stability* in the presence of antagonistic interests of the elements of the social system (decreasing the level of conflict within the social system);
(C) ensuring *effective governance*. Since the social system is a distributed system, the task of management is to ensure the coordinated functioning of all its subsystems. Efficiency conditions include the ability to manage and the availability of funds necessary for management (on the part of managers), as well as the consent to be ruled in conditions of social stratification (on the part of the governed).

The characteristic features of agrarian societies are the following:

- Resource base is constituted by land and its fertility.
- The volume of the resource is the area of cultivated land.
- The method of processing the resource base is the manual labor of peasants.
- The threshold of the production function F is small and can be overcome as a result of combining the efforts of several peasants in their agricultural work (see Fig. 1, curve 2 in Chapter "Modeling Social Self-Organization and Historical Dynamics. A General Approach" [Akaev et al., 2023a, this volume]);
- The production function F has saturation (see Fig. 1, curve 2 from the previous chapter), the value of which depends on the fertility of land and the technologies used for its cultivation.
- The consumption of peasants is limited mainly to agricultural products;
- The main social strata include the dependent peasants and landowners (elite).
- Problems (A), (B), and (C) in an agrarian society are solved as follows:

(A) The main source of wealth is agricultural production, while land resource is limited. The surplus product of the peasants is forcibly extracted by the elite, as a result of which property is sharply stratified, and a bimodal economic structure of society (ESS) is formed (see Fig. 2a in the previous chapter). This structure is stable due to the limited common resource (everything that is produced is spent on consumption by both peasants and the elite). Gross agricultural production can be increased in the following ways:
    (a) increasing the area of cultivated land by:
        - Plowing forests, draining swamps, artificial irrigation, etc. (restrictions include the limited amount of available reserves of the territory and labor intensity).
        - Conquests (restrictions include the resistance of the enemy and the costs of managing the conquered territories).

(b) Increasing the number of working-age population (restrictions include a decrease in the specific area of cultivated land per family, that is, the income of one family).

(c) Increasing the productivity of land (influence on the production function *F*):

- The threshold of function F can be overcome by peasants uniting their efforts to cultivate the land, that is, by the compatibility (coherence) of individual actions at the micro level of the family and community (as a result, there is demand for mutual assistance and collectivist psychology in an agrarian society).
- An increase in the scale of production is achieved due to the compatibility (coherence) of individual actions at the macro level of public works for melioration, irrigation, etc., organized by the state (as a result, the state is of great economic importance in the river civilizations of the ancient East, such as Egypt, Mesopotamia, China).

  When production and consumption become mismatched, a new equilibrium is achieved due to:
- Reducing consumption (hence the popularity of the ethics of asceticism).
- Decrease in population (due to migration, civil wars, etc.).

(B) Socio-psychological stability of an agrarian society, challenged by sharp property stratification, is ensured by:

- The formation of the corresponding ideological sphere: collectivist morality, religions that sanctify the existing order.
- The development of a system of patronage, clans, castes, implying and encouraging assistance to the poor by the rich.

(C) The effectiveness of governance in an agrarian society is ensured by:

- The development of measures that ensure non-economic coercion (this requires significant costs; therefore, the size of agrarian states is largely determined by the available resources for non-economic management and defense against the neighbors);
- Authoritarian (directive) method of governance (the elite opposes the producers and imposes their will on them).

In agrarian societies, the demographic problem comes to the fore. As the mechanization of labor is low, the economic stability of peasant farms is largely determined by the number of workers. Therefore, peasant families usually have many children. However, after several generations, the focus on large families, which ensures the local economic sustainability of individual farms, leads to a society-wide sociodemographic crisis due to the inevitable decline in the size of the peasants' plots (as the available land fund is limited). The decline in per capita agricultural production results in "crises of underproduction" (in contrast to "crises of overproduction" in industrial societies). This results in social cataclysms, food riots, mass migrations, civil wars, i.e., full-scale sociodemographic collapses.

Acknowledgements This work was done with the support of MSU Program of Development, Project No 23A-SCH05-03.

References

Akaev, A., Malkov, S., Bilyuga, S., Malkov, A., & Korotayev, A. (2023a). Modeling social self-organization and historical dynamics. A general approach. In V. Sadovnichy et al. (Eds.), *Reconsidering the limits to growth. A report to the Russian Association of the Club of Rome* (pp. 253–307). Springer. https://doi.org/10.1007/978-3-031-34999-7_15

Akaev, A., Malkov, S., Davydova, O., Kovaleva, N., Malkov, A., Grinin, L., & Korotayev, A. (2023b). Modeling social self-organization and historical dynamics. Industrial society. In V. Sadovnichy et al. (Eds.), *Reconsidering the limits to growth. A report to the Russian Association of the Club of Rome* (pp. 337–385). Springer. https://doi.org/10.1007/978-3-031-34999-7_17

Akimov, A. (1999). On the threshold of the seventh billion. Population and society. *Information Bulletin of the Center for Human Demography and Ecology of the Institute of Economic Forecasting of the Russian Academy of Sciences, 39*, 47–62.

Arab-Ogly, E. (1978). *Demographic and ecological forecasts. Criticism of modern bourgeois concepts*. Statistics.

Artzrouni, M., & Komlos, J. (1985). Population growth through history and the escape from the Malthusian trap: A homeostatic simulation model. *Genus, 41*(3–4), 21–39.

Baskin, K. (2020). Religion as the process driving social evolution. *Social Evolution & History, 19*(1), 43–72. https://doi.org/10.30884/seh/2020.01.03

Chu, C. Y. C., & Lee, R. D. (1994). Famine, revolt, and the dynastic cycle: Population dynamics in historic China. *Journal of Population Economics, 7*(4), 351–378.

Durand, J. D. (1974). *Historical estimates of world population: An evaluation*. University of Pennsylvania Population Center.

Fleitmann, D., Haldon, J., Bradley, R. S., Burns, S. J., Cheng, H., Edwards, R. L., et al. (2022). Droughts and societal change: The environmental context for the emergence of Islam in late antique Arabia. *Science, 376*(6599), 1317–1321. https://doi.org/10.1126/science.abg4044

Goldstone, J. A. (2016). *Revolution and rebellion in the early modern world: Population change and state breakdown in England, France, Turkey, and China, 1600–1850* (2nd ed.). Routledge.

Grinin, L., & Grinin, A. (2015). *From bifaces to nanorobots. The world is on the way to the era of self-governing systems (history of technologies and a description of their future)*. Uchitel.

Grinin, L., & Grinin, A. (2016). *The cybernetic revolution and the forthcoming epoch of self-regulating systems*. Uchitel.

Grinin, L., & Grinin, A. (2023). Technology. Limitless possibilities, effective control. In V. Sadovnichy et al. (Eds.), *Reconsidering the limits to growth. A report to the Russian Association of the Club of Rome* (pp. 139–154). Springer. https://doi.org/10.1007/978-3-031-34999-7_8

Grinin, L., Grinin, A., & Korotayev, A. (2020). A quantitative analysis of worldwide long-term technology growth: From 40000 BCE to the early 22nd century. *Technological Forecasting and Social Change, 155*, 119955. https://doi.org/10.1016/j.techfore.2020.119955

Grinin, L., & Korotayev, A. (2012). *Tsikly, krizisy, lovushki sovremennoy Mir-Sistemy. Issledovaniye kondrat'yevskikh, zhyuglyarovskikh i vekovykh tsiklov, global'nykh krizisov, mal'tuzianskikh i postmal'tuzianskikh lovushek*. LKI.

Grinin, L., & Korotayev, A. (2015). *Great divergence and great convergence. A global perspective*. Springer.

Grinin, L., Malkov, S., Gusev, V., & Korotayev, A. (2009). Nekotoryye vozmozhnyye napravleniya razvitiya teorii sotsial'no-demograficheskikh tsiklov i matematicheskiye modeli vykhoda iz mal'tuzianskoy lovushki. In L. Grinin, A. Korotayev, & S. Malkov (Eds.), *Istoriya i Matematika: Protsessy i modeli* (pp. 134–210). LIBROKOM/URSS.

Gushchina, Y., & Malkov, S. (2014). K voprosu o modelirovanii demograficheskikh tsiklov i o vykhode iz mal'tuzianskoy lovushki. In L. Grinin & A. Korotayev (Eds.), *Istoriya i matematika: aspekty demograficheskikh i sotsial'no-ekonomicheskikh protsessov* (pp. 29–42). Uchitel.

Ilyin, I., Issaev, L., & Malkov, S. (2020). A methodology for analyzing and forecasting sociopolitical destabilization. *Biosystems, 198*, 1–8. https://doi.org/10.1016/j.biosystems.2020.104229

Kapitza, S. (2006). Ob uskorenii istoricheskogo vremeni. *Istoriya i Matematika, 1*, 12–30.

Kapitza, S. (2009). *Obshchaya teoriya rosta chelovechestva. Kak ros i kuda idet mir cheloveka.* Alpina non-fikshn.

Kirdina, S. (2004). *X i Y-ekonomiki: Institutsional'nyy analiz*. Nauka.

Kirdina, S. (2014a). Institutional matrices theory, or X&Y theory: The main provisions and applications. *Journal of Institutional Studies, 6*(3), 13–33.

Kirdina, S. (2014b). *Institutsional'nyye matritsy i razvitiye Rossii: vvedeniye v X-Y-teoriyu* (3rd ed.). Nestor-Istoriya.

Klimenko, V. (2009). *Klimat: neprochitannaya glava istorii*. MEI.

Kögel, T., & Prskawetz, A. (2001). Agricultural productivity growth and escape from the Malthusian trap. *Journal of Economic Growth, 6*, 337–357. https://doi.org/10.1023/A:1012742531003

Komlos, J., & Artzrouni, M. (1990). Mathematical investigations of the escape from the Malthusian trap. *Mathematical Population Studies, 2*, 269–287.

Komlos, J., & Nefedov, S. (2002). A compact macromodel of pre-industrial population growth. *Historical Methods: A Journal of Quantitative and Interdisciplinary History, 35*(2), 92–94. https://doi.org/10.1080/01615440209604133

Korotayev, A. (1995a). Sabaean cultural-political area: Some general trends of evolution. *Proceedings of the Seminar for Arabian Studies, 23*, 49–60.

Korotayev, A. (1995b). *Ancient Yemen: Some general trends of evolution of the Sabaic language and Sabaean culture*. Oxford University Press.

Korotayev, A. (2017). Malthusian cycle. In F. M. Moghaddam (Ed.), *The SAGE Encyclopedia of political Behavior* (pp. 457–462). SAGE.

Korotayev, A. (2020a). Evolution of Sociopolitical institutions in north-East Yemen (the 1st millennium BCE–the 2nd millennium CE). In D. M. Bondarenko, S. A. Kowalewski, & D. B. Small (Eds.), *The evolution of social institutions. Interdisciplinary perspectives* (pp. 161–184). Springer. https://doi.org/10.1007/978-3-030-51437-2_7

Korotayev, A. (2020b). The 21st century singularity in the big history perspective. A re-analysis. In A. Korotayev & D. LePoire (Eds.), *The 21st century singularity and global futures. A big history perspective* (pp. 19–75). Springer. https://doi.org/10.1007/978-3-030-33730-8_2

Korotayev, A. (2021). From chiefdom to tribe? Trends in the evolution of political Systems in North-Eastern Yemen in the last two millennia. *Nouvelles Chroniques du manuscrit au Yémen, 32*, 93–115.

Korotayev, A., & Grinin, L. (2013). Urbanization and political development of the world system. *Entelequia. Revista interdisciplinar, 15*, 197–254.

Korotayev, A., Grinin, L., & Grinin, A. (2021). Mathematical model of interaction between civilization center and tribal periphery: A description. *Social Evolution & History, 20*(2), 50–78. https://doi.org/10.30884/seh/2021.02.03

Korotayev, A., Grinin, L., & Grinin, A. (2022). Mathematical model of interaction between civilization center and tribal periphery: An analysis. *Social Evolution & History, 21*(1), 67–100. https://doi.org/10.30884/seh/2022.01.03

Korotayev, A., & Khaltourina, D. (2006). *Introduction to social macrodynamics: Secular cycles and millennial trends in Africa*. KomKniga/URSS.

Korotayev, A., Klimenko, V., & Proussakov, D. (1999). Origins of Islam: Political-anthropological and environmental context. *Acta Orientalia Academiae Scientiarum Hungaricae, 52*(3/4), 243–276.

Korotayev, A., & Komarova, N. (2004). A new mathematical model of pre-industrial demographic cycle. In M. G. Dmitriev & A. P. Petrov (Eds.), *Mathematical Modeling of social and economic dynamics* (pp. 157–163). Russian State Social University.

Korotayev, A., Komarova, N., & Khaltourina, D. (2007). *Zakony istorii: Vekovyye tsikly i tysyacheletniye trendy. Demografiya, ekonomika, voyny*. KomKniga.

Korotayev, A., Malkov, A., & Khaltourina, D. (2006). *Introduction to social macrodynamics: Secular cycles and millennial trends*. KomKniga/URSS.

Korotayev, A., Malkov, S., & Grinin, L. (2014). A trap at the escape from the trap? Some demographic structural factors of political instability in modernizing social systems. *History of Mathematics, 4*, 201–267.

Korotayev, A., & Zinkina, J. (2014). How to optimize fertility and prevent humanitarian catastrophes in tropical Africa. *African Studies in Russia, 6*, 94–107.

Korotayev, A., & Zinkina, J. (2015). East Africa in the Malthusian trap? *Journal of Developing Societies, 31*(3), 385–420. https://doi.org/10.1177/2F0169796X15590322

Korotayev, A., & Zinkina, J. (2022). Egypt's 2011 revolution. A demographic structural analysis. In J. A. Goldstone, L. Grinin, & A. Korotayev (Eds.), *Handbook of revolutions in the 21st century: The new waves of revolutions, and the causes and effects of disruptive political change* (pp. 651–683). Springer. https://doi.org/10.1007/978-3-030-86468-2_25

Korotayev, A., Zinkina, J., & Andreev, A. (2016a). Secular cycles and millennial trends. *Cliodynamics, 7*(2), 204–216. https://doi.org/10.21237/C7CLIO7230484

Korotayev, A., Zinkina, J., Goldstone, J., & Shulgin, S. (2016b). Explaining current fertility dynamics in tropical Africa from an anthropological perspective: A cross-cultural investigation. *Cross-Cultural Research, 50*(3), 251–280. https://doi.org/10.1177/1069397116644158

Korotayev, A., Zinkina, J., Kobzeva, S., Bogevolnov, J., Khaltourina, D., Malkov, A., & Malkov, S. (2011). A trap at the escape from the trap? Demographic-structural factors of political instability in modern Africa and West Asia. *Cliodynamics: The Journal of Quantitative History and Cultural Evolution, 2*(2), 276–303. https://doi.org/10.21237/C7clio22217

Kovaleva, N. (2023). Ecology. Life in the "unstable biosphere". In V. Sadovnichy et al. (Eds.), *Reconsidering the limits to growth. A report to the Russian Association of the Club of Rome* (pp. 71–95). Springer. https://doi.org/10.1007/978-3-031-34999-7_5

Lee, F. (2007). Climatic change and Chinese population growth. *Human Geography, 24*(3), 347–364.

Lee, H. F., & Zhang, D. D. (2010). Changes in climate and secular population cycles in China, 1000 CE to 1911. *Climate Research, 42*(3), 235–246.

Livi-Bacci, M. (2012). *A concise history of world population*. Wiley-Blackwell.

Malkov, S. (2009). *Sotsial'naya samoorganizatsiya i istoricheskiy protsess: Vozmozhnosti matematicheskogo modelirovaniya*. LIBROKOM.

Malkov, S., Grinin, L., Grinin, A., Musieva, J., & Korotayev, A. (2023). Modeling social self-organization and historical dynamics. Global phase transitions. In V. Sadovnichy et al. (Eds.), *Reconsidering the limits to growth. A report to the Russian Association of the Club of Rome* (pp. 387–417). Springer. https://doi.org/10.1007/978-3-031-34999-7_18

Malthus, T. (1798). *An essay on the principle of population, as it affects the future improvement of society with remarks on the speculations of Mr. Godwin, M. Condorcet, and other writers*. St. Paul's Church-yard.

McEvedy, C., & Jones, R. (1978). *Atlas of world population history*. Facts on File.

Nefedov, S. A. (2004). A model of demographic cycles in a traditional society: The case of ancient China. *Social Evolution & History, 3*(1), 69–80.

Nefedov, S. A. (2013). Modeling Malthusian dynamics in pre-industrial societies. *Cliodynamics: The Journal of Quantitative History and Cultural Evolution, 4*(2), 229–240. https://doi.org/10.21237/C7clio4221335

Nefedov, S. A. (2014a). Modeling Malthusian dynamics in pre-industrial societies. *Social Evolution & History, 13*(1), 172–184.

Nefedov, S. A. (2014b). Modeling Malthusian dynamics in pre-industrial societies: Mathematical Modeling. *History of Mathematics, 4*, 190–220.

Parker, G. (2013). *Global crisis: War, climate change, & catastrophe in the seventeenth century*. Yale University Press.

Shcherbakova, Y. (2019). Stareniye naseleniya mira po otsenkam OON 2019 goda. *Demoskop Weekly*, 837–838. http://demoscope.ru/weekly/2019/0837/barom01.php
Steinmann, G., & Komlos, J. (1988). Population growth and economic development in the very long run: A simulation model of three revolutions. *Mathematical Social Sciences, 16*(1), 49–63. https://doi.org/10.1016/0165-4896(88)90004-2
Steinmann, G., Prskawetz, A., & Feichtinger, G. (1998). A model on the escape from the Malthusian trap. *Journal of Population Economics, 11*, 535–550. https://doi.org/10.1007/s001480050083
Thompson, W. R., & Zakhirova, L. (2021). *Climate change in the Middle East and North Africa: 15,000 years of crises, setbacks, and adaptation*. Routledge.
Turchin, P. (2018). *Historical dynamics: Why states rise and fall* (2nd ed.). Princeton University Press.
Turchin, P., & Korotayev, A. (2006). Population density and warfare: A reconsideration. *Social Evolution and History, 5*(2), 121–158.
Turchin, P., & Nefedov, S. (2009). *Secular Cycles*. Princeton University Press.
Valentey, D. (Ed.). (1974). *Marksistsko-leninskaya teoriya narodonaseleniya* (2nd ed.) Mysl.
Vishnevsky, A. (1973). *Demograficheskaya revolyutsiya. Voprosy filosofii, 2*, 53–64.
Usher, D. (1989). The dynastic cycle and the stationary state. *The American Economic Review, 79*, 1031–1044.
Wood, J. W. (1998). A theory of pre-industrial population dynamics: Demography, economy, and Well-being in Malthusian systems. *Current Anthropology, 39*, 99–135.
Young, L. D. (2022). Testing tilly: Does war really make states? *Social Evolution & History, 21*(1), 175–199. https://doi.org/10.30884/seh/2022.01.07
Zhang, D. D., Brecke, P., Lee, H. F., He, Y. Q., & Zhang, J. (2007). Global climate change, war, and population decline in recent human history. *Proceedings of the National Academy of Sciences, 104*(49), 19214–19219. https://doi.org/10.1073/pnas.0703073104
Zinkina, J., & Korotayev, A. (2014a). Explosive population growth in tropical Africa: Crucial omission in development forecasts (emerging risks and way out). *World Futures, 70*(4), 271–305. https://doi.org/10.1080/02604027.2014.894868
Zinkina, J., & Korotayev, A. (2014b). Projecting Mozambique's demographic futures. *Journal of Futures Studies, 19*(2), 21–40.

Sergey Malkov heads the Center for Long-Term Forecasting and Strategic Planning at Lomonosov Moscow State University; he also works as a Senior Research Professor at the Institute of Economics of the Russian Academy of Sciences, and at the Centre for Stability and Risk Analysis at HSE University in Moscow. He is the author of over 300 scholarly publications, including such monographs as *Social self-organization and historical process: possibilities of mathematical modeling* (URSS, 2009), *Modeling and forecasting of world dynamics* (ISPI RAS, 2012, together with Victor Sadovnichy and others), or *The twenty-first Century Singularity and Global Futures. A Big History Perspective* (Springer, 2020, together with David J. LePoire and others).

Natalia Kovaleva heads the Laboratory of Ecological Soil Science of the Faculty of Soil Science of Lomonosov Moscow State University. She is also a visiting professor at Kazan Federal (Volga Region) University. Natalia Kovaleva has authored and co-authored over 700 scientific publications including such monographs as *Recreational territories and ecological well-being of the population* (1996), *Soil science: history, sociology, methodology* (2005), *Soil in the modern world* (2008), *Soils in the biosphere and human life: monograph (2012), Protected areas – the basis of environmental safety* (2018), *Soil biomarkers* (2020).

Leonid Grinin is Senior Research Professor at the Centre for Stability and Risk Analysis at HSE University in Moscow, as well as Senior Research Professor of the Institute of Oriental Studies, Russian Academy of Sciences. He is co-editor of the international journals *Social Evolution &*

History and the *Journal of Globalization Studies*, as well as co-editor of the international almanacs *Evolution*, *History & Mathematics*, and *Kondratieff Waves*. He is author of more than 600 publications, including 33 monographs in Russian, English, and Chinese. His current research focuses on comparative political studies, theory of revolution, political anthropology, global economy, global history, historical sociology, and futurology.

Andrey Korotayev heads the Centre for Stability and Risk Analysis at HSE University in Moscow, Russia. He is also Senior Research Professor at the Institute of Oriental Studies and Institute for African Studies, Russian Academy of Sciences, as well as at the Faculty of Global Studies of the Lomonosov Moscow State University and the Russian Presidential Academy of National Economy and Public Administration. He has authored or co-authored over 650 scholarly publications, including such monographs as *Ancient Yemen* (Oxford University Press, 1995), *Introduction to Social Macrodynamics: Compact Macromodels of the World System Growth* (URSS, 2006; with Daria Khaltorina and Artemy Malkov), *Great Divergence and Great Convergence. A Global Perspective* (Springer, 2015; with Leonid Grinin), *Islamism, Arab Spring, and the Future of Democracy. World System and World Values Perspectives* (Springer, 2019; with Leonid Grinin and Arno Tausch).

Modeling Social Self-Organization and Historical Dynamics. Industrial Society

Askar Akaev, Sergey Malkov, Olga Davydova, Natalia Kovaleva, Artemy Malkov, Leonid Grinin, and Andrey Korotayev

Abstract

This chapter presents the results of using the set of basic models described in the chapter "Modeling Social Self-Organization and Historical Dynamics. A General Approach" to model the dynamics of an industrial society. Both interactions in

A. Akaev
Lomonosov Moscow State University, Moscow, Russia

HSE University, Moscow, Russia

S. Malkov
Lomonosov Moscow State University, Moscow, Russia

HSE University, Moscow, Russia

Institute of Economics, Russian Academy of Sciences, Moscow, Russia
e-mail: s@malkov.org

O. Davydova
Lomonosov Moscow State University, Moscow, Russia

N. Kovaleva
Soil Science Faculty, Lomonosov Moscow State University, Moscow, Russia

A. Malkov
Moscow Institute of Physics and Technology, Moscow, Russia
e-mail: as@malkov.org

L. Grinin
HSE University, Moscow, Russia

Institute of Oriental Studies, Russian Academy of Sciences, Moscow, Russia

A. Korotayev (✉)
Lomonosov Moscow State University, Moscow, Russia

HSE University, Moscow, Russia

Institute of Oriental Studies, Russian Academy of Sciences, Moscow, Russia

V. Sadovnichy et al. (eds.), *Reconsidering the Limits to Growth*, World-Systems Evolution and Global Futures, https://doi.org/10.1007/978-3-031-34999-7_17

the "society–nature" system and social interactions within society between the main social groups (including economic and political interactions) are considered. It is shown that industrial society is characterized by the continuous growth of the main demographic and economic characteristics due to the high pace of technological development and the expansion of the resource base associated with this. Based on the modeling of social interactions, the main structural and functional features of an industrial society are identified, leading to the formation of social structures of the so-called Y-type (which are characterized by a market economy, republican forms of government, and individualism in the socio-psychological sphere). It is shown that a necessary condition for the effective functioning of Y-type social structures is economic growth, in the absence of which they experience crisis and transformation into structures of other types.

1 Modeling of Interactions in the "Society-Nature" System

1.1 Simulation of Dm-Tl-En Dynamics

A. The most important typological features inherent in industrial societies (approximately from the mid-eighteenth century to the early 1970s) include the following:

- The main type of production activity of an industrial society is industrial production (agricultural products make up a small share of GDP).
- The driving force behind economic development is technological progress, primarily the replacement of manual labor with machine labor.
- Due to technological progress, periodically arising resource restrictions are usually removed by introducing new resources into circulation or their more efficient use.
- Market-oriented specialized industrial (semi-industrial) production prevails. The main production unit is an enterprise using machines and mechanisms, as well as hired labor to maintain machines and mechanisms. The main purpose of production is to sell the manufactured products on the market and make profit.
- Demographic dynamics is characterized by the dependence shown in Fig. 4a in Sect. 1.3 of chapter "Modeling Social Self-Organization and Historical Dynamics. A General Approach" (Akaev et al., 2023a) with a gradual transition in high-income countries to the dependence shown in Fig. 4b in Sect. 1.3 of chapter "Modeling Social Self-Organization and Historical Dynamics. A General Approach" (Akaev et al., 2023a).

Thus, the basic equations for a typical industrial society take the following form.
A.1. Generalized equation of demographic dynamics.

The basic equation (1),[1] reflecting the nature of the demographic dynamics characteristic of a classical industrial society and depicted in Fig. 4a,[2] can be written as follows:

$$dN/dt \approx r \cdot N \cdot (1 - x_0/x), \tag{1}$$

where N is the population, x is the GDP per capita, x_0 is the level of the "subsistence minimum" (physical survival level), and r is a constant coefficient. It means that when the level of well-being x exceeds the critical value x_0, society escapes from the "Malthusian trap"[3] and population growth accelerates due to a decrease in mortality (especially infant mortality).

A.2. Generalized equation of economic dynamics.

For an industrial society, industrial products represent the largest proportion of production, while the share of agricultural products intended for consumption (primarily food) is consistently falling. Urbanization (i.e., the number of people living in cities with high population density) is on the rise. Accordingly, the costs of transforming the external environment in order to create comfortable/acceptable conditions for life (I_2) are growing. At the same time, an increase in production is absorbed by an increase in consumption (the desire for consumption is higher than the possibilities of production, which ensures a continuous increase in demand). Equation (3)[4] takes the following form:

$$dx/dt \approx (F(R^-, I_1) - I_1 \ - I_2)/N - Q \approx (F(R^-, I_1) - I_1 - I_2)/N - q \cdot x, \tag{2}$$

where $q{\cdot}x$ is the consumption (it is assumed that consumption is proportional to savings: the more material goods x, the greater their consumption; q is the proportionality coefficient).

As mentioned above, it is necessary to distinguish between situations where (1) there are no resource constraints (resources are available in abundance) and (2) when resources are scarce.

Until the mid-twentieth century, industrial development essentially took place within the framework of the first situation, since, due to technological progress, periodically arising resource restrictions were removed by introducing new resources into circulation, as well as due to their more efficient use.

---

[1] From chapter "Modeling Social Self-Organization and Historical Dynamics. A General Approach" (Akaev et al., 2023a, this volume).

[2] From chapter "Modeling Social Self-Organization and Historical Dynamics. A General Approach" (Akaev et al., 2023a, this volume).

[3] On the issues connected with the "Malthusian trap" (see, e.g., Malthus, 1798; Artzrouni & Komlos, 1985; Steinmann & Komlos, 1988; Kögel & Prskawetz, 2001; Komlos & Artzrouni, 1990; Steinmann et al., 1998; Wood, 1998; Komlos & Nefedov, 2002; Korotayev & Khaltourina, 2006; Korotayev et al. 2006a, 2006b, 2011, 2014; Grinin et al., 2009; Zinkina & Korotayev, 2014a, 2014b; Korotayev et al., 2016; Korotayev & Zinkina, 2014, 2015, 2022; Korotayev, 2017).

[4] From chapter "Modeling Social Self-Organization and Historical Dynamics. A General Approach" (Akaev et al., 2023a, this volume).

Accordingly, the production function in this case is written in the form (4),[5] and the main factors of production are the number of producers (labor force) and the amount of means of labor associated with them. To reflect this situation in macroeconomic models, the Cobb-Douglas production function is usually used. In a modified form, it can be written as follows:

$$F = A_R(T) \cdot V = A \cdot N^u \cdot K^v \tag{3}$$

where it is assumed that the amount of the labor force is proportional to N (i.e., the share of workers in the total population is stable), A is the general factor productivity (depending primarily on the level of technological development), K is the production capital (depending on investment I_1), u and v are exponents reflecting the diminishing returns of production factors ($u < 1$, $v < 1$) and characterizing the ratio between production factors N and K. This ratio changes over time: manual labor prevailed in the agrarian era (so, $u \to 1$, $v \to 0$). As manual labor is increasingly replaced by machine labor, the value of u decreases, while the value of v increases.

Thus, Eq. (2) becomes

$$dx/dt \approx (A \cdot N^u K^v - I_1 - I_2)/N - q \cdot x \ \approx A \cdot N^u \cdot K^v/N - q \cdot x - \alpha_1 \cdot x - \alpha_2 \cdot x, \tag{4}$$

where it is assumed that the values of I_1 and I_2 in the absence of severe resource constraints grow in proportion to the growth of x: $I_1/N \approx \alpha_1 \cdot x$, $I_2/N \approx \alpha_2 \cdot x$ (coefficients α_1 and α_2 may change over time). Given that $A \cdot N^u \cdot K^v/N$ is GDP per capita and that, within the framework of the basic model, GDP per capita is a measure of technological progress, Eq. (4) can be written as

$$dx/dt \approx S - q \cdot x - \alpha_1 \cdot x - \alpha_2 \cdot x. \tag{5}$$

A.3. Generalized equation of **technological** dynamics.

In the industrial era, technological progress becomes an important factor in economic dynamics (the total factor productivity A, as well as the value of K in Eqs. (3) and (4) critically depend on it). To describe the technological dynamics, the basic equation (9)[6] is used, in which the values of c and S_0 are determined on the basis of statistical data.

A.4. Generalized equation of **natural resource** consumption.

In an industrial society, industrial production uses a wide variety of resources. As already mentioned, a feature of industrial development in the period up to the middle of the twentieth century was that thanks to technological progress, the periodically arising resource restrictions were removed by introducing new resources into

---

[5]From chapter "Modeling Social Self-Organization and Historical Dynamics. A General Approach" (Akaev et al., 2023a, this volume).

[6]From chapter "Modeling Social Self-Organization and Historical Dynamics. A General Approach" (Akaev et al., 2023a, this volume).

circulation, as well as through their more efficient use. The long-term consequences of the depletion of single-use resources (non-renewable and renewable) were not taken into account.

For the main reusable resource, land, Eq. (14)[7] can be used. The increase in the area of agricultural land occurred with the help of technical means through anthropogenic changes in natural landscapes (for example, plowing of virgin lands). The decrease in the area of agricultural land occurred mainly due to the growth of urbanization and the replacement of agricultural land by settlements, as well as due to the desertification of former pastures, soil salinization, etc. (see Sect. 1.2). Equation (15)[8] can be used to analyze changes in the quality of land resources.

A.5. Generalized equation of accumulation of **waste products**.

The long-term consequences of the accumulation of waste products were not taken into account until the 1970s; accordingly, the value of I_2 did not depend on them.

A.6. Generalized equation for **changes in the characteristics of the external environment** U_j as a result of anthropogenic impact.

The long-term consequences of changes in characteristics of the external environment as a result of anthropogenic impact were not taken into account until the 1970, so the value of I_2 did not depend on them.

B. The main equations that determine the economic and demographic dynamics of an industrial society in the period up to the 1970s are Eqs. (1), (5), (9):[9]

$$dN/dt \approx r \cdot N \cdot (1 - x_0/x), \tag{6}$$

$$dx/dt \approx S - q \cdot x - \alpha_1 \cdot x - \alpha_2 \cdot x, \tag{7}$$

$$dS/dt \approx c \cdot N \cdot (S - S_0). \tag{8}$$

The characteristic time of the economic equation (7) is the production cycle (a year or less), while for the demographic equation (6), it is the generational cycle (25–30 years). The characteristic time of the technological equation (8) during the formation of industrial society exceeds the time of the generational cycle (technological growth was still slow and was only gaining momentum, primarily in Britain, where conditions for this emerged), and by the mid-twentieth century, it became less than the generational cycle (the rate of technological change outpaces the rate of demographic growth). In this regard, it appears appropriate to divide the development of industrial society into two phases:

---

7 From chapter "Modeling Social Self-Organization and Historical Dynamics. A General Approach" (Akaev et al., 2023a, this volume).

8 From chapter "Modeling Social Self-Organization and Historical Dynamics. A General Approach" (Akaev et al., 2023a, this volume).

9 From chapter "Modeling Social Self-Organization and Historical Dynamics. A General Approach" (Akaev et al., 2023a, this volume).

Fig. 1 Population and economic growth in Britain, 1700–1860 (1700 level = 100). *Source*: Clark (2007)

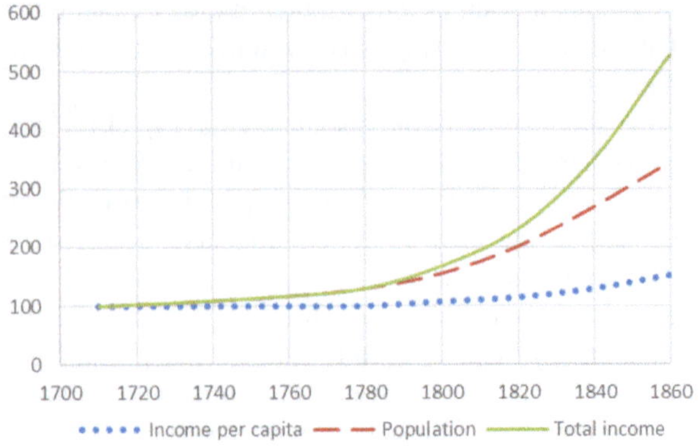

- The first phase is the eighteenth century (the period of predominant manufactory production and the beginning of the Industrial Revolution[10] in Britain).
- The second phase comprised the nineteenth century and lasted till the 1960s (industrialization of the economies of the leading countries, the formation of capitalism as the leading socio-economic system).

The economic and demographic dynamics in Britain in the eighteenth and nineteenth centuries is shown in Fig. 1:

It can be seen that economic growth in Britain began in the mid-eighteenth century, but until 1800 it only led to population growth[11] with a constant per capita income, that is, society continued to remain in the "Malthusian trap" (this is the first phase of the development of industrial society). The growth of per capita income (with continued population growth) begins in 1800 (this is the second phase of the development of industrial society).

Figure 2 shows the results of calculations with model (6)–(8).

The calculation results (Fig. 2) correlate very well with the statistical data (Fig. 1): up to the time point 200 (conventional year 1800), GDP/capita shows almost no growth, only demographic growth is observed, and only then the growth of GDP/capita follows. The presence of the *first* phase, when the result of the emerging technological growth is only population growth, and not the growth of average incomes (i.e., the "Malthusian trap" persists), indicates the threshold of the transition from an agrarian to an industrial society. Apparently, the presence of this threshold did not allow China in the Sung era to move to an industrial-type society, although

[10]For more detail on the Industrial Revolution see, e.g., Cipolla (1976), Allen (2009, 2011), Goldstone (2009), Grinin and Korotayev (2015), Komlosy, (2021).

[11]This is the so-called first phase of demographic transition when economic development leads to explosive acceleration of population growth due to the decreasing mortality (especially, infant mortality)—[see, e.g., Chesnais, 1992; Caldwell et al., 2006; Reher, 2011; Dyson, 2010; Livi-Bacci, 2012, as well as chapter "Demography. Toward Optimization of Demographic Processes" (Korotayev et al., 2023a, this volume)].

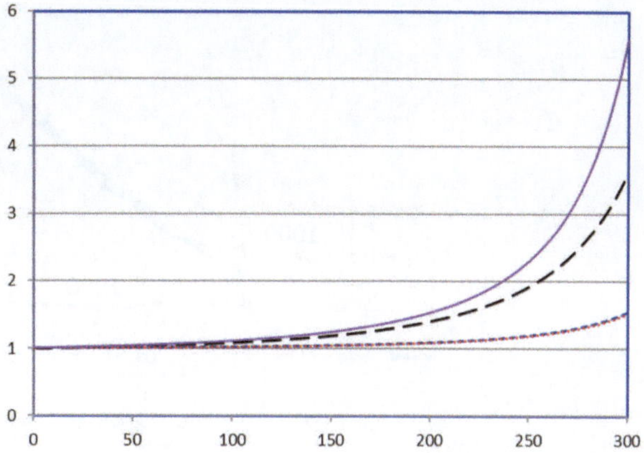

Fig. 2 Calculation results for model (6)–(8) (solid line is GDP, dashed line is population, dotted line is GDP/capita; divisions on the *x*-axis approximately correspond to half-year, 0 corresponds to 1700, 300 to 1860)

there were prerequisites for this in the form of clear technological progress (see, e.g., McNeill, 2013). The advantages of Britain, that overcame this threshold, were

- The absence of external military threats in the eighteenth century (to some extent due to the insular position of Britain).
- A favorable internal institutional environment after the "Glorious Revolution" of the late seventeenth century.
- Income from foreign, including colonial, trade ("positive sum game" due to the influx of resources from the outside world).
- The possibility of "dumping" the excess population in the colonies (e.g., Allen, 2009; Goldstone, 2009).

All this created the conditions for Britain to cross the threshold in the eighteenth century and to further accelerate the innovation process, which became self-sustaining (due to the presence of positive feedbacks).

The transition to the second phase is the formation of a society with a high degree of urbanization [chapter "Global Aging – an Integral Problem of the Future. How to Turn a Problem into a Development Driver?" (Grinin et al., 2023a, this volume)] and the predominance of industrial production over agricultural production in GDP. In the first phase, the "living wage" x_0 corresponded to the "living wage" of an agrarian society.[12] In the second phase, the value of the "living wage" begins to increase, since in an urbanized society[13] life support in cities requires additional resources for housing and communal services, construction, water and energy supply, transport, etc. In this regard, the problem arises of estimating the growth of x_0 in the second phase of the development of industrial society.

[12] Around 500 of 1990 international dollars annually, according to Maddison's (2001) estimation.

[13] In Britain, the proportion of the urban population reached 50% by 1850 (Clark, 2007).

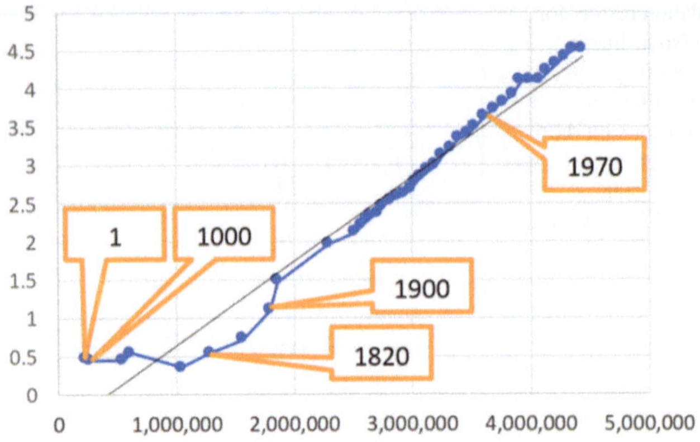

Fig. 3 Relationship between the world average value of GDP/capita (*y*-axis, thousands of 2011 international dollars at PPP) and the population of the Earth (*x*-axis, thousand people) in the period from 1 AD to 1980. *Data source:* Maddison Project (2020)

As noted above, in the system of Eqs. (6)–(8), the variable x is a "fast" variable with respect to the variables N and S; therefore, according to the Tikhonov theorem, the differential equation (7) can be replaced with an algebraic equation:

$$S - q \cdot x - \alpha_1 \cdot x - \alpha_2 \cdot x \approx 0, \tag{9}$$

from which we infer:

$$x \cdot (q + \alpha_1 + \alpha_2) \approx S. \tag{10}$$

Accordingly, Eq. (8) can be written as follows:

$$dS/dt \approx c \cdot S \cdot N \cdot (1 - k \cdot x_0/x), \tag{11}$$

and Eq. (6) can be written as follows:

$$dN/dt \approx r \cdot N \cdot (1 - S_0/(k \cdot S)), \tag{12}$$

where k is a coefficient close to unity; $S_0(t)$ is the minimum value of GDP per capita that is necessary to ensure the "living wage" in the period of time t. Not long ago, this level of GDP/capita was typical for the countries of the World System periphery. If we talk about the World System as a whole, then the dynamics of change in $S_0(t)$ reflects the dynamics of the growth of the average world GDP/capita, since the main contribution to this indicator is made by the countries of the World System periphery, where the main part of the world population lives.[14] Figure 3 is a graph showing

[14] At the same time, the value of GDP/capita in high income countries can exceed the world average by several times.

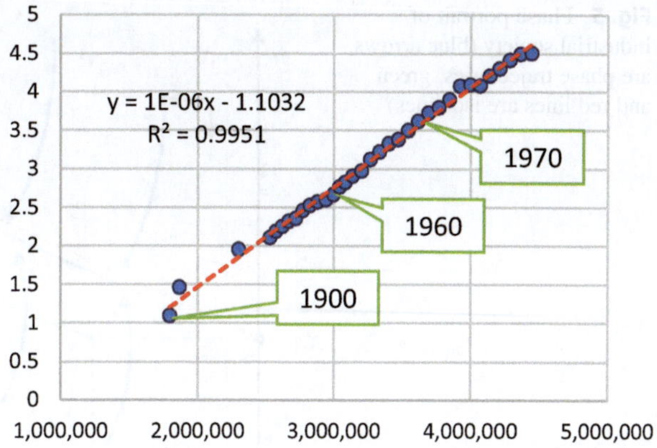

Fig. 4 Relationship between the world average value of GDP/capita (*y*-axis, thousands of 2011 international dollars at PPP) and the world population (*x*-axis, thousand people) in the period from 1900 to 1980. *Data source:* Maddison Project (2020)

the relationship between the world average GDP/capita and the world population in the period from 1000 to 1980 (calculated with data from the Maddison Project, 2020).

Until the early nineteenth century, the world average value of GDP/capita practically did not change and was equal to approximately 500 of 1990 international dollars, which reflects the constancy of x_0 under the conditions of the "Malthusian trap". In the second phase of the development of an industrial society, the world average value of GDP/capita began to grow, and since the beginning of the twentieth century, it has grown strictly in proportion to the population (see Fig. 4, calculated with data from Maddison Project, 2020):

Bearing in mind the close relationship between the world average value of GDP/capita and the world population, the expression reflecting the growth of the value of x_0 in the second phase of the development of industrial society (that is, since the beginning of the nineteenth century) can be written as follows:

$$x_0(t) \approx x_0 \cdot (1 + b \cdot (N(t) - N(1800))/N(1800)), \qquad t \geq 1800, \qquad (13)$$

where x_0 corresponds to the "subsistence minimum" characteristic of an agrarian society.

The phase portrait of systems (6)–(8) and (13) in the x and N coordinates has the following form (see Fig. 5):

Unlike the phase portrait of agrarian society [see Fig. 1 in chapter "Modeling Social Self-Organization and Historical Dynamics. Agrarian Society" (Malkov et al., 2023b, this volume)], the phase portrait of industrial society (see Fig. 5 above) does not have a point of stable equilibrium. This is a consequence of the fact that, in contrast to agrarian societies, where there was an inverse negative relationship between per capita x and N [the larger N, the lower the per capita income due to the limited land resource, see Eq. (6) in chapter "Modeling Social Self-Organization and Historical Dynamics. Agrarian Society" (Malkov et al., 2023b, this volume)], in industrial societies, there is a positive relationship between per capita x and N (the larger N, the more inventors and inventions that increase labor productivity and,

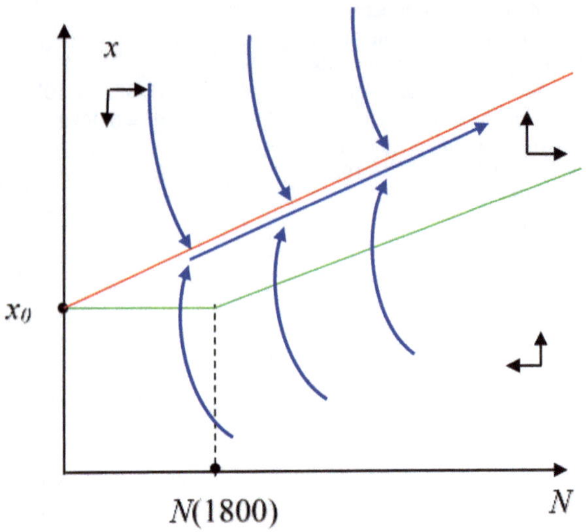

Fig. 5 Phase portrait of industrial society (blue arrows are phase trajectories, green and red lines are isoclines)

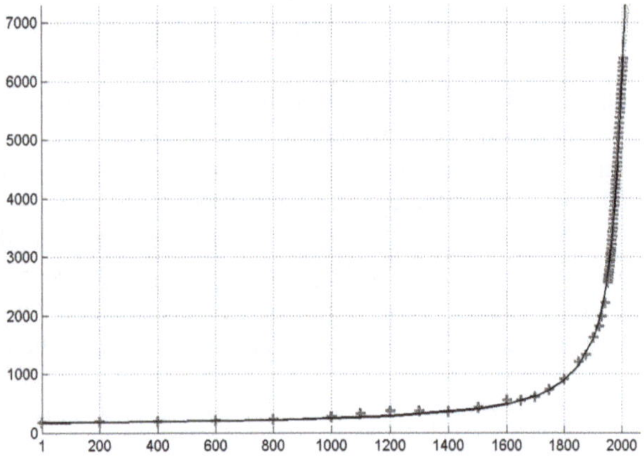

Fig. 6 Dynamics of the world population (y-axis, million people), 1–2000 CE. *Data sources*: Maddison Project (2020)

ultimately, per capita income). Accordingly, the "society of restrictions", which was agrarian society, is turning into the "society of growth". At the same time, the positive feedback observed in Eqs. (6)–(8) increases over time, which leads not just to exponential growth, but to explosion-like hyperbolic growth, which is confirmed by statistical data on world demographics and economy up to the 1960s (see Fig. 6).

A specific feature of hyperbolic growth is that it ends with a *singularity*, when at a certain point in time the indicator in question becomes infinite. So, in 1960 Heinz von Foerster and his colleagues calculated the singularity point for the growth of the world population on the basis of processing the statistical data for 1. It turned out that the hyperbolic trend they calculated reaches infinite values on November 13, 2026 (von Foerster et al., 1960; see also, e.g., Kapitza, 1992, 1996, 1999, 2003, 2006, 2010; Korotayev, 2020).

However, real growth cannot be infinite. Therefore, causes and mechanisms must inevitably arise that transfer the system from the mode of unlimited growth to a more restrained mode of development. It is also clear that society must change in a significant way. We are living in this transitional period.

1.2 Simulation of En-El-Na Dynamics

In the industrial era, due to the development of industry and the rapid growth of the world population, the influence of human activity on nature intensified sharply. Until the 1970s, the problem of limited natural resources was not taken seriously, since new deposits of minerals were constantly discovered, methods of their extraction were improved, and new types of resources were developed. Environmental problems also did not attract global attention, since they were mostly local in nature and were seriously perceived only in certain regions. Meanwhile, these problems were becoming more and more serious and were profoundly analyzed in the reports to the Club of Rome (e.g., Meadows et al., 1972, 2004; Pestel, 1989; Randers, 2012; Wijkman & Rockström, 2012; Bardi, 2014, 2017; Martin, 2015; Randers et al., 2018; von Weizsäcker & Wijkman, 2018; Álvarez Pereira, 2021; Perissi & Bardi, 2021, among others).

Among other issues, those reports analyze the situation with the consumption of non-renewable (single use) resources such as coal, oil, metal ores, etc. Their consumption is described by Eq. (12) in chapter "Modeling Social Self-Organization and Historical Dynamics. A General Approach" (Akaev et al., 2023a, this volume). Less attention has been paid to reusable resources such as agricultural land. Meanwhile, the problem of degradation of this type of resource became increasingly urgent. So, while over the past 10,000 years, the area of loss of fertile land in the world as a whole has amounted to 2000 million ha, over the past 300 years (industrial society), it has amounted to 700 million ha. That is, the average annual loss rate in the last 10,000 years was about 0.2 million ha, while in the last 300 years, it has been ten times higher, 2.3 million ha [chapter "Ecology. Life in the "Unstable Biosphere" (Kovaleva, 2023), this volume).

The interaction of humankind with the natural environment changed during the industrial era. In the agrarian era, the impact of human activity on the natural environment was slow and was mainly of a localized nature,[15] while the reverse effect of natural (primarily climatic) changes on human society was much more significant. In the industrial era, people became less dependent on weather conditions and began to penetrate into unfavorable natural zones more actively (for example, in the regions of the Far North). So, they also began to actively influence changes in nature and climate. An example of this is the impact of

[15] As a rule, it took place in the areas of intensive agriculture and animal husbandry.

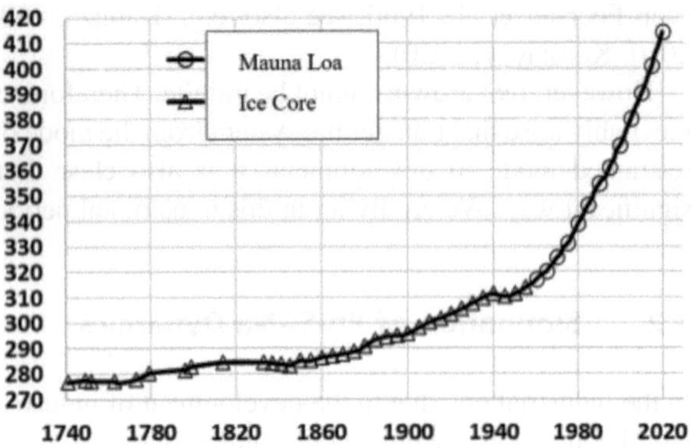

Fig. 7 The increase in CO_2 concentration (million^{-1}) in the atmosphere according to the ice core of Antarctica 1745–1973 and measurements at the Mauna Loa station 1959–2020. *Data sources*: Boden et al. (1994), NOAA (2021), Our World in Data (2021), Lindsey (2021)

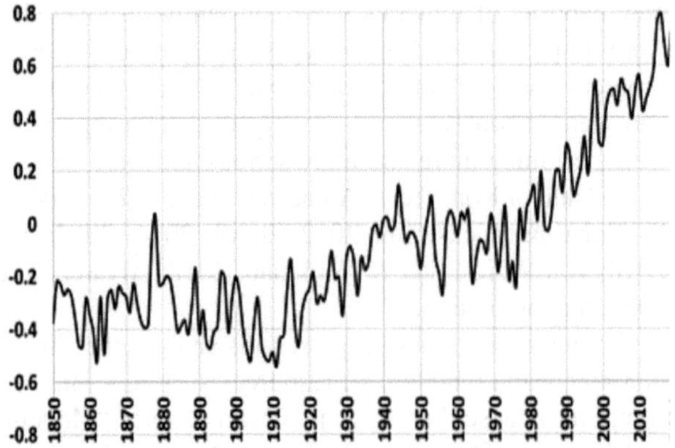

Fig. 8 Anomalies of surface global temperature of the atmosphere (°C) in 1856–2020. *Data sources*: Jones et al. (1994), Our World in Data (2021), Met Office Hadley Centre (2022)

human activity on the increase in the temperature of the Earth atmosphere, which began in the industrial age.

On this topic, five theses are formulated in the chapter "Climate and Energy. Scenarios of Energy Transition and Global Temperature Changes Based on Current Technologies and Trend" (Akaev & Davydova, 2023, this volume), which have been convincingly confirmed by evidence over the past decades:

1. Starting around 1850, atmospheric concentrations of CO_2 have increased significantly, as shown in Fig. 7, from 280 ppm, typical of interglacials over the last 400,000 years, to 420 ppm today.
2. The reason for this change is human economic activity: first of all, the burning of fossil fuels, and secondly, deforestation.
3. CO_2 affects the climate by changing the radiation balance of the Earth: an increase in the concentration of this gas leads to an increase in the surface temperature of the atmosphere. It is very likely that doubling its concentration (from 280 to 560 ppm) will increase the average global temperature by 3±1 °C.
4. In the twentieth century, the climate experienced significant warming (by about 0.8 °C globally, and by 1 °C in Europe). The temperatures of the past 10 years

have been the highest on a global scale since records began in the nineteenth century, as shown in Fig. 8.

5. The main part of this warming ≈0.6 °C is due to the increase in the concentration of CO_2 and other anthropogenic gases; a smaller part has natural causes, such as the fluctuations in solar activity.

Air temperature is the most important climatic factor that determines the nature and conditions of human life and economic activity. Changes in atmospheric temperature lead to changes in the intensity of biological processes on land and in the ocean and cause disturbances in the established biogeochemical cycles. The consequences of global warming can be quite devastating (e.g., Budyko, 1980; Rahmstorf & Schellnhuber, 2007; Randers et al., 2018; Akaev & Davydova, 2020; Gills & Morgan, 2022). Therefore, it is necessary to use every opportunity to reduce the concentration of CO_2 in the atmosphere. The most accessible and effective means on this path will be transition to green energy sources, and constant focus on increasing the absorptive capacity of natural carbon sinks; it is necessary to plant forests and reclaim agricultural land.

Thus, in the industrial era, interaction in the "society-nature" system moved to a qualitatively new level and requires serious attention at the global level, and not just at the level of individual regions.

2 Stability of Social Interactions

2.1 Socio-economic Stability

2.1.1. Modeling Economic and Non-economic Interactions of the Main Social Groups

Equations (21) from Sect. 1.3 of chapter "Modeling Social Self-Organization and Historical Dynamics. A General Approach" (Akaev et al., 2023a, this volume), adapted to the considered historical epoch, are the key ones for the mathematical description of socio-economic processes in industrial society.[16] Let us analyze the basic features of the provision for socio-economic stability in the society of this type.

1. Formalization of basic economic processes.

The main social groups in industrial society are entrepreneurs and wage workers. Entrepreneurs are the owners of the means of production and the organizers of production. Their income comes from sales of manufactured products. These incomes are spent on their own consumption, on the payroll of the wage workers, and on investments (costs for subsequent production cycles). Wage workers work at

[16] For simplicity, a closed system is considered: there are no external actors (trading partners, neighboring states).

the enterprises of businesspersons, receive wages proportional to the output, and spend them on their own consumption.

The basic economic processes that reflect the economic interaction of these social groups can be described using a system of differential equations that reflect the dynamics of material (or cash[17]) flows in the system [see the basic model of socio-economic interactions (33) in Sect. 3 of chapter "Modeling Social Self-Organization and Historical Dynamics. A General Approach" (Akaev et al., 2023a, this volume)]. In a generalized form, these equations for a society consisting of entrepreneurs and wage workers have the following form.

The equation for the dynamics of the material resources of employees $X_1(t)$ has the form:[18]

$$dX_1(t)/dt = (\text{wages}) - (\text{consumption}) \tag{14}$$

The equation for the dynamics of material resources of entrepreneurs $X_2(t)$ has the form:

$$\begin{aligned} dX_2(t)/dt = &(\text{income from manufactured products}) \\ &- (\text{salary for employees}) - (\text{consumption}) - (\text{investment}) \end{aligned} \tag{15}$$

or

$$\begin{cases} \dfrac{dX_1}{dt} = W - Q_1 \\ \dfrac{dX_2}{dt} = F - W - Q_2 \ - I \end{cases}, \tag{16}$$

where F is the production function, Q_1 is the consumption expenditure of wage workers, Q_2 is the consumption expenditure of entrepreneurs, W is the wages paid to workers, and I is an investment. Production in industrial society depends on the labor of workers and the productivity of this labor, which, in turn, depends on the technical equipment (material factor), the efficiency of the production organization (organizational factor), the quality and coherence of the personnel work (socio-psychological factor). As noted above, in model (16), the population is assumed to be constant, that is, $N_1 = \text{const}$, $N_2 = \text{const}$, where N_1 and N_2 are the number of workers and entrepreneurs, respectively.

Under fairly general assumptions about the nature of the dependence of the functions on the right-hand sides of the equations on the variables X_1 and X_2, system (16) can be written as follows:

[17] Taking into account the purchasing power of money, that is, taking into account its depreciation due to inflation.

[18] To simplify the description, it is assumed that on the considered time scales, the change in the population size can be neglected.

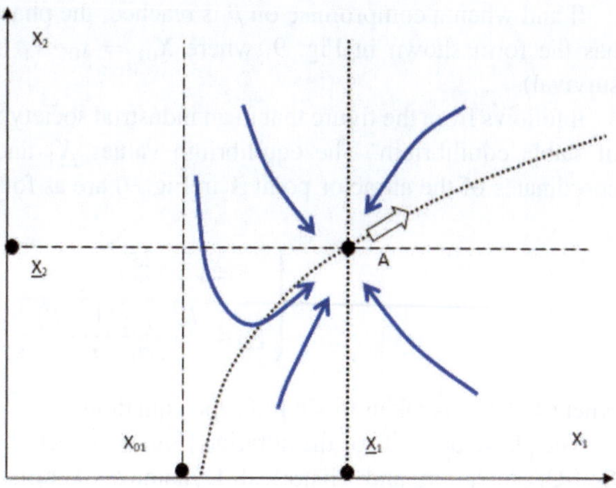

Fig. 9 Phase portrait of the system of Eq. (17) describing the industrial-type social systems (arrows are phase trajectories, dotted lines are isoclines, point *A* at the intersection of isoclines is a stable state of equilibrium)

$$\begin{cases} \dfrac{dX_1}{dt} = \beta \cdot F' - \alpha_1 \cdot X_1 = B \cdot X_1^k \cdot \beta - \alpha_1 \cdot X_1 \\ \dfrac{dX_2}{dt} = F' - \beta \cdot F' - \alpha_2 \cdot X_2 = B \cdot X_1^k \cdot (1 - \beta) - \alpha_2 \cdot X_2 \end{cases}, \tag{17}$$

Where F' is a modified production function (taking into account investment costs, that is, $F' = F - I$), while, in accordance with the above, it is assumed that $F' = BX_1{}^k$, where B is the labor productivity, the multiplier $X_1{}^k$ characterizes the "labor" factor, taking into account diminishing returns ($k < 1$); β is the wage coefficient (characterizes the share of wages in the cost of manufactured products); α_1 and α_2 are coefficients that characterize the share of funds of the wage workers and the entrepreneurs that they spend on consumption.

2. Basic contradictions.

The basic contradictions are *economic* (arising primarily about the payroll of workers). In relation to model (17), these are contradictions over the value of the wage coefficient β: entrepreneurs want a decrease in the value of β, and employees want it to increase. This divergence of interests leads to the fact that employees fight for an increase in wages, while entrepreneurs are opposed to this. As a result of social interaction (confrontation), a temporary equilibrium is established (and the value of β corresponding to it), fixing the achieved compromise for the period until the next change in the situation.

3. The current state of the social system, the mechanisms for its maintenance and stabilization.

If and when a compromise on β is reached, the phase portrait of the system (17) has the form shown in Fig. 9, where $X_{01} = x_{01} \cdot N_1$ (x_{01} is the level of physical survival).

It follows from the figure that in an industrial society with a fixed β, there is a state of stable equilibrium. The equilibrium values $\underline{X}_1$ and $\underline{X}_2$ (corresponding to the coordinates of the attractor point **A** in Fig. 9) are as follows:

$$\begin{cases} \underline{X}_1 = \left(\frac{\beta \cdot B}{\alpha_1}\right)^2 \\ \underline{X}_2 = \dfrac{B^2 \cdot \beta \cdot (1-\beta)}{\alpha_1 \cdot \alpha_2} \end{cases}, \tag{18}$$

where $k = 0.5$ is taken to simplify the equations.

The phase portraits of the agrarian [Fig. 5 in Sect. 2 of chapter "Modeling Social Self-Organization and Historical Dynamics. Agrarian Society" (Malkov et al., 2023b)] and industrial (Fig. 9) societies are very similar. The fundamental difference is that in agrarian society, due to the invariance of F, the position of the equilibrium point **A** at given values of β and γ does not change over time, while in industrial society, it gradually changes (shifts towards increasing values of $\underline{X}_1$ и $\underline{X}_2$) due to a progressive increase in labor productivity [cp. chapter "Modeling Social Self-Organization and Historical Dynamics. Agrarian Society" (Malkov et al., 2023b, this volume)].

An analysis of expressions (18) with a change in the value of β shows that with high values of β, the ratio of $\underline{X}_1$ and $\underline{X}_2$ shifts in favor of workers, and with a decrease in β, it shifts in favor of entrepreneurs. At the same time, very low values of β are unprofitable for entrepreneurs since the impoverishment of employees reduces the quality of their work. In addition, the reduction of $\underline{X}_1$ to the level of physical survival X_{01} entails a sharp increase in social instability due to the growing discontent of workers. This, in turn, reduces the current value of labor productivity B (due to the high level of social conflict, protests, strikes, etc.) and leads to a general decrease in production volumes, which devalues the decrease in the value of β for entrepreneurs. In this regard, there is an optimal value of β for entrepreneurs [in terms of achieving the maximum value of $\underline{X}_2$ in Eq. (18)], which at $k = 0.5$ is equal to

$$\beta_{\text{opt}} = 0.5. \tag{19}$$

The maximum value of $\underline{X}_2$ in this case is equal to

$$\underline{X}_2 = \frac{\mathrm{B}^2}{4 \cdot \alpha_1 \cdot \alpha_2}. \tag{20}$$

Thus, the current state of the social system is established as a compromise between entrepreneurs and workers about the value of β (i.e., about the size of wages). However, the nature of establishing this compromise in the nineteenth and twentieth centuries was different. In the nineteenth century and the early twentieth century, entrepreneurs (capitalists) sought to minimize the value of β in order to reduce their production costs (which included the wages paid to workers) and increase profits. The resistance of workers to lower wages took the form of class struggle (strikes and social movements). This forced the entrepreneurs to make certain concessions and move to social state policies (see, e.g., Zinkina & Andreev, 2020); as a result, a compromise was established in terms of the value of β. Since the "Great Depression" of the 1930s, it has become clear to entrepreneurs that it is more profitable not to reduce β, but to set it at the level of β_{opt}, especially since by that time the workers had become the main consumers of manufactured products and provided solvent demand. For its part, high demand made it possible to increase production and stimulated technological development and growth in labor productivity (e.g., Greenberg et al., 2008).

The tools for achieving and maintaining the achieved compromise were as follows:

- In the sphere of governance, they included the formation of a legal system (including labor legislation regulating the conflicts and disputes between entrepreneurs and workers) and organizational forms of interaction between entrepreneurs and workers (trade unions, etc.). The legal system fixed the "rules of the game" and established punishment for their violation (due to this, the utilitarian component of psychological impact a^G was actualized; on the utilitarian component, see Ilyin et al., 2020).
- In the socio-psychological sphere, they included the legitimization of the legal system through the introduction of democratic procedures (elections, referendums, etc.), which made it look like all opinions were taken into account and motivated the population to comply with the laws [due to this, the deontological component of psychological impact a^F was actualized; on the deontological component, see (Ilyin et al., 2020)]. In this case, religion is no longer required; moreover, the conservatism of religion and tradition reduces the flexibility of the procedure for establishing a compromise, so industrial society is not religious, the legitimation of compromises is carried out not by religion, but by ideology (liberal values, democracy, human rights, etc.).

4. Possible reasons for changing the current state.

The reasons for changing the current equilibrium (current values of β) can be:

- Economic crisis, slowdown in growth, which leads to a decrease in profits (unprofitability of production) and encourages entrepreneurs to reduce wages (reduce β). In times of severe crises, when the equilibrium value $\underline{X}_1$ decreases to the level of physical survival X_{01}, the initiators of the upward change in β are workers.
- In an era of economic growth, on the wave of positive expectations from workers, an initiative is possible to change β upwards.

5. Assessment of the risk of social system destabilization.

If there are factors (causes) that violate the previously existing balance, the following is possible:

either a transition to new values of β (and to new equilibrium values of $\underline{X}_1$ and $\underline{X}_2$) on the basis of a new social compromise (on the basis of legitimized legal procedures for resolving labor disputes);

or social system destabilization, since a change in β leads to $\underline{X}_1$ and $\underline{X}_2$ going beyond the range of permissible values,[19] as a result of which the stable attractor shown in Fig. 9 disappears and the system ceases to exist in its former form.[20] In real life, social desynchronization is expressed in the form of strikes, mass demonstrations, riots (on the part of workers) and retaliatory repressive measures (on the part of entrepreneurs). As a result, a new value of β and a new system of its legitimation are established (for example, the suspension of the constitution and the declaration of a state of emergency).[21]

2.1.2. Stability of Social Interactions

In this chapter, we use a mathematical model to analyze the impact of competitive interactions on the evolution of the economic system in industrial society. The purpose of mathematical modeling is to determine the conditions under which the sustainable functioning of the economic system is possible in the presence of market competition between producers. At the microeconomic level, competition between different firms is considered in many works (see, e.g., Fama & Laffer, 1972; Matsubayashi & Yamada, 2008; Bourreau et al., 2011), some of which use mathematical modeling. Mathematical modeling for the analysis of competitive relations at the mesoeconomic level is used much less frequently, but it is this level that is of greatest interest to us, since it becomes possible to analyze the situation as a whole, minimize the number of exogenously specified parameters, and identify internal mechanisms of self-organization of the economic system in various external conditions.

The model presented below is of a basic nature. The task of the model is to identify the most general patterns of evolution of a competitive market economy in

[19] Permissible values are restricted by the level of physical survival for workers, and zero profit for entrepreneurs.

[20] The reason for the destabilization may also be a decrease in real income growth rates in relation to the expected ones (e.g., Davies, 1962; Korotayev & Shishkina, 2020).

[21] On the so called Marxian trap see, e.g., Grinin (2013a, 2013b, 2022).

various conditions.[22] To facilitate the analysis, we will consider the simplest model case, namely, a closed[23] economic system consisting of two subsystems that produce the same type of goods, and two groups of households that work in these subsystems, receive wages, and buy products produced by the subsystems with the money they earn. Since the goods are of the same type, households can buy products of either subsystem in accordance with their preferences, while the effective demand of households is limited by their wages. The model describes the dynamics of funds of these economic actors: M_{Y1} is funds of the first subsystem, M_{Y2} is funds of the second subsystem, M_{H1} is funds of the first group of households, M_{H2} is funds of the second group of households.

As a starting point for the analysis, we consider the situation when subsystems produce identical products that enjoy the same consumer demand. In this case, the system of equations describing the dynamics of cash can be written as follows:

$$dM_{Y1}/dt = (k_1 \cdot M_{H1} + k_2 \cdot M_{H2}) \cdot (F_1/(F_1 + F_2)) - p \cdot h_1 \cdot F_1 \tag{21}$$

$$dM_{H1}/dt = p \cdot h_1 \cdot F_1 - k_1 \cdot M_{H1} \tag{22}$$

$$dM_{Y2}/dt = (k_1 \cdot M_{H1} + k_2 \cdot M_{H2}) \cdot (F_2/(F_1 + F_2)) - p \cdot h_2 \cdot F_2 \tag{23}$$

$$dM_{H2}/dt = p \cdot h_2 \cdot F_2 - k_2 \cdot M_{H2} \tag{24}$$

$$dp/dt = a \cdot ((k_1 \cdot M_{H1} + k_2 \cdot M_{H2})/(F_1 + F_2 - k_{Y1} \cdot M_{Y1}/p - k_{Y2} \cdot M_{Y2}/p) - p) \tag{25}$$

$$F_i = f_i \cdot (k_{Yi} \cdot M_{Yi}/p)^{ci} \tag{26}$$

$$M_{H1} + M_{H2} + M_{Y1} + M_{Y2} = M \tag{27}$$

Equations (21) and (23) describe the dynamics of funds of the first and the second production subsystems, respectively. The right-hand parts of the equations reflect the balance of income and expenses at the considered point in time. It is assumed that income is the money that the subsystems receive by selling goods they produced to households (it is assumed that the flow of this money is proportional to the amount of consumer products produced by the subsystem F_i and the total funds of households; k_1 and k_2 are proportionality coefficients). Subsystem expenditures are salaries paid to their group of households for their work in production (it is assumed that wages are piecework, so they are proportional to the quantity of products produced and are

[22] See also Grinin (2022), as well as chapter "High Income and Low Income Countries. Toward a Common Goal at Different Speeds" (Grinin et al., 2023b, this volume) and chapter "Economics. Optimizing Growth" (Grinin et al., 2023c, this volume).

[23] The condition of closedness is fundamental. It means that all commodity and cash flows are localized within the considered economic system, that is, the condition of system autonomy is fulfilled.

indexed to actual inflation; h_1 and h_2 are proportionality coefficients; p is the price level).

The second and fourth equations describe the dynamics of funds of the first and the second groups of households, respectively. The right-hand parts of the equations reflect the balance of income and expenses at the considered point in time. Income is considered to consist of the wages received from the subsystems, and expenses are the money spent on the purchase of consumer goods produced by the subsystems. Subsystem incomes are household expenditures, and subsystem expenditures are household incomes.

The fifth equation describes the dynamics of the price level. It is assumed that it varies depending on the balance of money demand for consumer products and the quantitative supply of these products. The quantity of consumer output is defined as all manufactured output minus industrial output (needed to support the production process).

The sixth equation is the production function. It is assumed that the total amount of products produced by subsystems is proportional to their monetary investment in production capital (equal to $k_{Y1}{\cdot}M_{Y1}$ and $k_{Y2}{\cdot}M_{Y2}$, respectively) taking into account inflation and the effect of diminishing returns (c_1 and c_2 are diminishing returns coefficients).

The seventh equation reflects the closedness of the economic system (M is the total amount of money in the system) and is essentially the law of conservation of money in the system.

To simplify the analysis, we will assume that the rate of price establishment is significantly higher than the rate of change of M_i (which, as a rule, is well performed in efficiently functioning market economies). Then, according to Tikhonov's theorem, Eq. (25) can be written as

$$k_1 \cdot M_{H1}/p + k_2 \cdot M_{H2}/p - (F_1 + F_2 - k_{Y1} \cdot M_{Y1}/p - k_{Y2} \cdot M_{Y2}/p) \approx 0 \tag{28}$$

To simplify the analysis, let us assume that $k_i = k$. Then, it follows from Eqs. (27) and (28) that

$$p \approx k \cdot M/(F_1 + F_2) \tag{29}$$

Accordingly, it follows from Eqs. (21), (23), and (29) that

$$\begin{aligned} dM_{Y1}/dt &\approx k \cdot (M - M_{Y1} - M_{Y2}) \cdot (F_1/(F_1 + F_2)) - k \cdot M \cdot h_1 \cdot F_1) \cdot (F_1/(F_1 + F_2)) \\ &= k \cdot (M \cdot (1 - h_1) - M_{Y1} - M_{Y2}) \cdot (F_1/(F_1 + F_2)) \end{aligned} \tag{30}$$

$$\begin{aligned} dM_{Y2}/dt &\approx k \cdot (M - M_{Y1} - M_{Y2}) \cdot (F_2/(F_1 + F_2)) - k \cdot M \cdot h_2 \cdot (F_2/(F_1 + F_2)) \\ &= k \cdot (M \cdot (1 - h_2) - M_{Y1} - M_{Y2}) \cdot (F_2/(F_1 + F_2)) \end{aligned} \tag{31}$$

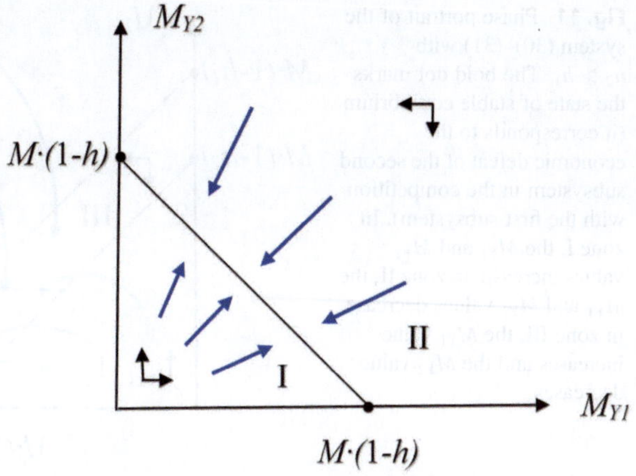

Fig. 10 Phase portrait of system (30)–(31) with $h_i = h$. Arrows indicate phase trajectories. In zone I, the values of M_{Y1} and M_{Y2} increase; in zone II, the values of M_{Y1} and M_{Y2} decrease

Consider a situation where the amount of money in the economic system is unchanged, that is, there is no money emission. This means that the condition is satisfied:

$$M(t) = const. \tag{32}$$

To construct a phase portrait of system (30) and (31), it is necessary to determine the expressions for the isoclines. They can be obtained by equating the right-hand sides of Eqs. (30) and (31) to zero. As a result, we get:

$$M \cdot (1 - h_1) - M_{Y1} - M_{Y2} = 0 \tag{33}$$

$$M \cdot (1 - h_2) - M_{Y1} - M_{Y2} = 0 \tag{34}$$

If the costs of production per unit are the same for both subsystems, that is, $h_i = h$, then the phase portrait has the following form (see Fig. 10):

A characteristic feature of this phase portrait is that the isoclines (33) and (34) coincide; this means that the system has an infinite number of equilibrium states, which include all points of the combined isoclines. This, in turn, means that the phase trajectories are pulled together to the isoclines, where they slow down. Thus, as a result of competitive interaction, a simple reproduction mode is eventually established, in which the situation stabilizes, the values of M_{Y1} and M_{Y2} stop changing, incomes are equal to expenses, net profit is zero, while each of the subsystems occupies its own economic niche without suppressing each other. This is an important result showing that equilibrium states are possible in a closed system under conditions of competition between producers, and there can be many such states. However, the considered situation (the final equilibrium of competing subsystems) is possible only if the equality of the values h_i is strictly satisfied.

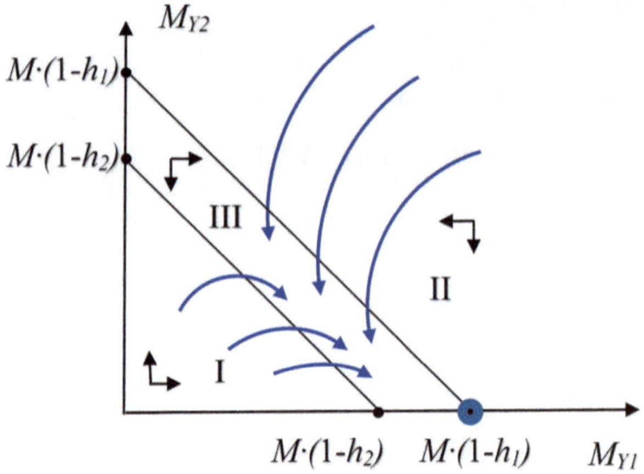

Fig. 11 Phase portrait of the system (30)–(31) with $h_2 > h_1$. The bold dot marks the state of stable equilibrium (it corresponds to the economic defeat of the second subsystem in the competition with the first subsystem). In zone I, the M_{Y1} and M_{Y2} values increase, in zone II, the M_{Y1} and M_{Y2} values decrease, in zone III, the M_{Y1} value increases and the M_{Y2} value decreases

If the costs of production per unit for one of the subsystems are higher than for the other, for example, $h_2 > h_1$, the phase portrait of system (30)–(31) changes significantly and takes the form:

With $h_2 \neq h_1$, the isoclines cease to coincide and zone III appears in the phase portrait, where the value of M_{Y1} increases and the value of M_{Y2} decreases. At the same time, the possibility of stable coexistence of two subsystems disappears, and the subsystem with higher costs ultimately suffers an economic defeat: competition leads to the monopoly of the first subsystem.

A similar situation arises if the first subsystem starts producing products that are more attractive to consumers, and consumer demand shifts towards this subsystem. If the additional demand for the products of the first subsystem is δ, then the system of Eqs. (21)–(27) is transformed into the following form:

$$dM_{Y1}/dt = (k_1 \cdot M_{H1} + k_2 \cdot M_{H2}) \cdot (F_1 \cdot (1+\delta)/(F_1 + F_2)) - p \cdot h_1 \cdot F_1 \quad (35)$$

$$dM_{H1}/dt = p \cdot h_1 \cdot F_1 - k_1 \cdot M_{H1} \quad (36)$$

$$dM_{Y2}/dt = (k_1 \cdot M_{H1} + k_2 \cdot M_{H2}) \cdot (F_2/(F_1 + F_2) - F_1 \cdot \delta/(F_1 + F_2)) - p \cdot h_2 \cdot F_2 \quad (37)$$

$$dM_{H2}/dt = p \cdot h_2 \cdot F_2 - k_2 \cdot M_{H2} \quad (38)$$

$$dp/dt = a \cdot p \cdot (k_1 \cdot M_{H1}/p + k_2 \cdot M_{H2}/p - (F_1 + F_2 - k_{Y1} \cdot M_{Y1}/p - k_{Y2} \cdot M_{Y2}/p)) \quad (39)$$

$$F_i = f_i \cdot (k_{Yi} \cdot M_{Yi}/p)^{c_i} \quad (40)$$

$$M_{H1} + M_{H2} + M_{Y1} + M_{Y2} = M \quad (41)$$

Equation (37) accounts for the fact that an increase in demand for the products of the first subsystem, due to the closed nature of the economic system, automatically leads to a decrease in effective demand for the products of the second subsystem.

It can be easily shown that the phase portrait of system (35)–(41) is similar to the one shown in Fig. 11; accordingly, even a slight advantage of the products of the first subsystem over the products of the second subsystem ultimately leads to the bankruptcy of the latter.[24]

Thus, the situation of competitive interaction characterized by the equations under consideration is *fundamentally unstable* and, if condition (32) is met (i.e., in a "zero-sum game"), it inevitably leads to *monopolism* due to the presence of positive feedbacks, as a result of which those who, for one reason or another, have achieved an advantage in the competitive struggle, receive in the future even more opportunities to strengthen their positions.[25]

However, if we move from the "zero-sum game" to the "positive-sum game", that is, to a situation where the total resource in the system is continuously growing (in our case, this means the growth of $M(t)$ as a result of the constant emission "pumping" of money into the economic system[26]), the emergence of monopoly can be avoided. In this case, the functioning of the economic system in the competitive mode can continue indefinitely. In the language of phase portraits, this can be explained as follows.

Consider the situation depicted in Fig. 11, where it is assumed that $h_2 > h_1$. We will further assume that the initial state is in zone I, where both M_{Y1} and M_{Y2} are growing. If the "zero-sum game" condition (32) were satisfied, the trajectory of the system would reach isocline (34) and turn in the direction of the attractor indicated by the thick dot, which means the suppression of the second subsystem by the first subsystem. However, under the conditions of the "*game with a positive sum*", when the value of $M(t)$ is constantly increasing, isocline (34) continuously moves away from the zero point, expanding zone I. Accordingly, despite the growth of M_{Y1}, and M_{Y2}, the phase trajectory does not leave zone I and does not move into zone III, in which the first subsystem fatally suppresses the second and eventually becomes a monopolist.

The expansion of zone I gives the second subsystem a chance to turn the tide and try to get ahead. To do this, it needs additional financial resources that it could use to gain advantages in the competition with the first subsystem. Additional financial resources are provided by a bank loan, and benefits can be obtained through the development and implementation of innovations that increase labor productivity

[24] The situation is similar if the first subsystem seeks to gain an advantage through price dumping.

[25] In real situations, to prevent monopoly, antimonopoly laws and special regulations are adopted, but these are external regulation measures. We are interested in the dynamics of competitive interactions in their "pure" form.

[26] This may be a loan issue to enterprises and/or the population, or an increase in the money supply due to an increase in government spending or due to a positive balance of international trade, etc.

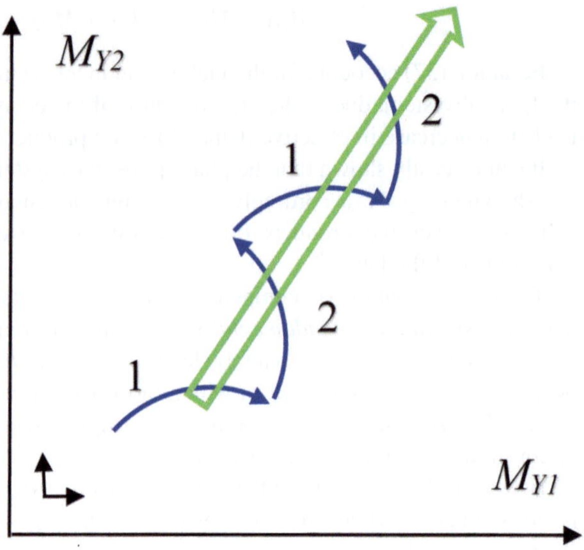

Fig. 12 The phase trajectory in the case when either the first or the second subsystem alternately takes the lead due to the periodic introduction of technological innovations to gain an advantage in the competitive struggle: (1)—sections of the phase trajectory when the first subsystem receives a technological advantage; (2)—sections of the trajectory when the second subsystem receives a technological advantage. The bold arrow indicates the overall growth of the economic system

and/or improve the quality and attractiveness of manufactured goods. If the second system succeeds in this, it outpaces the first system which, in its turn, is forced to catch up. And so on (see Fig. 12).

Thus, a necessary condition for competition not to lead to monopolism is the continuing growth of the system, as well as the constant creation of conditions for a "positive sum game". That is why liberal market economies are so concerned about maintaining high growth rates. Economic growth in competitive market systems is not so much the desire to ensure the material well-being of citizens (as is often stated for propaganda purposes), but the desire to ensure the sustainability of the functioning of a competitive economy. The presence of competition and the threat of bankruptcy force economic agents to make every effort to maintain their business, stimulate economic activity, and compel them to introduce innovations,[27] which, in turn, creates conditions for the continuation of overall economic growth.[28] The system becomes self-consistent: the "positive-sum game" maintains the stability of competitive interactions, and competition, in turn, stimulates activity and economic growth, creating conditions for the "positive-sum game".

[27] This "being forced to innovate" is the reason why in the industrial age the value of coefficient c in the equation of technological dynamics (9) in chapter "Modeling Social Self-Organization and Historical Dynamics. A General Approach" (Akaev et al., 2023a, this volume) becomes higher than in the agrarian age.

[28] In relation to the competitive market economy, we can only talk about dynamic equilibrium, similar to the equilibrium of a riding cyclist, who, in order not to fall, must pedal with all the might.

The need to ensure a "positive sum game" dramatically increases the role of the financial system in an industrial society, since it helps to increase the money supply. At the same time, in an industrial society based on competition and private property, the financial system uses its privileged position to strengthen its economic and political power.

At the same time, it is important to understand that money emission alone is not enough to ensure economic growth: real growth is possible only with the expansion of production and the necessary resource base.[29] When it is depleted, growth stops, despite the injection of money into the economy (these injections only lead to higher inflation[30]). In this case, competition leads to monopoly and/or to a distributive system, where relations in society are strictly regulated (an example of this is the concept of "inclusive capitalism" promoted by Schwab & Malleret, 2020).

2.2 Sociopolitical Stability

In industrial society, there is no dominant material resource (similar to land in agrarian society) [chapter "Modeling Social Self-Organization and Historical Dynamics. Agrarian Society" (Malkov et al., 2023b, this volume)]. Moreover, technological innovations constantly change the resource field, this leads to a periodic change in economic leaders, and after them, political leaders. In the "positive sum game" with economic competition, the establishment of monopoly becomes difficult. At the same time, there emerges an intentional resistance to the tendency towards monopolism by following the strategy of "uniting the weak against the strong" strategy (in other words, conscious opposition to the strengthening of any particular economic or political actor), institutionalizing the principle of separation of powers, the rule of law, and the independence of judges.[31] That is, the regime of "checks and balances", shown in Fig. 19 in Sect. 3.2.2 of chapter "Modeling Social Self-Organization and Historical Dynamics. A General Approach" (Akaev et al., 2023a, this volume), emerges and consolidates.

In the context of the "multi-actor" political and economic life and the growing interdependence of the main social groups, it is an adaptive democratic rather than directive authoritarian governance system[32] that becomes more effective and efficient, allowing for a more rapid social compromise and "synchronization" of society

[29] Western countries, starting from the eighteenth century, supported the growth of the resource base and the "positive sum game" for their economies, in particular, through the policies of colonialism and neo-colonialism.

[30] Growth can also be hampered by increased costs, for example, to address the growing environmental problems.

[31] In this sense, the history of Britain in the second half of the seventeenth to early eighteenth centuries is indicative, when, after half a century of civil wars, new rules of political struggle were established, same for all.

[32] Shown in Fig. 13 in Sect. 3.2.1 of chapter "Modeling Social Self-Organization and Historical Dynamics. A General Approach" (Akaev et al., 2023a, this volume).

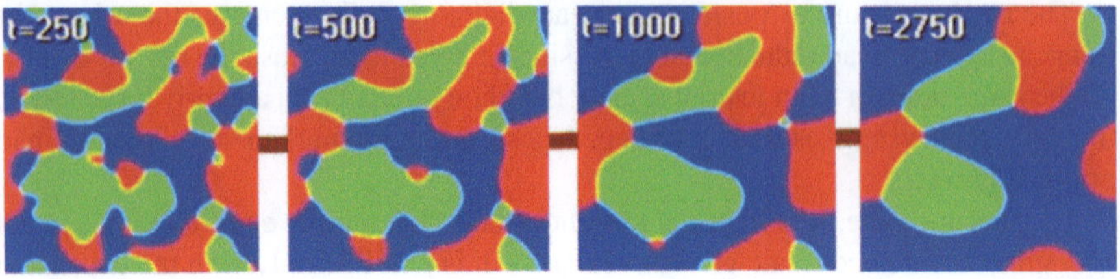

Fig. 13 Dynamics of cluster enlargement when modeling competitive interactions with the use of model (45) presented in chapter "Modeling Social Self-Organization and Historical Dynamics. A General Approach" (Akaev et al., 2023a, this volume) (the numbers in the upper left corner of each frame correspond to a respective cycle of calculation)

in the event of changes in external and internal circumstances. This system is beginning to be called "democratic" because it is based on procedures for coordinating interests and finding compromises, which, until a significant change in circumstances, become mandatory for everyone. As is shown in Sect. 3.2.1 of chapter "Modeling Social Self-Organization and Historical Dynamics. A General Approach" (Akaev et al., 2023a, this volume), this system is stable and reproducible, but a necessary condition for this in an internally competitive society is the presence of opportunities for a "positive sum game" (see Sect. 2.1.2).

2.3 Inter-state Interactions

Unlike agrarian society, where subsistence farming and local economic systems dominated, industrial society is based on the division of labor and is interested in sales markets and economic ties. Thus, in industrial society there is a natural tendency to enlarge and homogenize social systems. These processes correspond to stage II of the dynamics described by basic model (44) in Sect. 3.2.3 of chapter "Modeling Social Self-Organization and Historical Dynamics. A General Approach" (Akaev et al., 2023a, this volume), which reflects the enlargement of territorial state entities (clusters) that are in competitive relations with each other. The simulation results of this process are presented in Fig. 13 [they correspond to frames 4–7 of Fig. 20 in Sect. 3.2 of chapter "Modeling Social Self-Organization and Historical Dynamics. A General Approach" (Akaev et al., 2023a, this volume)].

In essence, the frames of calculation cycles in Fig. 13 can serve as an illustration of the process of formation and development of nation-states in Europe in modern times. For comparison, Fig. 14 shows how the political map of France changed during the transition from feudal fragmentation to the formation of a single national state.

As a result of this process, states grew in their territorial sizes and became more homogeneous: with uniform laws, a single financial system, a single official language, and a single education system. A new concept of "nation" emerged, which united local social groups, previously divided along ethnic, class and territorial lines,

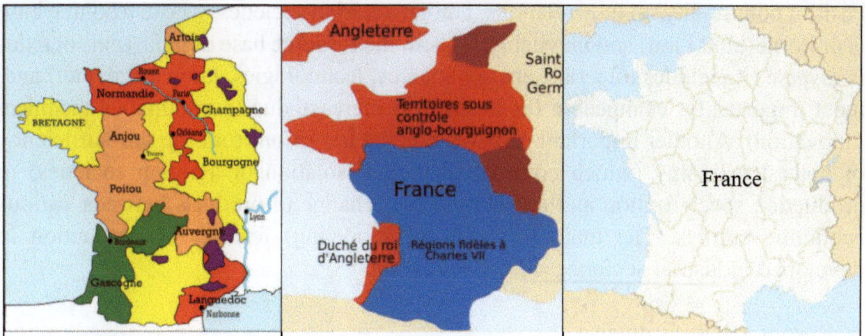

13th century (https://commons.wikimedia.org/wiki/File:Carte_France_1270.svg)	**15th century** (https://commons.wikimedia.org/wiki/File:Guerre_de_cent_ans_(1435).svg)	**Late 18th century** (https://commons.wikimedia.org/wiki/File:France_Regions_et_departements.png)

Fig. 14 Changes in the political map of France from the thirteenth to the early nineteenth century

into the "friend-friend" system. The interaction "friend or foe" shifted to the level of relations between nation-states.

2.4 Patterns of Social Self-Organization

The features of social interactions inherent in industrial society, which were discussed above in Sects. 2.1–2.3, allow us to speak of self-organization mechanisms characteristic of this society, which result in the formation and stable reproduction of social structures of a certain type. In works (Kirdina, 2004, 2014a, 2014b; Malkov, 2009), they are called Y-type social structures [see chapter "Socio-Political Transformations. A Difficult Path to Cybernetic Society" (Grinin et al., 2023d, this volume) for detail].

To ensure the stability of the social system to external and internal destabilizing influences, three main problems must be solved [see Sect. 1.1 of chapter "Modeling Social Self-Organization and Historical Dynamics. A General Approach" (Akaev et al., 2023a, this volume)]:

(A) Providing for the material needs of members of society.
(B) Pnsuring socio-psychological stability against the backdrop of different interests of the social system elements.
(C) Pnsuring effective governance.

The economic basis of developed industrial societies is the production of industrial products and their market redistribution among the members of society on the basis of commodity–money relations. Unlike agrarian societies, whose resource base is uniform (land fund) and inevitably limited, the resource base of industrial societies is diverse (it includes minerals, energy sources, technologies, knowledge, etc.) and, most importantly, changeable (as it tends to increase due to continuous technical innovation). Another important difference is the transition from the self-sufficiency of natural economy (which contributed to the isolation of agrarian societies) to production specialization and active market exchange of products between various economic entities. The main (in terms of modeling) features of production in developed industrial societies are the following:

- Resource base is diverse.
- The volume of the resource changes (due to the emergence of new technologies).
- The method of processing the resource base is the mechanized labor of workers (machines plus workers).
- The threshold of production function *F* is high: significant investments are needed to organize production [see curve 3 of Fig. 1 in chapter "Modeling Social Self-Organization and Historical Dynamics. A General Approach" (Akaev et al., 2023a, this volume)].
- The saturation of function *F* is weak [see curve 3 of Fig. 1 in chapter "Modeling Social Self-Organization and Historical Dynamics. A General Approach" (Akaev et al., 2023a, this volume)]: it is determined by the applied technologies, which are constantly being improved.
- Consumption does not have saturation, since it is determined mainly by durable goods intended to meet social needs (comfort, prestige).
- Social structure is complex and changeable;[33] the main social strata are workers (direct producers of material and intellectual products) and entrepreneurs (owners of the means of production, employers).

Consequences of production features:

- There is no resource limit (or it seems that there is not), it is constantly moving forward due to the development of technology. Supply outstrips demand, the problem is not to produce, but to sell the products and justify the investment. Competition stimulates the development of technologies and the search for new resources, that is, there is positive feedback that determines the dynamism of society and the lack of stability.
- Unlike peasants in agrarian society, hired workers are not self-sufficient, they depend on the employer for getting jobs and wages. In industrial society, there is interdependence: hired workers are interested in the availability of work (that is, in entrepreneurs), while entrepreneurs are interested in the purchasing power of

[33] The social structure of a feudal society is perhaps even more complex than an industrial one due to many gradations and ranks.

workers, since they consume the goods produced. The determining factor is not the use of force, but rather economic coercion.

– The threshold of production function F in an industrial society is overcome with the help of specially created technologies. Employment is carried out individually, employees are rather weakly connected with the production team. Hence, individualism develops even among hired workers. Due to the fact that urbanization is growing, and in cities, there are smaller families than in rural areas, as well as with the development of health care technologies (including the family planning ones), proliferation of female education and a number of other factors (e.g., Chesnais, 1992; Caldwell et al., 2006; Livi-Bacci, 2012) the birth rate is gradually starting to decline. There is a demographic transition (e.g., Kapitza, 2006, 2010; Korotayev et al., 2006a, 2015; Dyson, 2010; Reher, 2011; Erçetin et al., 2019): population growth in industrialized countries stops. Having few children removes the threat of demographic crises and, at the same time, leads to an increase in the psychology of individualism.

The ways of solving problems (**A**), (**B**) and (**C**) in a developed industrial society are as follows:

(A) The resources are not limited (or this seems to be so) due to the permanent emergence of new technologies and the introduction of new types of resources. In this society, it is possible to achieve a local supply–demand balance, but it is unstable and constantly disturbed due to the expansion of production, leading to crises of overproduction. Thus, while an agrarian society is characterized by crises of underproduction (sociodemographic crises) due to a limited resource base, in an industrial society, the situation is reversed: due to the constant growth of labor productivity and the expansion of the resource base, it is characterized by periodic crises of overproduction.

Under these conditions, entrepreneurs can increase their income by:

(a) Reducing the costs by lowering the wages of workers. But then the purchasing power of the population decreases, a crisis of overproduction arises, and entrepreneurial income falls. Thus, in a developed industrial society, this strategy is not optimal.
(b) Increasing labor productivity (attracting more demand for their products by lowering prices without loss of profitability). This path requires a permanent search for new resources and technologies and stimulates competition.

It is impossible to overcome the threshold of function F only by combining the efforts of workers in production. It requires considerable expenditures on equipment, raw materials, workers' wages, etc. This is done by entrepreneurs, using the credit system existing in the society.

In order for manufactured products to be bought, effective demand is needed. Accordingly, the main buyers (workers) must be relatively wealthy. Therefore, bimodal economic structure of society (ESS) (i.e., a sharp polarization of property) in an industrial society is not beneficial. Unimodal ESS is more efficient [see Fig. 2b in chapter "Modeling Social Self-Organization and Historical Dynamics. A General Approach" (Akaev et al., 2023a, this volume)], where the so-called middle class prevails. It is profitable for entrepreneurs to pay high wages to workers, but on the condition that they spend the money they received on buying the goods, so that entrepreneurs receive income. Under such a system, entrepreneurs get richer not because workers get poorer, but because the ESS curve (see Fig. 2b in chapter "Modeling Social Self-Organization and Historical Dynamics. A General Approach" (Akaev et al., 2023a, this volume)] shifts entirely to the right. This is possible only if the resource in the system is constantly growing. If the growth of the resource stops, due to the properties of v [see Sect. 1 of chapter "Modeling Social Self-Organization and Historical Dynamics. A General Approach" (Akaev et al., 2023a, this volume)], the unimodal ESS quickly transforms into a bimodal one.

(B) With a unimodal ESS, there is no need for religious or ideological mechanisms to ensure socio-psychological stability.[34] Something else becomes necessary. As has been mentioned above, the economic stability of industrial society is possible with consistent effective demand. This means that not only must the workers have money, but they must be willing to buy. The desire to buy becomes a factor in ensuring the stability of production and society as a whole. Therefore, the society's propaganda machine is aimed at arousing desires ("take everything from life"), at forming a "consumer society" (in contrast to agrarian society, in which asceticism is preached). This results in de-ideologization and pluralism, on the one hand, and switching consciousness from spiritual values to material ones, primitivization, dominance of mass culture, and so on. The constantly heated desire to buy unwinds a spiral of demand, and with it an ever-increasing cost of resources. That is, positive feedback is realized, making industrial society increasingly resource-consuming.

In the socio-psychological sphere, there is a decrease in the "friend – foe" confrontation that was characteristic of pre-industrial societies. Tolerance increases due to the need to establish and maintain numerous contacts with different people in the course of production and trade activities.

(C) The authoritarian (directive) governance model does not correspond to the nature of the developed industrial society. The constant emergence of new technological leaders and industries that provide super-profits does not allow the production elite to stabilize, constantly changing the balance of power.

[34] On the religious foundations of the entrepreneurial ideology of early capitalism and the role of Protestantism in the emergence of capitalism (see Weber, 2004 [1905]; Korotayev et al., 2006a: 87–91; Grinin et al., 2022; Grinin, 2022).

There is a constant "social mixing", as a result of which positive feedbacks in the field of redistribution [see Sect. 1 of chapter "Modeling Social Self-Organization and Historical Dynamics. A General Approach" (Akaev et al., 2023a, this volume) on property *v*] do not lead to rigid hierarchization of the elite, which was characteristic of agrarian society. The absence of a single center of power, the ongoing competition of various political and economic groups lead to the fact that for the type of society under consideration, an adaptive (democratic) model of governance with constant adjustment of the status quo in power structures through periodic elections becomes more adequate.

Comparison of ways to solve problems (**A**), (**B**) and (**C**) in industrial and agrarian societies shows that changing conditions and methods of resource use lead to a fundamental change in strategies for ensuring socio-economic sustainability. What was natural and useful in agrarian societies becomes unnatural and harmful in industrial ones. The psychology of asceticism is replaced by the psychology of consumerism, the threat of overpopulation is replaced by the threat of depopulation, and so on [see Sect. 3 of chapter "Modeling Social Self-Organization and Historical Dynamics. Global Phase Transitions" for more detail].

3 Modernization Processes and Their Modeling

The process of transformation of traditional agrarian societies [see chapter "Modeling Social Self-Organization and Historical Dynamics. Agrarian Society" (Malkov et al., 2023b, this volume)] into industrial societies (described in the present chapter) is called the process of socio-economic and political modernization. In the history of each country, this process takes a fairly long period of time and has different forms. But with all the variety of these forms, modernization as a historical phenomenon has well-defined internal patterns and features that can be described using mathematical modeling methods (Apter, 1965; Bendix, 1967; Black, 1966; Levy, 1966, 1967; Eisenstadt, 1966, 1978; Smelser, 1967; Collins, 1968; Rostow, 1971; Tipps, 1973; Huntington, 1968, 1986; Przeworski & Limongi, 1997; Christian, 2011; Galor, 2011; Grinin 2013a, 2013b, 2022; Korotayev et al., 2015; Grinin et al., 2015; Korotayev et al., 2020).

3.1 Features of Socio-economic Modernization

The processes of modernization in the nineteenth and twentieth centuries encompassed all countries of the world, but proceeded differently in different countries, with different speed and intensity. This can be clearly seen in Fig. 15, which shows graphs of demographic growth and GDP growth per capita in a number of countries in the period from 1820 to 2008.

The logic of the modernization process can be described as follows.

A society that has embarked on the path of modernization is initially traditional, based on agricultural production using manual labor. The population lives

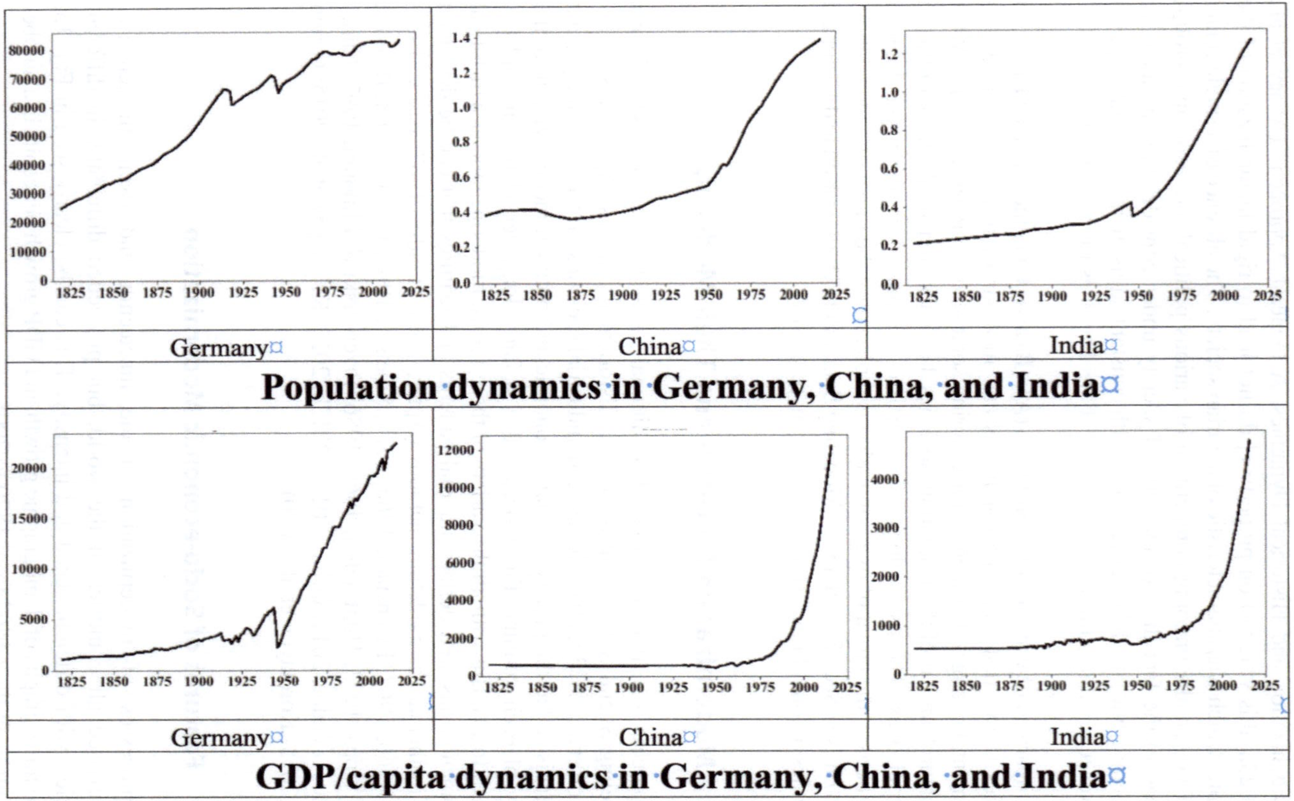

Fig. 15 Dynamics of population (thousand people) and GDP per capita (international 1990 dollars at PPP) in a number of countries of the world in 1820–2008. *Data source*: Maddison Project (2020)

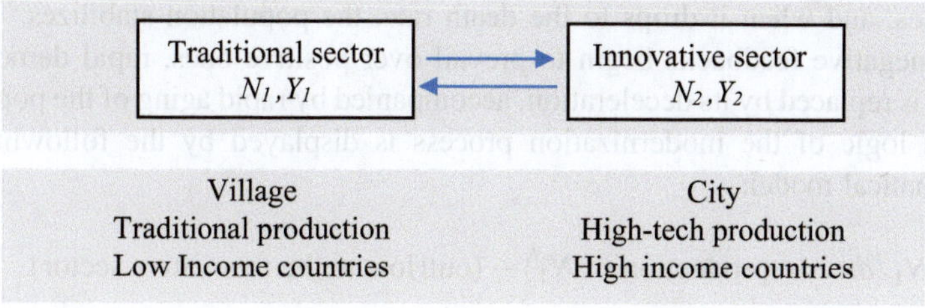

Fig. 16 Scheme of interaction between the traditional and innovative sectors of the society in the process of modernization (*N* is population, *Y* is the level of production)

predominantly in rural areas. Due to the limited resource base, such a society is in the Malthusian trap, the population is limited by the demographic capacity of the territory.

Modernization begins with the emergence of an innovative sector in a traditional society (see Fig. 16), which develops on the basis of new technologies.

The innovation sector can emerge as a result of both internal[35] and external[36] causes. Centers for the development of the innovation sector are located in cities and industrial towns, where people from rural areas begin to move to, supplying the innovation sector with cheap labor. The reason for this migration is that in the traditional sector the level of productivity (and hence the standard of living) is lower than in the innovative sector. The growth of living standards contributes to a decrease in mortality (especially in infant mortality), which (against the background of still high fertility) leads to the acceleration of population growth [chapter "Demography. Toward Optimization of Demographic Processes" (Korotayev et al., 2023a, this volume)]. If, at the same time, labor productivity continues to grow, then the Malthusian trap does not emerge and demographic growth turns into a "population explosion". This is the initial phase of modernization. During this phase, as a rule, strong social and sociodemographic disproportions arise (in particular, the so-called youth bulge[37] is formed), often giving rise to social instability and political crises. But at the same time, positive feedbacks prevail, which lead to rapid growth of both the population and the innovative sector of the economy.

The final phase of modernization begins when the population becomes predominantly urban, the innovation sector prevails and is already spreading to agricultural production. At the same time, along with the growth of the material well-being of the population, the family model shifts to having fewer children. The birth rate

---

[35] An example of this is the development of England since the sixteenth century (e.g., Allen, 2009; Grinin & Korotayev, 2015).

[36] External causes are usually due to the influence of the more high income countries.

[37] See, e.g., Moller (1968), Choucri (1974), Mesquida & Wiener (1996, 1999), Goldstone (2001, 2002, 2016), Urdal (2004, 2006, 2008), Korotayev et al. (2011, 2014, 2021, 2023b), Goldstone et al. (2012), Weber (2019), Romanov et al. (2021), Sawyer et al. (2022).

decreases, and when it drops to the death rate, the population stabilizes.[38] In this phase, negative feedbacks begin to prevail over positive ones, rapid demographic growth is replaced by its deceleration, accompanied by rapid aging of the population.

This logic of the modernization process is displayed by the following basic mathematical model:

$$dN_1/dt = (\text{reproduction of } N_1) - (\text{outflow to the innovative sector}) \approx a_1(t) \cdot N_1 - b(t) \cdot N_1 \cdot N_2 \quad (42)$$

$$dN_2/dt = (\text{reproduction of } N_2) + (\text{inflow from the traditional sector}) \approx a_2(t) \cdot N_2 + b(t) \cdot N_1 \cdot N_2 \quad (43)$$

$$y = (GDP)/(population) = \frac{Y}{N_1 + N_2}, \quad (44)$$

where Y is the gross domestic product (GDP) of the society; y is GDP per capita; N_1 is the number of the population in the traditional sector of the economy; N_2 is the number of the population in the innovative sector of the economy; a_1, a_2 are variable coefficients characterizing the reproduction of the population in the traditional (N_1) and innovative sectors (N_2); b is a variable coefficient characterizing the intensity of migration from the traditional sector to the innovative one (the migration term in the form of the product of N_1 and N_2 reflects the fact that the migration rate is proportional to the number of people in the traditional sector (from which migration occurs), on the one hand, and the "capacity" of the innovative sector to receive migrants, which in the demographic sense is equal to N_2, on the other hand). The dynamics of GDP can be described by the following expression:

$$Y = Y_1 + Y_2 = y_1(t) \cdot N_1 + y_2(t) \cdot N_2, \quad (45)$$

where Y_1 and Y_2 are the GDP produced in the traditional and innovative sectors, respectively; $y_1(t)$ and $y_2(t)$ are GDP per capita in the traditional and innovative sectors, respectively, with $y_1(t) < y_2(t)$.

In the process of modernization, the ratio of $y_1(t)$ and $y_2(t)$ changes. This change can be characterized by the function $g(t)$:

$$g(t) = (y_2(t) - y_1(t))/y_1(t), \quad (46)$$

where the beginning of the modernization process is taken as the reference point of time t.

In the process of modernization, the value of $g(t)$ grows (however, this growth slows down at the end of the modernization process, since the returns from the introduced technologies gradually come to saturation. For high-income countries,

[38] This process can develop further through the so-called second demographic transition (e.g., Lesthaeghe, 2020) and lead to a general decrease in the population.

the growth of $g(t)$ is determined by the logic of the development of the scientific sphere and the rate of introduction of new technologies into production. For low-income countries, the growth of $g(t)$ is determined mainly by the diffusion of innovations from high-income countries.

Equations (42)–(46) represent a basic modernization model. For its further specification, it is necessary to define the form of functions $a_1(t)$, $a_2(t)$, $b(t)$, $y_1(t)$, $g(t)$ included in these equations as time-varying coefficients. Since our goal is an attempt to create a general model of modernization processes based on empirical data, it is desirable that these coefficients would not be functions of time t, but functions of model variables N_1 and N_2, so that the system of equations would be autonomous.[39] From this point of view, it is expedient to describe the change in model coefficients depending on the change in the value of N_2/N_1, which characterizes the ratio of the innovative and traditional sectors in society. Let us consider each of the indicated coefficients successively.

The demographic reproduction coefficient a_1 during the modernization process grows from zero to some constant value (primarily due to a decrease in mortality, especially infant and child mortality in rural areas). The demographic reproduction rate a_2 during the modernization process gradually decreases to zero (due to the decline in the birth rate, especially in urban areas). Since, within the framework of the model, the change in a_1 and a_2 depends on the phase at which society is being modernized (which, in turn, is characterized by the value of N_2/N_1, reflecting the degree of the population involvement in the innovation sector), the coefficients a_1 and a_2 can be represented as functions of N_2/N_1, for example, in the form:

$$a_1 = \frac{a_{11} \cdot \frac{N_2}{N_1}}{\frac{N_2}{N_1} + a_{12}}, \tag{47}$$

$$a_2 = \frac{a_{21}}{\frac{N_2}{N_1} + a_{22}}, \tag{48}$$

where a_{ij} are parameters that take into account the specifics of demographic processes in the society in question.

A typical view of the relationships a_1 and a_2 corresponding to Eqs. (47) and (48) is shown in Fig. 17.

The value of the migration coefficient b depends on how much the values y_1 and y_2 differ in the innovative and traditional sectors, to what extent the innovative sector is more attractive than the traditional one. In the simplest case, we will assume that the value of b is directly proportional to the value of g, that is, the more the value of y_1 exceeds the value of y_2, the more intense the migration flow is from the traditional sector to the innovative one:

[39] An autonomous system of differential equations is a special case when the argument t is not explicitly included in the functions defining the system.

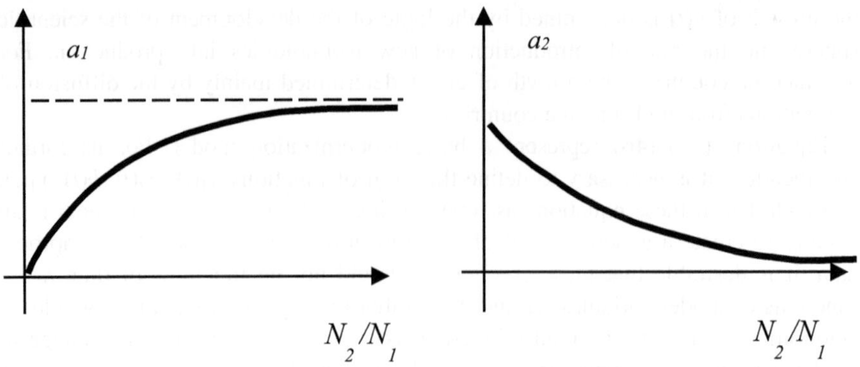

Fig. 17 A typical form of dependencies a_1 and a_2 characterizing the reproduction of population in the traditional and innovative sectors of society during modernization

$$b = b_0 \cdot g, \tag{49}$$

where b_0 is the proportionality coefficient.

At the same time, the value of g increases during the modernization process. Within the framework of the model, we will assume that g grows from zero to some constant value (because the returns from the introduced technologies gradually come to saturation):

$$g = \frac{g_0 \cdot \frac{N_2}{N_1}}{\frac{N_2}{N_1} + k}, \tag{50}$$

where g and k are coefficients. The typical form of relationship $g(N_2/N_1)$ is similar to the form of relationship $a_1(N_2/N_1)$ (see Fig. 17).

In accordance with Eq. (45), the GDP of the country under consideration, taking into account both sectors, is described by the expression:

$$Y = Y_1 + Y_2 = y_1 \cdot N_1 + y_2 \cdot N_2 = y_1 \cdot (N_1 + N_2 \cdot (g + 1)). \tag{51}$$

The population is equal to:

$$N = N_1 + N_2. \tag{52}$$

Equations (42)–(52) represent a modernization model where all values are functions of N_2/N_1.

It should be noted that, although the model was substantiated on the example when migration was understood as the movement of rural residents from villages to cities, the term "migration" actually has a much broader meaning. For example, migration from the traditional to the innovative sector could be the transition of the

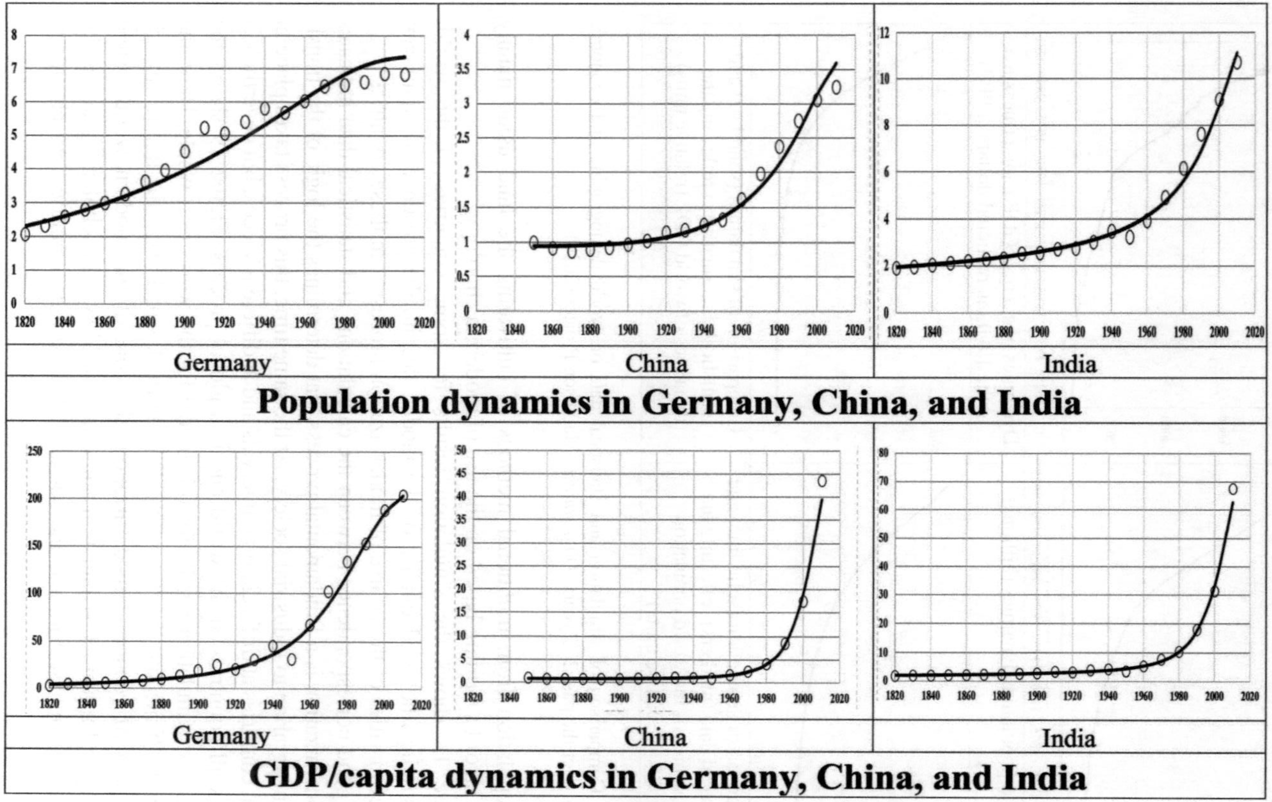

Fig. 18 Comparison of simulation results (solid line) and statistical data (markers) reflecting the dynamics of population and GDP/capita in a number of countries of the world (data are given in relative units). *Data source*: Maddison Project (2020)

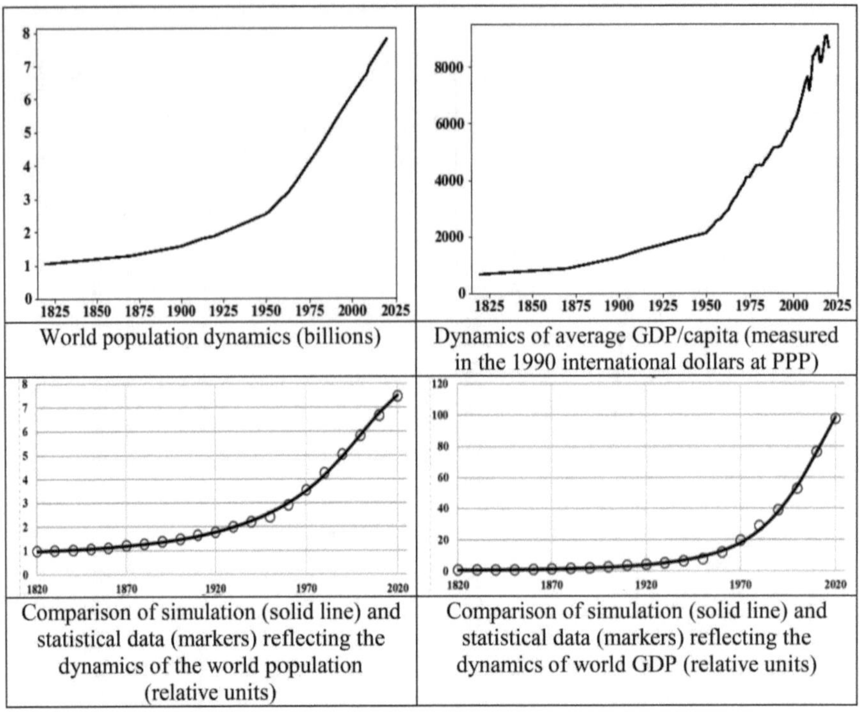

Fig. 19 Comparison of simulation and statistical data on world demographic and economic dynamics over the past 200 years. *Data source*: Maddison Project (2020)

rural population from traditional methods of cultivating the land using mainly manual labor to farming based on the use of modern technology.

Figure 18 shows the simulation results for some countries of the world, demonstrating the good quality of the model, despite the presence of significant local peculiarities of the course of modernization processes in these countries.

The good correspondence between the calculations and statistical data indicates that modernization is a fundamental process that determines the logic of the transition from agrarian to industrial society. At the same time, this process takes place in different countries at different time intervals and at different speeds, but nevertheless it covers all countries in the world and is a global process. Figure 19 shows the statistics and simulation results for this model for the world as a whole. Here, the so-called high-income countries (OECD countries), which were the first to embark on the path of economic modernization (the core of the modern world-system according to Immanuel Wallerstein and other world-system scholars[40]), act as the global innovation sector, and the so-called low-income countries, which started

[40] See, e.g., Wallerstein (1974, 2004), Arrighi and Silver (1999), Chase-Dunn and Hall (1997), Chase-Dunn and Anderson (2005), Chase-Dunn and Lerro (2016).

economic modernization later under the influence of high-income countries represent the traditional sector (periphery of the modern world-system according to Wallerstein and other world-systems scholars).

3.2 Features of Socio-political Modernization

As shown above, the process of modernization is a gradual contraction of the traditional sector and expansion of the innovative sector of society (see Fig. 16 above). The problem here, among other things, is that these two sectors of society are characterized by different principles of social self-organization.[41] Moreover, what is considered right in the innovation sector is not approved and even condemned by members of the traditional sector, and vice versa: while the innovation sector welcomes the desire for novelty, and its ideal is individual freedom, in a traditional society, it is rather the adherence to traditions and collectivism that are welcomed. This difference in behavioral strategies and value systems of members of the traditional and innovative sectors gives rise to the social instability of society in the period of modernization: it is difficult for members of these two sectors to "synchronize" in their actions, and it is difficult to manage such a society (see, e.g., Huntington, 1968, 1986). In the language of the basic model of political stabilization/destabilization [see (Malkov & Bilyuga, 2015) and Sect. 3.2.1 of chapter "Modeling Social Self-Organization and Historical Dynamics. A General Approach" (Akaev et al., 2023a)], this means an increase in the value of randomization coefficient λ. A consequence of this is the instability and complexity of modernization transitions: during these periods, society becomes vulnerable to socio-political destabilization. As a rule, it is during these periods that social and various "color" revolutions take place (Korotayev et al., 2011, 2014; Grinin, 2022).

The modernization process can also be hampered by the so-called development traps (Aiyar et al., 2013; Grinin, 2013a; Malkov et al., 2018; Korotayev et al., 2019). Development traps are traps in which countries that have embarked on the path of economic modernization fall, objectively having growth potential, but not realizing their potential due to implicit internal or external constraints. Such traps are characteristic of the economies of low-income countries, their social, political, and other spheres. Together, development traps create a formidable barrier that can separate the low-income countries from the high-income countries for long periods of time.

The *traps of economic development* include the trap of resource-rich countries, the middle-income trap, the trap of innovative backwardness, and financial traps.

---

[41] See Sect. 2.4 of chapter "Modeling Social Self-Organization and Historical Dynamics. Agrarian Society" (Malkov et al., 2023b, this volume), Sect. 2.4 of chapter "Modeling Social Self-Organization and Historical Dynamics. Industrial Society" (Akaev et al., 2023b, this volume), and Sect. 3 of chapter "Modeling Social Self-Organization and Historical Dynamics. Global Phase Transitions" (Malkov et al., 2023a, this volume).

Table 1 Development traps specific to low-income countries

Trap	The essence of the trap
Traps of economic development:	
(A) Resource-richness trap (typical for low-income countries specializing in the export of commodities)	Orientation to an external buyer of raw materials (at world prices) stimulates the policy of reducing domestic costs in order to reduce the cost of production and increase its competitiveness in foreign markets. This pushes producers to reduce wages, which leads to (a) a contraction in domestic demand and degradation of the domestic manufacturing industry, (b) dependence of the economy on foreign technologies, (c) a decrease in the well-being of the population in relation to the well-being of the population of industrialized countries.
(B) Middle-income trap (typical for low-income countries specializing in the export of manufactured goods)	If a low-income country is oriented towards the export of manufacturing products based on foreign technologies, the trap lies in the fact that: (a) economic growth in the low-income country depends on external demand and on the inflow of foreign investments; (b) the development of domestic demand is hampered by the need to keep wages low (for the sake of reducing costs and ensuring the price competitiveness of manufactured products); (c) the poor cannot get high-quality education, a low level of education hinders the development of science and technology, which reinforces technological dependence on the high-income countries.[a]
(C) The innovation backwardness trap	Significantly higher salaries in science and R&D in high-income countries attract scientists and talented young people from low-income countries ("brain drain"). The constant outflow of talent keeps low-income countries lagging behind in the development of advanced technologies and perpetuates their raw materials curse (dependence on Western technologies).
(D) Financial traps	The reasons for the weakness of the financial and economic systems of low-income countries, in particular, are the following: (1) the competitiveness of export-oriented low income countries can increase due to the depreciation of their national currencies. But the currency devaluation leads to a decrease in the ability to purchase modern foreign equipment and technologies, and to capital outflow; (2) low-income countries, as a rule, are not interested in the depreciation of the reserve currency, since it is the basis of their foreign exchange reserves; (3) the emission of the national currency is carried out not depending on the real needs of the economy, but depending on the inflow of reserve currency into the country through foreign trade operations; (4) inflation targeting by means of monetary policy suppresses economic growth.
Social development traps:	
(A) Demographic trap	(1) Economic growth improves living conditions in low-income countries, resulting in reduced infant mortality. This leads to an increase in population growth, which in turn impedes economic growth (since GDP per capita in the end remains virtually unchanged).

(continued)

Table 1 (continued)

Trap	The essence of the trap
	(2) Decreasing infant mortality in the process of economic growth leads to a change in the age structure of the population, and, particularly, to the emergence of a "youth bulge" (an abnormal increase in the proportion of young people). Without social lifts, these young people can become a "combustible material" for anti-government actions and a factor of instability.
(B) Property stratification trap	The transition to a liberal market economy inevitably leads to property stratification, which then leads to social tension and instability [See chapter "Economics. Optimizing Growth" (Grinin et al., 2023c, this volume), chapter "Socio-Political Transformations. A Difficult Path to Cybernetic Society" (Grinin et al., 2023d, this volume), and chapter "High Income and Low income countries. Toward a Common Goal at Different Speeds" (Grinin et al., 2023b, this volume)].
(C) Social stratification trap	Property stratification leads to the fragmentation of society into social groups with different interests and values, which increases social tension.
Traps of political development:	
(A) The trap of institutional transitions	Western countries are characterized by a Y-institutional structure of society, which is based on the principles of competition and individualism, while low-income countries are mostly characterized by an X-institutional structure based on the principles of collectivism (the features of X- and Y-institutional structures are described in the Note). The modernization transition from X- to Y-structure is associated with a decrease in the stability of society and the transition to a vulnerable unstable state (X-structures were abandoned, and Y-structures did not give the expected result).
(B) De-ideologization trap	Pluralism (as the ideal of the Y-system) leads to the de-ideologization of society, loss of common goals and values, disorientation, desynchronization of actions and goals.
(C) Elite evolution trap	(1) In low-income countries, market reforms are commonly paralleled by a reduction in the role of the state in the economy, and privatization of large state assets. This leads to a sharp enrichment of part of the elite, its oligarchization and strengthening of its influence on power. (2) In modernizing societies, the elite, as a rule, professes individualism (Y-system), while the people endorse collectivism (X-system). This results in a cognitive dissonance, misunderstanding, mutual rejection, and internal division in society. (3) In export-oriented low-income countries, the economy is focused not on domestic, but on external demand. This makes the financial and economic elite cosmopolitan, not nationally oriented (denationalization of the elite occurs).
(D) Authoritarianism trap	(1) Reforms require concentration of resources and strict governance; these are the reasons for the effectiveness of authoritarianism. But authoritarianism is "not democratic", so there is always an excuse for Western sanctions.

(continued)

Table 1 (continued)

Trap	The essence of the trap
	(2) The authoritarian form of government complicates the transfer of power, which leads to the conservation of regimes, a reluctance to give up power, and to the accumulation of political problems. This, in turn, inevitably leads to protest moods in society, and to political destabilization.
(E) Imitation democracy trap	An attempt to transition to democracy in countries that do not have democratic traditions and experience in achieving political compromises leads to intensified confrontation between various political groups, to increased radicalism and social instability.
(F) Political instability trap	"Democracy" and "freedoms" in a transitional society provoke acts of defiance by radicals. At the same time, the "democratic" government is not allowed to suppress the "people's will". Therefore, protest actions, without meeting adequate opposition, can ultimately destabilize the situation in the country and lead to a "color revolution".

Note: The combination "distributive economy—directive (hierarchical) governance system—primacy of collectivism in social consciousness" is characteristic of institutional X-structures. Social structures of this type are formed if security tasks (ensuring survival in difficult natural and social conditions with a shortage of resources) are priority (vital) for society. If the priorities are the tasks of economic development in conditions of resource sufficiency (or redundancy), society naturally acquires the features of the so-called Y-structure, which is characterized by the institutional combination of a liberal market economy, adaptive (democratic) governance system, and the primacy of individualism in social consciousness. See Sect. 3 of chapter "Modeling Social Self-Organization and Historical Dynamics. Global Phase Transitions" (Malkov et al., 2023a) for detail on X- and Y-structures

[a]However, for a number of countries, especially East Asian ones, including China, this trap did not work (see Grinin, 2013a, 2013b; Grinin et al., 2015)

Social development traps include demographic trap, inequality trap, corruption trap, social instability trap (due to high expectations of the population during reforms).

The traps of political development include the trap of institutional transitions (breaking down the old mechanisms for ensuring socio-economic and political stability when the new ones are inoperable), the trap of authoritarianism, the trap of imitation democracy, and others.

A brief description of the traps is given in Table 1.

Acknowledgements This work was done with the support of MSU Program of Development, Project No 23A-SCH05-03.

References

Aiyar, M. S., Duval, M. R. A., Puy, M. D., Wu, M. Y., & Zhang, M. L. (2013). *Growth slowdowns and the middle-income trap*. International Monetary Fund.

Akaev, A., & Davydova, O. (2020). The Paris Agreement on climate is coming into force: Will the great energy transition take place? *Herald of the Russian Academy of Sciences, 90*(5), 588–599.
Akaev, A., & Davydova, O. (2023). Climate and energy. Scenarios of energy transition and global temperature changes based on current technologies and trends. In: V. Sadovnichy et al., (Eds.) *Reconsidering the limits to growth. A report to the Russian Association of the Club of Rome* (pp. 53–70). Springer. https://doi.org/10.1007/978-3-031-34999-7_4
Akaev, A., Malkov, S., Bilyuga, S., Malkov, A., & Korotayev, A. (2023a). Modeling social self-organization and historical dynamics. A general approach. In: V. Sadovnichy et al., (Eds.) *Reconsidering the limits to growth. A report to the Russian Association of the Club of Rome.* (pp. 253–307). Springer. https://doi.org/10.1007/978-3-031-34999-7_15
Akaev, A., Malkov, S., Davydova, O., Kovaleva, N., Malkov, A., Grinin, L., & Korotayev, A. (2023b). Modeling social self-organization and historical dynamics. Industrial society. In: V. Sadovnichy et al. (Eds.), *Reconsidering the limits to growth. A Report to the Russian Association of the Club of Rome* (pp. 337–385). Springer. https://doi.org/10.1007/978-3-031-34999-7_17
Allen, R. C. (2009). *The British industrial revolution in global perspective*. Cambridge University Press.
Allen, R. C. (2011). *Global economic history: A very short introduction*. Oxford University Press.
Álvarez Pereira, C. (2021). *Learning new ways of becoming human. A report to The Club of Rome*. The Club of Rome.
Apter, D. E. (1965). *The politics of modernization*. University of Chicago Press.
Arrighi, G., & Silver, B. J. (1999). *Chaos and governance in the modern world system*. University of Minnesota Press.
Artzrouni, M., & Komlos, J. (1985). Population growth through history and the escape from the Malthusian trap: A homeostatic simulation model. *Genus, 41*(3–4), 21–39.
Bardi, U. (2017). *The Seneca effect. Why growth is slow but collapse is rapid. A report to The Club of Rome*. Springer. https://doi.org/10.1007/978-3-319-57207-9
Bardi, U. (2014). *Extracted: How the quest for mineral wealth is plundering the planet. A report to The Club of Rome*. Chelsea Green Publishing.
Bendix, R. (1967). Tradition and modernity reconsidered. *Comparative Studies in Society and History, 9*(3), 292–346.
Black, C. (1966). *The dynamics of modernization: A study in comparative history*. Harper Torchbooks.
Boden, T. A., Kaiser, D. P., Sepanski, R. J., & Stoss, F. W. (1994). *TREND's 93. Compendium of data on global change*. Carbon Dioxide Informational Analysis Center, Oak Ridge National Laboratory.
Bourreau, M., Hombert, J., Pouyet, J., & Schutz, N. (2011). Upstream competition between vertically integrated firms. *The Journal of Industrial Economics, 59*(4), 677–713.
Budyko, M. (1980). *Klimat v proshlom i budushchem*. Gidrometeoizdat.
Caldwell, J. C., Caldwell, B. K., Caldwell, P., McDonald, P. F., & Schindlmayr, T. (2006). *Demographic transition theory*. Springer. https://doi.org/10.1007/978-1-4020-4498-4
Chase-Dunn, C., & Anderson, E. (Eds.). (2005). *The historical evolution of world-systems*. Springer. https://doi.org/10.1057/9781403980526
Chase-Dunn, C., & Hall, T. D. (1997). *Rise and demise: Comparing world-systems*. Westview Press.
Chase-Dunn, C., & Lerro, B. (2016). *Social change: Globalization from the Stone Age to the present*. Routledge.
Chesnais, J. C. (1992). *The demographic transition: Stages, patterns, and economic implications*. Clarendon Press.
Choucri, N. (1974). *Population dynamics and international violence: Propositions, insights, and evidence*. Lexington Books.
Christian, D. (2011). *Maps of time: An introduction to big history*. University of California Press.
Cipolla, C. M. (Ed.). (1976). *The Industrial Revolution, 1700–1914*. Harvester Press.

Clark, G. (2007). *A farewell to alms: A economic history of the world*. Princeton University Press.
Collins, R. (1968). A comparative approach to political sociology. In R. Bendix et al. (Eds.), *State and society: A reader in comparative political sociology* (pp. 42–67). Little & Brown.
Davies, J. C. (1962). Toward a theory of revolution. *American Sociological Review, 27*(1), 5–19.
Dyson, T. (2010). *Population and development. The demographic transition*. Zed Books.
Eisenstadt, S. (1966). *Modernization: protest and change*. Prentice-Hall.
Eisenstadt, S. (1978). *Revolution and the transformation of societies: A comparative study of civilizations*. Free Press.
Erçetin, Ş. Ş., Ray, N., & Sen, S. (Eds.). (2019). *Economic growth and demographic transition in third world nations: A chaos and complexity theory perspective*. CRC Press.
Fama, E. F., & Laffer, A. B. (1972). The number of firms and competition. *The American Economic Review, 62*(4), 670–674.
von Foerster, H., Mora, P. M., & Amiot, L. W. (1960). Doomsday: Friday, 13 November, A.D. 2026: At this date human population will approach infinity if it grows as it has grown in the last two millennia. *Science, 132*(3436), 1291–1295. https://doi.org/10.1126/science.132.3436.1291
Galor, O. (2011). *Unified growth theory*. Princeton University Press.
Gills, B., & Morgan, J. (Eds.). (2022). *Economics and climate emergency*. Routledge.
Goldstone, J. A. (2001). Demography, environment, and security: An overview. In M. Weiner & S. S. Russell (Eds.), *Demography and national security* (pp. 38–61). Berghahn Books.
Goldstone, J. (2002). Population and security: How demographic change can lead to violent conflict. *Journal of International Affairs, 56*(1), 3–21.
Goldstone, J. A. (2009). *Why Europe? The rise of the west in world history. 1500–1850*. McGraw-Hill.
Goldstone, J. A. (2016). *Revolution and rebellion in the early modern world: Population change and state breakdown in England, France, Turkey, and China, 1600–1850* (2nd ed.). Routledge.
Goldstone, J. A., Kaufmann, E. P., Toft, M. D., & (Eds.). (2012). *Political demography: How population changes are reshaping international security and national politics*. Oxford University Press.
Greenberg, B., Watts, L., Greenwald, R. A., Reavley, G., George, A. L., Beekman, S., et al. (2008). *Social history of the United States*. ABC-CLIO.
Grinin, L. (2013a). State and socio-political crises in the process of modernization. *Social Evolution & History, 12*(2), 35–76.
Grinin, L. (2013b). The tiger and the dragon. Development models and perspectives of India and China. *Journal of Globalization Studies, 4*(1), 5–31.
Grinin, L. (2022). Revolutions and modernization traps. In J. A. Goldstone, L. Grinin, & A. Korotayev (Eds.), *Handbook of revolutions in the 21st century: The new waves of revolutions, and the causes and effects of disruptive political change* (pp. 219–238). Springer. https://doi.org/10.1007/978-3-030-86468-2_8
Grinin, L., Grinin, A., & Korotayev, A. (2023a). Global aging – An integral problem of the future. How to turn a problem into a development driver? In: V. Sadovnichy et al. (Eds.), *Reconsidering the limits to growth. A report to the Russian Association of the Club of Rome* (pp. 117–135). Springer. https://doi.org/10.1007/978-3-031-34999-7_7
Grinin, L., Grinin A., & Malkov, S. (2023c). Economics. Optimizing growth. In: V. Sadovnichy et al. (Eds.), *Reconsidering the limits to growth. A report to the Russian Association of the Club of Rome* (pp. 155–168). Springer. https://doi.org/10.1007/978-3-031-34999-7_9
Grinin, L., Grinin, A., & Malkov, S. (2023d). Socio-political transformations. A difficult path to cybernetic society. In: V. Sadovnichy et al. (Eds.), *Reconsidering the limits to growth. A report to the Russian Association of the Club of Rome* (pp. 169–189). Springer. https://doi.org/10.1007/978-3-031-34999-7_10
Grinin, L., & Korotayev, A. (2015). *Great divergence and great convergence. A global perspective*. Springer. https://doi.org/10.1007/978-3-319-17780-9

Grinin, L., Korotayev, A., & Tsirel, A. (2015). Will the explosive growth of China continue? *Technological Forecasting and Social Change, 95*(2015), 294–308.
Grinin, L., Malkov, S., Grinin, A., & Korotayev, A. (2022). Will capitalism die? Reflections on the capitalism of the past, present and future. *Sotsiologicheskiy Zhurnal = Sociological Journal, 28*(2), 100–130. https://doi.org/10.19181/socjour.2022.28.2.8988
Grinin, L., Malkov, S., Gusev, V., & Korotayev, A. (2009). Nekotoryye vozmozhnyye napravleniya razvitiya teorii sotsial′no-demograficheskikh tsiklov i matematicheskiye modeli vykhoda iz "mal′tuzianskoy lovushki". In L. Grinin, A. Korotayev, & S. Malkov (Eds.), *Istoriya i Matematika: Protsessy i modeli* (pp. 134–210). LIBROKOM/URSS.
Grinin, L., Malkov, S., & Korotayev, A. (2023b). High income and low income countries. Toward a common goal at different speeds. In: V. Sadovnichy et al. (Eds.), *Reconsidering the limits to growth. A report to the Russian Association of the Club of Rome* (pp. 207–224). Springer. https://doi.org/10.1007/978-3-031-34999-7_12
Huntington, S. P. (1968). *Political order in changing societies*. Yale University Press.
Huntington, S. (1986). Revolution and political order. In J. A. Goldstone (Ed.), *Revolutions: Theoretical, comparative, and historical studies* (pp. 39–47). Harcourt Brace Jovanovich.
Ilyin, I., Issaev, L., & Malkov, S. (2020). A methodology for analyzing and forecasting sociopolitical destabilization. *BioSystems, 198*(104229), 1–8. https://doi.org/10.1016/j.biosystems.2020.104229
Jones, P. D., Parker, D. E., Osborn, T. J., & Briffa, K. R. (1994). Global temperature anomalies in 1856-1999. In T. A. Boden, D. P. Kaiser, R. J. Sepanski, & F. W. Stoss (Eds.), *TREND's 93. Compendium of data on global change*. Carbon Dioxide Informational Analysis Center, Oak Ridge National Laboratory.
Kapitza, S. (1992). A mathematical model for global population growth. *Matematicheskoe modelirovanie, 4*(6), 65–79.
Kapitza, S. (1996). Phenomenological theory of world population growth. *Physics Uspekhi, 39*(1), 57–71. https://doi.org/10.1070/PU1996v039n01ABEH000127
Kapitza, S. (1999). *An essay on the demographic imperative: Non-linear theory of growth of humankind*. The Edwin Mellen Press.
Kapitza, S. (2003). The statistical theory of global population growth. In *Formal descriptions of developing systems* (pp. 11–35). Springer. https://doi.org/10.1007/978-94-010-0064-2_2
Kapitza, S. (2006). *Global population blow-up and after. A report to the Club of Rome*. Global Marshall Plan Initiative.
Kapitza, S. (2010). On the theory of global population growth. *Physics Uspekhi, 53*(12), 1287–1296. https://doi.org/10.3367/UFNe.0180.201012g.1337
Kirdina, S. (2004). *X i Y-ekonomiki: Institutsional′nyy analiz*. Nauka.
Kirdina, S. (2014a). Institutional matrices theory, or X&Y theory: The main provisions and applications. *Journal of Institutional Studies, 6*(3), 13–33.
Kirdina, S. (2014b). *Institutsionalnyye matritsy i razvitiye Rossii: vvedeniye v X-Y-teoriyu* (3rd ed.). Nestor-Istoriya.
Kögel, T., & Prskawetz, A. (2001). Agricultural productivity growth and escape from the Malthusian trap. *Journal of Economic Growth, 6*, 337–357. https://doi.org/10.1023/A:1012742531003
Komlosy, A. (2021). Entanglements of catching-up: Rethinking 'industrial revolution' from a global perspective. *Journal of Globalization Studies, 12*(1), 79–100. https://doi.org/10.30884/jogs/2021.01.05
Komlos, J., & Artzrouni, M. (1990). Mathematical investigations of the escape from the Malthusian trap. *Mathematical Population Studies, 2*, 269–287.
Komlos, J., & Nefedov, S. (2002). A compact macromodel of pre-industrial population growth. *Historical Methods: A Journal of Quantitative and Interdisciplinary History, 35*(2), 92–94. https://doi.org/10.1080/01615440209604133
Korotayev, A. (2017). Malthusian cycle. In F. M. Moghaddam (Ed.), *The SAGE encyclopedia of political behavior* (pp. 457–462). SAGE.

Korotayev, A. (2020). The 21st century singularity in the big history perspective. A re-analysis. In A. Korotayev & D. LePoire (Eds.), *The 21st century Singularity and global futures. A big history perspective* (pp. 19–75). Springer. https://doi.org/10.1007/978-3-030-33730-8_2

Korotayev, A., Goldstone, J. A., & Zinkina, J. (2015). Phases of global demographic transition correlate with phases of the Great Divergence and Great Convergence. *Technological Forecasting and Social Change, 95*, 163–169. https://doi.org/10.1016/j.techfore.2015.01.017

Korotayev, A., & Khaltourina, D. (2006). *Introduction to social macrodynamics: Secular cycles and millennial trends in Africa*. KomKniga/URSS.

Korotayev, A., Malkov, A., & Khaltourina, D. (2006a). *Introduction to social macrodynamics: Compact macromodels of the world system growth*. KomKniga/URSS.

Korotayev, A., Malkov, A., & Khaltourina, D. (2006b). *Introduction to social macrodynamics: Secular cycles and millennial trends*. KomKniga/URSS.

Korotayev, A., Malkov, S., & Grinin, L. (2014). A trap at the escape from the trap? Some demographic structural factors of political instability in modernizing social systems. *History & mathematics, 4*, 201–267.

Korotayev, A., Malkov, S., & Musieva, J. (2023a). Demography. Toward optimization of demographic processes. In: V. Sadovnichy et al. (Eds.), *Reconsidering the limits to growth. A report to the Russian Association of the Club of Rome* (pp. 97–116). Springer. https://doi.org/10.1007/978-3-031-34999-7_6

Korotayev, A., Romanov, D., Zinkina, J., & Slav, M. (2023b). Urban youth and terrorism: A quantitative analysis (Are youth bulges relevant anymore?). *Political Studies Review, 21*(3), 550–574. https://doi.org/10.1177/14789299221075908

Korotayev, A., Sawyer, P., Gladyshev, M., Romanov, D., & Shishkina, A. (2021). Some sociodemographic factors of the intensity of anti-government demonstrations: Youth bulges, urbanization, and protests. *Sociologiceskoe Obozrenie = Russian Sociological Review, 20*(3), 98–128. https://doi.org/10.17323/1728-192x-2021-3-98-128

Korotayev, A., & Shishkina, A. (2020). Relative deprivation as a factor of sociopolitical destabilization: Toward a quantitative comparative analysis of the Arab spring events. *Cross-Cultural Research, 54*(2–3), 296–318. https://doi.org/10.1177/1069397119882364

Korotayev, A., Slinko, E., Meshcherina, K., & Zinkina, J. (2020). Variation of human values and modernization: Preliminary results. *Cross-Cultural Research, 54*(2–3), 238–272. https://doi.org/10.1177/1069397119874781

Korotayev, A., Tsirel, S., & Bilyuga, S. (2019). Corruption, human values and attempts of violent changes of government in countries with various levels of per capita GDP: A comparative quantitative analysis. *Sravnitelnaya Politika-Comparative Politics, 10*(1), 98–123. https://doi.org/10.24411/2221-3279-2019-10007

Korotayev, A., & Zinkina, J. (2014). How to optimize fertility and prevent humanitarian catastrophes in Tropical Africa. *African Studies in Russia, 6*, 94–107.

Korotayev, A., & Zinkina, J. (2015). East Africa in the Malthusian trap? *Journal of Developing Societies, 31*(3), 385–420. https://doi.org/10.1177/0169796X15590322

Korotayev, A., & Zinkina, J. (2022). Egypt's 2011 revolution. A demographic structural analysis. In J. A. Goldstone, L. Grinin, & A. Korotayev (Eds.), *Handbook of revolutions in the 21st century: The new waves of revolutions, and the causes and effects of disruptive political change* (pp. 651–683). Springer. https://doi.org/10.1007/978-3-030-86468-2_25

Korotayev, A., Zinkina, J., Goldstone, J., & Shulgin, S. (2016). Explaining current fertility dynamics in Tropical Africa from an anthropological perspective: A cross-cultural investigation. *Cross-Cultural Research, 50*(3), 251–280. https://doi.org/10.1177/1069397116644158

Korotayev, A., Zinkina, J., Kobzeva, S., Bogevolnov, J., Khaltourina, D., Malkov, A., & Malkov, S. (2011). A trap at the escape from the trap? Demographic-structural factors of political instability in modern Africa and West Asia. *Cliodynamics, 2*(2), 276–303. https://doi.org/10.21237/C7clio22217

Kovaleva, N. (2023). Ecology. Life in the "unstable biosphere". In V. Sadovnichy et al. (Eds.), Reconsidering the limits to growth. A report to the Russian Association of the Club of Rome (pp. 71–95). Springer. https://doi.org/10.1007/978-3-031-34999-7_5

Lesthaeghe, R. (2020). The second demographic transition, 1986–2020: Sub-replacement fertility and rising cohabitation—A global update. *Genus, 76*(1), 1–38. https://doi.org/10.1186/s41118-020-00077-4

Levy, M., Jr. (1966). *Modernization and the structure of societies*. Princeton University Press.

Levy, M., Jr. (1967). Social patterns (structures) and problems of modernization. In W. E. Moore & R. M. Cook (Eds.), *Readings on social change* (pp. 189–208). Prentice House.

Lindsey, R. (2021). *Climate change: Atmospheric carbon dioxide*. https://www.climate.gov/news-features/understanding-climate/climate-change-atmospheric-carbon-dioxide

Livi-Bacci, M. (2012). *A concise history of world population*. Wiley-Blackwell.

Maddison, A. (2001). *Monitoring the world economy: A millennial perspective*. OECD.

Maddison Project. (2020). *Maddison Project Database 2020*. Groningen Growth and Development Centre. https://www.rug.nl/ggdc/historicaldevelopment/maddison/releases/maddison-project-database-2020

Malkov, S. (2009). *Sotsia'naya samoorganizatsiya i istoricheskiy protsess: Vozmozhnosti matematicheskogo modelirovaniya*. LIBROKOM.

Malkov, S., Bilyuga, S., & Davydova, O. (2018). The use of "development traps" in interstate confrontation. The middle-income trap: Modeling and analysis. *Strategic Stability, 1*(82), 47–56.

Malkov, S., & Bilyuga, S. (2015). Model of stability/destabilization of political systems. *Information Wars, 1*(33), 7–18.

Malkov, S., Grinin, L., Grinin, A., Musieva, J., & Korotayev, A. (2023a). Modeling social self-organization and historical dynamics. Global phase transitions. In: V. Sadovnichy et al. (Eds.), *Reconsidering the limits to growth. A report to the Russian Association of the Club of Rome* (pp. 387–417). Springer. https://doi.org/10.1007/978-3-031-34999-7_18

Malkov, S., Kovaleva, N., Grinin, L., & Korotayev, A. (2023b). Modeling social self-organization and historical dynamics. Agrarian society. In: V. Sadovnichy et al. (Eds.), *Reconsidering the limits to growth. A report to the Russian Association of the Club of Rome* (pp. 309–335). Springer. https://doi.org/10.1007/978-3-031-34999-7_16

Malthus, T. (1798). *An essay on the principle of population, as it affects the future improvement of society with remarks on the speculations of Mr. Godwin, M. Condorcet, and other writers*. St. Paul's Church-yard.

Martin, C. (2015). *On the edge. A report to The Club of Rome*. Ingram.

Matsubayashi, N., & Yamada, Y. (2008). A note on price and quality competition between asymmetric firms. *European Journal of Operational Research, 187*(2), 571–581.

McNeill, W. H. (2013). *The pursuit of power: Technology, armed force, and society since AD 1000*. University of Chicago Press.

Meadows, D. H., Meadows, D. L., Randers, J., & Behrens, W. W. (1972). *The limits to growth: A report for the Club of Rome's project on the predicament of mankind*. Universe Books.

Meadows, D. H., Randers, J., & Meadows, D. L. (2004). *Limits to growth: The 30-year update. A report to The Club of Rome*. Earthscan.

Mesquida, C. G., & Wiener, N. I. (1996). Human collective aggression: A behavioral ecology perspective. *Ethology and Sociobiology, 17*(4), 247–262.

Mesquida, C. G., & Weiner, N. I. (1999). Male age composition and severity of conflicts. *Politics and the Life Sciences, 18*, 113–117. https://doi.org/10.1017/S0730938400021158

Met Office Hadley Centre. (2022). *Met Office Hadley Centre observations datasets, HadCRUT4*. https://www.metoffice.gov.uk/hadobs/hadcrut4/data/current/download.html

Moller, H. (1968). Youth as a force in the modern world. *Comparative Studies in Society and History, 10*(3), 237–260. https://doi.org/10.2307/177801

NOAA. (2021). *Global Monitoring Laboratory*. Earth System Research Laboratories. https://www.esrl.noaa.gov/gmd/ccgg/trends/data.html

Our World in Data. (2021). *CO_2 and greenhouse gas emissions.* https://ourworldindata.org/co2-and-other-greenhouse-gas-emissions#how-have-global-co2-emissions-changed-over-time
Perissi, I., & Bardi, U. (2021). *The empty sea. The future of the blue economy. A report to The Club of Rome*. Springer. https://doi.org/10.1007/978-3-030-51898-1
Pestel, E. (1989). *Beyond the limit to growth. A report to The Club of Rome*. The Club of Rome.
Przeworski, A., & Limongi, F. (1997). Modernization: Theories and facts. *World Politics, 49*, 155–183. https://doi.org/10.1353/wp.1997.0004
Rahmstorf, S., & Schellnhuber, H. J. (2007). *Der Klimavandel. Diagnose, prognose, therapie*. Verlag C.H. Beck OHG.
Randers, J. (2012). *2052. A global forecast for the next forty years. A report to The Club of Rome*. Chelsea Green Publishing.
Randers, J., Rockström, J., Stoknes, P. E., Golüke, U., Collste, D., & Cornell, S. (2018). *Transformation is feasible. A report to The Club of Rome*. Stockholm Resilience Centre.
Reher, D. (2011). Economic and social implications of the demographic transition. *Population and Development Review, 37*, 11–33. https://doi.org/10.1111/j.1728-4457.2011.00376.x
Romanov, D., Meshcherina, K., & Korotayev, A. (2021). The share of youth in the total population as a factor of intensity of non-violent protests: A quantitative analysis. *Polis. Political Studies, 3*, 166–181. https://doi.org/10.17976/jpps/2021.03.11
Rostow, W. (1971). *The stages of economic growth: A non-communist manifesto*. Cambridge University Press.
Sawyer, P., Romanov, D., Slav, M., & Korotayev, A. (2022). Urbanization, the youth, and protest: A cross-national analysis. *Cross-Cultural Research, 56*(2–3), 125–149. https://doi.org/10.1177/10693971211059762
Schwab, K., & Malleret, T. (2020). *COVID-19: The great reset*. Forum Publishing.
Smelser, N. J. (1967). Processes of social change. In N. J. Smelser (Ed.), *Sociology* (pp. 674–728). Wiley.
Steinmann, G., & Komlos, J. (1988). Population growth and economic development in the very long run: A simulation model of three revolutions. *Mathematical Social Sciences, 16*(1), 49–63. https://doi.org/10.1016/0165-4896(88)90004-2
Steinmann, G., Prskawetz, A., & Feichtinger, G. (1998). A model on the escape from the Malthusian trap. *Journal of Population Economics, 11*, 535–550. https://doi.org/10.1007/s001480050083
Tipps, D. C. (1973). Modernization theory and the comparative study of societies: A critical perspective. *Comparative Studies in Society and History, 15*(2), 199–226.
Urdal, H. (2004). *The devil in the demographics: The effect of youth bulges on domestic armed conflict, 1950–2000*. World Bank.
Urdal, H. (2006). A clash of generations? Youth bulges and political violence. *International Studies Quarterly, 50*(3), 607–629. https://doi.org/10.1111/j.1468-2478.2006.00416.x
Urdal, H. (2008). Population, resources, and political violence: A subnational study of India, 1956–2002. *Journal of Conflict Resolution, 52*(4), 590–617. https://doi.org/10.1177/0022002708316741
von Weizsäcker, E. U., & Wijkman, A. (2018). *Come on! Capitalism, short-termism, population and the destruction of the planet*. Springer.
Wallerstein, I. (1974). *The modern world-system*. Academic Press.
Wallerstein, I. (2004). *World-systems analysis: An introduction*. Duke University Press.
Weber, H. (2019). Age structure and political violence: A re-assessment of the "youth bulge" hypothesis. *International Interactions, 45*(1), 80–112. https://doi.org/10.1080/03050629.2019.1522310
Weber, M. (2004 [1905]). *Die protestantische Ethik und der Geist des Kapitalismus*. CH Beck.
Wijkman, A., & Rockström, J. (2012). *Bankrupting nature. A report to The Club of Rome*. Routlege.

Wood, J. W. (1998). A theory of pre-industrial population dynamics: Demography, economy, and well-being in Malthusian systems. *Current Anthropology, 39*, 99–135.

Zinkina, J., & Andreev, A. (2020). The earliest history of the evolution of social states in nineteenth-century Europe. *Social Evolution & History, 19*(1), 135–146. https://doi.org/10.30884/seh/2020.01.07

Askar Akaev is the First President of the Kyrgyz Republic and Foreign Full Member of the Russian Academy of Sciences. He works as Chief Research Professor in the Prigogine Institute for Mathematical Investigations of Complex Systems at Lomonosov Moscow State University. He is also Academic Supervisor of the Centre for Stability and Risk Analysis in HSE University, Moscow, Russia. Askar Akaev has authored and co-authored over 300 scientific publications, including such monographs as *Economic Transition from the Point of View of a Physicist. A Mathematical Model of Emerging Economy* (2000), or *Quality of Education, Effectiveness of R&D, and Economic Growth. Quantitative Analysis and Mathematical Modeling* (2016), etc.

Sergey Malkov heads the Center for Long-Term Forecasting and Strategic Planning at Lomonosov Moscow State University; he also works as a Senior Research Professor at the Institute of Economics, Russian Academy of Sciences, and at the Centre for Stability and Risk Analysis at HSE University in Moscow. He is the author of over 300 scholarly publications.

Olga Davydova is a programmer and a specialist in data analysis, development of mathematical models, implementation of models in Python, identification of parameters and verification of models. She has co-authored more than 25 academic publications.

Natalia Kovaleva heads the Laboratory of Ecological Soil Science of the Faculty of Soil Science of Lomonosov Moscow State University. She is also a visiting professor at Kazan Federal (Volga Region) University. Natalia Kovaleva has authored and co-authored over 700 scientific publications.

Artemy Malkov is a Russian mathematician, specialist in mathematical modeling of socio-historical processes, spatial historical dynamics, lecturer at Moscow Institute of Physics and Technology. He is the author or co-author of more than 100 scientific publications.

Leonid Grinin is Senior Research Professor at the Centre for Stability and Risk Analysis at HSE University in Moscow, as well as Senior Research Professor of the Institute of Oriental Studies, Russian Academy of Sciences. He is the author of more than 600 publications, including more than 33 monographs in Russian, English, and Chinese.

Andrey Korotayev heads the Centre for Stability and Risk Analysiss at HSE University in Moscow, Russia. He is also Senior Research Professor at the Institute of Oriental Studies and Institute for African Studies, Russian Academy of Sciences, as well as at the Faculty of Global Studies, Lomonosov Moscow State University and the Russian Presidential Academy of National Economy and Public Administration, Moscow, Russia. He has authored or co-authored over 650 scholarly publications.

Modeling Social Self-organization and Historical Dynamics: Global Phase Transitions

Sergey Malkov, Leonid Grinin, Anton Grinin, Jameelah Musieva, and Andrey Korotayev

Abstract

This chapter presents the results of using the set of basic models outlined in Chapter "Modeling Social Self-Organization and Historical Dynamics. A General Approach" to model transitional eras, when, due to technological revolutions, there is a sharp acceleration in the pace of economic (and, after that, demographic) development, a "society of stability/stagnation" is transformed into a "growth society," the dominance of X-type social structures is replaced by the dominance of Y-type structures. Examples of such transitional eras are the Axial Age (in the first millennium BCE) and the industrial era (the last 200 years). Malkov et al. consider characteristic features of transitional epochs, analyze the logic of the evolution of economic and social processes, and model these mathematically. It is shown that in the initial phases of such eras, there is a hyperbolic growth of the main demographic and economic characteristics, and at the final stage, there is a sharp slowdown, accompanied by strong social instability and a reverse transition from the dominance of Y-type social structures to the dominance of X-type structures.

S. Malkov
Lomonosov Moscow State University, Moscow, Russia

HSE University, Moscow, Moscow, Russia

Institute of Economics, Russian Academy of Sciences, Moscow, Russia
e-mail: s@malkov.org

L. Grinin
HSE University, Moscow, Moscow, Russia

A. Grinin · J. Musieva
HSE University, Moscow, Moscow, Russia

Faculty of Global Studies, Lomonosov Moscow State University, Moscow, Russia

A. Korotayev (✉)
HSE University, Moscow; Lomonosov Moscow State University; Institute of Oriental Studies, Russian Academy of Sciences, Moscow, Russia

V. Sadovnichy et al. (eds.), *Reconsidering the Limits to Growth*, World-Systems Evolution and Global Futures, https://doi.org/10.1007/978-3-031-34999-7_18

1 The Issue of Acceleration of Historical Development

A. In recent years, research and popular literature have given considerable attention to the topic of accelerating historical development[1] and the phenomenon of "singularity," observed in the analysis of various indicators related to the evolution of nature and society.[2] Various authors write about evolutionary, demographic, and technological singularities. In 2020, Springer published a collective monograph on this topic (Korotayev & LePoire, 2020). The essence of the problem is that mathematical extrapolation into the future of some global processes carried out by various researchers yields a startling result. It turns out that these processes are developing with acceleration, but not just with exponential acceleration, which is common for many natural and socio-economic processes, but with acceleration according to a hyperbolic law that has a singularity point, and this hyperbola goes to infinity in the coming next years or decades. So, Figs. 1 and 2 present data on the population of the Earth from 3000 BCE to 2000 CE and world GDP (gross domestic product) from the beginning of the Common Era to 2008 CE.

In both cases, the data fit hyperbolic curves surprisingly well. Heinz von Foerster and his colleagues, who first made such an approximation for the population of the Earth, even called their article with the results of the calculation "Doomsday: Friday, November 13, 2026" (von Foerster et al., 1960), where the date 11/13/2026 means the point of singularity.

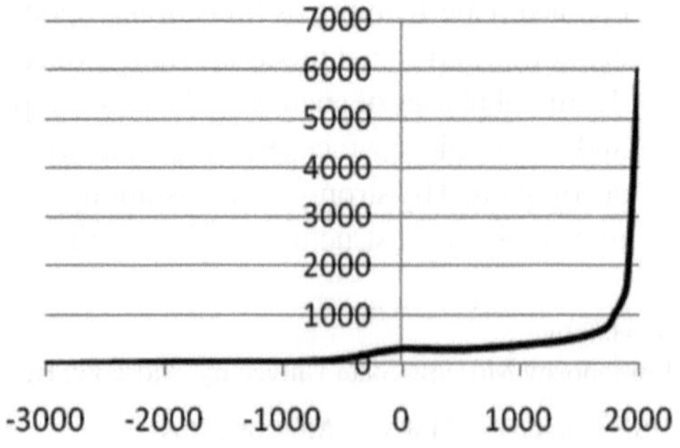

Fig. 1 Dynamics of the world population, million people (x-axis is years). *Data source* Maddison Project (2020)

[1] On the speed of historical and technological progress, see, e.g., Grinin (2007), Grinin and Grinin (2016), Grinin et al. (2020).

[2] Ulam (1958), p. 5; von Foerster et al. (1960), von Hoerner (1975), Taagepera (1976, 1979), Kapitza (1992, 1996, 1999a, 1999b, 2003, 2006, 2010), Kremer (1993), Belavin et al. (1998), Tsirel (2004, 2008, 2009), Kurzweil (2001, 2005), Nazaretyan (2005, 2015, 2017, 2018, 2020), Panov (2005, 2011, 2017, 2020), Korotayev (2005, 2006a, 2006b, 2007, 2013, 2020a, 2020b), Korotayev and Khaltourina (2006), Korotayev et al. (2006a, 2006b, 2016), Markov and Korotayev (2007, 2008), Markov et al. (2010), Korotayev and Grinin (2013), Dobrolyubov (2016), Korotayev and Malkov (2016), Korotayev and Zinkina (2017), Podlazov (2017), Fomin (2020), Last (2020), LePoire and Devezas (2020), Malkov (2020), Panov et al. (2020), Widdowson (2020).

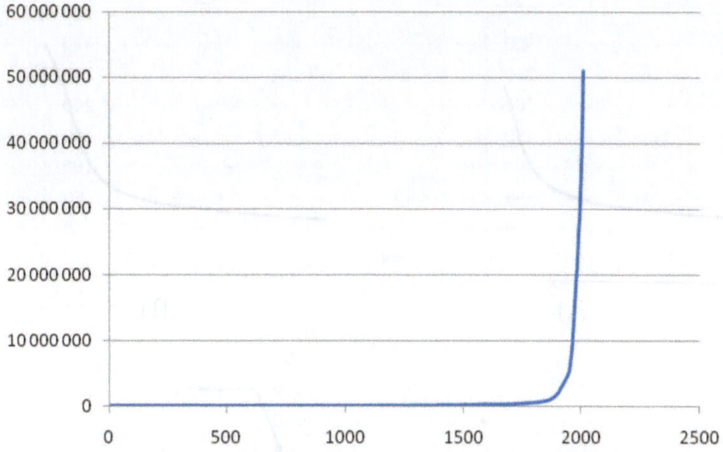

Fig. 2 Dynamics of world GDP, thousand 1990 international dollars at PPP (x-axis is years). *Data source* Maddison Project (2020)

The question arises: what is the reason for such a surprising result of the approximation, is the occurrence of singularity points in the analysis of global processes an artifact, or is it a reflection of a real pattern?

B. A more detailed examination of the historical data series makes it clear that demographic processes comparable in scale to what is currently observed have already occurred in human history. We are talking about (1) the Neolithic (Agricultural) Revolution (~ 10 thousand years ago, the transition from a society of hunter-gatherers to a society of agriculturalists),[3] (2) the "Urban Revolution" (~ 5 thousand years ago, the emergence of the first cities, states, writing systems, metallurgy, plow agriculture and so on),[4] (3) the "Axial Age" (~ 2.5 thousand years ago, the widespread use of iron tools and related technologies, accompanied by breakthroughs in cultural, political, and social evolution).[5] Each time, large-scale technological transformations were the cause of changes in society, as a result of which labor productivity and human capabilities in various spheres of life increased sharply, and the resource base expanded, which in turn led to a sharp increase in the human population. There is little demographic data on the first two transformations, so quantitative modeling is not an option. It is interesting to take a closer look at the demographic dynamics of the "Axial Age," for which there are more reliable and accurate data. It is already

[3] See, e.g., Childe (1934, 1944), Watts (1977), Harris and Hillman (1989), Eshed et al. (2004), Grinin and Grinin (2016), Grinin et al. (2023), as well as chapter "Technology. Limitless Possibilities, Effective Control" (Grinin & Grinin, 2023b, this volume).

[4] E.g., Frank and Gills (1993), Childe (1948, 1952), Bondarenko and Korotayev (2003), Grinin and Korotayev (2006, 2009a, 2012a), Grinin (2009), Grinin et al. (2004), Korotayev and Grinin (2013).

[5] E.g., Jaspers (1953), Eisenstadt (1982), Bellah (2005), Grinin and Korotayev (2009b), Bellah and Joas (2012), Panov et al. (2020).

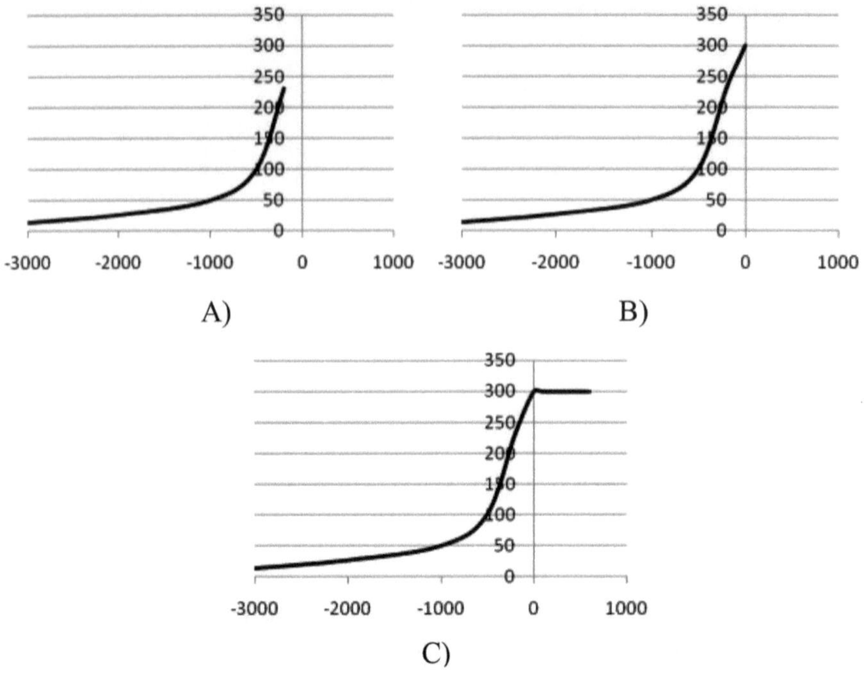

Fig. 3 The population of the Earth by 200 BCE (A), by 1 CE (B), and by 600 CE (C), million people (the x-axis is time in years) (See Malkov, 2020 for data sources)

clear from Fig. 1 that from the eighth to the first centuries BCE, there was some population growth. The scale of this growth seems insignificant when compared with the population explosion of the last two centuries. But the situation will change dramatically when viewed through the lens of the contemporaries of those events. Figure 3 shows the demographic dynamics seen by contemporaries in 200 BCE, in 1 CE, and in 600 CE.

By 600 CE, the transitional demographic growth had already ended (with the transition from antiquity to the Middle Ages taking place). By 1 CE, a slowdown in demographic growth became apparent (i.e., the time of the transformation of the Roman Republic into the Roman Empire and the unification of the Mediterranean into a single political and economic space). In 200 BCE, the transition process was still in full swing, and the demographic curve was perfectly approximated by a hyperbola with a singularity point at 1 CE and with the exponent $m = 1.00$ (see Fig. 4).

C. If we look at the demographic history of humankind as a whole, then even with the proviso that the accuracy of demographic data decreases significantly as we move away from the present by hundreds and thousands of years, a picture of extremely uneven development is revealed (see Fig. 5).

Periods of demographic growth alternate with periods of relative stability (or even demographic decline). Moreover, during periods of growth, population

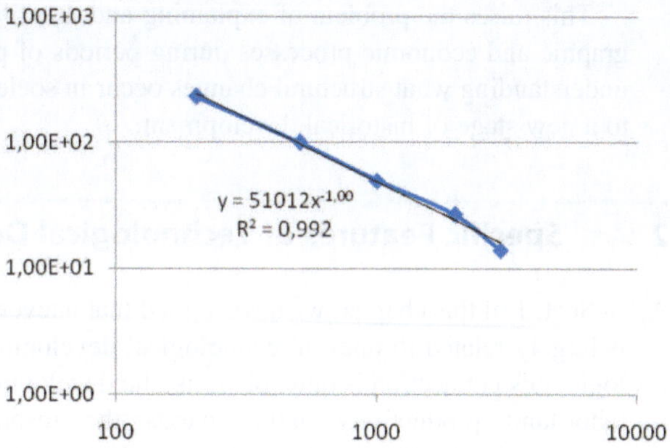

Fig. 4 Hyperbolic approximation of the world population from 3000 BCE to 200 BCE

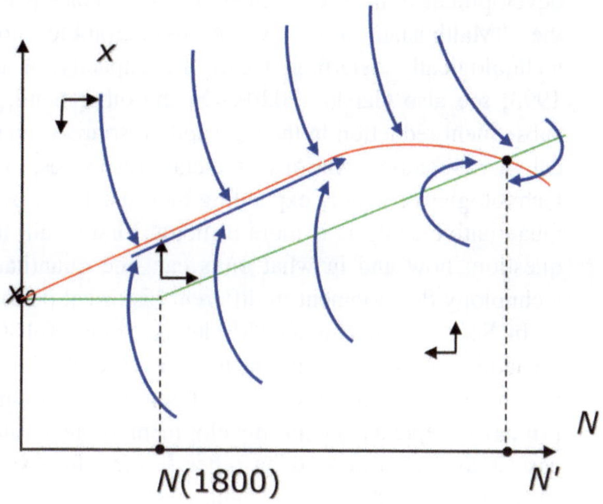

Fig. 5 The dynamics of the world population in the period from 15 thousand years BCE to 2000 CE (million people) with superimposed exponential (thin dotted line) and hyperbolic (dashed line) approximation (See Malkov, 2020 for data sources)

increase occurs not so much according to the exponential, but rather the hyperbolic law (which gives rise to the scientific discourse about the "singularity" [see, e.g., Korotayev, 2020a, 2020b; LePoire, 2020; LePoire & Korotayev, 2020; Malkov, 2020; Panov et al., 2020]). The hyperbolic nature of demographic growth is evidence of the cumulative interaction of various factors that mutually reinforce each other in these historical periods.

Each time, however, the period of rapid demographic growth ended, and a qualitatively new historical epoch began. Thus, periods of sharp demographic growth essentially marked a phase transition from the old to the new economic, demographic, and political reality (see Fried, 1967a, 1967b; Service, 1975; Carneiro, 1970; Claessen & van de Velde, 1985; Chase-Dunn & Hall, 1997; Grinin, 2003, 2004a, 2004b, 2007, 2008; Korotayev, 2006a, 2006b; Korotayev et al., 2006b: 147–162; Grinin & Korotayev, 2009a; LePoire, 2015, 2020; Zinkina et al., 2016; Malkov, 2020).

This raises the problem of explaining and describing the dynamics of demographic and economic processes during periods of phase transitions, as well as understanding what structural changes occur in society after the phase transition to a new stage of historical development.

2 Specific Features of Technological Development

A. In Sect. 1 of this chapter, we have argued that uneven demographic development is largely related to uneven technological development. At the level of general logic, this connection is quite obvious: the development of technology increases labor/land productivity and expands the resource base, which creates opportunities for the population to grow. In turn, the inhibition of technological development limits the available resource base, resulting in society falling into the "Malthusian trap," when demographic growth is limited by the technologically determined carrying capacity of the territory (e.g., Kremer, 1993; see also Garske, 2020). On the other hand, population growth and the subsequent reduction in the per capita resources available at a given technological level causes various societal responses, including developing new technologies or widely expanding their use (e.g., Boserup, 1965).
Quantitative analysis is more difficult. First of all, it is necessary to answer the question: how and in what units can one quantitatively measure the level of technology development in different historical periods?

In Sect. 1.3 of chapter "Modeling Social Self-Organization and Historical Dynamics. A General Approach," the value of GDP/capita is used as a quantitative indicator of the level of technology development *S*, since this indicator primarily depends on the development of technologies (but in each specific case it may also depend on other factors: for example, climate, soil fertility, etc.). For the industrial era, this method of measuring *S* is quite natural and convenient, since the methodology for estimating the value of GDP in a market economy is well developed. However, for an Agrarian society based on a subsistence economy, only indicative and/or indirect estimates of GDP are possible; the less reliable, the earlier the historical era.[6]

For the source of information about the development of technologies in the early periods, one can use the data from historical and archaeological sources about inventions and technologies that appeared and were introduced into economic life in various regions of the world. Data on inventions from 2.5 million years BCE to 2003 CE are collected in the Bunch-Hellemans database (Bunch & Hellemans, 2004). The dynamics of the number of inventions as a cumulative

[6] For example, it is very difficult to evaluate the GDP in a hunter-gatherer society.

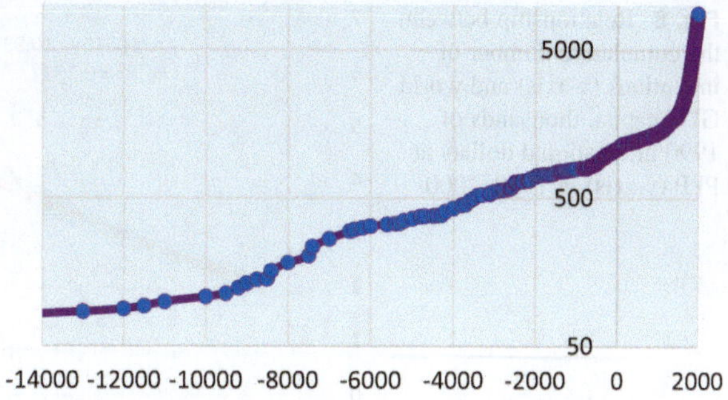

Fig. 6 Cumulative number of inventions over the last 16,000 years, logarithmic scale. *Data source* Bunch and Hellemans (2004)

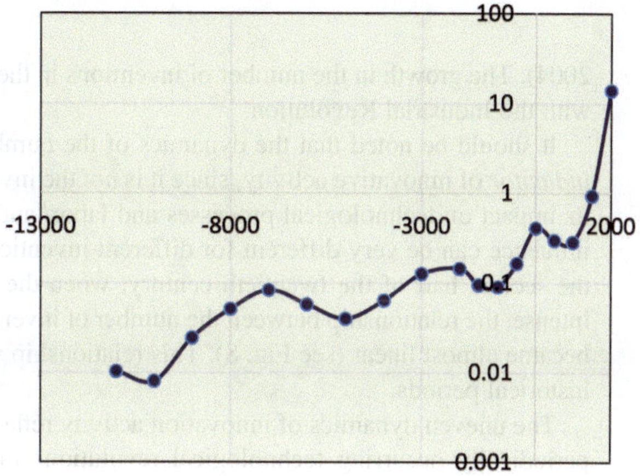

Fig. 7 Average rate of technological innovation activity (number of major inventions per year) from the 12th millennium BCE to 2000 CE

total in the period from the 14th millennium BCE to 2003 CE is shown[7] in Fig. 6.

The dynamics of inventions are very uneven and resemble the graph of the population dynamics in the same period [see Fig. 3 in Sect. 1 (this chapter)]. The uneven growth in the number of inventions can be clearly seen in Fig. 7, where the average number of inventions per year is calculated based on the data of the same source (averaging was carried out over an interval of 500 years).

The growth in the number of inventions in the 10th–7th millennia BCE is associated with the Agrarian (Neolithic) Revolution; the growth in the number of inventions in the 4th–3rd millennia BCE is associated with the Urban Revolution and the transition to the Bronze Age, while the growth in the number of inventions in the 1st millennium BCE is associated with the Axial Revolution in connection with the mass diffusion of iron tools and accompanying major inventions, e.g., water- and windmills, new generations of simple machines, ships, plows, wheeled transportation devices, and so on (Bunch & Hellemans,

[7] When constructing a graph, the inventions of the period before 1 CE were taken into account with weight coefficients reflecting their significance in the subsequent history of humankind (weight coefficients were determined through expert procedures).

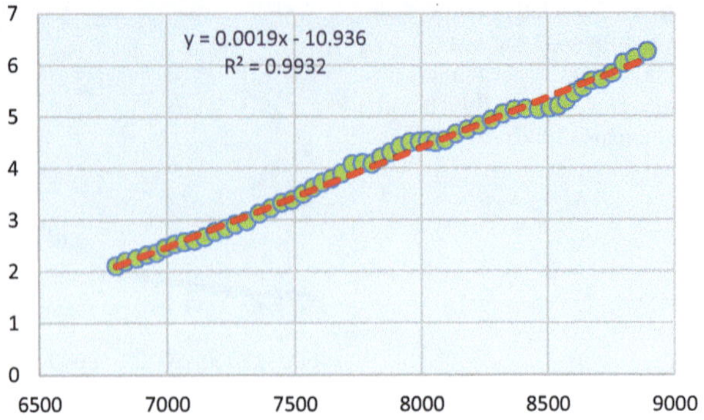

Fig. 8 Relationship between the cumulative number of inventions (*x*-axis) and world GDP/capita, thousands of 1990 international dollars at PPP (*y*-axis) in 1950–2000

2004). The growth in the number of inventions in the last 500 years is associated with the Industrial Revolution.

It should be noted that the dynamics of the number of inventions is only an *indicator* of innovative activity, since it is not the invention itself that matters but its impact on technological processes and labor/land productivity growth. Such influence can be very different for different inventions. However, in general, in the second half of the twentieth century, when the flow of inventions became intense, the relationship between the number of inventions and world GDP/capita became almost linear (see Fig. 8). This relationship was not so clear in previous historical periods.

The uneven dynamics of innovation activity reflected in Fig. 7 can be seen as periodically occurring technological revolutions that significantly change the nature of labor and its productivity. Of all the diverse technological and industrial changes that have taken place in history, three revolutions have had the most profound and comprehensive consequences for society:

1. Agrarian (or agricultural) – the Neolithic Revolution. It resulted in the transition to systematic food production and, on this basis, to the complex social division of labor. This revolution is also associated with the use of new energy sources (animal power) and materials.
2. The Industrial Revolution, as a result of which the main production was concentrated in industry and began to rely on machines and mechanisms. The significance of this revolution is not only in the replacement of manual labor with machines and biological energy with water and steam but also in the fact that it opens up the process of labor-saving in a broad sense (and not only in the field of physical labor but also in accounting, control, management, exchange, credit, transfer of information).
3. The Cybernetic Revolution began with the appearance of powerful information technologies, the introduction of new materials and types of energy, and automation. Its final phase will see the transition to the widespread use of AI-based self-regulating and self-governing systems in various fields of activity that can run without the need for human intervention. The Cybernetic

Revolution is still ongoing and is expected to make huge strides in improving human health, quality of life, and ability to influence and control the human body (Grinin & Grinin, 2023a).

The important role of these revolutions is due to the fact that they created the technological basis for the development and change in the principles of production that determined the course of human history. These production principles include:

- hunter-gatherer;
- agricultural-handicraft;
- industrial-commercial;
- scientific-cybernetic (still in the nascent phases of development).

The cycle of development of each principle of production can be represented in six phases (see, e.g., Grinin, 2007; Grinin et al., 2020). Its first three phases correlate with three phases of the corresponding production revolution:

1. Production revolution begins. A new, still undeveloped and incomplete production principle is being formed, based on the development and implementation of new technologies.
2. The phase of primary modernization of production principle, its diffusion, and strengthening.
3. Production revolution is complete. The principle of production acquires some developed characteristics.

The last three phases of the development of the production principle characterize its mature features:

4. The principle of production becomes mature and expands. New technologies spread widely across territories and sectors, bringing the principle of production to mature forms. Some transformations occur in the socio-economic sphere.
5. The principle of production becomes totally dominant. Technologies intensify, bringing opportunities to the limit beyond which crisis phenomena arise.
6. The phase of nonsystemic phenomena (the preparatory period for the transition to a new principle of production). Intensification leads to the emergence of nonsystemic elements that prepare the birth of a new production principle. (When, under favorable circumstances, these elements can form a system, in some societies, the transition will begin to a new principle of production, and the cycle will repeat itself.)

The features of the principles of production determined the specifics of socio-economic processes and influenced the institutional development of society. In this regard, it is expedient to speak of *Agrarian, industrial and cybernetic* societies as the most important stages of historical development.

B. An important marker of the technological changes taking place in the world is the dynamics of the urban population (see Fig. 9). Indeed, at all times cities have accumulated people who create and use advanced technologies, such as artisans,

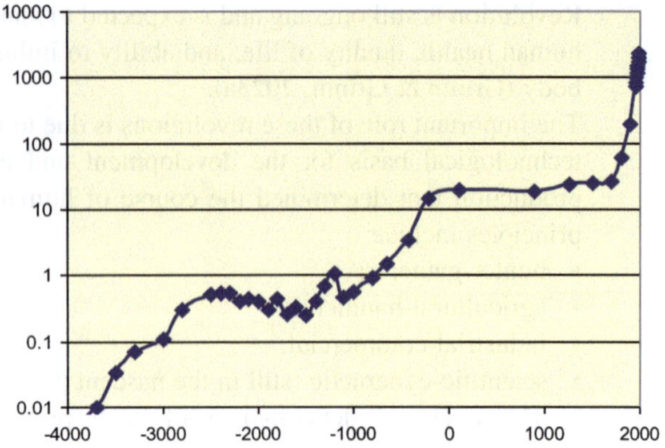

Fig. 9 Change in the world urban population on a logarithmic scale, millions (taking into account cities with a population of more than 10,000 people), over the past six thousand years. *Data source* Korotayev (2006b)

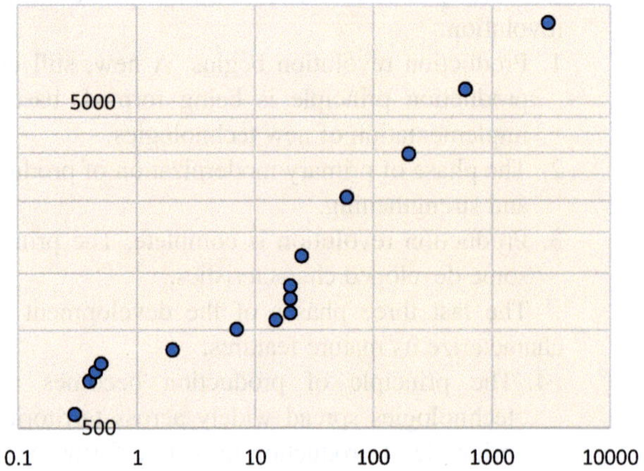

Fig. 10 Correlation between the urban population (*x*-axis, millions) and the cumulative number of inventions (*y*-axis) over the past 5 thousand years, double logarithmic scale

merchants, or industrial workers. Therefore, periods of growth in innovative (inventive) activity (Fig. 7) and periods of urban population growth (Fig. 9) correspond to each other.

This correlation can be clearly seen in Fig. 10:

As can be seen in the above graphs, technological development and demographic growth in the history of humankind were generally synchronous but extremely uneven: eras of rapid growth were replaced by eras of stabilization/stagnation. At the same time, each successive epoch of stabilization/stagnation was characterized by a higher population and a higher share of the urban population (see Fig. 11).

The alternation of demographic and technological growth eras and demographic and technological stabilization/stagnation influenced the course of social self-organization processes and the change in the nature of these processes (Korotayev et al., 2023, this volume).

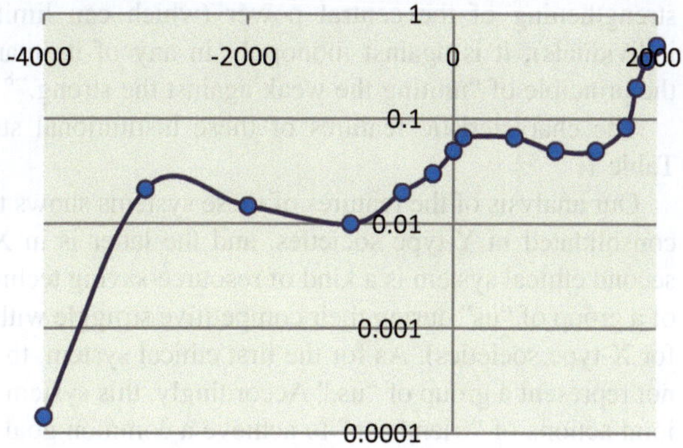

Fig. 11 Change in the share of the urban population in the world population over the past five thousand years. *Data source* Korotayev (2006b)

3 Features of Macrosocial Dynamics

Section 2 of the present chapter shows that the alternation of periods of demographic and technological growth and stabilization/stagnation is an important feature of the macrohistorical process (if we consider the world as a single developing system). In the eras of technological growth, the resource base expanded, and labor productivity increased (an example of such an era is the 200 years since the beginning of the Industrial Revolution), while in the eras of technological stagnation, the growth of the resource base stopped, the situation of the "Malthusian trap" arose, and the struggle for limited resources intensified (an example of such era is the Middle Ages). Changing the conditions in which society finds itself affects the change in the nature of social self-organization and the institutional structures that are formed on its basis.

Thus, given a limited resource base and serious external threats, ensuring survival in difficult natural and social conditions is a vital priority for society. Accordingly, as a result of self-organization, society acquires the features of the so-called X-structure [see Sect. 2.4 of chapter "Modeling Social Self-Organization and Historical Dynamics. Agrarian Society" (Malkov et al., 2023, this volume)], which is characterized by the institutional combination "distributive economy – directive (hierarchical) governance system – the primacy of collectivism in the public mind." The essence of this structure is the unification of common efforts to combat an external threat. This structure objectively contributes to strengthening the central power and can be characterized as "uniting the weak around the strong."

If, however, the priorities for society are the tasks of economic development in conditions of resource sufficiency (or abundance), then society naturally acquires the features of the so-called Y-structure [see Sect. 2.4 of chapter "Modeling Social Self-Organization and Historical Dynamics. Agrarian Society" (Malkov et al., 2023, this volume)], which is characterized by the institutional combination of "liberal-type market economy – adaptive (democratic) system of governance – the primacy of individualism in the public mind." Such a society is not interested in an excessive strengthening of the central power (which can limit the freedom of action of individuals); it is against monopoly in any of its manifestations and implements the principle of "uniting the weak against the strong."[8]

The characteristic features of these institutional structures are summarized in Table 1.

Our analysis of the features of these systems shows that the former is shaped and consolidated in Y-type societies, and the latter is in X-type societies. Indeed, the second ethical system is a kind of resource-saving technology for the social survival of a group of "us" during their competitive struggle with "them" (which is important for X-type societies). As for the first ethical system, the competing social actors do not represent a group of "us." Accordingly, this system is not a means of organizing joint actions of "friends/us" to achieve a common goal but a technology for cutting off actions that are undesirable for society through a normatively fixed legal system (this is important for Y-type societies).

The difference between the ethical systems and behavior patterns in X- and Y-type societies leads to misunderstanding and rejection of one society by another.

It must be borne in mind that we are talking about a shift in social self-organization toward strengthening either X- or Y-structures when external conditions change. Real social systems always contain both X- and Y-elements, but either X- or Y-institutions dominate and shape the image of society as a whole, so we can speak of X-type societies and Y-type societies, respectively.

X-type societies prevail during periods of demographic and technological stabilization/stagnation, while Y-type societies are formed and strengthened during periods of demographic and technological growth (see Fig. 12).

The problem is that, as follows from Table 1, the characteristics of X- and Y-structures are fundamentally different from each other, and the principles of self-organization are often opposite: that is, what is good for X-structure is, as a rule, bad for Y-structure, and vice versa. Therefore, the process of transformation of institutional structures in transitional eras is associated with a decrease in socio-political stability and is often accompanied by social instability and revolutions.

4 Simulation of Global Phase Transitions

Throughout history, periods of demographic and technological growth alternate with periods of demographic and technological stabilization/stagnation. An example of social evolution in the phase of demographic and technological growth is industrial society (see Chapter "Modeling Social Self-Organization and Historical Dynamics. Industrial Society" [Akaev et al., 2023b], this volume). An example of social evolution in the phase of demographic and technological stabilization/stagnation is

[8]On X and Y structures, see also chapter "Economics. Optimizing Growth" (Grinin et al., 2023a, this volume) and chapter "Socio-Political Transformations. A Difficult Path to Cybernetic Society" (Grinin et al., 2023b, this volume),

Table 1 Distinctive features of X- and Y-types of social structures

Feature	X-structure	Y-structure
Institutional Features	1. Regulated economy 2. Directive centralized control system (vertical hierarchies) 3. The primacy of collectivism in the socio-psychological sphere	1. Liberal market economy 2. Adaptive (democratic) control system (horizontal networks) 3. The primacy of individualism in the socio-psychological sphere
Formation conditions	• serious external threats; • lack of resources (zero-sum game)	• absence of serious external threats; • abundance of resources (positive sum game)
Nature of competition[a]	Social competition (the strongest society survives), the "friend or foe" principle	Competition of individuals (survival of the strongest individual), the "foe-foe" principle
Goal	Security (society survival)	Development (increasing individual well-being)
Way to achieve the goal	The weak uniting around the strong (strong central government)	The weak uniting against the strong (weak central government)
Priorities	• cooperation as a principle; • ensuring the unity of society; • improved governance	• competition as a principle; • initiation of pluralism, economic activity
Ethical system[b]	"Declaration of the good" (ideological unity)	"Prohibition of the evil" (freedom of action within the law)
Threats to the system	• loss of unity of society; • decrease in the efficiency of power, bureaucracy, corruption	• monopoly in business and politics; • wealth inequality
Object of protection	Social organization (state)[c]	Individual rights and freedoms

[a]In X-type societies, there is a pronounced division into "friends" and "foes." "Friends" are united by a common goal of ensuring collective survival in the face of external threats. Permanent confrontation with the "foes" requires an effective central government capable of formulating common tasks, organizing joint activities of "friends," and protecting them from outside encroachments. The mobilization functions of the central government require it to make decisions quickly and imply the unconditional implementation of the decisions. Under these conditions, a rigid directive system of social governance is being formed [the basic scheme of which is shown in Fig. 12 in Sect. 3.2 of chapter "Modeling Social Self-Organization and Historical Dynamics. A General Approach" (Akaev et al., 2023a, this volume)]. In contrast, there is no rigid division in Y-type societies into "friends" and "foes." Each actor pursues his or her own interest, counting on his or her strength. The interaction of actors is regulated by formal rules (laws), and the central government controls the compliance with them. Accounting for the interests of society is carried out through the election of governance bodies. The actors' economic independence limits the possibility of pressure on them by the authorities. On the contrary, society can influence the government through re-elections. Under these conditions, an adaptive democratic system of social management is being formed [the basic scheme of which is shown in Fig. 13 in Sect. 3.2 of chapter "Modeling Social Self-Organization and Historical Dynamics. A General Approach" (Akaev et al., 2023a, this volume)].

[b]This refers to the ethical system according to Vladimir Lefebvre. In his work (Lefebvre, 1982) he theoretically substantiates the existence of ethical systems of two opposite types. The first of them is based on the prohibition of the evil, the second is based on the declaration of the good. Both systems are internally logical and consistent, but in practical situations they lead to opposite patterns of behavior. So, while in the first system the desire to agree with a partner is approved, in the second it is considered correct to subordinate him or her to one's own will.

[c]Naturally, we are talking only about periods of a centralized state, and not about periods of decentralization

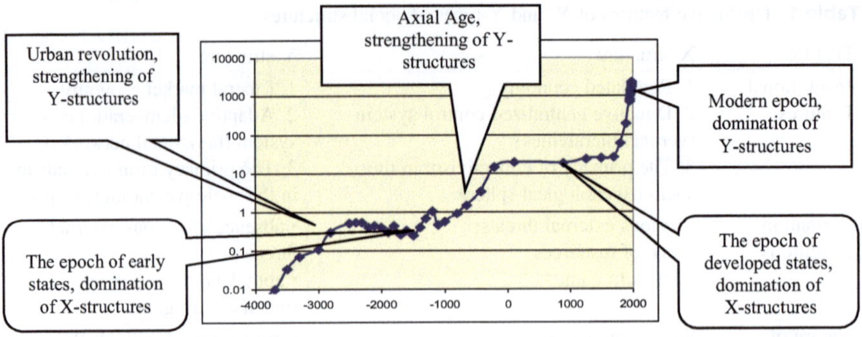

Fig. 12 Relationship between the processes of social self-organization and the phases of historical development (the graph reflects the change in the world urban population in millions for cities with a population of more than 10,000 people, logarithmic scale)

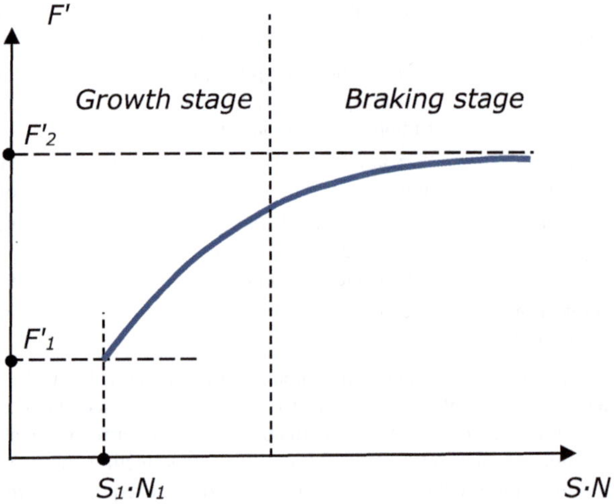

Fig. 13 Function F' described by expression (5)

the typical Agrarian society (see Chapter "Modeling Social Self-Organization and Historical Dynamics. Agrarian Society" [Malkov et al., 2023] this volume).[9] It is important to understand how and why transitions occur from the demographic and technological growth phase to the phase of demographic and technological stabilization/stagnation and vice versa.

[9]The term "typical" is used here because in the pre-industrial period, within the framework of the agrarian macrophase of historical development, there were periods/phases of demographic and technological growth due to fundamental technological shifts ("Urban Revolution", due to the technological breakthroughs of the 4th millennium BCE; "Axial Age Revolution", due to the transition to iron metallurgy and accompanying technological breakthroughs). The term "typical agrarian society" refers to an agrarian society beyond the time limits of the indicated periods of demographic and technological growth.

4.1 Dynamics of Dm-Tl-En in the Eras of Change

Let us apply the basic model "demography –technology – economy" (Dm-Tl-En) to the analysis of global phase transitions. Let the society initially be in a state of the "Malthusian trap," when there is no demographic growth due to a limited resource base [this situation was modeled in Sect. 1.1 of chapter "Modeling Social Self-Organization and Historical Dynamics. A General Approach" (Akaev et al., 2023a, this volume)], using Eqs. (1), (6), (7) from that chapter with constant value of F' – production function characterizing the production of final consumer products). Let the development of technologies, that expand the resource base and ensure growth of the production function, begin at time t_1. Thus, the process of demographic and technological growth is launched [this situation is modeled in Sect. 1.1 of chapter "Modeling Social Self-Organization and Historical Dynamics. Industrial Society" (Akaev et al., 2023b, this volume)] using Eqs. (6)–(8) from that chapter. We assume that economic growth begins to slow down after some time due to the occurrence of certain constraints (for example, resource or environmental), and the production function is eventually fixed at unchanged level F'_2 ($F'_2 \gg F'(t_1)$).

The system of equations during the period of growth corresponds to Eqs. (6)–(8) from Sect. 1.1 of chapter "Modeling Social Self-Organization and Historical Dynamics. Industrial Society" (Akaev et al., 2023b, this volume), and during subsequent deceleration and stabilization transforms into Eqs. (1), (6), (7) from Sect. 1.1 of chapter "Modeling Social Self-Organization and Historical Dynamics. Agrarian Society" (Malkov et al., 2023, this volume), has the following form[10]:

$$dN/dt \approx r \cdot N \cdot (1 - x_0(t)/x), \tag{1}$$

$$dS/dt \approx c \cdot N \cdot S \cdot (1 - x_0(t)/x), \tag{2}$$

$$dx/dt \approx F'/N - q \cdot x, \tag{3}$$

$$x_0(t) \approx x_0 \cdot (1 + b \cdot (N - N_1)/N_1), \tag{4}$$

where subscript $_1$ means that the value of the variable corresponds to the initial time t_1. Equation (4) reflects the growth of the x_0 value as the population grows and society becomes more complex [see Sect. 1.1 of chapter "Modeling Social Self-Organization and Historical Dynamics. Industrial Society" (Akaev et al., 2023b, this volume)]. To account for the growth of function F' at the beginning of the phase transition and its subsequent deceleration due to the occurrence of resource, environmental, and other limitations, the following expression can be used (see Fig. 13):

[10] As before, it is taken into account that the speed of economic processes (the characteristic time of production cycles is months and years) significantly exceeds the rate of demographic and technological changes (the characteristic time is decades), and therefore Tikhonov's theorem can be applied to replace the economic differential equation with an algebraic one.

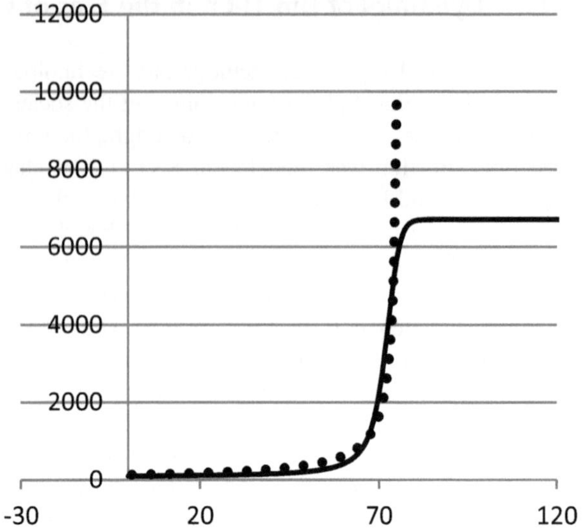

Fig. 14 Typical results of calculating the dynamics of population N (solid line) during the phase transition (values on x- and y-axes are in conventional units). The dotted line is a hyperbolic curve

$$F' \approx (S_1 \cdot N_1)^2/(N_1 \cdot S_1 + F'_2) + S \cdot N \cdot F'_2/(N \cdot S + F'_2) \tag{5}$$

A typical solution of the system of Eqs. (1)–(5) is shown in Fig. 14:
The simulation results lead us to the following conclusions:

- the global phase transition described by the system of Eqs. (1)–(5) is characterized by rapid demographic and technological growth, as well as rapid deceleration in growth at the final stage of this transition (such dynamics actually took place at the end of the Axial Age, see Fig. 3);
- during the phase transition at the stage of growth, the demographic dynamics is approximated with high accuracy by a hyperbolic curve (dashed line in Fig. 14), which was pointed out by Heinz von Foerster in 1960 when analyzing statistical data (von Foerster et al., 1960). The hyperbolic nature of growth is due to strong positive feedback between the variables N and S: the growth of N accelerates the growth of S (see Eq. (2)), and the growth of S, in its turn, accelerates the growth of N, increasing the value of x in Eq. (1).

Thus, modeling shows that singularity is not an artifact but a regular phenomenon in the dynamics of global phase transition due to the synergistic mutual influence of demographic and technological growth (see Korotayev, 2006a, 2020a; LePoire, 2015, 2020; LePoire & Korotayev, 2020; Malkov, 2020 for detail). Summarizing, we can say that if we observe hyperbolic trends in the dynamics of the indicators of complex systems, these are markers that we are dealing with a phase transition that actualizes positive feedback in the processes determining the running of these systems.

The question arises as to under what conditions positive feedback loops are actualized, and the explosive growth process starts. Modeling shows that the main condition is the following: a simultaneous increase in the value of the coefficient c in Eq. (2), which describes the dynamics of technology development, and an increase in the return on innovation (that is, the growth of S and the expansion of the resource base).

An increase in the value of the coefficient c depends on many factors, one of the most important being the social demand for innovation. In a traditional society, the demand for innovation, as a rule, is small; moreover, innovators who offer something new and unusual are treated with caution and apprehension. Priority is given to the use of traditional social and material technologies[11] tested over time and proven to be effective in the past. The demand for innovation, as a rule, arises in a crisis, when for one reason or another, usual decision-making algorithms cease to produce the desired result (these can be natural, social, and other crises). It is also important that the demand for innovation would not fade after overcoming the crisis. For this, it requires a shift in social consciousness, for example, a new ideology (religion) that would encourage innovative activity (among other things, Protestant ethics [Weber, 2004/1905] became such a social technology in modern times in Europe).

The fulfillment of the second condition for the transition to growth – an increase in the return on innovation – is possible if these innovations are effective and provide a sharp increase in labor/land productivity, introduce new types of resources into circulation, and increase production efficiency.[12] Naturally, in the initial period, these effects can only be local (geographically tied to the place where the innovation first gave a positive result), and only then, as success is consolidated, can they spread to peripheral regions. The modernization model describing this process is presented in Sect. 3 of chapter "Modeling Social Self-Organization and Historical Dynamics. Industrial Society" (Akaev et al., 2023b, this volume). The expansion of the region covered by the modernization processes includes several phases.

Stage 1: Traditional society.

This phase is characterized by a low level of economic growth with the economy based on the use of traditional technologies. Socio-economic and political processes are cyclical, and there is a dominance of X-type institutional structures (see Table 1 in Sect. 3 of the present chapter).

Stage 2: Phase transition influenced by technological and social innovations.

The emergence and development of a set of technological and social innovations that dramatically increase labor and land productivity in a number of societies disrupts the relative balance of power in the world system and launches a process

[11] The cautious attitude towards innovation in traditional societies is due, among other things, to the fact that these societies, as a rule, are in the "Malthusian trap", when a significant part of society is concerned about physical survival and is not ready for drastic changes that upset the fragile balance.

[12] This, in turn, stimulates further innovative search, encourages innovative activity, raises the social status of innovators, and creates a favorable environment for development in society.

of modernization that gradually expands and eventually embraces the entire system. This process has the following subphases:

Subphase 2.1. The world system is differentiated into core countries (which include societies that have mastered and introduced innovations) and periphery (which includes societies that still use traditional technologies). In societies of the core, Y-institutional structures based on internal competition[13] are being formed. Societies of the periphery are still dominated by X-institutional structures based on cooperation among "us" and opposition to "them" (see Table 1 in Sect. 3 of the present chapter).

Subphase 2.2. Great Divergence (the explosive growth of a gap in per capita incomes between the countries of the world-system core and the countries of the world-system periphery)[14]; the countries of the core establish a system of economic and political rules, a certain type of globalization is implemented in the interests of the countries of the core (in order to use the resources of the periphery by the core countries to stimulate the development of the latter).

Subphase 2.3. Modernization process develops in the countries of the periphery due to the diffusion of innovations and advanced technologies from the core countries.

Subphase 2.4. Great Convergence: the countries of the periphery start catching up with the countries of the core as a result of modernization of their economies; this is accompanied by slowdown and critical phenomena in the core countries.[15]

Subphase 2.5. Structural world-system crisis: the loss of leading positions by the core countries followed by the destruction of the system of rules and the dominant ideology established by them. As a consequence, chaos ensues in economic and political interactions, as well as progressive regionalization, and the formation of new competing political blocs (see, e.g., Grinin & Korotayev, 2011b, 2015).

Stage 3: New society.

At this stage, the competition between the X- and Y-systems is removed due to transformations in the ideological and institutional spheres, the differentiation into the countries of the core and the countries of the periphery is eliminated, a new type of society and new geopolitical blocs arise.

This raises the second question as to why the period of growth eventually ends. The influence of the following factors is important here:

- the depletion of the capabilities of the basic technologies that started the growth process (see, e.g., LePoire & Korotayev, 2020);

[13] At the beginning of the Axial Age, the leader in the formation of Y-institutions in the Circum-Mediterranean region was Ancient Greece, at the beginning of the Modern period this role was played by Britain.

[14] See e.g., Goldstone (2009, 2021), Grinin and Korotayev (2015), Korotayev et al. (2015), Pomeranz (2021).

[15] See e.g., Korotayev et al. (2011, 2020), Korotayev and de Munck (2013), Korotayev and Zinkina (2014), Akaev (2015), Grinin and Korotayev (2015), Korotayev et al. (2015), Grinin et al. (2016).

- an increase in population gives rise to problems that previously seemed insignificant (for example, environmental problems), but now require serious attention and resource costs;
- the stocks of the main natural resources, which were intensively used during the period of growth, are being depleted;
- an increase in the size and density of the population requires a continuous increase in the cost of infrastructure and livelihoods, which increases the burden on the economy;
- an increase in the density and overcrowding of the population contributes to the emergence of new diseases and epidemics;
- an increase in the scale of social systems and an increase in the complexity of managing them requires a significant restructuring of the entire system of social interactions, which usually happens very painfully, increases internal conflict, and leads to social crises.

Until now, under the influence of these and other factors, periods of growth ended with the opening of a new historical era, during which society adapted to a new reality, to a world changed by new technologies. This was accompanied by the emergence of new ideologies (religions) and types of statehood, changes in mentality, emergence of new social strata, consolidation of social systems and their internal connectivity (transport, economic, informational, mental). Society 'digested' the changes that occurred as a result of the phase transition and accumulated the potential for the next round of evolution.

It remains to be understood why the intervals between successive "epochs of change" are decreasing: the Neolithic Revolution took place 10 thousand years ago, the "urban revolution" – 5 thousand years ago, the "Axial Age" – 2.5 thousand years ago, the Industrial Revolution – 200 years ago (we use such a list of global transformations because global population explosions accompanied them, see Fig. 15, as well as Korotayev, 2020a). We can try to answer this question based on Taagepera – Kremer equation (Taagepera, 1976, 1979; Kremer, 1993; Korotayev, 2020b)[16], which reflects the fact that the speed of innovation is proportional to the current technological level of development:

$$dS/dt \approx c \cdot N \cdot S, \tag{6}$$

where S is the number of technological innovations; thus, it turns out that an increase in the number of innovations leads to an increase in technological development and, as we already know, to an increase in the population (and hence to an increase in the number of innovators, which make up a very small but fairly stable percentage of the population). That is, there is positive feedback that accelerates technological growth

[16]See also Taagepera (1976, 1979), Tsirel (2004), Korotayev (2005, 2007), Podlazov (2017).

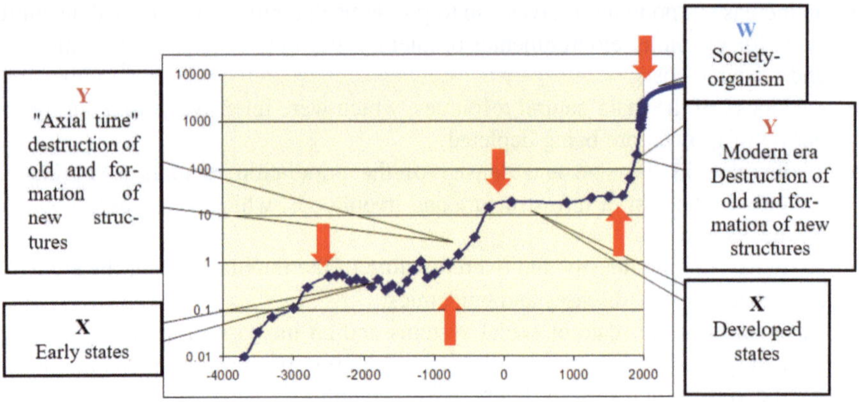

Fig. 15 Correlation between demographic, social, and spiritual development in world history (solid line – dynamics of the world urban population, millions, logarithmic scale, see Fig. 9). Arrows mark turning points in the formation of ideological currents (see commentary in the text)

and reduces the time between the next "technological revolutions," which is actually observed in human history.[17]

4.2 Macrosocial Dynamics in the "Era of Change"

So, as shown above, at the macro level, human history is a series of alterations of periods of intensive economic and demographic growth in connection with technological transformations and periods of economic and demographic stability (and, at times, stagnation). It is clear that the processes of social self-organization during periods of intensive growth and during periods of stability differ significantly from each other, which leads to the formation of different social and institutional structures in these societies: growth conditions contribute to the formation of the so-called Y-structures, while stability results in the formation of the so-called X-structures. A description of X- and Y-institutional structures is given above in Sect. 3 of this chapter.

What is important for us here is that in conditions of limited resources and serious external threats, the economic system acquires a distributive character, which is sanctified by tradition and ideology (usually in the form of religion). In complex Agrarian societies, traditions, and religions impose severe restrictions and self-restraints on the members of society ("thou shalt not kill," "thou shalt not steal,"

[17] Interestingly, the effect of technological acceleration is very reminiscent of the well-known effect of accelerating speciation in biology, which has been observed over the past 500 million years and, apparently, is due to the fact that earlier species change their habitat by their vital activity, making it more diverse, saturating it with organic matter, generating new biological niches, thereby creating favorable conditions for the emergence of new species (see, e.g., Markov & Korotayev, 2007, 2008).

"thou shalt not commit adultery," etc.), which have a stabilizing effect on the social system (see, e.g., Turchin et al., 2022a, 2022b). Violation of these norms in conditions of limited resources leads to destabilization and weakening of society; therefore, it is condemned and punished.

With a growing resource, the need for rigid regulation of society is no longer necessary. Moreover, existing traditions and norms become a burden, as they hinder introducing and disseminating innovations. The concept of "freedom" acquires a status of the most important value; the former restrictions and self-restraints are gradually rejected as outdated and no longer needed. The rejection of the old canons leads to the formation of new institutional structures that are more adequate to the growth conditions, encouraging individual activity and competition (Y-structures). However, this reduces the stability of society, and if (and when) the growth of the resource begins to slow down, the "society of freedom" may get destabilized and disintegrate.

Consider how the transitions from periods of stability[18] to periods of growth[19] and vice versa took place:

- during periods of stability/stagnation, society is in the conditions of a "zero-sum game."[20] X-structures dominate. Tradition prevails, ensuring the balance of society with the environment. Ideologically, tradition is sanctified by religion, which establishes a system of restrictions and self-restraints. Accordingly, the observance of ideological (religious) norms is of great importance in ensuring the stability of society; innovations are not welcome;
- with a sharp change in the external environment (for example, due to climate change), the old norms and self-limitations no longer give the required result in terms of stabilizing the society. Society destabilizes, the search for a way out of the crisis begins, and there is a demand for innovation, respectively. Society begins to abandon the restrictions and self-restraints;
- if the search for innovations aimed at overcoming the crisis turns out successful[21] and the introduction of new technologies expands the resource base, then society

[18]Periods of stability can be characterized in terms of game theory as periods of dominance of the "zero-sum game" in the interactions of social actors: under conditions of a limited resource base, victory in the competitive struggle of one of the actors leads to the death (physical, economic, political) of the defeated. Therefore, within society, competition was condemned (so as not to lead to its destabilization), while solidarity and mutual assistance were encouraged. However, in relation to other societies, antagonistic relations and the "image of the enemy" were common.

[19]Periods of growth can be characterized in terms of game theory as periods of dominance of the "positive sum game" in the interactions of social actors: under conditions of a growing resource base, the victory of one of the actors does not lead to the death (physical, economic, political) of the defeated, but stimulates the latter be more competitive. Competition generates economic activity leading to a further expansion of the resource base. Accordingly, it is not condemned, but welcomed, which leads to the formation of an internally competitive society.

[20]In game theory, a "zero-sum game" is a game in which one player's gain equals the other's loss.

[21]This happens very rarely. History, on the contrary, is replete with examples when states and entire civilizations fell into decay and disintegrated under the influence of adverse climatic and environmental changes.

enters the phase of economic growth. This creates a positive feedback loop: economic success stimulates further rejection of traditional norms and self-restraints. Society begins to encourage initiative, the desire for novelty, and freedom of action. An internally competitive society with a Y-type[22] institutional structure is being formed, focused on a "positive sum game."[23] Internal competition stimulates individual activity and the search for additional resources, which in turn supports and enhances economic growth;
- however, after some time, the potential of the technological revolution launched by an innovative impulse reaches its saturation and is depleted, the resource base stops growing, and the "positive sum game" stops. In the transition to the "zero-sum game," the Y-society loses its effectiveness: with limited resources, the orientation toward competition as a principle of relationships leads to increased internal conflict and destabilization of society (to a war of all against all). A new system of norms is needed that restrains competition, transforming competitive relations into cooperative ones and a new system of self-restraints.[24] There is a need to revive the X-system, in which collective interests prevail over individual ones, and distributive principles replace free competition.

Thus, the following sequence of events takes place. The traditional X-society, having fallen into a crisis situation, is forced to change, and if these changes are successful and launch an innovative process that expands the resource base, then economic growth begins, in which the social X-structure begins to transform into a Y-structure. The Y-structure remains effective as long as the "positive sum game" is maintained. When the innovation momentum dries up and the resource base ceases to grow, the Y-structure is again transformed into an X-structure focused on the "zero-sum game." The circle closes according to the scheme: X → Y → X → .... (see Fig. 15).

Three points are important here.

Firstly, each technological revolution increased the unevenness of development and launched another round of globalization (see Zinkina et al., 2019; Grinin & Korotayev, 2012b, 2014b; Carneiro et al., 2017), since the leading societies that mastered new technologies earlier than others pulled ahead in economic and political relations and began external expansion (in terms of trade, military sphere, ideology)

---

[22] Such societies included the city-states of Ancient Mesopotamia at the beginning of the "urban revolution", the poleis of Ancient Greece at the beginning of the "Axial Age", England at the beginning of the Modern period. Further on, the locally emerged institutional Y-structures begin to spread and scale. Thus, in the Axial Age many achievements of the poleis of Ancient Greece were borrowed and used by the Ancient Romans when creating their empire. In the Modern era, the achievements of Britain were borrowed and used by the United States when creating "Pax Americana".

[23] In game theory, a "positive sum game" is a game in which all players can benefit (but perhaps to varying extent).

[24] This was one of the reasons why, at the end of the Roman Empire, Christianity became the state religion.

outside their territories. As a result, the consolidation of socio-economic systems took place:

- the Urban Revolution saw the transition from fragmented structures of tribes, chiefdoms, and their analogues to early states and early state analogues (e.g., Claessen & Skalník, 1978; Service, 1975; Bondarenko & Korotayev, 2003; Grinin et al., 2004; Grinin & Korotayev, 2009b, 2011a, 2012a);
- the Axial Age saw the process of the formation of vast empires that absorbed many early states and their analogues (e.g., Taagepera, 1968, 1976, 1978, 1979, 1997; Diakonoff, 1999; Chase-Dunn et al., 2010; Turchin, 2016; Korotayev, 2020a);
- the modern era witnessed the process of economic and political globalization in the full sense of the word with the formation of the truly global world system (e.g., Wallerstein, 1974, 2004; Modelski & Thompson, 1996; Modelski et al., 2007; Korotayev & Malkov, 2012; Sheffield et al., 2013; Grinin & Korotayev, 2014a; Chase-Dunn & Lerro, 2016; Zinkina et al., 2019; Grinin, 2021).

In this regard, the X-structures that are formed after each technological revolution are not a repetition of the previous X-structures but new social phenomena, larger and more complex in their composition and internal connections.

Secondly, since X- and Y-type societies are radically different in their ideological and value orientations, the transition from X-society to Y-society, as well as the transition from Y-society to X-society, is impossible without a revolution in people's minds, without the emergence of new mass ideologies (religions) that change value orientations, ideas about good and evil, about right and wrong behavior (e.g., Korotayev, 2004; Zinkina et al., 2022). Such ideologies (religions) emerged and diffused both at the beginning of the "epochs of change" (providing the transformation of X-society into Y-society) and at their end (providing the transformation of Y-society into X-society). In Fig. 15, ideologies (religions) of the first type are indicated by upward arrows, and ideologies (religions) of the second type are indicated by downward arrows.[25] Accordingly, in the current transitional era, the emergence of new ideologies of the second type is very likely.

Thirdly, the chain: X (early states of the Bronze Age) → Y ("Axial Age") → X (developed states of the Middle Ages) → Y (modern capitalism) → (?) most likely will not end with a return to social X-structures. One should expect the emergence of a fundamentally new "postsingular" society. We will conditionally call it a "W-society."

[25] So, for example, the religion that contributed to the transformation of the ancient Greek X-society into the Y-society was the religion of the Olympian gods with its physicality and very competitive spirit. The religion that contributed to the transformation of the ancient Roman Y-society (Roman Republic) into the X-society (the Roman Empire of the Dominate period) was the Christian religion, which proclaimed the virtue of "love thy enemy." The confessions that contributed to the transformation of European medieval X-society into Y-society of modern times were certain versions of Protestantism with their spirit of individualism and active competitiveness.

Acknowledgment This work was done with the support of MSU Program of Development, Project No 23A-SCH05-03.

References

Akaev, A. (2015). *Ot epokhi Velikoy divergentsii k epokhe Velikoy konvergentsii: Matematicheskoye modelirovaniye i prognozirovaniye dolgosrochnogo tekhnologicheskogo i ekonomicheskogo razvitiya mirovoy dinamiki*. LENAND/URSS.

Akaev, A., Malkov, S., Bilyuga, S., Malkov, A., Musieva, J., & Korotayev, A. (2023a). Modeling social self-organization and historical dynamics. A general approach. In V. Sadovnichy et al. (Eds.), *Reconsidering the limits to growth. A report to the Russian Association of the Club of Rome* (pp. 309–335). Springer. https://doi.org/10.1007/978-3-031-34999-7_16

Akaev, A., Malkov, S., Davydova, O., Kovaleva, N., Malkov, A., Grinin, A., & Korotayev, A. (2023b). Modeling social self-organization and historical dynamics. Industrial society. In V. Sadovnichy et al. (Eds.), *Reconsidering the limits to growth. A report to the Russian Association of the Club of Rome* (pp. 337–385). Springer. https://doi.org/10.1007/978-3-031-34999-7_17

Belavin, V., Kapitsa, S., & Kurdyumov, S. (1998). Mathematical model of global demographic processes with regard to spatial distribution. *Journal of Computational Mathematics and Mathematical Physics, 38*(6), 885–902.

Bellah, R. N. (2005). What is axial about the Axial Age? *Archives Européennes de Sociologie/European Journal of Sociology/Europäisches Archiv für Soziologie, 46*(1), 69–89.

Bellah, R. N., & Joas, H. (2012). *The Axial Age and its consequences*. Harvard University Press.

Bondarenko, D. M., & Korotayev, A. V. (2003). "Early state" in cross-cultural perspective: A statistical reanalysis of Henri J. M. Claessen's database. *Cross-Cultural Research, 37*(1), 105–132. https://doi.org/10.1177/1069397102238924

Boserup, E. (1965). *The conditions of agricultural growth: The economics of agrarian change under population pressure*. Routledge.

Bunch, B., & Hellemans, A. (2004). *The history of science and technology*. Houghton Mifflin Company.

Carneiro, R., Grinin, L., & Korotayev, A. (2017). *Chiefdoms: Yesterday and today*. Eliot Werner Publications.

Carneiro, R. L. (1970). A Theory of the origin of the state: Traditional theories of state origins are considered and rejected in favor of a new ecological hypothesis. *Science, 169*(3947), 733–738. https://doi.org/10.1126/science.169.3947.733

Chase-Dunn, C., & Hall, T. D. (1997). *Rise and demise: Comparing world-systems*. Westview Press.

Chase-Dunn, C., & Lerro, B. (2016). *Social change: Globalization from the Stone Age to the present*. Routledge.

Chase-Dunn, C., Niemeyer, R., Alvarez, A., Inoue, H., & Love, J. (2010). Cycles of rise and fall, upsweeps and collapses: Changes in the scale of settlements and polities since the Bronze Age. In L. Grinin, P. Herrmann, A. Korotayev, & A. Tausch (Eds.), *History and mathematics: Processes and models of global dynamics* (pp. 64–91). Uchitel.

Childe, V. G. (1934). *New light on the most ancient East*. Routledge & Kegan.

Childe, V. G. (1944). *Progress and archaeology*. Watts & Company.

Childe, V. G. (1948). *What happened in history*. Penguin Books.

Childe, V. G. (1952). *New light on the most ancient East* (4th ed.). Routledge & Paul.

Claessen, H. J. M., & Skalník, P. (Eds.). (1978). *The early state*. Mouton.

Claessen, H. J. M., & van de Velde, P. (1985). *Development and decline. The evolution of sociopolitical organization*. Bergin & Garvey.

Diakonoff, I. (1999). *The paths of history*. Cambridge University Press.

Dobrolyubov, S. (2016). Globalnoye obshchestvo kak tochka singulyarnosti i fazovogo perekhoda k novomu tipu sotsial'noy evolyutsii. *Evolyutsiya, 8*, 229–262.

Eisenstadt, S. N. (1982). The axial age: The emergence of transcendental visions and the rise of clerics. *European Journal of Sociology/Archives Européennes de Sociologie, 23*(2), 294–314.

Eshed, V., Gopher, A., Gage, T. B., & Hershkovitz, I. (2004). Has the transition to agriculture reshaped the demographic structure of prehistoric populations? New evidence from the Levant. *American Journal of Physical Anthropology, 124*(4), 315–329. https://doi.org/10.1002/ajpa.10332

Fomin, A. (2020). Hyperbolic evolution from biosphere to technosphere. In A. Korotayev & D. LePoire (Eds.), *The 21st century singularity and global futures. A Big History perspective* (pp. 105–118). Springer. https://doi.org/10.1007/978-3-030-33730-8_5

Frank, A. G., & Gills, B. K. (Eds.). (1993). *The world system: Five hundred years of five thousand*. Routledge.

Fried, M. H. (1967a). *The evolution of political society. An essay in political anthropology*. Random House.

Fried, M. H. (1967b). On the concepts of 'tribe' and 'tribal society'. In J. Helm (Ed.), *Essays on the problem of tribe* (pp. 3–20). American Ethnological Society.

Garske, J. P. (2020). Climate change, Malthusian Catastrophe, and a global rule of law. *Journal of Globalization Studies, 10*(1), 3–15. https://doi.org/10.30884/jogs/2020.01.01

Goldstone, J. A. (2009). *Why Europe? The rise of the West in world history. 1500–1850*. McGraw-Hill.

Goldstone, J. A. (2021). Dating the great divergence. *Journal of Global History, 16*(2), 266–285. https://doi.org/10.1017/S1740022820000406

Grinin, L. (2003). The early state and its analogues. *Social Evolution and History, 2*(1), 131–176.

Grinin, L. (2004a). Democracy and early state. *Social Evolution and History, 3*(2), 93–149.

Grinin, L. (2004b). Early state and democracy. In L. E. Grinin, R. L. Carneiro, D. M. Bondarenko, & A. V. Korotayev (Eds.), *The early state, its alternatives and analogues* (pp. 419–463). Uchitel.

Grinin, L. (2007). Production revolutions and periodization of history: A comparative and theoretic-mathematical approach. *Social Evolution & History, 6*(2), 75–120.

Grinin, L. (2008). Early state, developed state, mature state: Statehood evolutionary sequence. *Social Evolution & History, 7*(1), 67–81.

Grinin, L. (2009). The pathways of politogenesis and models of the early state formation. *Social Evolution, 8*(1), 92–132.

Grinin, L. (2021). How globalization shifts the global economic and political balance. *Journal of Chinese Philosophy, 48*(2), 222–234.

Grinin, L., Carneiro, R. L., Bondarenko, D., Kradin, N., & Korotayev, A. (Eds.). (2004). *The early state, its alternatives and analogues*. Uchitel.

Grinin, L., & Grinin, A. (2016). *The cybernetic revolution and the forthcoming epoch of self-regulating systems*. Uchitel.

Grinin, L., & Grinin, A. (2023a). Analyzing social self-organization and historical dynamics. Future cybernetic W-society: Socio-political aspects. In V. Sadovnichy et al. (Eds.), *Reconsidering the limits to growth. A Report to the Russian Association of the Club of Rome* (pp. 491–519). Springer. https://doi.org/10.1007/978-3-031-34999-7_21

Grinin, L., & Grinin, A. (2023b). Technologies: Limitless possibilities and effective control. In V. Sadovnichy et al. (Eds.), *Reconsidering the limits to growth. A Report to the Russian Association of the Club of Rome* (pp. 139–154). Springer. https://doi.org/10.1007/978-3-031-34999-7_8

Grinin, L., Grinin, A., & Korotayev, A. (2020). A quantitative analysis of worldwide long-term technology growth: From 40,000 BCE to the early 22nd century. *Technological Forecasting and Social Change, 155*, 1–19. https://doi.org/10.1016/j.techfore.2020.119955

Grinin, L., Grinin, A., & Korotayev, A. (2023). Global aging – An integral problem of the future. How to turn a problem into a development driver? In V. Sadovnichy et al. (Eds.), *Reconsidering the limits to growth. A report to the Russian Association of the Club of Rome* (pp. 117–135). Springer. https://doi.org/10.1007/978-3-031-34999-7_7

Grinin, L., Grinin, A., & Malkov, S. (2023a). Economics. Optimizing growth. In V. Sadovnichy et al. (Eds.), *Reconsidering the limits to growth. A report to the Russian Association of the Club of Rome* (pp. 155–168). Springer. https://doi.org/10.1007/978-3-031-34999-7_9
Grinin, L., Grinin, A., & Malkov, S. (2023b). Socio-political transformations. A difficult path to cybernetic society. In V. Sadovnichy et al. (Eds.), *Reconsidering the limits to growth. A report to the Russian Association of the Club of Rome* (pp. 169–189). Springer. https://doi.org/10.1007/978-3-031-34999-7_10
Grinin, L., & Korotayev, A. (2006). Political development of the world system: A formal quantitative analysis. In P. Turchin, L. E. Grinin, V. C. de Munck, & A. V. Korotayev (Eds.), *History & mathematics: Historical dynamics and development of complex societies* (pp. 63–114). KomKniga/URSS.
Grinin, L., & Korotayev, A. (2009a). Social macroevolution: growth of the world system integrity and a system of phase transitions. *World Futures, 65*(7), 477–506.
Grinin, L., & Korotayev, A. (2009b). The epoch of the initial politogenesis. *Social Evolution and History, 8*(1), 52–91.
Grinin, L., & Korotayev, A. (2011a). Chiefdoms and their analogues: alternatives of social evolution at the societal level of medium cultural complexity. *Social Evolution & History, 10*(1), 276–335.
Grinin, L., & Korotayev, A. (2011b). The coming epoch of new coalitions: Possible scenarios of the near future. *World Futures, 67*(8), 531–563. https://doi.org/10.1080/02604027.2011.625749
Grinin, L., & Korotayev, A. (2012a). Emergence of chiefdoms and states: A spectrum of opinions. *Social Evolution & History, 11*(2), 191–204.
Grinin, L., & Korotayev, A. (2012b). The Afroeurasian world-system: genesis, transformations, characteristics. In S. J. Babones & C. Chase-Dunn (Eds.), *Routledge handbook of world-systems analysis* (pp. 30–41). Routledge.
Grinin, L., & Korotayev, A. (2014a). Globalization shuffles cards of the world pack: In which direction is the global economic-political balance shifting? *World Futures, 70*(8), 515–545. https://doi.org/10.1080/02604027.2014.982474
Grinin, L., & Korotayev, A. (2014b). Origins of globalization in the framework of the Afroeurasian world-system history. In T. D. Hall (Ed.), *Comparing globalizations. Historical and world-systems approaches* (pp. 37–70). Springer. https://doi.org/10.1007/978-3-319-68219-8_3
Grinin, L., & Korotayev, A. (2015). *Great divergence and great convergence. A global perspective*. Springer. https://doi.org/10.1007/978-3-319-17780-9
Grinin, L., Korotayev, A., & Tausch, A. (2016). *Economic cycles, crises, and the global periphery*. Springer. https://doi.org/10.1007/978-3-319-41262-7
Harris, D., & Hillman, G. (Eds.). (1989). An evolutionary continuum of people – plant interaction. In *Foraging and farming. The evolution of plant exploitation*. Unwin Hyman..
Jaspers, K. (1953). *The origin and goal of history*. Yale University Press.
Kapitza, S. (1992). A mathematical model for global population growth. *Matematicheskoe modelirovanie, 4*(6), 65–79.
Kapitza, S. (1996). Phenomenological theory of world population growth. *Physics Uspekhi, 39*(1), 57–71. https://doi.org/10.1070/PU1996v039n01ABEH000127
Kapitza, S. (1999a). *Obshchaya teoriya rosta chelovechestva. Skol'ko lyudey zhilo, zhivet i budet zhit' na Zemle*. Nauka.
Kapitza, S. (1999b). *An essay on the demographic imperative: Non-linear theory of growth of humankind*. The Edwin Mellen Press.
Kapitza, S. (2003). The statistical theory of global population growth. In J. Nation, I. Trofimova, J. D. Rand, & W. Sulis (Eds.), *Formal descriptions of developing systems* (pp. 11–35). Springer. https://doi.org/10.1007/978-94-010-0064-2_2
Kapitza, S. (2006). *Global population blow-up and after. A report to the Club of Rome*. Global Marshall Plan Initiative.
Kapitza, S. (2010). On the theory of global population growth. *Physics Uspekhi, 53*(12), 1287–1296. https://doi.org/10.3367/UFNe.0180.201012g.1337

Korotayev, A. (2004). *World religions and social evolution of the Old World Oikumene civilizations: A cross-cultural perspective*. The Edwin Mellen Press.
Korotayev, A. (2005). A compact macromodel of World System evolution. *Journal of World-Systems Research, 11*(1), 79–93. https://doi.org/10.5195/jwsr.2005.401
Korotayev, A. (2006a). The World System history periodization and mathematical models of socio-historical processes. *History & Mathematics, 1*, 39–98.
Korotayev, A. (2006b). The World System urbanization dynamics: a quantitative analysis. *History & Mathematics, 2*, 44–62.
Korotayev, A. (2007). Compact mathematical models of world system development, and how they can help us to clarify our understanding of globalization processes. In G. Modelski, T. Devezas, & W. Thompson (Eds.), *Globalization as evolutionary process: Modeling global change* (pp. 133–160). Routledge.
Korotayev, A. (2013). Globalization and mathematical modeling of global evolution. *Evolution, 3*, 69–83.
Korotayev, A. (2020a). How singular is the 21st century Singularity? In A. Korotayev & D. LePoire (Eds.), *The 21st century Singularity and global futures. A Big History perspective* (pp. 571–595). Springer. https://doi.org/10.1007/978-3-030-33730-8_26
Korotayev, A. (2020b). The 21st century Singularity in the Big History perspective. A re-analysis. In A. Korotayev & D. LePoire (Eds.), *The 21st century Singularity and global futures. A Big History perspective* (pp. 19–75). Springer. https://doi.org/10.1007/978-3-030-33730-8_2
Korotayev, A., Bilyuga, S., & Shishkina, A. (2020). Which countries generate Kondratieff waves in global GDP growth rate dynamics in the contemporary world? *Journal of Globalization Studies, 11*(1), 33–63. https://doi.org/10.30884/jogs/2020.01.03
Korotayev, A., & de Munck, V. (2013). Advances in development reverse inequality trends. *Journal of Globalization Studies, 4*(1), 105–124.
Korotayev, A., Goldstone, J. A., & Zinkina, J. (2015). Phases of global demographic transition correlate with phases of the Great Divergence and Great Convergence. *Technological Forecasting and Social Change, 95*, 163–169. https://doi.org/10.1016/j.techfore.2015.01.017
Korotayev, A., & Grinin, L. (2013). Urbanization and political development of the world system. *Entelequia Revista Interdisciplinar, 15*, 197–254.
Korotayev, A., & Khaltourina, D. (2006). *Introduction to social macrodynamics: Secular cycles and millennial trends in Africa*. KomKniga/URSS.
Korotayev, A., & LePoire, D. J. (Eds.). (2020). *The 21st century Singularity and global futures. A Big History perspective*. Springer. https://doi.org/10.1007/978-3-030-33730-8
Korotayev, A., & Malkov, A. (2016). A compact mathematical model of the World System economic and demographic growth, 1 CE–1973 CE. *International Journal of Mathematical Models and Methods in Applied Sciences, 10*, 200–209.
Korotayev, A., Malkov, A., & Khaltourina, D. (2006a). *Introduction to social macrodynamics: Compact macromodels of the World System growth*. KomKniga/URSS.
Korotayev, A., Malkov, A., & Khaltourina, D. (2006b). *Introduction to social macrodynamics: Secular cycles and millennial trends*. KomKniga/URSS.
Korotayev, A., & Malkov, S. (2012). Mathematical models of the world-system development. In S. Babones & C. Chase-Dunn (Eds.), *Routledge handbook of world-systems analysis* (pp. 158–161). Routledge.
Korotayev, A., Malkov, S., & Musieva, J. (2023). Demography: Toward optimization of demographic processes. In V. Sadovnichy et al. (Eds.), *Reconsidering the limits to growth. A report to the Russian Association of the Club of Rome* (pp. 97–116). Springer. https://doi.org/10.1007/978-3-031-34999-7_6
Korotayev, A., & Zinkina, J. (2014). On the structure of the present-day convergence. *Campus-Wide Information Systems, 31*(2–3), 139–152. https://doi.org/10.1108/CWIS-11-2013-0064
Korotayev, A., & Zinkina, J. (2017). Systemic boundary issues in the light of mathematical modeling of world-system evolution. *Journal of Globalization Studies, 8*(1), 78–96.

Korotayev, A., Zinkina, J., & Andreev, A. (2016). Secular cycles and millennial trends. *Cliodynamics, 7*(2), 204–216. https://doi.org/10.21237/C7CLIO7230484

Korotayev, A., Zinkina, J., Bogevolnov, J., & Malkov, A. (2011). Global Unconditional Convergence among Larger Economies after 1998? *Journal of Globalization Studies, 2*(2), 25–62.

Kremer, M. (1993). Population growth and technological change: One million B.C. to 1990. *The Quarterly Journal of Economics, 108*, 681–716. https://doi.org/10.2307/2118405

Kurzweil, R. (2001). The law of accelerating returns. *KurzweilAI.net, 07.03.2001*. Retrieved from https://www.kurzweilai.net/the-law-of-accelerating-returns

Kurzweil, R. (2005). *The singularity is near: When humans transcend biology*. Viking Penguin.

Last, C. (2020). Global brain: foundations of a distributed singularity. In A. Korotayev & D. LePoire (Eds.), *The 21st century Singularity and global futures. A Big History perspective* (pp. 363–375). Springer. https://doi.org/10.1007/978-3-030-33730-8_16

Lefebvre, V. (1982). *Algebra of conscience. A comparative analysis of Western and Soviet ethical systems*. Springer. https://doi.org/10.1007/978-94-010-9051-3

LePoire, D. (2015). Interpreting Big History as complex adaptive system dynamics with nested logistic transitions in energy flow and organization. *Emergence: Complexity and Organization, 17*(1), 1–16.

LePoire, D. (2020). Exploring the singularity concept within Big History. In A. Korotayev & D. LePoire (Eds.), *The 21st century Singularity and global futures. A Big History perspective* (pp. 77–97). Springer. https://doi.org/10.1007/978-3-030-33730-8_3

LePoire, D., & Devezas, T. (2020). Near-term indications and models of a singularity trend. In A. Korotayev & D. LePoire (Eds.), *The 21st century Singularity and global futures. A Big History perspective* (pp. 213–224). Springer. https://doi.org/10.1007/978-3-030-33730-8_11

LePoire, D., & Korotayev, A. (2020). Conclusion. In A. Korotayev & D. LePoire (Eds.), *The 21st century Singularity and global futures. A Big History perspective* (pp. 599–620). Springer. https://doi.org/10.1007/978-3-030-33730-8_27

Maddison Project. (2020). *Maddison Project Database 2020*. Groningen Growth and Development Centre. Retrieved from https://www.rug.nl/ggdc/historicaldevelopment/maddison/releases/maddison-project-database-2020

Malkov, S. (2020). About the singularity in biological and social evolution. In A. Korotayev & D. LePoire (Eds.), *The 21st century Singularity and global futures. A Big History perspective* (pp. 517–534). Springer. https://doi.org/10.1007/978-3-030-33730-8_23

Malkov, S., Kovaleva, N., Grinin, L., & Korotayev, A. (2023). Modeling social self-organization and historical dynamics. Agrarian society. In V. Sadovnichy et al. (Eds.), *Reconsidering the limits to growth. A report to the Russian Association of the Club of Rome* (pp. 309–335). Springer. https://doi.org/10.1007/978-3-031-34999-7_16

Markov, A., Anisimov, V., & Korotayev, A. (2010). Relationship between genome size and organismal complexity in the lineage leading from prokaryotes to mammals. *Paleontological Journal, 44*(4), 363–373. https://doi.org/10.1134/S0031030110040015

Markov, A., & Korotayev, A. (2007). Phanerozoic marine biodiversity follows a hyperbolic trend. *Palaeoworld, 16*(4), 311–318. https://doi.org/10.1016/j.palwor.2007.01.002

Markov, A., & Korotayev, A. V. (2008). Hyperbolic growth of marine and continental biodiversity through the Phanerozoic and community evolution. *Zhurnal Obshchei Biologii, 69*(3), 175–194.

Modelski, G., Devezas, T., & Thompson, W. R. (Eds.). (2007). *Globalization as evolutionary process: Modeling global change*. Routledge.

Modelski, G., & Thompson, W. R. (1996). *Leading sectors and world powers: The coevolution of global politics and economics*. University of South Carolina Press.

Nazaretyan, A. (2005). Big (Universal) history paradigm: versions and approaches. *Social Evolution & History, 4*(1), 61–86.

Nazaretyan, A. (2015). Megahistory and its mysterious singularity. *Herald of the Russian Academy of Sciences, 85*(4), 352–361. https://doi.org/10.1134/S1019331615040061

Nazaretyan, A. (2017). Mega-history and the twenty-first century singularity puzzle. *Social Evolution & History, 16*(1), 31–52.

Nazaretyan, A. (2018). The polyfurcation century: Does the evolution on earth have a cosmological relevance? *Journal of Big History, 2*(1), 27–41. https://doi.org/10.22339/jbh.v2i1.2253

Nazaretyan, A. (2020). The 21st century's "mysterious singularity" in the light of the Big History. In A. Korotayev & D. LePoire (Eds.), *The 21st century Singularity and global futures. A Big History perspective* (pp. 345–362). Springer. https://doi.org/10.1007/978-3-030-33730-8_15

Panov, A. (2005). Scaling law of the biological evolution and the hypothesis of the self-consistent Galaxy origin of life. *Advances in Space Research, 36*(2), 220–225. https://doi.org/10.1016/j.asr.2005.03.001

Panov, A. (2011). Post-singular evolution and post-singular civilizations. *Evolution, 2*, 212–231.

Panov, A. (2017). Singularity of evolution and post-singular development. In B. Rodrigue, L. Grinin, & A. Korotayev (Eds.), *From Big Bang to galactic civilizations. A Big History anthology. Volume III: The ways that Big History works: Cosmos, life, society and our future* (pp. 370–402). Primus Books.

Panov, A. (2020). Singularity of evolution and post-singular development in the Big History perspective. In A. Korotayev & D. LePoire (Eds.), *The 21st century Singularity and global futures. A Big History perspective* (pp. 439–465). Springer. https://doi.org/10.1007/978-3-030-33730-8_20

Panov, A., LePoire, D., & Korotayev, A. (2020). The 21st century Singularity in the Big History perspective. An overview. In A. Korotayev & D. LePoire (Eds.), *The 21st century Singularity and global futures. A Big History perspective* (pp. 1–18). Springer. https://doi.org/10.1007/978-3-030-33730-8_1

Podlazov, A. (2017). A theory of the global demographic process. *Herald of the Russian Academy of Sciences, 87*(3), 256–266. https://doi.org/10.1134/S1019331617030054

Pomeranz, K. (2021). *The great divergence. China, Europe, and the making of the modern world economy* (2nd ed.). Princeton University Press. https://doi.org/10.1515/9780691217192

Service, E. (1975). *Origins of the state and civilization. The process of cultural evolution*. Norton.

Sheffield, J., Korotayev, A., & Grinin, L. (2013). *Globalization: Yesterday, today, and tomorrow*. Emergent Publications.

Taagapera, R. (1968). Growth curves of empires. *General Systems, 13*, 171–175.

Taagapera, R. (1978). Size and duration of empires: Growth-decline curves, 3000 to 600 B.C. *Social Science Research, 7*, 180–196.

Taagapera, R. (1979). Size and duration of empires: Growth-decline curves, 600 B.C. to 600 A.-D. *Social Science History, 3*, 115–138.

Taagapera, R. (1997). Expansion and contraction patterns of large polities: Context for Russia. *International Studies Quarterly, 41*(3), 475–504. https://doi.org/10.1111/0020-8833.00053

Taagepera, R. (1976). Crisis around 2005 AD? A technology-population interaction model. *General Systems, 21*, 137–138.

Taagepera, R. (1979). People, skills, and resources: An interaction model for world population growth. *Technological Forecasting and Social Change, 13*, 13–30. https://doi.org/10.1016/0040-1625(79)90003-9

Tsirel, S. (2004). On the possible reasons for the hyperexponential growth of the earth population. In M. Dmitriev & A. Petrov (Eds.), *Mathematical modeling of social and economic dynamics* (pp. 367–369). Russian State Social University.

Tsirel, S. (2008). Zametki ob istoricheskom vremeni i putyah istoricheskoj evolyutsii. *Istoriya i Matematika, 5*, 246–278.

Tsirel, S. (2009). Skorost evolutsii: pulsiruyushaya, zamedlayushayasya, uskoryayushchayasya. *Evolutsiya, 1*, 62–98.

Turchin, P. (2016). *Ultrasociety: How 10,000 years of war made humans the greatest cooperators on earth*. Beresta Books.

Turchin, P., Whitehouse, H., Larson, J., Cioni, E., Reddish, J., Hoyer, D., Savage, P. E., Covey, R. A., Baines, J., et al. (2022a). Big Gods and big science: Further reflections on theory, data, and analysis. *Religion, Brain & Behavior, 13*, 218. https://doi.org/10.1080/2153599X.2022.2065354

Turchin, P., Whitehouse, H., Larson, J., Cioni, E., Reddish, J., Hoyer, D., Savage, P. E., Covey, R. A., Baines, J., et al. (2022b). Explaining the rise of moralizing religions: A test of competing hypotheses using the Seshat Databank. *Religion, Brain & Behavior, 13*, 167. https://doi.org/10.1080/2153599X.2022.2065345

Ulam, S. (1958). Tribute to John von Neumann. *Bulletin of the American Mathematical Society, 64*(3), 1–49.

von Foerster, H., Mora, P. M., & Amiot, L. W. (1960). Doomsday: Friday, 13 November, A.D. 2026: At this date human population will approach infinity if it grows as it has grown in the last two millennia. *Science, 132*(3436), 1291–1295. https://doi.org/10.1126/science.132.3436.1291

von Hoerner, S. J. (1975). Population explosion and interstellar expansion. *Journal of the British Interplanetary Society, 28*, 691–712.

Wallerstein, I. (1974). *The modern world-system*. Academic Press.

Wallerstein, I. (2004). *World-systems analysis: An introduction*. Duke University Press.

Watts, R. C. A. (Ed.). (1977). *The origins of agriculture*. Mouton.

Weber, M. (2004 [1905]). *Die protestantische Ethik und der Geist des Kapitalismus*. CH Beck.

Widdowson, M. (2020). The 21st century Singularity: The role of perspective and perception. In A. Korotayev & D. LePoire (Eds.), *The 21st century Singularity and global futures. A Big History perspective* (pp. 489–516). Springer. https://doi.org/10.1007/978-3-030-33730-8_22

Zinkina, J., Grinin, L., Christian, D., Ilyin, I., Andreev, A., Aleshkovski, I., Shulgin, S., & Korotayev, A. (2019). *Big History of globalization. The emergence of a global world system*. Springer. https://doi.org/10.1007/978-3-030-05707-7

Zinkina, J., Korotayev, A., & Andreev, A. (2016). Circumscription theory of the origins of the state: A cross-cultural re-analysis. *Cliodynamics, 7*(2), 187–203. https://doi.org/10.21237/C7clio7232817

Zinkina, J., Ustiuzhanin, V., & Shulgin, S. (2022). Social macroevolution and changes in the human value systems. How has modernization affected human values. *Social Evolution & History, 21*(2), 91–105. https://doi.org/10.30884/seh/2022.02.06

Sergey Malkov heads the Center for Long-Term Forecasting and Strategic Planning at Lomonosov Moscow State University; he also works as a Senior Research Professor at the Institute of Economics of the Russian Academy of Sciences as well as at the Centre for Stability and Risk Analysis, HSE University, Moscow. He is the author of over 300 scholarly publications, including such monographs as *Social self-organization and historical process: possibilities of mathematical modeling* (URSS, 2009), *Modeling and forecasting of world dynamics* (ISPI RAS, 2012, together with Victor Sadovnichy and others), *A new theory of the reproduction of capital: development and practical application* (Nestor-Istoriya, 2016, together with Vladimir Mayevsky and others), *The 21st Century Singularity and Global Futures. A Big History Perspective* (Springer, 2020, together with David J. LePoire and others). In 2022 he was awarded with the Kondratieff Award by the Russian Academy of Sciences.

Leonid Grinin is a Senior Research Professor at the Centre for Stability and Risk Analysis at HSE University in Moscow, as well as a Senior Research Professor of the Institute of Oriental Studies, Russian Academy of Sciences. He is co-editor of the international journals *Social Evolution & History* and the *Journal of Globalization Studies*, as well as co-editor of the international almanacs *Evolution, History & Mathematics*, and *Kondratieff Waves*. He is the author of more than 600 publications, including 33 monographs in Russian, English, and Chinese. His current research focuses on comparative political studies, the theory of revolution, political anthropology, global economy, global history, historical sociology, and futurology. Among his monographs are *The Cybernetic Revolution and the Forthcoming Epoch of Self-Regulating Systems* (Uchitel, 2016, with Anton Grinin), *Big History of Globalization. The Emergence of a Global World System* (Springer 2019, in co-authorship with Julia Zinkina, David Christian, and others), *Islamism, Arab Spring, and* *the Future of Democracy. World System and World Values Perspectives* (Springer, 2019; with Andrey Korotayev and Arno Tausch). In 2004 he was awarded as a winner of the best publication of *Socis* journal during the year for the article "People of Publicity – a New Social Layer?". In 2012 he was awarded with the Gold Kondratieff Medal by the International N. D. Kondratieff Foundation. In 2016 he received the Belyaev Award for the book *From Bifaces to Nanorobots* (the book published in Russian in co-authorship with Anton Grinin).

Anton Grinin is a Russian scientist specializing in the study of modern technological trends and future studies, as well as an evolutionist and philosopher. He has a PhD in Biology and works as a Research Fellow for Lomonosov Moscow State University, Russia and the Centre for Stability and Risk Analysis at HSE University in Moscow. He is the author of more than 80 publications in Russian and English, including 2 monographs. Awards: Gold Kondratieff Young Scholars Medal, Alexander Belyaev Literature Award. Anton Grinin focuses on the issues of the technological revolution (the Cybernetic revolution) evolving in the current decades of the twenty-first century and the resultant profound transformations in economy and society as well as related economic, political, ethic-legal, and other risks, including political revolutions. He also investigates the correlation between the Cybernetic revolution and Kondratieff waves as well as between the Cybernetic revolution and global aging.

Jameelah Musieva is a Research Assistant at Lomonosov Moscow State University, Faculty of Global Studies, as well as a Research Fellow at the Centre for Stability and Risk Analysis at HSE University in Moscow. She is the author and co-author of more than 10 scholarly publications, including such articles as "The Concept of "Quality of Life"" (2019), "Methodology for Cross-Country Assessment of The Quality of Life Based on The LQI Index" (2019), "A Cross-Country Analysis of The Quality of Life" (2020), "Quality of Life in The BRICS Countries Based on The LQI Index" (2021).

Andrey Korotayev heads the Centre for Stability and Risk Analysis at HSE University in Moscow, Russia. He is also a Senior Research Professor at the Institute of Oriental Studies and Institute for African Studies, Russian Academy of Sciences, as well as at the Faculty of Global Studies of Lomonosov Moscow State University and at the Russian Presidential Academy of National Economy and Public Administration in Moscow. He has authored or co-authored over 650 scholarly publications, including such monographs as *Ancient Yemen* (Oxford University Press, 1995), *World Religions and Social Evolution of the Old World Oikumene Civilizations: A Cross-Cultural Perspective* (The Edwin Mellen Press, 2004), *Introduction to Social Macrodynamics: Compact Macromodels of the World System Growth* (URSS, 2006; with Daria Khaltourina and Artemy Malkov), *Introduction to Social Macrodynamics: Secular Cycles and Millennial Trends* (URSS, 2006; with Daria Khaltourina and Artemy Malkov), *Great Divergence and Great Convergence. A Global Perspective* (Springer, 2015; with Leonid Grinin), *Economic Cycles, Crises, and the Global Periphery* (Springer, 2016; with Leonid Grinin and Arno Tausch), *Islamism, Arab Spring, and the Future of Democracy. World System and World Values Perspectives* (Springer, 2019; with Leonid Grinin and Arno Tausch). He is a laureate of the Russian Science Support Foundation Award in 'The Best Economists of the Russian Academy of Sciences' Nomination (2006); in 2012, he was awarded with the Gold Kondratieff Medal by the International N. D. Kondratieff Foundation.

Modeling Social Self-Organization and Historical Dynamics: Modern Society and a Look into the Global Future. Cybernetic W-Society

Askar Akaev, Sergey Malkov, Leonid Grinin, Stanislav Bilyuga, Olga Davydova, Anton Grinin, Natalia Kovaleva, Artemy Malkov, Jameelah Musieva, and Andrey Korotayev

A. Akaev
Lomonosov Moscow State University, Moscow, Russia

HSE University, Moscow, Russia

S. Malkov
Lomonosov Moscow State University, Moscow, Russia

HSE University, Moscow, Russia

Institute of Economics, Russian Academy of Sciences, Moscow, Russia
e-mail: s@malkov.org

L. Grinin
HSE University, Moscow, Russia

Institute of Oriental Studies, Russian Academy of Sciences, Moscow, Russia

S. Bilyuga · O. Davydova
Lomonosov Moscow State University, Moscow, Russia
e-mail: bilyuga@fgp.msu.ru

A. Grinin · J. Musieva
HSE University, Moscow, Russia

Faculty of Global Studies, Lomonosov Moscow State University, Moscow, Russia

N. Kovaleva
Soil Science Faculty, Lomonosov Moscow State University, Moscow, Russia

A. Malkov
Moscow Institute of Physics and Technology, Moscow, Russia
e-mail: as@malkov.org

A. Korotayev (✉)
HSE University, Moscow, Russia

Institute of Oriental Studies, Russian Academy of Sciences, Moscow, Russia

Faculty of Global Studies, Lomonosov Moscow State University, Moscow, Russia

V. Sadovnichy et al. (eds.), *Reconsidering the Limits to Growth*, World-Systems Evolution and Global Futures, https://doi.org/10.1007/978-3-031-34999-7_19

Abstract

This chapter presents the results of the application of the set of basic models outlined in chapter "Modeling Social Self-Organization and Historical Dynamics. A General Approach" (Akaev et al., 2023, this volume) to model the current historical situation and to forecast its further development. Akaev et al. show that the modern historical period is a period of transition from the "epoch of growth," which followed the industrial revolution of the early nineteenth century, to the "era of deceleration," one of the strongest indicators of which is a rapid decrease in the growth rate of the Earth's population and its aging. They perform the analysis of the ongoing changes in all spheres of life. Akaev et al show the uniqueness of the ongoing changes and consider alternative options for further development.

1 Modern Society and a Look into the Global Future

In chapter "Modeling Social Self-Organization and Historical Dynamics. Global Phase Transitions" (Malkov et al., 2023a, this volume), it is shown that in recent decades the world has entered a period of cardinal transformations: a transition has begun to a new historical era, to a new type of society (we have conditionally called it "W-society"), the exact characteristics of which are not quite clear yet. Before proceeding to the analysis and modeling of trends of further historical development, consider the most important features of the current historical situation.

In the past few decades, there has been a fundamental change in the nature of the dynamics of world development characteristic of the industrial era. Figure 1 presents statistical data on the decade-averaged growth rates of the world population, world GDP, and GDP/capita over the past 200 years.

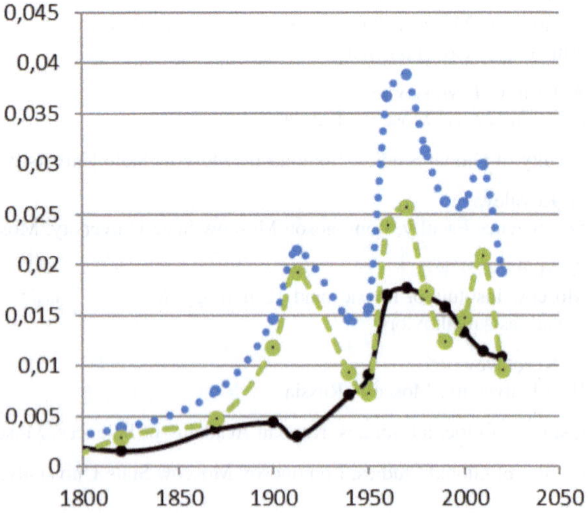

Fig. 1 Decade-averaged growth rates of the world population (solid line), world GDP (dotted line) and GDP/capita (dashed line) over the last two centuries. *Data source* Maddison Project (2020)

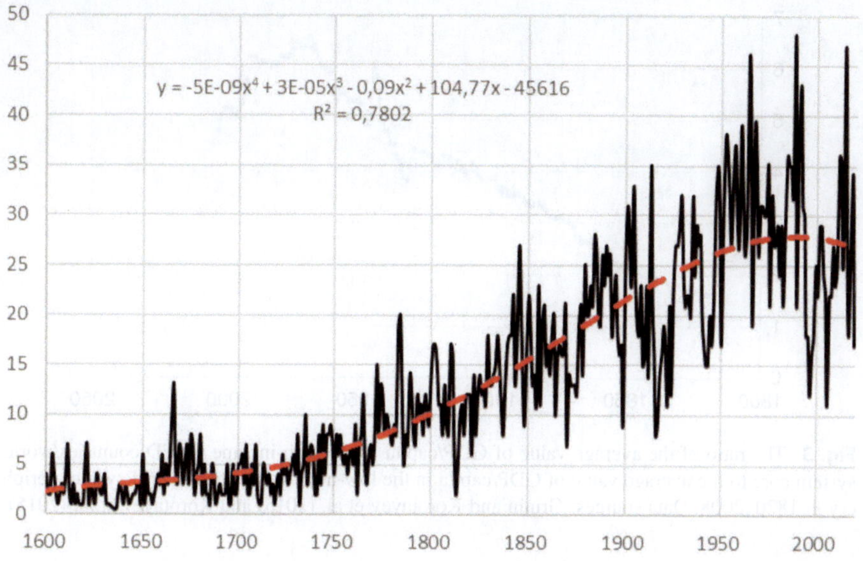

Fig. 2 The number of major inventions per year in 1600–2020. *Data sources* Bunch and Hellemans (2004) till 2003; the authors' own calculations afterward

It can be seen that all of these indicators experienced accelerated growth from the beginning of the nineteenth century, reached their historical maximum in the 1960s and early 1970s, and then began to decline.[1]

Figure 2 presents data on the dynamics of invention activity over the past four centuries.

Thus, invention activity, consistently increasing as from the mid-eighteenth century, reached its maximum in the 1960s and 1970s, and then began to slow down (see also Korotayev, 2020; LePoire & Korotayev, 2020; Panov et al., 2020).

Figure 3 presents data on the ratio of the average value of GDP/capita in the high-income OECD countries/world-system core to the average value of GDP/capita in the low income countries/world-system periphery.

Starting from the 1970s, the "Great Divergence" (i.e., the increasing gap between the GDP/capita of the leading countries of the industrial era and the GDP/capita of the rest of the world) was replaced by the "Great Convergence" (a decrease in this gap in GDP/capita).[2]

[1] See chapter "Demography. Toward Optimization of Demographic Processes" (Korotayev, Malkov, & Musieva, 2023, this volume) and chapter "Global Aging – An Integral Problem of the Future. How to Turn a Problem into a Development Driver?" (Grinin, Grinin & Korotayev 2023b, this volume) for more detail. See also, e.g., Korotayev, Malkov, and Khaltourina (Korotayev et al., 2006a; Korotayev et al., 2006b), Korotayev et al. (2020), Korotayev et al. (2020).

[2] See, e.g., Goldstone (2009, 2021); Korotayev, Zinkina, Bogevolnov, and Malkov (2011), Korotayev and de Munck (2013), Grinin and Korotayev (2014a, 2015), Korotayev and Zinkina (2014), Akaev (2015), Korotayev et al. (2015); Grinin et al. (2016), Korotayev et al. (2020),

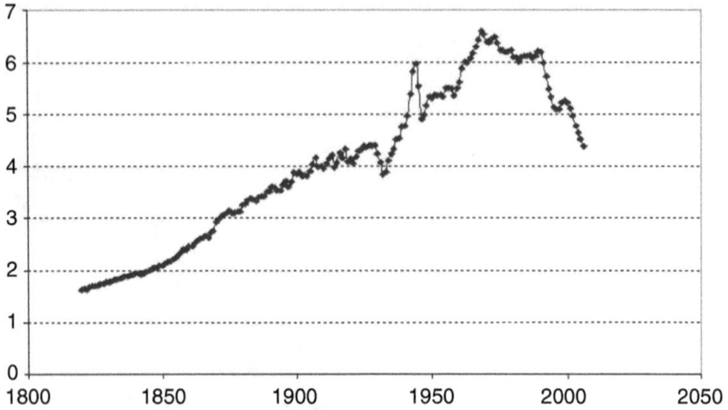

Fig. 3 The ratio of the average value of GDP/capita in the high-income OECD countries/world-system core to the average value of GDP/capita in the low-income countries/world-system periphery in 1820–2008. Data sources: Grinin and Korotayev et al. (2015) and Korotayev et al. (2015)

Thus, a change occurred in the development trends characteristic of the industrial era. It seems that since the 1970s, we have entered a new era that is reminiscent of what happened during the slowdown in past "eras of change" [see chapter "Modeling Social Self-Organization and Historical Dynamics. Global Phase Transitions" (Malkov et al., 2023a, this volume)]. Indeed, if we formally make a forecast based on the modernization model [see Sect. 3.1 of chapter "Modeling Social Self-Organization and Historical Dynamics. Industrial Society" (Akaev et al., 2023, this volume)], which describes well the world dynamics in the industrial era, we will get the result presented in Fig. 4.

A comparison of the model calculations (Fig. 4) with statistical data (Fig. 1) shows that although the model does not reflect the impact of the world wars and Kondratieff cycles[3] clearly visible in the statistical data, it well reflects the global trends. With these trends, the peak of economic and demographic growth rates,

Pomeranz (2021); see also chapter "High Income and Low Income Countries. Toward a Common Goal at Different Speeds" (Grinin, Malkov, and Korotayev 2023, this volume); about economic growth rates see also chapter "Economics. Optimizing Growth" (Grinin, Grinin, and Malkov 2023a, this volume) and chapter "Technology. Limitless Possibilities, Effective Control" (Grinin & Grinin, 2023b, this volume).

[3]On those cycles and their connections with the global dynamics see, e.g., Kondratieff (1926, 1935, 1984), Schumpeter (1939), Forrester (1978, 1981, 1985), Mensch (1979), Mandel (1980). Wallerstein (1984), Freeman (1987), Goldstein (1988), Berry (1991), Gruebler and Nakicenovic (1991), Tylecote (1992), Hirooka (2006), Korotayev, Zinkina, and Bogevolnov (2011), Akaev et al. (2012, 2017), Korotayev and Grinin (2012), Akaev and Sadovnichiy (2016), Grinin et al. (2016), Grinin and Grinin (2016b), Grable (2019). Note that the basic modernization model set out in Sect. 3.1 chapter "Modeling Social Self-Organization and Historical Dynamics. Industrial Society" (Akaev, Malkov, Davydova et al., 2023, this volume) has not been designed to model economic cycles.

Fig. 4 Decade-average growth rates of the world population (solid green line) and world GDP (dotted red line) calculated using the modernization model [see Sect. 3.1 of chapter "Modeling Social Self-Organization and Historical Dynamics. Industrial Society" (Akaev et al., 2023, this volume)]

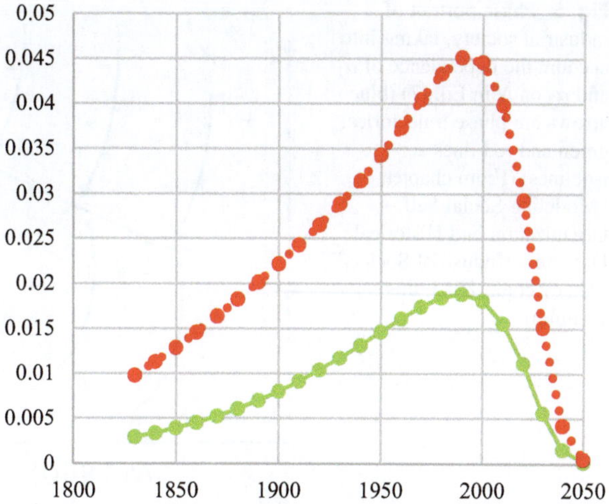

according to the model, was reached in the last quarter of the twentieth century (which corresponds to the statistical data), and then a rapid decline was bound to begin (which is also already clearly visible from the statistical data). At the same time, the model forecasts that, if global trends continue, a stationary regime (i.e., a regime with zero demographic and economic growth rates) will emerge by the middle of the twenty-first century. This is a very strong conclusion.[4]

This conclusion is quite close to the inertial forecast made in the first report to the Club of Rome (Meadows et al., 1972). One of the key prerequisites in this work was the thesis that as the population (N) grows, unit production costs will increase, the resource base will shrink, and environmental costs will increase. In terms of the industrial society model (6)–(8),[5] this means an increase in the coefficients α_1 and α_2 in Eq. (7)[6] as N grows. The corresponding situation is reflected in Fig. 5, which shows the phase portrait of industrial society with this premise.

When the dependence of α_1 and α_2 on N is taken into account, the isoclines intersect, and a stable attractor of the "focus" type appears, which was not in the phase portrait of industrial society when this dependence was absent in Eq. (7)[7] (see Fig. 5).[8] This situation is similar to what was typical for an agrarian

[4]On the other hand, chapter "Technology. Limitless Possibilities, Effective Control" (Grinin & Grinin, 2023b, this volume) suggests that in connection with the Cybernetic Revolution economic growth rates are likely to increase after 2030s (see also Grinin et al., 2020).

[5]Equations from chapter "Modeling Social Self-Organization and Historical Dynamics. Industrial Society" (Akaev, Malkov, Davydova et al., 2023, this volume).

[6]From chapter "Modeling Social Self-Organization and Historical Dynamics. Industrial Society" (Akaev, Malkov, Davydova et al., 2023, this volume).

[7]From the same chapter.

[8]From the same chapter.

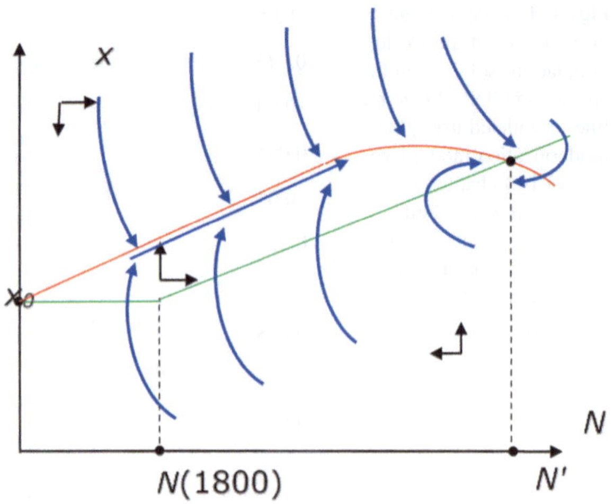

Fig. 5 Phase portrait of industrial society, taking into account the dependence of α_1 and α_2 on N in Eq. (7) (blue arrows are phase trajectories, green and red lines are isoclines) (From chapter "Modeling Social Self-Organization and Historical Dynamics. Industrial Society" (Akaev et al., 2023, this volume))

society,[9] in which the presence of a stable attractor hampered development and led to the "Malthusian trap" and the emergence of demographic cycles. The anticipation of the onset of this situation, inevitable when maintaining the Y-type socioeconomic system,[10] predetermined the pessimistic view of the members of the Club of Rome on the prospects for world development.

However, the situation is not so simple and straightforward. Over the past few decades, there have been serious qualitative changes in the world, which must be taken into account when analyzing and modeling options for future development. We are talking about the following changes that qualitatively distinguish the modern era from previous eras. These are fundamental changes; they are imperative. These changes will exert a principal influence on the formation of future society:

- the process of *globalization*, which was gradually proceeding over many centuries (see Modelski et al., 2007; Sheffield et al., 2013; Grinin & Korotayev, 2014b, 2014c; Korotayev & Zinkina, 2017; Zinkina et al., 2019; Grinin, 2021), has technically ended: the world has become unified, entangled in information, economic, and communication ties.[11] While the division into "us" and "them" (those who were not subject to moral norms adopted when interacting with

[9]See the phase portrait in Fig. 1 in chapter "Modeling Social Self-Organization and Historical Dynamics. Agrarian Society" (Malkov, Koroleva et al., 2023b, this volume).

[10]See Sect. 2.4 of chapter "Modeling Social Self-Organization and Historical Dynamics. Industrial Society" (Akaev, Malkov, Davydova et al., 2023, this volume).

[11]The process of deglobalization that has begun recently has no way of completely stopping the process of globalization. Deglobalization in our understanding is a kind of ebb, which will be replaced by a new, stronger flow of globalization (see Grinin & Korotayev, 2021; Grinin et al., 2021; see also Ilyin & Leonova, 2022).

"ours")[12] was typical earlier in interstate relations, now the task of forming common ideological principles of co-existence in the world comes to the fore;

- the world has become "*a closed circuit*," with its parts being *interdependent* on each other; global resource constraints and environmental problems have arisen,[13] the irreversible impact of human activity on nature and climate has been recognized[14];
- there is a slowdown in the growth of the world population; at the same time, the world population will age, and life expectancy will continue to increase. Under these conditions, the proportion of the elderly will increase, while the proportion of the young will fall. The age pyramid is changing from triangular to rectangular[15];
- digitalization and total computerization in all spheres of life will continue and accelerate. Information networks will cover the world's entire population, everyone will be under control, and human autonomy will disappear. Human beings will increasingly transfer the decision-making function to artificial intelligence and, accordingly, will increasingly depend on the algorithms embedded in it [see chapter "Technology. Limitless Possibilities, Effective Control" (Grinin & Grinin, 2023b, this volume)];
- the world will be more and more technogenic; however, while nature was earlier the object of technological transformations, now the objects of changes are people themselves (biotechnology, genetic engineering, chipping, artificial organs, cloning, etc.) and their consciousness (cognitive technologies, neuropsychological impact, psychotropic substances, etc.).

In addition, it is important to keep in mind that while until the beginning of the twenty-first century, the main part of the world's population lived in rural areas (see Fig. 6), now the situation has changed; the main part of the population lives in cities (56% in 2020), and the share of the urban population continues to grow.[16] People living in cities have greater material resources than rural residents, but at the same time, they are more dependent on the condition of infrastructure and on changing economic situations in their countries and the world; their behavior differs from the behavior of rural residents.

[12] See chapter "Modeling Social Self-Organization and Historical Dynamics. Agrarian Society" (Malkov, Kovaleva et al., 2023b, this volume).

[13] See chapter "Economics. Optimizing Growth" (Grinin, Grinin, & Malkov, 2023a, this volume).

[14] See chapter "Climate and Energy. Energy Transition Scenarios and Global Temperature Changes Based on Current Technologies and Trends" (Akaev & Davydova, 2023, this volume).

[15] See chapter "Demography. Toward Optimization of Demographic Processes" (Korotayev, Malkov & Musieva, 2023, this volume) and chapter "Global Aging – an Integral Problem of the Future. How to Turn a Problem into a Development Driver?" (Grinin, Grinin & Korotayev, 2023b, this volume).

[16] See chapter "Demography. Toward Optimization of Demographic Processes" (Korotayev, Malkov, & Musieva, 2023, this volume) for more detail.

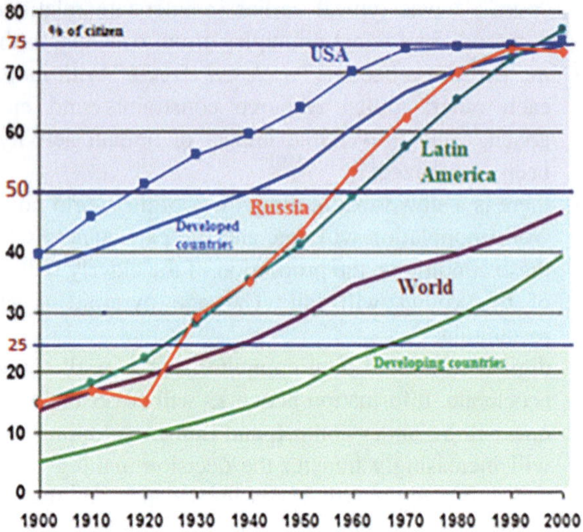

Fig. 6 Dynamics of the share of the urban population in the twentieth century for some countries, groups of countries, and continents

Another important point is that until the 1970s, during the second phase of industrial society, the value of GDP/capita grew in the countries of the world-system core due to accelerated economic growth. In contrast, the countries of the world-system periphery, despite the beginning of economic growth, remained in the "Malthusian trap." [17] Since the 1970s, the gap between GDP/capita in high-income countries and in the rest of the world began to gradually decrease (see Fig. 3) as low-income countries started their escape from the "Malthusian trap" and their consistent catch-up with high-income countries in terms of GDP/capita (see above). More and more people in the world have their basic material needs satisfied, and they have a certain surplus of money that can be spent at their discretion. Accordingly, this affects their behavior and their decision-making in various situations. Such a situation has emerged for the first time in the history of humankind (earlier, only a narrow stratum of the elite had a relative excess of material goods), and the impact of this situation on social processes must be taken into account when modeling—see Goldstone (1991), Grinin & Korotayev (2009, 2010, 2016), von Weizsäcker & Wijkman (2018).

Thus, there is good reason to think that the 1970s represent a certain milestone from which a new phase of historical development began. Major phase transitions in human history are closely related to the following revolutionary technological transformations:

- the Neolithic (Agrarian) Revolution represented the transition of humankind from a foraging economy (hunting and gathering) to a food-producing economy

[17]For the notion of "Malthusian trap" see chapter "Modeling Social Self-Organization and Historical Dynamics. Agrarian Society" (Malkov, Kovaleva et al., 2023b, this volume).

(agriculture and animal husbandry based on manual labor); it led to the formation of *agrarian* societies;
- the Industrial Revolution, when manual labor was replaced with machine work, led to the formation of *industrial* societies;
- the Digital (or Cybernetic) Revolution is based on the invention of high-speed computing devices and their introduction into all spheres of life, which allows the gradual replacing human intellectual labor with machine work and leads to the formation of a society whose functioning is decisively dependent on computer algorithms.

The Cybernetic Revolution is a fundamental transition from the industrial principle of production to the production of services and goods based on the widespread implementation of self-regulating systems, that is, systems that can not only function in the absence (or with minimal involvement) of humans but also independently make complex decisions (see chapter "Technology. Limitless Possibilities, Effective Control" (Grinin & Grinin, 2023b, this volume); Grinin, 2006; Grinin & Grinin, 2015a, 2015b, Grinin & Grinin, 2016a, 2016b; Grinin et al., 2017a, 2017b, 2020, 2021).[18]

In this regard, the emerging W-society, which differs both from X-society (typical of the agrarian era) and Y-society (typical of the industrial era), can be logically called cybernetic [for important features of cybernetic society, see chapters of the main text of the report[19] and chapter "Modeling Social Self-Organization and Historical Dynamics. Global Phase Transitions" (Malkov et al., 2023a, this volume)].[20]

2 Modeling the Interactions in the "Society-Nature" System

A. In cybernetic societies, all processes will become more interconnected than in previous eras. On the one hand, this is because the world population increased several times at the end of the twentieth century compared to the times of the Industrial Revolution. At the same time, human pressure on nature has multiplied due to the work of industrial enterprises; as a result, the rate of natural and

[18] We should point out that automatization which was considered as the major characteristic of the start of the Cybernetic Revolution (see, e.g., Lilley, 1976; Philipson, 1962; Bernal, 1965; Benson & Lloyd, 1983) appears to be only the initial level of future self-regulating systems.

[19] Chapter "Technology. Limitless Possibilities, Effective Control" (Grinin & Grinin, 2023b, this volume), chapter "Economics. Optimizing Growth" (Grinin, Grinin, & Malkov, 2023a, this volume), chapter "Socio-Political Transformations. A Difficult Path to Cybernetic Society" (Grinin, Grinin & Malkov, 2023b, this volume), and chapter "Global Aging – an Integral Problem of the Future. How to Turn a Problem into a Development Driver?" (Grinin, Grinin, & Korotayev, 2023b, this volume).

[20] It is important to note here that the term "cybernetic society" focuses on the technological foundation of future society, and the term "W-society" characterizes its institutional structure.

environmental changes can no longer be considered a slow process compared to the rate of demographic processes. On the other hand, the rate of technological change has increased: it has significantly exceeded the rate at which the generations of people change. Accordingly, the division of processes into the "fast" (demographic-economic) and the "slow" (natural) ones used in the analysis of agrarian and industrial societies becomes irrelevant. Therefore, the "nature-society" interactions must be considered within the framework of a single system of equations.

So, the basic Eqs. (1)–(8)[21] for a new type of society acquire the following form.

A.1 Generalized Equation of Demographic Dynamics
Fifty years ago (Meadows et al., 1972), it was believed that if special restrictive measures were not taken, the population would grow exponentially (that is, in accordance with the dynamics shown in Fig. 4a).[22] However, the historical experience of the late twentieth and early twenty-first centuries has shown that in an urbanized society, the family model is transformed from having many children to having few children, so the problem of high-income countries is not population growth, but depopulation. Different variants of population growth slowing down are shown in Fig. 4b–d[23],[24],[25], in chapter "Modeling Social Self-Organization and Historical Dynamics. A General Approach" (Akaev et al., 2023, this volume). The basic equations corresponding to these variants of demographic dynamics can be written as:

$$dN/dt \approx r \cdot N \cdot (1 - x_0/x) \cdot (x_0/x) \text{ (dynamics in Fig.4b)}, \tag{1}$$

$$dN/dt \approx r \cdot N \cdot (1 - x_1/x - x/x_2) \text{ (dynamics in Fig.4c)}, \tag{2}$$

$$dN/dt \approx r \cdot N \cdot (1 - x_1/x - x/x_2) \cdot (x_0/x)^2 \text{ (dynamics on Fig.4d)} \tag{3}$$

This leads to a gradual stabilization of the population (Eqs. 1 and 3) or its decrease (Eq. 2) due to fertility decline.

A.2 Generalized Equation of Economic **Dynamics**
Since the basis of the economy of both industrial and cybernetic societies is industrial production and the development of technologies, the general form of the economic dynamics equation remains the same [see Eq. (2) from chapter "Modeling

[21] From chapter "Modeling Social Self-Organization and Historical Dynamics. A General Approach" (Akaev, Malkov, Bilyuga et al., 2023, this volume).

[22] From chapter "Modeling Social Self-Organization and Historical Dynamics. A General Approach" (Akaev, Malkov, Bilyuga et al., 2023, this volume).

[23] From chapter "Modeling Social Self-Organization and Historical Dynamics. A General Approach" (Akaev, Malkov, Bilyuga et al., 2023, this volume).

[24] From chapter "Modeling Social Self-Organization and Historical Dynamics. A General Approach" (Akaev, Malkov, Bilyuga et al., 2023, this volume).

[25] From chapter "Modeling Social Self-Organization and Historical Dynamics. A General Approach" (Akaev, Malkov, Bilyuga et al., 2023, this volume).

Social Self-Organization and Historical Dynamics. Agrarian Society" (Malkov et al., 2023b, this volume)], the changes only affect the form of its terms.

As mentioned above, it is necessary to distinguish between situations (1) when there are no resource constraints (resources are available in abundance) and (2) when resources are scarce.

A special feature of cybernetic society is that technological progress makes it possible to remove the resource limitations (that appear from time to time) by introducing new resources and through their more efficient use.[26] However, unlike at the industrial phase of development, environmental problems are exacerbated, as well as the problems of other changes in nature (such as climate change), leading to a deterioration in living conditions.

Accordingly, keeping the form of Eq. (2) from chapter "Modeling Social Self-Organization and Historical Dynamics. Agrarian Society" (Malkov et al., 2023b, this volume], it is necessary to introduce into it the dependence of I_1 and I_2 on the state of the external environment, the quality of resources, etc.

A.3 Generalized Equation of *Technological* Dynamics

Technological progress remains the key driver of economic dynamics in the cybernetic age. To describe the technological dynamics, we use the basic Eq. (9) from chapter "Modeling Social Self-Organization and Historical Dynamics. A General Approach" (Akaev et al., 2023, this volume), in which the values of c and S_0 are set on the basis of the available statistical data and additional hypotheses.

A.4 Generalized Equation of *Natural Resource* Consumption

In industrial society, technological progress made it possible to remove the resource limitations (that arose from time to time) by introducing new resources and their more efficient use, while the cost of resource support for production became more and more expensive. We believe that the same situation will continue in cybernetic society. So, for the use of various categories of natural resources, Eqs. (12)–(13)[27] can be used, taking into account the specifics of resources. At the same time, based on the solution of these equations, the value of I_1 can be estimated, which is necessary to ensure sustainable crisis-free development. This value should then be used in the equation of economic dynamics.

A.5 Generalized Equation of the Accumulation of *Waste Products*

In the context of globalization, the accumulation of waste products has become a serious problem. For its analysis, Eqs. (16)[28] are used, taking into account the

[26] This is also typical for industrial society.

[27] Equations from chapter "Modeling Social Self-Organization and Historical Dynamics. A General Approach" (Akaev, Malkov, Bilyuga et al., 2023, this volume).

[28] Equation from chapter "Modeling Social Self-Organization and Historical Dynamics. A General Approach" (Akaev, Malkov, Bilyuga et al., 2023, this volume).

specifics of waste products. Based on the solution of these equations, the value of I_2 is estimated, which is necessary to ensure sustainable crisis-free development.

A.6 Generalized Equation for the *Changing Characteristics of the Environment* U_j Because of Anthropogenic Impact

The preservation of the environment characteristics requires additional costs I_2. Eq. (16)[29] and then Eq. (18)[30] are used to evaluate them.

B. Thus, the basic model of interactions in the "society-nature" system can be written as the following system of equations:

$$dN/dt \approx r \cdot N \cdot G(x_0, x), \tag{4}$$

$$dx/dt \approx (F(R^-, I_1) - I_1 - I_2)/N - Q = S - I_1/N - I_2/N - Q, \tag{5}$$

$$dS/dt \approx c \cdot N \cdot (S - S_0), \tag{6}$$

$$dR/dt = k^+{}_R \cdot q_R \cdot I_1 - F/A_R(S), \tag{7}$$

$$dR/dt = R^+{}_{np} + k^+{}_R \cdot q_R \cdot I_1 - R^-{}_{np} - k^-{}_R \cdot F, \tag{8}$$

$$dR/dt = k^+{}_R \cdot q_R \cdot I_1 - f^-{}_R(t), \tag{9}$$

$$dW_i/dt = k^+{}_{Wi} \cdot F - W_i{}^-{}_{np} - k^-{}_{Wi} \cdot q_{Wi} \cdot I_2 \tag{10}$$

The equation of technological progress (6) remains unchanged (compared to Eqs. (6)–(8),[31] which describe the dynamics of industrial society). The demographic Eq. (4) is written in a general form; formulas (1), (2), or (3) can be used in various calculations; the main thing is that all these options take into account the decline in the birth rate (the second phase of the demographic transition[32]) with an increase in the level of well-being x. In the economic Eq. (5), it becomes essential to quantitatively consider the impact of resource constraints and the need to solve environmental problems on the values of I_1 and I_2 (see Eqs. 7–10).

---

[29] Equation from chapter "Modeling Social Self-Organization and Historical Dynamics. A General Approach" (Akaev, Malkov, Bilyuga et al., 2023, this volume).

[30] Equations from chapter "Modeling Social Self-Organization and Historical Dynamics. A General Approach" (Akaev, Malkov, Bilyuga et al., 2023, this volume).

[31] From chapter "Modeling Social Self-Organization and Historical Dynamics. Industrial Society" (Akaev, Malkov, Davydova et al., 2023, this volume).

[32] The second phase of demographic transition (leading to a slowdown in population growth) took place in economically high income countries of the world already in the second phase of industrial society. However, on the global scale it manifested itself after the 1960s, which, among other things, led to the acceleration of economic growth in the countries of the World System semi-periphery and periphery [see chapter "Demography Toward Optimization of Demographic Processes" (Korotayev, Malkov, & Musieva, 2023, this volume), as well as Grinin & Korotayev, 2015; Korotayev et al., 2015, Korotayev et al., 2020].

Since an economically developed society uses a variety of resources, some of which are interchangeable, the quantitative accounting for the impact of the presence of resource constraints on the values of I_1 and I_2 is a complex task that requires the analysis and processing of a large amount of statistical data. At the level of the basic model designed to analyze the long-term trends in world dynamics, it appears appropriate to use a scenario approach based on general assumptions. As the basic scenario for the development of a postindustrial cybernetic society that makes extensive use of technological advances, it appears appropriate to accept a scenario in which humanity will eventually be able to enter a balanced/sustainable development mode, in which resource and environmental problems can be solved with appropriate technologies and social reforming.[33] Mathematically, this means that the right-hand sides of Eqs. (7)–(10) are approximately equal to zero under the conditions of balanced, sustainable development. Our analysis of the corresponding algebraic relations shows that this is possible if the values of I_1 and I_2 grow proportionally to F, that is, $I_1 \approx g_1 \cdot F$, $I_2 \approx g_2 \cdot F$. As $F = S \cdot N$, the economic Eq. (5) is transformed to the following form:

$$dx/dt \approx S - I_1/N - I_2/N - Q \approx S - g_1 \cdot S - g_2 \cdot S - q \cdot x, \tag{11}$$

where g_1 and g_2 are coefficients, the value of which can change over time.

Since in the system of Eqs. (4), (6), (11) the variable x is a "fast" variable with respect to the variables N and S, then, according to Tikhonov's theorem, the algebraic equation can replace the differential Eq. (11)

$$S - g_1 \cdot S - g_2 \cdot S - q \cdot x \approx 0, \tag{12}$$

whence it follows that:

$$x \approx S \cdot (1 - g_1 - g_2)/q. \tag{13}$$

Versions of the phase portrait of systems (4), (6), (11) in the x and N coordinates for different values of the parameters c, g_1, g_2, and q are shown in Fig. 7.

The specific features of the phase portrait of cybernetic society (Fig. 7) in comparison with the industrial one (Fig. 5)[34] are as follows:

- demographic growth ceases to be infinite, and the population eventually stabilizes (in Fig. 7, this is point N''). Moreover, this deceleration occurs naturally (with the second phase of the demographic transition that is currently taking place in most countries of the world and includes, among other things, a transition from a

[33] A necessary condition for solving problems of this type (environmental, energy-related, etc.) which affect everyone is the development of mechanisms for making collectively agreed decisions and the strengthening of centralized governance.

[34] From chapter "Modeling Social Self-Organization and Historical Dynamics. Industrial Society" (Akaev, Malkov, Davydova et al., 2023, this volume).

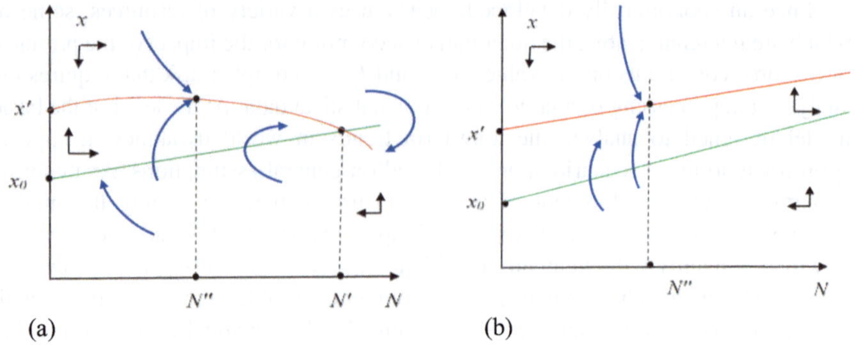

Fig. 7 Phase portrait of a cybernetic society. (**a**) – low value of c, high values of g_1, g_2, q. (**b**) – high value of c, moderate values of g_1, g_2, q (blue arrows are phase trajectories, green and red lines are isoclines, and the location of the red isocline depends on the dynamics of S)

family with many children to a family with few children). Unlike agrarian society [see Fig. 1 from the chapter "Modeling Social Self-Organization and Historical Dynamics. Agrarian Society" (Malkov et al., 2023b, this volume)], where the population limit was determined by the demographic capacity of the territory (ultimately, by the amount of land resources), in a cybernetic society, the value of N'' can vary widely (it can even significantly decrease if the demographic dynamics proceeds in accordance with Eq. (2)) and largely depends on the ongoing demographic policy;

- the dynamics of the level of well-being x does not depend on the size of the population [as it was in an agrarian society, see Fig. 1 from the chapter "Modeling Social Self-Organization and Historical Dynamics. Agrarian Society" (Malkov et al., 2023b, this volume)], but depends on the level of technological development. At the same time, the amount of effort spent on the restoration of resources, waste disposal, etc., plays a significant role (that is, the values of g_1 и g_2 in the Eq. (13));
- thus, the position of the state of equilibrium (that is, the point N'' on the phase portrait) is not fixed. It can vary significantly and depends on the ongoing socio-economic and demographic policy. In any case, the period of hyperbolic economic and demographic growth characteristic of the industrial era is coming to an end, and the nature of the further development trajectory largely depends on measures regulating the socio-economic and demographic processes;
- under these conditions, the centralized regulation of interactions in the "society-nature" system will inevitably strengthen. It is necessary to decide on the number of resources and funds allocated to solving the common economic and climate problems. The question is, who will do it and how? To what extent is the world community capable of concerted action in order to implement a favorable development option?

C. When considering interactions in the "society-nature" system, it is necessary to discuss the topic of climate change separately due to its high relevance.

Section 1.2 of chapter "Modeling Social Self-Organization and Historical Dynamics. Industrial Society" (Akaev et al., 2023 in this volume) shows that during the industrial age, and especially at the end of it, human activity became an essential factor influencing nature and climate. According to the IPCC, the Intergovernmental Panel on Climate Change, until 1976, climate variability did not exceed natural variations estimated over a period of about 1000 years. After that, the amplitude of temperature fluctuations exceeded this threshold, and the issue of anthropogenic global warming due to the increased greenhouse effect became relevant. Thus, the danger is not the greenhouse effect as such but its excess over the established environmental level. Indeed, because of the greenhouse effect, the average annual temperature in the surface atmosphere in the last millennia was approximately + 14 °C, making Earth's climate comfortable and suitable for life. Without the greenhouse effect, our entire planet would inevitably be covered with ice, since, in that case, the average temperature of the surface atmosphere would be −18 °C. The reason for the concern about global warming is that the natural mechanism is artificially amplified by human economic activity.[35]

The dynamics of temperature over the past millennium, according to the third report of the IPCC, is shown in Fig. 8 in Sect. 1.2 of chapter "Modeling Social Self-Organization and Historical Dynamics. Industrial Society" (Akaev et al., 2023, this volume), which clearly demonstrates an explosive increase in global average air temperature by about 0.8 °C compared to 1900. The contribution of anthropogenic factors (greenhouse gases and aerosols) to climate warming in the twentieth century is estimated at about 0.6 °C, according to the IPCC. The main anthropogenic cause of the increased concentration of greenhouse gases in the atmosphere is the growing energy consumption and industrial CO_2 emissions. Prior to the oil crisis of the 1970s, the energy demand increase was proportional to the square of the world population ($E \sim N^2$). It is necessary to move to a new paradigm of energy consumption in which it will grow in direct proportion to the population ($E \sim N$). The new paradigm is essentially aimed at practically implementing the Blue Card energy scenario (IEA, 2010) through the huge yet untapped potential for energy savings and efficiency improvements. The effect of technologies aimed at reducing CO_2 emissions should be considered at the stage of calculating the dynamics of carbon dioxide emissions into the atmosphere. As for hydrocarbon fuel replacement technologies, they should be taken into account when considering the structure of energy consumption by source.

Thus, the climate forecast calculation scheme includes the following steps:

[35] See chapter "Climate and Energy. Scenarios of Energy Transition and Global Temperature Changes Based on Current Technologies and Trends" (Akaev & Davydova, 2023, this volume) for detail.

1) developing a mathematical model to describe the process of transition to a new paradigm of energy consumption;
2) calculating various scenarios of population growth in the world as a whole and individual countries (demographic dynamics): $N(t)$;
3) calculating appropriate scenarios for the development of energy demand dynamics based on the new energy consumption paradigm: $E(N(t))$;
4) forecasting the structural dynamics of energy consumption by energy source type (coal, oil, gas, renewable energy sources, nuclear energy, hydropower);
5) forecasting the dynamics of demand for organic fossil fuels (coal, oil, gas);
6) forecasting the dynamics of changes in the structure of hydrocarbon energy sources (coal, oil, gas);
7) calculating the dynamics of CO_2 emissions into the atmosphere during the combustion of hydrocarbon fuels, taking into account structural changes in the consumption of organic fossil fuels (coal, oil, gas), as well as the use of carbon capture and storage (CCS) technologies:

$$W_{CO2} = W_{CO2}(E(N(t)), S);$$

8) calculating the dynamics of CO_2 accumulation in the atmosphere, taking into account the nonindustrial CO_2 emissions (due to deforestation and soil erosion) and the absorption of part of the emissions by oceans and land ecosystems;
9) forecasting the dynamics of climate change; calculating the changes in the average global temperature of the surface atmosphere:

$$\Delta T = f(W_{CO2}) = f(W_{CO2}(E(N(t)), S));$$

10) analyzing the results and developing practical recommendations.

The developed mathematical models, calculation results, and their analysis are presented in (Akaev & Davydova, 2020a; Akaev & Davydova, 2020b; Akaev & Davydova, 2021a, Akaev & Davydova, 2021b) and in chapter “Climate and Energy. Scenarios of Energy Transition and Global Temperature. Changes Based on Current Technologies and Trends” (Akaev & Davydova, 2023, this volume) of the main text of the report. Several possible scenarios are considered. It is shown that the energy transition from the currently dominant fossil hydrocarbons to the predominant use of renewable energy sources (RES) when the share of RES in the total energy balance exceeds 40%, may occur in the 2060s. Also, this transition is impossible without the dynamic development of nuclear energy as a low-carbon source for generating the base and peak electricity necessary to create a stable and sustainable energy system. It has been shown that to keep global warming at 1.5 °C by the end of this century, it will be necessary to use more chemical technology to capture, fix, and store carbon dioxide emitted while burning hydrocarbons in power stations and directly from the atmosphere. However, this option is still hindered by the cost of the technology. Solving these problems is possible, but it will depend to a decisive extent on how coordinated and purposeful the actions of the world community will be.

D. In the late 1960s, the growth rate of world GDP reached its maximum. At the same time, it was acknowledged that the growing environmental problems were not something secondary, that they were growing rapidly and, in a few decades, would become a serious headache for all humankind.

Perhaps the most serious problem in the field of ecology is the accelerating degradation of fertile land, the main agricultural resource. While over the past 10,000 years, the area of lost fertile lands totaled 2000 million hectares, it rocketed to 300 million hectares over the past 50 years. The average annual loss rate over the past 10,000 years was 0.2 million hectares, while over the past 50 years, it reached 6 million ha, up by 30 times. Large-scale plowing of land and the destruction of natural vegetation over large areas have led to a reduction in land resources in all countries as a result of the following processes:

1) Accelerating soil erosion, washout of the upper fertile layer, and accumulation of pedosediments in relief depressions.
2) Accumulation of alluvium (solid sediments from river waters) and increased flooding of river floodplains. The spread of water meadows observed today in floodplains, especially those of small rivers, is not a natural phenomenon; it has been caused by the anthropogenic activity of agrarian societies, which commonly settled in the river valleys.
3) In the humid areas of Western Europe with ancient agriculture, leaching, and eluviation (washing and depletion) have intensified up to the replacement of brown soils by podzolic soils.
4) Desertification resulting from overgrazing in arid regions has increased due to the destruction of vegetation, overconsolidation of soils, and activation of water and wind erosion. Almost all the broken drift sands in the world (sand deserts of Sahara, Arabia, Karakum, etc.) are the legacy of the former immoderate herding. The primary biological productivity of arid ecosystems is limited by the lack of water and can only be supported by the alienation of the current annual growth of plant biomass. If, as a result of overgrazing, not only annual growth is alienated but also the accumulated stock of phytomass, pasture digression inevitably begins, leading to desertification. The area of the Sahara is increasing by 1.5 million hectares per year – among other things, due to the absorption of oases by loose sands. In the history of agriculture, humankind has turned 1 billion hectares of productive land into deserts. Soils of subboreal and subtropical deserts, subtropical semi-deserts, subboreal, and subtropical steppes, and dry steppes involved in irrigated agriculture, pastoral animal husbandry, as well as construction and communications, are most at risk of desertification.
5) Salinization processes have increased due to high salinity of irrigation water on watersheds and/or due to rising groundwater levels in floodplains and deltas. In Iraq and Syria, more than 50% of all irrigated lands are subject to salinization, more than 30% are in Egypt, and more than 15% are in Iran. An increase in salt

content in the arable layer of irrigated soil by 1% reduces the yield by one-third; with 2-3% of salt, the harvest dies completely. Agrarian society left 50-60 million hectares of secondarily saline soils as a legacy.

6) Extraction of water for irrigation in the upper and middle reaches of the rivers results in the lowering of river levels in low water and the inability of the river to cope with the waste discharged into it. Water extraction near sea coasts leads to saline water intrusion into underground reservoirs. Water carriers of diseases (malaria, encephalitis, yellow fever, etc.) spread on irrigated fields.
7) Formation of "urban soils" in places of urban settlements has intensified; they are characterized by a high content of heavy metals and phosphorus in the soil, an alkaline reaction of the environment, an abundance of mechanical inclusions, etc.
8) Simplification of soil cover structures determines the loss of biodiversity by ecosystems and the destruction of natural habitats of animals.

As a result of agricultural development of territories, 8% of the land, or 1176 million hectares, are lands excluded from use in the course of history, as anthropogenic deserts or badlands represent them.

The only way out of the irreversible reduction in the area of fertile soils dubbed as the "quiet crisis of the planet" is to develop and implement organic farming technologies and soil protection technologies. However, these technologies aim not to obtain the maximum yield at any cost. The main result they are aimed at is increasing soil fertility with minimal costs for agricultural production. The main principles of the system of soil-protective and resource-saving agriculture include the following:

- plowing is not the most important component in growing crops;
- plant residues are a valuable product and should be kept on the soil surface as mulch;
- burning of crop residues (mulch) should be prohibited;
- permanent vegetation cover should be present;
- the development of biological processes that ensure high soil fertility should be emphasized;
- biological pest control should be active;
- water and wind erosion should be acknowledged as a symptom of inappropriate agricultural principles and practices used for a given field or ecosystem.

The implementation of these principles might ensure the restoration of degrading soils or, at least, will slow down the accelerated degradation as much as possible.

3 Transformations of Social Interactions

Section 1 has outlined the most important changes that will have a decisive impact on the reformatting of social relations and on the mechanisms of social self-organization:

1) globalization: the world has become unified, there is no division into "us" and existential "them" (accordingly, the task of shaping the principles of co-existence becomes urgent);
2) the world has become "a closed circuit": the possibility of external expansion has disappeared along with the cessation of the influx of external resources, the "positive sum game" has ended, orientation toward profit maximization is no longer possible;
3) stabilization of the world population, its aging;
4) transition to the information/cybernetic society; information technologies and artificial intelligence have penetrated all spheres of life;
5) the object of technological influence and change is not only nature (as it used to be in previous eras) but also people themselves (biotechnology, genetic engineering, chipping, artificial organs) and their consciousness.

As a result, the following social transformations are very likely:

- the end of the era of a liberal market economy based on profit maximization[36]; transition to a predominantly distributive society (how and by what methods distribution will be carried out is a separate issue);
- satisfying the basic material needs of the population. With technological progress, material goods cease to be scarce, the greatest value is placed not on the material objects but rather on information;
- total control over human behavior (through social networks, gadgets, and other systems with artificial intelligence and cognitive technologies); people become "transparent" [about negative consequences of this, see chapter "Technology. Limitless Possibilities, Effective Control" (Grinin & Grinin, 2023b, this volume)]. A number of works have already been devoted to the analysis of various aspects of this problem in the present and the future (e.g., Westin, 1966; Ashman et al., 2014; Cecere et al., 2015; Moustaka et al., 2019; Schwartz, 1999; Solove, 2008; Brammer et al., 2020; Alharbi, 2020);
- a fundamental change in the demographic structure of society as it is becoming older and, consequently, more conservative; transition from a "society of growth" to a "society of stability"[37];
- transition to self-regulating systems and decreasing of the human factor in production processes, deep automation of production processes, the involvement of people in human-machine systems, a decrease in the need for personal experience (lack of personal experience is compensated by the use of artificial

[36]This is a consequence of the decline in the overall rate of profit and the end of an era when the "positive sum game" was the rule rather than the exception.

[37]See chapter "Global Aging – an Integral Problem of the Future. How to Turn a Problem into a Development Driver?" (Grinin, Grinin & Korotayev, 2023b, this volume), chapter "Technology: Limitless Possibilities, Effective Control" (Grinin & Grinin, 2023b, this volume) and chapter "Economics. Optimizing Growth" (Grinin, Grinin & Malkov, 2023a, this volume), as well as Grinin et al. (2020), Korotayev et al. (2021), and Korotayev et al. (2023c).

intelligence), narrowing space of freedom of action, dependence on IT technologies in all spheres of life[38];

- as a consequence, an increase in the specialization of people's professional activities (see Grinin & Grinin, 2015a, 2015b), the emergence of a "new estate system."[39] The emergence of a "new class system" is also possible. Who will be the elite in the cybernetic W-society? This is still unclear. Of course, the political and financial elites will remain, although they are being transformed. New layers of the elite are likely to emerge. There are various options here. One of them, as we pointed out in chapter "Future Political Change: Toward a More Efficient World Order" (Grinin et al., 2023a, in this volume), is the formation of a layer of a sort of "expertocracy." At the same time, the elite in the distributive W-society will be those who control and redistribute resources (primarily information resources);
- the important role of ideology as a means of controlling people's behavior.
- Those transformations are very likely, and they determine the difference of the coming W-society from the X- and Y-societies that preceded it (see Table 1).

The institutional design and content of the transformations listed above (and hence the organization of the emerging society) can be different in different scenarios of future evolution.[40] In essence, what is currently being shaped is the path that the development of the postsingular W-society will take. There are some options here. Of these options, the one that will eventually be established is the one that will smooth out the processes of social self-organization in the beast way, ensuring the stability of the institutional structure formed within the framework of this option.

4 Socioeconomic Stability in the Cybernetic W-Society and Options for Its Formation

Consider the issues of ensuring socio-economic stability in future cybernetic W-society.

[38] On techno-medical systems in cybernetic society see chapter "Analysis of Social Self-Organization and Historical Dynamics. Future Cybernetic W-society: Socio-Political Aspects" (Grinin & Grinin, 2023a, this volume).

[39] See chapter "Analysis of Social Self-Organization and Historical Dynamics. Future Cybernetic W-society: Socio-Political Aspects" (Grinin & Grinin, 2023a, this volume), chapter "Global Aging – an Integral Problem of the Future. How to Turn a Problem into a Development Driver?" (Grinin, Grinin & Korotayev, 2023b, this volume), and chapter "Socio-Political Transformations. A Difficult Path to Cybernetic Society" (Grinin, Grinin, & Malkov, 2023b, this volume) for more detail.

[40] So, despite the fact that the Byzantine Empire, medieval monarchies, the USSR, and Germany of the 1930s are all X-type societies, their external appearance and internal content are very different from each other.

Table 1 Comparison of characteristics of preindustrial X-, industrial Y-, and postindustrial W-societies

	Preindustrial X-society	Industrial Y-society	Postindustrial W-society
Development of science and technology	Slow	Fast	Fast
Material resources	Land resources in a limited area (limited resource base)	Growth of the resource base due to scientific and technological progress	Growth of the resource base due to scientific and technological progress
Specific features of economy	Shortage of material goods, distribution of what is lacking	Periodic crises of overproduction, exchange of surplus (through trade)	Sufficiency of material goods, a provision according to individual needs
Population dynamics	Very weak growth (Malthusian trap), young people prevail	Fast growth, young people prevail	No growth, elderly people prevail
System closeness/ openness	Closeness (weak ties with the outside world)	Openness (there is a possibility of expansion to the outside world)	Closeness (the whole planet has been globalized; no possibility of the further expansion of the World System)
Specific features of governance	"subject-object" (means of influence: violence + ideology)	"subject-subject" (means of influence: material interest/dependence)	"subject-environment" (means of influence: influence on the mentality)
	Joining the elite is possible through access to power, to the management of redistribution. Power is more important than money.	Joining the elite is possible through access to cash flow management. Money is more important than power.	Joining the elite is possible through access to mentality management.
	Autocracy	Representative democracy	Options: digital autocracy or meritocracy with direct democratic network governance
Social structure	Hierarchically organized estates	Classes	Social groups united in networks
Means of ensuring social sustainability	Supression + ideology (to justify the legitimacy of power)	Money + contract (legal system)	Self-restraints + ideology (mind control)
Relations in society	The predominance of "friend-friend" relations within one's group and "friend-foe" in relation to everyone else. Clanship	The prevalence of "foe-foe" relations, competition, individualism	Options: either "friend or foe" (new estate system), or "friend or friend" (collectivism)

(continued)

Table 1 (continued)

	Preindustrial X-society	Industrial Y-society	Postindustrial W-society
Incentives	Spiritual, religious (a monk is worthier than a warrior or a merchant)	Material (the rich are worthier than the poor)	Mental

A. In future cybernetic W-society, due to the increasing role of age markers, as well as what we have called the economic type of alienation, former social divisions will become less significant. What will be the new social structure? Various options/ scenarios are also possible here. One such option, as we see it, is an increase in the fragmentation and multiplicity of social groups, which will result in a decrease in the factors of social cohesion. Therefore, when modeling socio-economic sustainability, it is no longer possible to divide society into a small number of social strata, the interaction of which with each other is decisive – in agrarian society, this was the interaction of farmers and landowners [see Sect. 2.1 of chapter "Modeling Social Self-Organization and Historical Dynamics. Agrarian Society" (Malkov et al., 2023b, this volume)], in industrial society this was/is the interaction of entrepreneurs and employees [see Sect. 2.1 of chapter "Modeling Social Self-Organization and Historical Dynamics. Industrial Society" (Akaev et al., 2023, this volume)]. It is necessary to use a multi-component interaction scheme involving N social actors (social groups), which is generally given in Sect. 3.1.1 of chapter "Modeling Social Self-Organization and Historical Dynamics. A General Approach" (Akaev et al., 2023, this volume), Eq. (21).

This raises the question: how will the "synchronization" of the actions of these social actors be ensured?

B. Another feature of future cybernetic W-society is the strengthening of distributive principles (in contrast to the dominance of liberal market principles in the out-going industrial society).[41] The question is, who will be the distributor, and on what principles will the distribution be carried out, and therefore, who will control the main processes?

C. There are a number of scenarios of the future global evolution; more precisely, there is a continuum of variants of the future social structure and order in different societies. Below we discuss only two polar versions of the future societal order.

C.1. The first option (if we summarize some common opinions) is the model for the formation of a hierarchical estate (caste) totalitarian society, controlled by an explicit or implicit "world government" (a global elite, which includes representatives of financial capital, transnational companies, Internet giants who

[41] Malkov (2021), using the basic model of economic competition, shows that when moving from a "positive-sum game" to a "zero-sum game", competitive interactions inevitably lead to the victory of one of the parties and to economic monopoly. Under these conditions, liberal market mechanisms are transformed into distributive ones, dependent on the monopolist. The question is who the distributor will be, and on what principles the distribution will be carried out. In the modern world, such monopolists can be states, transnational companies and financial capital (in the economic sphere), Internet giants (in the information sphere).

control resource and information flows). The basic material needs of the population are provided with an unconditional basic income, but the population is deprived of property and a real opportunity to influence the fundamentally important decisions made by the "world government."[42] At the same time, people's mentality is controlled (among other things, through chipping), which makes it possible to ensure the necessary "synchronization" of society. Digital technologies are used for total control over the behavior of the population. This is option No. 1 of future W-society (a phase portrait of such a society is shown in Fig. 7a, where population containment at a level not exceeding N'' is achieved, among other things, by gender policy, by encouraging child-free lifestyle etc.).

In essence, this is a sort of a new feudalism, the reincarnation of the X-system of a totalitarian type in the cybernetic era. But there is an essential difference from the feudalism of the Middle Ages. In medieval states, there was an internal competition of elites for power and external competition of states with each other, which, despite all the costs, served as an incentive for further development (including technological development). In option No. 1 of future W-society, in the context of globalization and monopoly in the field of governance and distribution of material goods, competition is suppressed, which will inevitably lead to stagnation and subsequent degradation.

C.2. An alternative to option No.1 can be option No. 2 of future W-society under the conditional name "World-Organism." This is a society where, in the presence of a deep professional specialization of the population, there is no discrimination, and governance is carried out on the basis of network technologies (network technologies are used to "synchronize" collective activities and implement direct democracy). An important role is played by ideology (in the form of ethical norms and self-restrictions that suppress antagonistic competition), which establishes a system of rules for developing coordinated decisions. At the same time, nonantagonistic competition is not suppressed; on the contrary, it is encouraged but channeled into the sphere of sports, science and technology, creative professions, etc. The activities of bureaucrats are controlled by society (this is ensured by information transparency and network platforms), which makes it possible to form a governing apparatus based on meritocratic principles. Of course, it is only a metaphor, but in a certain sense, such a society acts as a biological organism, where all organs act in concert without suppressing each other to achieve a common goal. The phase portrait of such a society is shown in Fig. 7b, where the population N'' is established naturally, and technological growth is not inhibited due to the encouragement of the creative activity of all social strata. In essence, option No. 2 of future W-society under the code name *"World-organism"* is the implementation of the ideas of the great Russian humanist scientists Vladimir Vernadsky (1998, 2012, 2013, 2015, 2018), Kondratieff (2002), Sorokin (1960, 1968), Moiseev (1982, 1990, 1996, 2022); Moiseev et al., 1985), Prigogine (1997, 2017); Prigogine et al., 1984),

[42] An analogue of "bread and circuses" for the plebs in ancient Rome.

Gvishiani (1977, 1980, 1984, 1986, 2010); Gvishiani et al., 1991), Kapitza (1992, 1996, 1999a, 1999b, 2003, 2006, 2010) and others.

D. These options are two extreme alternatives. They may not be implemented in their pure form, but what matters is which way the formation of future society will go: closer to option No. 1 or closer to option No. 2. The first path is the path when the antagonistic competitive principle in social relations prevails over the cooperative principle. The second way is when the achievement of common goals, cooperation, and not antagonistic competition, is at the forefront (but where nonantagonistic competition still plays an important role).

If the transition from industrial society (with its internal competition, the dominance of Y-institutions and individualism at the socio-psychological level)[43] to **option No. 1** of future W-society is, in a sense evolutionary, since it retains an antagonistic competitive mentality and is based on the desire of social actors to maximize the individual "utility function," then the transition to **option No. 2** faces fundamental difficulties. These difficulties are associated with the need to move from competitive to cooperative relations and require a restructuring of consciousness and decision-making principles at all levels of society. Such restructuring does not happen quickly, it takes time, measured in generations, and it needs a new ideology [see Sect. 4.2 of chapter "Modeling Social Self-Organization and Historical Dynamics. Global Phase Transitions" (Malkov et al., 2023a, this volume)].

As is known, the task of transforming mass consciousness from a competitive mentality to a mentality based on self-restraint and reducing mutual aggression was attempted in Europe by Christianity[44] with its well-known imperatives. In the modern era, such an ideology has not yet been formulated,[45] but it should take the form not of religion (since religions are canonical in nature and rigidly oppose the circle of their adherents to adherents of other religions), but *ethics*, generally accepted norms of morality and behavior.

An essential question is emerging – with the help of what social mechanisms will the fulfillment of moral imperatives be ensured and the "free rider effect" (that is, the desire of certain members of society to evade payment for common goods or generally accepted self-restraints) be stopped? In Y-society, the fight against violators of the legally established rules of conduct is based on a rigid system of punishments (administrative law, criminal law, etc.), but the rule "not caught is not a thief" really works.

In religious X-society, the rules of conduct are fixed in the sacred books, and the threat in case of their violation is the punishment of the Lord, which in the minds of

[43] See Sect. 3 of chapter "Modeling Social Self-Organization and Historical Dynamics. Global Phase Transitions" (Malkov et al., 2023a, this volume).

[44] Buddhism played the same role in South Asia, and Confucianism did it in East Asia.

[45] The communist ideology in its classical form is not suitable for this purpose, since it is a product of the industrial age and reflects its features. At present, attempts to formulate an ideology adequate to the modern era are multiplying, but the process of its formation has not yet been completed.

believers is a serious deterrent[46] (but this threat does not affect the minds of nonreligious people: "if there is no God, then everything is permitted" (F.M. Dostoevsky)). In cybernetic W-society, the information transparency of a person and the network organization of society becomes a means of identifying and suppressing the "free rider effect." At the same time, information transparency can be used both to create a "digital concentration camp" (in option No. 1 of W-society), to effectively control the work of officials on the part of citizens, to suppress corruption, and form a management apparatus based on meritocratic principles (in option No. 2 of W-society). How the possibilities of universal digitalization and information transparency will be used in future cybernetic W-society depends on society itself.

In this respect, one may say that the World System has found itself at a critical point which implies emerging short and midterm possibilities this criticality opens (that will be considered in more detail below). The extreme importance of self-organization, a key characteristic of living systems, brings the impossibility of determinism. Self-organization is a mechanism creating new forms out of critical points in ways that cannot be predicted ex-ante, and we are now just in one of such critical points. We are dealing here with important modeling limitations (in the classical sense of identifying what futures could be) and the potential for emergence of self-organized futures that cannot be anticipated. Hence, the possible trajectories, specified above, in no way exhaust all the possible trajectories, the World System may follow in the future.

5 Sociopolitical Sustainability

The above options for the formation of future W-society imply a different structure of political interactions.

Option No. 1 of cybernetic W-society governed by a global elite implies the establishment of a political monopoly [see Fig. 17 in chapter "Modeling Social Self-Organization and Historical Dynamics. A General Approach" (Akaev et al., 2023, this volume)] or hierarchical (vassal) relations in the political elite (see Fig. 18 in the same chapter). In this sense, it resembles the feudal system in the Middle Ages [see Sect. 2.2 of chapter "Modeling Social Self-Organization and Historical Dynamics. Agrarian Society" (Malkov et al., 2023b, this volume)].

In **option No. 2** of cybernetic W-society, the principle of antagonistic competition is excluded from political life. In this case, as is shown by the basic model of political competition [see Sect. 3.2.2 of chapter "Modeling Social Self-Organization and Historical Dynamics. A General Approach" (Akaev et al., 2023, this volume)], a typical situation is the establishment of equal relations between political actors without unilateral advantages for any of them (see Fig. 20 in the same chapter).

[46] See, e.g., Turchin et al. (2022a, 2022b).

Accordingly, in **option No. 1** of cybernetic W-society, a directive management scheme of a hierarchical (authoritarian) type is established [see Fig. 12 in chapter "Modeling Social Self-Organization and Historical Dynamics. A General Approach" (Akaev et al., 2023, this volume)] when key decisions are made at the top of the hierarchy by the "world government" and go down for their implementation.

In the case of W-society **option No. 2**, an adaptive (democratic) governance scheme is naturally established, as shown in Fig. 13 in Sect. 3.2.1 of chapter "Modeling Social Self-Organization and Historical Dynamics. A General Approach" (Akaev et al., 2023, this volume), which allows one to quickly find a social compromise and ensure the "synchronization" of society when external and internal circumstances change.

Both of these governance schemes, as shown (see Fig. 12 in the same chapter), are stable.

6 Interstate Interactions

In the cybernetic era, as a result of the development of globalization processes and the increasing connectivity of the world, the importance of establishing common technical standards, information exchange protocols, coordinated operation of information systems, international trade, etc., is objectively increasing. However, these processes in the variants of the formation options of future W-society considered above differ significantly from each other. These differences are well demonstrated by base model (45) in Sect. 3.2.3 of chapter "Modeling Social Self-Organization and Historical Dynamics. A General Approach" (Akaev et al., 2023, this volume). Figure 8 below shows the results of modeling the globalization process

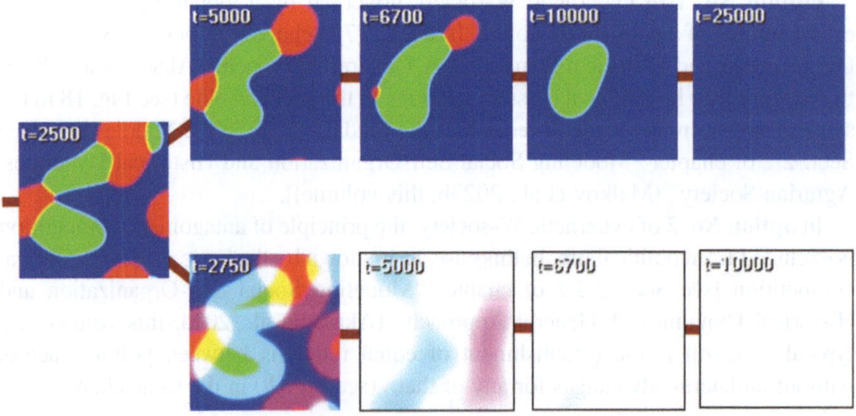

Fig. 8 Results of modeling the system dynamics described by Eq. (45) from Sect. 3.2.3 of chapter "Modeling Social Self-Organization and Historical Dynamics. A General Approach" (Akaev et al., 2023, this volume) (the numbers in the upper left corner of each frame corresponding to the calculation cycle)

corresponding to phase III of the evolution of the system (45) from chapter "Modeling Social Self-Organization and Historical Dynamics. A General Approach" (Akaev et al., 2023, this volume).

In relation to the case of interest to us, geopolitical actors, displayed in Fig. 8 by blue, green, and red clusters, correspond to civilizations; the nature of their interaction determines which path the formation of future W-society will take.

The top row[47] in Fig. 8 corresponds to the formation of option No. 1 of cybernetic W-society according to the *"clash of civilizations"* scenario [using Huntington's expression (Huntington, 1993, 1996)], as a result of which one of the civilizations begins to dominate.

The bottom row[48] in Fig. 8 corresponds to the formation of option No. 2 of W-society according to the scenario of Yury Yakovet's *"partnership of civilizations"* (Yakovets & Farah, 2014), following which there is an interaction of civilizations in the interests of developing common goals, mutual cultural enrichment, and solving common problems. The picture illustrates the fact that the mutual diffusion of cultures and the strengthening of partner interaction leads to a new qualitative state: blue, green, and red colors, when combined, give a white color in which they become integral components. White color is *common ethics*, general rules of cooperative interaction, which become natural for a single World-organism (unity of diversity).

An illustration of the transition from the modern world-system of competing states to an integrative global noospheric society to the World-organism (in the case of the implementation of option No. 2) is shown in Fig. 9.

7 Possible Scenarios of World Dynamics in the Twenty-First Century

Previous sections in this chapter discuss the main features of the ongoing global transformations. It is shown that the further course of events can vary. Below we analyze the most characteristic scenarios among the possible ones.

A. Of all the processes associated with the development of society in the modern era, the most inertial one is demographic dynamics. As is shown above, the most important feature of modern demographic dynamics is that in all regions of the world, the second phase of demographic transition is going on (or has already been

[47]The upper row of the calculation in Fig. 8 corresponds to the case when the competition between the actors is antagonistic, respectively, in Eqs. (45) from chapter "Modeling Social Self-Organization and Historical Dynamics. A General Approach" (Akaev, Malkov, Bilyuga et al., 2023, this volume) coefficients $c_{ij} < 0$.

[48]The bottom row of the calculation in Fig. 8 corresponds to the variant when there is no antagonistic competition between the actors with a simultaneous increase in the intensity of interaction (with an increase in the values of the coefficients D_i in Eqs. (45) from the same chapter).

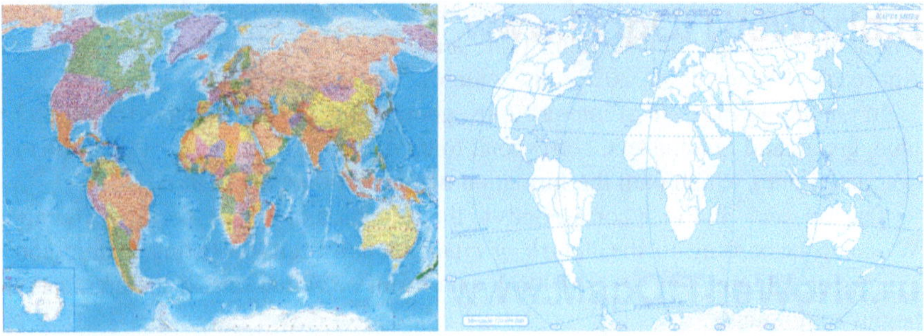

Modern political map Future political map

Fig. 9 Change in the political map of the world when option 2 of the World System transformation is implemented. It is implied that state borders on the political map of the world are the borders of clusters separating "us" and "them." In the case of an integrative society, there will no longer be such a division (everyone will be "us"). Only administrative boundaries will remain

completed), which implies a dramatic decrease in the birth rates[49] and a corresponding slowdown in population growth, or even population decline.[50] The only uncertain parameter is the speed at which the birth rates will decrease worldwide (at the same time, the demographic situation will develop differently in different countries and regions). This uncertainty determines a set of probable demographic scenarios, which were calculated on the basis of mathematical models, taking into account current trends and hypotheses about the further dynamics of demographic processes[51]:

the upper scenario implies continued growth of the world population in the twenty-first century with the stabilization of its numbers in the twenty-second century;
the medium scenario implies the stabilization of the world population by the end of the twenty-first century;
the lower scenario implies the stabilization of the world population with a subsequent decline in the second half of the twenty-first century.

These scenarios are presented in Fig. 10.

Calculations demonstrate the inevitability of a significant change of population in different regions and cultural communities (civilizations), which will significantly affect world processes during the twenty-first century (see Fig. 11).

[49]The decline in the birth rate is largely due to the processes of socio-economic modernization and the growth of the urban population, for which small family model is typical.

[50]See chapter "Demography. Toward Optimization of Demographic Processes" (Korotayev, Malkov & Musieva, 2023, this volume), chapter "Africa – the Continent of the Future. Challenges and Opportunities" (Grinin & Korotayev, 2023, this volume), and chapter "Modeling Social Self-Organization and Historical Dynamics. Africa's Futures" (Korotayev, Shulgin et al., 2023, this volume).

[51]For more detail see chapter "Demography. Toward Optimization of Demographic Processes" (Korotayev, Malkov & Musieva, 2023, this volume).

Fig. 10 Possible scenarios of world demographic dynamics up to 2100, thousands (1 – *the upper scenario*, 2 – *the medium scenario*, 3 – *the lower scenario*)

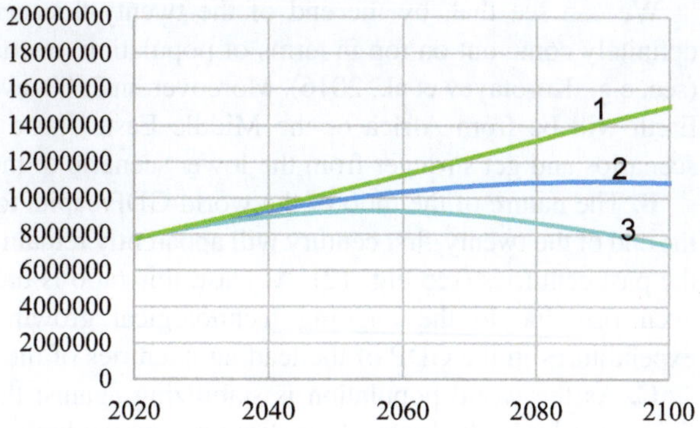

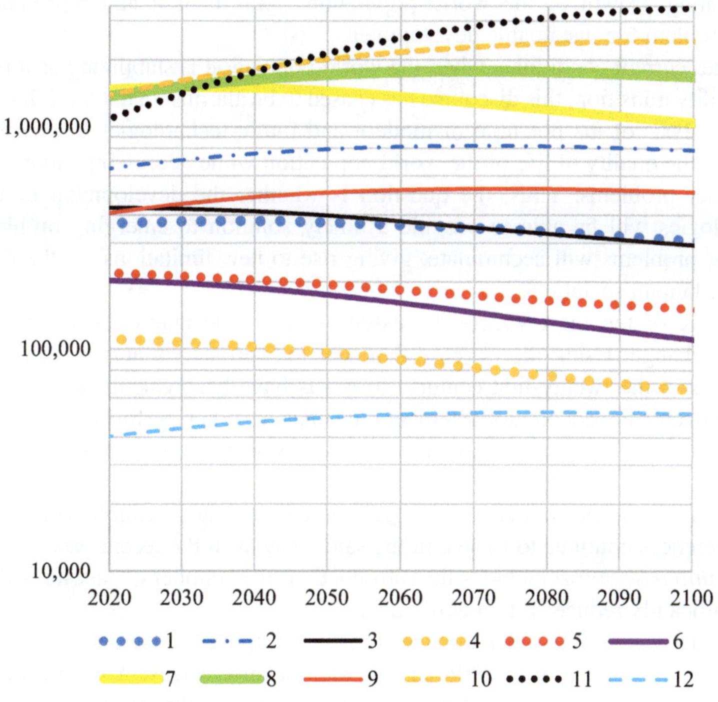

Fig. 11 Population dynamics in various regions of the world in accordance with the average scenario, thousands, logarithmic scale (1 = North America, 2 = Latin America, 3 = Western Europe, 4 = Eastern Europe, 5 = Russia and CIS countries, 6 = Japan and Korea, 7 = China, 8 = India, 9 = Southeast Asia, 10 = Middle East and North Africa, 11 = Sub-Saharan Africa, 12 = Australia, New Zealand, Oceania)

We can see that, by the end of the twenty-first century, Tropical Africa will definitely come out on top in terms of population, overtaking both China and India (see, e.g., Korotayev et al., 2016). Moreover, in 2100, every second inhabitant of the Earth will be from Africa or the Middle East. These changes are typical for all scenarios and get stronger from the lower scenario to the upper one (see Table 2).

B. The nature of the ratio of the world GDP/capita to the world population until the end of the twenty-first century will apparently remain the same as it has been over the past centuries (see Fig. 12). At least, this ratio is unlikely to deteriorate; it may even rise due to the ongoing technological growth and the increasing R&D expenditures in the GDP of the leading countries of the world.

C. As the world population is stabilizing against the background of the rising living standards, the focus of the discourse on the limits of growth is changing from what was initiated by the reports to the Club of Rome half a century ago. At that time, it was believed that with the escape from the "Malthusian trap," the accelerating growth of the world population[52] was natural and represented the main problem for sustainable development.

Under current conditions, when the world population is stabilizing as a result of the fertility transition, this discourse has ceased to be alarmist in nature. It has shifted to an analysis of the potential of modern and future technologies to preserve and improve the quality of life of the world population while addressing environmental and other problems. Thus, the question is whether the development of modern technologies will be able to provide a timely solution to emerging problems, or whether problems will accumulate, giving rise to new limitations for the development of human society.

D. One of the most widely discussed topics is the anthropogenic impact on climate change. Using the developed model for various variants of demographic dynamics in the twenty-first century, forecasts have been calculated to assess the anthropogenic impact on global temperature change. When performing calculations, three scenarios for the development of the energy sector were considered:

a conservative scenario assumes that government policies, technologies, and social preferences continue to evolve in the same way as in the recent past;

an ambitious scenario includes the introduction of a number of measures that will significantly reduce carbon emissions;

a Net Zero scenario assumes that the measures proposed in the ambitious scenario are complemented and reinforced by significant changes in the behavior and preferences of society, which further accelerates the reduction of carbon emissions – for more information on these scenarios, see chapter "Climate and Energy. Scenarios of Energy Transition and Global Temperature changes Based on Current Technologies and Trends" (Akaev & Davydova, 2023, this volume) and chapter "Modeling Social Self-Organization and Historical Dynamics.

[52]Indeed, in the1960s, the growth rate of the world population was the highest in the history of humankind (see Fig. 1).

Table 2 The share of the population of the region in the total population of the Earth

Year	Region											
	1 = North America	2 = Latin America	3 = Western Europe	4 = Eastern Europe	5 = Russia and CIS countries	6 = Japan and Korea	7 = China	8 = India	9 = Southeast Asia	10 = Middle East and North Africa	11 = Sub-Saharan Africa	12 = Australia, New Zealand, Oceania
2020	4.76%	8.37%	5.62%	1.44%	2.87%	2.66%	19.15%	17.74%	5.27%	17.83%	13.72%	0.52%
2100 upper scenario	2.49%	6.48%	2.26%	0.50%	1.15%	0.75%	6.76%	14.01%	4.22%	22.55%	38.43%	0.43%
2100 the medium scenario	2.83%	7.24%	2.70%	0.60%	1.40%	1.02%	9.61%	15.53%	4.68%	22.33%	31.58%	0.47%
2100 the lower scenario	3.29%	7.92%	3.24%	0.73%	1.59%	1.27%	11.33%	16.34%	4.97%	22.09%	26.75%	0.49%

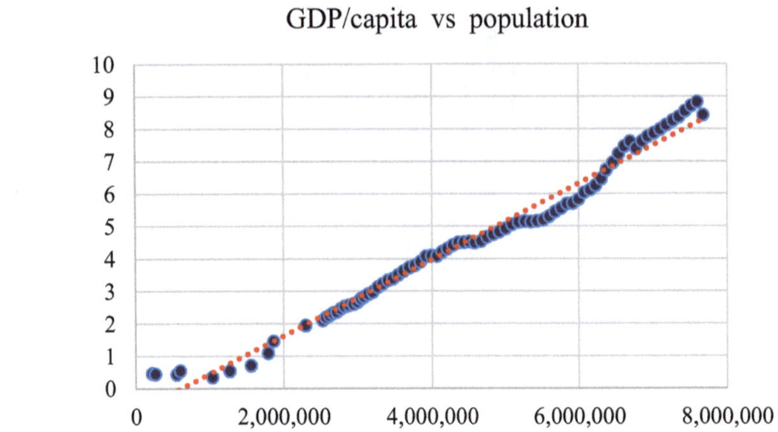

Fig. 12 The ratio of the world average value of GDP/capita (y-axis, thousands of international dollars 1990 at PPP) to the world population (x-axis, thousands), the beginning of the common era to 2020. Scatterplot with a fitted regression line. *Data source* Maddison Project (2020)

Industrial Society" (Akaev et al., 2023, this volume), as well as Akaev and Davydova (2020a), Akaev and Davydova (2020b), Akaev and Davydova (2021a, 2021b).

The calculation results are shown in Figs. 13, 14, and 15.

The calculations show that even in the upper demographic scenario, it is possible to keep the average global temperature of the surface atmosphere within acceptable limits, provided that modern technologies are used effectively.

A similar conclusion can be drawn in relation to global environmental problems (a number of important measures aimed at stabilizing the environmental situation are outlined in Sect. 1; see also Kiss, 2020).

E. Our analysis reveals that the most pressing modern problems are found not so much in the sphere of "society-nature" interactions (these problems can be solved with the necessary coordination of the efforts of the world community), but rather in the sphere of social and political interactions. It is important which way social transformations will take in the modern "era of change," and what society will be like after the end of the current transition period. In this historical period, the most important (but also the most difficult) task is social design, purposeful activity to shape future global society on the principles of cooperation and partnership rather than competition and rivalry (see Sect. 4 above). Some ideas on how to direct the development of society toward a favorable path of development are set out in the main text of the report [see chapter "Socio-Political Transformations. A

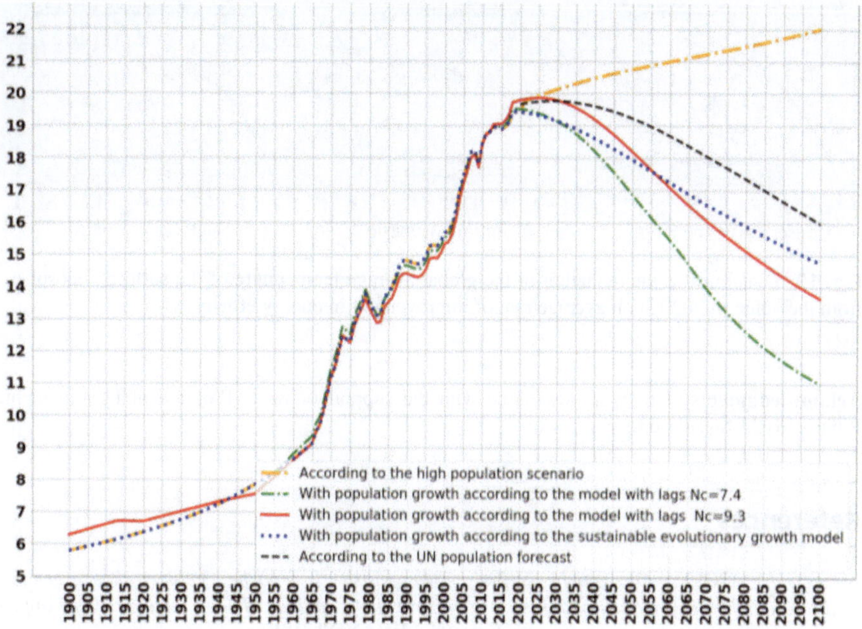

Fig. 13 Dynamics of real energy consumption at an average rate of energy efficiency improvement, billion tons of fuel equivalent per year

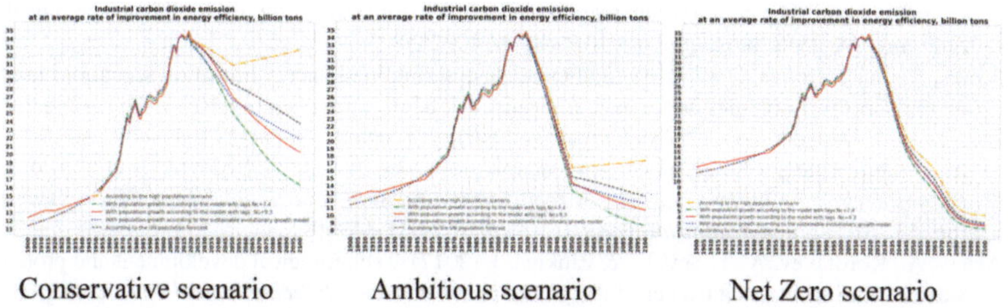

Fig. 14 Dynamics of industrial carbon dioxide emissions at an average rate of improvement in energy efficiency, billion tons

Difficult Path to Cybernetic Society" (Grinin et al., 2023b, this volume)] and in chapter "Analysis of Social Self-Organization and Historical Dynamics. Future Cybernetic W-society: Socio-Political Aspects" (Grinin & Grinin, 2023a, this volume) below.

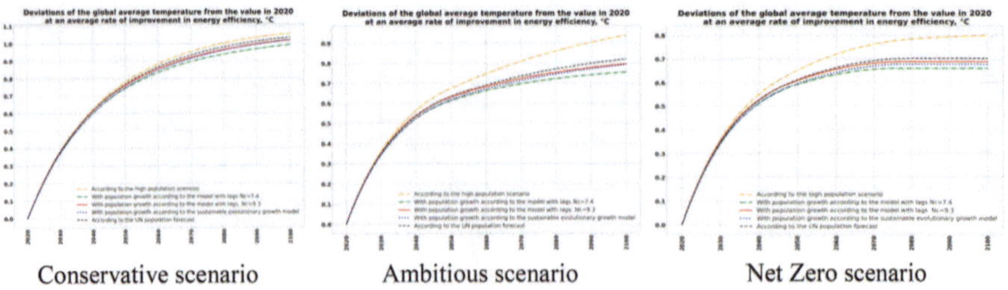

Fig. 15 Dynamics of the deviation of the global average temperature of the surface atmosphere from the value in 2020 at an average rate of improvement in energy efficiency

Acknowledgments This work was done with the support of MSU Program of Development, Project No 23-SCH05-03.

References

Akaev, A. (2015). *Ot epokhi Velikoy divergentsii k epokhe Velikoy konvergentsii: Matematicheskoye modelirovaniye i prognozirovaniye dolgosrochnogo tekhnologicheskogo i ekonomicheskogo razvitiya mirovoy dinamiki*. LENAND/URSS.

Akaev, A., & Davydova, O. (2020a). The Paris agreement on climate is coming into force: Will the great energy transition take place? *Herald of the Russian Academy of Sciences, 90*(5), 588–599.

Akaev, A., & Davydova, O. (2020b). Parizhskoye klimaticheskoye soglasheniye vstupa-yet v silu. Sostoitsya li Velikiy energeticheskiy perekhod? *Vestnik Rossiyskoy akademii nauk, 90*(10), 926–938.

Akaev, A., & Davydova, O. (2021a). A mathematical description of selected energy transition scenarios in the 21st century, intended to realize the main goals of the Paris climate agreement. *Energies, 14*, 2558. https://doi.org/10.3390/en14092558

Akaev, A., & Davydova, O. (2021b). Mathematical description of energy transition scenarios based on the latest technologies and trends. *Energies, 14*, 8360. https://doi.org/10.3390/en14248360

Akaev, A., & Davydova, O. (2023). Climate and energy. Scenarios of energy transition and global temperature changes based on current technologies and trends. In V. Sadovnichy et al. (Eds.), *Reconsidering the limits to growth. A report to the Russian Association of the Club of Rome* (pp. 53–70). Springer. https://doi.org/10.1007/978-3-031-34999-7_4

Akaev, A., Korotayev, A., Issaev, L., & Zinkina, J. (2017). Technological development and protest waves: Arab spring as a trigger of the global phase transition? *Technological Forecasting and Social Change, 116*, 316–321. https://doi.org/10.1016/j.techfore.2016.08.009

Akaev, A., Malkov, S., Bilyuga, S., Malkov, A., Musieva, J., & Korotayev, A. (2023a). Modeling social self-organization and historical dynamics. A general approach. In V. Sadovnichy et al. (Eds.), *Reconsidering the limits to growth. A report to the Russian Association of the Club of Rome* (pp. 253–307). Springer. https://doi.org/10.1007/978-3-031-34999-7_15

Akaev, A., Malkov, S., Davydova, O., Kovaleva, N., Malkov, A., Grinin, L., & Korotayev, A. (2023b). Modeling social selforganization and historical dynamics. Industrial Society. In V. Sadovnichy et al. (Eds.), *Reconsidering the limits to growth. A report to the Russian Association of the Club of Rome* (pp. 337–385). Springer. https://doi.org/10.1007/978-3-031-34999-7_17

Akaev, A., Sadovnichy, V., & Korotayev, A. (2012). On the dynamics of the world demographic transition and financial-economic crises forecasts. *The European Physical Journal Special Topics, 205*(1), 355–373. https://doi.org/10.1140/epjst/e2012-01578-2

Akaev, A. A., & Sadovnichiy, V. A. (2016). A closed dynamic model to describe and calculate the Kondratiev long wave of economic development. *Herald of the Russian Academy of Sciences, 86*(5), 371–383. https://doi.org/10.1134/S1019331616050014

Alharbi, F. S. (2020). Dealing with data breaches amidst changes in technology. *International Journal of Computer Science and Security (IJCSS), 14*(3), 108–115.

Ashman, H., Brailsford, T., Cristea, A. I., Sheng, Q. Z., Stewart, C., Toms, E. G., & Wade, V. (2014). The ethical and social implications of personalization technologies for e-learning. *Information & Management, 51*(6), 819–832. https://doi.org/10.1016/j.im.2014.04.003

Benson, I., & Lloyd, J. (1983). *New technology and industrial change: The impact of the scientific-technical revolution on labour and industry*. Kogan Page.

Bernal, J. D. (1965). *Science in history* (3rd ed.). Hawthorn Books.

Berry, B. J. L. (1991). *Long wave rhythms in economic development and political behavior*. Johns Hopkins University Press.

Brammer, S., Branicki, L., & Linnenluecke, M. (2020). COVID-19, societalization and the future of business in society. *Science of the Total Environment, 34*(4), 2–7. https://doi.org/10.5465/amp.2019.0053

Bunch, B., & Hellemans, A. (2004). *The history of science and technology*. Houghton Mifflin Company.

Cecere, G., Le Guel, F., & Soulié, N. (2015). Perceived internet privacy concerns on social networks in Europe. *Technological Forecasting and Social Change, 96*, 277–287. https://doi.org/10.1016/j.techfore.2015.01.021

Forrester, J. W. (1978). *Innovation and the economic long wave*. MIT.

Forrester, J. W. (1981). *The Kondratieff cycle and changing economic conditions*. MIT.

Forrester, J. W. (1985). Economic conditions ahead: Understanding the Kondratieff wave. *Futurist, 19*(3), 16–20.

Freeman, C. (1987). Technical innovation, diffusion, and long cycles of economic development. In T. Vasko (Ed.), *The long-wave debate* (pp. 295–309). Springer. https://doi.org/10.1007/978-3-662-10351-7_21

Goldstein, J. (1988). *Long cycles: Prosperity and war in the modern age*. Yale University Press.

Goldstone, J. A. (1991). *Revolution and rebellion in the early modern world*. University of California Press.

Goldstone, J. A. (2009). *Why Europe? The rise of the west in world history. 1500–1850*. McGraw-Hill.

Goldstone, J. A. (2021). Dating the great divergence. *Journal of Global History, 16*(2), 266–285. https://doi.org/10.1017/S1740022820000406

Grable, J. E. (2019). An introduction to Kondratieff waves. *Journal of Financial Service Professionals, 73*(3), 19–22.

Grinin, L. (2006). *Proizvoditelnyye sily i istoricheskiy protsess*. URSS.

Grinin, L. (2021). How globalization shifts the global economic and political balance. *Journal of Chinese Philosophy, 48*(2), 222–234.

Grinin, L., & Grinin, A. (2015a). Kiberneticheskaya revolyutsiya i shestoy tekhno-logicheskiy uklad. *Istoricheskaya psikhologiya i sotsiologiya istorii, 8*(1), 172–197.

Grinin, L., & Grinin, A. (2015b). *Ot rubil do nanorobotov. Mir na puti k epokhe samoupravlyayemykh sistem (istoriya tekhnologiy i opisaniye ikh budushchego)*. Uchitel.

Grinin, L., & Grinin, A. (2016a). *The cybernetic revolution and the forthcoming epoch of self-regulating systems*. Uchitel.

Grinin, L., & Grinin, A. (2016b). The sixth Kondratieff wave and the cybernetic revolution. In A. Korotayev, L. Grinin, I. Ilya, & P. Herrmann (Eds.), *Globalistics and globalization studies: Global transformations and global future* (pp. 337–556). Uchitel.

Grinin, L., & Grinin, A. (2023a). Analysis of social self-organization and historical dynamics. Future cybernetic W-society: Socio-political aspects. In: V. Sadovnichy et al. (Eds.), *Reconsidering the limits to growth. A report to the Russian Association of the Club of Rome* (pp. 521–530). Springer. https://doi.org/10.1007/978-3-030-*****-*_22

Grinin, L., & Grinin, A. (2023). Technology: Limitless possibilities, effective control. In V. Sadovnichy et al. (Eds.), *Reconsidering the limits to growth. A report to the Russian Association of the Club of Rome* (pp. 139–154). Springer. https://doi.org/10.1007/978-3-031-34999-7_8

Grinin, L., Grinin, A., & Korotayev, A. (2017a). Forthcoming Kondratieff wave, cybernetic revolution, and global aging. *Technological Forecasting and Social Change, 115*, 52–68. https://doi.org/10.1016/j.techfore.2016.09.017

Grinin, L., Grinin, A., & Korotayev, A. (2017b). The MANBRIC-technologies in the forthcoming technological revolution. In T. Devezas, J. Leitão, & A. Sarygulov (Eds.), *Industry 4.0 – Entrepreneurship and structural change in the new digital landscape: What is coming on along with the fourth industrial revolution* (pp. 243–261). Springer.

Grinin, L., Grinin, A., & Korotayev, A. (2020). A quantitative analysis of worldwide long-term technology growth: From 40,000 BCE to the early 22nd century. *Technological Forecasting and Social Change, 155*, 1–19. https://doi.org/10.1016/j.techfore.2020.119955

Grinin, L., Grinin, A., & Korotayev, A. (2021). COVID-19 pandemic as a trigger for the acceleration of the cybernetic revolution, transition from e-government to e-state, and change in social relations. *Technological Forecasting & Social Change, 175*, 1–16. https://doi.org/10.1016/j.techfore.2021.121348

Grinin, L., Grinin, A., & Korotayev, A. (2023a). Future political change: Toward a more efficient world order. In V. Sadovnichy et al. (Eds.), *Reconsidering the limits to growth. A report to the Russian Association of the Club of Rome* (pp. 191–206). Springer. https://doi.org/10.1007/978-3-031-34999-7_11

Grinin, L., Grinin, A., & Korotayev, A. (2023b). Global aging – An integral problem of the future. How to turn a problem into a development driver? In V. Sadovnichy et al. (Eds.), *Reconsidering the limits to growth. A report to the Russian Association of the Club of Rome* (pp. 117–135). Springer. https://doi.org/10.1007/978-3-031-34999-7_7

Grinin, L., Grinin, A., & Malkov, S. (2023c). Economics. Optimizing growth. In V. Sadovnichy et al. (Eds.), *Reconsidering the limits to growth. A report to the Russian Association of the Club of Rome* (pp. 155–168). Springer. https://doi.org/10.1007/978-3-031-34999-7_9

Grinin, L., Grinin, A., & Malkov, S. (2023d). Socio-political transformations. A difficult path to cybernetic society. In V. Sadovnichy et al. (Eds.), *Reconsidering the limits to growth. A report to the Russian Association of the Club of Rome* (pp. 169–189). Springer. https://doi.org/10.1007/978-3-031-34999-7_10

Grinin, L., & Korotayev, A. (2009). Urbanization and political instability: To the development of mathematical models of political processes. *Polis-Politicheskiye Issledovaniya, 4*, 34–52.

Grinin, L., & Korotayev, A. (2010). Will the global crisis lead to global transformations? The coming epoch of new coalitions. *Journal of Globalization Studies, 1*(2), 166–183.

Grinin, L., & Korotayev, A. (2014a). Globalization shuffles cards of the world pack: In which direction is the global economic-political balance shifting? *World Futures, 70*(8), 515–545. https://doi.org/10.1080/02604027.2014.982474

Grinin, L., & Korotayev, A. (2014b). Revolution vs democracy (revolution and counterrevolution in Egypt). *Polis-Politicheskiye issledovaniya, 3*, 139–158. https://doi.org/10.17976/jpps/2014.03.09

Grinin, L., & Korotayev, A. (2014c). Origins of globalization in the framework of the Afroeurasian world-system history. In T. D. Hall (Ed.), *Comparing globalizations. Historical and world-systems approaches* (pp. 37–70). Springer. https://doi.org/10.1007/978-3-319-68219-8_3

Grinin, L., & Korotayev, A. (2015). *Great divergence and great convergence. A global perspective*. Springer. https://doi.org/10.1007/978-3-319-17780-9

Grinin, L., & Korotayev, A. (2016). Global population ageing, the sixth Kondratieff wave, and the global financial system. *Journal of Globalization Studies, 7*(2), 11–31.

Grinin, L., & Korotayev, A. (2021). Seven weaknesses of the U.S. Donald trump and the future of American hegemony. *World Futures, 77*(1), 23–54. https://doi.org/10.1080/02604027.2020.1801309

Grinin, L., & Korotayev, A. (2023). Africa – The continent of the future. Challenges and opportunities. In V. Sadovnichy et al. (Eds.), *Reconsidering the limits to growth. A report to the Russian Association of the Club of Rome* (pp. 225–238). Springer. https://doi.org/10.1007/978-3-031-34999-7_13

Grinin, L., Korotayev, A., & Tausch, A. (2016). *Economic cycles, crises, and the global periphery*. Springer. https://doi.org/10.1007/978-3-319-41262-7

Grinin, L., Malkov, S., & Korotayev, A. (2023). High income and low income countries. Toward a common goal at different speeds. In V. Sadovnichy et al. (Eds.), *Reconsidering the limits to growth. A report to the Russian Association of the Club of Rome* (pp. 207–224). Springer. https://doi.org/10.1007/978-3-031-34999-7_12

Gruebler, A., & Nakicenovic, N. (1991). Long waves, diffusion, and substitution. *Review, 14*(2), 313–342.

Gvishiani, D. (1977). *Metodologicheskiye problemy modelirovaniya globalnogo razvitiya*. VNIISI.

Gvishiani, D. (1980). Methodological problems of global development modelling. In E. Velikhov, D. Gvishiani, & S. Mikulinsky (Eds.), *Science, technology and the future: Soviet scientists' analysis of the problems of and prospects for the development of science and technology and their role in society* (pp. 21–35). Pergamon Press.

Gvishiani, D. (Ed.). (1984). *Systems research: Methodological problems*. Pergamon Press.

Gvishiani, D. (1986) *Priroda modeley i modeley prirody*. Mysl.

Gvishiani, D. (2010). *Mosty v budushchee. Rimskiy klub, IFIAS, Venskiy sovet*. URSS.

Gvishiani, D., Velikhov, E., Leybin, V., et al. (1991). *Modelirovaniye protsessov mirovogo razvitiya i sotrudnichestva*. Nauka.

Hirooka, M. (2006). *Innovation dynamism and economic growth. A nonlinear perspective*. Edward Elgar.

Huntington, S. P. (1993). The clash of civilizations. *Foreign Affairs, 72*(3), 22–49.

Huntington, S. P. (1996). *The clash of civilizations and the remaking of world order*. Simon & Schuster.

Ilyin, I. V., & Leonova, O. G. (2022). 'Non-Global' globalization. New features of political globalization. *Journal of Globalization Studies, 13*(1), 20–36. https://doi.org/10.30884/jogs/2022.01.02

Kapitza, S. (1992). A mathematical model for global population growth. *Matematicheskoe Modelirovanie, 4*(6), 65–79.

Kapitza, S. (1996). Phenomenological theory of world population growth. *Physics Uspekhi, 39*(1), 57–71. https://doi.org/10.1070/PU1996v039n01ABEH000127

Kapitza, S. (1999a). *Obshchaya teoriya rosta chelovechestva. Skol'ko lyudey zhilo, zhivet i budet zhit' na Zemle*. Nauka.

Kapitza, S. (1999b). *An essay on the demographic imperative: Non-linear theory of growth of humankind*. The Edwin Mellen Press.

Kapitza, S. (2003). The statistical theory of global population growth. In *Formal descriptions of developing systems* (pp. 11–35). Springer. https://doi.org/10.1007/978-94-010-0064-2_2

Kapitza, S. (2006). *Global population blow-up and after. A report to the Club of Rome*. Global Marshall Plan Initiative.

Kapitza, S. (2010). On the theory of global population growth. *Physics Uspekhi, 53*(12), 1287–1296. https://doi.org/10.3367/UFNe.0180.201012g.1337

Kiss, E. (2020). The present dialogue between nature and society or the mutuality of nature preservation and society protection. *Journal of Globalization Studies, 11*(1), 86–101. https://doi.org/10.30884/jogs/2020.01.06

Kondratieff, N. (1926). Die langen Wellen der Konjunktur. *Archiv fuer Sozialwissenschaft und Sozialpolitik, 56*(3), 573–609.

Kondratieff, N. (1935). The long waves in economic life. *The Review of Economic Statistics, 17*(6), 105–115.

Kondratieff, N. (1984). The long wave cycle. .

Kondratieff, N. D. (2002 [1926]). *Osnovnyye tsikly kon"yunktury i teoriya prognoza*. Economika.

Korotayev, A. (2020). The 21st century Singularity in the Big History perspective. A re-analysis. In A. Korotayev & D. LePoire (Eds.), *The 21st century Singularity and global futures. A Big History perspective* (pp. 19–75). Springer. https://doi.org/10.1007/978-3-030-33730-8_2

Korotayev, A., Bilyuga, S., & Shishkina, A. (2020). Which countries generate Kondratieff waves in global GDP growth rate dynamics in the contemporary world? *Journal of Globalization Studies, 11*(1), 33–63. https://doi.org/10.30884/jogs/2020.01.03

Korotayev, A., Butovskaya, M., Shulghin, S., & Zinkina, J. (2021). Impact of global aging on the global values system. *Vek Globalizatsii, 4*(40), 69–80. https://doi.org/10.30884/vglob/2021.04.05

Korotayev, A., Butovskaya, M., Shulgin, S., & Zinkina, J. (2023c). How can the global ageing affect the global value system? An evolutionary perspective. *Social Evolution & History, 22*(1), 57–76. https://doi.org/10.30884/seh/2023.01.03

Korotayev, A., & de Munck, V. (2013). Advances in development reverse inequality trends. *Journal of Globalization Studies, 4*(1), 105–124.

Korotayev, A., Goldstone, J. A., & Zinkina, J. (2015). Phases of global demographic transition correlate with phases of the Great Divergence and Great Convergence. *Technological Forecasting and Social Change, 95*, 163–169. https://doi.org/10.1016/j.techfore.2015.01.017

Korotayev, A., & Grinin, L. (2012). Kondratieff waves in the world system perspective. *Kondratieff Waves, 1*, 23–64.

Korotayev, A., Malkov, A., & Khaltourina, D. (2006a). *Introduction to Social macrodynamics: Compact macromodels of the world system growth*. KomKniga/URSS.

Korotayev, A., Malkov, A., & Khaltourina, D. (2006b). *Introduction to social macrodynamics: Secular cycles and millennial trends*. KomKniga/URSS.

Korotayev, A., Malkov, S., & Musieva, J. (2023a) Demography. Toward optimization of demographic processes. In V. Sadovnichy et al. (Eds.), *Reconsidering the limits to growth. A report to the Russian Association of the Club of Rome* (pp. 97–116). Springer. https://doi.org/10.1007/978-3-031-34999-7_6

Korotayev, A., Shulgin, S., Ustyuzhanin, V., Zinkina, J., & Grinin, L. (2023b). Modeling social self-organization and historical dynamics. Africa's futures. In V. Sadovnichy et al. (Eds.), *Reconsidering the limits to growth. A report to the Russian Association of the Club of Rome* (pp. 491–519). Springer. https://doi.org/10.1007/978-3-031-34999-7_21

Korotayev, A., & Zinkina, J. (2014). On the structure of the present-day convergence. *Campus-Wide Information Systems, 31*(2–3), 139–152. https://doi.org/10.1108/CWIS-11-2013-0064

Korotayev, A., & Zinkina, J. (2017). Systemic boundary issues in the light of mathematical modeling of world-system evolution. *Journal of Globalization Studies, 8*(1), 78–96.

Korotayev, A., Zinkina, J., & Bogevolnov, J. (2011). Kondratieff waves in global invention activity (1900–2008). *Technological forecasting and social change, 78*(7), 1280–1284. https://doi.org/10.1016/j.techfore.2011.02.011

Korotayev, A., Zinkina, J., Bogevolnov, J., & Malkov, A. (2011). Global unconditional convergence among larger economies after 1998? *Journal of Globalization Studies, 2*(2), 25–62.

Korotayev, A., Zinkina, J., Goldstone, J., & Shulgin, S. (2016). Explaining current fertility dynamics in Tropical Africa from an anthropological perspective: A cross-cultural investigation. *Cross-Cultural Research, 50*(3), 251–280. https://doi.org/10.1177/1069397116644158

LePoire, D., & Korotayev, A. (2020). Conclusion. In A. Korotayev & D. LePoire (Eds.), *The 21st century singularity and global futures. A Big History perspective* (pp. 599–620). Springer. https://doi.org/10.1007/978-3-030-33730-8_27

Lilley, S. (1976). Technological progress and the industrial revolution, 1700–1914. In C. M. Cipolla (Ed.), *The industrial revolution, 1700–1914. The Fontana economic history of Europe* (pp. 187–254). The Harvest Press.

Malkov, S. (2021). Razmyshleniya na temu: Est li budushshee u kaitalizma? *Informatsionnye Voyny, 4*(60), 41–49.

Malkov, S., Grinin, L., Grinin, A., Musieva, J., & Korotayev, A. (2023a). Modeling social self-organization and historical dynamics. Global phase transitions. In V. Sadovnichy et al. (Eds.),

Reconsidering the limits to growth. A report to the Russian Association of the Club of Rome (pp. 387–417). Springer. https://doi.org/10.1007/978-3-031-34999-7_18
Malkov, S., Kovaleva, N., Grinin, L., & Korotayev, A. (2023b). Modeling social self-organization and historical dynamics. Agrarian society. In V. Sadovnichy et al. (Eds.), *Reconsidering the limits to growth. A report to the Russian Association of the Club of Rome* (pp. 309–335). Springer. https://doi.org/10.1007/978-3-031-34999-7_16
Mandel, E. (1980). *Long waves of capitalist development*. Cambridge University Press.
Meadows, D., Meadows, D., Randers, J., & Behrens, W. W. (1972). *The limits to growth. A report to the Club of Rome*. Potomac Associates.
Mensch, G. (1979). *Stalemate in technology – Innovations overcome the depression*. Ballinger.
Modelski, G., Devezas, T., & Thompson, W. R. (Eds.). (2007). *Globalization as evolutionary process: Modeling global change*. Routledge.
Moiseev, N. (1982). *Chelovek, okruzhayushchaya sreda, obshchestvo. Problema formalizatsii opisaniya*. Nauka.
Moiseev, N. (1990). *Chelovek i noosfera*. Molodaya gvardiya.
Moiseev, N. (1996). Limits of predictability for biospheric processes. In Y. A. Kravtsov & J. B. Kadtke (Eds.), *Predictability of complex dynamical systems* (pp. 169–185). Springer. https://doi.org/10.1007/978-3-642-80254-6_10
Moiseev, N. (2022). *How far it is to tomorrow…*. Birkhäuser. https://doi.org/10.1007/978-3-030-96651-5
Moiseev, N., Alexandrov, V., & Tarko, A. (1985). *Chelovek i biosfera: Opyt sistemnogo analiza i eksperimenty s modelyami*. Nauka.
Moustaka, V., Theodosiou, Z., Vakali, A., Kounoudes, A., & Anthopoulos, L. G. (2019). Enhancing social networking in smart cities: Privacy and security borderlines. *Technological Forecasting and Social Change, 142*, 285–300. https://doi.org/10.1016/j.techfore.2018.10.026
Panov, A., LePoire, D., & Korotayev, A. (2020). The 21st century singularity in the Big History perspective. An overview. In A. Korotayev & D. LePoire (Eds.), *The 21st century singularity and global futures. A Big History perspective* (pp. 1–18). Springer. https://doi.org/10.1007/978-3-030-33730-8_1
Philipson, M. (Ed.). (1962). *Automation: Implications for the future*. Vintage.
Pomeranz, K. (2021). *The great divergence. China, Europe, and the making of the modern world economy* (2nd ed.). Princeton University Press. https://doi.org/10.1515/9780691217192
Prigogine, I. (1997). *The end of certainty*. Free Press.
Prigogine, I. (2017). *Non-equilibrium statistical mechanics (Dover Books on Physics)*. Dover Publications.
Prigogine, I., Stengers, I., & Toffler, A. (1984). *Order out of chaos*. Bantam.
Schumpeter, J. A. (1939). *Business cycles*. McGraw-Hill.
Schwartz, P. M. (1999). Internet privacy and the state. *Connecticut Law Review, 32*, 815–829.
Sheffield, J., Grinin, L., & Korotayev, A. (Eds.). (2013). *Globalization: Yesterday, today, and tomorrow*. Emergent Publications.
Solove, D. J. (2008). *Understanding privacy*. Harvard University Press.
Sorokin, P. (1968). *Man and society in calamity – The effects of revolution, war, famine, pestilence upon human mind, behavior, social organization and cultural life*. E.P. Dutton & Co..
Sorokin, P. A. (1960). Mutual convergence of the United States and the USSR to the mixed sociocultural type. *International Journal of Comparative Sociology, 1*(2), 143–176.
Turchin, P., Whitehouse, H., Larson, J., Cioni, E., Reddish, J., Hoyer, D., Savage, P. E., Covey, R. A., Baines, J., et al. (2022a). Big Gods and big science: Further reflections on theory, data, and analysis. *Religion, Brain & Behavior*, 1–14. https://doi.org/10.1080/2153599X.2022.2065354
Turchin, P., Whitehouse, H., Larson, J., Cioni, E., Reddish, J., Hoyer, D., Savage, P. E., Covey, R. A., Baines, J., et al. (2022b). Explaining the rise of moralizing religions: A test of competing hypotheses using the Seshat Databank. *Religion, Brain & Behavior*, 1–28. https://doi.org/10.1080/2153599X.2022.2065345

Tylecote, A. (1992). *The long wave in the world economy*. Routledge.
Vernadsky, V. (1998). *The biosphere*. Copernicus.
Vernadsky, V. (2012). *Biosphere and noosphere*. Iris-Press.
Vernadsky, V. (2013). *Collected works: In 24 volumes*. Nauka.
Vernadsky, V. (2015). *The study of life and the new physics*. 21st Century Science & Technology.
Vernadsky, V. (2018). *Geochemistry and the biosphere: Essays*. Synergetic Press.
von Weizsäcker, E. U., & Wijkman, A. (2018). *Come on! Capitalism, short-termism, population and the destruction of the planet. A report to the Club of Rome*. Springer.
Wallerstein, I. (1984). Economic cycles and socialist policies. *Futures, 16*(6), 579–585.
Westin, A. F. (1966). Science, privacy, and freedom: Issues and proposals for the 1970s. Part I. The current impact of surveillance on privacy. *Columbia Law Review, 66*(6), 1003–1050.
Yakovets, Y., & Farah, S. (2014). *Dialogue and partnership of civilizations*. International Institute of Pitirim Sorokin-Nikolay Kondratieff.
Zinkina, J., Christian, D., Grinin, L., Ilyin, I., Andreev, A., Aleshkovski, I., Shulgin, S., & Korotayev, A. (2019). *A Big History of globalization: The emergence of a global world system*. Springer. https://doi.org/10.1007/978-3-030-05707-7

Askar Akaev is the First President of the Kyrgyz Republic and Foreign Full Member of the Russian Academy of Sciences. He works as Chief Research Professor in the Prigogine Institute for Mathematical Investigations of Complex Systems at Lomonosov Moscow State University. He is also the Academic Supervisor of the Centre for Stability and Risk Analysis at HSE University, Moscow, Russia. Askar Akaev has authored and co-authored over 300 scientific publications, including such monographs as *Coherent Optical Computers* (1977), *Optical Methods of Information Processing* (1983), *Holographic Memory* (1996), *Selected Lectures on Optical Computers* (1996), *Economic Transition from the Point of View of a Physicist. A Mathematical Model of Emerging Economy* (2000), *Modeling and Forecasting the World Dynamics* (2012), *Quality of Education, Effectiveness of R&D, and Economic Growth. Quantitative Analysis and Mathematical Modeling* (2016), etc.

Sergey Malkov heads the Center for Long-Term Forecasting and Strategic Planning at Lomonosov Moscow State University; he also works as a Senior Research Professor at the Institute of Economics of the Russian Academy of Sciences as well as at the Centre for Stability and Risk Analysis, HSE University, Moscow. He is the author of over 300 scholarly publications, including such monographs as *Social self-organization and historical process: possibilities of mathematical modeling* (URSS, 2009), *Modeling and forecasting of world dynamics* (ISPI RAS, 2012, together with Victor Sadovnichy and others), *A new theory of the reproduction of capital: development and practical application* (Nestor-Istoriya, 2016, together with Vladimir Mayevsky and others), *The 21st Century Singularity and Global Futures. A Big History Perspective* (Springer, 2020, together with David J. LePoire and others). In 2022 he was awarded with the Kondratieff Award by the Russian Academy of Sciences.

Leonid Grinin is a Senior Research Professor at the Centre for Stability and Risk Analysis at HSE University in Moscow, as well as Senior Research Professor of the Institute of Oriental Studies, Russian Academy of Sciences. He is co-editor of the international journals *Social Evolution & History* and the *Journal of Globalization Studies*, as well as co-editor of the international almanacs *Evolution, History & Mathematics*, and *Kondratieff Waves*. He is the author of more than 600 publications, including 33 monographs in Russian, English, and Chinese. His current research focuses on comparative political studies, the theory of revolution, political anthropology, global economy, global history, historical sociology, and futurology. Among his monographs are *The Cybernetic Revolution and the Forthcoming Epoch of Self-Regulating Systems* (Uchitel, 2016, with Anton Grinin), *Big History of Globalization. The Emergence of a Global World System* (Springer 2019, in co-authorship with Julia Zinkina, David Christian, and others), *Islamism, Arab Spring, and* *the Future of Democracy. World System and World Values Perspectives* (Springer, 2019; with Andrey Korotayev and Arno Tausch). In 2004, he was awarded as a winner of the best publication of *Socis* journal during the year for the article "People of Publicity – a New Social Layer?". In 2012 he was awarded with the Gold Kondratieff Medal by the International N. D. Kondratieff Foundation. In 2016 he received the Belyaev Award for the book *From Bifaces to Nanorobots* (the book published in Russian in co-authorship with Anton Grinin).

Stanislav Bilyuga has a Ph.D. degree in Political Sciences and is affiliated to the Faculty of Global Studies, Lomonosov Moscow State University. His research interests include modeling global trends and dynamics and forecasts of destabilization. Stanislav Bilyuga has authored and co-authored over 70 scientific publications concerning factors, trends, indices, and methods of research on sociopolitical destabilization of various countries.

Olga Davydova is a specialist in data analysis, development of mathematical models, implementation of models in Python, identification of parameters, and verification of models. She has co-authored more than 15 academic articles, including such papers as "Resistance to Socio-Political Instability as an Indicator of the Country's Successful Development" (2015), "Model for Assessing and Forecasting the Development of Human Capital" (2016), "Forecast of the Dynamics of World Imports by Commodity Groups" (2018), "The Paris Agreement on Climate Is Coming into Force: Will the Great Energy Transition Take Place?" (2020).

Anton Grinin is a Russian scientist specializing in the study of modern technological trends and future studies, as well as an evolutionist and philosopher. He has a Ph.D. in Biology and works as a Research Fellow for the Centre for Stability and Risk Analysis Laboratory at HSE University in Moscow and for Lomonosov Moscow State University, Russia. He is the author of more than 80 publications in Russian and English, including 2 monographs. Awards: Gold Kondratieff Young Scholars Medal, Alexander Belyaev Literature Award. Anton Grinin focuses on the issues of the technological revolution (the Cybernetic Revolution) evolving in the current decades of the twenty-first century and the resultant profound transformations in economy and society as well as related economic, political, ethic-legal, and other risks, including political revolutions. He also investigates the correlation between the Cybernetic Revolution and Kondratieff waves as well as between the Cybernetic Revolution and global ageing.

Natalia Kovaleva heads the Laboratory of Ecological Soil Science of the Faculty of Soil Science of Lomonosov Moscow State University. She is also a visiting professor at Kazan Federal (Volga Region) University. Natalia Kovaleva has authored and co-authored over 700 scientific publications including such monographs as *Recreational territories and ecological well-being of the population* (1996), *Soil science: history, sociology, methodology* (2005), *Soil in the modern world* (2008), *Soils in the biosphere and human life: monograph (2012), Protected areas – the basis of environmental safety* (2018), *Soil biomarkers* (2020).

Artemy Malkov is a Russian mathematician, specialist in mathematical modeling of socio-historical processes, spatial historical dynamics, lecturer at the Moscow Institute of Physics and Technology. He is the author or co-author of more than 100 scientific publications, including such monographs as *Introduction to Social Macrodynamics: Compact Macromodels of the World System Growth* (KomKniga/URSS, 2006; with Andrey Korotaev and Daria Khaltourina), *Introduction to Social Macrodynamics: Secular Cycles and Millennial Trends* (KomKniga/URSS, 2006; with Andrey Korotayev and Daria Khaltourina), *Laws of History: Mathematical Modeling of the Development of the World System. Demography, Economics, Culture* (LENAND/URSS, 2019, in Russian; with Andrey Korotayev and Daria Khaltourina).

Jameelah Musieva is a Research Assistant at Lomonosov Moscow State University, Faculty of Global Studies, and Research Fellow at the Centre for Stability and Risk Analysis at HSE University in Moscow. She is the author and co-author of more than 10 scholarly publications, including such articles as "The Concept of 'Quality of Life'" (2019), "Methodology for Cross-Country Assessment of The Quality of Life Based on The LQI Index" (2019), "A Cross-Country Analysis of The Quality of Life" (2020), "Quality of Life in The BRICS Countries Based on The LQI Index" (2021).

Andrey Korotayev heads the the Centre for Stability and Risk Analysis at HSE University in Moscow, Russia. He is also a Senior Research Professor at the Institute of Oriental Studies and Institute for African Studies, Russian Academy of Sciences, as well as at the Faculty of Global Studies of Lomonosov Moscow State University and at the Russian Presidential Academy of National Economy and Public Administration in Moscow. He has authored or co-authored over 650 scholarly publications, including such monographs as *Ancient Yemen* (Oxford University Press, 1995), *World Religions and Social Evolution of the Old World Oikumene Civilizations: A Cross-Cultural Perspective* (The Edwin Mellen Press, 2004), *Introduction to Social Macrodynamics: Compact Macromodels of the World System Growth* (URSS, 2006; with Daria Khaltourina and Artemy Malkov), *Introduction to Social Macrodynamics: Secular Cycles and Millennial Trends* (URSS, 2006; with Daria Khaltourina and Artemy Malkov), *Great Divergence and Great Convergence. A Global Perspective* (Springer, 2015; with Leonid Grinin), *Economic Cycles, Crises, and the Global Periphery* (Springer, 2016; with Leonid Grinin and Arno Tausch), *Islamism, Arab Spring, and the Future of Democracy. World System and World Values Perspectives* (Springer, 2019; with Leonid Grinin and Arno Tausch). He is a laureate of the Russian Science Support Foundation Award in 'The Best Economists of the Russian Academy of Sciences' Nomination (2006); in 2012, he was awarded with the Gold Kondratieff Medal by the International N. D. Kondratieff Foundation.

Modeling Social Self-Organization and Historical Dynamics: Africa's Futures

Andrey Korotayev, Sergey Shulgin, Vadim Ustyuzhanin, Julia Zinkina, and Leonid Grinin

Abstract

This chapter presents forecasts for the emergence of large-scale political and demographic collapses and for the economic growth of some countries in Sub-Saharan Africa (SSA), where the likelihood of armed civil conflicts and population impoverishment is the highest in the coming decades. Korotayev et al. apply several advanced mathematical models: (1) to forecast the risks of armed conflict, where population, median age, and education are the main explanatory

A. Korotayev (✉)
HSE University, Moscow, Russia

Institute for African Studies, Russian Academy of Sciences, Moscow, Russia

Lomonosov Moscow State University, Moscow, Russia

S. Shulgin
Russian Presidential Academy of National Economy and Public Administration (RANEPA), Moscow, Russia
e-mail: sergey@shulgin.ru

V. Ustyuzhanin
HSE University, Moscow, Russia

Russian Presidential Academy of National Economy and Public Administration (RANEPA), Moscow, Russia
e-mail: vvustiuzhanin@hse.ru

J. Zinkina
Russian Presidential Academy of National Economy and Public Administration (RANEPA), Moscow, Russia

Lomonosov Moscow State University, Moscow, Russia

L. Grinin
HSE University, Moscow, Russia

Institute of Oriental Studies, Russian Academy of Sciences, Moscow, Russia

V. Sadovnichy et al. (eds.), *Reconsidering the Limits to Growth*, World-Systems Evolution and Global Futures, https://doi.org/10.1007/978-3-031-34999-7_20

factors; and (2) to forecast economic growth, which is a function of the same variables and risks of large-scale armed civil conflicts. It is important to note that mathematical models consider the interaction of explanatory factors with each other, thereby creating feedback effects. Using these methods, Korotayev et al. calculate three possible scenarios for the development of each of the countries under consideration in the twenty-first century: (1) a pessimistic one, (2) an inertial one, and (3) an optimistic one assuming the achievement of sustainable development goals (SDGs) by 2030. The modeling results suggest that the Sahel could become the most disadvantaged region. The four countries of this region are characterized by: (1) a negligible difference between the inertial scenario and the pessimistic scenario; (2) extremely high risks of full-scale civil wars in the near future; and (3) reaching the level of middle-income countries only by the end of this century, even under the most optimistic scenario. Korotayev et al. conclude that the main way of mitigating the risks of sociodemographic collapses is rapid progress toward achieving the SDGs in the very near future, which seems impossible without adequate support from the world community.

1 Modeling the Sociodemographic Futures of Africa

1.1 Modeling the Sociodemographic Futures of Africa. General View

Currently, Tropical Africa is still characterized by very high fertility rates, which distinguish it from all the other regions of the world (see Fig. 1, see also Zinkina & Korotayev, 2014a, 2014b; Korotayev, Zinkina, et al., 2016; Nzimande &

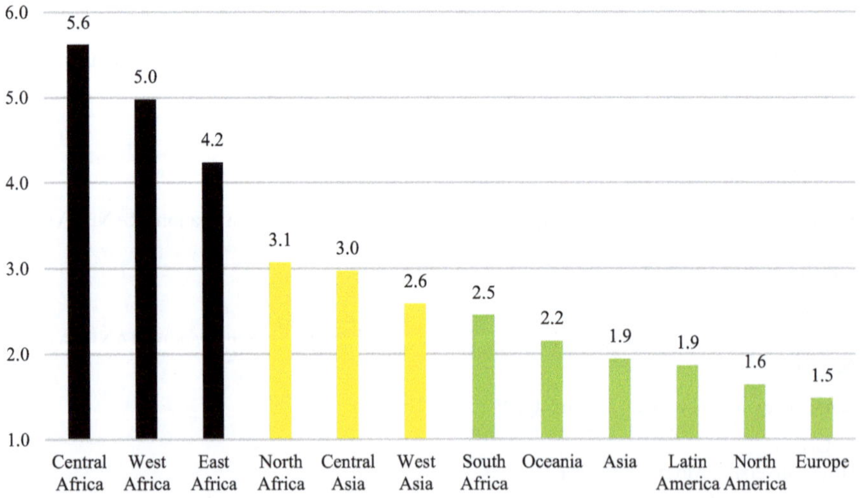

Fig. 1 Total fertility rates in various regions of the world in 2021. *Data source* UN Population Division database (2022)

Mugwendere, 2018; Kebede et al., 2019; Schoumaker, 2019; May & Rotenberg, 2020).

Given the very slow decline in fertility rates against the background of rather remarkable decline in mortality rates, it is hardly surprising that even the medium UN population forecasts imply impressive population growth in this region. For example, according to medium UN forecasts, the population of Kenya will catch up with the population of Russia by the end of this century, and the population of Uganda will even surpass it (see Fig. 2). The population of Tanzania will exceed the population of Russia by the middle of this century, and the population of Ethiopia will do this by the 2030s; by the late twenty-first century, Tanzania and Ethiopia are projected to overtake Russia in terms of population size by almost two and three times correspondingly.

The easiest way to compare the expected populations of two countries is to compare the sizes of child age groups within the structures of their current populations. Thus, Ethiopia can be expected to overtake Russia in terms of population size very soon, as the number of under-5 kids in Russia is currently less than 8 million, while Ethiopia has nearly 18 million of under-5 children (see Fig. 3).

The largest increase in population is forecasted for Nigeria; by 2100, its population is projected to be almost the same as the population of the whole of Europe (including Russia), rocketing to reach more than 540 million people (see Fig. 2). Figure 4 presents the population projections for Sub-Saharan countries with the largest expected population increase in absolute terms that can result in catastrophic consequences for the development of these countries (see, e.g., Abate, 2023). Projections of the population of Russia are presented for the sake of comparison.

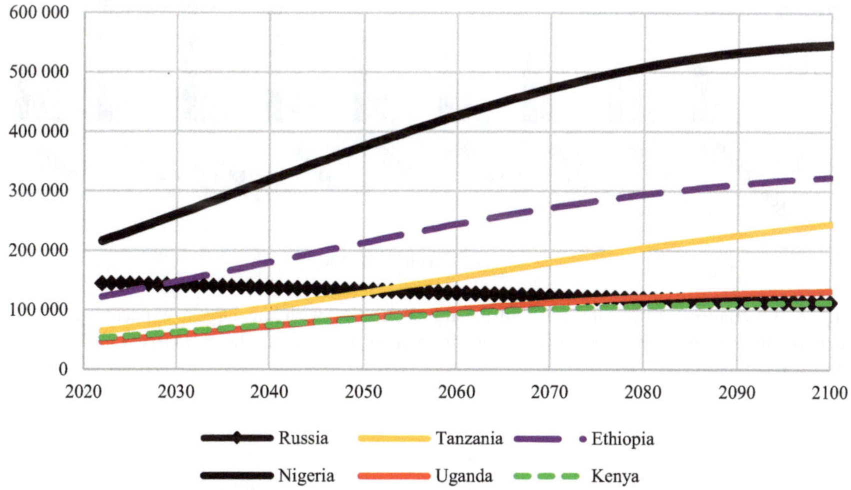

Fig. 2 Population projections for some African countries in comparison with Russia, UN forecasts from 2020 to 2100 (thousands). *Data source* UN Population Division database (2022)

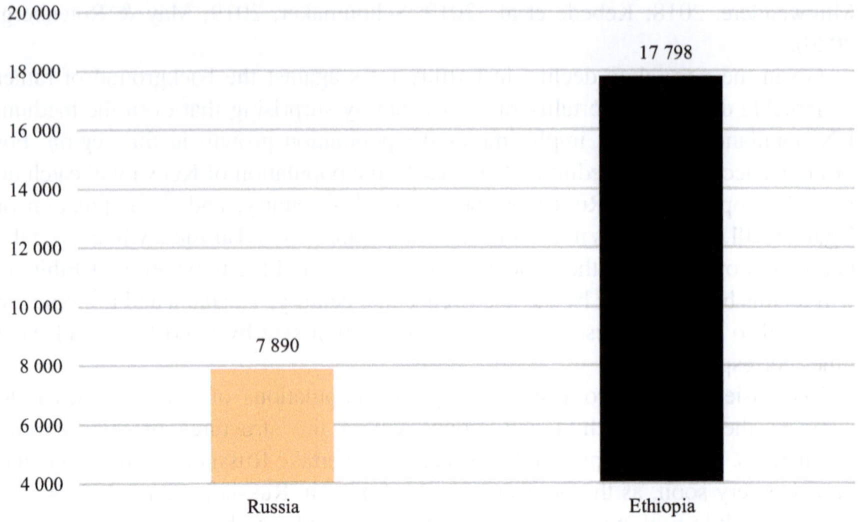

Fig. 3 The number of under-5 children in Russia and Ethiopia in 2021 (thousands). *Data source* UN Population Division database (2022)

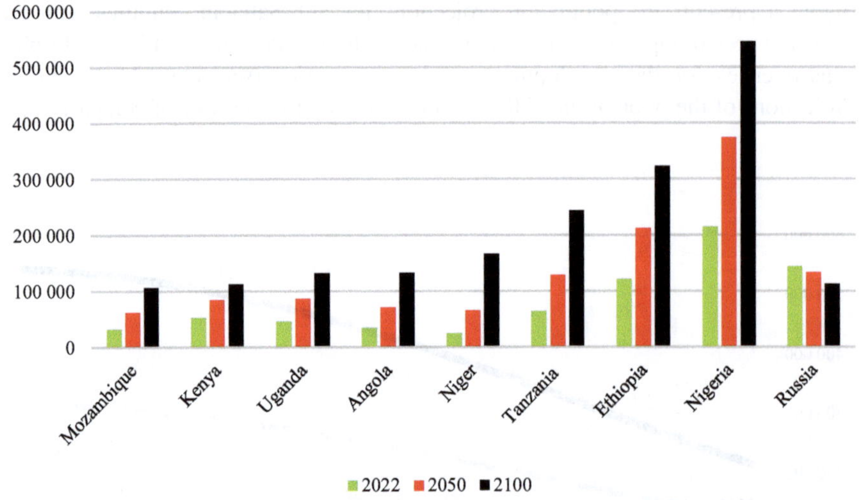

Fig. 4 Population projections for Sub-Saharan countries with the largest expected population increase in absolute terms and population projections for Russia. *Data source* UN Population Division database (2022)

Below, we consider several scenarios of the sociodemographic development of Tropical African countries, corresponding to different scenarios for the course of the modernization processes in this part of our planet. To model the demographic future of African countries, a mathematical model with the following cognitive scheme is used (see Fig. 5).

Fig. 5 Cognitive scheme of the influence of various factors on economic growth and their interaction with each other

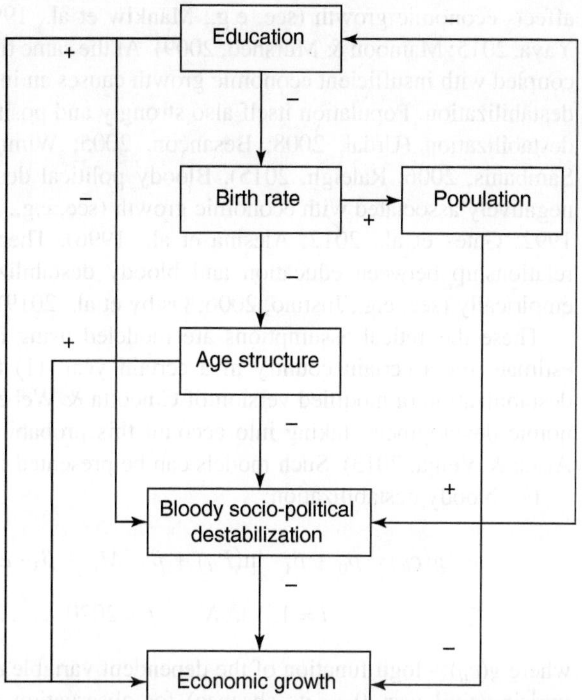

The expansion of formal education in the low-income and middle-income countries leads to a significant decrease in the birth rate (see, e.g., Zinkina & Korotayev, 2017; Kebede et al., 2019). In turn, fertility decline leads to an increase in the share of the working-age groups in the total population, creating the effect of the so-called "demographic bonus" (see, e.g., Bloom et al., 2003; Bloom & Williamson, 1998; Bloom et al., 2000; Omoruyi, 2021; Korotayev, Shulgin, et al., 2022c). Fertility decrease brings down the youth bulges and leads to population aging (an increase in the median age), which, in turn, leads to a decrease in the risks of bloody sociopolitical destabilization.[1] The spread of education leads to a reduction in the risks of bloody sociopolitical destabilization[2] and directly and positively

[1] See, e.g., Moller (1968), Choucri (1974), Mesquida and Wiener (1996, 1999), Goldstone (1991, 2001, 2002), Urdal (2004, 2006, 2008), Staveteig (2005), Korotayev et al. (2011, 2014), Goldstone et al. (2012), Farzanegan and Witthuhn (2017), Weber (2019), Cincotta and Weber (2021), Romanov et al. (2021), Korotayev, Sawyer, Gladyshev, et al. (2021); Korotayev, Romanov, et al. (2023), Sawyer et al. (2022).

[2] The diffusion of formal education in low- and middle-income countries, at the same time, leads to an increase in the risks of unarmed destabilization (e.g., Korotayev, Bilyuga, & Shishkina, 2017b, 2018; Korotayev et al., 2020; Korotayev, Sawyer, & Romanov, 2021; Sawyer & Korotayev, 2022) and, at the same time, to a decrease in the risks of armed bloody revolutions/civil wars (Urdal, 2008; Østby et al., 2019; Ustyuzhanin et al., 2022). However, it is bloody revolutions/civil wars that have a really strong negative impact on economic growth. Thus, education reduces namely those risks

affects economic growth (see, e.g., Mankiw et al., 1992; Lucas, 2002; Bonnal & Yaya, 2015; Mamoon & Murshed, 2009). At the same time, rapid population growth coupled with insufficient economic growth causes an increase in the risks of bloody destabilization. Population itself also strongly and positively influences the risks of destabilization (Urdal, 2008; Besançon, 2005; Wimmer et al., 2009; Hegre & Sambanis, 2006; Raleigh, 2015). Bloody political destabilization is strongly and negatively associated with economic growth (see, e.g., Aisen & Veiga, 2013; Fosu, 1992; Gates et al., 2012; Alesina et al., 1996). There is also a strong negative relationship between education and bloody destabilizations, which is supported empirically (see, e.g., Justino, 2006; Østby et al., 2019).

These theoretical assumptions are modeled using mathematical equations that estimate for a certain country in a certain year: (1) the probability of a bloody destabilization (a modified version of Cincotta & Weber, 2021 model) and (2) economic development, taking into account this probability (based on the model by Aisen & Veiga, 2013). Such models can be presented in the following form:

For bloody destabilization:

$$g(c_{it}) = \beta_0 + \beta_1 \cdot \ln(P_{it}) + \beta_2 \cdot M_{it} + \beta_3 \cdot E_{it} + \beta_4 \cdot \mu_t + \varepsilon_{it}$$

$$i = 1, \ldots, N \qquad t = 2020, \ldots, 2100$$

where $g(c_{it})$ – logit function of the dependent variable c_{it} (1 = armed revolutionary uprising/civil war, 0 = its absence) for observation i at time t; P_{it}, M_{it} *u* E_{it} – explaining variables of i observation at time t (population, median age, and education, respectively); β_{it} – estimated coefficients at the explanatory variables; μ_t – specific time effect; ε_{it} – the random errors of each observation i at time t.

To calculate the probability from the estimated $g(c_{it})$ it is necessary to transform it into exponential form. Then the equation of the estimated probability of event c_{it} will be:

$$P(c_{it} = 1) = \frac{e^{\beta_0 + \beta_1 \cdot \ln(P_{it}) + \beta_2 \cdot M_{it} + \beta_3 \cdot E_{it} + \mu_t + \varepsilon_{it}}}{1 + e^{\beta_0 + \beta_1 \cdot \ln(P_{it}) + \beta_2 \cdot M_{it} + \beta_3 \cdot E_{it} + \mu_t + \varepsilon_{it}}} \text{ or } \frac{e^{g(c_{it})}}{1 + e^{g(c_{it})}}$$

$$i = 1, \ldots, N \qquad t = 2020, \ldots, 2100$$

To evaluate economic growth, it makes sense to first model the overall level of economic development:

$$\ln(y_{it}) = \gamma_0 + \gamma_2 \cdot \ln(y_{i,t-1}) + \gamma_3 \cdot \Delta P_{it} + \gamma_4 \cdot M_{it} + \gamma_5 \cdot E_{it} + \varphi\left(\frac{e^{g(c_{it})}}{1 + e^{g(c_{it})}}\right) + \varepsilon_{it}$$

---

that negatively affect economic growth. Therefore, the impact of the spread of formal education through this channel on economic growth is rather positive, which is also confirmed in models of neoclassical economic growth.

$$i = 1, \ldots, N \qquad t = 2020, \ldots, 2100$$

where y_{it} – GDP per capita of observation i at time t; β_i – estimated coefficients at the explanatory variables; $y_{i,\ t\ -\ 1}$ – GDP per capita of observation i at time t-1 (lagged); $\Delta P_{it} = \frac{P_{it} - P_{i,t-1}}{P_{i,t-1}}$ – population delta; where P_{it} – population of observation i at time t, and $P_{i,\ t\ -\ 1}$ – population of observation i at time $t - 1$ (lagged); M_{it} – median age of observation i at time t; E_{it} – education of observation i at time t;ε_{it} – the random errors of each observation i at time t;$\varphi\left(\frac{e^{g(c_{it})}}{1+e^{g(c_{it})}}\right)$ – probability function of bloody destabilization of observation i at time t, which has the following form:

$$\varphi\left(\frac{e^{g(c_{it})}}{1+e^{g(c_{it})}}\right) = \begin{cases} \gamma_4 \cdot P(c_{it}), & P(c_{it}) \leq 0.25 \\ 2\gamma_4 \cdot P(c_{it}), & P(c_{it}) > 0.25 \text{ and } P(c_{it}) \leq 0.5 \\ 3\gamma_4 \cdot P(c_{it}), & P(c_{it}) > 0.5 \text{ and } P(c_{it}) \leq 0.75 \\ 4,5\gamma_4 \cdot P(c_{it}), & P(c_{it}) > 0.75 \end{cases}$$

$$i = 1, \ldots, N \qquad t = 2020, \ldots, 2100$$

where $P(c_{it})$ – estimated probability of bloody destabilization for observation i at time t.

Then economic growth can be modeled by following equitation:

$$\text{growth}_{it} = \frac{y_{it} - y_{i,\ t-1}}{y_{i,\ t-1}} \cdot 100\%$$

$$i = 1, \ldots, N \qquad t = 2020, \ldots, 2100$$

Based on this, the following scenarios are modeled:

1. The inertial scenario is based on the assumption that the trends of recent years will continue at the same pace in the respective countries (and in Sub-Saharan Africa [SSA] in general).
2. Pessimistic scenario of delayed modernization in SSA cannot be ruled out as there have already been precedents when, after a period of fairly rapid modernization (fertility decline proceeding, formal education expanding rapidly), many countries of the region experienced a noticeable slowdown. Thus, attempts to save on education within the framework of structural adjustment programs in the 1990s caused a delay in its spread that, in turn, led to remarkable slowdowns or, not infrequently, complete stalls of fertility decline (Schoumaker, 2019; Zinkina & Korotayev, 2017; Kebede et al., 2019).
3. Scenario of full achievement of the sustainable development goals (SDGs). When analyzing the impact of achieving the SDGs on fertility decline in Sub-Saharan countries, we rely on the calculations of the Institute of Health Metrics and Evaluation (IHME) (Vollset et al., 2020).

In terms of demographic characteristics, our inertial scenario corresponds to the reference scenario, and the pessimistic scenario corresponds to the scenario of slow achievement of the SDGs by the Institute for Health Metrics and Assessment (Vollset et al., 2020). These scenarios produce the following projections of population dynamics in Sub-Saharan Africa (see Fig. 6).

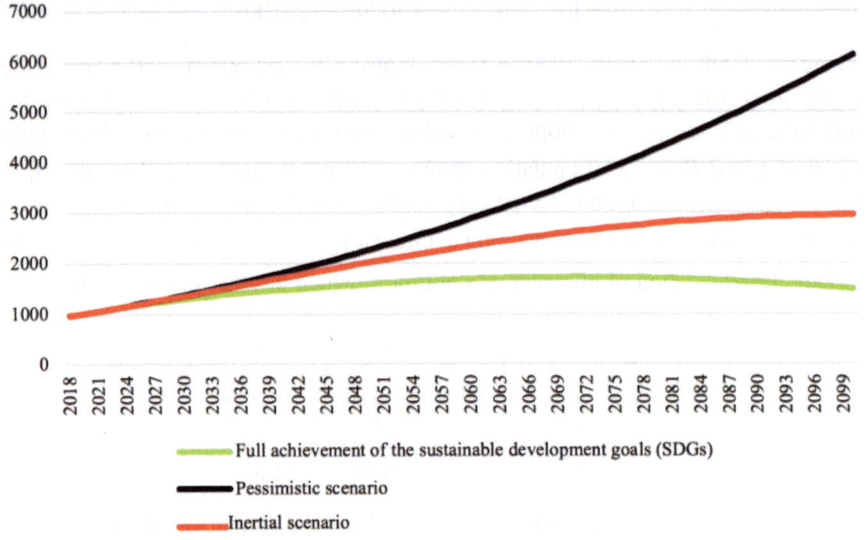

Fig. 6 Population growth scenarios (in millions) in Sub-Saharan Africa in the twenty-first century

Note that the pessimistic scenario (the upper black line in Fig. 6) is rather speculative. This scenario is fraught with the highest risks of sociodemographic and political-economic collapses in the not-too-distant future. Inertial scenario, though less threatening, also bears rather serious risks of collapses, although in the noticeably more distant future. The only scenario that can secure stable development of the SSA countries withstanding the risks of sociodemographic and political-economic collapses is the scenario of a fairly rapid and full achievement of the SDGs.[3]

1.2 Modeling the Sociodemographic Futures of Africa. Country Forecast Scenarios

Let us now consider the results of modeling the demographic, economic, and political future of some of the most important countries in Tropical Africa. We will start by presenting our results for **Angola**. To begin with, let us consider scenarios for the dynamics of the risks of armed revolutionary destabilization/civil wars (Fig. 7).

[3]Taking into account the fact that under any demographic scenario these countries follow the path of modernization, and this period is always more prone to processes of destabilization and revolutions, as well as due to immature statehood, the growth of nationalism (often in the form of ethnic tribalism), the risks of sociopolitical destabilization in SSA countries remain very high (Korotayev et al., 2011; Bondarenko, 2021; Goldstone et al., 2022; Grinin, 2022b; Grinin & Grinin, 2022).

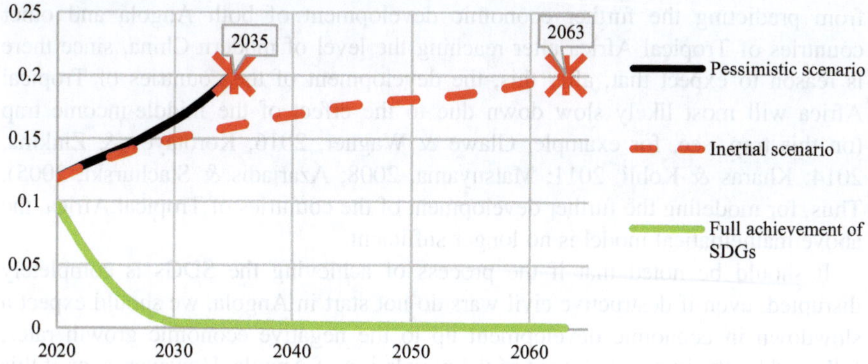

Fig. 7 Scenarios of risk dynamics of armed revolutionary destabilization/civil wars in Angola

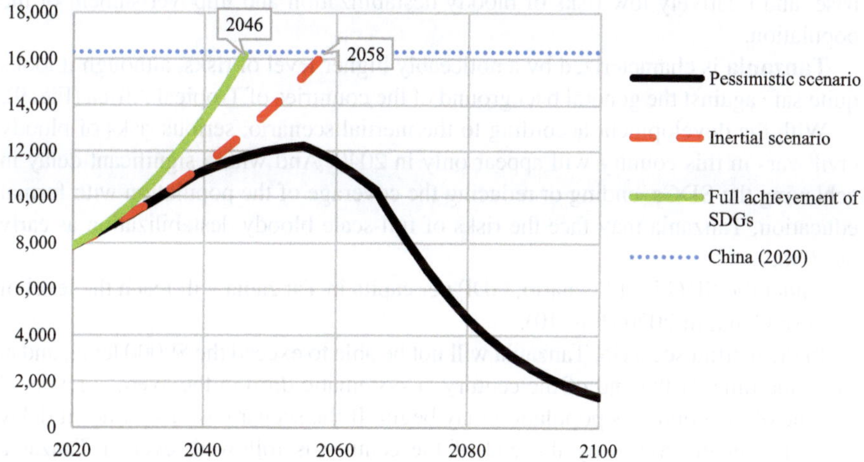

Fig. 8 Scenarios for GDP per capita dynamics in Angola. Here and throughout GDP per capita is estimated in constant 2011 international dollars at purchasing power parities

As will become clear from comparison with other analyzed countries of Tropical Africa, Angola is characterized by rather low risks of the outbreak of large-scale civil wars. Even if the inertial scenario is followed, noticeable risks of catastrophic civil wars in the country will appear only in the second half of this century. Only a significant delay in moving toward the achievement of the SDGs (primarily in the spread of modern education) can lead to the emergence of such risks in the next decade (a pessimistic scenario). At the same time, Angola will be able to reach the level of modern China in per capita GDP by 2058, even if it follows the inertial scenario (see Fig. 8).

However, if Angola follows the scenario of achieving the SDGs, it will be able to reach the level of modern China even before 2050. It should be noted that we refrain from predicting the further economic development of both Angola and other countries of Tropical Africa after reaching the level of modern China, since there is reason to expect that, after this, the development of the countries of Tropical Africa will most likely slow down due to the effect of the middle-income trap (on this trap, see, for example, Glawe & Wagner, 2016; Korotayev & Zinkina, 2014; Kharas & Kohli, 2011; Matsuyama, 2008; Azariadis & Stachurski, 2005). Thus, for modeling the further development of the countries of Tropical Africa, the above mathematical model is no longer sufficient.

It should be noted that if the process of achieving the SDGs is completely disrupted, even if destructive civil wars do not start in Angola, we should expect a slowdown in economic development up to the negative economic growth rates, followed by the impoverishment of the population of Angola. However, even if this scenario is followed, per capita GDP in Angola will drop to only $1,350 by 2100, which is still noticeably higher than this indicator, for example, in modern Niger ($985). This shows that Angola has a sufficient margin of safety, a good economic base, and relatively low risks of bloody destabilization and impoverishment of the population.

Tanzania is characterized by a noticeably higher level of risks, although it looks quite safe against the general background of the countries of Tropical Africa (Fig. 9).

With the development according to the inertial scenario, serious risks of bloody civil wars in this country will appear only in 2040. And with a significant delay in achieving the SDGs, ending or reducing the coverage of the population with formal education, Tanzania may face the risks of full-scale bloody destabilization as early as 2030.

Under the SDG Fast Scenario, GDP per capita in Tanzania will reach the level of modern China in 2070 (Fig. 10).

In the inertial scenario, Tanzania will not be able to exceed the $9000 level, and at the same time, at the end of the century, a systematic drop in the average levels of income of the country's population may begin. If the scenario of a significant delay in achieving the SDGs by the end of the century is followed, even if Tanzania

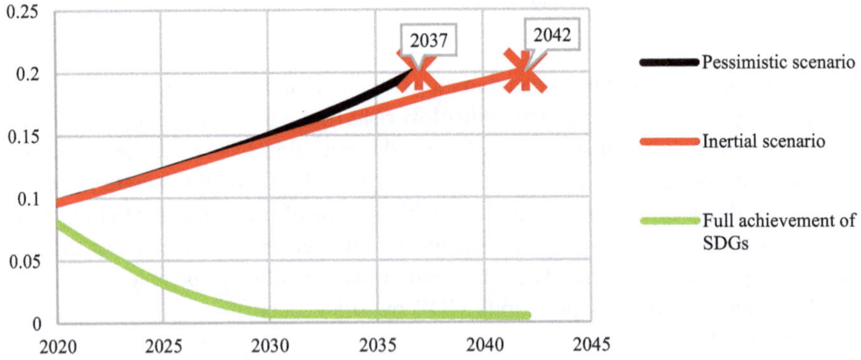

Fig. 9 Scenarios of risk dynamics of armed revolutionary destabilization/civil wars in Tanzania

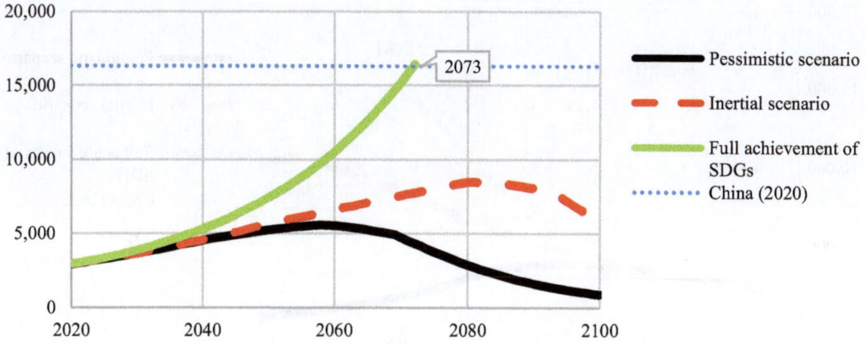

Fig. 10 GDP per capita scenarios in Tanzania

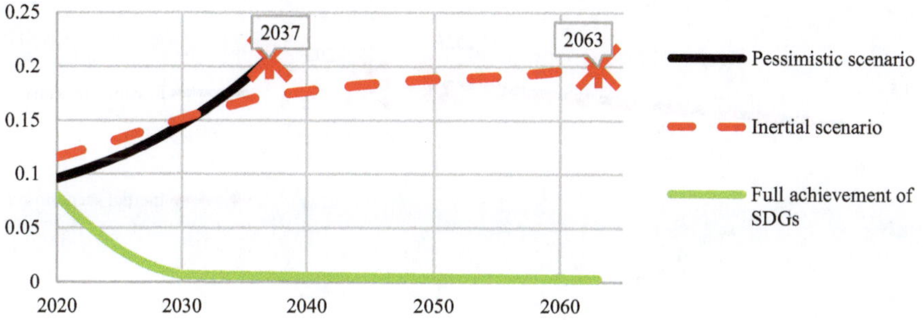

Fig. 11 Scenarios of risk dynamics of armed revolutionary destabilization/civil wars in Uganda

manages to avoid a large-scale political and demographic collapse, there will be a catastrophic impoverishment of the country's population with a drop in per capita GDP below $1000, which is less than the initial values in 2020.

Fairly similar scenarios can be traced for Uganda, where the risks of a full-scale civil war with development according to the inertial scenario are expected in the second half of this century, and with a significant delay in achieving the SDGs, in the 2030s (Fig. 11).

Uganda also reaches the level of the PRC in terms of per capita GDP under the scenario of rapid achievement of the SDGs in the second half of this century. It performs slightly better under the inertial scenario than Tanzania, but under a particularly significant SDG delay scenario, it is projected to be even more impoverished, with GDP per capita falling below $500 (see Figure 12).

Ethiopia is characterized by extremely high risks of armed revolutionary destabilization, full-scale civil wars and political and demographic collapses in the very coming years. At the same time, such risks are extremely high even in the case of development according to the inertial scenario. To prevent them, a noticeable acceleration in the achievement of the SDGs is necessary (see Fig. 13).

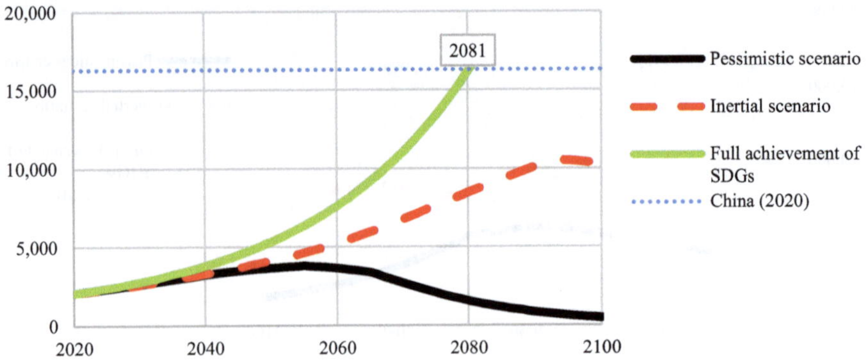

Fig. 12 Scenarios for GDP per capita in Uganda

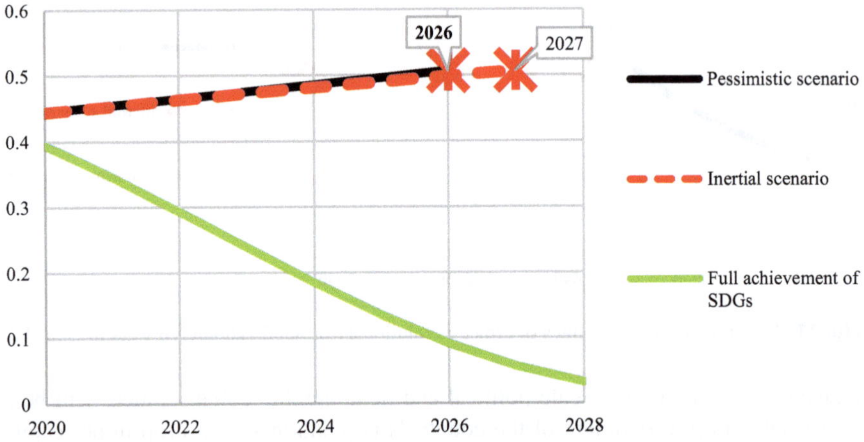

Fig. 13 Scenarios of risk dynamics of armed revolutionary destabilization/civil wars in Ethiopia

Accordingly, even the development according to the inertial scenario assumes a significant impoverishment of the Ethiopian population in the coming decades. However, an accelerated path to achieving the SDGs could allow Ethiopia to reach the level of modern China in the second half of this century (see Fig. 14).

Somewhat lower, but still very high, are the risks of political and demographic collapses in **Nigeria** as well (see Fig. 15).

However, Nigeria's economic prospect looks somewhat less ominous. Per capita GDP growth in Nigeria may continue until the middle of the century (albeit at a rather slow pace) even under the development according to the inertial scenario. In the case of development according to the optimistic scenario, Nigeria may approach the level of modern China by the middle of this century (see Fig. 16).

The risks of new full-scale civil wars are also extremely high in the **Democratic Republic of the Congo** (see Fig. 17).

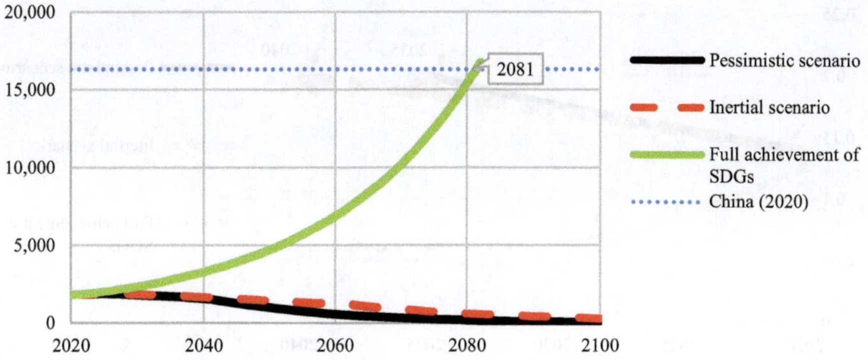

Fig. 14 GDP per capita scenarios in Ethiopia

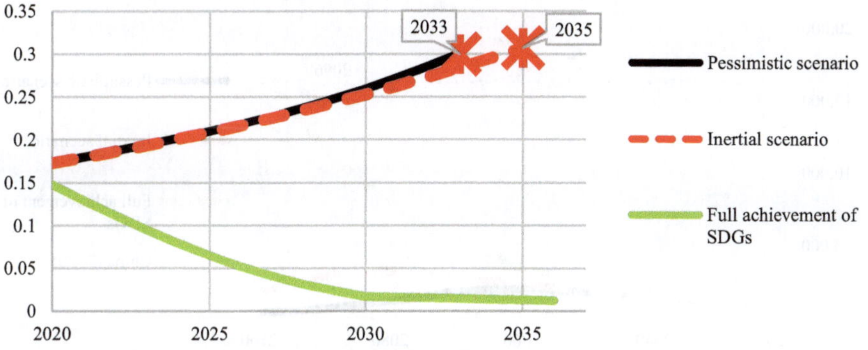

Fig. 15 Scenarios of risk dynamics of armed revolutionary destabilization/civil wars in Nigeria

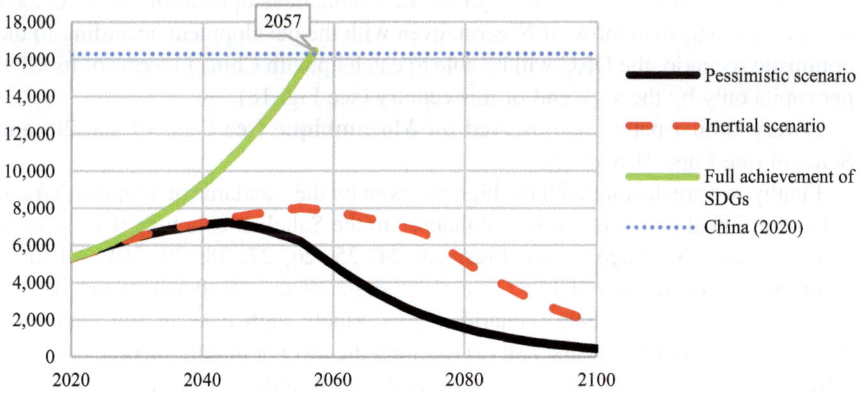

Fig. 16 GDP per capita scenarios in Nigeria

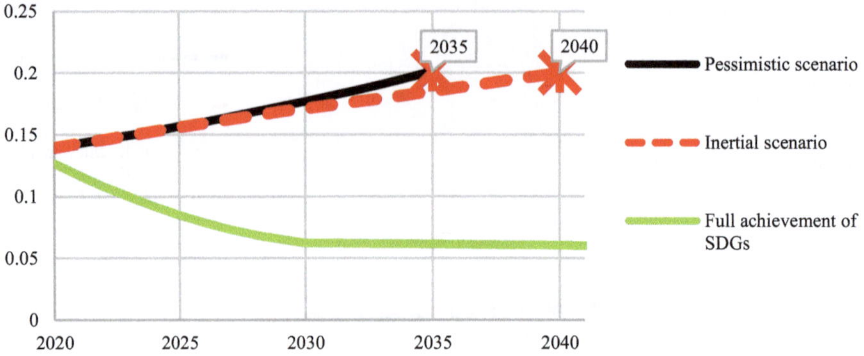

Fig. 17 Scenarios of risk dynamics of armed revolutionary destabilization/civil wars in the DRC

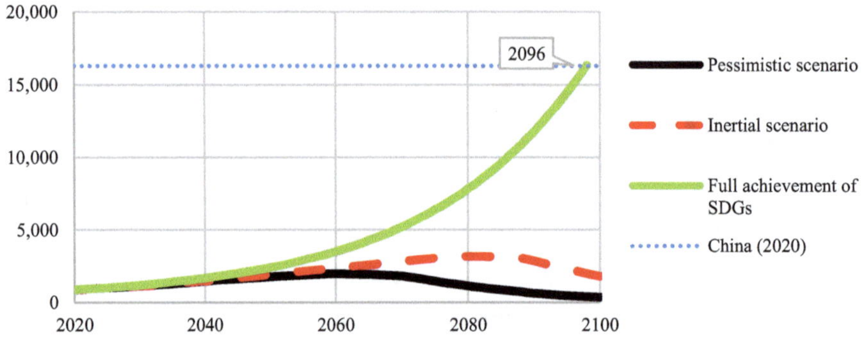

Fig. 18 GDP per capita scenarios in DR Congo

At the same time, the prospects for the economic development of the DRC look noticeably worse than those of Nigeria: even with the development according to the optimistic scenario, the DRC will be able to catch up with China in terms of its GDP per capita only by the very end of this century (see Fig. 18).

A very similar pattern is observed for **Mozambique** (see Figs. 19 and 20) and **Senegal** (see Figs. 21 and 22).

Finally, we are dealing with the highest (even by the standards of Tropical Africa) risks of political and demographic collapses in the Sahelian countries (Chad, Mali, Burkina Faso, and Niger – see Figs. 23, 24, 25, 26, 27, 28, 29, 30). All these countries are characterized by: (1) a very slight difference between the inertial scenario and the pessimistic scenario; (2) extremely high risks of full-scale civil wars in the very coming years; and (3) reaching the level of middle-income countries only by the end of this century, even under the most optimistic scenario.

This is largely due to the fact that all these four countries occupy the lowest lines in the global ranking of countries in terms of population coverage with modern education (see Fig. 31).

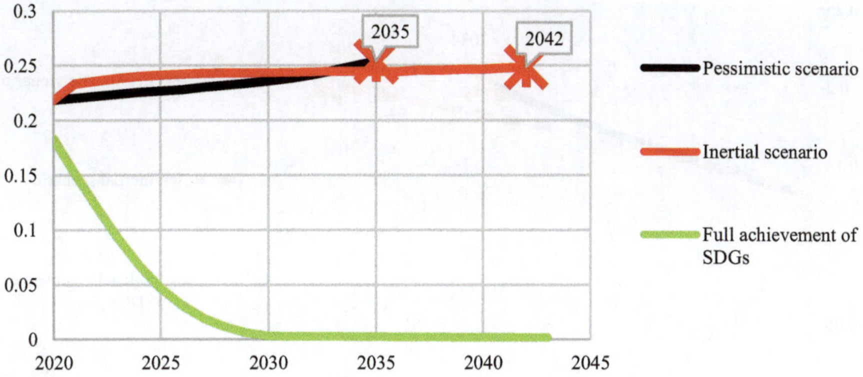

Fig. 19 Scenarios of risk dynamics of armed revolutionary destabilization/civil wars in Mozambique

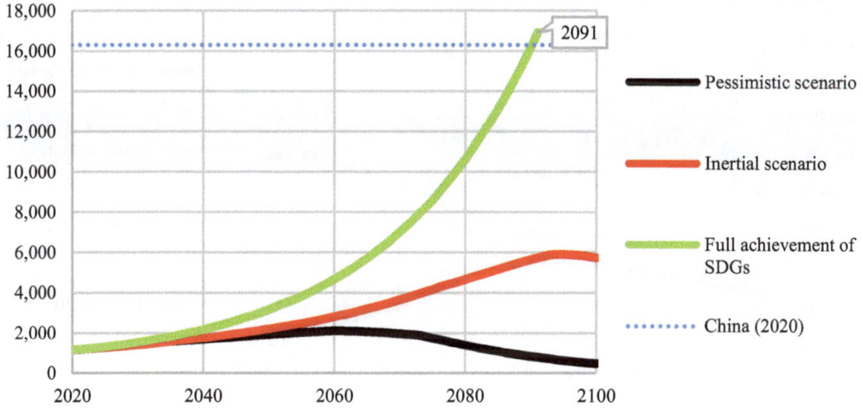

Fig. 20 GDP per capita scenarios in Mozambique

Meanwhile, as mentioned above, a very important factor in whether or not revolutionary actions in a country will take an armed form is the coverage of the population with formal education. In other words, armed revolutionary uprisings are most likely in countries with a high proportion of illiterate or semi-literate populations (Grinin & Korotayev, 2009, 2014; Korotayev, Bilyuga, & Shishkina, 2017b; Korotayev et al., 2020; Ustyuzhanin et al., 2022; Collier, 2004; Barakat & Urdal, 2009; Machado et al., 2011; Brancati, 2014; Butcher & Svensson, 2016; Kostelka & Rovny, 2019; Korotayev, Sawyer, & Romanov, 2021; Sawyer & Korotayev, 2022). The Sahelian countries belong to this category.

Of special concern is that the militant formations of Islamist radicals quite intentionally choose schools, teachers, and pupils as one of their main targets, in a deliberate attempt to reduce the coverage of the population with modern education.

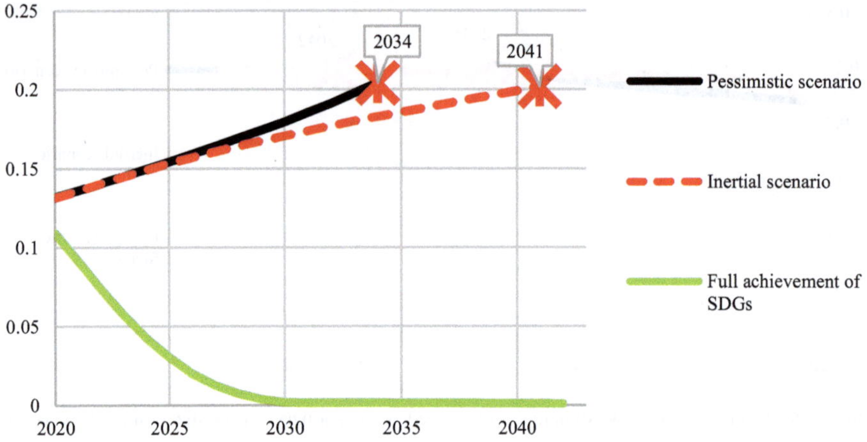

Fig. 21 Scenarios of risk dynamics of armed revolutionary destabilization/civil wars in Senegal

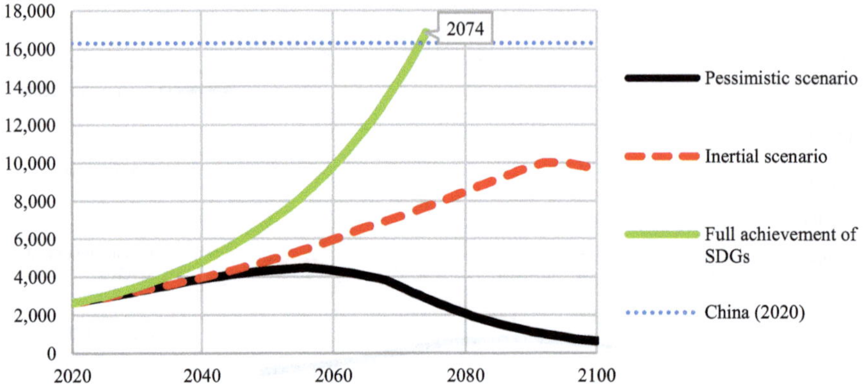

Fig. 22 GDP per capita scenarios in Senegal

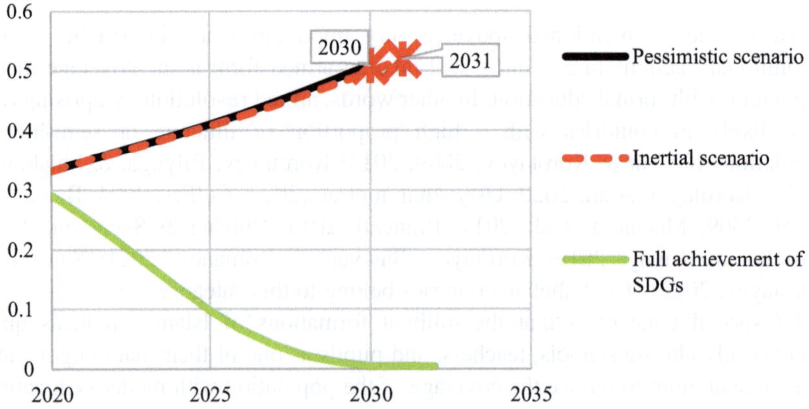

Fig. 23 Scenarios of risk dynamics of armed revolutionary destabilization/civil wars in Chad

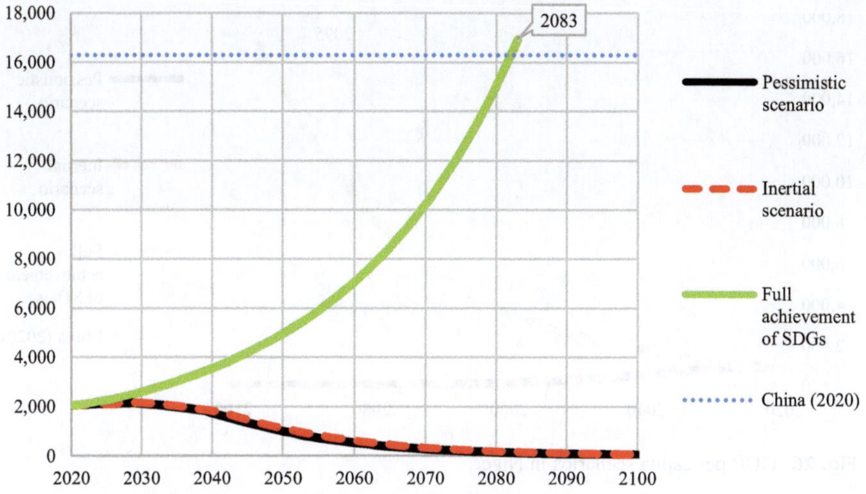

Fig. 24 GDP per capita scenarios in Chad

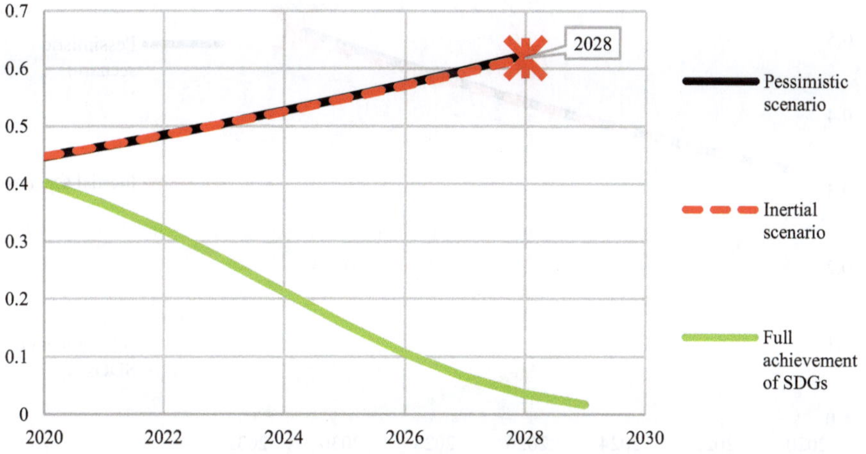

Fig. 25 Scenarios of risk dynamics of armed revolutionary destabilization/civil wars in Niger

Thus, speaking about the events of 2020 in Burkina Faso, Rida Lammouri notes that Islamist militant groups "continued to select schools as important targets for their attacks . . . In its May 2020 report, Human Rights Watch (HRW) counted 126 attacks and armed threats against teachers, students, and schools by jihadist groups in 2020, in addition to 222 education workers who were victims of the attacks. Consequently, the Burkinabe government has closed around 2500 schools, depriving about 350,000 students of education" (Lyammouri, 2020, pp. 2–3). And one of the most important

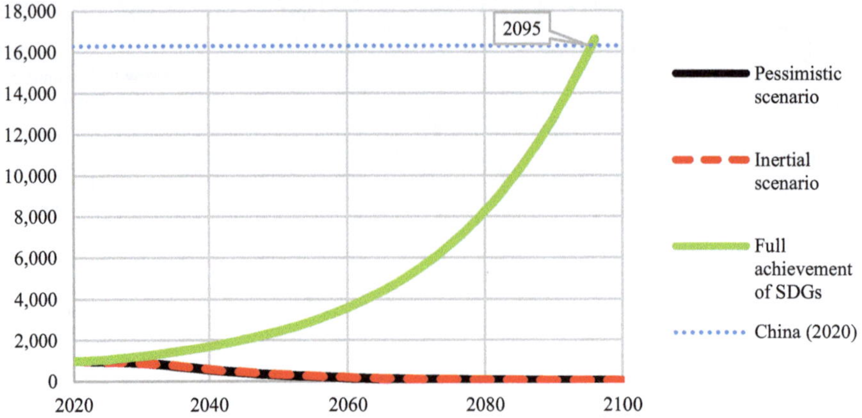

Fig. 26 GDP per capita scenarios in Niger

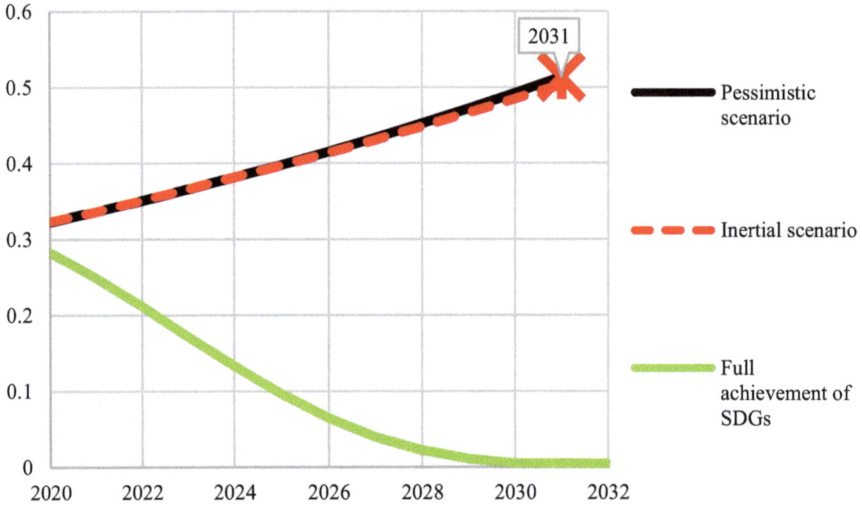

Fig. 27 Scenarios of risk dynamics of armed revolutionary destabilization/civil wars in Mali

radical Islamist groups in the Sahel, Boko Haram, has even used the slogan of combating the spread of modern education as its self-name (see, e.g., Walker, 2012).

Thus, a dangerous feedback loop is being formed in the Sahel, leading to a vicious circle and a development trap, when the low coverage of the population with modern education stimulates the growth of armed revolutionary Islamist activity, and the growth of the latter, in turn, leads to a decrease in the coverage of the population with education, which can lead not only to further growth of armed activity of the Islamists but also to a delay in the decline in the birth rate (or even to

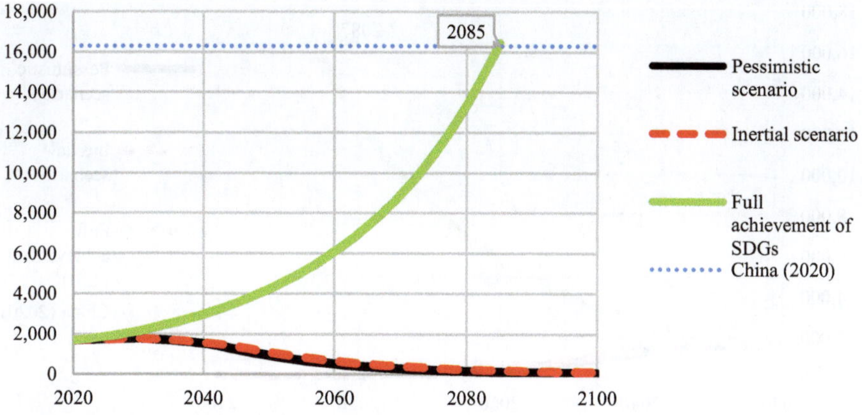

Fig. 28 GDP per capita scenarios in Mali

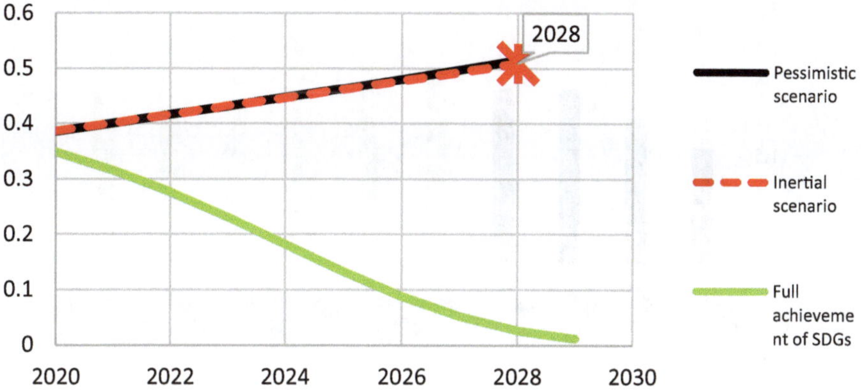

Fig. 29 Scenarios of risk dynamics of armed revolutionary destabilization/civil wars in Burkina Faso

some fertility increase[4]), to an increase in the population growth rate, to a slowdown in economic growth, to a fall in per capita GDP, to the preservation or even growth of "youth bulges," which will lead to an even more pronounced increase in the armed activities of radical Islamists. The countries of the Sahel need to break out of this trap as soon as possible, which, apparently, is no longer possible without the most serious support on the part of the world community, in general, and the African Union, in particular (see, e.g., Mustapha, 2020).

[4] See, e.g., Zinkina and Korotayev (2014a, 2014b, 2017)).

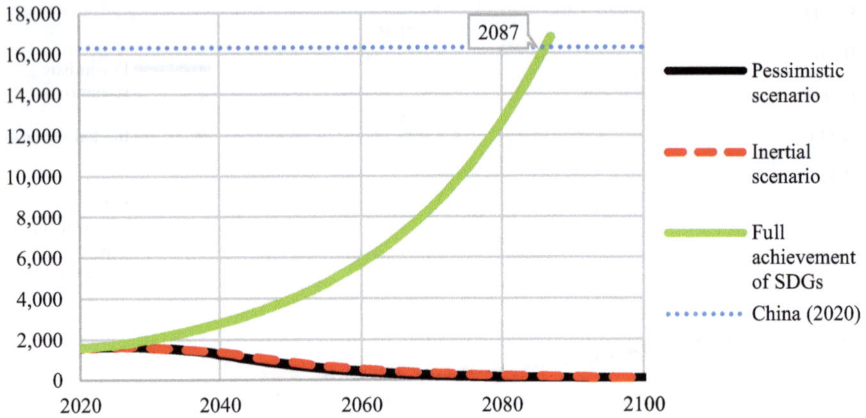

Fig. 30 GDP per capita scenarios for Burkina Faso

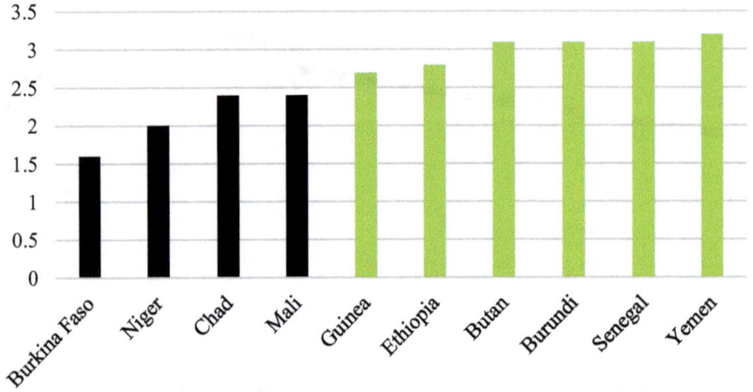

Fig. 31 10 countries in the world with the lowest mean years of schooling in 2018. Solid black bars indicate the countries of the G5 Sahel. *Source* Korotayev and Ustyuzhanin (2021)), based on the UNDP database

2 Brief Sociopolitical Forecasts of the Futures of Africa in a World-Systems Perspective

Thus, on the one hand, African countries are ahead of the path of active and dynamic development in the demographic, economic, political, cultural, and other spheres. But on the other hand, this path of development seems to be very thorny, associated with a high risk of destabilization events, possible kickbacks, and casualties.

The presented analysis well confirms the main idea expressed in the title of chapter "Africa – the continent of the future: challenges and opportunities" (Grinin & Korotayev, 2023, this volume).

As we see, on the one hand, Africa is a big problem for the modern World System in terms of poorly controlled population growth, which can lead to serious environmental, climatic, social, food, and other problems that, in one way or another, will concern the entire World System. Therefore, the leading powers are interested in seeing the demographic development of Africa proceed in a more manageable way. But on the other hand, in the context of a declining birth rate, African labor resources can partially help solve the problem of labor shortages. Thus, developed countries need to develop a common program of influence on African countries in a direction more acceptable to humankind so that this program does not infringe on the interests of African states. Meanwhile, not enough attention has been paid to Africa so far. Also, the world community faces an important task of trying to use the African resource as efficiently as possible to solve global problems with the maximum benefit for the Africans themselves.

Rapid demographic growth and an increase in the proportion of young people, taking into account rapid urbanization, as is clear, will inevitably exacerbate the already acute problem of unemployment in African countries. However, in the world-system aspect, this can contribute to the development of both regions with an elderly population and Africa itself. In the context of the strong development of remote work [see chapter "Global Aging – an Integral Problem of the Future. How to Turn a Problem into a Development Driver?" (Grinin, Grinin, & Korotayev, 2023b, this volume) for more detail], more Africans will be connected with economically developed countries. In this regard, developed countries face a very important task in providing more assistance than is being done today in the development of the education system and the possible recruitment of specialists for remote work. This is a project for decades. And this is one of the ways to reduce unemployment in Africa, raise living standards in these countries, and bring European and African cultures closer together. There are also possible ways to bring European pension systems closer to the governments of African countries (for more detail, see Grinin & Korotayev, 2010, 2016).

Taking into account global aging, which in the coming decades will greatly affect the situation in Europe, East Asia and other regions [chapter "Global Aging – an Integral Problem of the Future. How to Turn a Problem into a Development Driver?" (Grinin, Grinin, & Korotayev, 2023b, this volume)], young societies in Africa – with all their future problems and instability – will in a number of aspects positively influence the overall picture of the world's population.

Rapid population growth in Africa against the background of declining fertility is creating and will continue to create a huge demographic dividend in many African countries (see, e.g., Bloom et al., 2013; Groth & May, 2017; Jimenez & Pate, 2017; Hasan et al., 2019; Korotayev, Shulgin, et al., 2022c). At present, of course, a poorly literate working-age population (especially females) does not represent a very great value for the world economy. However, as the level and quality of education increases, these cohorts can become a very powerful resource for the deficient working-age population of aging developed countries[5] as well as a source of youth for the education systems of these countries. Already today, a large number of African students study in European, American and Asian universities, a large number of specialists from Africa work in one way or another for these economies.

Although aging in African countries will begin to be felt substantially only in the second half of the twenty-first century, however, it is necessary to forecast that African countries will most likely meet this process completely unprepared if they do not begin to prepare for it in advance (which is doubtful, in view of the more acute and numerous problems, in front of them). Meanwhile, although we have noted that almost all societies are severely underprepared for the consequences of global aging [chapter "Global Aging – an Integral Problem of the Future. How to Turn a Problem into a Development Driver?" (Grinin, Grinin, & Korotayev, 2023b, this volume)], African countries are prepared for this to a much lesser extent than, for example, European ones, since in African countries the pension and other social security systems are in a very weak state, one can say that they are only being formed. It may be that by the time the problems of aging rise to their full height, the world will already have formed models of relations in society that contribute to the institutionalization of aging [see chapter "Socio-Political Transformations. A Difficult Path to Cybernetic Society" (Grinin, Grinin, & Malkov, 2023, this volume), chapter "Analyzing Social Self-Organization and Historical Dynamics. Future Cybernetic W-Society: Socio-Political Aspects" (Grinin & Grinin, 2023, this volume) for more detail].

We would like to end this chapter with a thesis formulation of some conclusions, from which it follows that African countries will generate destabilization processes for a long time (although, of course, the scenario and the course of specific events will greatly depend on a number of circumstances) for many reasons. At the same time, the correct policy in the field of demography, education (especially for women), as well as in the ethnonational and economic spheres, as has been emphasized above in this chapter, can reduce in a very significant way the risks of such destabilizing events and their severity. It is worth noting here that the larger the population in the country, especially the young, the more acute conflicts and problems can be, so reducing the birth rate is the right way to reduce the risks and severity of destabilization problems.

Our general idea, which explains the inevitability under any circumstances of the emergence of destabilization processes both in the near and in the distant future of African countries, is the existence of a close correlation between immaturity/backwardness/underdevelopment and future instability events. And this correlation is as follows. As society matures and transitions to more mature relations, ideology, social, national, and political forces and conditions for certain processes appear in

[5] See chapter "Future Political Change. Toward a More Efficient World Order" (Grinin, Grinin, & Korotayev, 2023a, this volume) and chapter "Global Ageing – an Integral Problem of the Future. How to Turn a Problem into a Development Driver?" (Grinin, Grinin, & Korotayev, 2023b, this volume).

it, which can simultaneously influence destabilization. We mean the following factors: the desire for a nation-state, for democracy and social justice, the desire to get their share of benefits and power from certain elites, the struggle for resources and privileges, the strengthening of contradictions between the regions of the country, and many others.[6] All this can lead to internal confrontation and, as a result, to possible destabilization. Modernization continues in the Sahel and many other countries in Africa (see, e.g., Bondarenko, 2021; Prokopenko, 2021; Korotayev et al., 2022a, b, c; 2023). As a result, an increasingly complex conglomeration of the modern and the archaic in all spheres of production and life is being formed in these societies (Bondarenko, 2021). The archaic features will, of course, disappear, but over time. And with rapid modernization, a tough conflict arises between the advanced and the archaic, which in itself is a source of instability. Let us now list the reasons for this in more detail.

First, the overall impact of the World System reconfiguration processes will be felt (see Grinin & Korotayev, 2012, 2015, 2022; Grinin, 2022a). And this impact, other things being equal, is the more noticeable and stronger, the more unstable the sociopolitical and ethno-confessional situation in a particular region is. Africa is an extremely unstable and conflict zone, so various global processes will inevitably be reflected in it in the form of certain destabilizing events.

Second, Africa, including the Sahel zone and North Africa, as well as its other zones and countries (for example, the Democratic Republic of the Congo), is becoming a territory of increasing geopolitical rivalry between major powers.

Third, under external pressure, a synergistic effect can arise if there is an internal weakness and instability of such states and societies.

Fourth, the weakness of society in such cases is very often associated with:

a) the point that in the recent past, it was an artificial political formation whose boundaries were drawn for political reasons by external political actors;
b) there are no strong traditions of statehood; society is not ethnically monolithic, but divided, while conflicts between different ethnic groups, cultures, confessions, territories, tribes or clans are very significant, and also easily flare up.

Fifth, most countries of the Afrasian instability macrozone[7] correspond to one or more of the listed parameters to a greater or lesser extent, not to mention the fact that in the Sahel and other regions of Africa there are so-called failed states (e.g., Korotayev, Medvedev, & Zinkina, 2022), unable to resist criminal and terrorist

[6] See, e.g., Korotayev et al. (2011, 2015)), Grinin (2012, 2013, 2022a)), Korotayev et al. (2014)), Korotayev, Bilyuga, and Shishkina (2016, 2017a, 2017b, 2018)), Korotayev, Vaskin, and Bilyuga (2017)), Slinko et al. (2017)), Korotayev, Vaskin, et al. (2018), Vaskin et al. (2018)), Korotayev and Shishkina (2020), Korotayev, Vaskin, and Tsirel (2021), Slav et al. (2021), Korotayev, Grinin, et al. (2022).

[7] On the Afrasian instability macrozone see, e.g., Korotayev, Issaev, et al. (2016), Korotayev et al. (2019), Medvedev et al. (2022), and Korotayev and Khokhlova (2023).

groups to protect their own population (see above). Accordingly, it is quite possible to expect various problems in such countries.

Thus, it can be assumed that for a number of global, regional, cultural, religious, and country reasons, many African societies will experience increased risks of destabilization for quite a long time. At the same time, a set of destabilization phenomena will often be present in societies, that is, several components of destabilization (revolutions, civil wars, confessional confrontations, interventions, terrorism, etc.) will simultaneously interact, hybridizing, which, unfortunately, can open unstable eras (like the one that we could observe in the recent decades in the Democratic Republic of the Congo). And the period of overcoming these shortcomings and maturation of societies will also be a period of extremely high risk of destabilization processes.

However, we believe that African societies have a bright future ahead and, in general, Africa will increasingly contribute to the world-system development.

Acknowledgment This work was done with the support of MSU Program of Development, Project No 23-SCH05-03.

References

Abate, D. W. (2023). The global perspective of the dilemma of cooperation and complex relationships between the upper and lower Nile Riparian States after 2011. *Journal of Globalization Studies, 14*(1), 29–42. https://doi.org/10.30884/jogs/2023.01.03

Aisen, A., & Veiga, F. J. (2013). How does political instability affect economic growth? *European Journal of Political Economy, 29*, 151–167.

Alesina, A., Özler, S., Roubini, N., & Swagel, P. (1996). Political instability and economic growth. *Journal of Economic Growth, 1*(2), 189–211.

Azariadis, C., & Stachurski, J. (2005). Poverty traps. In S. N. Durlauf & P. Aghion (Eds.), *Handbook of economic growth* (pp. 295–384). Elsevier.

Barakat, B., & Urdal, H. (2009). *Breaking the waves? Does education mediate the relationship between youth bulges and political violence?* World Bank.

Besançon, M. L. (2005). Relative resources: Inequality in ethnic wars, revolutions, and genocides. *Journal of Peace Research, 42*(4), 393–415.

Bloom, D. E., Canning, D., & Malaney, P. (2000). Demographic change and economic growth in Asia. *Population and Development Review, 26*(suppl), 257–290.

Bloom, D. E., Canning, D., & Sevilla, J. (2003). *The demographic dividend: A new perspective on the economic consequences of population change.* RAND.

Bloom, D.E., Humair, S., Rosenberg, L., Sevilla, J.P., & Trussell, J. (2013). A demographic dividend for sub-Saharan Africa: Source, magnitude, and realization.

Bloom, D. E., & Williamson, J. G. (1998). Demographic transitions and economic miracles in emerging Asia. *World Bank Economic Review, 12*(3), 419–455.

Bondarenko, D. M. (2021). Nation-building in post-colonial states: Historical past and present-day realities. *Journal of Globalization Studies, 12*(1), 18–37. https://doi.org/10.30884/jogs/2021.01.02

Bonnal, M., & Yaya, M. E. (2015). Political institutions, trade openness, and economic growth: New evidence. *Emerging Markets Finance and Trade, 51*, 1276–1291. https://doi.org/10.1080/1540496x.2015.1011514

Brancati, D. (2014). Pocketbook protests: Explaining the emergence of pro-democracy protests worldwide. *Comparative Political Studies, 47*(11), 1503–1530.

Butcher, C., & Svensson, I. (2016). Manufacturing dissent: Modernization and the onset of major nonviolent resistance campaigns. *Journal of Conflict Resolution, 60*(2), 311–339.
Choucri, N. (1974). *Population dynamics and international violence: Propositions, insights, and evidence*. Lexington Books.
Cincotta, R., & Weber, H. (2021). Youthful age structures and the risks of revolutionary and separatist conflicts. In A. Goerres & P. Vanhuysse (Eds.), *Global political demography: Comparative analyses of the politics of population change in all world regions* (pp. 57–92). Palgrave Macmillan.
Collier, P. (2004). Greed and grievance in civil war. *Oxford Economic Papers, 56*(4), 563–595.
Farzanegan, M. R., & Witthuhn, S. (2017). Corruption and political stability: Does the youth bulge matter? *European Journal of Political Economy, 49*, 47–70. https://doi.org/10.1016/j.ejpoleco.2016.12.007
Fosu, A. K. (1992). Political instability and economic growth: Evidence from sub-Saharan Africa. *Economic Development and Cultural Change, 40*, 829–841. https://doi.org/10.1086/451979
Gates, S., Hegre, H., Nygård, H. M., & Strand, H. (2012). Development consequences of armed conflict. *World Development, 40*, 1713–1722. https://doi.org/10.1016/j.worlddev.2012.04.031
Glawe, L., & Wagner, H. (2016). The middle-income trap: Definitions, theories and countries concerned – A literature survey. *Comparative Economic Studies, 58*(4), 507–538.
Goldstone, J. A. (1991). *Revolution and rebellion in the early modern world*. University of California Press.
Goldstone, J. A. (2001). Demography, environment, and security: An overview. In M. Weiner & S. S. Russell (Eds.), *Demography and national security* (pp. 38–61). Berghahn Books.
Goldstone, J. A. (2002). Population and security: How demographic change can lead to violent conflict. *Journal of International Affairs, 56*(1), 3–21.
Goldstone, J. A., Grinin, L., & Korotayev, A. (2022). Conclusion. How many revolutions will we see in the 21st century? In J. A. Goldstone, L. Grinin, & A. Korotayev (Eds.), *Handbook of revolutions in the 21st century: The new waves of revolutions, and the causes and effects of disruptive political change* (pp. 1037–1061). Springer. https://doi.org/10.1007/978-3-030-86468-2_41
Goldstone, J. A., Kaufmann, E. P., & Toft, M. D. (2012). *Political demography: How population changes are reshaping international security and national politics*. Oxford University Press.
Grinin, L. (2012). State and socio-political crises in the process of modernization. *Cliodynamics, 3*(1), 124–157.
Grinin, L. (2013). State and socio-political crises in the process of modernization. *Social Evolution & History, 12*(2), 35–76.
Grinin, L. (2022a). On revolutionary situations, stages of revolution, and some other aspects of the theory of revolution. In J. A. Goldstone, L. Grinin, & A. Korotayev (Eds.), *Handbook of revolutions in the 21st century: The new waves of revolutions, and the causes and effects of disruptive political change* (pp. 69–104). Springer. https://doi.org/10.1007/978-3-030-86468-2_3
Grinin, L. (2022b). Revolutions and modernization traps. In J. A. Goldstone, L. Grinin, & A. Korotayev (Eds.), *Handbook of revolutions in the 21st century: The new waves of revolutions, and the causes and effects of disruptive political change* (pp. 219–238). Springer. https://doi.org/10.1007/978-3-030-86468-2_8
Grinin, L., & Grinin, A. (2022). Conclusion. New wave of Middle Eastern revolutionary events in the World System context. In L. Issaev & A. Korotayev (Eds.), *New wave of revolutions in the MENA region. A comparative perspective*. Springer.
Grinin, L., & Grinin, A. (2023). Analyzing social self-organization and historical dynamics. Future cybernetic W-society: socio-political aspects. In: V. Sadovnichy et al. (Eds.), *Reconsidering the limits to growth. A Report to the Russian Association of the Club of Rome* (pp. 491–519). Springer. https://doi.org/10.1007/978-3-031-34999-7_21
Grinin, L., Grinin, A., & Korotayev, A. (2023a). Future political change. Toward a more efficient world order. In: V. Sadovnichy et al. (Eds.), *Reconsidering the limits to growth. A Report to the*

Russian Association of the Club of Rome (pp. 191–206). Springer. https://doi.org/10.1007/978-3-031-34999-7_11

Grinin, L., Grinin, A., & Korotayev, A. (2023b). Global aging – an integral problem of the future. How to turn a problem into a development driver? In: V. Sadovnichy et al. (Eds.), *Reconsidering the limits to growth. A report to the Russian Association of the Club of Rome* (pp. 117–135). Springer. https://doi.org/10.1007/978-3-031-34999-7_7

Grinin, L., Grinin, A., & Malkov, S. (2023). Socio-political transformations. A difficult path to cybernetic society. In: V. Sadovnichy et al. (Eds.), *Reconsidering the limits to growth. A Report to the Russian Association of the Club of Rome* (pp. 169–189). Springer. https://doi.org/10.1007/978-3-031-34999-7_10

Grinin, L., & Korotayev, A. (2009). Urbanization and political instability: To the development of mathematical models of political processes. *Polis-Politicheskiye Issledovaniya, 4*, 34–52.

Grinin, L., & Korotayev, A. (2010). Will the global crisis lead to global transformations? The coming epoch of new coalitions. *Journal of Globalization Studies, 1*(2), 166–183.

Grinin, L., & Korotayev, A. (2012). Does "Arab Spring" mean the beginning of world system reconfiguration? *World Futures, 68*(7), 471–505. https://doi.org/10.1080/02604027.2012.697836

Grinin, L., & Korotayev, A. (2014). Revolution vs democracy (revolution and counterrevolution in Egypt). *Polis-Politicheskiye Issledovaniya, 3*, 139–158. https://doi.org/10.17976/jpps/2014.03.09

Grinin, L., & Korotayev, A. (2015). *Great divergence and great convergence. A global perspective*. Springer. https://doi.org/10.1007/978-3-319-17780-9

Grinin, L., & Korotayev, A. (2016). Global Population Ageing, the sixth Kondratieff wave, and the global financial system. *Journal of Globalization Studies, 7*(2), 11–31.

Grinin, L., & Korotayev, A. (2022). The future of revolutions in the 21st century and the world system reconfiguration. *World Futures, 79*, 69. https://doi.org/10.1080/02604027.2022.2050342

Grinin, L., & Korotayev, A. (2023). Africa – the continent of the future: Challenges and opportunities. In: V. Sadovnichy et al. (Eds.), *Reconsidering the limits to growth. A Report to the Russian Association of the Club of Rome* (pp. 225–238). Springer. https://doi.org/10.1007/978-3-031-34999-7_13

Groth, H., & May, J. F. (Eds.). (2017). *Africa's population: In search of a demographic dividend*. Springer. https://doi.org/10.1007/978-3-319-46889-1

Hasan, R., Loevinsohn, B., Moucheraud, C., Ahmed, S. A., Osorio-Rodarte, I., Suzuki, E., Pradhan, E., Madhavan, S., Troiano, S., Sexton, M., Mustapha, F. A., Odutolu, O., & Okunola, O. (2019). *Nigeria's demographic dividend? Policy note in support of Nigeria's ERGP 2017–2020*. World Bank.

Hegre, H., & Sambanis, N. (2006). Sensitivity analysis of empirical results on civil war onset. *Journal of conflict resolution, 50*(4), 508–535.

Jimenez, E., & Pate, M. A. (2017). Reaping a demographic dividend in Africa's largest country: Nigeria. In H. Groth & J. F. May (Eds.), *Africa's population: In search of a demographic dividend* (pp. 33–52). Springer.

Justino, P. (2006). *On the links between violent conflict and chronic poverty: How much do we really know?* Institute of Development Studies, University of Sussex.

Kebede, E., Goujon, A., & Lutz, W. (2019). Stalls in Africa's fertility decline partly result from disruptions in female education. *Proceedings of the National Academy of Sciences, 116*(8), 2891–2896. https://doi.org/10.1073/pnas.1717288116

Kharas, H., & Kohli, H. (2011). What is the middle-income trap, why do countries fall into it, and how can it be avoided? *Global Journal of Emerging Market Economies, 3*(3), 281–289.

Korotayev, A., Bilyuga, S., & Shishkina, A. (2016). GDP per capita, protest intensity and regime type: A quantitative analysis. *Sravnitelnaya Politika-Comparative Politics, 7*(4), 72–94. https://doi.org/10.18611/2221-3279-2016-7-4(25)-72-94

Korotayev, A., Bilyuga, S., & Shishkina, A. (2017a). Correlation between GDP per capita and protest intensity: a quantitative analysis. *Polis – Politicheskiye Issledovaniya, 2*, 155–169. https://doi.org/10.17976/jpps/2017a.02.1

Korotayev, A., Bilyuga, S., & Shishkina, A. (2017b). GDP per capita, intensity of anti-government demonstrations and level of education. Cross-national analysis. *Politeia-Journal of Political Theory, Political Philosophy and Sociology of Politics, 84*(1), 127–143. https://doi.org/10.30570/2078-5089-2017b-84-1-127-143

Korotayev, A., Bilyuga, S., & Shishkina, A. (2018). GDP per capita and protest activity: A quantitative reanalysis. *Cross-Cultural Research, 52*(4), 406–440. https://doi.org/10.1177/1069397117732328

Korotayev, A., Grinin, L., Medvedev, I., & Slav, M. (2022a). Political regime types and revolutionary destabilization risks in the twenty-first century. *Russian Sociological Review, 21*(4), 9–65. https://doi.org/10.17323/1728-192x-2022-2-9-65

Korotayev, A., Medvedev, I., & Zinkina, J. (2022b). Global systems for sociopolitical instability forecasting and their efficiency. A comparative analysis. *Comparative Sociology, 21*(1), 64–104. https://doi.org/10.1163/15691330-bja10050

Korotayev, A., Shulgin, S., Zinkina, J., & Slav, M. (2022c). Estimates of the possible economic effect of the demographic dividend for sub-Saharan Africa for the period up to 2036. *Vostok (Oriens), 2*, 108–123. https://doi.org/10.31857/S086919080019128-7

Korotayev, A., Romanov, D., Zinkina, J., & Slav, M. (2023). Urban youth and terrorism: A quantitative analysis (are youth bulges relevant anymore?). *Political Studies Review, 21*(3), 550–574. https://doi.org/10.1177/14789299221075908

Korotayev, A., Issaev, L., Shishkina, A., Rudenko, M., & Ivanov, E. (2016). A Afrasian instability zone and its historical background. *Social Evolution and History, 15*(2), 120–140.

Korotayev, A., Issaev, L., & Zinkina, J. (2015). Center-periphery dissonance as a possible factor of the revolutionary wave of 2013–2014: A cross-national analysis. *Cross-Cultural Research, 49*(5), 461–488. https://doi.org/10.1177/1069397115595374

Korotayev, A., & Khokhlova, A. (2023). MENA sociopolitical destabilization dimensions in comparative global perspective. *Journal of Globalization Studies, 14*(1), 67–94. https://doi.org/10.30884/jogs/2023.01.05

Korotayev, A., Malkov, S., & Grinin, L. (2014). A trap at the escape from the trap? Some demographic structural factors of political instability in modernizing social systems. *History & Mathematics, 4*, 201–267.

Korotayev, A., Meshcherina, K., Slinko, E., & Shishkina, A. (2019). Value orientations of the Afrasian zone of instability: Gender dimensions. *Vostok (Oriens), 1*, 122–154. https://doi.org/10.31857/S086919080003963-6

Korotayev, A., Sawyer, P., Gladyshev, M., Romanov, D., & Shishkina, A. (2021). Some sociodemographic factors of the intensity of anti-government demonstrations: Youth bulges, urbanization, and protests. *Sociologiceskoe Obozrenie = Russian Sociological Review, 20*(3), 98–128. https://doi.org/10.17323/1728-192x-2021-3-98-128

Korotayev, A., Sawyer, P., Grinin, L., Romanov, D., & Shishkina, A. (2020). Socio-economic development and anti-government protests in light of a new quantitative analysis of global databases. *Sotsiologicheskiy Zhurnal, 26*(4), 61–78. https://doi.org/10.19181/socjour.2020.26.4.7642

Korotayev, A., Sawyer, P., & Romanov, D. (2021). Socio-economic development and protests. A quantitative reanalysis. *Comparative Sociology, 20*(2), 195–222. https://doi.org/10.1163/15691330-bja10030

Korotayev, A., & Shishkina, A. (2020). Relative deprivation as a factor of sociopolitical destabilization: Toward a quantitative comparative analysis of the Arab spring events. *Cross-Cultural Research, 54*(2–3), 296–318. https://doi.org/10.1177/1069397119882364

Korotayev, A., & Ustyuzhanin, V. (2021). O strukturno-demograficheskikh faktorakh vooruzhennykh islamistskikh revolyutsionnykh vystupleniy v stranakh gruppy G5 Sahel. *Systemnyi monitoring globalnyh i regionalnyh riskov, 12*, 451–474.

Korotayev, A., Vaskin, I., & Bilyuga, S. (2017). Olson-Huntington hypothesis on a bell-shaped relationship between the level of economic development and sociopolitical destabilization: A quantitative analysis. *Sociologiceskoe Obozrenie = Russian Socioloogical Review, 16*(1), 9–49. https://doi.org/10.17323/1728-192X2017-1-9-49

Korotayev, A., Vaskin, I., Bilyuga, S., & Ilyin, I. (2018). Economic development and sociopolitical destabilization: A re-analysis. *Cliodynamics, 9*(1), 59–118. https://doi.org/10.21237/c7clio9137314

Korotayev, A., Vaskin, I., & Tsirel, S. (2021). Economic growth, education, and terrorism: A re-analysis. *Terrorism and Political Violence, 33*(3), 572–595. https://doi.org/10.1080/09546553.2018.1559835

Korotayev, A., & Zinkina, J. (2014). On the structure of the present-day convergence. *Campus-Wide Information Systems, 31*(2–3), 139–152. https://doi.org/10.1108/CWIS-11-2013-0064

Korotayev, A., Zinkina, J., Goldstone, J., & Shulgin, S. (2016). Explaining current fertility dynamics in Tropical Africa from an anthropological perspective: A cross-cultural investigation. *Cross-Cultural Research, 50*(3), 251–280. https://doi.org/10.1177/1069397116644158

Korotayev, A., Zinkina, J., Kobzeva, S., Bogevolnov, J., Khaltourina, D., Malkov, A., & Malkov, S. (2011). A trap at the escape from the trap? Demographic-structural factors of political instability in modern Africa and West Asia. *Cliodynamics, 2*(2), 276–303. https://doi.org/10.21237/C7clio22217

Kostelka, F., & Rovny, J. (2019). It's not the left: Ideology and protest participation in old and new democracies. *Comparative Political Studies, 52*(11), 1677–1712. https://doi.org/10.1177/0010414019830717

Lucas, R. E. (2002). *Lectures on economic growth.* Harvard University Press.

Lyammouri, R. (2020). *Burkina Faso elections, another box to check.* Policy Center for the New South. Retrieved from https://www.policycenter.ma/sites/default/files/PB_20-84_Rida%20Lyammouri_n.pdf

Machado, F., Scartascini, C., & Tommasi, M. (2011). Political institutions and street protests in Latin America. *Journal of Conflict Resolution, 55*(3), 340–365. https://doi.org/10.1177/0022002711400864

Mamoon, D., & Murshed, S. M. (2009). Want economic growth with good quality institutions? Spend on education. *Education Economics, 17*, 445–468. https://doi.org/10.1080/09645290801931782

Mankiw, N. G., Romer, D., & Weil, D. N. (1992). A contribution to the empirics of economic growth. *The Quarterly Journal of Economics, 107*(2), 407–437. https://doi.org/10.2307/2118477

Matsuyama, K. (2008). Poverty traps. In S. N. Durlauf & L. Blume (Eds.), *The new Palgrave dictionary of economics* (pp. 5073–5077). Palgrave Macmillan. https://doi.org/10.1007/978-1-349-58802-2_1317

May, J. F., & Rotenberg, S. (2020). A call for better integrated policies to accelerate the fertility decline in sub-Saharan Africa. *Studies in Family Planning, 51*(2), 193–204.

Medvedev, I., Ustyuzhanin, V., Zinkina, J., & Korotayev, A. (2022). Machine learning for ranking factors of global and regional protest destabilization with a special focus on Afrasian instability macrozone. *Comparative Sociology, 21*(6), 604–645. https://doi.org/10.1163/15691330-bja10062

Mesquida, C. G., & Weiner, N. I. (1999). Male age composition and severity of conflicts. *Politics and the Life Sciences, 18*, 113–117. https://doi.org/10.1017/S0730938400021158

Mesquida, C. G., & Wiener, N. I. (1996). Human collective aggression: A behavioral ecology perspective. *Ethology and Sociobiology, 17*(4), 247–262.

Moller, H. (1968). Youth as a force in the modern world. *Comparative Studies in Society and History, 10*(3), 237–260. https://doi.org/10.2307/177801

Mustapha, A. H. (2020). Nigeria and the African Union (2002–2019). *Journal of Globalization Studies, 11*(2), 35–54. https://doi.org/10.30884/jogs/2020.02.03

Nzimande, N., & Mugwendere, T. (2018). Stalls in Zimbabwe fertility: Exploring determinants of recent fertility transition. *Southern African Journal of Demography, 18*(1), 59–110.

Omoruyi, E. M. M. (2021). Harnessing the demographic dividend in Africa through lessons from East Asia's experience. *Journal of Comparative Asian Development, 18*(2), 1–38.

Østby, G., Urdal, H., & Dupuy, K. (2019). Does education lead to pacification? A systematic review of statistical studies on education and political violence. *Review of Educational Research, 89*(1), 46–92. https://doi.org/10.3102/0034654318800236

Prokopenko, L. Y. (2021). Development of the Institution of the First Ladyship in Africa. *Journal of Globalization Studies, 12*(1), 38–60. https://doi.org/10.30884/jogs/2021.01.03

Raleigh, C. (2015). Urban violence patterns across African states. *International Studies Review, 17*(1), 90–106.

Romanov, D., Meshcherina, K., & Korotayev, A. (2021). The share of youth in the total population as a factor of intensity of non-violent protests: A quantitative analysis. *Polis. Political Studies, 3*, 166–181. https://doi.org/10.17976/jpps/2021.03.11

Sawyer, P., & Korotayev, A. (2022). Formal education and contentious politics: The case of violent and non-violent protest. *Political Studies Review, 20*(3), 366–389. https://doi.org/10.1177/1478929921998210

Sawyer, P., Romanov, D., Slav, M., & Korotayev, A. (2022). Urbanization, the youth, and protest: A cross-national analysis. *Cross-Cultural Research, 56*(2–3), 125–149. https://doi.org/10.1177/10693971211059762

Schoumaker, B. (2019). Stalls in fertility transitions in sub-Saharan Africa: Revisiting the evidence. *Studies in family planning, 50*(3), 257–278. https://doi.org/10.1111/sifp.12098

Slav, M., Smyslovskikh, E., Novikov, V., Kolesnikov, I., & Korotayev, A. (2021). Deprivation, instability, and propensity to attack: how urbanization influences terrorism. *International Interactions, 47*(6), 1100–1130. https://doi.org/10.1080/03050629.2021.1924703

Slinko, E., Bilyuga, S., Zinkina, J., & Korotayev, A. (2017). Regime type and political destabilization in cross-national perspective: A re-analysis. *Cross-Cultural Research, 51*(1), 26–50. https://doi.org/10.1177/1069397116676485

Staveteig, S. (2005). The young and the restless: Population age structure and civil war. *Environmental Change and Security Program Report, 11*, 12–19.

UN Population Division. (2022). *United Nations Population Division Database*. United Nations, Department of Economics and Social Affairs, Population Division. Retrieved from http://www.un.org/esa/population

Urdal, H. (2004). *The devil in the demographics: The effect of youth bulges on domestic armed conflict, 1950–2000*. World Bank.

Urdal, H. (2006). A clash of generations? Youth bulges and political violence. *International Studies Quarterly, 50*(3), 607–629. https://doi.org/10.1111/j.1468-2478.2006.00416.x

Urdal, H. (2008). Population, resources, and political violence: A subnational study of India, 1956–2002. *Journal of Conflict Resolution, 52*(4), 590–617. https://doi.org/10.1177/0022002708316741

Ustyuzhanin, V., Grinin, L., Medvedev, I., & Korotayev, A. (2022). Education and revolutions. Why do some revolutions take up arms while others do not? *Politeia-Journal of Political Theory, Political Philosophy and Sociology of Politics, 104*(1), 50–71. https://doi.org/10.30570/2078-5089-2022-104-1-50-71

Vaskin, I., Tsirel, S., & Korotayev, A. (2018). Economic growth, education and terrorism: experience in quantitative analysis. *Sotsiologicheskiy Zhurnal, 24*(2), 28–65. https://doi.org/10.19181/socjour.2018.24.2.5844

Vollset, S. E., Goren, E., Yuan, C.-W., Cao, J., Smith, A. E., Hsiao, T., Bisignano, C., Azhar, G. S., Castro, E., Chalek, J., Dolgert, A. J., Frank, T., Fukutaki, K., Hay, S. I., Lozano, R., Mokdad, A. H., Nandakumar, V., Pierce, M., Pletcher, M., Robalik, T., Steuben, K. M., Wunrow, H. Y., Zlavog, B. S., & Murray, C. J. L. (2020). Fertility, mortality, migration, and population scenarios for 195 countries and territories from 2017 to 2100: a forecasting analysis for the

Global Burden of Disease Study. *The Lancet, 396*, 1285–1306. https://doi.org/10.1016/s0140-6736(20)30677-2
Walker, A. (2012). *What is Boko Haram?* US Institute of Peace.
Weber, H. (2019). Age structure and political violence: A re-assessment of the "youth bulge" hypothesis. *International Interactions, 45*(1), 80–112. https://doi.org/10.1080/03050629.2019.1522310
Wimmer, A., Cederman, L. E., & Min, B. (2009). Ethnic politics and armed conflict: A configurational analysis of a new global data set. *American Sociological Review, 74*(2), 316–337.
Zinkina, J., & Korotayev, A. (2014a). Explosive population growth in Tropical Africa: Crucial omission in development forecasts (emerging risks and way out). *World Futures, 70*(4), 271–305. https://doi.org/10.1080/02604027.2014.894868
Zinkina, J., & Korotayev, A. (2014b). Projecting Mozambique's demographic futures. *Journal of Futures Studies, 19*(2), 21–40.
Zinkina, J., & Korotayev, A. (2017). *Sotsial'no-demograficheskoye razvitiye stran Tropicheskoy Afriki: Klyuchevyye faktory riska, modifitsiruyemyye upravlyayushchiye parametry, rekomendatsii*. Lenand/URSS.

Andrey Korotayev heads the Centre for Stability and Risk Analysis at HSE University in Moscow, Russia. He is also a Senior Research Professor at the Institute of Oriental Studies and Institute for African Studies, Russian Academy of Sciences, as well as at the Faculty of Global Studies of Lomonosov Moscow State University and the Russian Presidential Academy of National Economy and Public Administration. He has authored or co-authored over 650 scholarly publications.

Sergey Shulgin is a Vice-Director of the International Research Laboratory for Demography and Human Capital at the Russian Presidential Academy of National Economy and Public Administration (RANEPA), Moscow, Russia. He has published over 70 research articles.

Vadim Ustyuzhanin is a Junior Research Fellow at the Centre for Stability and Risk Analysis at HSE University, as well as at the International Research Laboratory for Demography and Human Capital at the Russian Presidential Academy of National Economy and Public Administration (RANEPA), Moscow, Russia.

Julia Zinkina is a Research Professor at the International Research Laboratory for Demography and Human Capital of the Russian Presidential Academy of National Economy and Public Administration (RANEPA), Moscow and at the Faculty of Global Studies of Lomonosov Moscow State University. Julia Zinkina is the author of more than 100 scholarly publications.

Leonid Grinin is a Senior Research Professor at the Centre for Stability and Risk Analysis at HSE University in Moscow, as well as a Senior Research Professor at the Institute of Oriental Studies, Russian Academy of Sciences. He is the author of more than 600 publications, including 30 monographs in Russian, English, and Chinese.

Analyzing Social Self-Organization and Historical Dynamics. Future Cybernetic W-Society: Sociopolitical Aspects

Leonid Grinin and Anton Grinin

Abstract

Leonid Grinin and Anton Grinin look at the forthcoming decades up to the late twenty-first century, focusing on two important processes of development. The first of them is technological changes connected with a new technological wave or, in the authors' terms, the final phase of the Cybernetic Revolution. This technological revolution will result in the widespread introduction of self-regulating systems. The second one is the process of global aging, which will influence more and more all spheres and relations in future society. As a result of the two processes, a society will emerge, which Leonid Grinin and Anton Grinin call "cybernetic society" because its formation will be inextricably linked with the completion of the Cybernetic Revolution and its consequences. It will be society, which will have fundamentally new characteristics, and consequently, it will be very different from all previous societies. On the one hand, it will be closely connected with very deep technological changes in the management and regulation of our life at all levels. On the other hand, it is inextricably linked with the process of global aging because aging becomes an integral part of society. As it develops and the institutions adapt to it, no sphere of society will remain unchanged, including technology, economy, consumption, social sphere, ideology, etc. In short, cybernetic society is an elderly society with the institutionalization of age differences, relying almost entirely on smart (cybernetic) technologies and self-regulating systems. Accordingly, this society will have its strengths and

L. Grinin (✉)
HSE University, Moscow, Russia

Institute of Oriental Studies, Russian Academy of Sciences, Moscow, Russia

A. Grinin
HSE University, Moscow, Russia

Faculty of Global Studies, Lomonosov Moscow State University, Moscow, Russia

V. Sadovnichy et al. (eds.), *Reconsidering the Limits to Growth*, World-Systems Evolution and Global Futures, https://doi.org/10.1007/978-3-031-34999-7_21

weaknesses, successes, and even triumphs, but it will not be free from some serious problems. Cybernetic society must go through a difficult path of social frictions, conflicts, and optimizations before many things become institutionalized. Techno-social regulation has important advantages, but it also includes violence against the will and freedom of people. Therefore, a long way is required to optimize regulations to consider a variety of needs, desires, and problems. Even short-term regulation with the help of technology during the COVID-19 pandemic has shown how painful it could be for societies.

In several chapters of this volume, we discuss the possible social changes in a future society.[1] Sometimes the forecast horizon is very long there; it even considers some possible changes that might fully occur in the first third of the twenty-second century. Here we look at some nearer future (the coming decades up to the late twenty-first century), focusing on two important processes (vectors) of development. The first is technological change, connected with a new technological wave, the essence of which is described in chapter "Technology. Limitless Possibilities and Effective Control" (Grinin & Grinin, 2023, this volume). This is the final phase of the Cybernetic Revolution with the widespread introduction of self-regulating and self-managing systems. The second one is the process of global aging, which will be influencing more and more all the spheres and relations in a future society. It is discussed in chapter "Global Aging – an Integral Problem of the Future. How to Turn a Problem into a Development Driver?" (Grinin et al., 2023b, this volume).[2]

As a result of the two processes described, society will emerge, which we call Cybernetic W-society. Cybernetic, because its formation will be inextricably linked with the completion of the Cybernetic Revolution and its subsequent influences. W-society because it will be society, which will be very different from all the previous social systems, as well as modern societies. Its social structure, as well as the management systems of this society, will be very different from the authoritarian-hierarchical structures characteristic of late feudal, East Asian, and socialist industrial societies. They will also be very different from systems with an organic combination of vertical and horizontal structures, characteristic of market capitalist societies with a democratic political system (generally still dominant, although greatly transformed in modern high-income countries). We denote authoritarian-hierarchical structures as "X-structures," whereas market capitalist societies with a

[1] See chapter "Socio-Political Transformations. A Difficult Path to Cybernetic Society" (Grinin et al., 2023d, this volume) and chapter "Modeling Social Self-Organization and Historical Dynamics. A Look into the Future: Cybernetic W-society" (Akaev, Malkov, Grinin et al., 2023, this volume). See also Grinin (2006), Grinin and Grinin (2015a, 2015b), Grinin and Grinin (2016), Grinin, Grinin, and Korotayev (2017a, 2017b, 2020, 2021b).

[2] See also Grinin (2006), Grinin and Grinin (2015a, 2015b, 2016), Grinin, Grinin, and Korotayev (2017a, 2017b, 2020, 2021b).

democratic political system are called "Y-structure." As future Cybernetic W-society will be qualitatively different from both types and will have fundamentally new characteristics, we call it "W-society."[3] We have already discussed Cybernetic W-society in chapter "Socio-Political Transformations. A Difficult Path to Cybernetic Society" (Grinin et al., 2023d, this volume) above. However, there are many important characteristics of Cybernetic W-society that have not been analyzed yet.

Below is an analysis of some important (from our point of view) problems that will accompany the formation of this Cybernetic W-society.

1 Cybernetic W-Society as a Result of the Cybernetic Revolution

Every major technological change and technological revolution inevitably (albeit with a noticeable time lag) changes society, its socio-professional characteristics, and somewhat later, its cultural, ideological, and even political life. The initial phases of the Cybernetic Revolution have significantly changed society, which today is markedly different from what it was in the mid-twentieth century. It has changed in mental, cultural, and demographical aspects, as well as in terms of family and sexual relations – not to mention finance, economic, and consumer relations, communications, the way we express our thoughts and emotions, or the way we receive and comprehend information. Important changes have also taken place in the political sphere, although, as has been shown in chapter "Future Political Change. Toward a More Efficient World Order" (Grinin et al., 2023a, this volume), this sphere lags far behind the economic one in its transformations, which leads to rather strong tensions, crises, and conflicts (see Grinin, 2021, 2022; Grinin & Grinin, 2015c; Grinin & Korotayev, 2014a, 2014b; Grinin & Korotayev, 2015b; Grinin et al., 2021b).

Accordingly, the final phase of the Cybernetic Revolution (which, according to our forecasts, will take place in the 2030–2070s), as well as the mature phases of the scientific-cybernetic production principle, should also greatly change future society. It is worth mentioning that we call this revolution "cybernetic" because (a) cybernetics is the science of managing complex systems; (b) the Cybernetic

[3]The characteristics of X-structures and Y-structures have been described in detail in a number of chapters of Methodological and Mathematical Appendices. See chapter "Modeling Social Self-Organization and Historical Dynamics. Agrarian Society" (Malkov, Kovaleva et al., 2023, this volume), chapter "Modeling Social Self-Organization and Historical Dynamics. Industrial Society" (Akaev, Malkov, Davydova et al., 2023, this volume), chapter "Modeling Social Self-Organization and Historical Dynamics. Global Phase Transitions" (Malkov, Grinin et al., 2023, this volume), and chapter "Modeling Social Self-Organization and Historical Dynamics. Modern Society and a Look into the Global Future: Cybernetic W-Society" (Akaev, Malkov, Grinin et al., 2023, this volume). See also chapter "Socio-Political Transformations. A Difficult Path to Cybernetic Society" (Grinin et al., 2023d, this volume).

Revolution will bring fundamental changes not just in technology but also in management processes in all spheres of life due to the wide spread of self-regulating and self-managing systems.[4] In other words, we will see tremendous changes in the nature and methods of managing and controlling processes in the broadest sense of the word – from production to everyday life, from medicine to social behavior, from planning to consumption, from consumption to business,[5] from transportation to cooking, from elder care to communications. Moreover, revolutionary changes will affect social, economic, administrative, and even political regulation.

Cybernetic society is a society that will take shape (and is already taking shape) as a result of the completion of the Cybernetic Revolution and mature phases of the scientific-cybernetic production principle (see Grinin & Grinin, 2015c for details). On the one hand, it will be closely connected to very deep technological changes in the management and regulation of our life at all levels. On the other hand, it is inextricably linked with the process of global aging because aging is becoming an integral part of society. As it develops and the institutions adapt to it, no sphere of society will remain unchanged, including technology, economy, consumption, social sphere, ideology, etc.[6]

As a result of the Cybernetic Revolution, and the subsequent phases of the scientific-cybernetic production principle, society has undergone and will undergo very serious changes in terms of demography: from a population explosion at the beginning of the production principle (in the 1950s–1960s) to stabilization or even a reduction in the population in the late twenty-first century, and in other respects. See these developmental phase changes in Table 1 below, which shows the major demographic transformations in their combination with production revolutions, production principles and types of population reproduction.

Thus, cybernetic society will be (a) super-technologized; (b) socio-technologically regulated at all levels; (c) a society where, as a result of the aging process, division into age categories will become socially much more significant

---

[4]See for detail chapter "Technology. Limitless Possibilities, Effective Control" (Grinin & Grinin, 2023, this volume).

[5]Today, no one is surprised that the main exchange trading is carried out by automated programs. And this has greatly changed the work of exchanges.

[6]On various aspects of the aging process and its modern and future impact on society and its different spheres as well as geopolitics see, e.g., Clark and Spengler (1980), Lee et al. (1988), Bös and Von Weizsäcker (1989), Peterson (1999), Fukuyama (2002), Jackson et al. (2008, 2013), Jackson (2021), Lee et al. (2010), Lee and Mason (2011), UNFPA (2012), UN Population Division (2013, 2019, 2020), Powell and Khan (2013), Popova and Yanik (2014), Goldstone (2015), Goldstone et al. (2015), Orlická (2015), Bloom and Luca (2016), Maestas et al. (2016), McDaniel and Zimmer (2016), van Dullemen (2017), Bengtson (2018), Fichtner (2018), Randers et al. (2018), Yaşar and Yaşar (2018), Barsukov (2019), Hsu and Lo (2019), Kapelyushnikov (2019a, 2019b), Shcherbakova (2019), Mitchell and Walker (2020), Jackson (2021), Medici (2021). Here we try to explore global aging as an integral problem of the future (Grinin & Korotayev, 2015a, 2016; Grinin, Korotayev & Grinin, 2016; Grinin et al., 2017a, Grinin et al., 2020, 2021b; WHO, 2021; Vollset et al., 2020; Gallagher, 2020; Alkema et al., 2011; Kaneda et al., 2016; Lutz et al., 2018; Raftery et al., 2012).

Table 1 Largest world-system demographic transformations (LDTs)

N	Name	Description	Type of population reproduction	Correspondence to the stages of the production principle	Date[a]	Part of the World-System
1	Neolithic	Transition from medium-high fertility among hunter-gatherers to a model of high fertility among primitive farmers with transition from medium-high to high mortality[b]; although generally small, but more or less constant natural increase appeared. But this is a model of high fertility with a limited ecological niche	Slow growth	Initial and modernization phase of Agrarian Revolution	10–5 thousand years ago	Different regions of Eurasia and North Africa[c]
2	Civilizational	Development of the established model with a high birth rate and the achievement[d] of a somewhat lower mortality on the basis of a capacious ecological niche, which made it possible to increase natural growth and consequently the population growth; as well as improve the quality of life	Slow growth	Final phase of Agrarian Revolution and mature stages of craft-agrarian production principle	5000–500 years ago	In the majority of Afrasian world-system and partly in American world-system
3	Anti-Malthusian	Transition to a model of reduced mortality while maintaining a relatively high birth rate; this provided a high natural increase	Fast growth	Final phase of Industrial Revolution and mature stages of industrial production principles	1700s–1930s	In Europe, North-America, and several other places
4	Third World	Transition to a model of greatly reduced mortality while maintaining a high birth rate; it gave rise to an explosive global population growth	Fast growth	Final stage of industrial production principle and initial phase of Cybernetic Revolution	1890–1980	In low-income countries
5	Adaptation to Aging	Transition to a model of an elderly society with relatively low mortality and very low birth rates, minimal growth and an increase in the share of the elderly and old population	Qualitative changes	Initial, modernization and final phase of Cybernetic Revolution	1960–2050	In the majority of the World System countries

(continued)

Table 1 (continued)

N	Name	Description	Type of population reproduction	Correspondence to the stages of the production principle	Date[a]	Part of the World-System
6	Techno-bio-medical	Transition to a model of total technological support for health and quality of biological life with the stabilization (or reduction) of the population, with a very large proportion of the elderly population	Qualitative changes	Final phase of Cybernetic Revolution, the mature stages of cybernetic production principle, probably beyond it	2060s–2220s	In the World System

[a]Dates are given before the beginning of the next largest world-system demographic transformation while previous world-system demographic transformations are continued in different parts of the World System, coexisting with the new one
[b]At the same time, the birth rate increased significantly more than the death rate
[c] = Afro-Eurasian world-system
[d]In a number of cases
[e]Dates are given before the beginning of the next world-system demographic transformation while previous largest demographic transformations are continued in different parts of the World System, coexisting with the new one

than it is today (i.e., age will become a much more important social marker than it is today).

In short, *cybernetic society is an elderly society with the institutionalization of age differences, relying almost entirely on smart (cybernetic) technologies and self-regulating systems.*

Accordingly, this society will have its strengths and weaknesses, successes, and even triumphs, but it will not be free from some serious problems, like any society that preceded it. Information society has experienced and continues to experience conflicts and tensions and has been forced to introduce many regulations.[7] Cybernetic society must go through a difficult path of social frictions, conflicts, and optimizations before many things become institutionalized. Techno-social regulation has important advantages, but it also includes control over the will and freedom of people. Therefore, a long way is required to optimize regulations and adapt to them in order to take into account a variety of needs, desires, and problems. Even short-term regulation with the help of technology during the COVID-19 pandemic has shown how painful it can be for societies (about the influence of COVID-19 see, e.g., Grinin, 2020; Irshad, 2020; Chumakov & Yurchenko, 2021; Hussain, 2021; Widdowson, 2021; Komlosy, 2023).

First, we will analyze how the aging process will affect and transform society. We must not forget, however, that the aging process itself is due to the increase in life expectancy and can only develop if there is a corresponding or even accelerating development of technologies. In the last section, we will show the direct impact of technology on changes in the organization of the state, political, and social life.

2 Aging and Age-Related Institutionalization in Cybernetic W-Society

Institutions play a huge role in society, being integral systems that mediate between various spheres of life. Institutions arise and develop along with the development of society and changes in its economic, social, and political model. It is essential *that when some important processes are constantly developing, they must inevitably lead to institutionalization*, that is, either create new institutions or greatly change the existing ones, for example, by strengthening institutional relations with the development of various organizations, movements, etc. Let us take an example. Since the beginning of the twentieth century, the women's movement has intensified in Western countries; its role has increased with the world wars, which "called" many millions of women into production. It has also increased with the growth of women's education, women's suffrage, and then very much with the development of contraception, the emergence of family planning, and the decline of the fertility rate.

[7] In particular, the growing contradiction between the values of privacy and constitutional rights to keep secrets (correspondence, personal information), on the one hand, and increasing information transparency, which allows tracking every person, on the other hand.

This dramatically expanded the opportunities for women to actively engage in all spheres of life. As a result, since the 1950s, but especially for the last three decades, the division into men and women, which, of course, has always been very important in any society, has risen to a new level in social and legal terms. There have been many formal acts and informal settings that strongly influence support for women, provide opportunities for legal action, and also require quotas for women on top positions. The latter case shows a shift from the idea of meritocracy, that is, occupying a position according to personal qualities and merit. The situation is similar to some other groups.

These examples give an idea of how age can be institutionalized in future cybernetic society. We are talking about the institutionalization of age, since the role of division into age groups, strata or "age classes" (to use the ethnographic term) will greatly increase.[8] But it will increase, of course, as a result of aging, since an increase in the proportion of the elderly is likely to lead to the fact that they will be opposed to the young and the middle-aged. The role of age in the social and political context will be discussed below. In this section, we will talk in general about what institutions related to aging exist today, and how the institutionalization of age (aging) can develop.

After all, both existing institutions and those that will develop or form later will mediate the links between the age levels of population groups, on the one hand, and the economy, social, legal, technological, cultural and ideological spheres, on the other.

Modern Institutions Associated with Aging In modern society, in fact, there are few institutions that are aimed at helping the elderly to adapt in society, at "healthy aging" (see below) or "dignified aging" for those who cannot serve themselves. This is, firstly, the pension system. But it is far from developed in many countries; moreover, a number of countries have established a so-called solidarity system, where the size of the pension depends little on the contributions made by the individual himself. Accordingly, in the event of financial crises and inflation, the size of pensions is set by the state, and this size may be far from what is necessary for a normal standard of living. But even in societies where pensions are mainly dependent on contributions, current and future pensioners remain hostages of the financial well-being of the state, and the growing share of pensioners makes the

[8] Age groups are one of the most important lines of stratification in some societies of hunter-gatherers and simple pastoralists and farmers, where the most important age transitions were institutionalized. Societies were structurally divided into several age classes, each with its specific rights and responsibilities (see, e.g., Ember et al., 2018). Systems of such institutionally expressed associations or their traces are found among a significant number of peoples of the world, but especially in Africa, particularly in East Africa, where in many societies this institution was very pronounced and functioned as very relevant until very recently (see Kalinovskaya, 1976). According to some researchers, such classes normatively performed a function that **ensures** the social division of labor, including in the military sphere (see, e.g. Kalinovskaya & Markov, 1987, pp. 64–65). On the role of age groups and classes in social development and state formation see Grinin (2011), p. 111.

default on financial obligations increasingly real (on the problems of pension systems see Kapelyushnikov, 2019a, 2019b; Bloom & Luca, 2016; von Weizsäcker & Wijkman, 2018; Randers, 2012; Grinin & Korotayev, 2010, 2016).

Apart from pension systems, other institutions include: (a) a system of social assistance for the elderly through special bodies; (b) health insurance or free medical care (but this is not exclusive to retirees); (c) assisted-living facilities, which are common mainly in high-income countries. However, they are generally available to a small part of the elderly people because of their high cost. If the costs are low and/or equal to the size of a pension, they are more like shelters for homeless senior citizens. But, most elderly people are not particularly willing to move even to comfortable assisted-living facilities, and many of them, in fact, are sent there to die. In addition, the COVID pandemic has shown that such houses are potentially dangerous in terms of the spread of contagious diseases and this can lead to high mortality of their residents (see, for example, Yakushev et al., 2020).

The established institutional forms of support for the elderly were created in the early to mid-twentieth century, when there were much fewer elderly people than today. That is, they were created under completely different demographic and economic conditions, without considering the long-term problems they could generate (Kapelyushnikov, 2019a, 2019b). *Therefore, the aging process requires the development of institutions*.

In fact, the family remains today the main institution that deals with the issue of care for the bulk of the disabled and the integration of relatively healthy elderly people into social activities [for example, housework, childcare, etc. (for example, UN Population Division, 2015)]. However, today we are witnessing a family crisis associated with various reasons, including the fact that many parents, especially in the high-income countries, are trying to let their children start their own lives early and to live separately. At the same time, many elderly people tend not to want to take care of kids and help their children and grandchildren. This model has many advantages, but the weakening of family ties and individualism within the family is not well suited to integrate older people who need care into families. Thus, the main institution that for centuries has been carrying out (for better or for worse) the socialization of the elderly is weakening and will continue to weaken even further, all the more so given the declining birth rate and the increasing number of childless couples and single households.

Challenges and Conflicts on the Path to Adaptation to Aging and Institutionalization of Age Above, we have discussed modern institutions that help the elderly adapt to society. *But there are practically no institutions that would help society adapt to an increase in the number of the elderly*. This is a fundamental idea for understanding of the future challenges of global aging.

On the one hand, as we have already said, the main institution – the family – is weakening. On the other hand, an attempt to completely replace the family with the state, which seemed quite acceptable earlier, today demonstrates a very high probability of a crisis in the near future, since the burden on the state (in the form of rising pensions and social security expenses, medical and other types of care) becomes overwhelming. After all, young cohorts entering the economy are smaller than those that are leaving it due to age. More competent demographic policies might probably keep the birth rate at higher levels [see chapter "Demography. Toward Optimization of Demographic Processes" (Korotayev et al., 2023, this volume)], but still the general trend is toward technological changes that lead to higher life expectancy and unprecedented opportunities for family planning.

Thus, not only the pension system, tax system, healthcare system, labor, and capital market institutions (Kapelyushnikov, 2019a, 2019b, p. 45) but all subsystems of society, including the legal sphere, ideology, and others, must adjust to the situation with a high and constantly growing proportion of the elderly. They are lagging far behind, though some progress has been achieved. Some important UN and WHO policy papers have now been developed [for example, Healthy Aging (UN Population Division, 2015)]. "Healthy aging" is defined as developing and maintaining functional ability that ensures well-being in old age. The concept of well-being brings together all the components and elements of life that are valuable for people (UN Population Division, 2015). Thus, there is a kind of Declaration of the Rights of the Elderly. Similar to the declaration of human rights, it will serve as a guideline in forming institutions. However, it is important to realize that moving toward a more fundamental consideration of the empowerment and rights of older people will inevitably have a strong structural impact on the whole society.

We fully agree that all countries face a serious challenge in preparing their healthcare and social care systems to perform as effectively as possible in the face of this demographic shift (WHO, 2021). However, let us repeat that, despite the fact that population aging has accelerated significantly compared to the previous period, and it will accelerate for a long time to come, in general, the world (and the majority of societies, if not all of them) is clearly insufficiently prepared for the situation when society becomes old. This is all the more alarming since, in absolute terms, the number of older people will grow especially rapidly in low- and middle-income countries (WHO, 2021), which are very poorly prepared for this challenge.

Directions of Institutionalization The main directions of institutionalization, in our view, will be related to:

a) de jure or de facto different age-specific rights. Of course, such a difference in rights exists today, firstly, in relation to minors and the rights of people reaching a certain age to receive a pension. Other than this, legally, adult people are formally equal in rights. However, we believe that in the future the number of rights or restrictions associated with age will increase greatly. This will not necessarily be expressed legally, but may take place in the form of stable public perceptions, judicial precedents, and the like. For example, in most countries, constitutions or constitutional laws declare equal rights for women and men in all spheres. But in fact, as a result of the prevailing judicial practice in family, marriage, and sexual relations, women, in some sense, can have even more rights than men (e.g., in respect of rights of children in the case of divorce, aliments, right on earlier pension age in some countries, etc.; see also above about quotas).[9] We are talking about this without any value judgment goals, just to show how the institutionalization of rights and restrictions (duties) in connection with age division would take place. The restrictions imposed on the elderly (65+) due to COVID-19 also show how rather severe restrictions can be imposed on people according to their age. Accordingly, there can be a lot of combinations of rights, restrictions, and obligations, depending on society and its cultural and historical characteristics, the social position, the economic situation and much more;

b) self-organization of age groups of citizens in various forms ranging from clubs and mutual aid societies to world organizations; from providing mutual assistance to lobbying interests;
c) the structuring of the electorate by age categories and, accordingly, the formation of political and legal programs;
d) the strengthening of the ideological orientation when discussing the rights and obligations of age groups (including all kinds of campaigns like what we see in the feminist movement);
e) the inclusion of the age aspect in all types of planning and activities (e.g., in the planning of buildings, in legal acts and laws, in requirements for employers, and landlords). Here, to a certain extent, the institutionalization of disabled people today can serve as a guideline for the development of the institutionalization of age tomorrow[10];
f) the expansion of judicial and legal activity to protect the age rights of people.

In general, one can assume a rather difficult and conflicting path of adaptation to aging. Accordingly, the problem of ageism, which is already acute today, will manifest itself in ever-new aspects and directions, greatly changing society.[11] We assume that ageism will go through several stages in its development, including quite conflicting ones. But the laws, rules, and institutions gradually worked out by experience will lead to a situation where a system will arise for taking into account the interests of different age groups, fixing their rights, duties, and restrictions, up to quotas in political representation and holding posts (as is observed today in relation to quotas for women and people of color).

Please note that we are talking not only about the institutionalization of the elderly cohorts but the institutionalization of various age groups based on the change in the age composition of society (since ageism can be both in relation to the elderly

[9] However, with regard to equal pay for equal work, although women have achieved a great deal, and somewhere "overdid it," but on the whole, the advantages still remain with men. It is fair that the concept of popular psychology, that it is a man who should be able to provide financial support for his family, prevails.

[10] Here, to a certain extent, the institutionalization of people with disabilities can serve as a guideline.

[11] About the problem of ageism today see chapter "Global Aging – an Integral Problem of the Future. How to Turn a Problem into a Development Driver?" (Grinin et al., 2023b, this volume).

and the young, see below). How this will happen in practice *will depend on the activity of these groups, the specifics of the situation, and the intensity of the social and age struggle.* In general, we can assume a rather difficult and conflict-prone path. However, gradually, in the course of adjusting, adapting to aging, resolving problems and conflicts, more adequate institutions should develop in society, including those related to the institutionalization of ageism (see below), and the optimization of attitudes toward age.

Let us finish this topic with a remark on special settlements. We believe that *nursing homes are unlikely to be the most important institution associated with global aging in the future* (and COVID-19 has clearly confirmed their serious danger to their inhabitants; see above). They are, in fact, still a legacy of previous ideas about charity for the needy.

At the same time, special settlements for the elderly began to appear, but so far, they are common only in a few countries.[12] We do not rule out that there are prospects for such an institution. Such settlements contribute to the formation of the necessary institutions and the involvement of older people in active life, although age segregation also has negative sides, such as separating the elderly population from the mainstream of life and strengthening its conservatism.

3 Aging, Ageism, and Stratification of Cybernetic Society

Aging and Changing Socioeconomic Relations Each level of technology, which we call a production principle [see Chapter "Technology. Limitless Possibilities and Effective Control" (Grinin & Grinin, 2023, this volume) Grinin, 2006; Grinin & Grinin, 2015a, 2015b, 2016; Grinin et al., 2017a, 2017b, 2020, 2021a, 2021b on the production principles], corresponds to a special level, or type of relations.[13] And although there may be very large variations in economic and social relations in each society, some important characteristics of this set of relations can be traced in any society using the appropriate level of technology. We called the main characteristics of such a complex of relations corresponding to the production principle *the type of alienation of labor and personality* (abbreviated as the type of alienation), using the Hegel-Marx term "alienation."

[12]These settlements are special cities and towns where people mostly live ordinary lives, although there may be nursing homes on their territory. The beginning of such a movement was laid by Sun City in Arizona in the early 1960s. Today there are about 500 such communities in the United States, but they are all similar to the Sun City. The main feature of the Sun City is the availability of all kinds of infrastructure and employment centers for retirees. Now this village in Arizona, called Sun City, has about 24,000 households, there are about 40,000 retirees who create thousands of jobs.

[13]Each production principle signifies a transition within the World System to completely new, more effective production systems, which ultimately restructure the entire spectrum of economic activity and its relations. In other words, we are talking not only about new technologies and economic management, but also about a (fundamental) complete and worldwide change in the way of doing things.

The cybernetic production principle corresponds to the *economic type of alienation.*[14] How does it differ from the semi-economic type of alienation characteristic of the industrial production principle? The main difference lies in the fact that in modern society, there are many social guarantees and types of insurance from various problems and misfortunes that can happen to a person (unemployment, accidents, disability, etc.), as well as developed pension and social security systems. What is the economic type of alienation? On the one hand, thanks to a gigantic amount of resources, society turned out to be able to support the weak. Therefore, *each person, in conditions of various freedoms, simply because she or he is a citizen or resident of a certain country, can claim social guarantees and rights (to education, medicine, protection, assistance in old age or disability, etc.).* However, on the other hand, *the economic type of alienation is an opportunity to better demonstrate one's skills, qualifications, talent, and receive many times more than the social minimum.* Consequently, labor coercion factors are becoming weaker, while incentives for activating people's talent and energy are growing (for more details, see Grinin, 2006; Grinin & Grinin, 2015b). The economic type of alienation has evolved throughout the Cybernetic Revolution [see above and chapter "Technology. Limitless Possibilities and Effective Control" (Grinin & Grinin, 2023, this volume) about it] and will continue to evolve. The lack of labor resources significantly contributes to the development of the economic type of alienation because the pressure of unemployment is greatly weakened, so the hard economic reasons for coercion to labor are weakening. In the future, the introduction of some kind of guaranteed income cannot be ruled out.

Thus, *cybernetic society will correspond to the economic type of alienation in its full maturity.*

Very importantly, the above features of cybernetic society will contribute to the movement toward this maturity. Firstly, the development of self-regulating systems will partially solve the problems of labor shortages, first in simple labor (especially in services) and then in increasingly complex labor. And also, quite likely, it will lead to the possibility of a shorter workweek. Secondly, for many reasons, the development of this type of alienation will be greatly influenced by aging, as older people have a guaranteed income in the form of pensions and savings. In addition, unlike young people, they have already acquired a lot of basic necessities, so they will need less borrowing, etc., that is, they will be less dependent on employers. Thus, *the greater the proportion of elderly and old people in the population employed in the economy, the more noticeable will be the features of the economic type of alienation in the future society.*

[14]The hunter-gatherer production principle corresponded to the pre-economic type of alienation; craft-agrarian production principle corresponded to the non-economic type; trade-industrial production principle corresponded to semi-economic type (for more on the types of alienation see Grinin (2006) and Grinin and Grinin (2015c).

The Growing Role of Age Differences We believe that, in general, social division in cybernetic society will be increasingly linked to age differences. It is very interesting that since the second half of the twentieth century the role of class (economic) differences has been decreasing in the social struggle, whereas the differences associated with gender, age, and other anthropological characteristics have become noticeably greater.[15] It is worth mentioning that from the very beginning of the Cybernetic Revolution, that is, in the 1950s–1970s, women and youth movements,[16] as well as movements related to gender identity (LGBTQ+), became more active. The struggle for the rights of racial and ethnic minorities also became more active. Thus, while at the beginning of the Cybernetic Revolution the problem of the relationship between the generations of parents and children was the most important point, it can be assumed that in the final phase of the Cybernetic Revolution (in the 2030s–2070s), the elderly will come to the fore. The problem of generational conflict will arise and take a different form: the relationship between parents and children, on the one hand, and grandparents, on the other.

As we have said above, the process of institutionalization of age categories will intensify. Our idea is that the role of generations in the social division of cybernetic society will grow, and, accordingly, the positions of different generations should be more clearly taken into account in various relations, institutions, and, possibly, laws. Otherwise, the trends of global aging, the introduction of new technologies and the ability to transform the biological nature of human beings (see Grinin & Grinin, 2015b; Grinin et al., 2017a) will create certain ethical and legal difficulties, even more serious than today's gender problem. In any case, age categories will give rise to new social minorities and other problems.[17]

Institutionalization of Ageism In any society, there is a problem of ageism, since people of different ages coexist in it. Ageism is understood as discrimination or other infringement of the rights and opportunities of people on the basis of age in economic, legal, domestic, or psychological forms. Today, the problem of ageism has started to be addressed, but with the emphasis being mainly done on the infringement of the rights of the elderly, and the disabled. It is believed that every second person in the world holds ageist views (WHO, 2021). However, ageism can also occur in relation to the young. For example, they can experience problems trying to get hired, have lower wages, encounter limited opportunities for social mobility due to the dominance of the elderly in high positions, etc. Therefore, ageism can occur in relation to different age groups and from different age groups. As for ageism in relation to the elderly in modern society, it can intensify (including the

[15] Here, modern society began to suddenly echo some hunter-gatherer societies, where property differences were not so important, while gender and age differences were very important (Grinin, 2006; Grinin & Grinin, 2015c).

[16] As we know, youth riots shook the Western world especially in 1968.

[17] We do not rule out that if age segregation develops in settlements (see above), then social divisions like modern fraternities, that is, age territorial associations, as well as all sorts of horizontal ties between such communities, will become important.

self-ageism, that is, elderly people underestimating their own capabilities) as a result of the fact that life expectancy has increased very significantly, while the ideas about appropriate rights, duties, and opportunities of each age have largely remained at the same level. Meanwhile, people who were considered old 80 years ago (for example, at the age of 65) today can be seen as being in a late stage of maturity. It is worth noting that there will be big changes in the sense of age in the future.

It is fundamentally important to realize that the problem of ageism will now persist in society for a long time due to global aging, and we must learn how to live with it. Of course, we must try to smooth it out, e.g., by encouraging older people to increase their self-esteem and also trying to awaken respect for older ages in younger cohorts. But it is impossible to solve this problem fully so that it disappears. *In connection with the growing importance of the problem of age, it will manifest itself in new aspects and directions, greatly changing the society*. And since, as has already been mentioned, the importance of age categories in society will increase, in our opinion, it is more correct to institutionalize some aspects of these categories than to try to provide everyone with equal rights, because the latter may be unrealistic. We think that ageism will go through several stages in its development, including quite conflicting and contradictory ones, and the laws, rules, and compromises gradually developed by experience will lead to a new stage, where a system will take into account the interests of different age groups. Thus, *ageism can be gradually institutionalized in a certain and systematic consideration of the interests of different age groups, fixing their rights, duties and restrictions, up to, perhaps, quotas of political representation, holding positions, etc.*

We also believe that in cybernetic society, due to the increasing social significance of age differences, self-organization around age lines may increase. Already today, for example, self-help groups are springing up, where the "younger elderly" or stronger people help the "older elderly" or people whose abilities are limited. There are examples of this in countries with different levels of development (UN Population Division, 2015). Accordingly, different variants of such social self-organization or institutionalization around the age line are possible. Other variants of this type of social self-organization are also possible.

4 The Impact of Technology on the Social and Political Life of Future Cybernetic Society

As we have already pointed out, in the course of the Cybernetic Revolution, production, economy, and other spheres of life will see many self-regulating systems emerging [see chapter "Technology. Limitless Possibilities and Effective Control" (Grinin & Grinin, 2023, this volume); Grinin, 2006; Grinin & Grinin, 2015a, 2015b, 2016; Grinin et al., 2017a, 2017b, 2020; Grinin et al., 2016]. Among such systems, a special role will be played by socio-technical self-regulating systems (SSSs), which, with the help of artificial intelligence, will regulate various social and administrative relations. SSSs perform social and administrative functions (i.e., control, verification, distribution, security, checking, identification, issue of documents, rating, and other functions) using a set of technologies with little or no involvement of officials and specialists. Thus, they can be used by authorities at different levels and by the state as a whole (as well as by the administration of service centers where the authorities consider such regulation necessary: airports, crowded places, etc.). That is why the development of SSSs in one way or another, pushes society toward the formation of electronic state.

What is an electronic state? Traditionally (according to the United Nations e-Government Development Database), e-government has been defined as the use of ICTs to improve the efficiency of government institutions and deliver government services online (UNeGovDD, 2013). However, the concept of e-state in this respect is something more advanced than e-government. We understand e-state as a state with a significantly reduced number of state oversight bodies, mainly based on SSS technology (see Grinin et al., 2021a). This can also affect democratic procedures (see below).

The development of the SSS is currently proceeding rapidly. In some cases, they are beginning to be used to impose certain behaviors on people. Most notable in this regard is China, where officials have developed a Social Credit System for individuals, businesses, and the government to monitor and assess their reliability, for example, to control nonresidents (Kuznetsova & Mashkina, 2020). Many projects of the Social Credit System have been implemented only as regional pilot programs, although there are plans to expand them throughout the country (Chin & Wong, 2016; Creemers, 2018), they can become a model for other countries (Síthigh & Siems, 2019). 2020 was, in some ways, a turning point in that regard. For example, the use of facial recognition systems [including faces in masks (Ramanathan et al., 2021)] increased by about 7% and was expected to grow another 12% by 2022 (Technavio, 2020). A well-known vpn company reported having location tracking systems already in 42 messaging apps with at least 187 million downloads (Sean, 2021), and their number increased rapidly during the pandemic (Stanley & Granick, 2020). At the moment, the Chinese authorities use very strong anti-covid measures by means of SSS, which include drones.

The pandemic and the emerging need for control of anti-epidemic measures have significantly accelerated the development of such SSSs. In particular, e-pass systems, QR codes, COVID-immunity passports, movement control, punishment system, accounting systems, business control systems, and many other systems appeared as a result of blockages and can be considered as prototypes of the SSSs (Ram & Gray, 2020; Wang et al., 2020; Hasan et al., 2020; WHO, 2020; Barnes, 2021; Grinin et al., 2021a). Thus, the coronavirus pandemic has shown that the introduction of SSSs can be a rapid process that will unfold over 10–20 years. In the future, it can be expected that the development of socio-technical self-regulating systems will go in different directions, continuing to develop in the field of medicine. For example, there may be systems for continuous monitoring of the health status and critical parameters not only of patients in hospitals but also of people staying at home (Grinin & Grinin, 2016; Grinin et al., 2021a), including the chronically ill and the elderly.

Since such SSSs can regulate the entire society, it is already extremely important today to clearly regulate the possibilities for manipulating personal data and using them for social regulation. It is necessary to take into account the existence of different types of behavior that have the right to exist in society. Meanwhile, the SSSs can lead to total unification, thereby suppressing the individuality and special abilities of people.

The formation of SSSs will lead to serious changes in the political and social spheres:

a) as a result of the spread of SSSs, the number of officials will decrease, and interaction between citizens and state bodies bypassing local and intermediary bodies will become possible; in general, this will greatly affect the number of people employed in public administration;
b) the state will become cheap, but at the same time, digital control will increase, which can cause strong behavioral stratification (social sphere) and protests;
c) such development will lead to the formation of an e-state, in which it is quite possible that a transition from representative to direct democracy will take place. Direct democracy is good for getting feedback from the population quickly, but it can undermine the stability of governance (especially in large or highly divided societies) because of the rapid and frequent changes in public mood. Moreover, it will become difficult to make long-term decisions. It should be noted that direct democracy can increase the influence of the growing older part of society;
d) we also believe that the widespread introduction of SSSs can accelerate the change in social relations and increase confrontation in society because the use of self-regulating systems in governance will lead to increased technological and political control. This is all the more dangerous because, although such systems have not yet become widespread in private life, they will be.

Thus, both the introduction of socio-technical self-regulating systems and the development of the aging process must be adjusted and institutionalized sooner or later, then enshrined in laws or stable relations in such a way as to optimize interests and minimize dangerous consequences. If such an institutionalization does not take place, then the following will intensify: (a) social tensions; (b) the negative impact of the growing proportion of the elderly on democratic institutions, which, as mentioned above, will be more noticeable in direct democracy.

As a result, the aging process can turn a democracy into a gerontocracy. Consequently, a crisis in democratic government as a whole is quite likely, and even the very foundations of democracy may be undermined (Peterson, 1999). A drift toward gerontocracy is possible, that is, to a situation where the elderly population predominates among voters, which can lead to conflicts between generations and a change in political psychology. Obviously, as life expectancy increases and the proportion of young people in the structure of society decreases, the proportion of older people will also increase (Fukuyama, 2002; Grinin et al., 2020). And since the older generation is more conservative in their habits, this can affect the choice of policies and many other aspects that can put the younger and middle generations at a disadvantage.

Thus, in the last third of the twenty-first century, the society that we have called cybernetic will form in full maturity. Technology in it will regulate much more relationships than today and will maintain a very high life expectancy and health of the population, especially of the increasingly important elderly part of the population. The process of aging will give this society special dimensions and characteristics associated with age gradations.

5 Subsystems of Cybernetic Society

Let us consider the expected shape of cybernetic society in its subsystems (see Fig. 1):

1) in the technological sphere, active development of self-regulating systems is expected, which will allow: (a) ensuring an increase in life expectancy and an improvement in health-related (or biological) quality of life, which will intensify the process of global aging; (b) changing political, administrative, managerial, social and economic relations;
2) in the demographic sphere, we expect: (a) stabilization of the population (or movement toward it) on a global scale[18]; (b) an increase in the proportion of the elderly population in some societies, which will lead to a noticeable increase in the social and political significance of these age groups;
3) in the economic sphere, we expect that: (a) the aging of the population will lead to certain conservatism in needs and demand (Grinin et al., 2020). This will noticeably reduce consumerism and help rebuild the supply, demand, and consumption model (Grinin et al. 2023c). In particular, the movement from the acquisition of goods toward the improvement of the biological quality of life and healthy lifestyle will intensify; (b) capitalism as an economic structure will persist,[19] while market relations will change toward greater stability and predictability (that is, market volatility will decrease) due to the fact that self-managing systems will be leading technologies, and they essentially require certain stability;
4) in politics and administration, the development of self-regulating systems will entail very significant, one might say revolutionary, changes in the field of governance because many social and administrative relations will be regulated

[18]See chapter "Demography. Toward Optimization of Demographic Processes" (Korotayev et al., 2023, this volume),

[19]For the evolution of capitalism, see chapter "High Income and Low Income Countries. Towards a Common Goal at Different Speeds" (Grinin et al., 2023e, this volume) and chapter "Modeling Social Self-Organization and Historical Dynamics. Global Phase Transitions" (Malkov et al., 2023, this volume).

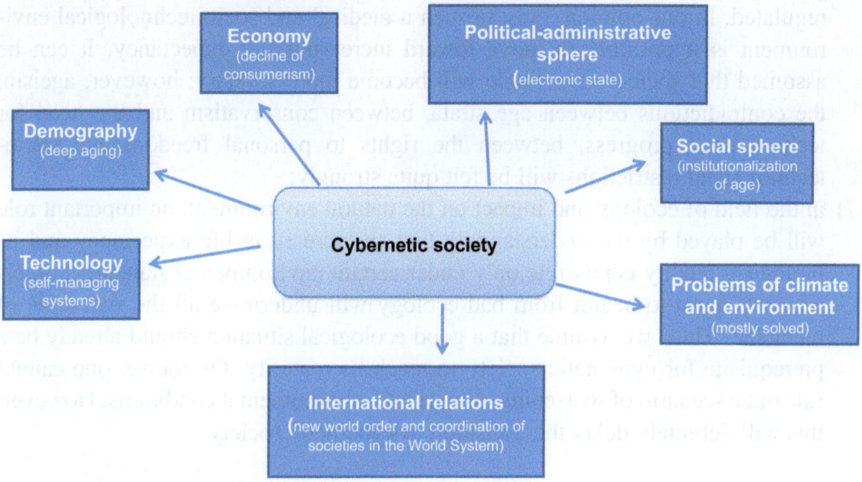

Fig. 1 Subsystems of cybernetic society

by these systems. This will take place both at the level of individual administrative units and cities (so-called smart cities), and at the level of the state as a whole. As a result, what we have called the "electronic state" (e-state) will emerge. This will lead to a sharp reduction in the number of bureaucrats, as well as to the "cheapening" of the state, and to a reduction in the vices associated with management (corruption, red tape, etc.). Also, a transition to direct democracy through permanent electronic voting is likely. But this will also create a number of problems, which we have mentioned above. We anticipate this will result in a special kind of political regime, like democratic authoritarianism. Finally, the role of the elderly in politics will increase dramatically and, accordingly, age will become an important line of political agitation;

5) in international relations, as a result of changes in individual societies, as well as the emergence of a new world order, the coordination of societies will increase.[20] It is also very likely that countries with older populations will become more peaceful, with less ambition to wage war (Jackson et al., 2008);
6) in the social sphere, it is expected that: (a) age will become a very important social marker; (b) regulation using the self-regulating systems will cover society from top to bottom, which can cause social tensions; (c) social tensions between different age cohorts are very likely, as a result of which the institutionalization of age is expected to occur; (d) the importance of the relationship between the individual and technology will increase, and a kind of socio-technological sphere will arise, in which people will be forced to stay almost constantly. This will affect, first of all, the elderly and the disabled, whose lives will be significantly

[20] See chapter "Future Political Change. Toward a More Efficient World Order" (Grinin et al., 2023a, this volume).

regulated. In our opinion, only in such a medical and socio-technological environment is it possible to move toward increasing life expectancy. It can be assumed that society as a whole will become more solidary; however, ageism, the contradictions between age strata, between conservatism and the need for technological progress, between the rights to personal freedom and socio-technological restrictions will be felt quite strongly;

7) in the field of ecology and impact on the natural environment, an important role will be played by the understanding that an increase in life expectancy and its biological quality is possible only under certain environmental standards; otherwise the harm to health from bad ecology will undermine all the successes of medicine. Thus, we assume that a good ecological situation should already be a prerequisite for cybernetic society to reach its maturity. Of course, one cannot rule out a scenario of worsening climate and environmental conditions. However, this will definitely delay the transition to cybernetic society.

Thus, we believe that the role of age groups in the social division of cybernetic society will expand, and the positions of different generations will be explicitly taken into account in various relations and laws. But the path to this will be uneasy, often conflicting, and affect almost every person's interests. However, we think that due to the completion of the Cybernetic Revolution and complex socio-demographic transformations, in the last third or quarter of the twenty-first century, cybernetic society will fully mature in many countries.

Technology will regulate many more relationships in it than it does today, and technology will support very high life expectancy and the health of the population, especially its increasingly important elderly part. And the aging process will give this society special dimensions and characteristics associated with age gradations.

6 Additional Notes

Above, we have mentioned that future Cybernetic W-society will be different from all previous societies; it will have fundamentally new characteristics. For a better understanding of this important point, we present several tables.

As we said above, Table 1 shows the largest demographic transformations (LDTs) in their combination with types of population reproduction (TPRs), production revolutions (Agrarian, Industrial, Cybernetic) and production principles (craft-agrarian, trade-industrial, and scientific-cybernetic). For these revolutions and production principles, see above.[21] Changes in the World System are also shown.

[21] For detail see chapter "Technology. Limitless Possibilities, Effective Control" (Grinin & Grinin, 2023, this volume); see also Grinin and Grinin (2015c, 2016), Grinin and Korotayev (2015b), Grinin et al. (2020).

We identify six largest world-system demographic transformations (LDTs) and four types of population reproduction (TPRs) in the historical process.[22] It is worth making clarifications about the correlation between LDT and TPR. Each production principle corresponds to its own type of population reproduction (see Table 2 below). However, the latter is not formed immediately, but, as it were, in two stages due to the fact that technological development in the production principle has a staged development. Therefore, for each type of population reproduction (TPR), we identified two major demographic transformations (see column 4 in Table 1). However, the first, i.e., the hunting-gatherer production principle, is not associated with any largest demographic transformation (LDT). Therefore, the remaining three TPRs (corresponding to craft-agrarian, trade-industrial, and scientific-cybernetic production principles) quite logically correspond to the six largest demographic transformations. Their characteristics are systematized in Table 1 ("The largest world-systemic demographic transformations.")

The first two largest demographic transformations (LDT) relate to the craft-agrarian production principle (one took place in the process of the Agrarian Revolution, the second began in the final phase of the Agrarian Revolution and ended already in the mature stages of the craft-agrarian production principle). The third LDT occurred during the period of the trade-industrial production principle, but the fourth demographic transformation took place at the end of the industrial production principle and the start of the Cybernetic Revolution (it swept the low-income countries, as we have already said above).

Finally, during the period of the first two phases of the Cybernetic Revolution (the 1950s–2020s), the fifth LDT is actively going on and the features of a new type of population reproduction (TPR) are being formed.

And at the final phase of the Cybernetic Revolution, we assume the beginning of the transition to a new LDT, which will take place already at the mature phases of the scientific-cybernetic production principle. So, for the formation of future cybernetic society, we expect a new (future) largest demographic transformation, which we denote as "techno-bio-medical" (see line 6 in Table 1). Also, the fourth type of population reproduction (TPR) will be formed (see line 4 in Table 2) although many of its features have already been identified (see chapter "Modeling Social Self-Organization and Historical Dynamics. Modern Society and a Look into the Global Future: Cybernetic W-Society" (Akaev et al., 2023, this volume).

The dynamics of demographic and social transformations and characteristics during the scientific-cybernetic production principle are shown in Table 3.

The main characteristics of all four types of population reproduction (TPRs) are described in Table 2 ("Production principle and type of population reproduction").

---

[22] About some details of demographic transformations see chapter "Global Aging – an Integral Problem of the Future. How to Turn a Problem into a Development Driver?" (Grinin et al., 2023b, this volume), chapter "Socio-Political Transformations. A Difficult Path to Cybernetic Society" (Grinin et al., 2023d, this volume), chapter "Modeling Social Self-Organization and Historical Dynamics. Modern Society and a Look into the Global Future: Cybernetic W-Society" (Akaev, Malkov, Grinin et al., 2023, this volume), and chapter "Demography. Toward Optimization of Demographic Processes" (Korotayev et al., 2023).

Table 2 Production principles and types of population reproduction

N	Production Principle	Type of Population Reproduction (TPR)	Birth rate	Death rate	Natural increase	Life expectancy	Settlement type	Quality of biological life[a]
1	Hunter-gatherer	Simple reproduction (natural)	Medium-high[b]	Medium-high for all population groups[c]	Very low	Medium-low	No permanent settlements	Low
2	Craft-agrarian	Slow growth (socio-natural)	Very high[d]	High (especially for children)[e]	Low with a tendency to slow acceleration	Average low; but higher for adults	Rural settlements, but the number of dwellers is gradually growing	Low (while improving in some aspects)
3	Trade-industrial	Fast growth (socio-quantitative)	High	Medium-low	High	Medium-high	Cities	Medium
4.	Scientific-cybernetic	Qualitative changes (socio-qualitative)	Low	Low	Low	High and very high	De-urbanization	High

[a]Includes a certain comfort, the development of medicine and assistance, including the one for the elderly and the infirm
[b]25–30 life births per thousand per year
[c]25–30 deaths per thousand per year
[d]More than 45 life births per thousand per year
[e]More than 40 deaths per thousand per year

Table 3 Demographic changes during the scientific-cybernetic principle of production

Phase	Dates	Techno-social changes	Demographic changes	Limitations	Population at the start of the phase
Initial phase of the Cybernetic Revolution	1950s–1990s	Powerful growth of science, technology, and medicine. Major political and social changes in the World System; liberation of the colonies; the rise of social policies	In low-income countries, due to the reduction in mortality, very rapid population growth began; in high-income countries, after a period of rising birth rates, there was a transition from a decrease in the birth rate to the level of simple reproduction	Huge gap in the level of development of the core and periphery of the World System; advances in medicine, the rise of women's education and rights, and economic policy led to declining birth rates in the developed world	2.5 billion
Modernization phase of the Cybernetic Revolution	1990s–2030s	Mass diffusion of ICT and AI, growth of education in the low-income countries	Transition to TPR of high life expectancy, population aging, and low population aging in most parts of the World System	In terms of reducing the birth rate, traditional and religious ideologies and practices are a limitation	5.27 billion[a]
Final phase of the Cybernetic Revolution	2030s–2070s	The emergence of major breakthroughs in medicine, the impact of aging on society; the struggle for a new world order	Global aging and adaptation to it; possible completion of the demographic transition in Africa	Traditional and religious ideologies in Africa and parts of Islamic countries	8.37–8.65 billion (forecast)[b]
Mature phases of the scientific-cybernetic production principle	2070s–2120s	The growth of conservatism, the final formation of new social institutions associated with aging	Noticeable increase in life expectancy; growth of adaptability to aging and disability due to the mass dissemination of techno-bio-medical technologies among the population[c]	The conservatism of the aging society; dangers of excessive techno-medical power; unpredictable dangers	10–12 billion (according to different forecasts)[d]

[a]Akimov (1999), Shcherbakova (2019)
[b]UN Population Division (2022)
[c]Among all population groups, but especially among the elderly, disabled and incapacitated (requiring constant care)
[d]See chapter "Demography. Toward Optimization of Demographic Processes" (Korotayev et al., 2023, this volume)

The formation of TPR took place during the entire production principle and could even partially involve the next production principle (as can be seen by the example of the trade-industrial and scientific-cybernetic production principles). It is very important that the type of population reproduction associated with the cybernetic production principle changes and will change the previous trajectory of the development of demographic trends dramatically, replacing quantitative growth with qualitative changes in the life of the population and each person. Therefore, an important conclusion can be drawn: in contrast to the rate of technological growth, which accelerates over time from phase to phase (see Grinin et al., 2020), the rate of demographic growth from phase to phase has begun to slow down since the 1970s.

References

Akaev, A., Malkov, S., Davydova, O., Kovaleva, N., Malkov, A., Grinin, L., & Korotayev, A. (2023). Modeling social self-organization and historical dynamics: Industrial society. In: V. Sadovnichy et al. (Eds.), *Reconsidering the limits to growth. A report to the Russian Association of the Club of Rome* (pp. 337–385). Springer. https://doi.org/10.1007/978-3-031-34999-7_17

Akaev, A., Malkov, S., Grinin, L., Bilyuga, S., Davydova, O., Grinin, A., Kovaleva, N., Malkov, A., Musieva, J., & Korotayev, A. (2023). Modeling social self-organization and historical dynamics. Modern society and a look into the global future: Cybernetic W-society. In: V. Sadovnichy et al. (Eds.), *Reconsidering the limits to growth. A report to the Russian Association of the Club of Rome* (pp. 419–460). Springer. https://doi.org/10.1007/978-3-031-34999-7_19

Akimov, A. (1999). Na poroge sedmogo milliarda. *Naseleniye i Obshchestvo, 39*, 1–4.

Alkema, L., Raftery, A. E., Gerland, P., Clark, S. J., Pelletier, F., Buettner, T., & Heilig, G. K. (2011). Probabilistic projections of the total fertility rate for all countries. *Demography, 48*(3), 815–839. https://doi.org/10.1007/s13524-011-0040-5

Barnes, R. (2021). Microsoft, Oracle and other tech giants team up on Covid-19 vaccine "passports". *Tech Compsmag*. News - Tech News, Latest News, Business News and world News. Retrieved from https://www.compsmag.com/news/technology/microsoft-oracle-and-other-tech-giants-team-up-on-covid-19-vaccine-passports/

Barsukov, V. N. (2019). From the demographic dividend to population aging: World trends in the system-wide transition. *Economic and Social Changes: Facts, Trends, Forecast, 12*(4), 167–182. https://doi.org/10.15838/esc.2019.4.64.11

Bengtson, V. (Ed.). (2018). *Global aging and challenges to families*. Routledge.

Bloom, D. E., & Luca, D. L. (2016). The global demography of aging: Facts, explanation, future. *Handbook of Economics of Population Aging, 1A*, 3–56.

Bös, D., & Von Weizsäcker, R. K. (1989). Economic consequences of an aging population. *European Economic Review, 33*(2–3), 345–354.

Chin, J., & Wong, G. (2016). China's new tool for social control: A credit rating for everything. *The Wall Street Journal, 28*. Retrieved from https://www.wsj.com/articles/chinas-new-tool-for-social-control-a-credit-rating-for-everything-1480351590

Chumakov, A. N., & Yurchenko, P. S. (2021). Global pandemic in terms of objective causes and subjective speculations. *Journal of Globalization Studies, 12*(2), 157–168. https://doi.org/10.30884/jogs/2021.02.08

Clark, R. J., & Spengler, J. J. (1980). *The economics of individual population aging*. Cambridge University Press.

Creemers, R. (2018). China's social credit system: An evolving practice of control. *Rochester, Social Science Research Network.* https://doi.org/10.2139/ssrn.3175792

Ember, C., Ember, M., & Peregrine, P. (2018). *Anthropology* (15th ed.). Pearson Publishing.

Fichtner, J. J. (2018). Global aging and public finance. *Business Economics, 53*(2), 72–78.

Fukuyama, F. (2002). *Our post-human future: consequences of the bio-technology revolution.* Farrar, Straus, and Giroux.

Gallagher, J. (2020). *Fertility rate: 'Jaw-dropping' global crash in children being born.* Retrieved from https://www.bbc.com/news/health-53409521

Goldstone, J. A. (2015). Population aging and global economic growth. In J. A. Goldstone, L. Grinin, & A. Korotayev (Eds.), *History & mathematics: Demography & aging* (pp. 147–155). Uchitel.

Goldstone, J. A., Grinin, L., & Korotayev, A. (2015). Research into global aging and its consequences. In J. A. Goldstone, L. Grinin, & A. Korotayev (Eds.), *History & mathematics: Political demography & global aging* (pp. 5–9). Uchitel.

Grinin, L. (2006). *Proizvoditel'nyye sily i istoricheskiy protsess.* URSS.

Grinin, L. (2011). *Gosudarstvo i istoricheskiy protsess. Epokha formirovaniya rannego gosudarstva: Obshchiy kontekst sotsial'noy evolyutsii pri obrazovanii gosudarstva.* LKI.

Grinin, L. E. (2020). How can COVID-19 change geopolitics and economy? *Journal of Globalization Studies, 11*(2), 121–134. https://doi.org/10.30884/jogs/2020.02.09

Grinin, L. (2021). How globalization shifts the global economic and political balance. *Journal of Chinese Philosophy, 48*(2), 222–234.

Grinin, L. (2022). Revolutions of the 21st century as a factor of the world system reconfiguration. In J. A. Goldstone, L. Grinin, & A. Korotayev (Eds.), *Handbook of revolutions in the 21st century: The new waves of revolutions, and the causes and effects of disruptive political change* (pp. 975–1000). Springer. https://doi.org/10.1007/978-3-030-86468-2_38

Grinin, L., & Grinin, A. (2015a). Kiberneticheskaya revolyutsiya i shestoy tekhno-logicheskiy uklad. *Istoricheskaya Psikhologiya i Sotsiologiya Istorii, 8*(1), 172–197.

Grinin, L., & Grinin, A. (2015b). The cybernetic revolution and historical process. *Social Evolution and History, 14*(1), 125–184.

Grinin, L., & Grinin, A. (2015c). *Ot rubil do nanorobotov. Mir na puti k epokhe samoupravlyayemykh sistem (istoriya tekhnologiy i opisaniye ikh budushchego).* Uchitel.

Grinin, L., & Grinin, A. (2016). *The cybernetic revolution and the forthcoming epoch of self-regulating systems.* Uchitel.

Grinin, L., & Grinin, A. (2023). Technology. Limitless possibilities and effective control. In: V. Sadovnichy et al. (Eds.), Reconsidering the limits to growth. A report to the Russian Association of the Club of Rome (pp. 139–154). Springer. https://doi.org/10.1007/978-3-031-34999-7_8

Grinin, L., Grinin, A., & Korotayev, A. (2017a). Forthcoming Kondratieff wave, cybernetic revolution, and global aging. *Technological Forecasting and Social Change, 115*, 52–68. https://doi.org/10.1016/j.techfore.2016.09.017

Grinin, L., Grinin, A., & Korotayev, A. (2017b). The MANBRIC-technologies in the forthcoming technological revolution. In T. Devezas, J. Leitão, & A. Sarygulov (Eds.), *Industry 4.0 – entrepreneurship and structural change in the new digital landscape: What is coming on along with the fourth industrial revolution* (pp. 243–261). Springer. https://doi.org/10.1007/978-3-319-49604-7_13

Grinin, L., Grinin, A., & Korotayev, A. (2020). A quantitative analysis of worldwide long-term technology growth: From 40,000 BCE to the early 22nd century. *Technological Forecasting and Social Change, 155*(June), 119955. https://doi.org/10.1016/j.techfore.2020.119955

Grinin, L., Grinin, A., & Korotayev, A. (2021a). COVID-19 pandemic as a trigger for the acceleration of the cybernetic revolution, transition from e-government to e-state, and change in social relations. *Technological Forecasting and Social Change, 175*, 1–15. https://doi.org/10.1016/j.techfore.2021.121348

Grinin, L., Grinin, A., & Korotayev, A. (2021b). Global trends and forecasts of the 21st century. *World Futures, 77*(5), 335–370. https://doi.org/10.1080/02604027.2021.1949939

Grinin, L., Grinin, A., & Korotayev, A. (2023a). Future political change. Toward a more efficient world order. In: V. Sadovnichy et al. (Eds.), *Reconsidering the limits to growth. A report to the Russian Association of the Club of Rome* (pp. 191–206). Springer. https://doi.org/10.1007/978-3-031-34999-7_11

Grinin, L., Grinin, A., & Korotayev, A. (2023b). Global aging: An integral problem of the future. How to turn a problem into a development driver? In: V. Sadovnichy et al. (Eds.), *Reconsidering the limits to growth. A report to the Russian Association of the Club of Rome* (pp. 117–135). Springer. https://doi.org/10.1007/978-3-031-34999-7_7

Grinin, L., Grinin, A., & Malkov, S. (2023c). Economics: Optimizing growth. In: V. Sadovnichy et al. (Eds.), *Reconsidering the limits to growth. A report to the Russian Association of the Club of Rome* (pp. 155–168). Springer. https://doi.org/10.1007/978-3-031-34999-7_9

Grinin, L., & Grinin, A. Malkov, S. (2023d). Socio-political transformations: A difficult path to cybernetic society. In: V. Sadovnichy et al. (Eds.), *Reconsidering the limits to growth. A report to the Russian Association of the Club of Rome* (pp. 169–189). Springer. https://doi.org/10.1007/978-3-031-34999-7_10

Grinin, L., Malkov, S., & Korotayev, A. (2023e). Developed and low income countries. Towards a common goal at different speeds. In: V. Sadovnichy et al. (Eds.), *Reconsidering the limits to growth. A report to the Russian Association of the Club of Rome* (pp. 169–189). Springer. https://doi.org/10.1007/978-3-031-34999-7_10

Grinin, L., & Korotayev, A. (2010). Will the global crisis lead to global transformations? The coming epoch of new coalitions. *Journal of Globalization Studies, 1*(2), 166–183.

Grinin, L., & Korotayev, A. (2014a). Revolution vs democracy (revolution and counterrevolution in Egypt). *Polis-Politicheskiye Issledovaniya, 3*, 139–158. https://doi.org/10.17976/jpps/2014.03.09

Grinin, L., & Korotayev, A. (2014b). Globalization shuffles cards of the world pack: In which direction is the global economic-political balance shifting? *World Futures, 70*(8), 515–545. https://doi.org/10.1080/02604027.2014.982474

Grinin, L., & Korotayev, A. (2015a). Global'noye stareniye naseleniya, shestoy tekhnologicheskiy uklad i mirovaya finansovaya Sistema. In L. Grinin & A. Korotayev (Eds.), *Istoriya i matematika: futurologicheskiye i metodologicheskiye aspekty* (pp. 31–56). Uchitel.

Grinin, L., & Korotayev, A. (2015b). *Great divergence and great convergence. A global perspective*. Springer. https://doi.org/10.1007/978-3-319-17780-9

Grinin, L., & Korotayev, A. (2016). Global population aging, the sixth Kondratieff wave, and the global financial system. *Journal of Globalization Studies, 7*(2), 11–31.

Grinin, L., Korotayev, A., & Grinin, A. (2016). Global'noye stareniye naseleniya, shestoy kondrat'yevskiy tsikl i kiberneticheskaya revolyutsiya. *Problemy teorii i praktiki upravleniya, 7*, 54–59.

Grinin, L., Korotayev, A., & Tausch, A. (2016). *Economic cycles, crises, and the global periphery*. Springer. https://doi.org/10.1007/978-3-319-41262-7

Hussain, M. (2021). The changing dynamics of liberal international order: Covid-19 and its global implications. *Journal of Globalization Studies, 12*(2), 143–156. https://doi.org/10.30884/jogs/2021.02.07

Hasan, H. R., Salah, K., Jayaraman, R., Arshad, J., Yaqoob, I., Omar, M., & Ellahham, S. (2020). *Blockchain-based solution for COVID-19 digital medical passports and immunity certificates*. Retrieved from https://doi.org/10.1109/ACCESS.2020.3043350

Hsu, Y. H., & Lo, H. C. (2019). The impacts of population aging on saving, capital formation, and economic growth. *American Journal of Industrial and Business Management, 9*(12), 2231–2246. https://doi.org/10.4236/ajibm.2019.912148

Irshad, M. S. (2020). Covid-19 and Global 'At-Risk Community': From benefit-sharing to risk-sharing of economic crisis. *Journal of Globalization Studies, 11*(2), 135–145. https://doi.org/10.30884/jogs/2020.02.10

Jackson, R. (2021). *The macro challenges of population aging*. The Concord Coalition.

Jackson, R., Howe, N., Strauss, R., & Nakasima, K. (2008). *The graying of the great powers. Demography and geopolitics in the 21st century*. Center for Strategic & International Studies.

Jackson, R., Macaranas, R., & Peter, T. (2013). *U.S. development policy in an aging world. New challenges and new priorities for a new demographic era. A research report from project on U.S. Leadership*. CSIS (Center for Strategic and International Studies).

Kalinovskaya, K. (1976). *Age groups of the peoples of East Africa. Structure and functions*. Nauka.

Kalinovskaya, K., & Markov, G. (1987). Obshchestvennoye razdeleniye truda u skotovodcheskikh narodov Azii i Afriki. *Vestnik MGU. Seriya «Istoriya», 6*, 56–69.

Kaneda, T., Greenbaum, C., & Patierno, K. (2016) 2018 World population data sheet. *Population Reference Bureau*. Retrieved from https://www.prb.org/2016-world-population-data-sheet/

Kapelyushnikov, R. (2019a). *Fenomen stareniya naseleniya: ekonomicheskiye effekty*. Institut ekonomiki RAN.

Kapelyushnikov, R. (2019b). Fenomen stareniya naseleniya: ekonomicheskiye effekty. *Ekonomicheskaya Politika, 14*(2), 8–63.

Komlosy, A. (2023). Turn of an Era. Corona, big data and the cybernetic future. *Journal of Globalization Studies, 14*(1), 190–192. https://doi.org/10.30884/jogs/2023.01.12

Korotayev, A., Malkov, S., & Musieva, J. (2023). Demography. Toward optimization of demographic processes. In: V. Sadovnichy et al. (Eds.), *Reconsidering the limits to growth. A report to the Russian Association of the Club of Rome* (pp. 97–116). Springer. https://doi.org/10.1007/978-3-031-34999-7_6

Kuznetsova, V., & Mashkina, O. (2020). Trudovaya migratsiya i problemy obucheniya detey rabochikh-migrantov v Kitaye. *Istoriya I Sovremennost', 2*(36), 46–69.

Lee, R., & Mason, A. (2011). *Population aging and the generational economy: A global perspective*. Edward Elgar.

Lee, R. D., Arthur, W. B., & Rodgers, G. (1988). *Economics of changing age distributions in high income countries*. Oxford University Press.

Lee, R. D., Mason, A., & Cotlear, D. (2010). *Some economic consequences of global aging. A discussion note for the World Bank*. World Bank.

Lutz, W., Goujon, A., Kc, S., Stonawski, M., & Stilianakis, N. (2018). *Demographic and human capital scenarios for the 21st century: 2018 assessment for 201 countries*. Publications Office of the European Union.

Maestas, N., Mullen, K.J., & Powell, D. (2016). The effect of population aging on economic growth. The labor force and productivity. .

Malkov, S., Grinin, L., Grinin, A., Musieva, J., & Korotayev, A. (2023a). Modeling social self-organization and historical dynamics. Global phase transitions. In: V. Sadovnichy et al. (Eds.), *Reconsidering the limits to growth. A report to the Russian Association of the Club of Rome* (pp. 387–417). Springer. https://doi.org/10.1007/978-3-031-34999-7_18

Malkov, S., Kovaleva, N., Grinin, L., & Korotayev, A. (2023b). Modeling social self-organization and historical dynamics. Agrarian society. In: V. Sadovnichy et al. (Eds.), *Reconsidering the limits to growth. A report to the Russian Association of the Club of Rome* (pp. 309–335). Springer. https://doi.org/10.1007/978-3-031-34999-7_16

McDaniel, S. A., & Zimmer, Z. (2016). *Global aging in the twenty-first century: Challenges, opportunities and implications*. Routledge.

Medici, A. C. (2021). *Health sector challenges and policies in the context of aging populations*. United Nations, Department of Economics and Social Affairs, Population Division.

Mitchell, E., & Walker, R. (2020). Global aging: successes, challenges and opportunities. *British Journal of Hospital Medicine, 81*(2), 1–9. https://doi.org/10.12968/hmed.2019.0377

Orlická, E. (2015). Impact of population aging and elderly poverty on macroeconomic aggregates. *Procedia Economics and Finance, 30*, 598–605. Retrieved from https://reader.elsevier.com/reader/sd/pii/S2212567115012721?token=288EB7A0AC3FDB18A990AC8D7EDC3CB6B011C420BEB3125FB6F4BB002FA87B096BE4D58E889373F940046194EEDCC0E8&originRegion=eu-west-1&originCreation=20220203133449

Peterson, P. G. (1999). Gray down: The global aging crisis. *Foreign Affairs, 78*(1), 42–55.

Popova, S., & Yanik, A. (2014). Problemy global'nogo stareniya naseleniya: Analiz dokumentov i strategii OON. *Mezhdunarodnoye pravo i mezhdunarodnyye organizatsii, 3*, 429–443.
Powell, J. L., & Khan, H. T. (2013). Aging and globalization: A global analysis. *Journal of Globalization Studies, 4*(1), 137–146.
Raftery, A. E., Li, N., Ševčíková, H., Gerland, P., & Heilig, G. K. (2012). Bayesian probabilistic population projections for all counties. *Proceedings of the National Academy of Sciences, 109*(35), 13915–13921. https://doi.org/10.1073/pnas.1211452109
Ram, N., & Gray, D. (2020). Mass surveillance in the age of COVID-19. *Journal of Law and the Biosciences, 7*(lsaa023). https://doi.org/10.1093/jlb/lsaa023
Ramanathan, S. S. K., Basha, R. F. K., & Banu, A. (2021). A novel face recognition technology to enhance health and safety measures in hospitals using SBC in pandemic prone areas. *Materials Today: Proceedings, 45*, 2584–2588. https://doi.org/10.1016/j.matpr.2020.11.336
Randers, J. (2012). *2052: A global forecast for the next forty years. A report to the Club of Rome*. Chelsea Green Publishing.
Randers, J., Rockström, J., Stoknes, P. E., Golüke, U., Collste, D., & Cornell, S. (2018). *Transformation is feasible. A report to the Club of Rome*. Stockholm Resilience Centre.
Sean. (2021). We analyzed 450 apps and found location trackers in every one. *ExpressVPN*. Home of internet privacy. Retrieved from https://www.expressvpn.com/blog/digital-security-lab-location-trackers-smartphone-apps-research/
Shcherbakova, Y. (2019). Stareniye naseleniya mira po otsenkam OON 2019 goda. *Demoskop Weekly*, pp. 837–838. Retrieved from http://demoscope.ru/weekly/2019/0837/barom01.php
Síthigh, D. M., & Siems, M. (2019). The Chinese social credit system: A model for other countries? *The Modern Law Review, 82*(6), 1034–1071. https://doi.org/10.1111/1468-2230.12462
Stanley, J., & Granick, J. S. (2020). *The limits of location tracking in an epidemic*. ACLU.
Technavio. (2020). *Global 3D facial recognition systems market 2018–2022*. Retrieved from https://www.technavio.com/report/global-3d-facial-recognition-systems-market-analysis-share-2018
UN Population Division. (2013). *World population aging 2013*. United Nations, Department of Economics and Social Affairs, Population Division.
UN Population Division. (2015). *World population aging 2015*. United Nations, Department of Economics and Social Affairs, Population Division.
UN Population Division. (2019). *World population prospects 2019: volume I: Comprehensive tables*. United Nations: UN Population Division.
UN Population Division. (2020). *World population aging 2020*. United Nations, Department of Economics and Social Affairs, Population Division.
UN Population Division. (2022). *World population prospects 2022*. United Nations, Department of Economics and Social Affairs, Population Division.
UNeGovDD. (2013). *E-government. The United Nations e-Government development database*. Retrieved from https://publicadministration.un.org/egovkb/en-us/About/UNeGovDD-Framework
UNFPA. (2012). *Aging in the twenty-first century: A celebration and a challenge*. UNFPA and HelpAge International. Retrieved from https://www.unfpa.org/publications/aging-twenty-first-century
van Dullemen, C. E. (2017). The politics of aging, global challenges and national solutions. *Journal of Aging and Neuropsychology., 01*(2017), 1–4. https://doi.org/10.29011/JANP-104.100004
Vollset, S. E., Goren, E., Yuan, C.-W., Cao, J., Smith, A. E., Hsiao, T., Bisignano, C., Azhar, G. S., Castro, E., Chalek, J., Dolgert, A. J., Frank, T., Fukutaki, K., Hay, S. I., Lozano, R., Mokdad, A. H., Nandakumar, V., Pierce, M., Pletcher, M., Robalik, T., Steuben, K. M., Wunrow, H. Y., Zlavog, B. S., & Murray, C. J. L. (2020). Fertility, mortality, migration, and population scenarios for 195 countries and territories from 2017 to 2100: A forecasting analysis for the Global Burden of Disease Study. *The Lancet, 396*(10258), 17–23. https://doi.org/10.1016/S0140-6736(20)30677-2

von Weizsäcker, E. U., & Wijkman, A. (2018). *Come on! Capitalism, short-termism, population and the destruction of the planet. A report to the Club of Rome*. Springer. https://doi.org/10.1007/978-1-4939-7419-1

Wang, S., Ding, S., & Xiong, L. (2020). A new system for surveillance and digital contact tracing for COVID-19: Spatiotemporal reporting over network and GPS. *JMIR mHealth and uHealth, 8*(6), e19457.

WHO – World Health Organization. (2020). *"Immunity passports" in the context of COVID-19*. Retrieved from https://www.who.int/news-room/commentaries/detail/immunity-passports-in-the-context-of-covid-19

WHO – World Health Organization. (2021). *Ageism is a global challenge*. Retrieved from https://www.who.int/news/item/18-03-2021-ageism-is-a-global-challenge-un

Widdowson, M. (2021). From Covid-19 to zero-gravity: Complex crises and production revolutions. *Journal of Globalization Studies, 12*(1), 119–146. https://doi.org/10.30884/jogs/2021.01.07

Yakushev, E., Sinyavskaya, O., & Voron, O. (2020). *COVID-19 i smertnost' v domakh prestarelykh v mire i v Rossii*. Institut sotsial'noy politiki NIU VSHE.

Yaşar, E., & Yaşar, M. (2018). Global wealth and aging population. In R. Yilmaz & G. Löschnigg (Eds.), *Studies on Balkan and near Eastern Social Sciences* (Vol. 2, pp. 47–59). Peter Lang.

Leonid Grinin is Senior Research Professor at Centre for Stability and Risk Analysis at HSE University in Moscow, as well as Senior Research Professor of the Institute of Oriental Studies, Russian Academy of Sciences. He is co-editor of the international journals *Social Evolution & History* and the *Journal of Globalization Studies*, as well as co-editor of the international yearbooks *Evolution, History & Mathematics*, and *Kondratieff Waves*. He is author of more than 600 publications, including 33 monographs in Russian, English, and Chinese. His current research focuses on comparative political studies, theory of revolution, political anthropology, global economy, global history, historical sociology, and futurology. Among his monographs are *Great Divergence and Great Convergence. A Global Perspective* (Springer, 2015; with Andrey Korotayev); *The Cybernetic Revolution and the Forthcoming Epoch of Self-Regulating Systems* (Uchitel, 2016, with Anton Grinin), *Economic Cycles, Crises, and the Global Periphery* (Springer, 2016; with Andrey Korotayev and Arno Tausch); *Big History of Globalization. The Emergence of a Global World System* (Springer 2019, in co-authorship with Julia Zinkina, David Christian, and others), *Islamism, Arab Spring, and the Future of Democracy. World System and World Values Perspectives* (Springer, 2019; with Andrey Korotayev and Arno Tausch). In 2004 he was awarded as a winner of the best publication of *Socis* journal during the year for the article "People of Publicity – a New Social Layer?". In 2012 he was awarded with the Gold Kondratieff Medal by the International N. D. Kondratieff Foundation. In 2016 he received the Belyaev Award for the book *From Bifaces to Nanorobots* (the book published in Russian in co-authorship with Anton Grinin).

Anton Grinin is a Russian scientist specializing in the study of modern technological trends and future studies, as well as evolutionist and philosopher. He has a PhD in Biology and works as a Research Fellow for Lomonosov Moscow State University, Russia as well as Centre for Stability and Risk Analysis, HSE University, Moscow. He is the author of more than 100 publications in Russian and English, including 3 monographs. Awards: Gold Kondratieff Young Scholars Medal and Alexander Belyaev Literature Award. Anton Grinin focuses on the issues of the technological revolution (the Cybernetic Revolution) evolving in the current decades of the twenty-first century and the resultant profound transformations in economy and society as well as related economic, political, ethic-legal, and other risks, including political revolutions. He also investigates the correlation between the Cybernetic Revolution and Kondratieff waves as well as between the Cybernetic Revolution and global aging.

Analyzing Social Self-Organization and Historical Dynamics: Life Quality Index

Sergey Malkov, Stanislav Bilyuga, and Jameelah Musieva

Abstract

It is quite natural that in the current global transition from the "era of growth" to the "era of stability," the importance of qualitative changes in all spheres of life is increasing. It is indicators of the quality of life in different countries of the world that play an important role among the indicators characterizing these changes. Quality of life depends on both material, social, and psychological factors. Taking into account the existing approaches to determining the quality of life, Malkov et al. make an attempt to create an integrated index that considers both subjective (psychological) and objective aspects of its assessment. The notion that the higher the quality of life, the more basic material, social, and spiritual needs are met, is used as a methodological basis for the construction of the index. When constructing the index, Malkov et al use Maslow's theory of the "pyramid" of needs (the first, basic, hierarchical level within the "pyramid" is physiological needs, the last level is the need for self-actualization). For each type of needs, indicators characterizing the degree of their satisfaction (or, on the contrary, dissatisfaction) have been selected. The chapter contains a description of the methodology used and some calculation results.

S. Malkov (✉)
Lomonosov Moscow State University, Moscow, Russia

HSE University, Moscow, Russia

Institute of Economics, Russian Academy of Sciences, Moscow, Russia

S. Bilyuga
Faculty of Global Studies, Lomonosov Moscow State University, Moscow, Russia
e-mail: bilyuga@fgp.msu.ru

J. Musieva
HSE University, Moscow, Russia

Faculty of Global Studies, Lomonosov Moscow State University, Moscow, Russia

V. Sadovnichy et al. (eds.), *Reconsidering the Limits to Growth*, World-Systems Evolution and Global Futures, https://doi.org/10.1007/978-3-031-34999-7_22

As has been repeatedly noted above [see the main text of the report, as well as chapter "Modeling Social Self-Organization and Historical Dynamics. Modern Society and a Look into the Global Future: Cybernetic W-Society" (Akaev et al., 2023, this volume)] and chapter "Modeling Social Self-Organization and Historical Dynamics. Global Phase Transitions" (Malkov et al., 2023, this volume)], currently, there is a global transition from the "era of growth" to the "era of stability" (though this is a long way). Accordingly, if until now, when analyzing world dynamics, the main attention has been paid to indicators reflecting the growth rates of economic, demographic, and other characteristics (e.g., GDP growth rates), now indicators reflecting qualitative characteristics, in particular, *the quality of life*, come to the fore.[1]

The quantitative assessment and comparative analysis of *the quality of life* of the population in different countries of the world are an urgent and very difficult task. This is due to the fact that there is still no generally accepted understanding of what this concept includes. A description of the existing approaches to assessing *the quality of life* is provided in some of our previous publications (Malkov et al., 2019; Musieva, 2019). Based on the results of the analysis, it was revealed that the methods used and the indices calculated on their basis do not provide a full assessment of *the quality of life* of the population in various countries. These include, for example, the life quality indices calculated by The Economist analysts, the International Living Legatum Prosperity Index, and the well-known *Human Development Index* (HDI). Each index is unique and requires special consideration, but a common disadvantage for all of them is insufficient consideration (as in the case of The Economist, Legatum Prosperity Index) or the absence (in other cases) of indicators that take into account the cultural aspects of people's lives, their spiritual, and psychological condition. For this reason, there is a paradoxical situation: according to the happiness index, which is compiled on the basis of sociological surveys of the population of various countries of the world (Veenhoven, 1995, 2020), citizens of island or other low-income countries are in the first positions (Vanuatu, Costa Rica), while high-income countries with strong economic indicators lag behind, although they are leaders in the indices mentioned above. This means that in order to reveal the completeness of the idea of quality of life, it is necessary to find additional parameters that will eliminate the existing shortcomings in the estimates.

Taking into account the existing approaches to determining the quality of life, an attempt has been made to create an integrated index that takes into account both subjective (psychological) and objective aspects of its assessment. The creation of

[1] The GDP indicator is often subjected to generally fair criticism (see, e.g., Easterlin, 2010; Costanza et al., 2014; McElwee & Daly, 2014; Goldsmith, 2019; Kapoor & Debroy, 2019), however sometimes criticism is mainly intended to hide the fact that the potential of high income states has weakened (see Grinin & Grinin, 2021a, 2021b, 2021c for detail). On economic growth and its perspectives see chapter "Economics. Optimizing Growth" (Grinin et al., 2023a, this volume). Work on the formation of indicators which can replace GDP is underway (see, e.g., Ahmad et al., 1991; ul Haq, 1995; Banerjee & Duflo, 2019; Masood, 2021, Colman, 2022).

this index is largely based on the work of Malkov and Lyubovikova (2014). The notion that the higher the quality of life, the more basic material, social, and spiritual needs are met has been used as a methodological basis for the construction of the index. To determine the fundamental needs, Abraham Maslow's (1970) theory is used, on the basis of which the following hierarchy/"pyramid" of needs is identified:

1. Physiological needs
2. The need for security
3. Love, communication, belonging to a social group
4. Respect, recognition, prestige
5. Self-expression, creativity
6. Self-actualization

For each type of needs, we have selected indicators characterizing the degree of their satisfaction (or, on the contrary, dissatisfaction).

The first level of the "pyramid" is **physiological needs**, which consist of the needs for food, water, reproduction, etc., i.e., those whose satisfaction is necessary for each person in the first place. To characterize this group of needs, indicators of *infant mortality* and *life expectancy* were chosen. As important indicators, they reflect the possibility of full-fledged human existence. *Infant mortality* was chosen as an indicator due to the fact that the survival of newborn children directly depends on the conditions and quality of life in the country of residence of their parents, as well as on the state of health care. *Life expectancy at birth* is one of the basic characteristics in modern socioeconomic and demographic studies. Life expectancy at birth is the average number of years a newborn should live if, at each age, the conditions for maintaining their life were hypothetically the same as they were for the corresponding age group in the year of his birth. It characterizes various social subsystems – the health care system, the social security system, etc., and also evaluates the effectiveness of the administrative apparatus of the state and its social policy. There is a direct relationship between life expectancy and the life quality index, while the infant mortality rate is inversely proportional to the life quality index.[2] The source of data for these indicators is the World Bank database, within the framework of the indicators of the Sustainable Development Goals.

To characterize the second level of the "pyramid" – **the need for security** – the following indicators are used: *the number of premeditated murders* and *mortality from environmental pollution*. These indicators reflect the presence of threats to human life both from other people's direct actions and environmental pollution. Ensuring security is one of the fundamental tasks of the state, without which human existence is also impossible. In this case, it implies the presence of security from

[2] On relationship between life expectancy and quality of life see also chapter "Global Ageing – an Integral Problem of the Future. How to Turn a Problem into a Development Driver?" (Grinin et al., 2023, this volume) and chapter "Socio-Political Transformations. A Difficult Path to Cybernetic Society" (Grinin et al., 2023b, this volume).

deliberate killings and from a lethal action of environmental pollution. Since both indicators reflect the level of insecurity in the country, there is an inverse relationship between them and the quality-of-life index. The source of statistical data is the World Bank database within the framework of the Sustainable Development Goals indicators and data from the American Institute for Health Metrics and Evaluation (IHME).

The third level of the "pyramid" of needs includes the need for love and for communication. Needs of this kind are best realized in the family, i.e., when people marry and start their own families where there is love, care, joint activities and, of course, a sense of belonging to the family. In most cases, divorced or single individuals do not feel satisfied with their lifestyle and seek to start a family. In view of this point, the indicator of *registered marriages and divorces* was chosen to characterize this need. The source of the statistical data is the UN Database.

The fourth level of Abraham Maslow's hierarchy of needs includes the need for respect, for personal achievements, and for recognition from others. The quality of life is directly related to the satisfaction of this need. However, both in international and national statistics, it is difficult to find an indicator that would characterize the satisfaction of the need for respect. As mentioned above, recognition can be achieved through activities in different categories. To characterize the success/failure of a person's labor activity, the following indicators were chosen: *the intensity of doing business and the unemployment rate*. Each specific person needs to have confidence in the future, in the fulfillment of their abilities, and in the ability to provide for themselves and their families financially. The fulfillment of this need can be characterized by the indicator of the *intensity of doing business*. On the other hand, the possibility of meeting this need largely depends on the state of the labor market, an indicator of which is the unemployment rate. Unemployment is a negative factor, so there is an inverse relationship between its level and the quality-of-life index. The sources of statistical data are the World Bank database within the framework of the indicators of the Sustainable Development Goals.

At the fifth level of Abraham Maslow's "pyramid" there is a **need for self-expression and for creativity**. An indicator of how many people in a society are engaged in creative activities is *the number of researchers*, as well as *the mean years of schooling*, since the ability to engage in creative activities depends on the level of education. Usually, indexes use only the number of years of education, which cannot fully reflect its quality. The latter seems to be a rather difficult task based on the available statistical data, since either there are no relevant indicators or data are not available for all countries. Determining the number of researchers makes it possible to identify, in general, the interest in knowledge among citizens, as well as the desire of the state to develop the field of science. Thus, these indicators together will make it possible to assess both the level of education in the states and its quality. The main source for this indicator was the statistics of the World Bank within the framework of the Sustainable Development Goals indicators and the UNESCO database.

At the last level of the hierarchy of needs, we placed the need for self-actualization. An indicator of the security of this need is the extent to which people are satisfied with their lives and consider themselves to be happy. This is an integrative and subjective assessment. It is reflected by the so-called *happiness index*, determined on the basis of sociological surveys of representative samples of the population in various countries of the world (Veenhoven, 2020). It characterizes satisfaction with personal life, social relations, environment, and state of affairs in the country. Another integrative indicator reflecting the subjective perception of the quality of life is the suicide rate. This is an anti-indicator reflecting the state of the social climate in the country. This opinion is confirmed by the results of an empirical study by Emile Durkheim (Durkheim, 1897/1951), who was one of the first to establish the relationship between modernization and the growth of deviant behavior. The source of statistical data for determining these indicators is the happiness survey data posted on the World Happiness Database website [this is a database developed at the Erasmus University Rotterdam and based on surveys of the population of more than 150 states (Veenhoven, 2020)], as well as data on the number of suicides from the World Bank database.

If the indicators proposed for calculating the life quality index are presented in data sources in absolute terms, then they are reduced to a specific form (per 100,000 people), making it possible to compare countries by the value of each of the selected indicators.

The process of forming an integral index of quality of life consists of the following successive stages:

A) unification of the values of indicators. Before calculating the life quality index, it becomes necessary to unify the values of the indicators, because they differ in measurement units and the spread of numerical values. Between the value of individual indicators and the quality of life, both direct and inverse relationships can be observed, i.e. for some indicators, the lowest value of the indicator will correspond to the best quality of life, for others – the highest. The unification of indicators is a transformation of their values, as a result of which their new values will be located in the interval from 0 to 1. Thus, "1" will be equal to the best value of the indicator in relation to the quality of life in the sample of countries under consideration, and zero will equal the worst. Taking into account the fact that the values of indicators and the life quality index can be directly and inversely proportional to each other, their unification is carried out using the following formula:
if the best indicator corresponds to its maximum value:

$$\tilde{x} = \frac{x - x_{\min}}{x_{\max} - x_{\min}},$$

if the best indicator corresponds to its minimum value:

$$\tilde{x} = \frac{x_{\max} - x}{x_{\max} - x_{\min}}$$

where $\widetilde{x}$ – unified data; x – initial data (the value of the considered indicator for the ith country); x_{max} – the maximum value of the indicator for the considered sample of countries; x_{min} – the minimum value of the indicator for the considered sample of countries.

Unification of the measurement scale of indicators allows solving the problem of quantitative incompatibility of changes in different variables, as well as the problem of multidirectional measurements "for the better";

B) formation of an integral index of quality of life based on multiplicative convolution. The integral life quality index is formed on the basis of a multiplicative convolution, which involves the multiplication of all indicators. It should be noted here that often the development of the life quality index in other works was an additive convolution in which particular indicators were added up. However, the experience of using such indices has shown that additive convolution has certain disadvantages. In particular, according to the meaning of such a convolution, the satisfaction of each need can be fully compensated by any other need (for example, a low level of security can be fully compensated by a high level of satisfaction of physiological needs), which is not true. When using an additive convolution, it turns out that a country that has the worst scores on a number of indicators can compensate for this by having the best scores on other variables (for example, if the country has an extremely high level of pollution, but the material well-being of the population is at a high level, then the index will show that the quality of life in the country is high). Multiplicative convolution is devoid of such shortcomings. When using it, only the country, all indicators of which are at a high level, will have a high quality of life.

Due to the fact that the life quality index is a multiplicative convolution, it is advisable to transfer the values of the indicators from the interval (0; 1) to the interval (2; 3) (by simultaneously increasing the values of all unified indicators by two units) and then to calculate the logarithm of the result.[3]

In accordance with the above, the calculation of the integral index of quality of life was carried out according to the following formula:

$$LQI = \widetilde{x}_1^{n_1} \times \widetilde{x}_2^{n_2} \times \ldots \times \widetilde{x}_k^{n_k}$$

where life quality index (LQI) is the value of the life quality index in the country under consideration;k – the number of indicators that determine the level of quality of life;n_k – the value of the indicator weight;$\widetilde{x}_k$ – the unified value of the indicator after calculating its logarithm.

The important problem is to determine the weight (significance) of each indicator when calculating the life quality index. Usually, these weights are determined on the

[3]In principle, other methods of transforming unified data are also possible, but as further studies have shown, they have little effect on the final comparative assessments of the quality of life in different countries.

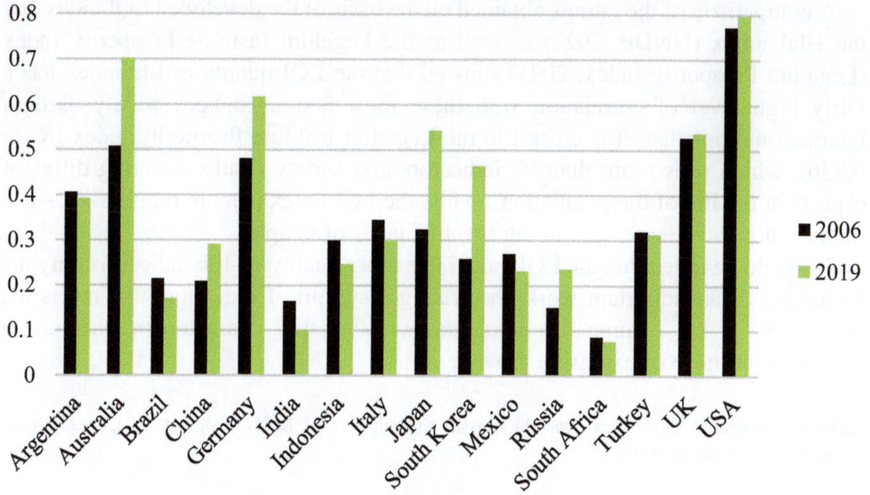

Fig. 1 LQI values in 2006 and 2019 for some G20 countries

basis of sociological research or by expert opinion. In this version of the integral index, it was assumed that the satisfaction of all the needs of Abraham Maslow's "pyramid" is equally important for people, regardless of the country in which they live. Taking into account the fact that each need in the integral index is represented by the same number of indicators (namely, two indicators), each indicator was assigned a unit weight.

Below are some results of the calculation of the life quality index (LQI) according to the method described. Figure 1 shows the results of calculating the LQI index for a number of countries that are included in the Group of Twenty (G20), which consists of the largest economies in the world and includes both high income and low income countries. It should be noted that, according to the logic of its structure, the index is designed to reflect comparative characteristics in relation to the countries included in the index rather than absolute value of the quality of life in a given country.

Figure 1 shows that the LQI value in the period under review was growing in such countries as Australia, China, Germany, Japan, South Korea, Russia, the United Kingdom, and the United States, while the rest of the countries presented in the histogram show a relative decrease in the value of the index.

Calculations of the LQI index allow ranking the countries of the world in terms of quality of life. The index provides the possibility to find out exactly what changes in indicators led to an increase or decrease in the overall rating of the country in question. For example, the rise of Russia's rating in the life quality index from 46th to 38th place in 13 years (from 2006 to 2019) is associated with an increase in such indicators as life expectancy and happiness index, as well as with a decrease in such indicators as the unemployment rate, the number of deliberate and premediated murders, or the number of suicides. Almost similar indicators caused the growth of the index value in 2019 in other countries.

A comparison of the ratings obtained on the basis of the developed LQI index and the HDI index (UNDP, 2022) as well as the Legatum Institute Prosperity Index (Legatum Prosperity Index, 2021) showed that the LQI quality-of-life index has a fairly high level of correlation with these most famous indices widely used in international practice. It is closest to the Legatum Institute Prosperity Index ($R^2 = 0.736$), which uses more than 90 indicators and survey results covering different aspects of the life of the population[4], while the LQI index is more transparent, since its indicators are clearly tied to the specific needs of people.

It should be noted that the LQI index, like other quality-of-life indices, is only the beginning of an important work on creating a system of indicators that meets the challenges of the coming historical era, when quality characteristics become a priority over those of extensive growth.

Acknowledgment This work was done with the support of MSU Program of Development, Project No 23A-SCH05-03.

References

Ahmad, E., Drèze, J., Hills, J., & Sen, A. (1991). Social security in developing countries.

Akaev, A., Malkov, S., Grinin, L., Bilyuga, S., Davydova, O., Grinin, A., Kovaleva, N., Malkov, A., Musieva, J., & Korotayev, A. (2023). Modeling social self-organization and historical dynamics. Modern society and a look into the global futures: Cybernetic W-society. In: V. Sadovnichy et al. (Eds.), *Reconsidering the limits to growth. A Report to the Russian Association of the Club of Rome* (pp. 419–460). Springer. https://doi.org/10.1007/978-3-031-34999-7_19

Banerjee, A. V., & Duflo, E. (2019). *Good economics for hard times*. Allen Lane.

Colman, R. (2022). *What really counts: The case for a sustainable and equitable economy*. Columbia University Press.

Costanza, R., Kubiszewski, I., Giovannini, E., Lovins, H., McGlade, J., Pickett, K. E., Ragnarsdóttir, K. V., Roberts, D., De Vogli, R., & Wilkinson, R. (2014). Development: Time to leave GDP behind. *Nature News, 505*(7483), 283–285. https://doi.org/10.1038/505283a

Durkheim, E. (1897/1951). *Suicide: A study in sociology*. Routledge.

Easterlin, R. A. (2010). Well-being, front and center: A note on the Sarkozy report. *Population and Development Review, 36*(1), 119–124. https://doi.org/10.1111/j.1728-4457.2010.00320.x

Goldsmith, C. (2019). Why GDP is no longer the most effective measure of economic success. World Finance 25.07.2019. Retreived from https://www.worldfinance.com/strategy/why-gdp-is-no-longer-the-most-effective-measure-of-economic-success

Grinin, L., & Grinin, A. (2021a). Analyzing the global problems of the twenty-first century. A review and forecast based on the Report to the Club of Rome 'Come On!'. *Journal of Globalization Studies, 12*(2), 181–195. https://doi.org/10.30884/jogs/2021.02.10

Grinin, L., & Grinin, A. (2021b). Reflections on the economic growth and future. Part 1. GDP growth and "the Decline of the West". *Philosophy and society, 3*, 5–34. https://doi.org/10.30884/jfio/2021.03.01

[4]Unlike the Legatum Prosperity Index, the HDI index includes only three indicators, which, in our opinion, is clearly not enough to assess the quality of life.

Grinin, L., & Grinin, A. (2021c). Reflections on the economic growth and future. Part 2. Globalization and economic growth. *Philosophy and society, 4*, 5–38. https://doi.org/10.30884/jfio/2021.04.01

Grinin, L., Grinin, A., & Korotayev, A. (2023). Global aging – An integral problem of the future. How to turn a problem into a development driver? In: V. Sadovnichy et al. (Eds.), *Reconsidering the limits to growth. A report to the Russian Association of the Club of Rome* (pp. 117–135). Springer. https://doi.org/10.1007/978-3-030-*****-*_7

Grinin, L., Grinin, A., & Malkov, S. (2023a). Economics. Optimizing growth. In: V. Sadovnichy et al. (Eds.), *Reconsidering the limits to growth. A report to the Russian Association of the Club of Rome* (pp. 155–168). Springer. https://doi.org/10.1007/978-3-031-34999-7_9

Grinin, L., Grinin, A., & Malkov, S. (2023b). Socio-political transformations. A difficult path to cybernetic society. In: V. Sadovnichy et al. (Eds.), *Reconsidering the limits to growth. A report to the Russian Association of the Club of Rome* (pp. 169–189). Springer. https://doi.org/10.1007/978-3-031-34999-7_10

Kapoor, A., & Debroy, B. (2019). GDP is not a measure of human well-being. *Harvard Business Review, 10*. Retrieved from https://hbr.org/2019/10/gdp-is-not-a-measure-of-human-well-being

Legatum Prosperity Index. (2021). *Key findings of the Legatum Prosperity Index*. Retrieved from https://www.prosperity.com/feed/key-findings-2021-prosperity-index

Malkov, S., Bilyuga, S., & Musieva, J. (2019). Metodika mezhstranovoy otsenki kachestva zhizni na osnove indeksa LQI. *Informatsionnyye Voyny, 4*(52), 38–44.

Malkov, S., Grinin, L., Grinin, A., Musieva, J., & Korotayev, A. (2023). Modeling social self-organization and historical dynamics. Global phase transitions. In: V. Sadovnichy et al. (Eds.), *Reconsidering the limits to growth. A report to the Russian Association of the Club of Rome* (pp. 387–417). Springer. https://doi.org/10.1007/978-3-031-34999-7_18

Malkov, S., & Lyubovikova, D. (2014). Indeks kachestva zhizni kak pokazatel′ razvitiya obshchestva i gosudarstva. *Informatsionnyye Voyny, 2*(30), 48–52.

Maslow, A. H. (1970). *Motivation and personality* (2nd ed.). Harper and Row.

Masood, E. (2021). *GDP: The world's most powerful formula and why it must now change*. Icon Books.

McElwee, S., & Daly, L. (2014). Why we should abolish the GDP. *Washington Post, 05.06.2014*. Retrieved from https://www.washingtonpost.com/posteverything/wp/2014/06/05/why-we-should-abolish-the-gdp/

Musieva, J. (2019). Ponyatiye "kachestvo zhizni" i podkhody k yego opredeleniyu. *Informatsionnyye Voyny, 3*(51), 16–20.

ul Haq, M. (1995). *Reflections on human development*. Oxford University Press.

UNDP (United Nations Development Programme). (2022). *2022 Special report on human security*. UN.

Veenhoven, R. (1995). World database of happiness. *Social Indicators Research, 34*, 299–313.

Veenhoven, R. (2020). *Happiness in nations. World database of happiness*. Erasmus University Rotterdam. Retrieved from http://worlddatabaseofhappiness.eur.nl/hap_nat/nat_fp.php?mode=1

Sergey Malkov heads the Center for Long-Term Forecasting and Strategic Planning at Lomonosov Moscow State University; he also works as a Senior Research Professor at the Institute of Economics of the Russian Academy of Sciences and at the Centre for Stability and Risk Analysis of HSE University in Moscow. He is the author of over 300 scholarly publications, including such monographs as *Social self-organization and historical process: possibilities of mathematical modeling* (URSS, 2009), *Modeling and forecasting of world dynamics* (ISPI RAS, 2012, together with Victor Sadovnichy and others), *A new theory of the reproduction of capital: development and practical application* (Nestor-Istoriya, 2016, together with Vladimir Mayevsky and others), *The 21st Century Singularity and Global Futures. A Big History Perspective* (Springer, 2020, together with David J. LePoire and others). In 2022 he was awarded with the Kondratieff Award by the Russian Academy of Sciences.

Stanislav Bilyuga has a Ph.D. degree in Political Sciences and is affiliated to the Faculty of Global Studies, Lomonosov Moscow State University. His research interests include the modeling of global trends and dynamics and forecasts of destabilization. He has authored and co-authored over 70 scientific publications concerning factors, trends, indices, and methods of research on sociopolitical destabilization of various countries.

Jameelah Musieva is a research assistant at Lomonosov Moscow State University, Faculty of Global Studies as well as a research fellow at the Centre for Stability and Risk Analysis, HSE University, Moscow. She is the author/co-author of more than 10 scholarly publications, including such articles as "The Concept of 'Quality of Life'" (2019), "Methodology for Cross-Country Assessment of The Quality of Life Based on The LQI Index" (2019), "A Cross-Country Analysis of The Quality of Life" (2020), "Quality of Life in The BRICS Countries Based on The LQI Index" (2021).

Conclusion: Reconsidering the Limits – Suggestions (Come On!)

Victor Sadovnichy, Askar Akaev, Ilya Ilyin, Sergey Malkov, Leonid Grinin, Dmitry Ivanov, and Andrey Korotayev

Abstract

This concluding chapter provides summary of main ideas expressed in the previous chapters of the present report to the Club of Rome, which are: (1) population growth has ceased to be the main cause of many other global problems; the issue has shifted to other dimensions and requires optimization of demographic processes, rather than general limitation of population growth; (2) modern

V. Sadovnichy · I. Ilyin
Lomonosov Moscow State University, Moscow, Russia
e-mail: info@rector.msu.ru; dekanat@fgp.msu.ru

A. Akaev
Lomonosov Moscow State University, Moscow, Russia

HSE University, Moscow, Russia

S. Malkov
Lomonosov Moscow State University, Moscow, Russia

HSE University, Moscow, Russia

Institute of Economics, Russian Academy of Sciences, Moscow, Russia
e-mail: s@malkov.org

L. Grinin
HSE University, Moscow, Russia

Institute of Oriental Studies, Russian Academy of Sciences, Moscow, Russia

D. Ivanov
Volga State Technological University, Yoshkar-Ola, Russia
e-mail: ivanovdv@volgatech.net

A. Korotayev (✉)
Lomonosov Moscow State University, Moscow, Russia

HSE University, Moscow, Russia

Institute for African Studies, Russian Academy of Sciences, Moscow, Russia

V. Sadovnichy et al. (eds.), *Reconsidering the Limits to Growth*, World-Systems Evolution and Global Futures, https://doi.org/10.1007/978-3-031-34999-7_23

political system is obviously in a transitional period to a fundamentally new condition based on new principles, that will allow to overcome major contradictions between countries and regions and find new solutions to global problems; (3) the problem of resource depletion could be to a very considerable extent solved through the development and diffusion of fundamentally new technologies, through the movement toward resource and energy saving in the framework of the ongoing Cybernetic Revolution, which is associated with the saving of resources and labor; (4) aging of the global population now becomes one of the most important issues of the world development and requires special attention to minimize the emergent problems and at the same time utilizing advantages of increasing life expectancy; (5) the problem of inequality both within countries and between them requires a stronger social policy and a general restructuring of socio-economic relations in combination with measures preserving the potential for economic growth; (6) technological progress together with global aging will lead societies to a new societal type – cybernetic W-society, characterized by its reliance on AI-based smart (cybernetic) technologies and self-regulating systems, as well as the institutionalization of age differences; (7) a transition to a new global socio-economic system ("world-organism") is expected, from the modern system, from competing states to a more coordinated global management, and, in the long term, to a closer association based on cooperative principles and aimed at solving common problems, a move to a more solidary society; (8) the inevitable transformation of the current social system requires critical reflection on the previous experience, taking into account the best achievements of the past, which could be realized with the help of media. Thus, generally speaking, this report's key message is that the time of historical bifurcations is coming, and we need to prepare for it to make the future more beneficial to us and humanity as a whole.

As mentioned in the Introductory Chapter[1], in our report to the Club of Rome, the concept of limits is understood in a broad sense, that is, the limits should be interpreted as very strong challenges, which, however, can eventually be made less relevant.

We have revealed the problems and presented forecasts and possible solutions to these problems. Now, returning to the topic of our report, we would like to say that we are cautious optimists about the future. Even though the problems described by us and in other reports to the Club of Rome (Perissi & Bardi, 2021; Berg, 2019, Randers et al., 2018, von Weizsäcker & Wijkman, 2018; Lovins et al., 2018; Wijkman & Skånberg, 2017; Bardi, 2017; etc.; see also Grinin & Grinin, 2021) are very complex, although today we are going through a very turbulent and, one might even say, acute crisis in the development of our human civilization, we believe that in general many problems will be solved to some extent.

[1] See chapter "Introduction. Hoping for the Future" (Sadovnichy et al., 2023b).

In this section, based on the optimal development scenarios for the coming decades, we would like to summarize how to move to a positive development scenario, or, in other words, how to reconsider the limits.

So, what limits (challenges) are we talking about?

1 Population Growth Limit

The Club of Rome traditionally considers population growth as a critically important problem that determines all other ones (ecology, climate, and lack of resources). However, as we have shown in chapter "Future Society and Transition to It" (Sadovnichy et al., 2023a, this volume) and in chapter "Demography. Toward the Optimization of Demographic Processes" (Korotayev et al., 2023, this volume), the problem of rapid population growth in its original, neo-Malthusian sense has left an agenda in most countries of the world. On the contrary, in many high and middle-income countries, demographic growth is no longer the problem, while depopulation is (when the birth rate does not make up for population decline). The rest of the countries (with the exception of Tropical Africa so far[2]) are moving in the same direction. Therefore, we consider the problem of population growth a global problem (because population growth in Africa and some other countries will affect the whole world), but not a catastrophic one.

In this sense, the world has already mitigated the problem of limiting population growth, but now we have to reconsider the threat of depopulation; otherwise, we will lose many nations. We also have to find the optimum for population stabilization in Africa. In other words, we need to consider the problem of optimizing the demographic processes rather than limiting population growth.

2 The Limit of Systemic Restrictions of the Existing Type of Society, State, and World Relations

Firstly, the main political systems of modernity (sovereign states) fit the industrial society very well and not the modern (cybernetic) society. Hence, the concept of a world order headed by a hegemon (the most powerful state in the military and economic sense) is clearly outdated. Therefore, we must see the new world order not as yet another change of the world hegemon (from the United States to another power) but as a fundamentally new construction based on new principles, thanks to which it will be possible to mitigate the contradictions between countries and regions and open up new possibilities for solutions to social problems; it might be even more appropriate to call it a "world consensual system" rather than a "world order" (see

[2] See chapter "Africa – the Continent of the Future. Challenges and Opportunities" (Grinin & Korotayev, 2023a, this volume) and chapter "Modeling Social Self-Organization and Historical Dynamics. Africa's Future" (Korotayev, Shulgin et al., 2023, this volume).

chapter "Future Political Change. Toward a More Efficient World Order" (Grinin et al., 2023a, this volume).

Secondly, the World System is currently going through a global phase transition to a fundamentally new condition, which in any case will be significantly different from the previous (let us call it "industrial") condition. This transition will probably continue for another century, going through several phases that will have strong differences from each other.[3]

As a result, we have to reconsider the limits of the modern market capitalist society with its individualism and move to a more solidary society.[4] In this new society, consumerism will weaken, along with the desire for growing consumption and, accordingly, the transformation of the entire economic system as less focused on provoking an increase in consumption and profit.

Thus, some systemic restrictions of modern society will be partially overcome with the transition of the World System to a new condition.

3 The Limit of Resources

The problem of limited resources manifests itself remarkably from time to time. Of course, there are still many virgin fields of limited resources in Africa [see chapter "Africa – the Continent of the Future. Challenges and Opportunities" (Grinin & Korotayev, 2023a, this volume)]. However, in any case, it is necessary to realize the presence of fundamental limitations (including the ones related to the environment, climate, energy, and resources) within the modern resource- and energy-intensive type of development. The world needs a transition to a new technological development model associated with the Cybernetic Revolution [see chapter "Technology. Limitless Possibilities and Effective Control" (Grinin & Grinin, 2023b, this volume)] and the creation of self-regulating and self-managing systems that will lead an active search for energy- and resource-saving, environmentally friendly production and social technologies. On the basis of such technological development, it will be possible to solve the problem of limitations and form new directions of development. At the same time, the development of technologies will be much less spontaneous

[3] See chapter "Modeling Social Self-Organization and Historical Dynamics. Modern Society and a Look into the Global Futures: Cybernetic W-Society" (Akaev et al., 2023, this volume), chapter "Analyzing Social Self-Organization and Historical Dynamics. Future Cybernetic W-Society: Sociopolitical Aspects" (Grinin & Grinin, 2023a, this volume), and chapter "Socio-Political Transformations. A Difficult Path to Cybernetic Society" (Grinin, Grinin, & Malkov, 2023b, this volume).

[4] See chapter "Socio-Political Transformations. A Difficult Path to Cybernetic Society" (Grinin, Grinin, & Malkov, 2023b, this volume), chapter "High Income and Low Income Countries. Towards a Common Goal at Different Speeds" (Grinin & Korotayev, 2023b, this volume), and chapter "Analyzing Social Self-Organization and Historical Dynamics. Future Cybernetic W-Society: Socio-Political Aspects" (Grinin & Grinin, 2023a, this volume).

than before (previously, it was largely a result of economic competition and military rivalry).

A gradual transition is expected from prioritizing "growth" to prioritizing "optimization," which combines the level of economic growth necessary for a healthy society with the sustainability of long-term development.[5]

Thus, some of the resource constraints of modern society can be reconsidered through the development and diffusion of fundamentally new technologies through the movement toward resource and energy saving, since the Cybernetic Revolution is inherently associated with saving resources and labor.[6]

Our scenario calculations [see chapter "Climate and Energy. Scenarios of Energy Transition and Global Temperature Changes Based on Current Technologies and Trends" (Akaev & Davydova, 2023, this volume)] show that the problem of global warming can be solved on the basis of modern technologies, including the use of nuclear power plants, hydrogen, and, in the future, thermonuclear energy, with coordinated actions of the world community.

4 The Limits of the Current Life Expectancy Values

The Cybernetic Revolution should expand these limits by 10–20 years, as at its first phases, the main driver is the development of medicine and related industries.[7] Although population aging is uneven, by the end of the century it will engulf all countries of the world. Aging will affect all areas of life and we must prepare for that. It is necessary to institutionalize age characteristics so that the future society can minimize the problems that arise in connection with aging, and at the same time use all the advantages of greatly increased life expectancy [we are talking about this in chapter "Modeling Social Self-Organization and Historical Dynamics. Global Phase

[5] On the ways to optimize growth rates see chapter "Economics. Optimizing Growth" (Grinin, Grinin, & Malkov, 2023a, this volume), chapter "Demography. Toward Optimization of Demographic Processes" (Korotayev, Malkov, & Musieva, 2023, this volume), chapter "Modeling Social Self-Organization and Historical Dynamics. Modern Society and a Look into the Global Futures: Cybernetic W-Society" (Akaev et al., 2023, this volume), chapter "High Income and Low Income Countries. Toward a Common Goal at Different Speeds" (Grinin & Korotayev, 2023b, this volume), chapter "Future Political Change. Toward a More Efficient World Order" (Grinin et al., 2023a, this volume), and chapter "Africa – the Continent of the Future. Challenges and Opportunities" (Grinin & Korotayev, 2023a, this volume).

[6] On new resources see chapter "Economics. Optimizing Growth" (Grinin, Grinin, & Malkov, 2023a, this volume) and chapter "Technology. Limitless Possibilities and Effective Control" (Grinin & Grinin, 2023b, this volume).

[7] See chapter "Global Ageing – an Integral Problem of the Future. How to Turn a Problem into a Development Driver?" (Grinin, Grinin, & Korotayev, 2023b, this volume), chapter "Technology. Limitless Possibilities and Effective Control" (Grinin & Grinin, 2023b, this volume), chapter "Socio-Political Transformations. A Difficult Path to Cybernetic Society" (Grinin, Grinin, & Malkov 2023b, this volume), and chapter "Modeling Social Self-Organization and Historical Dynamics. Global Phase Transitions" (Malkov et al., 2023, this volume).

Transitions" (Malkov et al., 2023, this volume)]. As a result of the Cybernetic Revolution, a Cybernetic W-society will emerge; it will be (a) supertechnological; (b) socio-technologically regulated at all levels; (c) a society where, as a result of the aging process, division into age categories will become socially much more significant than today (i.e. age will become a much more important social marker than at present).[8]

We would present the next two limits as problems.

5 The Limit/Problem of Inequality in the Standard of Living among Different Societies (High-Income, Middle-Income, and Low-Income Ones) and among Different Social Strata within the Same Society

It is necessary to simultaneously solve two opposite and equally important tasks: to provide opportunities for the low-income countries to raise the economy and standards of living, bringing them closer to the level of the high income countries [see chapter "High Income and Low Income Countries. Towards a Common Goal at Different Speeds" (Grinin & Korotayev, 2023b, this volume)] and to ensure progress toward climate conservation [see chapter "Climate and Energy. Scenarios of Energy Transition and Global Temperature Changes Based on Current Technologies and Trends" (Akaev & Davydova, 2023)], reducing the unconstrained consumption of resources, and ultimately ensure the sustainability of society in the second half of the century.

To solve the problem of inequality within the framework of one society, a stronger social policy is needed, along with a general restructuring of socio-economic relations.

6 The Limit/Problem of the Lack of Unity in Societies

A transition to a new global socio-economic system is expected, from the modern World System of competing states to a more integrated supranational coordinating governance and, in the long term, to a closer association (world organism) based on cooperative principles to solve common problems.

The transition to the new system will not be smooth. During the phase transition that has begun, various options for transitional systems will arise. Even an attempt to create a global techno-totalitarian society ("digital concentration camp") is possible.

The transition to the new society will be gradual. The following steps are expected:

[8]For more details see chapter "Socio-political transformations: a difficult path to a cybernetic society" (Grinin, Grinin, & Malkov, 2023b, this volume) and chapter "Analyzing Social Self-Organization and Historical Dynamics. Future Cybernetic W-Society: Socio-Political Aspects" (Grinin & Grinin, 2023b, this volume).

- the next 10–15 years: a way out of global political confrontation and crisis;
- by 2050: formation of a new world order/"world consensual system" and a possible convergence of the levels of development of different world regions through the "Great Convergence" processes (which can facilitate the dialogue and partnership of supranational entities and civilizations [Grinin & Korotayev, 2015]);
- by 2100: the formation of new principles and institutions that would allow the world to move toward a real unity based on cooperative principles;
- by 2130: a possible formation of a new global system of the world (world organism) based on cooperative principles.

7 The Limit of the Destruction of the Old During the Transition to the New

The rejection of the old during the transition to the new is inevitable, but in the process of transition to a new state of the world, it is important to preserve all the best and smooth working practices from the present and the near past. Therefore, we should talk about constructing a new society based on critically reflecting on the previous eras and taking into account their best achievements (including the best achievements of capitalism, socialism, etc.).[9]

To solve this task along with the two tasks discussed above, it is necessary to de-monopolize the media throughout the world and ban attempts by any political or other forces to monopolize these media. Since the media have become a super-powerful, global, and extremely strong force influencing the formation of public opinion and ideology, checks and balances are needed in this area, both at the national and international levels.

Thus, in conclusion, we can say that the essence of change (as shown in the report, as well as by statistical data and mathematical modeling) is that humankind is moving to a fundamentally new phase of historical development. The old economic and social technologies (liberal capitalism, state socialism) no longer work as before. The time of historical bifurcations is coming.

Under these conditions, an extremely important (but also the most difficult) task is social design, purposeful activity to shape the future global society on the principles of cooperation and partnership. This will allow solving other global problems related to the environment, climate, resources, etc., that is, overcoming

[9] Schumpeter's "creative destruction" (Schumpeter, 1942) should be replaced here with "preserving destruction". See our understanding of "preserving destruction" in chapter "Economics. Optimizing Growth" (Grinin, Grinin, & Malkov, 2023a, this volume), chapter "High Income and Low Income Countries. Toward a Common Goal at Different Speeds" (Grinin & Korotayev, 2023b, this volume), chapter "Socio-Political Transformations. A Difficult Path to Cybernetic Society" (Grinin, Grinin, & Malkov, 2023b, this volume), and chapter "Modeling Social Self-Organization and Historical Dynamics. Modern Society and a Look into the Global Futures: Cybernetic W-Society" (Akaev et al., 2023, this volume).

the limits hindering the realization of humankind's potential and narrowing its prospects for the future.

We have proposed alternative projects for a more solidary and humane society in the future, and we invite everyone to discuss them and cooperate in working on the best projects for the future society. And once again, we repeat the idea that the more actively we fight for the future, as we would like to see it, the more likely it will become closer to our ideal, although we must simultaneously remain realistic.

References

Akaev, A., & Davydova, O. (2023). Climate and energy. Scenarios of energy transition and global temperature changes based on current technologies and trends. In V. Sadovnichy et al. (Eds.), *Reconsidering the limits to growth. A report to the Russian Association of the Club of Rome* (pp. 53–70). Springer. https://doi.org/10.1007/978-3-031-34999-7_4

Akaev, A., Malkov, S., Grinin, L., Bilyuga, S., Davydova, O., Grinin, A., Kovaleva, N., Malkov, A., Musieva, J., & Korotayev, A. (2023). Modeling social self-organization and historical dynamics. Modern society and a look into the global futures: Cybernetic W-society. In V. Sadovnichy et al. (Eds.), *Reconsidering the limits to growth. A report to the Russian Association of the Club of Rome* (pp. 419–460). Springer. https://doi.org/10.1007/978-3-031-34999-7_19

Bardi, U. (2017). *The Seneca effect. Why growth is slow but collapse is rapid. A report to the Club of Rome*. Springer. https://doi.org/10.1007/978-3-319-57207-9

Berg, C. (2019). *Sustainable action: Overcoming the barriers. A report to the Club of Rome*. Routledge.

Grinin, A. L., & Grinin, L. E. (2021). Analyzing the global problems of the twenty-first century. A review and forecast based on the report to the Club of Rome 'Come On!'. *Journal of Globalization Studies, 12*(2), 181–195. https://doi.org/10.30884/jogs/2021.02.10

Grinin, L., & Grinin, A. (2023a). Analyzing social self-organization and historical dynamics. Future cybernetic W-society: Socio-political aspects. In V. Sadovnichy et al. (Eds.), *Reconsidering the limits to growth. A report to the Russian Association of the Club of Rome* (pp. 491–519). Springer. https://doi.org/10.1007/978-3-031-34999-7_21

Grinin, L., & Grinin, A. (2023b). Technology: Limitless possibilities and effective control. In V. Sadovnichy et al. (Eds.), *Reconsidering the limits to growth. A report to the Russian Association of the Club of Rome* (pp. 139–154). Springer. https://doi.org/10.1007/978-3-031-34999-7_8

Grinin, L., Grinin, A., & Korotayev, A. (2023a). Future political change. Toward a more efficient world order. In V. Sadovnichy et al. (Eds.), *Reconsidering the limits to growth. A report to the Russian Association of the Club of Rome* (pp. 191–206). Springer. https://doi.org/10.1007/978-3-031-34999-7_11

Grinin, L., Grinin, A., & Korotayev, A. (2023b). Global ageing – An integral problem of the future. How to turn a problem into a development driver? In V. Sadovnichy et al. (Eds.), *Reconsidering the limits to growth. A report to the Russian Association of the Club of Rome* (pp. 117–135). Springer. https://doi.org/10.1007/978-3-031-34999-7_7

Grinin, L., Grinin, A., & Malkov, S. (2023a). Economics. Optimizing growth. In V. Sadovnichy et al. (Eds.), *Reconsidering the limits to growth. A report to the Russian Association of the Club of Rome* (pp. 155–168). Springer. https://doi.org/10.1007/978-3-031-34999-7_9

Grinin, L., Grinin, A., & Malkov, S. (2023b). Socio-political transformations. A difficult path to cybernetic society. In V. Sadovnichy et al. (Eds.), *Reconsidering the limits to growth. A report to the Russian Association of the Club of Rome* (pp. 169–189). Springer. https://doi.org/10.1007/978-3-031-34999-7_10

Grinin, L., & Korotayev, A. (2015). *Great divergence and great convergence. A global perspective*. Springer. https://doi.org/10.1007/978-3-319-17780-9

Grinin, L., & Korotayev, A. (2023a). Africa – The continent of the future. Challenges and opportunities. In V. Sadovnichy et al. (Eds.), *Reconsidering the limits to growth. A report to the Russian Association of the Club of Rome* (pp. 225–238). Springer. https://doi.org/10.1007/978-3-031-34999-7_13

Grinin, L., & Korotayev, A. (2023b). High income and low income countries. Towards a common goal at different speeds. In V. Sadovnichy et al. (Eds.), *Reconsidering the limits to growth. A report to the Russian Association of the Club of Rome* (pp. 207–224). Springer. https://doi.org/10.1007/978-3-031-34999-7_12

Korotayev, A., Malkov, S., & Musieva, J. (2023). Demography. Toward the optimization of demographic processes. In V. Sadovnichy et al. (Eds.), *Reconsidering the limits to growth. A report to the Russian Association of the Club of Rome* (pp. 97–116). Springer. https://doi.org/10.1007/978-3-031-34999-7_6

Korotayev, A., Shulgin, S., Ustyuzhanin, V., Zinkina, J., & Grinin, L. (2023). Modeling social self-organization and historical dynamics. Africa's futures. In V. Sadovnichy et al. (Eds.), *Reconsidering the limits to growth. A report to the Russian Association of the Club of Rome* (pp. 461–490). Springer. https://doi.org/10.1007/978-3-031-34999-7_20

Lovins, L. H., Wallis, S., Wijkman, A., & Fullerton, J. (2018). *A finer future. Creating an economy in service to life. A report to the Club of Rome*. New Society Publishers.

Malkov, S., Grinin, L., Grinin, A., Musieva, J., & Korotayev, A. (2023). Modeling social self-organization and historical dynamics. Global phase transitions. In V. Sadovnichy et al. (Eds.), *Reconsidering the limits to growth. A report to the Russian Association of the Club of Rome* (pp. 387–417). Springer. https://doi.org/10.1007/978-3-031-34999-7_18

Perissi, I., & Bardi, U. (2021). *The empty sea. The future of the blue economy. A report to the Club of Rome*. Springer. https://doi.org/10.1007/978-3-030-51898-1

Randers, J., Rockström, J., Stoknes, P. E., Golüke, U., Collste, D., & Cornell, S. (2018). *Transformation is feasible. A report to the Club of Rome*. Stockholm Resilience Centre.

Sadovnichy, V., Akaev, A., Ilyin, I., Malkov, S., Grinin, L., Andreev, A., & Korotayev, A. (2023a). Future society and transition to it. In V. Sadovnichy et al. (Eds.), *Reconsidering the limits to growth. A report to the Russian Association of the Club of Rome* (pp. 35–50). Springer. https://doi.org/10.1007/978-3-031-34999-7_3

Sadovnichy, V., Akaev, A., Ilyin, I., Malkov, S., Grinin, L., Sayamov, Y., & Korotayev, A. (2023b). Introduction. Hoping for the future. In V. Sadovnichy et al. (Eds.), *Reconsidering the limits to growth. A report to the Russian Association of the Club of Rome* (pp. 1–14). Springer. https://doi.org/10.1007/978-3-031-34999-7_1

Schumpeter, J. (1942). *Capitalism, socialism, and democracy*. Harper & Bros.

von Weizsäcker, E.U., & Wijkman, A. (2018) *Come on! Capitalism, short-termism, population and the destruction of the planet. A report to the Club of Rome*. Springer. https://doi.org/10.1007/978-1-4939-7419-1

Wijkman, A., & Skånberg, K. (2017). *The circular economy and benefits for society. Jobs and climate clear winners in an economy based on renewable energy and resource efficiency. A report to the Club of Rome*. Club of Rome.

Viktor Sadovnichy is a Doctor of Physics and Mathematics, Professor, full member of the Russian Academy of Sciences, and rector of Lomonosov Moscow State University since 1992. Field of scientific research: mathematics, mechanics, and computer science. Sadovnichy's research relates to the theory of differential operators, functional analysis, and mathematical modeling of long-term social and economic dynamics. He has obtained important theoretical and applied results related to the issues of dynamic simulation of controlled flights, in particular, of spacecrafts and aircrafts. This work formed the basis for the creation of the first simulator of zero-gravity in the Earth conditions in the world.

Askar Akaev is the First President of the Kyrgyz Republic and Foreign Full Member of the Russian Academy of Sciences. He works as Chief Research Professor in the Prigogine Institute for Mathematical Investigations of Complex Systems at the Moscow State University. He is also Academic Supervisor of the Centre for Stability and Risk Analysis at HSE University, Moscow, Russia. Askar Akaev has authored and co-authored over 300 scientific publications, including such monographs as Coherent Optical Computers (1977), Optical Methods of Information Processing (1983), Holographic Memory (1996), Selected Lectures on Optical Computers (1996), Economic Transition from the Point of View of a Physicist. A Mathematical Model of Emerging Economy (2000), Modeling and Forecasting the World Dynamics (2012), Quality of Education, Effectiveness of R&D, and Economic Growth. Quantitative Analysis and Mathematical Modeling (2016), etc.

Ilya Ilyin is a well-known Russian researcher, Doctor of Political Sciences, Professor, Head of the Department of Global Studies, Dean of the Faculty of Global Processes at Moscow State University named after M.V. Lomonosov, author of more than 350 scientific papers. In addition, he is the Chairman of the Presidium of the Society for Global Studies, President of the International Public Organization for the Promotion of Public Diplomacy, Scientific, Educational and Youth Cooperation "Eurasian Commonwealth", Chief Scientific Secretary of the International Academy of Global Studies, Chairman of the Russian Committee and Vice President of the Intergovernmental Council for the UNESCO Program "Management of Social transformations – MOST," honorary worker in the field of state youth policy and education, editor-in-chief of the journal *Bulletin of Moscow University*. Series 27: *Global Studies and Geopolitics*.

Sergey Malkov heads the Center for Long-Term Forecasting and Strategic Planning at the Moscow State University; he also works as a Senior Research Professor at the Centre for Stability and Risk Analysis at HSE University in Moscow. He is the author of over 300 scholarly publications, including such monographs as Social self-organization and historical process: possibilities of mathematical modeling (URSS, 2009), Modeling and forecasting of world dynamics (ISPI RAS, 2012, together with V. A. Sadovnichy and others), A new theory of the reproduction of capital: development and practical application (Nestor-Istoriya, 2016, together with V. I. Mayevsky and others), The 21st Century Singularity and Global Futures. A Big History Perspective (Springer, 2020, together with David J. LePoire and others).

Leonid Grinin is a Senior Research Professor at the Centre for Stability and Risk Analysis at HSE University in Moscow, as well as Senior Research Professor of the Institute of Oriental Studies, Russian Academy of Sciences. He is co-editor of the international journals *Social Evolution & History* and the *Journal of Globalization Studies* and the international yearbooks *Evolution, History & Mathematics*, and *Kondratieff Waves*. He is the author of more than 600 publications, including 33 monographs in Russian, English, and Chinese.

Dmitry Ivanov is a Russian mathematician, specialist in the field of location and telecommunication systems, corresponding member of the Russian Academy of Sciences (2016). Vice-Rector of the Volga State Technological University, Yoshkar-Ola. He is a specialist in the field of location and telecommunication systems. Author and co-author of over 200 scientific publications, including 3 monographs and 12 copyright certificates and patents. Member of the Bureau of the Scientific Council for Radio Wave Propagation of the Russian Academy of Sciences.

Andrey Korotayev heads the Centre for Stability and Risk Analysis at HSE University in Moscow, Russia. He is also a Senior Research Professor at the Institute of Oriental Studies and Institute for African Studies, Russian Academy of Sciences, as well as at the Faculty of Global Studies of the Lomonosov Moscow State University and the Russian Presidential Academy of National Economy and Public Administration. He has authored or co-authored over 650 scholarly publications.

MIX
Papier aus verantwortungsvollen Quellen
Paper from responsible sources
FSC® C105338
FSC
www.fsc.org

If you have any concerns about our products,
you can contact us on
ProductSafety@springernature.com

In case Publisher is established outside the EU,
the EU authorized representative is:
Springer Nature Customer Service Center GmbH
Europaplatz 3, 69115 Heidelberg, Germany

Printed by Libri Plureos GmbH
in Hamburg, Germany